EVOLUTION

EVOLUTION

Carl T. Bergstrom

University of Washington

Lee Alan Dugatkin

University of Louisville

W. W. NORTON & COMPANY

NEW YORK • LONDON

W. W. Norton & Company has been independent since its founding in 1923, when William Warder Norton and Mary D. Herter Norton first published lectures delivered at the People's Institute, the adult education division of New York City's Cooper Union. The Nortons soon expanded their program beyond the Institute, publishing books by celebrated academics from America and abroad. By mid-century, the two major pillars of Norton's publishing program—trade books and college texts—were firmly established. In the 1950s, the Norton family transferred control of the company to its employees, and today—with a staff of four hundred and a comparable number of trade, college, and professional titles published each year—W. W. Norton & Company stands as the largest and oldest publishing house owned wholly by its employees.

Editor: Betsy Twitchell
Developmental editors: Sandy Lifland, Andrew Sobel
Project editor: Carla L. Talmadge
Editorial assistant: Cait Callahan
Production manager: Chris Granville
Managing editor, College: Marian Johnson
Design director: Rubina Yeh
Book designer: Lissi Sigillo
Photo editor: Stephanie Romeo
Photo researcher: Dena Digilio Betz
Media editor: Patrick Shriner
Associate editor, emedia: Callinda Taylor
Media assistant: Carson Russell
Marketing manager: John Kresse
Composition: Preparé, Inc.
Illustration studio: Precision Graphics
Manufacturing: Quad Graphics

Library of Congress Cataloging-in-Publication Data

Bergstrom, Carl T.
  Evolution / Carl T. Bergstrom, Lee Alan Dugatkin.—1st ed.
    p. cm.
  Includes bibliographical references and index.
  ISBN 978-0-393-92592-0 (hardcover)—ISBN 978-0-393-91341-5 (pbk.)
1. Evolution (Biology) I. Dugatkin, Lee Alan, 1962– II. Title.
  QH366.2.B483 2012
  576.8—dc23

                        2011036572

W. W. Norton & Company, Inc., 500 Fifth Avenue, New York, NY 10110
www.wwnorton.com
W. W. Norton & Company Ltd., Castle House, 75/76 Wells Street, London W1T 3QT

1 2 3 4 5 6 7 8 9 0

BRIEF CONTENTS

CONTENTS

PART I
Foundations of Evolutionary Biology

PART II
Evolutionary Genetics

PART III
The History of Life

CHAPTER 11 The Origin and Evolution of Early Life   378

CHAPTER 12 Major Transitions   404

PART IV
Evolutionary Interactions

CHAPTER 16 **The Evolution of Sex**   538

CHAPTER 17 Sexual Selection   562

CHAPTER 18 The Evolution of Sociality   584

CHAPTER 19 **Coevolution** 620

CHAPTER 20 **Evolution and Medicine** 646

ABOUT THE AUTHORS

Carl T. Bergstrom is a professor in the Department of Biology at the University of Washington in Seattle and a member of the external faculty at the Santa Fe Institute. Dr. Bergstrom's research uses mathematical, computational, and statistical models to understand how information flows through biological and social systems. His recent projects include contributions to the game theory of communication and deception, applications of information theory to the study of evolution by natural selection, development of mathematical techniques for mapping and comprehending large network datasets, and a number of more applied studies in disease evolution, including analysis of antibiotic-resistant bacteria in hospital settings and models of the interaction between ecology and evolution in novel emerging pathogens such as those causing SARS and H5N1 avian influenza. At the University of Washington, Dr. Bergstrom teaches undergraduate courses on evolutionary biology and on the applications of evolutionary biology to medical problems.

Lee Alan Dugatkin is a professor and Distinguished University Scholar in the Department of Biology at the University of Louisville. His main area of research is the evolution of social behavior. He is currently studying the evolution of cooperation, aggression, antibiotic resistance, risk-taking behavior, and the interaction between genetic and cultural evolution. Dr. Dugatkin is the author of over 145 articles on evolution and behavior in journals such as *Nature* and the *Proceedings of the National Academy of Sciences* and of several trade monographs on the evolution of cooperation and the history of science. He is also the author of *Principles of Animal Behavior*, Second Edition.

PREFACE

We envy the student taking a class in evolutionary biology today. It is an exciting time to study science, and the science of evolutionary biology is more vibrant now than ever. In part, this is due to the new tools now available to scientists—tools that just a generation ago were the stuff of dreams. But there is more to it than that. Evolutionary biology is increasingly important in integrating the biological sciences. Evolutionary biologists are now collaborating in new, dynamic ways with researchers in many disciplines, and in so doing, they are bringing together a diverse set of perspectives—from areas like phylogenetics, population genetics, the study of adaptation, molecular genetics, and developmental biology, to name just a few. The result is a much deeper understanding of the history and diversity of life on Earth over the last 4 billion years or so. Our job as the authors of this book is to capture the exciting work that has gone into this effort, and to present it in a rigorous and engaging fashion.

To achieve this goal, we draw on our dual roles as researchers in and teachers of evolutionary biology. We each run active labs abuzz with the excitement that surrounds the science of evolution; we both lecture about evolution to students at our own universities and to audiences around the world. And we are each enthusiasts about the history of science in general and the history of evolutionary biology in particular. The successful strategies we've developed for communicating with these diverse audiences have informed the tone, emphases, and features in this textbook in a way that we hope will excite the scientific imaginations of students and instructors alike.

First and foremost, we relish the fact that *all science* is about testing hypotheses. Hypothesis-driven science has proven to be the most powerful approach ever devised for understanding the nature of the physical world we live in. We convey this through the abundant use of examples in which evolutionary biologists generate and test hypotheses. In this way, students can gain an intimate understanding of how theoretical ideas translate into testable predictions, and how this process of hypothesis testing leads to refinements of theory.

More generally, we understand that learning is an interactive process, and that such a process is greatly facilitated by the use of *stories*. And so, in each chapter, we make use of the natural human inclination to acquire and process information in the form of stories. Within the field of evolutionary biology are fascinating stories on many levels: stories of individual scientists and how they came to their discoveries, stories of how human thought has changed over the centuries, stories of how major evolutionary innovations arose in the history of life, stories of how individual species have changed over millennia through biological evolution or, as in the case of many microbes, how a population can change dramatically in a matter of weeks.

To tell these stories, we reference the primary literature as would any working scientist, acknowledging that the ideas in this book come from the endeavors of individual researchers—many of whom are active today—rather than from some accepted body of traditional wisdom that lies beyond challenge. But rather than re-using the same stable of illustrative examples that appear in text after text, a large majority of the examples used to illustrate principles in this book appear here for the first time in a biology textbook, and a great fraction of these are drawn from papers published within the past decade. Through the lens of current research, students can see how the scientific understanding of evolutionary biology is ever-changing, and that built into science is a system that allows each assumption to be challenged and refined or even rejected based on a preponderance of evidence.

Because models play a fundamental role in much of the current research in evolutionary biology today, we will devote considerable attention to simple conceptual models of evolutionary processes. Often such models can be profitably expressed through the language of mathematics, and one of our principal aims in the text is to help students become comfortable with this approach. We have found that one of the most important things that students learn in college-level physics or economics classes is how to formulate questions about the real world in the language of mathematical models, and how to answer these questions appropriately using mathematical analysis. We believe that this should be a critical component of a college education in the biological sciences as well. But we also recognize that students enter this course with varying degrees of mathematical preparedness. Thus, we have placed the more advanced concepts in boxes in an effort to offer instructors maximum flexibility in integrating mathematical models into their course.

So that students will gain a firm understanding of the essential foundations of evolutionary reasoning, we introduce several fundamental components of evolutionary thought in Chapter 1, and emphasize them throughout this textbook. These include:

- **Phylogenetics**. All living things on the planet today—and indeed all life that has ever existed—are linked by a shared evolutionary history that evolutionary biologists represent using phylogenetic trees. Thus, in order to understand evolutionary relationships, whether between two HIV strains or among the three major domains of life, students must learn to think in terms of phylogenetic relationships. We consider it critical that any textbook on evolution seamlessly integrates phylogenetic thinking throughout, and we have done so here. If students walk away remembering just one thing about this book—though of course we hope they walk away remembering much, much more—it will be the importance of phylogenetic thinking.

- **Population thinking**. Evolutionary change occurs in populations, but in our experience students are often more comfortable thinking at the level of the individual, as one would in a physiology course, for example. In this book, we demonstrate how to think at the population level as well, paying careful attention to the properties of populations: population composition, variation among individuals within and between populations, change in the properties of a population over time, and so forth. This population-level perspective, particularly as it relates to the process of natural selection, permeates our book. Because we know that some students initially struggle to master this type of population-level thinking, we devote considerable space to teaching this skill.

- **Natural selection**. Evolution is often defined as "descent with modification." As a population geneticist (CTB) and a behavioral evolutionary biologist (LAD), we both study the processes responsible for such "modification." We convey the importance of this topic to students by teaching them how the process of natural selection has shaped the diversity of life on this planet, and how other processes—most notably genetic drift—have also contributed to the myriad forms of life around us.

Features

This textbook integrates the big themes in evolutionary biology—phylogenetics and population thinking—in a way that is both current and accessible. Extensive, in-depth, current research examples, an emphasis on problem solving, and a stunning art program engage students, helping them understand fundamental concepts and processes. Major features:

- Extensive coverage of **phylogenetics**, which is introduced in Chapter 1 through the examination of a few engaging examples that demonstrate the power of phylogenetic thinking. Soon after, in Chapter 4, Phylogeny and Evolutionary History, and Chapter 5, Inferring Phylogeny, students are taught how to interpret and then build trees that generate testable hypotheses about evolutionary history and compare the relatedness of living organisms. This strong foundation in phylogenetic reasoning is then integrated into the text and art in virtually every chapter that follows.

- Examination of fundamental concepts through the lens of phylogenetics and population thinking are reinforced by **current research examples**, many of which are drawn from research done in the last decade. From Chapter 4's in-depth examination of Bryan Fry's study of venomousness in snakes to find the evolutionary origins of snake venom, to Chapter 11's coverage of Natasha Paul and Gerald Joyce's research on self-replicating RNA to find clues to the RNA world, to Chapter 19's discussion of Toby Kier's work on the mutualism between a soybean legume and a rhizobial bacterium to examine coevolution and the response to cheaters, the excitement of current research is captured throughout.

- Significant coverage of **contemporary topics** such as genomics, evo–devo, and molecular evolution, including full chapters on the following subjects: Genome Evolution (Chapter 10), Evolution and Development (Chapter 13), Coevolution (Chapter 19), and Evolution and Medicine (Chapter 20).

- An in-depth focus on **a few research studies** in each chapter promote a more complete understanding of how evolutionary biologists come to understand specific concepts. The examples were carefully chosen to offer a **balance of classic and contemporary studies** that most fully illustrate the concept being discussed.

- A beautiful and information-rich **art program** was carefully developed to promote understanding of key concepts described in the text by both engaging students visually and providing them with just the right amount of detail. The art includes distinctive figures that help students in the following ways:

  1. **Phylogenetic relationships** are made clear through the many phylogenetic trees that appear in virtually every chapter. Many of these trees also include in-figure captions, photographs, and line art that enrich students' understanding of the concept or example.

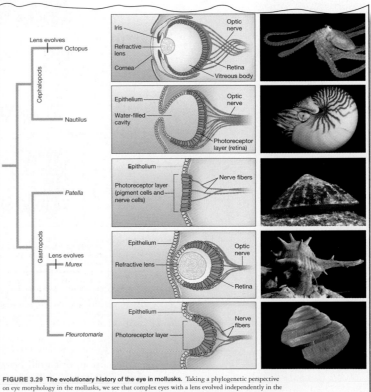

FIGURE 3.29 The evolutionary history of the eye in mollusks. Taking a phylogenetic perspective on eye morphology in the mollusks, we see that complex eyes with a lens evolved independently in the cephalopods and in the gastropods (Oakley and Pankey 2008). From top to bottom: The octopus eye uses a lens to focus light on the retina, much as does the vertebrate eye. The nautilus eye functions like a pinhole camera, casting a sharp image on the retina at the expense of a loss in brightness due to its small aperture. The limpet *Patella* has only a light-sensitive patch that can distinguish between light and dark. The predatory snail *Murex* uses a simple lens to focus incoming light. The snail *Pleurotomaria* has an indented eye cup that can detect the direction of a light source. Phylogeny is inspired by Oakley and Pankey (2008) and informed by Ponder and Lindberg (1997).

2. **Research-style data graphics** are presented much like they appear in the primary literature, but with carefully developed labels and in-figure captions that teach students to interpret and analyze the image or graph visually.

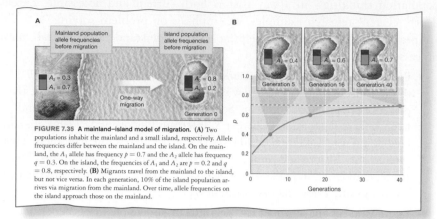

FIGURE 7.35 **A mainland–island model of migration.** **(A)** Two populations inhabit the mainland and a small island, respectively. Allele frequencies differ between the mainland and the island. On the mainland, the A_1 allele has frequency $p = 0.7$ and the A_2 allele has frequency $q = 0.3$. On the island, the frequencies of A_1 and A_2 are $p = 0.2$ and $q = 0.8$, respectively. **(B)** Migrants travel from the mainland to the island, but not vice versa. In each generation, 10% of the island population arrives via migration from the mainland. Over time, allele frequencies on the island approach those on the mainland.

3. Diagrams of **experimental processes** encourage students to visualize not just the outcome of a research study, but the specifics of how the experiment was constructed so that they can better understand the meaning behind the data.

FIGURE 3.3 **Phenotype depends on the effects of both genotype and environment.** Here we see how the height of a yarrow plant (*Achillea millefolium*) depends on its genotype and the altitude at which it is raised, as shown by populations of yarrow plants grown in gardens at three sites that were at different altitudes: high, medium, and low elevation. For example, the green screen behind the plants in genotype 1 shows that these plants grow tall at high and low elevations, but are short at medium elevation. The blue screen behind the plants in genotype 4 shows that the plants respond very differently to elevation. This genotype grows tallest at medium elevation and shorter at high and low elevations. Adapted from Clausen, Keck, and Hiesey (1940, 1948).

- Clear and accessible coverage of **quantitative methods**, the most difficult of which are in optional boxes. This teaches students how to formulate questions about evolutionary processes and relationships the ways researchers do—in the language of quantitative models.
- High-quality **problem sets** in the end-of-chapter material and online provide students with extensive practice in formulating and solving problems.

Resources for Instructors

Instructor's Resource Disc

This disc includes the following presentation features that can be used in your lectures:

- Labeled and unlabeled versions of every figure and photograph in the textbook, offered in JPEG and PowerPoint formats.
- All of the process animations for offline use.

Downloadable Instructor's Resources

These include content for use in both the classroom and online:

- Book art in zipped JPG and PowerPoint formats.
- Free, customizable coursepacks in a variety of formats.
- Test Bank in Examview, Word RTF, and PDF formats.
- Instructor's Manual in PDF format.

For more information and to view samples, visit **wwnorton.com/instructors**.

Test Bank

The Test Bank has been developed using the Norton Assessment Guidelines and provides a quality bank of 1000 items consisting of multiple-choice and short answer/essay questions. Each chapter of the Test Bank consists of three question types classified according to Norton's taxonomy of knowledge types:

- Factual questions test students' basic understanding of facts and concepts.
- Applied questions require students to apply knowledge in the solution of a problem.
- Conceptual questions require students to engage in qualitative reasoning and to explain why things are as they are.

Questions are further classified by section and difficulty, making it easy to construct tests and quizzes that are meaningful and diagnostic.

Instructor's Manual

This helpful online resource for instructors consists of detailed chapter outlines and guides to key readings in the text for every chapter. The manual also includes brief guides to accessing and using online simulations, including EvoBeaker.

Coursepacks

At no cost to instructors or students, Norton coursepacks are available in a variety of formats, including all versions of Blackboard and WebCT. With a simple download from our instructor's website, an adopter can bring high-quality digital media into a new or existing online course (with no additional student passwords or logins required). Content includes chapter-based assignments, test banks, quizzes, and animations from the StudySpace student website.

Resources for Students

StudySpace

Students rely on effective and well-designed online resources to help them succeed in their courses. StudySpace is a free and open website that shows students what they need to know, what they still need to review, and provides them with an organized study plan for mastering the material. Resources for students on StudySpace:

- Process animations based on topics nominated by more than 100 reviewers and based directly on powerful and information-rich art in the textbook. These animations bring to life fundamental concepts such as genetic drift, phylogenetics, the Hardy–Weinberg equilibrium, and mutation. In addition to being available online to students, offline access is available on the Instructor's Resource Disc.

- Quiz+ Diagnostic Quizzes provide a customized study plan based on students' right and wrong quiz answers by offering specific page references and links to the ebook and other online learning tools, including the animations.

- Detailed Study Plans guide students through the core concepts for each chapter and help them utilize all the online resources for the chapter.

- Flashcards for vocabulary terms help students master new terminology.

- A news feed links to current articles on the web relating to evolutionary biology.

Visit StudySpace at **wwnorton.com/studyspace**.

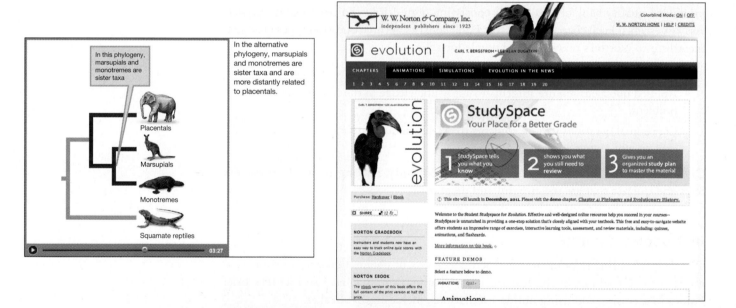

Ebook

Same great content, one-third the price. An affordable and convenient alternative to the print book, Norton ebooks retain the content and design of the print book and allow students to highlight and take notes with ease, print chapters as needed, search the text, and more. Norton ebooks are available online and as downloadable PDFs. They can be purchased directly from **nortonebooks.com** or with a registration folder that can be sold in the bookstore.

Chapter Select ebook chapters are available for just $4 each, $30 minimum purchase. Visit **nortonebooks.com** for pricing and details.

Acknowledgments

Creating this textbook has been a labor of love, and there are many people whose extraordinary commitment we'd like to acknowledge. First, we'd like to thank our Norton editors, Jack Repcheck, for seeing something special in this project and signing it in the first place, and Michael Wright, for bringing us together as coauthors, thus forging a relationship that has been both personally and professionally rewarding for both of us. Thanks also to editor Leo Weigman for his contributions. And thanks especially to our current editor, Betsy Twitchell. Without her insightful feedback and careful attention to every aspect of the development and production of the book, you would not be reading these words right now.

Sandy Lifland expertly developed our manuscript in each draft with a critical eye and thorough hand, for which she has our deepest thanks. Thanks also to Andrew Sobel, whose contributions as developmental editor for the chapters in Part II improved our coverage of population genetics immeasurably. We are grateful to our project editor, Carla Talmadge, for her exceptional attention to detail, tireless commitment to staying on schedule, and ability to synthesize the innumerable moving parts of this project. And we thank Marian Johnson, Norton's managing editor, for her help through the years in coordinating the complex process of turning a manuscript into a book. Chris Granville oversaw the final assembly into the beautiful book you hold in your hands; for this he has our thanks.

Thank you to our excellent copy editor, Brian Rose, who was assisted by Connie Parks, to our proofreader, Justine Cullinan, and to Dena Digilio Betz, Junenoire Mitchell, Stephanie Romeo, and Trish Marx for finding all the remarkable photographs you see in this book. For the truly stunning figures throughout this text, we thank the team at Precision Graphics. Rubina Yeh and Lissi Sigillo are responsible for the attractive design. We are grateful to editorial assistants Cait Callahan and Jennifer Cantelmi, who managed the enormous amount of information flowing between the members of the team and executed the ambitious review program we've benefited so much from. We also thank associate editors Matthew Freeman and Callinda Taylor for their hard work producing the instructor and student resources accompanying the text. Thank you to media editors Rob Bellinger and Patrick Shriner, who did a great job producing the innovative digital media that add so much value to our book for instructors and students.

We are grateful for the tireless advocacy of marketing manager John Kresse, director of marketing Steve Dunn, director of sales Michael Wright, and every one of Norton's extraordinary sales people, who will ensure our book reaches as wide an audience as possible. Finally, our deepest thanks to Drake McFeely, Roby Harrington, and Julia Reidhead for their unfaltering commitment to this project through all its twists and turns.

Publishing this book would not have been possible without the involvement of our reviewers. At each stage of development, their thoughtful feedback has helped us make this book more accurate, complete, and fun to read. For this, they have our deepest thanks. We are especially grateful to our accuracy reviewers: Sara Via, University of Maryland, who reviewed the entire book, Shannon Hedtke, University of Texas at Austin, Tamra Mendelson, University of Maryland, Baltimore County, Neil Sabine, Indiana University East, and Martin Tracey, Florida International University. We thank the following reviewers for their comments on various chapters of the book:

Byron Adams, Brigham Young University

Ron Aiken, Mount Allison University

Jonathan Armbruster, Auburn University

Peter Armbruster, Georgetown University

Ricardo Azevedo, University of Houston

Eric Baack, University of British Columbia

Felix J. Baerlocher, Mount Allison University

Christopher Beck, Emory University

Peter Bednekoff, Eastern Michigan University

Alison Bell, University of Illinois at Urbana-Champaign

Giacomo Bernardi, University of California, Santa Cruz

Annalisa Berta, San Diego State University

Edmund Brodie, University of Virginia

Sibyl Rae Bucheli, Sam Houston State University

Mark Buchheim, University of Tulsa

Christina Burch, University of North Carolina

Ashley Carter, California State University, Long Beach

Teresa Crease, University of Guelph

Charles D. Criscione, Texas A&M University

Patrick Danley, Baylor University

Margaret Docker, University of Manitoba

Thomas Dowling, Arizona State University

Victor Fet, Marshall University

David H. A. Fitch, New York University

Susan Foster, Clark University

Ronald Frank, Missouri University of Science and Technology

Robert Friedman, University of South Carolina

James Fry, University of Rochester

Jessica Garb, University of Massachusetts, Lowell

David Garbary, St. Francis Xavier University

Nicole Gerardo, Emory University

Jennifer Gleason, University of Kansas

David Gray, California State University, Northridge

Linda Green, Georgia Institute of Technology

Matthew Gruwell, Penn State Erie, The Behrend College

Shannon Hedtke, University of Texas at Austin

Michael Henshaw, Grand Valley State University

Chad Hoefler, Arcadia University

Eric A. Hoffman, University of Central Florida

James Hogue, California State University, Northridge

Dale Holen, Penn State Worthington, Scranton

Brett Holland, California State University, Sacramento

Timothy Holtsford, University of Missouri

Lisa Horth, Old Dominion University

Christopher Horvath, Illinois State University

Anne Houde, Lake Forest College

Laurence Hurst, University of Bath

David Innes, Memorial University of Newfoundland

Rebecca Jabbour, Saint Mary's College of California

Jerry Johnson, Brigham Young University

Mark Johnston, Dalhousie University

David Kass, Eastern Michigan University

Nicole Kime, Edgewood College

Charles Knight, California Polytechnic State University

Eliot Krause, Seton Hall University

Patrick J. Lewis, Sam Houston State University

Dale Lockwood, Colorado State University

Therese Markow, University of California, San Diego

Gary McCracken, University of Tennessee, Knoxville

Donald McFarlane, Scripps College

Douglas Meikle, Miami University

Tamra Mendelson, University of Maryland, Baltimore County

Mirjana Milosevic-Brockett, Georgia Institute of Technology

Yolanda Morbey, University of Western Ontario

Serena Moseman-Valtierra, Boston College

Laurence Mueller, University of California, Irvine

Maurine Neiman, University of Iowa

Mark Nielsen, University of Dayton

Juliet Noor, Duke University

David Orange, California State University, Sacramento

Kevin Padian, University of California, Berkeley

Matthew Parker, Binghamton University, State University of New York

Chris Parkinson, University of Central Florida

Leslee Parr, San José State University

Andrew Peters, University of Wisconsin, Madison

Raymond Pierotti, University of Kansas

Patricia Princehouse, Case Western Reserve University

Jayanti Ray-Mukherjee, Florida International University

Sean Rice, Texas Tech University

Antonis Rokas, Vanderbilt University

Michael Rosenberg, Arizona State University

Stephen Rothstein, University of California, Santa Barbara

Eric Routman, San Francisco State University

Harry Roy, Rensselaer Polytechnic Institute

Neil Sabine, Indiana University East

Joel Sachs, University of California, Riverside

Dietmar Schwarz, Western Washington University

Douglas Scofield, Umeå Plant Sciences Center

David Scott, South Carolina State University

Jon Seger, University of Utah

Matthew Shawkey, University of Akron

Rebecca Simmons, University of North Dakota

Kelly Smith, Clemson University

Nancy G. Solomon, Miami University

Maureen Stanton, University of California, Davis

William Starmer, Syracuse University

Stephen Stearns, Yale University

Don Stewart, Acadia University

J. Todd Streelman, Georgia Institute of Technology

Gerald Svendsen, Ohio University

Daniel Thompson, University of Nevada, Las Vegas

Jeffrey Townsend, Yale University

Martin Tracey, Florida International University

Priscilla Tucker, University of Michigan

J. Albert Uy, University of Miami

Sara Via, University of Maryland

Peter Waddell, Purdue University

Yufeng Wang, University of Texas at San Antonio

Christopher S. Willett, University of North Carolina at Chapel Hill

Barry Williams, Michigan State University

Roger Williams, Winthrop University

Christopher Witt, University of New Mexico

Lorne Wolfe, Georgia Southern University

Rebecca Zufall, University of Houston

Writing this book would not have been possible without our families, whom we thank for the enthusiasm, support, patience, and love that they provided throughout the entire process. Carl thanks his wife Holly for accommodating the many disruptions to family life that are imposed by a project of this scope, and thanks his children Helen and Teddy for their patience when he was working on the book and their excitement and wonder when he shared the stories and photographs therein. Carl also thanks his sister, Karen Hausdoerffer, who helped extensively in writing and revising Chapters 1 and 2. Lee would like to thank his wife Dana for her patience when asked "Hon, can you just proofread this chapter one more time?" and his son Aaron for going to Bats, Reds, and Yankees games with him so he could clear his head. Lee would also like to thank "2R," but he can't say why.

To the reader: Thank you as well! We greatly appreciate your consideration and selection of this book as your introduction to evolutionary biology. We welcome your comments.

Carl T. Bergstrom
cbergst@u.washington.edu
Lee Alan Dugatkin
Lee.dugatkin@louisville.edu

PART I

Foundations of Evolutionary Biology

Giant tortoises from inside the Alcedo volcano on Isabela Island. This island is part of the Galapagos Archipelago that Darwin visited while on board the HMS *Beagle*.

1

An Overview of Evolutionary Biology

◀ The process of evolution is responsible for the incredible diversity of living forms that cover the planet, including the spiral fronds of this uluhe fern (*Dicranopteris linearis*) in Hawaii Volcanoes National Park.

In his classic book, *The Structure of Scientific Revolutions*, Thomas Kuhn argued that major advances in science are rare, and that true scientific revolutions involve fundamental changes in the way we think (Kuhn 1962). Once such a revolution takes place, the world is never seen or understood in the same way. When early astronomers and physicists demonstrated that Earth was not at the center of the universe, what Kuhn described as a "paradigm shift" occurred. A similar paradigm shift occurred in biology when Charles Darwin laid out his theory of evolution by natural selection.

In *On the Origin of Species*, published in 1859, Darwin presented two revolutionary ideas. Each had been suggested independently by others before, but never had they been brought together with the conceptual brilliance and the naturalist's eye of Charles Darwin (Chapter 2). After decades of observations, collecting data from near and far, reading incessantly, and synthesizing and resynthesizing theories from a number of different disciplines, Darwin recognized that the wide diversity of life

3

we see around us has descended from previously existing species, which share a common ancestor from further back in time. Second, Darwin realized that the often exquisite fit of species to their environments is primarily a result of **natural selection**, a gradual process in which forms that are better suited to their environment increase in frequency in a population over sufficiently long periods of time. As we will see throughout this book, "sufficiently long" can range from a matter of days to tens of thousands of years, depending on the strength of natural selection and the rate of reproduction of the organisms we are studying. Together, these two ideas proposed by Darwin suggest that the entire organic world—much of everything we see, feel, smell, taste, and touch—is the result of evolutionary changes that have taken place over time.

Once the theory of evolution by natural selection was developed, scientists had at their disposal a natural—as opposed to a supernatural—explanation for the diversity of life on the planet, as well as an explanation for why the vast majority of life-forms that have ever existed are now extinct. More than that, they had a theory that could be used to explain the similarities and differences among all the creatures on Earth, both past and present.

Paradigm shifts have wide-ranging effects, and that was certainly the case for Darwin's theory—so much so that the renowned geneticist Theodosius Dobzhansky wrote, "nothing in biology makes sense except in the light of evolution" (Dobzhansky 1973, p. 125). Without evolutionary theory, biology is composed of a large number of very important, but disparate, subdisciplines. With evolution as its theoretical and conceptual foundation, however, biological science shares a common framework that allows us to understand both the commonalities and differences among living forms; it allows us to make sense of the way that living things function now and to understand how they came to be.

The study of physics is fundamental to understanding our universe, because it allows us to reconstruct the grand story of how the universe came to be as it is, and it lets us understand how the universe operates today. The study of evolution is similarly fundamental in that it allows us to reconstruct the grand story of how all living things came to be, and how they (and we) function.

As you will see as you work your way through this book, the characteristics of the organisms you are studying have been shaped by the evolutionary process. Whether you are interested in anatomy, physiology, behavior, molecular biology, genetics, development, or any other area of biology, a solid understanding of evolution is indispensable.

In this chapter, we will:

- Provide a brief introduction to evolution and natural selection, including examples related to (1) artificial selection, (2) antibiotic resistance, (3) conservation biology, (4) molecular genetics and evolution in primates, and (5) sperm competition in fowl.

- Give an overview of empirical and theoretical approaches to the study of evolution.

- Discuss a more detailed example of the way that empirical and theoretical approaches interact by looking at the evolution of sex ratios.

1.1 A Brief Introduction to Evolution and Natural Selection

The science of evolutionary biology reads like a great detective story in the sense that it unravels a great mystery. Indeed, evolutionary biologists *are* detectives—as are all scientists—but they are much more than that. The discipline of evolutionary biology allows us not only to infer the relationships among all life that has ever lived, and to track the diversity of life across vast stretches of time, but also to test hypotheses through a rigorous combination of observation and experimental manipulations. These observations and experiments may involve examining fossils or living organisms; they may use, among other things, anatomical, molecular genetic, developmental, and behavioral data; they may involve analyzing data from DNA sequences to population composition (Figure 1.1).

At its core, evolutionary biology is the study of the origin, maintenance, and diversity of life on Earth over approximately the last 3.5 billion years. To fully understand the **evolution** of a species, we need to know the ancestral species from which it *descended* and we need to know what sort of *modifications* have occurred along the way. Darwin referred to this entire process as **descent with modification**.

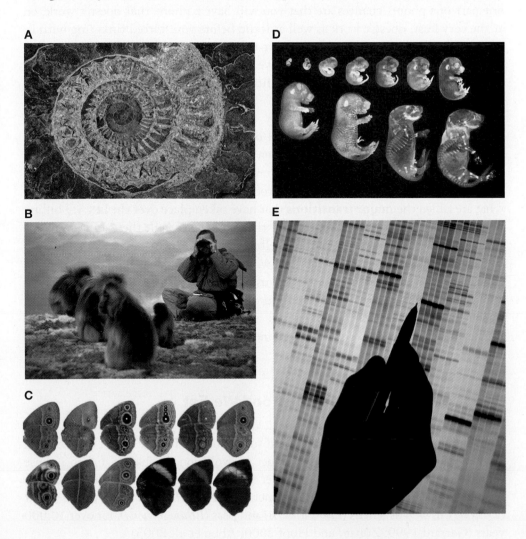

FIGURE 1.1 Sources of data for testing models of evolution. A few examples of the sources of data that evolutionary biologists use to test their hypotheses. **(A)** Data from the fossil record, as shown by this fossil ammonite found in Dorset, England, **(B)** behavioral data, as shown by observing the behavior of gelada baboons in Ethiopia, **(C)** morphological data, as shown by this display of wing color patterns on *Bicyclus anynana* butterfly wings, **(D)** embryological data, as shown by the magnetic resonance imaging of developing mouse embryos between day 9.5 and day 19, when the mouse is born, and **(E)** molecular genetic data, as shown by this DNA sequence film.

If the species in question is *Homo sapiens,* for example, we need to understand the primate species from which it descended (as well as other species closely related to this ancestral species), and the changes that occurred over the period in which *H. sapiens* evolved. We use the same reasoning if the species in question is the malaria parasite (*Plasmodium falciparum*) or corn (*Zea mays*). That is, we try to discern the ancestral history of the species in question, and at the same time, we attempt to track the modifications that have occurred in that species. Again, we aim to understand the process of descent with modification.

One of the most important processes responsible for the modifications that occur over time is natural selection. We will discuss natural selection and other evolutionary processes in greater detail in coming chapters. For the time being, we can summarize natural selection as follows. Genetic **mutations**, or changes to the DNA sequence, arise continually and change the **phenotype**—the observable, measurable characteristics—of organisms. These mutations can increase fitness, decrease fitness, or have no effect on fitness, where **fitness** is measured in terms of relative survival rates and reproductive success. Many mutations will disrupt processes that are already fine-tuned, and thus they will have harmful effects on fitness. By analogy, consider tinkering with a telephone. If you randomly change one part of a phone, chances are that you will have a phone that doesn't work, or, at the very least, doesn't work as well as it did before you started tinkering with it. But, by chance, some mutations will turn out to be advantageous in the sense that the individuals who carry them may have more surviving offspring than average. Such genetic changes that improve the fitness of individuals will tend to increase in frequency over time.

The result is evolutionary change by natural selection. The accumulation of advantageous genetic changes, amassed over long periods of time, can produce dramatic effects within a population, even to the extent of producing new species, genera, families, and higher taxonomic orders. Indeed, as we will see many times throughout the course of this book, the process of natural selection is fundamental in what are called the **major transitions** that have taken place over the last 3.5 billion years of life on Earth—the evolution of the prokaryotic cell, the evolution of the eukaryotic cell, the evolution of multicellularity, and so on.

Repeatedly throughout this book, we will examine the power of natural selection in shaping the life that we see around us. We begin with some of the practical applications of understanding evolution via natural selection. The examples in this section, as well as all the examples we discuss in this chapter, are meant to illustrate some of the major concepts, methods, and tools that biologists use to understand evolution.

Evolutionary Change and the Food We Eat

The next time you sit down for a meal, take a look at the items on your plate. Whether you're enjoying a home-cooked supper or fast-food takeout, the food you are eating is almost certainly the product of evolutionary change due to intense **selective breeding** over time (Denison et al. 2003) (Figure 1.2). Indeed, humans have been selectively breeding grains, such as barley (*Hordeum vulgare*) and wheat (*Triticum*), as well as lentils (*Lens culinaris*) and peas (*Pisum sativum*), for over 10,000 years (Garrard 1999; Zohary and Hopf 2000; Abbo et al. 2003).

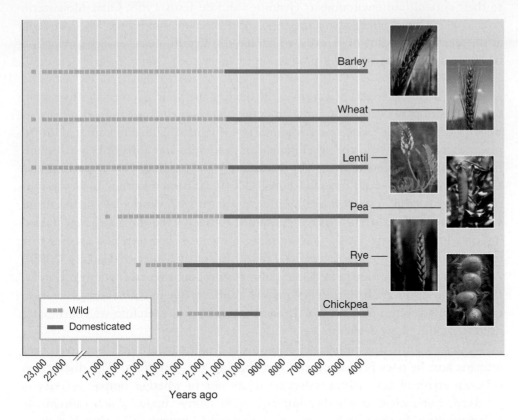

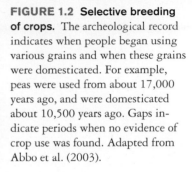

FIGURE 1.2 Selective breeding of crops. The archeological record indicates when people began using various grains and when these grains were domesticated. For example, peas were used from about 17,000 years ago, and were domesticated about 10,500 years ago. Gaps indicate periods when no evidence of crop use was found. Adapted from Abbo et al. (2003).

The process of human-directed selective breeding, known as **artificial selection**, is straightforward. In the case of crops, in each generation the best plants—for example, those that are the hardiest, quickest growing, and best tasting—are chosen as the parental stock for the next generation (Figure 1.3).

Artificial selection by humans is thus a counterpart to natural selection. With natural selection, traits that are associated with increased survival and reproduction increase in frequency. With artificial selection, humans choose which individuals get to reproduce, and in so doing, we select traits that are in some way beneficial to us. Such selective breeding can produce dramatic results. For example, the productivity of wheat (*Triticum aestivum*), rice (*Oryza sativa*), and corn (*Zea mays*) have doubled since 1930; much of that increase is due to selection for genetic crop strains better adapted

FIGURE 1.3 The process of artificial selection. Darwin used strawberries as an example of artificial selection, writing, "As soon, however, as gardeners picked out individual [strawberry] plants with slightly larger, earlier, or better fruit, and raised seedlings from them, and again picked out the best seedlings and bred from them, then, there appeared (aided by some crossing with distinct species) those many admirable varieties of the strawberry which have been raised during the last thirty or forty years" (Darwin 1859, pp. 41–42).

to their agricultural environments (Jennings and de Jesus 1968; Ortiz-Monasterio et al. 1997; Duvick and Cassmann 1999). And the same holds true when we look at the selective breeding of animals, which has resulted in increased egg production by chickens and increased milk production by dairy cows (Muir 1996; Rocha et al. 1998).

Even as artificial selection improves the quality and yield of crops and livestock, other evolutionary changes have detrimental effects on the human food supply, as we see with pesticide resistance. Although 10–35% of all U.S. crops are still lost to insect damage each year, the development of pesticides was a major breakthrough in reducing crop pests and thereby increasing crop productivity (Pimentel and Lehman 1991; National Research Council 2000). Natural selection, however, will tend to favor crop pests that are most resistant to such pesticides—as occurred when diamondback moths evolved resistance to one of the most frequently used insecticides of the late 1980s—creating an "arms race" between pest species that feed on crops, and humans determined to get rid of such species (Ceccatti 2009). As resistant pests increase in frequency, humans produce ever stronger insecticides. Because evolutionary change occurs quickly in insects due to their short generation times, humans often lose this particular arms race, and therefore we continually need to produce new pesticides.

Why do we call this natural selection instead of artificial selection, given that humans are the ones producing and distributing the pesticides? The distinction between artificial and natural selection refers not to whether human activity is involved, but rather to whether humans deliberately choose which individuals will reproduce. In the case of increasing grain yields, humans actively select those varieties with higher yield; in the case of increasing pesticide resistance, humans produce the pesticides but do not deliberately choose pesticide-resistant strains of insects for further reproduction. Indeed, what we want—pests easily killed by our pesticides—is just the opposite.

Unfortunately, a problem similar to that of resistance to pesticides unfolds when we look at another manufactured product: antibiotics.

Evolutionary Change and Pharmaceuticals

One theme that we will return to repeatedly throughout this book is the manner in which research in evolutionary biology can inform our understanding of disease and help us to design ever more effective responses to disease. For example, the discovery and development of antibiotic drugs for preventing or treating bacterial infections was one of the major medical developments of the twentieth century. But ever since humans first began using antibiotics, medical practitioners have had to deal with bacteria that are resistant to these drugs. The first modern antibiotic, penicillin, was introduced clinically in 1943; within a single year, penicillin resistance was observed, and within 5 years it had become common in a number of bacterial species. Since then, numerous new antibiotics have been developed and introduced to the market, only to lose their effectiveness within a matter of years as bacteria evolved resistance to the drug (Lacey 1973; Piddock et al. 1998; CDC 2007) (Table 1.1). The evolution of **antibiotic resistance** is the result of natural selection, and can be understood only in the context of evolutionary biology.

Bacteria reproduce at an astounding rate—in some cases, as frequently as once every 20 minutes. They reach enormous population sizes—a single gram of feces

TABLE 1.1		
Bacteria Have Rapidly Evolved Resistance to Clinical Antibiotics		
Antibiotic	Date Introduced	Date Resistance Observed
Penicillin	1943	1945
Chloramphenicol	1949	1950
Erythromycin	1952	1956
Methicillin	1960	1961
Cephalothin (first-generation cephalosporin)	1964	1966
Vancomycin	1958[a]	1986
Second- and third-generation cephalosporins	1979, 1981	1987
Carbapenems	1985	1987
Linezolid	2000	2002

[a]Vancomycin was first released in 1958; however, it was not widely used until the early 1980s.
From Bergstrom and Feldgarden (2008).

can contain 100 billion bacteria—which offer plentiful opportunities for mutations that provide resistance to arise. Antibiotics impose very strong natural selection for resistant strains. For all of these reasons, bacteria can evolve extremely rapidly, and when exposed to antibiotics, this is precisely what they do (Genereux and Bergstrom 2005).

Imagine that we are watching the evolution of resistance to the antibiotic ciprofloxacin. Ciprofloxacin is often prescribed for severe cases of food poisoning by the bacterium *Campylobacter jejuni*. This bacterium is common in the intestines of livestock, where it causes no symptoms in animals, but it can cause acute food poisoning in humans who acquire it by eating contaminated meat. At the start of the process, the gut of a single human patient houses millions or even billions of *Campylobacter* cells that are exposed to ciprofloxacin. Early on, the antibiotic may be deadly to these cells. But with vast numbers of bacterial cells exposed to the antibiotic, and with each cell dividing quickly, it is only a matter of time (weeks, months, perhaps years) before a mutation appears that creates a strain of *Campylobacter* cells that are somewhat resistant to our antibiotic. In a patient being treated with ciprofloxacin, this new strain can outcompete the susceptible strain, and the resistant *Campylobacter* will eventually become the dominant form. The process then starts anew, and soon another genetic change occurs, producing a strain of *Campylobacter* that is even more resistant to the antibiotic, and that strain quickly takes over. Repeating this process over and over results in a strain of *Campylobacter* that is highly resistant to the antibiotic (Figure 1.4).

FIGURE 1.4 Ciprofloxacin use selects for resistant *Campylobacter*. (A) Prior to antibiotic treatment, most *Campylobacter* are ciprofloxacin-sensitive (meaning that they cannot grow in the presence of ciprofloxacin), but a few resistant variants may also be produced by mutation. (B) Ciprofloxacin treatment kills or halts the growth of the sensitive strains, but the resistant strains survive. (C) In the presence of ciprofloxacin, resistant *Campylobacter* take over the population.

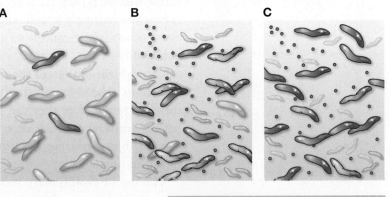

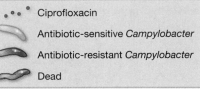

Ciprofloxacin

Antibiotic-sensitive *Campylobacter*

Antibiotic-resistant *Campylobacter*

Dead

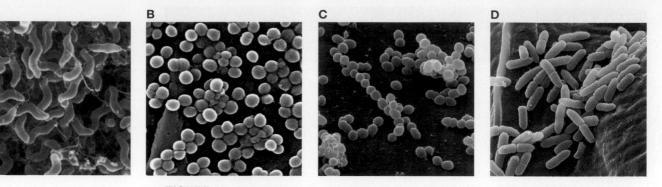

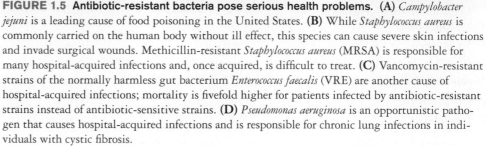

FIGURE 1.5 **Antibiotic-resistant bacteria pose serious health problems.** **(A)** *Campylobacter jejuni* is a leading cause of food poisoning in the United States. **(B)** While *Staphylococcus aureus* is commonly carried on the human body without ill effect, this species can cause severe skin infections and invade surgical wounds. Methicillin-resistant *Staphylococcus aureus* (MRSA) is responsible for many hospital-acquired infections and, once acquired, is difficult to treat. **(C)** Vancomycin-resistant strains of the normally harmless gut bacterium *Enterococcus faecalis* (VRE) are another cause of hospital-acquired infections; mortality is fivefold higher for patients infected by antibiotic-resistant strains instead of antibiotic-sensitive strains. **(D)** *Pseudomonas aeruginosa* is an opportunistic pathogen that causes hospital-acquired infections and is responsible for chronic lung infections in individuals with cystic fibrosis.

While *Campylobacter* is rarely life threatening, its consequences are certainly dramatic when considered in aggregate: antibiotic-resistant strains of *Campylobacter* are estimated to be responsible annually for nearly 500,000 more days of diarrhea in the United States than would occur in the absence of *Campylobacter* (Travers and Barza 2002). Other antibiotic-resistant bacteria such as *Staphylococcus aureus*, *Enterococcus* species, and *Pseudomonas aeruginosa* pose an even more significant threat (Figure 1.5). Today, antibiotic-resistant strains of these and other bacteria are largely responsible for an epidemic of hospital-acquired infections that kill an estimated 90,000 people per year in the United States—more than AIDS, influenza, and breast cancer combined (Bergstrom and Feldgarden 2008).

The study of evolutionary biology allows us to understand how these antibiotic-resistant bacteria evolve; this understanding in turn helps us deal with the health threat that such bacteria pose. In the course of drug development, pharmaceutical companies routinely screen potential new antibiotics by exposing bacteria to a wide range of antibiotic concentrations and making sure that antibiotic resistance does not readily evolve. Physicians often prescribe antibiotics in combination because drug combinations retard the rate at which antibiotic resistance evolves; even if the mutations needed for resistance to one drug should arise, the other drug will kill these bacteria before they can spread. Livestock producers have significantly (but not yet entirely) cut back on antibiotic use in agriculture now that we understand how antibiotic-resistant strains of bacteria can evolve in farm animals and then spread to humans.

The use of evolutionary models to address questions relevant to disease is not limited to the case of antibiotic resistance. In subsequent chapters, we will see other examples in which ideas and experiments from evolutionary biology have contributed to a fundamental understanding of influenza, sexually transmitted diseases such as AIDS, and many other infectious diseases. Evolutionary biology has likewise contributed to our understanding of chronic ailments such as diabetes and obesity, and even to an understanding of the phenomenon of aging itself. In some instances, such as that of antibiotic resistance, we can use our understanding of natural selection to design and construct models and experiments relevant to the study of disease; in other instances, we will examine how understanding patterns of common descent can achieve the same ends. But in all cases, our underlying approach will be evolutionary.

Evolution and Conservation Biology

All living things have descended from a common ancestor, and over eons the descendants of this common ancestor have diversified to yield the myriad forms that we observe in the world today (Chapter 4). We can view all species that live or ever have lived as forming a vast branching tree of relationships known as the **tree of life** (Figure 1.6). Such a tree captures the historical relationships among life-forms and is known as a **phylogenetic tree**. Often, each tip of a phylogenetic tree represents a species that is currently living in the world today or a species that has gone extinct; branch points represent divergence events—events associated with

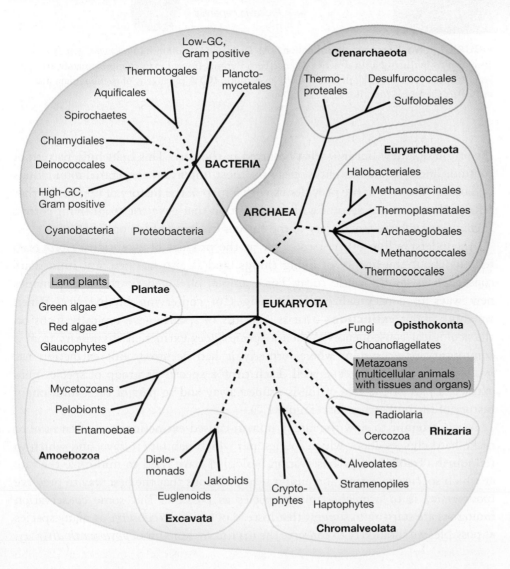

FIGURE 1.6 The tree of life. Bacteria, Archaea, and Eukaryota represent the main branches on the tree of life. These three groups are often referred to as "domains." We depict hypothetical relationships as dashed lines. Metazoan species—animals composed of cells differentiated into tissues and organs—including humans, are in the purple box. This tree is based on molecular genetic (ribosomal RNA) data. Adapted from Delsuc et al. (2005).

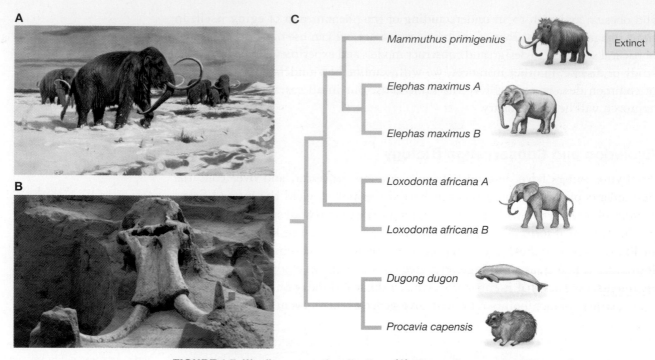

FIGURE 1.7 Woolly mammoth extinction. (A) The woolly mammoth (*Mammuthus primigenius*) once roamed northern North America and northern Eurasia. It went extinct approximately 10,000 years ago. **(B)** A fossilized skull of the woolly mammoth. **(C)** A phylogeny of paenungulata mammals. Part C adapted from Rogaev et al. (2006).

the origin of a new lineage—that occurred in the past. This branching pattern of common ancestry and descent is one of the most important conceptual foundations of biology. The tree of life provides us with a map of the history of life, a map that reflects the process of descent with modification that gave rise to all living forms. It connects evolutionary history to the current diversity of life on Earth.

An understanding of the tree of life as the product of evolutionary processes tells us about the history of living things, and it also has practical immediate consequences for the world today. For example, phylogenetic thinking provides new ways to conceptualize the challenges of conserving biodiversity. When we think about **extinction**—that is, the loss of species—we typically focus on the ecological consequences: When a species goes extinct, it disappears from a community or ecosystem where formerly it had occurred. But extinction has evolutionary consequences as well. Each time a species or group of species goes extinct, a part of the tree of life is pruned away and so part of the evolutionary history of life on Earth is lost (Figure 1.7).

As we attempt to slow the rate of human-caused extinctions, we often have to make hard choices about which species and which habitats to save, and which to relinquish. Traditionally, conservation biologists have tried to minimize the rate at which species go extinct, because it seems obvious that the best way to preserve biodiversity is to protect as many species as possible. But some conservation biologists are starting to suggest that instead of trying to conserve as many species as possible, we should try to conserve the maximum amount of *phylogenetic diversity*.

That is, we should conserve as much as possible of the evolutionary history represented by currently living species (Mace et al. 2003).

For example, in Figure 1.8, the extinction of the three species E, F, and I results in the loss of three twigs (indicated in red) at the tips of the tree, but nothing more. By contrast, extinction of the two species B and C results in the loss of a major branch (indicated in blue) of the phylogenetic tree. If we are interested in conserving **phylogenetic diversity**, the latter is a greater loss.

Appealing as it may sound to base conservation goals on phylogenetic diversity, our conservation agenda should probably not focus exclusively on preserving evolutionary history. For example, you might reasonably argue that, rather than focusing on history, it is important to save a population in which evolution is occurring rapidly and new species are being formed. While you might think that such a population would also be a major contributor to phylogenetic diversity, the opposite is often true. Species in areas where rapid diversification is occurring will be relative newcomers—new twigs on the tree of life—and it will be unlikely that all twigs on a major branch of the tree will perish at once. So, if we wish *only* to preserve phylogentic diversity, we need not be as concerned with areas where rapid **speciation** is occurring (Figure 1.9).

Consider the cichlid fishes of Lake Victoria in Africa. The approach of conserving phylogenetic diversity might suggest that we should be relatively unconcerned about species losses among these fishes. Why? Because Lake Victoria has been a hotbed of speciation for cichlids. Cichlid evolution there represents one of the most spectacular **evolutionary radiations** in recent history—400 new species of cichlid fish have evolved in the mere 14,000 years since Lake Victoria filled with water.

In the 1960s, the Nile perch (*Lates niloticus*)—a large, voracious predator of cichlids—was introduced into Lake Victoria as a game fish that would provide food for the local human populations (Figure 1.10). Since this introduction, approximately half of the native cichlids have gone extinct. Because the now-extinct species were phylogenetically very closely related to one another (that is, they shared a recent common ancestor), and because the combined evolutionary histories of all 200 lost species total only 3 million years or so, this is a very minor loss, where phylogenetic diversity is used as a measure. That is, phylogenetic diversity, as measured today, would not be reduced significantly by such extinctions.

On the other hand, perhaps we should also be interested in conserving the species and habitats within which evolution is operating most rapidly, as that is where the most change will occur in the future. If so, one could scarcely imagine a worse set of species to lose to extinction than the Lake Victoria cichlids.

The point here is not that one particular evolutionary model is best suited to solve all problems in conservation biology. Rather, the point is that when making decisions regarding biodiversity, conservation biologists could not even address these issues or have this important debate until they started thinking about evolutionary processes and consequences (Nee and May 1997). Thinking in this way also helps us put the current wave of human-caused extinctions into context. As Georgina Mace and her colleagues point out, "The tree of life is currently being pruned by extinction very much more rapidly than it is growing" (Mace et al. 2003, p. 1708).

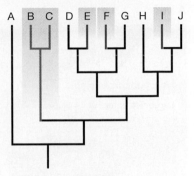

FIGURE 1.8 Extinction and twigs on phylogenetic trees. Assuming that species D, G, H, and J survive regardless, the extinction of species B and C (in blue) prunes this phylogenetic tree more severely than does the extinction of species E, F, and I (in red).

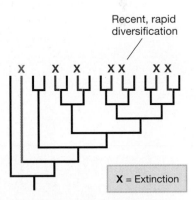

FIGURE 1.9 Hot zones, extinction, and evolutionary history. If we try to preserve evolutionary history, the loss of the single blue species which represents the only species on its branch of our tree—would produce a deeper cut than the loss of all the red species in a hot zone where speciation is occurring rapidly.

A

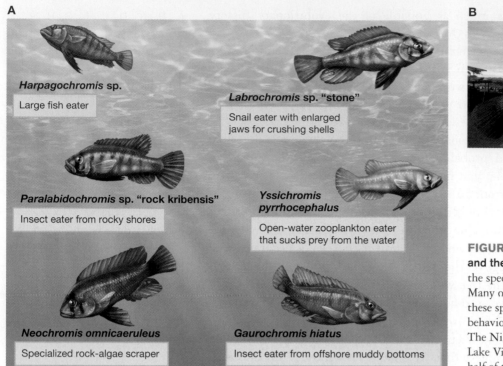

B

FIGURE 1.10 The cichlids of Lake Victoria, and their introduced predator. (A) A few of the species of cichlids found in Lake Victoria. Many of the modifications that distinguish these species involve feeding morphology and behavior. Adapted from Spinney (2010). **(B)** The Nile perch, a predator introduced into Lake Victoria. Since its introduction, about half of the native cichlids have gone extinct.

1.2 Empirical and Theoretical Approaches to the Study of Evolution

We have seen how evolutionary principles can be applied to a variety of subjects, but what *approaches* do evolutionary biologists employ in their quest to understand why things are the way they are? Any field of scientific endeavor requires us to generate and test alternative hypotheses. Indeed, the scientific process is all about postulating a series of testable hypotheses, ruling out alternatives, and honing in on the hypotheses that seem to best represent what is happening in nature (Mayr 1982, 1983). In generating and testing hypotheses, evolutionary biologists use a combination of empirical and theoretical approaches.

Empirical Approaches

The majority of this book focuses on empirical research. As we will see, empirical work in evolutionary biology can take many forms, but it almost always falls under one of two categories—observations or manipulations. Observational experiments entail gathering data without attempting to manipulate or control the system being studied. Examples include: (1) studying the fossil record to test predictions from evolutionary biology, as well as to generate new predictions, (2) inferring evolutionary history from genetic sequences, and (3) measuring behaviors occurring in a natural population of organisms (we will examine all of these in later chapters). Observational studies like these make up a powerful form of scientific research, and they have been used to test a myriad of evolutionary hypotheses.

Another approach is to design controlled manipulative experiments to test a specific hypothesis. Manipulative experiments allow a scientist to directly assess how changes in one component of a system influence the other components. This allows the scientist to examine not only correlations among data, but also causality—that is, what causes what. Ideally, manipulative experiments alter only one variable at a time, so that the investigator can ascertain what changes yield what results.

In order to examine how empirical studies in evolution work, we will consider two examples: (1) a comparison of the human and chimp genomes, and what this can teach us about primate evolution, and (2) natural selection and how it has shaped sexual behavior in birds—specifically, sperm allocation strategies in chickens.

Molecular Genetics and Evolution in Chimps and Humans

More than 100 years ago, Darwin and his colleague, Thomas Henry Huxley, hypothesized that humans share a common ancestor with the great apes (chimpanzees, gorillas, and orangutans). Their evidence was primarily based on **comparative anatomy**. Darwin, and especially Huxley, made inferences about the evolutionary history of humans by comparing the anatomical similarities and differences seen between humans and other primates in such traits as tooth and jaw shape, bone structure of the hands and feet, mode of locomotion, and brain size and structure (Figure 1.11).

If Darwin and Huxley's hypothesis is correct—if the great apes are our closest living relatives—modern molecular genetics should also find evidence for such relationships. Indeed, it does. Evidence from molecular genetics provides strong support for Darwin and Huxley's hypothesis, with chimpanzees and bonobos (pygmy chimps) as our closest living relatives. Humans and chimps, for example, differ by one set of chromosomes, with humans having 23 pairs, and chimps 24 pairs. When high-resolution pictures are taken of human and chimpanzee chromosomes, and the size and structure of chromosomes in both chimpanzees and humans are compared, researchers can see that human chromosome 2 is the

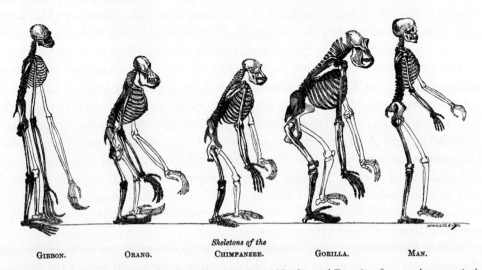

Skeletons of the

GIBBON.   ORANG.   CHIMPANZEE.   GORILLA.   MAN.

FIGURE 1.11 Huxley, Darwin, and primate evolution. Huxley and Darwin often used anatomical comparisons to infer the evolutionary history of humans and other primates. This example is from Huxley's *Evidence as to Man's Place in Nature* (originally published in 1863).

FIGURE 1.12 Primate chromosomes. From left to right for each set of chromosomes: The chromosomes of **(A)** humans, **(B)** chimpanzees, **(C)** gorillas, and **(D)** orangutans. Humans have one fewer pair of chromosomes as a result of the fusion of chromosomes 2p and 2q in chimpanzees (the second and third strands in the chromosome 2 panel).

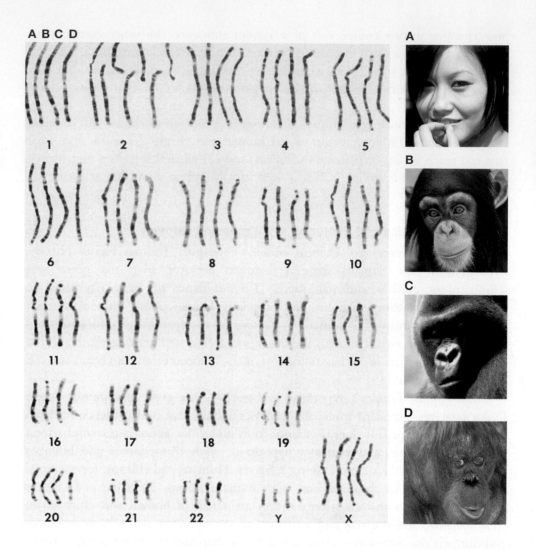

result of a fusion of two chromosomes at some point in human evolutionary history (Yunis and Prakash 1982) (Figure 1.12). Subsequent molecular genetic analyses, in which the DNA sequences from chromosome 2 in both chimps and humans were lined up and compared—nucleotide by nucleotide—has shown researchers the exact location where the chromosomal fusion occurred (Fan et al. 2002).

The entire genomes of both the chimpanzee and the human have now been mapped out in great detail. This allows us to make unprecedented molecular genetic comparisons to examine questions of primate evolution (Mikkelsen et al. 2005; Khaitovich et al. 2006). Tarjei Mikkelsen and his colleagues in the Chimpanzee Sequencing and Analysis Consortium mapped out approximately 95% of the chimpanzee genome (from eight chimpanzees) and compared that with the genome of a small set of humans. A whole genome comparison of DNA nucleotides found that humans and chimps differ by about 1.3%, although comparisons of specific sections of the genome show that DNA **sequence divergence** is greater in some areas and lower in others (Figure 1.13).

When Mikkelsen's group compared 13,454 pairs of genes in humans and chimpanzees, they began by calculating how much we would expect the human and chimp genomes to diverge due to the accumulation of **neutral mutations**—that is, changes that would have no effect on fitness. This served as a baseline value

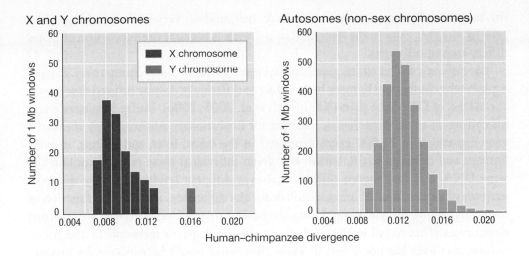

FIGURE 1.13 Human–chimp divergence rates. Human–chimp divergence rates across 1 Mb (1 million nucleotide) areas of the human and chimp genomes. Divergence is generally low, but varies across locations. Adapted from Mikkelsen et al. (2005).

that accounted for differences between the human and chimp genomes that were not due to natural selection.

Once neutral genetic differences were accounted for, Mikkelsen and his colleagues could search for evidence of divergence between chimps and humans that was due to natural selection. To do this, they examined whether some genes changed at higher rates than expected for neutral genes. When they found such alleles, Mikkelsen and his team could often correlate these increased rates of divergence with known functions of the alleles in question. This type of analysis found evidence for rapid evolutionary changes as a result of natural selection. These included genetic changes in humans associated with resistance to a bacterium that causes tuberculosis and a protozoan that causes malaria.

Mikkelsen and his colleagues used the same approach outlined above to compare *clusters of genes* in humans and chimps. That is, they again calculated the rate of divergence between humans and chimps expected due to neutral genetic change, and then they searched for evidence of divergence that is above that rate as evidence for natural selection; however, this time they used this approach for clusters of genes rather than single genes. In addition, they compared the rates at which clusters of genes have been evolving in both humans and chimps compared to other mammals whose genomes have been sequenced. This analysis revealed that natural selection has been acting strongly on both human and chimp genes in gene clusters associated with both survival and reproduction. Gene clusters associated with resistance to disease are evolving rapidly, as are gene clusters linked with reproductive traits such as sperm production, and production of various proteins during pregnancy. Understanding such evolutionary changes has implications for many medical issues, including maternal health and male infertility.

If chimpanzee and human genomes differ by only about 1.3% at the level of DNA base pairs, then how can we explain the dramatic differences in appearance and behavior between humans and chimps? Because these genomes have only recently been sequenced, and the amount of data cataloging tens of thousands of genes is astronomically large, we are just beginning to answer this sort of question. Progress is already evident, as researchers have found that important differences between humans and chimps may stem from the *expression* of genes. To understand the power of **gene expression**—which genes are turned on and off, and the timing of when they are turned on and off—remember that every cell in your body has the same set of genes, but skin cells look, feel, and do very different things than cells in muscles, cells in the liver, and so on. This is because the expression of genes differs among cell types.

The different way in which genes are expressed in humans and chimps may, in part, explain why chimps and humans look and act so differently, despite limited divergence at the level of DNA base pairs (Khaitovich et al. 2005, 2006). Philip Khaitovich and his colleagues at the Max Planck Institute for Evolutionary Anthropology measured gene expression in 21,000 genes expressed in the heart, liver, and kidney, in both humans and chimps. Again, within any given individual these cells all contain the same DNA, but they express different genes at different levels. It is these different patterns of expression that are responsible for the differences in form and function of liver cells, brain cells, and so on. Using the basic statistical approach that we outlined above researchers found evidence that suggests that gene expression in the heart, kidney, and liver has not diverged more than what would be expected by chance. On the other hand, a much higher divergence rate, and much stronger evidence for natural selection, was found when Khaitovich and his colleagues compared gene expression in the cells of human and chimp testes. Divergence in gene expression in the testes is likely a result of the very different **mating systems**—the way in which reproductive behaviors are structured in a population—seen in humans and chimps (Harcourt et al. 1981; Kappeler and van Schaik 2004).

Khaitovich and his team (2006) also examined gene expression in the brains of humans and chimps. Here the results were surprising. Given the evolution of language and other cognitively sophisticated traits in humans, we might expect high divergence in gene expression in the brains of humans and chimps (Dorus et al. 2004). Yet, this was not the case. Indeed, divergence in gene expression between the brains of humans and chimps was quite small compared to differences in gene expression in other organs. There is, however, some subtle evidence that natural selection has operated on gene expression in the brain during human evolution. Although the divergence in gene expression in the brains of humans and chimps was low, much of the difference that did exist appears to be due to natural selection on humans, not chimps, suggesting selection for brain function in humans relative to other primates. When Khaitovich and his team compared gene expression in *both* humans and chimps to gene expression in other mammalian species, they found evidence that, although there were relatively few changes in gene expression in the brains of humans versus chimpanzees, the changes that had occurred were large in magnitude and were more often due to changes in the human brain than the chimp brain. This result highlights a question that has been central to evolutionary biology since the time of Darwin, and that we are just starting to answer using studies in evolutionary genomics: Does major evolutionary change occur as a result of a large number of mutations with modest effects, or a small number of mutations that have large effects?

Sperm Allocation in Chickens

While genetic and molecular studies can reveal a great deal about the evolutionary process, we need not restrict our analysis of evolutionary questions to the molecular genetic level, as in the case of our comparison of human and chimp genomes. Evolutionary processes can be studied at levels far removed from the nucleotides that make up DNA. Indeed, much work on evolution by natural selection has examined behavioral traits—traits that are sometimes very difficult to trace back

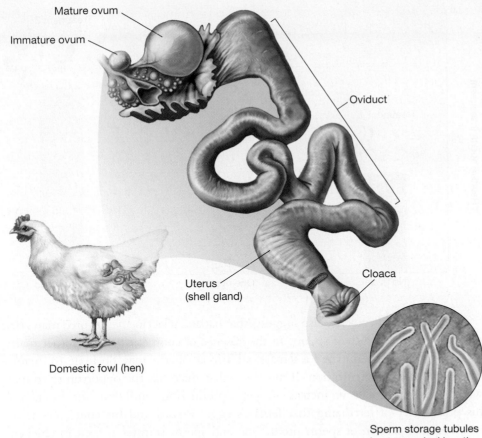

Mature ovum

Immature ovum

Oviduct

Uterus (shell gland)

Cloaca

Domestic fowl (hen)

Sperm storage tubules in uterovaginal junction

FIGURE 1.14 The reproductive system of chickens. Sperm storage occurs in the sperm storage tubule (SST) at the uterovaginal junction. Females can store sperm from multiple matings, and only a small proportion of sperm make it into the SSTs. Adapted from Birkhead and Møller (1992).

to the action of a particular gene, or set of genes. To see this, let's examine natural selection and one component of sexual behavior in birds.

In chickens (*Gallus gallus*), females mate with many males and can store male sperm; males also mate with many females (Figure 1.14). As a result, sperm from many males often compete for access to a female's eggs. This raises an interesting evolutionary question: namely, what sorts of male behavior has natural selection favored so as to maximize the probability that their sperm will be successful in fertilizing females? To answer that question, we will focus on sperm allocation strategies—the amount of sperm transferred during a particular act of mating (Parker 1970; Smith 1984; Birkhead and Møller 1998; Simmons 2001; Wedell et al. 2002).

Tommaso Pizzari and his colleagues examined how chickens allocate sperm by assigning male birds to one of three experimental treatments (Gage 2003; Pizzari et al. 2003). In one treatment, a lone male was placed together with a female. In a second treatment, a male and female were again placed together, but now an additional male competitor was present. In the last treatment, three additional male competitors were placed together with the original male subject and the female subject (Figure 1.15).

Because **sperm competition** within the female reproductive tract should increase when the number of males competing for mating opportunities increases, Pizzari and colleagues hypothesized that males—especially, aggressive dominant males—would respond to increased competition by transferring more sperm per copulation. Their results indicate that this is indeed the case: As the number of male competitors increased, the quantity of sperm a male used to inseminate a female increased.

FIGURE 1.15 Pizzari's sperm competition experiment. An outline of the experimental protocol used to examine sperm allocation strategies in chickens. In Treatment 1, a single male and female were placed together. In Treatment 2, a second male was added, and in Treatment 3, three additional males were added. Average sperm investment of dominant males increased as a function of the number of competitors a male had. The vertical axis is scaled against the largest sperm output produced by a given male. Adapted from Pizzari et al. (2003).

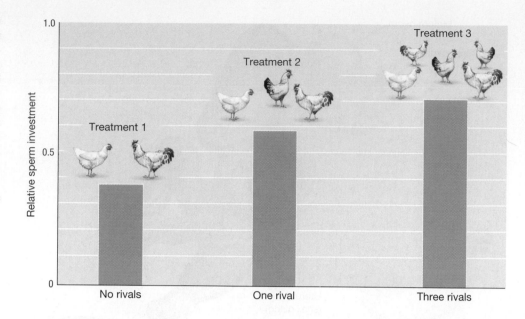

A second line of evidence also suggests that natural selection has favored males that strategically allocate their sperm. In the absence of competition from other males, once a male has inseminated a female, additional sperm may do little to increase the probability of fertilization. If, however, that male has the opportunity to mate with a different female, an increase in sperm production could dramatically increase his probability of fertilizing that female's eggs. Pizzari and his colleagues tested whether this pattern of sperm production and sperm transfer is seen in chickens. They found that in the absence of competition, a male inseminates a female (let's call her H1) with fewer and fewer sperm per copulation. But this is not simply a case of a male running out of sperm. If the same male is suddenly placed together with a new female (H2), the number of sperm transferred during copulations with H2 rises dramatically, as compared to sperm levels during the last copulations with H1. Taken together, the Pizzari studies indicate that natural selection has favored males that strategically allocate sperm so as to maximize the number of eggs they fertilize.

These two studies offer just a glimpse of how researchers investigate the evolutionary process and its consequences. There are literally tens of thousands of observational and experimental studies of evolution in the literature. For the time being, however, let us move on to the next tool in the evolutionary biologist's toolbox—theoretical approaches.

Theoretical Approaches

In evolutionary biology, theory plays an important role in shaping and furthering the research agenda of the field. Theoretical biology often, but not always, involves creating mathematical models of biological systems (Godfrey-Smith 2006). In evolutionary biology, as in science more broadly, mathematical models are used for many different purposes. At the most general level, models help us understand how complicated systems work. A good model does this, in part, by making assumptions that allow us to focus on only the critical details of a system, so we can understand how that system operates. Once we do this, we can use our model to make predictions and inferences.

Throughout the sciences, one of the most common uses of models is to make predictions and plan for the future. When we check a weather report, we are relying on a set of models of weather patterns to help us predict what the weather will be like tomorrow and to enable us to make sensible decisions about what to wear and whether to bring along an umbrella. Models from evolutionary biology can be used similarly. For example, when conservation biologists design captive breeding plans for highly endangered species, they use population genetic models (Chapters 7–9) to ensure that they are able to preserve sufficient genetic variation for the species to remain viable.

Another common use of models is to make inferences. Models of processes that we understand in detail help us use observable patterns to infer information that is more difficult to observe directly. When a policeman clocks the speed of a motorist using a radar gun, he is not measuring speed directly. Rather, he is measuring the Doppler shift in radio waves emitted by the gun, as they bounce off of the target automobile. The radar gun then uses a simple mathematical model to compute a motorist's speed from the observed Doppler shift. When evolutionary biologists estimate fitness by measuring the change in allele frequencies over time, they are doing something similar: They are using a mathematical model to connect the observable changes in gene frequencies to the less easily observed differences in fitness (we discuss this in more depth in Chapter 7). Similarly, whenever we infer phylogenetic trees from genetic data, we are applying a model of how genetic sequences change over time to observed gene sequences, in order to make inferences about evolutionary history (Chapter 5).

For a better sense of how evolutionary biologists employ models in their work, let's consider the evolution of the **sex ratio**.

Why Is There an Even Sex Ratio?

In this section, we will use a model to address a simple but far-reaching question: Why do so many species—humans included—exhibit an approximately even sex ratio of one male to one female? Is it a consequence of natural selection? While we are so accustomed to this aspect of the world that it may be hard to imagine things any other way, it is not at all obvious that natural selection should favor a so-called even sex ratio. For example, in most species a single male can fertilize numerous females, and often males provide nothing toward the care of the offspring. Why could there not be an excess of females in these species, so that the sex ratio differs from one male to one female? For mammals, one answer to this question lies in the mechanics of our **chromosomal sex determination**. Females have two X chromosomes. Males have an X and a Y chromosome; during meiosis, these segregate evenly to produce 50% X-bearing sperm and 50% Y-bearing sperm. As a result, roughly half of the fertilized embryos are XX females and half are XY males, producing an even sex ratio in zygotes. But the evolutionary question for us is this: Why has this sort of system evolved instead of some system that produces a different sex ratio? And why do species with other sex determination systems also commonly exhibit sex ratios near 1:1?

Using a simple model of sex ratios, first hypothesized by Darwin and then fully developed by R. A. Fisher in 1930, we can see clearly why natural selection usually favors an even sex ratio (Fisher 1930). The key to developing a useful model of this type is to find a way to express all of the important features relevant to the problem, while removing as many unimportant details as possible. The challenge and the art of modeling are determining what features one needs to retain, and what details may be safely omitted.

Let us look at what Fisher chose to include in his model, and what he chose to omit. He envisioned a sexually reproducing species, but he did not specify the details of its diet, habitat, life span, and so forth. At a first approximation, these are likely irrelevant to the sex ratio problem that he was trying to answer; after all, most species have an even sex ratio irrespective of their diet, habitat, and life span. He then assumed that sex ratio is under genetic control. This is an important assumption; if sex ratio were not under genetic control, it could not evolve by natural selection. Fisher assumed that parents can influence the sex ratio of their offspring. He also assumed that the fitness of a male depends on the frequency of males in the population, and similarly the fitness of a female depends on the frequency of females. And finally, Fisher realized that when parents influence the sex ratio of their offspring, their actions are manifested not in the survival of their offspring, but rather in the reproductive success of their offspring. This is because by altering the sex ratio of their offspring, individuals are not affecting the *number of young* they produce, just the proportion of males versus females. Thus, we cannot measure the benefits in the first generation by directly counting the number of offspring. Instead, we have to measure the benefits in the second generation by counting the number of surviving grandchildren.

Given this imagined population and method for assessing fitness, evolutionary biologist William D. Hamilton clearly summarized Fisher's basic conceptual argument:

1) Suppose male births are less common than female.
2) A newborn male then has better mating prospects than a newborn female, and therefore can expect to have more offspring.
3) Therefore parents genetically disposed to produce males tend to have more than average numbers of grandchildren born to them.
4) Therefore the genes for male-producing tendencies spread, and male births become commoner.
5) As the 1:1 sex ratio is approached, the advantage associated with producing males dies away.
6) The same reasoning holds if females are substituted for males throughout. Therefore 1:1 is the equilibrium ratio. (Hamilton 1967, p. 477)

This is a purely conceptual way to think about the evolution of an even sex ratio, and the logic Hamilton invokes is powerful. But we can also construct a simple mathematical model to check our intuition. We present such a model in Box 1.1.

Testing the Sex Ratio Model—A Rapid Change of Sex Ratio

As we discussed, models allow us to simplify a complex reality and thereby make useful predictions about what should happen under specific circumstances. We can then test such models through observational or manipulative experiments. One of the predictions that Fisher's sex ratio model makes is that if the sex ratio should deviate from 1:1, natural selection will strongly favor genetic changes that restore an even ratio. Thus, when the sex ratio becomes unbalanced, we expect a rapid return to a 1:1 ratio. This prediction was put to the test in a species of butterflies on the adjacent Samoan islands of Upolu and Savaii (Charlat et al. 2007). In 2001, 99% of the blue moon butterflies (*Hypolimnas bolina*) on Upolu and Savaii were

BOX 1.1 A Mathematical Model of the Sex Ratio

Imagine a population of sexually reproducing organisms. Let us suppose that there are m adult males and f adult females in this population. For simplicity, assume that these individuals live for a single year, reproduce at the end of their lifetimes, and then die. Let N be the number of offspring produced annually in this population. Our model, so far, contains just three variables: m, f, and N.

Regardless of the sex ratio, each offspring in our population has a mother and a father. This may seem obvious, but for our purposes it tells us something important—the total reproductive success of the males in the population must be equal to the total reproductive success of the females in the population. In other words, the total reproduction, N, is shared among f females and m males. On average, each male therefore has N/m offspring and each female has N/f offspring.

Suppose a parent produces offspring such that a fraction k are sons and the remaining fraction, $1 - k$, are daughters. How many grandoffspring will that parent have? On average, that parent will have

$$k\frac{N}{m} + (1 - k)\frac{N}{f}$$

grandoffspring per child. The first term in this expression represents the number of grandoffspring produced by male offspring, and the second represents the number of grandoffspring produced by female offspring.

When there are more females than males—that is, when $f > m$—parents with high k values will have more grandchildren than parents with lower k values. Under this condition, natural selection favors parents who produce more males, and the sex ratio moves toward 1:1. Conversely, when $f < m$, parents with low k values will have more grandchildren than parents with higher k values. Now selection favors parents who produce more females, and again, the sex ratio moves toward 1:1. What this model demonstrates is that, as Fisher surmised, natural selection drives the sex ratio to an even 1:1 ratio.

A numerical example helps illustrate the model. Imagine a population with more males than females: there are $m = 25$ males and $f = 20$ females, and they produce a total of $N = 100$ offspring. In this case, the average number of offspring produced by a male will be $100/25 = 4$, whereas the average number of offspring produced by a female parent will be $100/20 = 5$. Now suppose that a parent produces half sons and half daughters ($k = 1/2$). The average number of grandoffspring will be $0.5(4) + 0.5(5) = 4.5$ grandoffspring per child produced. Suppose instead that a parent produces all daughters ($k = 0$). Now the average number of offspring per child will be $0(4) + 1(5) = 5$. Thus, in a population with an excess of males, parents will have more grandoffspring when they produce extra daughters. As time passes, the excess of daughters means that their average fitness declines, favoring parents who produce extra sons.

---

female, and only 1% were male (Figure 1.16A). This extreme sex ratio bias was the result of male mortality due to a widespread infection by the *Wolbachia* bacterium. This infection has the curious effect of killing most males as embryos, while leaving females unharmed.

Fisher's model predicts that if there were to arise a genetic variant of the blue moon butterfly that produced as many males as females, despite infection by *Wolbachia*, this variant would spread rapidly. As this variant spread, the sex ratio would approach 1:1. This is exactly what happened on Upolu. Sometime after 2001, such a mutant arose on Upolu (or arrived from another island). By 2006, the sex ratio among *Hypolimnas* butterflies on the island had returned to approximately 1:1. Even though female butterflies on the island were still infected with the same variety of *Wolbachia* as in 2001, they now produced as many surviving males as females. On the nearby island of Savaii, sex ratios were returning to 1:1 as well. In one population, on the side of Savaii that was nearest to Upolu, males actually outnumbered females among the offspring followed by the experimenters. Only on the far side of the island of Savaii was the sex ratio still strongly female biased. Sex ratio theory predicts a return to 1:1 on this side of the island as well, once migrants capable of generating the 1:1 sex ratio arrive and spread there (Figure 1.16B).

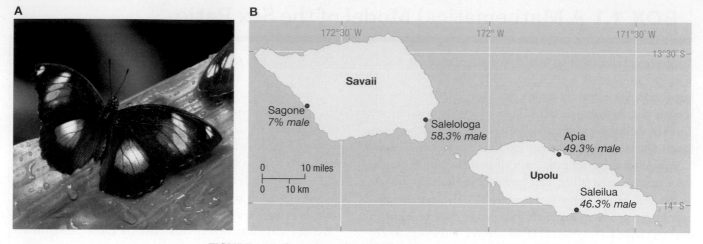

FIGURE 1.16 Sex ratios of butterflies on the Samoan islands. (A) The blue moon butterfly, *Hypolimnas bolina*, was used to test sex ratio theory on the Samoan islands of Upolu and Savaii. **(B)** Sex ratios of the blue moon butterfly across different islands. Although butterflies remain infected with *Wolbachia*, rapid evolution in the butterflies returned the sex ratio back to close to even in all sites except Sagone.

How do we know that the shift back to an even sex ratio was the result of genetic changes in the butterfly, and not the bacterium? To test this, Sylvian Charlat and colleagues extracted the *Wolbachia* bacterium from the offspring of Samoan females who produced an even sex ratio (Charlat et al. 2007). They introduced these bacteria into captive blue moon butterflies from the island of Moorea (near Tahiti). After they were infected by the Samoan *Wolbachia* strain, the Moorean butterflies produced only female offspring. That is, the tendency for Samoan *Wolbachia* to kill male butterfly embryos had not changed. From this, Charlat was able to conclude that evolutionary change had occurred in the Samoan butterflies, and not the Samoan *Wolbachia*, and that this involved the evolution of a gene that allowed the butterfly to suppress the male-killing effect of *Wolbachia*. This drastic change in sex ratio—from 99% female to approximately 50% female in only 5 years, or 10 generations for the butterflies—illustrates both the predictive power of Fisher's model and the speed with which evolution by natural selection can change the characteristics of a population.

Theory and Experiment

In the case of sex ratio evolution, Fisher's mathematical theory was the impetus for many subsequent experiments. Yet, Fisher developed his model in part because so much observational data suggested that the 1:1 sex ratio was common in nature, and he wanted to understand why that was. This raises a series of general questions: Is there any natural ordering when it comes to empirical and theoretical approaches? Does theory come before or after empirical work? The answer is, "It depends." Good theory can either precede or postdate data collecting and hypothesis testing. On some occasions, an observation or experiment will suggest to a researcher that a model should be developed. On the other hand, theory can also precede, encourage, and facilitate experimental research. Regardless of whether theoretical work predates or postdates empirical work, a powerful feedback loop typically emerges wherein advances in one area—either theoretical or empirical—lead to advances in the other area.

In this chapter, we have skimmed the surface in terms of understanding how evolution operates. In order to understand the details of evolutionary biology, however, we need to examine the historical context in which the discipline developed. And so in Chapter 2, we will explore some of the ideas that existed before Darwin revolutionized the study of biology, before proceeding to treat Darwin's insights.

SUMMARY

1. Charles Darwin's theory of evolution by natural selection produced a paradigm shift in the life sciences.

2. In *On the Origin of Species*, Darwin presented two revolutionary ideas: (a) the wide diversity of life we see around us has descended from previously existing species, which share common ancestry, and (b) the present forms of these species are a result primarily of natural selection, a process in which forms that are better suited to their environment increase in frequency over time.

3. Evolutionary biologists infer the causes of ancient events, and develop and test hypotheses through a combination of observation and experimental manipulations.

4. Artificial selection by humans is the counterpart to natural selection. Humans select which individuals get to reproduce by choosing those that possess traits that are beneficial to us, which changes the phenotype of domesticated varieties over time.

5. Practical applications of understanding evolution via natural selection include, but are not limited to, controlling resistance to insecticides and antibiotics, as well as using evolutionary principles to address problems in conservation biology and the medical sciences.

6. All species that have ever lived form a vast branching tree of evolutionary relationships known as the tree of life.

7. Theory plays an important role in shaping and furthering the research agenda in evolutionary biology. Models can be employed both to make predictions and to use observable patterns to infer information that is more difficult to observe directly.

KEY TERMS

antibiotic resistance (p. 8)

artificial selection (p. 7)

chromosomal sex determination (p. 21)

comparative anatomy (p. 15)

descent with modification (p. 5)

evolution (p. 5)

evolutionary radiation (p. 13)

extinction (p. 12)

fitness (p. 6)

gene expression (p. 17)

major transitions (p. 6)

mating systems (p. 18)

mutation (p. 6)

natural selection (p. 4)

neutral mutations (p. 16)

phenotype (p. 6)

phylogenetic diversity (p. 13)

phylogenetic tree (p. 11)

selective breeding (p. 6)

sequence divergence (p. 16)

sex ratio (p. 21)

speciation (p. 13)

sperm competition (p. 19)

tree of life (p. 11)

REVIEW QUESTIONS

1. Can you think of another paradigm shift that has occurred in biology in the last 100 years? Make your case for why this shift has fundamentally changed the way that scientists see the world and the sorts of questions they ask.

2. How has artificial selection been used to shape the looks and behavior of farm animals such as the cow, and family pets such as the domesticated dog?

3. In addition to the arms race that we discussed with respect to pesticide resistance and antibiotic resistance, can you describe another such evolutionary arms race that has practical applications? Describe this arms race. Hint: Think about viral or fungal infections.

4. Fisher's sex ratio model predicts a 1:1 female:male sex ratio. But this model assumes that the cost to a parent for producing and providing for a female offspring is equal to the cost to a parent for producing and providing for a male offspring. Suppose that this is not the case.
   a. Consider a case where each male offspring is twice as expensive to produce and raise to maturity as each female offspring. Build a mathematical model to predict the expected sex ratio (female:male) in such a population.
   b. Suppose that in a population you are studying you find that the sex ratio is different from what you predicted in your model. Using simple proportions of male and female offspring, show how, over the course of a few generations, natural selection should move the sex ratio to the value you predicted in your model.
   c. What are some of the ecological or behavioral reasons that male offspring might be more expensive to produce and raise to maturity than female offspring? Why might female offspring be more expensive?

5. We have seen how a comparison of the chimp and human genomes has shed light on evolutionary (as well as medical) questions. Make the case that, as researchers sequence more and more representative genomes from across the tree of life, we can use phylogenetic methods to better reconstruct the evolutionary history of our own species.

6. How does the sex ratio study on butterflies demonstrate that important work in science often involves a sound theoretical base, good observational and experimental skills, and often a bit of *luck*?

7. We touched on how evolutionary thinking can affect the way that conservation biologists think about extinction. Can you think of other ways that evolutionary thinking might impact studies in conservation biology?

8. Sometimes Dobzhansky's quote "nothing in biology makes sense except in the light of evolution" is misinterpreted to mean that evolutionary biology is the most important subdiscipline in biology. Why would this interpretation be a mistake? What do you think was the key point Dobzhansky was trying to make?

9. Take a look at Figure 1.1 in which we show some of the types of data that evolutionary biologists collect to both generate and test hypotheses. Can you think of other types of data that would be useful for evolutionary biologists to collect? Make a short list and justify each point on that list.

10. Good theory can either precede or postdate data collecting and hypothesis testing. Describe an example from a discipline of your choice in which an observation or experiment prompted the development of a model, and a second example in which already existing theory encouraged experimental research.

SUGGESTED READINGS

Birkhead, T. R., and T. Pizzari. 2002. Postcopulatory sexual selection. *Nature Reviews Genetics* 3: 262–273. This paper will give you a better understanding of the sperm competition and sperm allocation work we discussed in this chapter.

Engelstädter, J., and G. D. D. Hurst. 2009. The ecology and evolution of microbes that manipulate host reproduction. *Annual Review of Ecology, Evolution and Systematics* 40: 127–149. A detailed review on issues we discussed in the *Wolbachia*/blue moon butterfly sex ratio example.

Huxley, T. H. 1863. *Evidence of Man's Place in Nature*. D. Appleton, New York. Huxley—Darwin's close friend—presented evidence for human evolution in this book.

Kuhn, T. 1962. *The Structure of Scientific Revolutions*. University of Chicago Press, Chicago. In this volume, a classic in the philosophy of science, Kuhn outlines the idea of a paradigm shift.

Varki, A., D. H. Geschwind, and E. E. Eichler. 2008. Explaining human uniqueness: genome interactions with environment, behaviour and culture. *Nature Reviews Genetics* 9: 749–763. An interesting discussion of how to understand what molecular genetic comparisons tell us (and don't tell us) about similarities and differences between humans and other primates.

 Visit StudySpace at wwnorton.com/studyspace.

a party ... a salt mine there ... are there ... 200 ft ... Terrapin could ... There ... in ... b had ... one was not ... at that been that ... to manage ... men ... away their ... I industriously collected insects & reptiles from this Island. — It be very interesting to find to what district the organized beings ... of this Archipelago be attached. ... I ascended highest hill ... 2000 ... grass

2

Early Evolutionary Ideas and Darwin's Insight

◀ Some of the Galápagos finch species that so fascinated Darwin on his voyage aboard the HMS *Beagle*. These museum specimens are arrayed on a copy of Darwin's research journal.

Long before the science of evolutionary biology was born, people contemplated both the origin of life and the fact that organisms often seem so well suited for the environments in which they live. More than two millennia ago, the Greek philosopher Empedocles (ca. 492–432 B.C.) proposed a theory in which body parts arose independently from the ground, describing organisms

> where many heads grew up without necks, and arms were wandering about naked, bereft of shoulders, and eyes roamed about alone with no foreheads. (Empedocles, Book II, 244, in Fairbanks 1898, p. 189)

These unattached parts then wandered Earth before reassorting, sometimes into monstrous combinations such as creatures with two faces and animals with human heads, and sometimes into the well-proportioned forms that we observe in the animal world. When we read of such theories, we need to be careful not to fall into the trap of judging them based on what we know today. At the time, Empedocles was making a serious attempt to understand the origin of animals. He *might* have been correct, he just wasn't; but most ideas turn out to be wrong over the long run.

Empedocles' ideas did more than suggest how animal life originated: They also provided an explanation for why organisms seem to be so well adapted to their environments. Empedocles argued that if individuals were assembled from parts that were unable to function together to reproduce, they became extinct. Without turning to supernatural intervention, Empedocles proposed a theory that explained not only why we observe an incredible diversity of living forms, but also why the component parts of each species tend to be well suited to one another and to the species' habitats.

Empedocles and his ideas remind us that science has a rich and deep history. Sir Isaac Newton, the great physicist and mathematician, wrote in 1676 that if he had seen farther than others, it was only "by standing on the shoulders of giants." Therein lies the tremendous power of the scientific approach. On the one hand, scholars can build on decades, or even centuries, of previous work without needing to reinvent every step themselves. On the other hand, each of these previous discoveries or theories remains continually open to challenge, revision, and reinterpretation based on new evidence. Like all other great scientific ideas, Darwin's theory of evolution by natural selection did not arise in a vacuum. Instead, the idea of natural selection as a process in which forms that are better suited to their environment increase in frequency in a population emerged from a rich philosophical and scientific tradition that came before it.

Given that many theories from this pre-Darwinian tradition have since been discredited, why should a contemporary biologist study these ideas about evolution? Why pause in assessing the view from our time to look back at the giants that came before us?

We study the past to improve our work in the present. We hone our own scientific thinking by following the reasoning that led to both correct and incorrect conclusions, and we come to appreciate the intellectual risks that sparked the theories that we now take for granted. We learn from the giants that came before us to be flexible in our current thinking. Exploring the debates underlying our assumptions reminds us to question our understanding and to approach contemporary problems from new angles.

And so, before investigating Darwin's theory and the developments in biology that have followed from it, we will examine the ideas about the nature of the biological world that preceded the publication of *On the Origin of Species* in 1859. The first part of this chapter will serve as an introduction to how *pre-Darwinian thinkers* answered the big questions about life and biology, including these:

- What separates science from mythology?
- How should scientists reach conclusions about the natural world?
- How does the natural world change, and over what length of time?
- Why is the world filled with an astonishing diversity of living forms, instead of a few basic types?
- Where do species come from?
- Why are organisms well suited to the environments in which they live?

Once we have tackled these questions, in the second part of the chapter we will introduce Darwin's ideas on the evolutionary process.

We will begin by briefly addressing what separates science from mythology, and we will discuss what sorts of explanations scientists can pursue.

2.1 The Nature of Science: Natural versus Supernatural Explanations

Throughout recorded history, every human culture has cultivated a set of creation myths that purport to explain—literally or metaphorically—how the world came to be the way that it is. These mythologies address universal questions that stimulate the human imagination and gratify our need for explanations of our place in the world. Prior to the sixth or seventh century B.C., these creation myths provided the only answers that humankind had to the grand questions of our existence (Armstrong 2005). This approach to knowledge through mythmaking began to change with the early Greek philosophers.

FIGURE 2.1 Anaximander (ca. 610–546 B.C.). Anaximander proposed a mechanistic view of the Earth and heavens. The philosopher is illustrated in the 1493 history of the world, *The Nuremberg Chronicle.*

Methodological Naturalism

Although the early Greeks had their own creation myths, philosophers such as Anaximander (ca. 610–546 B.C.) (Figure 2.1) were among the first to develop a philosophy of a natural world driven by physical laws to replace a supernatural world driven by divine action. They sought to explain the world around them according to fixed laws of nature, rather than by the operation of divine whim.

At a time when heavenly bodies were regarded as divine personages, Anaximander provided a mechanistic rather than divine conception of the moon, sun, and stars. He suggested that just like the earthly structures of our common experience, the celestial bodies were physical objects (Figure 2.2). Earth, he proposed, was a cylindrical disk. The sun and the moon rotated around it as if on wagon wheels. Beyond the sun and the moon, tiny holes in the firmament let through the light from a vast dome of fire; these pinpoints of light were the stars. Anaximander got the details wrong, but given the state of scientific knowledge at the time, this is to be expected. The important component of these ideas is that Anaximander and some of the Greek philosophers who followed him developed explanations based on natural, rather than supernatural, phenomena.

The strategy of trying to explain the world based solely on natural phenomena is fundamental to the scientific method, and is at the heart of modern evolutionary biology. It is sometimes called **methodological naturalism**. We call it "naturalism" because of the focus on the natural rather than the supernatural. We use the adjective "methodological" because this strategy provides a method or procedure for seeking scientific explanations of the world. Although philosophers began using methodological naturalism as early as 600 B.C., this approach would not be solidified or universally embraced until the eighteenth century.

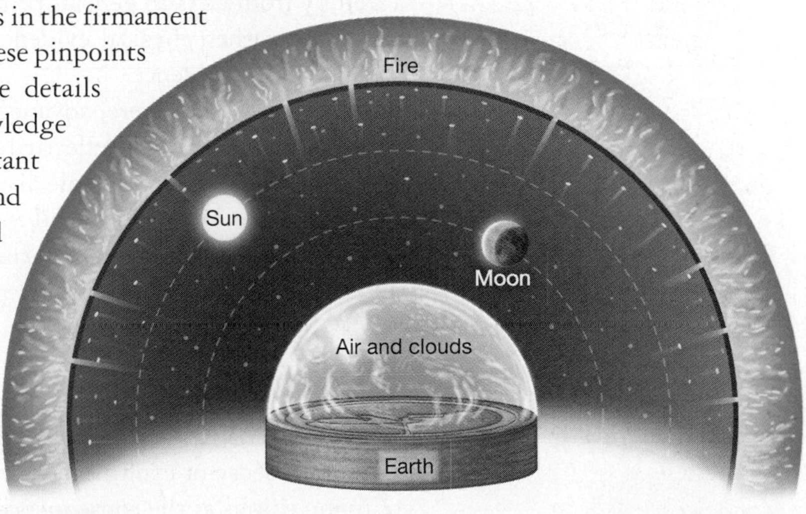

FIGURE 2.2 Anaximander's cosmology. In Anaximander's cosmology, Earth is a disk surrounded by vast wheels on which the sun and moon rotate and a dome of fire from behind which the stars glow.

Hypothesis Testing and Logic

Although they were able to make the shift from supernatural to natural explanations, the early Greek philosophers failed to exploit one of the greatest advantages of methodological naturalism: hypothesis testing. If we explain a phenomenon based

FIGURE 2.3 Aristotle (ca. 384–322 B.C.). The Greek philosopher Aristotle wrote, "We must not accept a general principle from logic only, but must prove its application to each fact; for it is in facts that we must seek general principles, and these must always accord with the facts."

on natural processes, we can then test our explanation, because we can observe and often manipulate these processes. By contrast, we have no way to observe, let alone manipulate, the supernatural, and thus we cannot test supernatural explanations. However, the early Greeks formulated **hypotheses** without refining them through testing. This lack of verification for ideas would begin to change with the great philosopher Aristotle (ca. 384–322 B.C.) (Figure 2.3).

Unlike those before him, Aristotle recognized the significance of testing one's hypotheses. In his *Natural History of Animals*, Aristotle was clear that "we must not accept a general principle from logic only, but must prove its application to each fact; for it is in facts that we must seek general principles, and these must always accord with the facts" (Aristotle, Book 1, p. 6, cited in Osborn 1894). In other words, principles must agree with the facts. If not, we need to rethink our principles and start over. This sort of approach is well accepted by modern evolutionary biologists, and for this we can thank Aristotle and those who followed in his footsteps. Of course, this approach did not take hold overnight, and even Aristotle did not always follow the practice he preached. In the very same volume where he advocated checking principles against the facts, Aristotle incorrectly asserted that men have more teeth than women. Philosopher Bertrand Russell famously remarked that "Aristotle maintained that women have fewer teeth than men; although he was twice married, it never occurred to him to verify this statement by examining his wives' mouths" (Russell 1952, p. 7).

After Aristotle, one advance in scientific methodology came through the use of logic. Application of logical and mathematical laws allowed thinkers to move carefully from facts to general principles. In modern evolutionary theory, not only must one gather physical evidence, but one must formulate and test hypotheses based on such evidence.

Profound as they were, advances in methodological naturalism and logic alone would not prepare the intellectual framework necessary for eventual breakthroughs in evolutionary theory. People also needed to become accustomed to the idea of a world that was both *ancient* and *ever-changing*. In the next section, we will examine historical conceptions of the nature of change, of the timescale for such changes, and of the sources of evidence for past changes.

2.2 Time and a Changing World

Darwin's theory of evolution by natural selection explains the form and diversity of living things as the consequence of gradual change over vast periods of time. As we will see in this section, Darwin was not the first to propose this idea, but the notion of change and huge expanses of time arrived late in the history of Western thought. This view was not the dominant one during most of Western history.

The view of the world as unchanging seems counterintuitive to anyone who has watched a storm roll in, a child grow up, or a candle burn. Yet, some Greek philosophers claimed that everything that exists has always existed and will always exist. The material world was permanent, unalterable, and unmoving. Even Aristotle, although he recognized change over small timescales, thought of the world as static and unchanging over longer periods of time. On the other hand, Empedocles (Figure 2.4) recognized that, historically, plant life preceded animal life, and Xenophanes (570–470 B.C.) studied fossils in sedimentary rocks in the mountains and concluded that at one time the rocks must have been under water.

The ideas of both Empedocles and Xenophanes implied that important changes in the biological world had occurred. What sorts of changes had occurred, however, remained contentious for nearly 2000 years. Indeed, until the work of French natural historians Buffon and Cuvier in the eighteenth century, the idea that species had gone extinct was thought of as an absurd challenge to the notion of a flawless Creator.

Even if philosophers accept and study the importance of change, a full theory of evolution by natural selection cannot exist without an understanding of the vast expanses of time over which some changes take place. That would not come for almost 2000 years following these early conjectures by the Greeks. Along the way, in the late Middle Ages, the written records of the Bible provided a starting place for estimating the age of the Earth. Following similar endeavors by scholars before him, James Ussher (1581–1656), a seventeenth-century Anglican archbishop in Northern Ireland, performed complex calculations based on the Old Testament, and he concluded that the universe had been created on October 23, 4004 B.C. Though the precision of the date may sound ludicrous today, Ussher's attempt to date the creation of the world was part of a serious research tradition at the time (Gould 1991). Famous scientific contemporaries of Ussher made similar attempts—for example, Isaac Newton dated creation at 3998 B.C.

At the same time that Archbishop Ussher was making his calculations, a radical shift was taking place in the way that other scholars viewed time and history. Inspired by the vastness of space made clear with the invention of the telescope and the discovery of countless stars beyond those visible to the naked eye, thinkers looked to an equally vast expanse of time.

Scientists began to suggest that both the universe and Earth were much, much older than the thousands of years suggested by a literal interpretation of the Old Testament. In the latter part of the eighteenth century, Georges-Louis Leclerc, comte de Buffon (1707–1788), a French naturalist and writer, used physical laws about the rate at which objects as large as Earth both heat up and cool down to calculate the age of the Earth at between 75,000 and 2 to 3 million years (Buffon 1778; Roger 1997). Around the same time, James Hutton (1726–1797), a Scottish geologist, naturalist, and chemist, argued that geological evidence—the way that rock strata were aligned, the processes of erosion and sedimentation, and the fossil data—suggested that the world was inconceivably old (Hutton 1795; Repcheck 2003). Once the idea of a changing world and vast stretches of time became established, the question became this: How can we fully employ the power of observation and experimentation to understand change over immense periods of time? To do so, we require explanations that not only appeal to natural processes, but, more specifically, that appeal to natural processes that are ongoing and observable, or otherwise accessible to us. Historically, the method to do this emerged first in the field of geology, and from there migrated to the biological sciences. To see how, we need to examine the work of Scottish geologist Charles Lyell (1797–1875) (Figure 2.5).

Building on ideas first proposed by Hutton, Lyell aimed to explain Earth's geological features by appealing to the same geological processes currently observable, operating over very long periods of time. From this, Lyell came up

FIGURE 2.4 Empedocles (ca. 492–432 B.C.). Empedocles recognized that plant life came before animal life.

A

B

FIGURE 2.5 Charles Lyell (1797–1875) and uniformitarianism. (A) Lyell's theory of uniformitarianism helped pave the way for modern evolutionary thinking about the vast expanse of time. (B) Uniformitarianism posits that the slow process of erosion (left), when carried out over long stretches of time, can produce massive canyons (right).

with the title of his famous book, *Principles of Geology, Being an Attempt to Explain the Former Changes of the Earth's Surface, by Reference to Causes Now in Operation* (Lyell 1830). As we will see shortly, this approach, known as **uniformitarianism**, had a strong influence on Charles Darwin.

Uniformitarianism explained the geological features of Earth in a radically different way than did **catastrophism**, the common theory of the time. According to catastrophism, Earth's major geological features arose through sudden catastrophic, large-scale events, rather than through slow gradual change. Moreover, catastrophism posited that these catastrophic events often involve different forces than those that are currently operating. The shift from catastrophism to uniformitarianism was an important development not only for geology, but also for science as a whole.

Science attempts to relate natural processes to observable patterns. In the extreme catastrophic view, these processes are not themselves observable or subject to manipulative experiments, and they are not expected to occur again in the future, making it hard—but not impossible—to test hypotheses about how observed patterns have been generated. In the uniformitarian view, all of the processes that have generated the current geological patterns we see around us can themselves be observed in operation at present, providing scientists with much more power to test hypotheses.

While Lyell's work related directly to geology, his concept of change over time would also influence evolutionary biology. Darwin read Lyell's *Principles of Geology* while serving as captain's companion and ship's naturalist on the HMS *Beagle*, and he was profoundly affected by Lyell's ideas (Recker 1990). Prior to publishing *On the Origin of Species*, Darwin wrote three books on geology, each of which drew heavily on Lyell's work on uniformitarian change. And, as we will see later in this chapter, in many ways Darwin's ideas on the gradual changes associated with evolution by natural selection are biological interpretations of Lyell's uniformitarianist ideas on geological processes. The diversity of life on Earth, Darwin proposed, can be explained by mechanisms that are in operation today, acting over very long periods of time.

By the time Darwin began his work, the approach to scientific inquiry had changed from mythmaking and supernatural explanations to naturalism—a method built on an increasingly sophisticated system of hypothesis testing and reason. By explaining the dramatic features of Earth's geography through uniformitarianism, Lyell conceived the world as changing across enormous expanses of time. In the next section, when we explore theories of how new species come into existence, we will see that both uniformitarianism and the concept of deep time (vast periods of time) were essential in understanding the origins of the diversity of organisms on Earth.

2.3 The Origins of Life and Its Diversity

In addition to taking the first steps toward the scientific method and hypothesizing about events from the past, the Greek philosophers also developed a keen appreciation for the study of natural history. Again, Aristotle's contributions were exceptional. With Aristotle's books *Physics* and *Natural History of Animals*, the field of **natural history** was born—an enterprise that would be important for the development of any theory of the astonishing diversity of life, whether that theory was evolutionary or not (Schneider 1862).

Aristotle distinguished among 500 species of birds, mammals, and fishes, and he wrote entire tracts on the anatomy and movement of animals. He also proposed a taxonomy of nature—a classification system of life—that led from polyps to the existence of humans. This would later be called "the great chain of being," or *scala naturae.* According to this linear classification system, each species occupied a link in a chain of ever-increasing complexity. This concept influenced Western thinkers for over 2000 years. While this view of nature contributed to the sense of the diversity of life, it was missing two critical concepts that were necessary for the development of evolutionary biology: shared degrees of complexity and the potential to change. On the *scala naturae*, every organism represented a specific and unique link in the chain, and each link represented a different level of complexity, which meant that different organisms could not share comparable degrees of complexity. Likewise, in this view, each specific link on the chain of being would remain forever fixed—precluding the possibility that organisms might change. Both of these misconceptions would have to be overcome before evolutionary biology could emerge as a science.

In addition to cataloging the details of natural history, the ancient Greeks also turned their attention to the problem of how life got started, and how all of the diverse living forms around them arose. As we learned at the start of the chapter in our discussion of Empedocles, without the ability to directly observe life arising and diversity being generated, and without a broad conceptual framework for the diversity of the life they saw, the Greeks resorted to speculative accounts of how this process may have occurred. While these speculations represented progress in the sense that they involved natural rather than supernatural explanations, many of the specific mechanisms that the Greeks proposed seem bizarre today. The commonality among almost all of their suggestions is that they relied on **spontaneous generation**—the idea that complex life-forms arise, repeatedly, without external stimuli, from nonliving matter.

Ideas on spontaneous generation existed before the Greeks and persisted for more than 2000 years after the Greeks. In Egypt, for example, people thought that frogs were created spontaneously from mud. This is because when the Nile River flooded every year, it transformed dry mudflats into wet mud, and simultaneously, hundreds of frogs appeared. The Egyptians therefore believed that the frogs must have spontaneously been created from the mud. Similarly, many medieval European farmers believed that mice were generated from moldy grain, and many urban residents believed that sewage created rats.

Finally, in 1668, in an early example of a modern experiment, Francesco Redi (1626–1697), an Italian physician and naturalist, addressed the following question: Are flies spontaneously generated from meat carcasses? It seemed as if they were, because when meat rotted, flies appeared. So, Redi placed raw meat in a series of jars. Covering some (for a control group) and leaving other jars uncovered or partially uncovered, Redi determined that flies only arise from the maggot offspring of other flies, and that maggots cannot spontaneously generate from meat (Figure 2.6). Redi's experiment prompted his contemporaries to question whether any organism could appear from a nonliving substance, or whether an organism

FIGURE 2.6 Redi's experiment. Redi's experiment demonstrated that maggots did not arise through spontaneous generation. Uncovered jars with meat have fly eggs and maggots. When the jars are covered, and flies cannot enter and lay eggs on the meat, no eggs or maggots are found.

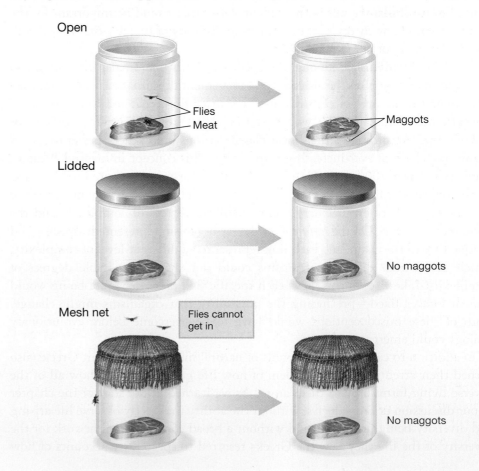

Open

Flies
Meat

Maggots

Lidded

No maggots

Mesh net

Flies cannot get in

No maggots

must come from parents. In spite of this experiment, spontaneous generation persisted as a theory, in part because the new technology of the microscope showed organisms like bacteria and fungi appearing on substances like spoiled broth without any clear parental source.

The late eighteenth and early nineteenth centuries brought new theories to explain the origins of life and the diversity of species. Erasmus Darwin (1731–1802) (Figure 2.7), an English physician, philosopher, and the grandfather of Charles Darwin, was one of the first to propose the idea of evolutionary change in his book *Zoonomia* (Darwin 1796; King-Hele 1998).

Erasmus Darwin argued that all life developed from what he called a "single living filament" (Darwin 1796). For Erasmus Darwin, this living filament had been modified in endless ways, over millions of years, to produce the life that he saw around him. He also hypothesized that man had initially walked on four limbs and, even more remarkably, that humans had descended from another primate species. This was a radical idea at the time. In addition, Erasmus Darwin understood the **struggle for existence**—the notion that organisms are in a constant struggle to obtain resources and to use these resources to produce more offspring than those around them can produce. Despite Erasmus Darwin's insights, he came up short of a full-blown theory of evolution of new species by natural selection, for at least two reasons: (1) with a few notable exceptions, he failed to connect the struggle for existence, which he described over and over again, to the evolutionary changes that such a struggle would produce (Krause 1879), and (2) he believed in the widely accepted, but incorrect, idea that new traits acquired *during the lifetime of an organism* could be passed down to progeny. We will return to this "inheritance of acquired characteristics" below, in our discussion of its most famous proponent, Jean-Baptiste Lamarck.

After Erasmus Darwin, Robert Chambers (1802–1871), a Scottish geologist, writer, and publisher (Figure 2.8), presented a more formally developed and widely influential theory on how new species originate from existing species in his 1845 book, *Vestiges of the Natural History of Creation* (Chambers 1845).

In the section of his book on what today we would call evolution, Chambers highlighted two critical points: (1) the composition of species has changed over time, and (2) this change is slow, gradual, and unlinked to catastrophes (Mayr 1982). From these ideas, Chambers outlined his *principle of progressive development*, in which he hypothesized that new species arise from old species: "The simplest and most primitive type . . . gave birth to the type next above it . . . and so on to the very highest, the stages of advance being in all cases very small—namely, from one species only to another; so that the phenomenon has always been of a simple and modest character" (Chambers 1845, p. 222).

One aspect of *Vestiges* that often goes unnoticed is that Chambers thought not in terms of individuals so much as **populations**—groups of individuals of the same species that are found within a defined area and, if they are a sexual species, interbreed with one another. In the parlance of modern evolutionary biology, we would say that, over time, populations evolve; individuals do not. Chambers recognized this, although he didn't phrase it in the terms we use today.

Robert Chambers and his *Vestiges* profoundly influenced a broad range of readers. *Vestiges* was widely read by scientists and laypeople alike, including a young Abraham Lincoln, who quickly became "a warm advocate of the doctrine" (Herndon and Weik 1893). *Vestiges* would eventually sell an astonishing 100,000 copies (Secord 2000).

FIGURE 2.7 Erasmus Darwin (1731–1802). Charles' grandfather raised the idea of evolutionary change in his book *Zoonomia*.

FIGURE 2.8 Robert Chambers (1802–1871). Chambers authored *Vestiges of the Natural History of Creation*.

For all its success, the greatest deficit in Chambers' book was the lack of a *theory* to explain *why* new species come into being. That is, there was nothing akin to the theory of natural selection that Darwin would propose some 15 years later.

2.4 Organisms Are Well-Suited to Their Environment

While *Vestiges* presented the idea of new species gradually arising from existing species, the book did not consider the enormous influence of the environment on these slow changes. Any observer of nature will notice the remarkable degree of fit between the structure of organisms and their environments. The mammals of cold climates have thick coats and layers of insulating fat; swimming animals have shapes that allow them to move efficiently through the water; desert plants have thick waxy cuticles and low surface area that help them avoid water loss. How do we explain this seemingly marvelous fit? Prior to Darwin's work, philosophers and scientists entertained a diverse array of answers to this question.

Paley's Natural Theology

FIGURE 2.9 William Paley (1743–1805). Paley discussed the exquisite fit of organism to environment by using an analogy in which, just as a watch requires a watchmaker, so too living organisms require a conscious designer.

For William Paley (1743–1805), an English naturalist and theologian, the fit of diverse species to their environments resulted from God's planning. In his textbook, *Natural Theology*, Paley discussed the famous metaphor of God as watchmaker (Paley 1802) (Figure 2.9). If a single part of the clockwork within a watch were shaped differently or placed elsewhere, the watch would fail to function, Paley observed. Because living creatures are even more complex than watches, they could not have come to perfectly fit their habitats through chance, Paley argued, just as it is virtually impossible for a fully working watch to come into being simply by chance arrangement of clockwork parts. Organisms, then, must have been intentionally designed by a benevolent deity in order to thrive in their environments.

Years later, Darwin would read and admire Paley's work, particularly his arguments on how the structures of organisms fit the functions they need to serve in order for individuals to survive. As we will see in greater detail in a moment, however, Darwin would disagree with Paley's explanation of the source of these adaptations. Darwin sought to explain adaptation by purely natural, rather than supernatural, causes.

Jean-Baptiste Lamarck and the Inheritance of Acquired Characteristics

FIGURE 2.10 Jean-Baptiste Lamarck (1744–1829). Lamarck developed a "transformation" theory for evolutionary change in his *Zoological Philosophy*.

With Jean-Baptiste Lamarck (1744–1829), we return fully to methodological naturalism as the explanation for species fitting their environments (Figure 2.10). Originally trained as a botanist at the French Jardin du Roi, Lamarck eventually became an animal systematist specializing in the study of invertebrates. His long-term studies of such organisms as mussels, which he compared to less complex fossil mussels, no doubt led him to think in terms of increasing complexity occurring in a group of organisms over time.

In his 1809 book, *Zoological Philosophy*, Lamarck rejected the idea that new species suddenly appeared following large-scale extinctions resulting from catastrophic events. Instead he advocated a transformationist theory, based on the idea that new, more complex species—humans being the most complex—had descended, gradually, from older, less complex species.

Lamarck is not remembered so much as a transformationist, but rather as the person who developed the first truly evolutionary theory for how such transformation came about through species adapting to their different environments. Actually, Lamarck outlined two mechanisms for evolutionary change, but here we will focus on his more famous one—the **inheritance of acquired characteristics**.

The idea behind the inheritance of acquired characteristics is that *during the lifetime of an organism*, the habits of the organism bring about changes in its structure, and such structural changes are passed down across generations (Lamarck 1809). Consider Lamarck's description of this process in birds (Figure 2.11):

> One may perceive that the bird of the shore, which does not at all like to swim, and which however needs to draw near to the water to find its prey, will be continually exposed to sinking in the mud. Desiring to avoid immersing its body in the liquid [it] acquires the habit of stretching and elongating its legs. The result of this for the generations of these birds that continue to live in this manner is that the individuals will find themselves elevated as on stilts, on naked long legs. (Lamarck 1801, cited in Burkhardt 1995, p. 172)

FIGURE 2.11 Lamarck, acquired characteristics, and shorebirds. Lamarck argued that the long legs of shorebirds such as this black-necked stilt (*Himantopus mexicanus*) are the result of birds stretching their legs as far as possible to avoid sinking in the mud. This stretching itself, Lamarck postulated, not only lengthened the legs of individuals doing the stretching, but their new trait of "longer legs" was then passed down to offspring.

Lamarck observed that we find long-legged birds in environments in which long legs are beneficial. Rather than crediting a watchmaker deity for this perfect fit, he hypothesized adaptations over time. Lamarck's hypothesis that traits acquired *during* the lifetime of an individual are passed on to its progeny was interesting, reasonable, and based on an idea that was universally accepted by scientists and nonscientists alike. After all, we are all aware of how our habits of life lead to changes in physiology; lifting weights, for example, leads to the development of increased muscle mass and lifting power. In the absence of evidence to the contrary, it is only a short leap from there to suppose that such changes could also be passed on to one's offspring. Today, however, we have plenty of evidence to the contrary. We know that acquired characteristics are not inherited, and we now ground our ideas of how traits are passed from generation to generation in the laws of genetics, which were formulated about 100 years after Lamarck (Chapter 6).

Lamarck's legacy, however, is not that he postulated the wrong processes for evolutionary change, but that he proposed a process in the first place, and that he connected it to environmental fit. As we will see, although Darwin did not completely reject the inheritance of acquired characteristics, his ideas on how and why evolutionary changes occur were quite different from those of Lamarck.

Patrick Matthew and Natural Selection

In the history of biology, we hear little about the developments in ideas of environmental adaptations in the 50 years between Lamarck's *Zoological Philosophy* (1809) and Charles Darwin's *On the Origin of Species* (1859). Yet, it was during this period that Patrick Matthew (1790–1874), a Scottish landowner and writer, proposed his own theory of evolution by natural selection, predating the ideas laid out in *On the Origin of Species* by more than a quarter of a century (Matthew 1831; Mayr 1982; Dempster 1996). In an obscure 1831 work entitled *On Naval Timber and Arboriculture*, Matthew put forth a theory very similar to Darwin's on the interaction between environment and evolutionary change. In the notes at the end of *On Naval Timber and Arboriculture*, in a section only tangentially related to the rest of book, Matthew outlined his ideas on both evolution and natural selection. He understood the idea that individuals best suited to their environment would be selected over others. The difference between this idea and Lamarck's theory is that Matthew relied on principles of survival of the fittest rather than the inheritance of acquired traits.

Matthew's discussion of environmental fit and natural selection—what he called "the circumstance-adaptive law"—is remarkably similar to what Darwin would discuss almost 30 years later. Matthew, for example, noted,

> The self regulating adaptive disposition of organized life may, in part, be traced to the extreme fecundity of Nature, who . . . has in all the varieties of her offspring, a prolific power much beyond (in many cases a thousandfold) what is necessary to fill up the vacancies caused by senile decay. As the field of existence is limited and pre-occupied, it is only the hardier, more robust, better suited to circumstance, individuals who are able to struggle forward to maturity . . . from the strict ordeal by which Nature tests their adaptation to her standard of perfection and fitness to continue their kind by reproduction, . . . the breed gradually acquiring the very best possible adaptation. (Matthew 1831, pp. 384–385)

Matthew outlines three important evolutionary ideas here: (1) resources are limited, and only so many offspring can survive to the age of reproduction, (2) individuals will differ in terms of traits that allow them to garner such resources, and (3) over time, this will lead to organisms that are well adapted to their environment.

Matthew's name is not readily associated with the theory of evolution by natural selection—despite the fact that on page 22 of the preface to the sixth edition of *The Origin of Species*, Darwin noted that Matthew presented "precisely the same view on the origin of species as that propounded by . . . myself . . . in the present volume." There are many reasons for this. Matthew's ideas were published in an obscure book that no one interested in biological diversity would have been likely to read, and even there his ideas were hidden in his notes and appendix section rather than presented as a unified theory. Moreover, Darwin discussed both natural selection and common descent, while Matthew mentioned only the former. Perhaps most importantly, Matthew presented scant evidence in support of his ideas. Darwin, on the other hand, spent 20 years gathering evidence for evolution by natural selection before publishing *On the Origin of Species*.

If we stop and take stock for a moment, what we have seen is that five major developments preceded and facilitated Darwin's *On the Origin of Species*. These changes involved moving: (1) from supernatural explanations to methodological naturalism, (2) from catastrophism to uniformitarianism, (3) from logic and pure reason to observation, testing, and refutation, (4) from an unchanging world to an evolving world, and (5) away from the idea of spontaneous generation to the idea that species come from other closely related species.

2.5 Darwin's Theory

We will begin our exploration of Darwin's contributions with a brief overview of the major ideas that he presented in *On the Origin of Species*. Darwin had two fundamental insights (he himself referred to them as "two great laws") about the process of evolution.

Darwin's Two Fundamental Insights

The first of Darwin's fundamental insights deals with the conditions of existence and the process of natural selection. Here, Darwin hypothesized that the environment selects on variation in the traits of individual organisms, because some variants are more successful than others at surviving and reproducing in their environment.

With this hypothesis, Darwin offered a mechanistic explanation both for how the characteristics of organisms change over time and for why organisms are well suited to their environments. That explanation was, of course, the process that Darwin dubbed "natural selection." The effect that a given variant of a trait has on survival and ultimately reproductive success depends on the environment in which an organism finds itself. As Darwin noted, once the "conditions of existence" are determined, "natural selection acts by either now adapting the varying parts of each being to its organic and inorganic conditions of life; or by having adapted them during past periods of time" (Darwin 1859, p. 206). Then, when Darwin spoke of the conditions of existence, he was describing the living (organic) and nonliving (inorganic) environment that sets the stage on which natural selection operates.

The second of Darwin's insights centers on the common ancestry of all living things. Here, Darwin hypothesized that all species have descended from one or a few common ancestors; species that share a recent common ancestor tend to resemble one another in many respects for the very reason that they share recent common ancestry. In short, Darwin hypothesized that new species do not arise through independent acts of creation or spontaneous generation, but rather from preexisting species. This process generates a branching pattern of ancestry relating all life.

These two insights are major themes not only within this chapter, but throughout the textbook, and we will go into much more detail about them in other chapters. For now, we will look at how Darwin arrived at these ideas, at how he collected evidence to support them, and at how he chose to present his challenging conclusions to his nineteenth-century contemporaries.

Publication of *On the Origin of Species*

On the Origin of Species begins as follows: "When on board H.M.S. 'Beagle,' as naturalist, I was much struck with certain facts in the distribution of the inhabitants of South America, and in the geological relations of the present to the past inhabitants of that continent. These facts . . . seemed to throw some light on the origin of species—that mystery of mysteries" (Darwin 1859, p. 1) (Figure 2.12). As we have seen, some of Darwin's predecessors talked of evolutionary change and even of processes akin to natural selection. Darwin's book, however, was the first to present

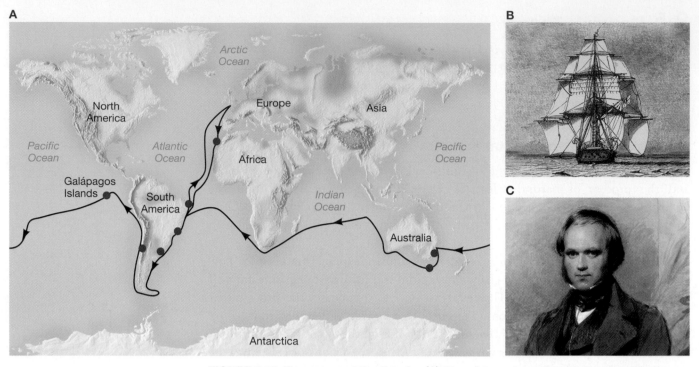

A

B

C

FIGURE 2.12 **The voyage of the *Beagle*.** **(A)** Map of the voyage of the HMS *Beagle*. **(B)** The HMS *Beagle* was a 10-gun brig of the British Royal Navy. **(C)** Portrait of a young Charles Darwin, ca. 1840.

a *complete theory* of evolution by natural selection, and to support that theory with an enormous body of evidence: evidence that included his observations of finches, tortoises, coral reefs, and so much more in the Galápagos.

Twenty-three years separated Darwin's return from his time on the HMS *Beagle* and the publication of *On the Origin of Species*. Darwin postponed releasing his work, partly because he knew that his ideas were revolutionary, and he wanted to have the strongest possible case before unveiling them to both the scientific world and the general public. But in the end, competition pressured Darwin into publishing. In 1858, as part of an ongoing correspondence with Alfred Russel Wallace (1823–1913), Darwin received a manuscript in which Wallace proposed a theory very similar to his own (Figure 2.13).

Wallace was a brilliant natural historian, geographer, and collector; he identified numerous new species of birds and insects, and his collections can be seen today in natural history museums around the world. Wallace had written a paper in 1855 in which he speculated on the origin of species; there he concluded from the similarity of geographically nearby species that new species must arise from preexisting ones (Wallace 1855). Wallace's concept of how species are formed led him to suggest the hierarchical branching relationship among species that is fundamental to our current understanding of the diversity of life.

It was during a bout with malaria on the Spice Islands, however, as he suffered from fever, that Wallace figured out the mechanism that drives species to change (Raby 2001). As he recollected, "I at once saw that the ever present variability of all living things would furnish that material from which, by the mere weeding out of those less

FIGURE 2.13 **Alfred Russel Wallace (1823–1913).** Wallace independently developed a theory of evolution by natural selection very similar to that of Darwin.

adapted to the actual conditions, the fittest alone would continue the race" (Wallace 1908, pp. 191–192). Darwin would call this process "natural selection."

When Wallace wrote to Darwin outlining these ideas on evolution, Darwin yielded to pressure from friends and colleagues, and publicized his own theories, first in the joint Darwin–Wallace paper that was read to the Linnaean Society in 1858 (with neither Darwin nor Wallace present), and later in longer form as *On the Origin of Species*. Wallace still holds a place in the pantheon of great evolutionary thinkers, but history primarily associates Darwin's name with the theory of evolution by natural selection. In large part, this is due to Wallace's professional generosity. While his theory closely resembled Darwin's, Wallace graciously agreed that Darwin deserved the credit. Darwin had worked for decades on developing the theory and had amassed huge amounts of data from many sources to provide evidence for his theory of evolution by natural selection.

In 1859, when Darwin finally published *On the Origin of Species*, he laid out his evidence and his argument carefully, cognizant of the criticism his ideas would draw. But before he could describe either his data or the process involved in generating a new species, Darwin needed to explain what defined an independent species. He did so cautiously, but in a strategically brilliant fashion.

Means of Modification and Pigeon Breeding

The opening chapter of *On the Origin of Species* may seem strange to the modern reader, with Darwin writing such things as:

> It is, therefore, of the highest importance to gain a clear insight into the means of modification. . . . At the commencement of my observations it seemed to me probable that a careful study of domesticated animals and of cultivated plants would offer the best chance of making out this obscure problem. (Darwin 1859, p. 4)

Indeed, Darwin writes at length about numerous domestication programs, with a particular emphasis on pigeon breeding (Figure 2.14).

FIGURE 2.14 Pigeon varieties. Darwin used pigeon breeding to explain artificial selection to the readers of *On the Origin of Species*. Here we see three domesticated pigeon varieties: **(A)** the carrier pigeon, **(B)** the beard pigeon, and **(C)** the pouter pigeon.

A

B

C

While this choice of subject matter appears unusual today, pigeon breeding was a popular pastime in Victorian England and would have been comfortingly familiar to Darwin's audience. With this example, Darwin set up an analogy that would help his readers of 1859 relate to the novel ideas in the rest of the book.

Darwin hoped to introduce readers to natural selection by first convincing them that the breeding programs that pigeon fanciers had developed—programs that had led to a wide range of extraordinary variation in pigeon color, flying habits, behavior, and so on—resembled the processes that led to differences within and between species in nature. In this section, Darwin aimed first to illustrate the processes by which he believed species change over time, and second to help his readers get beyond their preconceptions of species as eternal and immutable. We address these two aims in turn.

Artificial Selection

The process that pigeon breeders developed is an example of *artificial selection*, whereas the process leading to the wide variety of traits we see in nature is *natural selection*. In artificial selection, humans systematically breed certain varieties of an organism over others. For thousands of years, humans have been shaping animals and plants by this process. Ever since our ancestors *selected* some varieties of wheat, corn, and rice over others, and systematically planted such seeds, we have engaged in artificial selection. The same process describes our systematic breeding of certain types of dogs and our domesticated livestock.

Following Darwin, let us examine how artificial selection works in the context of pigeon breeding. Suppose that, like pigeon breeders in Victorian days, we want to produce a variety of pigeon with snow-white plumage. We would begin our artificial selection process by systematically allowing only those individuals in our population with the whitest plumage to breed. We would then continue this process generation after generation, in each generation sorting the birds based on plumage coloration, and allowing the whitest—those that are closest to the type we want to produce—to breed. If offspring resembled their parents in terms of plumage coloration, each generation of offspring would have whiter and whiter feathers. Eventually we would exhaust all genetic variation for plumage coloration and, so far as possible, we would have achieved our goal of a snow-white pigeon (Figure 2.15).

Generation 1   Generation 2   Generation 3   Generation N

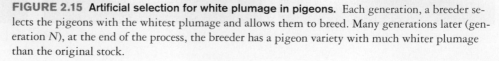

FIGURE 2.15 Artificial selection for white plumage in pigeons. Each generation, a breeder selects the pigeons with the whitest plumage and allows them to breed. Many generations later (generation *N*), at the end of the process, the breeder has a pigeon variety with much whiter plumage than the original stock.

Changing Species

While many of Darwin's contemporaries would have accepted the explanation of artificial selection as the mechanism producing new *varieties* of pigeons—new colors, new behaviors, and so on—the claim that this process could generate new *species* was much more controversial, as it implied that it would lead to original and new life-forms, an idea that was still widely unaccepted at the time. Therefore, in Chapter 2 of *On the Origin of Species*, Darwin seems almost obsessed with the definition of a variety versus a species, and with the problems in distinguishing between these two categories.

Darwin presents example after example in which one naturalist calls a group of organisms "species 1," while another classifies the same group as a "variety of species 2." In Darwin's eyes, the line between a variety and a species was arbitrary. Darwin saw species as merely "strongly marked and permanent varieties." Conversely, when he saw varieties, he viewed them as "leading to subspecies and then to species," and he often spoke of varieties as "incipient species"—species in the making.

Challenging the distinction between species and varieties was essential to Darwin's overarching argument. Pointing to examples in plant and animal breeding, Darwin could provide extensive evidence that new *varieties* often arise from a single stock, through a branching mechanism of descent. Having established that varieties are similar to species, Darwin could then claim that they probably both respond to similar processes, most notably, some process of selection (artificial or natural). As such, he could conclude that, like varieties, species change over time, and that new species arise from other species.

To explain how varieties were on the path to becoming new species, Darwin introduced the concept of descent with modification. For example, he hypothesized that if we want to understand how species 2 got to be what it is today, we need to recognize that it *descended* from another species—let's call it species 1—and that over evolutionary time, numerous *modifications* occurred. Darwin argued that these modifications resulted largely from the process he dubbed natural selection, a process analogous to the familiar technique of artificial selection that had been used by breeders for thousands of years.

Once Darwin had walked the reader of *On the Origin of Species* through the process of artificial selection and the concept of species as changing entities similar to varieties, he could move on to the details of natural selection.

2.6 Darwin on Natural Selection

As Darwin argued, the process of natural selection resembles that of artificial selection. The two important differences between the processes are the *selective agent* and the *traits* being selected. With artificial selection, the selective agent is the human breeder who chooses which traits to modify, and attempts to modify them in a way that is beneficial to the breeder. In the case of natural selection, we can think of nature as the selective agent, but it is important to understand that nature is not, in any sense, a conscious agent in the way that humans are.

With respect to what traits are selected, Darwin noted,

> Man can act only on external and visible characters; nature cares nothing for appearances, except in so far as they may be useful to any being. She can act on every internal organ, on every shade of constitutional difference, on the whole machinery of life. (Darwin 1859, p. 83)

That is, the process of natural selection favors any variant of a trait that increases the survival and reproductive success of an individual, even if the difference is not easily detected by a human observer or if the increase in reproductive success is small.

Darwin, Variation, and Examples of Natural Selection

Darwin hypothesized that evolution by natural selection was a gradual, but powerful, process. He argued that the process of natural selection acted on small differences between individuals. If one variety of a trait led to even a small reproductive advantage compared to other varieties, it would be favored by natural selection. These small differences could translate into major changes as they accumulated over evolutionary time.

For example, Darwin asked his reader to imagine the wolf that "preys on various animals, securing some by craft, some by strength, and some by fleetness" (Darwin 1859, p. 90). When prey animals are scarce, natural selection acts strongly on such wolf populations. Wolves that possess the traits that best suit them for hunting (speed, stealth, and so on) tend to survive longer and produce more offspring. These offspring in turn are likely to possess the traits that benefited their parents in the first place. The repetition of this process for generation after generation produces wolves that are very efficient hunters. "Slow though the process of selection may be," noted Darwin, the eventual outcome is a more effective wolf predator.

Darwin applied similar arguments to many examples in nature. Among these, he discussed the process of natural selection on plants that rely on insects for cross-fertilization. Darwin saw this case as more complicated than the case of the wolves, because insects often eat most of the plant's pollen. He argued that natural selection might nonetheless favor plant traits that foster more efficient insect pollination, because only a small amount of pollen is needed by the plant for fertilization (Figure 2.16). Darwin explained:

> . . . as pollen is formed for the sole object of fertilisation, its destruction appears a simple loss to the plant; yet if a little pollen were carried, at first occasionally and then habitually, by the pollen-devouring insects from flower to flower, and a cross thus

FIGURE 2.16 Plants and their pollinators. Darwin discussed the relationship between plants and the insects that cross-fertilized them as an example of how natural selection operates. Insects, such as the bee seen here, may eat some of the pollen produced by a plant, but if they move enough pollen from plant to plant, their actions may be in the plant's reproductive interests as well.

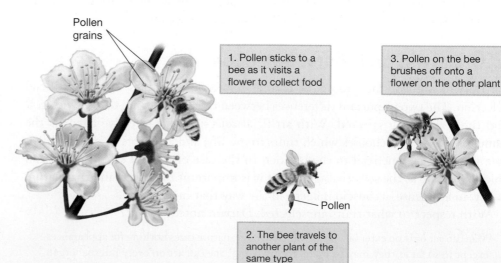

Pollen grains

1. Pollen sticks to a bee as it visits a flower to collect food

3. Pollen on the bee brushes off onto a flower on the other plant

Pollen

2. The bee travels to another plant of the same type

effected, although nine-tenths of the pollen were destroyed, it might still be a great gain to the plant; and those individuals which produced more and more pollen, and had larger and larger anthers, would be selected. (Darwin 1859, p. 92)

Once we see traits in terms of their effect on overall *reproductive success*—as Darwin did for wolves, insect-pollinated plants, and myriad other examples—the concept of natural selection becomes a powerful tool for understanding the world around us.

The Power of Natural Selection

Darwin's own writings demonstrate that he attributed enormous power to the process of natural selection. He ends the introductory chapter of *On the Origin of Species* by claiming, "I am convinced that natural selection has been the most important, but not the exclusive, means of modification" (Darwin 1859, p. 6). Darwin lays out his position in even more detail for the reader in a later passage:

> It may be said that natural selection is daily and hourly scrutinising, throughout the world, every variation, even the slightest; rejecting that which is bad, preserving and adding up all that is good; silently and insensibly working, whenever and wherever opportunity offers, at the improvement of each organic being in relation to its organic and inorganic conditions of life. We see nothing of these slow changes in progress, until the hand of time has marked the long lapse of ages. . . . (Darwin 1859, p. 84)

For Darwin, the process of natural selection operated 24 hours a day, every day, everywhere, over vast periods of time. Only a process of such magnitude could have shaped all the life that we see around us, and for that matter, all life that has ever lived. Any differences in reproductive success associated with varieties of a given trait will be acted on by natural selection. This includes differences so slight that even the most thorough and patient human investigator might struggle to detect them.

An analogy might help here: The process of natural selection acts as an editor, removing that which is not as well suited to its environment by increasing the frequency of that which is better suited. Changes take place constantly, but usually they will not manifest in measurable differences until the passing of eons. In later chapters, we will see that Darwin underestimated the potential rate of evolutionary change in some cases. Indeed, under certain conditions, the effects of the process of natural selection—particularly selection operating in species that reproduce very quickly—can be detected and measured in a span of years or even less.

Malthus and the Scope of Selection

Before his readers could accept the potency of evolutionary change, Darwin needed them to reconsider their beliefs about survival in the natural world. To do this, Darwin used an analogy. Just as selective breeders must discard numerous individuals bearing undesirable traits in order for artificial selection to work, "nature" must "discard" numerous individuals in order for natural selection to be effective. While it may seem obvious to us, in Darwin's time this concept ran against the prevailing notion of an orderly, efficient, and harmonious operation of nature.

To persuade his readers that his mechanism of natural selection could shape the natural world, Darwin first had to persuade them that nature was sufficiently

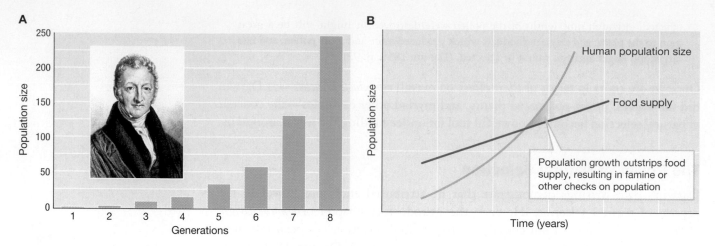

A

B

FIGURE 2.17 Malthus and population growth. Thomas Malthus argued that humans would outstrip the available resources necessary to sustain themselves, leading to population growth that would be checked by famine, war, and disease. Malthus' writings were influential in helping Darwin develop his ideas on natural selection. **(A)** Geometric population growth is shown in this graph. If each mother produces two replacements for herself, a single mother at time 0 gives rise to 2 additional mothers after a single generation. There will then be 4 mothers after 2 generations, 8 after 3 generations, 16 after 4 generations, and so forth. **(B)** Malthus argued that the human population was geometrically increasing (blue curve) and thus would inevitably outstrip its food supply (red curve), which he believed to be arithmetically increasing.

"wasteful" for selection to operate. That is, he needed to convince his readers that many individuals did not survive to the age of reproduction, and of those that did, only a fraction actually reproduced. Here Darwin drew on the ideas of Thomas Robert Malthus (1766–1834), an English political economist and demographer.

Malthus noticed that the human population, unless kept in check by war, famine, disease, or other causes, grows geometrically in time (Malthus 1798). Malthus contrasted the geometric growth of unconstrained human populations with the growth of food production, which he believed could increase at best arithmetically (Figure 2.17). As a result, Malthus surmised that humans would inevitably outstrip the available resources necessary to sustain themselves, and that population growth would inevitably be checked by famine, war, disease, or other forces.

Darwin recognized that Malthus' argument applies to animal and plant populations as well as to human populations. For these populations, food supply is usually not increasing at all, yet the power of reproduction would lead to a geometric increase in population size if growth were not checked by a struggle for existence. The difference between the potential growth and the maximum size allowed by the food supply denotes the number of individuals lost in the struggle for existence, and thus it represents the opportunity that natural selection has to sort populations based on even the smallest differences in form (Figure 2.18). Darwin neatly summarized this as follows:

> As many more individuals of each species are born than can possibly survive; and as, consequently, there is a frequently recurring struggle for existence, it follows that any being, if it vary however slightly in any manner profitable to itself, under the complex and sometimes varying conditions of life, will have a better chance of surviving, and thus be *naturally selected*. (Darwin 1859, p. 5)

Transformational and Variational Processes of Evolution

Darwin's mechanism of evolutionary change differs radically from previous concepts of evolution. Before Darwin, scientists had envisioned change as a **transformational process**, in which the properties of an ensemble change because every member of the ensemble itself changes. For example, a mountain range becomes less rugged and more rounded over geological timescales because *each* individual peak itself becomes more rounded.

Lamarck's theory of evolution was a transformational theory. According to Lamarck, the properties of a lineage (successive generations) of organisms shift over time because of changes that each member undergoes during its lifetime, and then passes along to its descendants. By contrast, Darwin's theory of evolutionary change was a variational one. In a **variational process** of evolution, the properties of an ensemble change, not because the individual elements change, but rather because of the action of some process *sorting* on preexisting variation within the ensemble (Levins and Lewontin 1987). For Darwin's theory, that process was the process of natural selection.

To see how such a sorting process operates, imagine sifting a bucket of soil with particles ranging in size from fine sand to small pebbles. After sifting, the soil particles remaining in the sifter will be considerably larger on average than those in the original soil mixture. This is not because of any change on the part of individual particles—no transformation in the size of soil particles has occurred—but rather it is because the sifter has sorted the members of the ensemble according to their characteristics (Figure 2.19).

This kind of sorting process is what takes place when we use artificial selection to change the characteristics of a breed of animals or plants. And just as a pigeon breeder sorts on variation when selecting breeding pairs so as to produce a snow-white pigeon, the conditions of existence sort on variation within the members of species. Natural selection favors those variants that survive and most successfully reproduce, passing on their characteristics to their offspring through the process of heredity.

To arrive at any theory of evolution, Darwin needed not only to establish that the process of natural selection involves "wasteful" deaths within populations, he also had to dispel the belief in an eternally unchanging world, as discussed earlier in this chapter. To arrive at a specifically *variational* theory of evolution, Darwin also had to reject the existing conception of nature that viewed any variation as

Population size

Time (years)

Plant or animal population size

Food supply

This difference reflects the zone in which natural selection by resource limitation occurs and in which less-fit forms are unable to survive

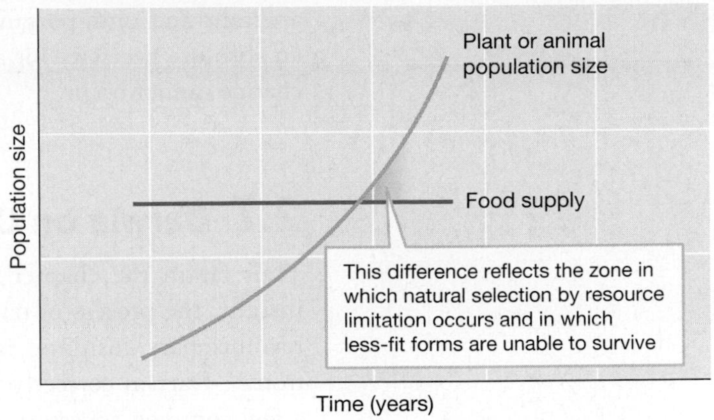

FIGURE 2.18 Darwin, Malthus, and natural selection. Darwin adapted Malthus' argument to natural populations of plants and animals. The food supply curve (red) is flatter here than in Figure 2.17. In that figure, the food supply curve also increased as a result of human innovations in food production.

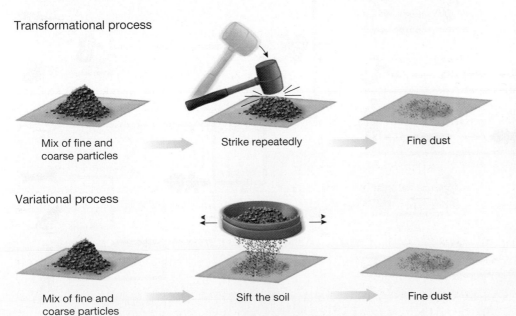

Transformational process

Mix of fine and coarse particles

Strike repeatedly

Fine dust

Variational process

Mix of fine and coarse particles

Sift the soil

Fine dust

FIGURE 2.19 Different processes of change. In a transformational process, the ensemble changes because each individual member changes. In a variational process, the ensemble changes because something sorts among the variants in the original ensemble. In this example, crushing the soil particles is a transformational process—the ensemble shifts toward smaller particles because the individual particles are reduced. Sifting the soil is a variational process—the ensemble shifts toward smaller particles because the larger particles are sorted out.

aberrant and unimportant, and instead place variation itself in the forefront, as an absolute necessity for a sorting process without which variational evolutionary change cannot occur.

2.7 Darwin on Common Ancestry

Thus far in the chapter we have concentrated on the details of Darwin's first insight, the process of natural selection. We now turn to the second of Darwin's revolutionary insights, his answer to the question: "Where do species come from?" Darwin correctly recognized that all living creatures derive from one or a few common ancestors, and that new species are formed when populations of a preexisting species diverge from one another.

The Tree of Life

In *On the Origin of Species*, Darwin explained that much as artificial selection can create multiple new varieties from a single domesticated variety, natural selection can generate multiple new species from a single ancestral species. Indeed, Darwin conjectured that the vast diversity of species that we see throughout the world has arisen from precisely this process.

FIGURE 2.20 Darwin's theory versus Lamarck's theory. In Lamarck's theory, species evolve independently and in parallel; in Darwin's theory, species are descended one from another to form a branching tree of life.

Darwin's explanation suggests that all living things are linked by a pattern of descent dramatically different from that implied by either special creation or Lamarck's theory of evolution (Figure 2.20). While these latter explanations

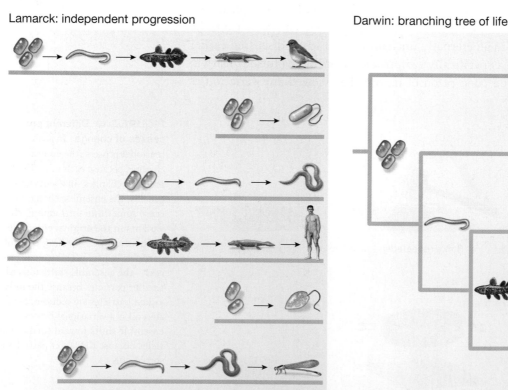

Lamarck: independent progression

Darwin: branching tree of life

envision species as a set of independent organisms, Darwin's theory links species according to their historical pattern of descent.

Darwin described the branching historical relationships among all living things using the metaphor of a tree of life (Figure 2.21). His eloquent depiction of the tree of life requires us to look at a lengthy quote, but this quotation is worth reproducing because of the profound implications of the tree of life metaphor:

> The affinities of all the beings of the same class have sometimes been represented by a great tree. I believe this simile largely speaks the truth. The green and budding twigs may represent existing species; and those produced during each former year may represent the long succession of extinct species. . . . The limbs divided into great branches, and these into lesser and lesser branches, were themselves once, when the tree was small, budding twigs. . . . Of the many twigs which flourished when the tree was a mere bush, only two or three, now grown into great branches, yet survive and bear all the other branches; so with the species which lived during long-past geological periods, very few now have living and modified descendants. From the first growth of the tree, many a limb and branch has decayed and dropped off; and these lost branches of various sizes may represent those whole orders, families, and genera which have now no living representatives, and which are known to us only from having been found in a fossil state. . . . As buds give rise by growth to fresh buds, and these, if vigorous, branch out and overtop on all sides many a feebler branch, so by generation I believe it has been with the great Tree of Life, which fills with its dead and broken branches the crust of the earth, and covers the surface with its ever branching and beautiful ramifications. (Darwin 1859, pp. 129–130)

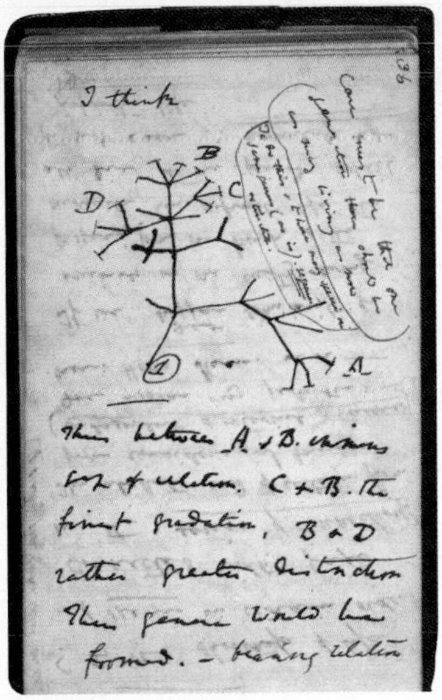

Darwin recognized the enormous importance of the branching relationships among species in this tree of life as a model for our understanding both of life's history and of the patterns of life's diversity. He chose to include only a single figure in *On the Origin of Species*, and this figure serves to illustrate his second insight—that of the branching historical relationships among all living things (Figure 2.22). Today, we refer to this type of figure as a phylogenetic tree.

FIGURE 2.21 An early phylogenetic tree from Darwin. From Darwin's notebook, one of his first sketches of the branching relationships among species.

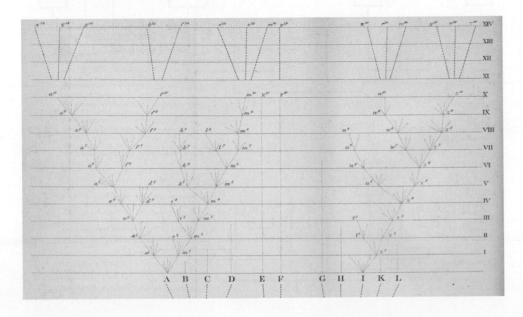

FIGURE 2.22 A phylogenetic tree from *On the Origin of Species*. Darwin included this diagram as the sole figure in *On the Origin of Species*. It illustrates the pattern of branching relationships among a number of initial populations (A–L) over vast periods of time (time moves forward as one moves up the vertical axis).

Groups within Groups

A major point in favor of the hypothesis of common ancestry with branching descent is that it explains hierarchical patterns of similarity that are observed in nature. By hierarchical patterns of similarity, we mean something like this: Different species of squirrels resemble each other more than they resemble a species of deer. And different species of deer resemble each other more than they resemble a species of squirrel. That is, species of squirrels *cluster* together because of their similarity to one another, and species of deer *cluster* together. At a different hierarchical level, species of squirrels and deer are more similar to one another than either is to a species of frog. And so, at this hierarchical level, species of squirrels and deer cluster together (as mammals), and species of frogs, toads, and salamanders cluster together (as amphibians). Finally, squirrels, deer, frogs, and toads are all more similar to one another (as vertebrates) than they are to species of octopus or squid (invertebrates).

In *On the Origin of Species*, Darwin argues that branching descent explains this hierarchical patterning seen in nature, saying that "the forms of life throughout the universe become divided into groups subordinate to groups" (Darwin 1859, p. 59). Neither special creation nor a theory such as Lamarck's can explain these groupings and subgroupings of organisms. But a process of branches dividing and subdividing naturally gives rise to a hierarchical structure of relationships—varieties nested within species within genera (a taxonomic group, intermediate in scale between species and families; the singular of genera is *genus*). Indeed, the modern field of **systematics**—the naming and classification of organisms—is based on the conceptual foundation of this hierarchical branching structure. As we will see in further detail in Chapter 4, systematists aim to classify organisms into hierarchically arrayed groups, or clades, of organisms, that have descended from a common ancestor (Figure 2.23).

FIGURE 2.23 Branching descent, clustering, and hierarchy. Darwin's view of branching descent explains both the clustering of species—indicated by the shaded grouping—in terms of similar form **(A)** and the hierarchical patterns of similarity **(B)** that we can discern when studying groups of species. In part B, some of the different clades are shown in different bracket colors, with the node representing the common ancestor of that entire clade in that same color. An X represents a lineage that has gone extinct.

A Clusters of species

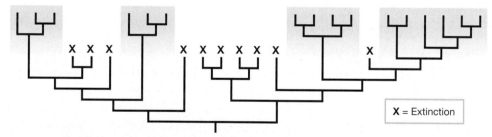

X = Extinction

B Hierarchical patterns of similarity

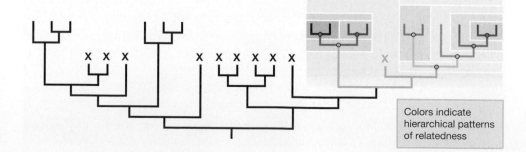

Colors indicate hierarchical patterns of relatedness

Darwin's view of common descent provides an explanation, not only for the hierarchy of organisms now studied by systematists, but also for the clustering of species: "No naturalist pretends that all the species of a genus are equally distinct from each other," Darwin told the reader of *On the Origin of Species* (Darwin 1859, p. 57). That is, we expect to see clusters at many levels, including that of the genus. Darwin reasoned that this clustering arose as a result of common ancestry. Groups of closely related species share common characteristics, in large part because they share common ancestry.

Common Descent and Biogeography

Both Wallace and Darwin traveled extensively across the globe, and in doing so, both were struck by the strong patterns that they observed in the geographic distribution of nature's diversity. In his 1855 paper that preceded Darwin's *On the Origin of Species* by 4 years, Wallace described his observations regarding the patterns of geographic distribution of "closely allied" species. He noted similar patterns in which closely related species occurred together throughout the fossil record. In short, Wallace found that highly similar species tend to be closely clustered in time and space, and from this observation he proposed that "Every species has come into existence coincident both in space and time with a pre-existing closely allied species" (Wallace 1855, p. 186).

Wallace recognized that this pattern of descent—new species coming into existence from previous species—implies the branching system of phylogenetic relationships that we have described in detail earlier in this section. He describes the groupings of species as the "complicated branching of the lines of affinity, as intricate as the twigs of a gnarled oak or the vascular system of the human body" (Wallace 1855, p. 187).

Darwin came to similar conclusions about the causes for groupings of species based on similar evidence. In *On the Origin of Species*, Darwin notes that similarities in "conditions of existence"—climate and physical conditions, for example—are insufficient to explain the geographic clustering of similar, closely related species. Instead, he thought that geographic features seemed to play an important role. He described the following pattern: Species separated by major geographic barriers to migration—mountain ranges, deserts, or large bodies of water—tend to be dissimilar even when the climate and physical conditions are similar on each side of the divide. Adjacent species that are not separated by geographic barriers tend to be similar to one another despite major differences in climate and habitat.

These geographic correlations supported Darwin's theory that each species arises only a single time in a single place, by descent with modification from a closely related species. Darwin then extrapolated from these patterns among groups of related species to suggest that in fact all living things have descended, with modification, from one or a few common ancestors. Darwin hypothesized that all living things—plants, protozoa, humans, birds, insects, and every other life-form—share a common origin. In the next few chapters, we will explore the overwhelming weight of evidence that has since accumulated in support of Darwin's conclusion. But first, we will consider some of the problems that troubled Darwin in his lifetime.

2.8 Problems with Darwin's Theory

In science, no grand theory is without its problems, especially in its early stages. The important issue is whether the researcher acknowledges such problems and generates new hypotheses, or simply ignores any inconsistencies. In *On the Origin of Species*, Darwin was not afraid to discuss many of the problems associated with his theory of evolution by natural selection.

Here we briefly touch on three of the major challenges that Darwin faced, and we provide pointers to where we will discuss some of these problems in greater detail in later chapters. Although not all of these challenges were resolved within Darwin's lifetime, today we have a good understanding of how to account for each of them. In Chapters 6 and 7, we will also show how another challenge Darwin faced—understanding how inheritance operated—was finally resolved.

FIGURE 2.24 Complex traits. One of the challenges that Darwin faced was to explain how natural selection could create complex traits such as **(A)** the vertebrate eye, **(B)** the mammary gland, or **(C)** the instincts for constructing the hexagonal cells of a honeycomb.

Problem 1: Accounting for Complex Structures with Multiple Intricate Parts

Darwin generally portrayed natural selection as a slow process acting on very small differences between individuals. It is relatively straightforward to see how this process could lead to gradual adjustments in the thickness of an otter's fur or the length of a badger's forelimb. But how might natural selection operate as a genuinely creative process? How might it generate complex structures such as the eye, the mammary gland, or the instincts needed to construct the hexagonal cells of a honeycomb (Figure 2.24)?

Darwin's critics seized on this issue. If natural selection operates by gradual increments, they reasoned, the eye must be preceded by half of an eye—and what good is half of an eye? These critics argued that complex traits would have no selective value until fully formed, and thus natural selection would not favor the intermediate steps necessary along the way. Darwin responded to this challenge with confidence; we will explore his explanation in depth in Chapter 3.

Problem 2: Explaining Traits and Organs of Seemingly Little Importance

At the opposite extreme, Darwin wondered how his theory could explain traits that appear to lack any biological function. If a trait does not contribute to survival and/or reproductive success, it will not be favored by natural selection, and yet it seemed as though such traits existed. Snakes have "limb buds" that appear to have no function, ruminants have incisor teeth that never break through their gums, and so on. How can these things be explained? We explore the answers in Chapter 4 (where we treat vestigial traits) and Chapter 8 (where we consider the neutral theory of evolution).

Problem 3: Why Doesn't Natural Selection Run Out of Variation to Sort On?

As we saw earlier in this chapter, Darwin's theory relied on a variational process of evolution rather than a transformational one. This posed a problem: In order for natural selection to operate, it must have variation to sort on—but the action of natural selection itself reduces the amount of variation in a population as less-fit variants are eliminated from that population. Thus, the fire of natural selection threatens to consume the variation that fuels it. How can we explain the persistence of variation? Why doesn't evolution just stop?

Adding to the scope of the problem, when Darwin wrote *On the Origin of Species*, biologists did not understand the basic principles of heredity. Mendel's laws were not known to Darwin; instead, like most of his contemporaries, Darwin envisioned inheritance as a blending of the hereditary determinants from each parent. Such a blending process also consumes variation. In Chapters 6 and 7, we will explore the sources of new variation, and in Chapter 9, we will see how scientists in the early part of the twentieth century reconciled the process of inheritance with Darwin's ideas about natural selection.

2.9 The Reaction to Darwin and Early History of the Modern Synthesis

While various religious leaders challenged almost all of the major conclusions that Darwin presented in *On the Origin of Species*, the scientific community exhibited a more mixed reaction (Mayr 1982). Early on, for example, British scientists almost universally embraced Darwin's ideas on evolution, but they rejected his theory of natural selection. That is, they accepted that evolutionary change, rather than special acts of creation, explained the world that we see around us, but they rejected the idea that the primary force generating evolutionary change was natural selection. A few British naturalists, including Alfred Russel Wallace, Henry Walter Bates (1825–1892), and Joseph Dalton Hooker (1817–1911), thought that natural selection was important in driving evolutionary change, but virtually all early experimental evolutionary biologists disagreed (Glick 1974).

In the 1880s, experimental work—primarily that of German geneticist and evolutionary biologist August Weismann (1834–1914), who demonstrated that traits acquired during the lifetime of an organism could not be inherited—dealt a death blow to previous theories of Lamarckian inheritance. Scientists were left with only two possible mechanisms of evolution. The processes were either natural selection acting in a slow and methodological way on small genetic differences, or **saltationism**, that is, "evolution via large, sudden changes from the existing norm" (Mayr 1982).

In his now-famous experiments of the 1850s and 1860s, Augustinian monk, plant breeder, and biologist Gregor Mendel (1822–1884) found that inherited factors that form the basis of traits come from both parents. His work on pea plants demonstrated that each parent plant has two copies of each gene, and that the two gene copies separate with equal probability into gametes (eggs, sperm, pollen, and so on). In Chapter 6, we will discuss Mendel's experiments in more detail.

Mendel's results remained virtually unnoticed until 1900, when three scientists (Hugo de Vries, Carl Correns, and Eric von Tschermak) independently rediscovered his work and made it available to the scientific world. Biologists began to explore how natural selection might operate when inherited material operated as Mendel suggested.

Evolutionary biologists fell into one of two camps. On one side was a group called the Mendelians. These scientists primarily worked in the lab, were trained more as physical than as biological scientists, and thought that the continuous variation in so many traits seen in nature was not primarily genetic in origin. This was because the Mendelian camp's original interpretation of Mendel's work allowed for discrete variation—for example, tall versus short—but not continuous variation in traits. Mendelians viewed evolution as a saltational process. In the other camp were the biometricians, including the English geneticist and statistician Karl Pearson (1857–1936). The biometricians were impressed by the amount of continuous variation—that is, extremely fine gradations of difference—that they saw all around them and thought natural selection was a slow, gradual process.

The differences between the Mendelians and the biometricians began to dissolve with experimental work in the 1930s and 1940s in what came to be called the **modern synthesis**, or the **evolutionary synthesis**. This synthesis included experimental work in genetics that demonstrated that:

- Genes are passed on from parents to offspring in an intact form, even if they are not expressed in the offspring's phenotype. That is, genes are particulate: they don't "blend" with other genes.

- One source of genetic variation is mutation.

- Genetic variants that generate large and small phenotypic differences are not qualitatively different from one another—the effects of large differences may be more pronounced, but genetic variation is generated and inherited in similar ways in both cases.

- Not all genetic mutations are harmful, so that positive changes can accrue over time—either slowly, or in some cases, more rapidly.

- Sexual reproduction is an important contributor to the production of massive amounts of genetic variation.

- Some traits are the result of the interaction of numerous genes, while some genes can affect more than one trait, helping to explain the evolution of complex traits without necessarily assuming some saltational (that is, large and sudden) change.

- Many (but not all) changes in the genotype affect the phenotype. Variation in the phenotype is the raw material for natural selection.

This work demonstrated that there was no conflict between what was being found in the new, burgeoning field of genetics and Darwin's idea that evolutionary change was primarily a slow process, driven by natural selection. Another crucial ingredient of the modern synthesis was the work of mathematical population geneticists such as Ronald Fisher (1890–1962), Sewall Wright (1889–1988), and

J. B. S. Haldane (1892–1964), who developed elegant models of how evolutionary processes lead to changes in gene frequencies, and how changes in gene frequencies map onto changes in the phenotypes of organisms.

The modern synthesis represented the collected efforts of systematists, geneticists, paleontologists, population biologists, population geneticists, and naturalists. Although often associated with the publication of British biologist Julian Huxley's (1887–1975) book, *Evolution: The Modern Synthesis*, this synthesis was not so much an event per se, as the result of a gradual accumulation of information that melded together to shape biology at the time (Huxley 1942). In addition to the work listed above, this synthesis involved an elegant combination of theoretical models and experimental manipulations, like that of German-American evolutionary biologist and ornithologist Ernst Mayr's (1904–2005) pathbreaking work on the process of speciation and its relationship to systematics (classifying organisms) (Mayr 1942). In essence, the evolutionary approach provided a framework for understanding both the fit of organisms to their environment and the diversity and history of life. We will discuss the major findings of the evolutionary synthesis in many subsequent chapters.

We have seen that midway into the nineteenth century, thinkers sought mechanistic, rather than supernatural, explanations for the physical world, and scientists valued experimentation, data gathering, and hypothesis testing. Theories in geology had created a sense of deep time and gradual, versus catastrophic, changes. Robert Chambers had suggested that new species might arise from existing species, Jean-Baptiste Lamarck had hypothesized that there were generational adaptations to environmental needs, and Patrick Matthew had presented a preliminary theory of natural selection. It was in this context that Charles Darwin developed his ideas. Having laid out both the basic elements of Darwin's theory and the problems facing that theory, we are now in a good position to examine the components of evolutionary change in subsequent chapters.

SUMMARY

1. Critical changes that set the stage for Darwin and Wallace to come up with their ideas on evolutionary change and natural selection included the shift from supernatural to natural explanations, the move from catastrophism to uniformitarianism, the use of logic and pure reason, the acceptance that the world—both the biotic and abiotic world—was constantly changing, and the rejection of the idea that life formed by spontaneous generation.

2. Scientists sought mechanistic rather than supernatural explanations for the features of the physical world; they valued experimentation, data gathering, and hypothesis testing.

3. Lyell's ideas in geology created a sense of deep time, Robert Chambers proposed that new species arose from existing species, Jean-Baptiste Lamarck hypothesized generational adaptations to environmental needs, and Patrick Matthew presented a preliminary theory of natural selection.

4. Darwin prepared his readers for his revolutionary ideas on natural selection by introducing them to the artificial selection programs breeders had long used.

5. Darwin's ideas on natural selection put variation at the forefront of evolutionary change. In this way, they differed dramatically from the transformational

evolutionary changes that Lamarck had suggested at the start of the nineteenth century.

6. Charles Darwin had two great insights: (a) natural selection occurs because populations are variable and because some individuals are more successful than others at sur- viving and reproducing in their environment, and (b) all species have descended from one or a few common ancestors; species that share a recent common ancestor tend to resemble one another in many respects for the very reason that they share recent common ancestry.

KEY TERMS

catastrophism (p. 34)

evolutionary synthesis (p. 56)

hypotheses (p. 32)

inheritance of acquired characteristics (p. 39)

methodological naturalism (p. 31)

modern synthesis (p. 56)

natural history (p. 35)

population (p. 37)

saltationism (p. 55)

spontaneous generation (p. 36)

struggle for existence (p. 37)

systematics (p. 52)

transformational process (p. 48)

uniformitarianism (p. 34)

variational process (p. 49)

REVIEW QUESTIONS

1. Make the following argument: The very fact that the Greek gods themselves were fallible, rather than in- fallible, made it easier for Greek civilization to first come up with a natural interpretation of biological phenomenon.

2. Explain why the linear hierarchy of Aristotle's *scala naturae* is incompatible with Darwin's phylogenetic view of biological diversity.

3. Why do you think the discovery that species go ex- tinct was important for the development of evolu- tionary ideas?

4. Write a short rejoinder to Paley's argument that, just as a watch must have a watchmaker who sets out to make that watch, all living forms too must have a conscious designer.

5. In this chapter, we used pigeon breeding as an exam- ple of artificial selection. Choose another example of artificial selection and describe a breeding program that would produce the desired aim of the breeder.

6. Robert bought a small iPod that held a small frac- tion of his full CD collection. It seemed like too much trouble to select his favorite CDs, so he sim- ply picked 50 of his discs at random, and put them on the iPod. Each month, he deleted any of the al-

bums that he didn't listen to over the past month; he added new ones, again selected randomly, in their place. At first, Robert thought the music on his iPod was so-so, but after a year, he thought the music it contained was really great. Is this a transformational or variational process of evolution? Explain.

7. Sarah had almost exactly the opposite experience as Robert in the previous question. She bought herself a cheap turntable and a stack of her favorite records on vinyl. Her problem was that each time she played a record, the cheap phonographic needle scratched and wore down the record—so that after a year, her music collection didn't sound nearly as good as when she first bought it. Is this a transformational or variational process of evolution? Explain.

8. It is well known that many lizard species have evolved the ability to detach their tails as a mecha- nism of escaping from the grasp of predators. In his *Natural History*, Pliny the Elder (23–79 A.D.) spins a similar tale about beavers (Healy 1991). He re- ports that beavers castrated themselves in order to escape hunters who pursued them for their testicles, which could be used to produce an analgesic medi- cation (Book 8, Chapter 47). Borrowing from Pliny, the Roman author Claudius Aelianus (ca. 175–ca.

235 A.D.) describes this behavior in detail in his encyclopedic series *On the Nature of Animals* (Johnson 1997). When pursued by hunters, he writes, the beaver "puts down its head and with its teeth cuts off its testicles and throws them in their path, as a prudent man who, falling into the hands of robbers, sacrifices all that he is carrying, to save his life, and forfeits his possessions by way of ransom." Of course, beavers do not actually do anything of the sort. Explain why Darwin would have considered it reasonable that lizards should drop their tails, but implausible that beavers should self-castrate even to spare their own lives.

9. Given the fact that most people in the 1850s were familiar with the sorts of breeding programs that were used to produce dog varieties and that Victorian Englishmen and Englishwomen were fascinated with pigeon breeding, why was it such a brilliant strategy for Darwin to open *On the Origin of Species* with a discussion of artificial selection?

10. Besides the process of natural selection, can you describe any other variational processes, either in biology, chemistry, or physics? Discuss similarities and differences between your example and the process of natural selection.

SUGGESTED READINGS

Burkhardt, R. W., ed. 1984. *The Zoological Philosophy of J.-B. Lamarck*. University of Chicago Press, Chicago. An edited volume on the work of Lamarck, with special reference to his theories of transformation.

Darwin, C. 1859. *On the Origin of Species*. John Murray, London. Here Darwin lays out his grand theory of descent with modification.

Glick, T. F., ed. 1988. *The Comparative Reception of Darwinism*. University of Chicago Press, Chicago. An excellent cross-cultural discussion of the reaction to Darwin's ideas.

Malthus, T. 1798. *An Essay on the Principle of Population, As It Affects the Future Improvement of Society*. J. Johnson, London. For insight into why overpopulation and competition play a role in Darwin's theory of natural selection.

Mayr, E. 1991. *One Long Argument*. Harvard University Press, Cambridge. A short, but wonderful book on Darwin and evolutionary biology.

Raby, P. 2001. *Alfred Russel Wallace: A Life*. Princeton University Press, Princeton. A biography of evolutionary biology's co-founder.

Visit StudySpace at wwnorton.com/studyspace.

3

Natural Selection

◀ A long-nosed horned frog (*Megophrys nasuta*) exquisitely matched to its environment in Sabah, Borneo.

Droughts rarely have good consequences for human populations—crops fail, people lack drinking water, livestock starve—but they can provide unique opportunities to test hypotheses generated in the natural sciences, including evolutionary biology. Let us take a look at an example.

Southern California is accustomed to fluctuations in rainfall because of El Niño cycles, but from 2000 to 2004 the area was hit by intense drought—even by Southern California standards. The droughts were so intense, in fact, that the governor of California declared a state of emergency in every year between 2000 and 2004. The drought hit animals hard. But animals are mobile, and they have the ability to respond with flexible behaviors. They can search out cooler, wetter refuges, for example. Plants can't.

The 2000–2004 drought also dramatically shortened the growing season of *Brassica rapa*, a species of mustard plant in Southern California. What does evolutionary theory predict the response to intense drought in populations of this plant species should be? If they can't move, what can we expect to see?

Evolutionary theory predicts that in such cases, natural selection should favor plants that flower earlier in their abbreviated growing seasons. It predicts this shift in flowering time because such a strategy should increase the reproductive success of plants that flower early, compared to plants that do not flower early. Steve Franks and his colleagues put this theory to the test, using an ingenious experimental approach (Franks et al. 2007).

Franks and his colleagues wanted to test the hypothesis that *B. rapa* plants from postdrought populations flowered earlier than plants from the same populations before the drought (Figure 3.1). It sounds simple enough in principle, but how could they do this? Obtaining plants from postdrought populations was easy enough—the researchers simply went out to the field in late 2004 and collected them. But all plants from predrought years were long gone—how could the researchers compare the flowering times of plants from postdrought times to plants present before 2000 but long since gone?

The researchers' solution as to how to compare predrought and postdrought populations tells us something about the importance of long-term studies and the collection of specimens in evolution and ecology. To gain a deep understanding of their system, Franks and his team had studied this population of *B. rapa* for many years, and they had collected seeds in 1997, just a few years before the drought. They could now directly compare these two seed stocks. To control for differences in age of the 1997 and 2004 seeds, they grew adult plants from each seed stock, and crossed those plants. In this way they obtained a supply of fresh seeds from 1997 parents, and a separate supply of fresh seeds from 2004 parents. They could grow seeds from the 1997 and 2004 populations under similar conditions, and test whether natural selection had affected flowering times as they predicted. They found that plants derived from the seeds of the 2004 parents flowered earlier, on average, than plants derived from the seeds of the 1997 parents. As predicted, flowering times had shortened from 1997 to 2004, presumably as a result of natural selection imposed by the drought.

The process of natural selection has played an essential role in driving the endless modifications that lead to the biological diversity of the living world. We have discussed this process in general terms, but we are now ready for a

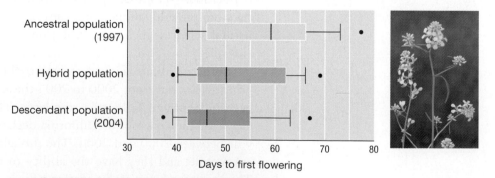

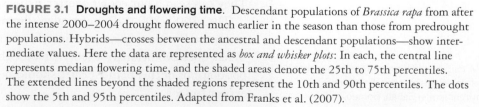

FIGURE 3.1 Droughts and flowering time. Descendant populations of *Brassica rapa* from after the intense 2000–2004 drought flowered much earlier in the season than those from predrought populations. Hybrids—crosses between the ancestral and descendant populations—show intermediate values. Here the data are represented as *box and whisker plots*: In each, the central line represents median flowering time, and the shaded areas denote the 25th to 75th percentiles. The extended lines beyond the shaded regions represent the 10th and 90th percentiles. The dots show the 5th and 95th percentiles. Adapted from Franks et al. (2007).

more detailed exploration of natural selection. We are also ready to move from Darwin's discoveries to the specific manifestation of his theory in contemporary evolutionary biology.

In this chapter, we will examine the following questions:

- What are the components of natural selection?
- What is an adaptation, and how do we study adaptations?
- How can natural selection be examined in the wild and in the laboratory?
- Why are there constraints on natural selection, and what are these constraints?
- How do complex traits originate?

3.1 The Components of Natural Selection

Though natural selection is the primary mechanism responsible for generating the exceptional diversity and complexity of all living forms, it is conceptually a very simple process. Natural selection is the inevitable consequence of three conditions being met (Figure 3.2):

1. **Variation**. Individuals in a population differ from one another.
2. **Inheritance**. Some of these differences are inherited by offspring from their parents.
3. **Differential reproductive success**. Individuals with certain traits are more successful than others at surviving and reproducing in their environment.

We will explore variation, inheritance, and differential reproductive success in detail later in this section, but before we do, let's examine why each is necessary and how together they lead to evolution by natural selection. In so doing, we should keep four points in mind.

First, mutation is one of the major sources generating the variation on which natural selection acts. While some mutations may be favored by natural selection, mutations *occur* at random with respect to the needs of the organism, independently of whether or not they would be favored by natural selection.

Second, when evolutionary biologists study the process of natural selection, they typically focus on how some *trait* of interest changes or remains constant over time. Researchers can study many different kinds of traits. They often examine a physical characteristic of an organism—for example, the color of a bird's plumage, the shape of a mammal's teeth, or the structure of a plant's flower. Other times researchers study behavioral traits, such as the elaborate dance of a lyrebird or the predator-avoidance behavior of the sea slug *Tritonia*. Sometimes the trait will simply be a genetic character: Which sequence of some particular gene does an individual have, or how many chromosomes does a species of grass have? Irrespective of the type of trait, most studies of natural selection begin by specifying which trait or traits are to be considered.

Third, natural selection does not directly sort on genotypes, but rather it sorts on phenotypic differences among the individuals in a population. Thus, to understand natural selection, we have to understand how the interplay between genotype and

FIGURE 3.2 The three components of natural selection. Evolution by natural selection occurs when there is variation, inheritance, and differential reproductive success among individuals in a population.

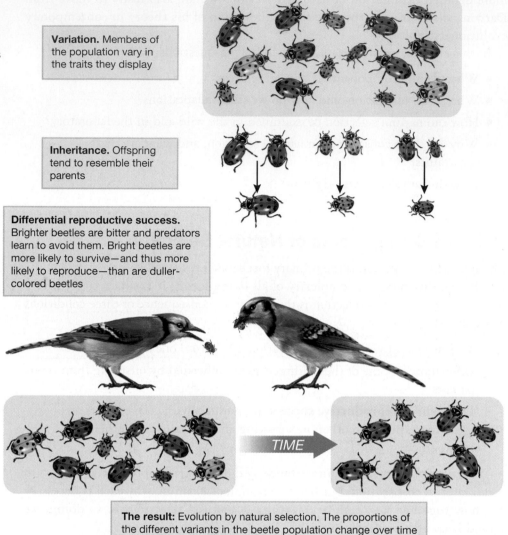

Variation. Members of the population vary in the traits they display

Inheritance. Offspring tend to resemble their parents

Differential reproductive success. Brighter beetles are bitter and predators learn to avoid them. Bright beetles are more likely to survive—and thus more likely to reproduce—than are duller-colored beetles

TIME

The result: Evolution by natural selection. The proportions of the different variants in the beetle population change over time

environment determines the phenotype. The key is to understand that a gene by itself does not code for a trait, but rather a gene codes for a trait *in the context of a particular set of environmental conditions*. For example, Figure 3.3 illustrates the way that elevation and genotype interact to determine the height of individuals in different populations of a yarrow plant (*Achillea millefolium*).

In other words, genes do not affect a trait in just one environment but rather produce what we call a **norm of reaction**. Each column in Figure 3.3 gives us the information we need to construct a norm of reaction for one particular genotype. For example, the column with green shading shows how the heights of plants of genotype 1 depend on the elevations at which they are grown. Genotype 1 doesn't produce "tall" or "short" plants. Rather, genotype 1 produces for the norm of reaction "tall at low and high elevations, short at medium elevation." Norms of reaction are often represented as functions or curves, as illustrated in Figure 3.4. Each genotype is represented by a single

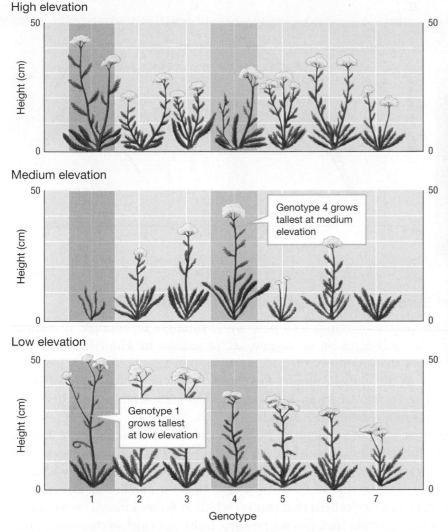

High elevation

Medium elevation

Genotype 4 grows
tallest at medium
elevation

Low elevation

Genotype 1
grows tallest
at low elevation

Genotype

FIGURE 3.3 Phenotype depends on the effects of both genotype and environment.
Here we see how the height of a yarrow plant (*Achillea millefolium*) depends on its genotype and the
altitude at which it is raised, as shown by populations of yarrow plants grown in gardens at three
sites that were at different altitudes: high, medium, and low elevation. For example, the green
screen behind the plants of genotype 1 shows that these plants grow tall at high and low elevations,
but are short at medium elevation. The blue screen behind the plants of genotype 4 shows that
these plants respond very differently to elevation. This genotype grows tallest at medium elevation
and shorter at high and low elevations. Adapted from Clausen, Keck, and Hiesey (1940, 1948).

curve, showing how expression of a genotype depends on the environmental
conditions. Environmental conditions are shown on the *x*-axis, and phenotypes
are shown on the *y*-axis. Such norms of reaction can be quite complex, with
a given genotype producing different phenotypes across an environmental
gradient, such as an altitudinal gradient.

Fourth, natural selection is a process by which the characteristics of a
population—not those of an individual—change over time. When we study
natural selection, we will typically do so with reference to one or more specified
populations of individuals.

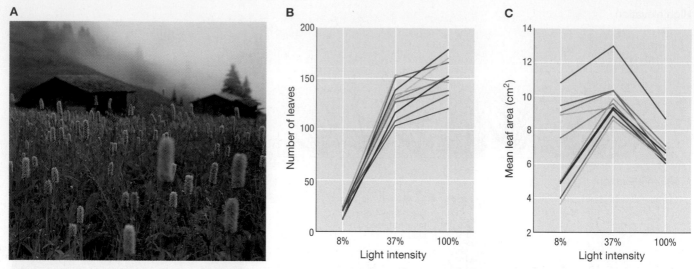

FIGURE 3.4 Norm of reaction curves for two traits. In the weedy annual plant *Persicaria maculosa* (**A**), the total number of leaves (**B**) and the mean leaf area (**C**) depend on the light intensity—ranging from full shade to full direct sunlight—that the plant experiences. Each curve for one specific genotype is called a norm of reaction. Here we see the norms of reaction for 10 different genotypes (each a different color), under light intensities of 8%, 37%, and 100% of available sunlight. Thus, the genotypes do not code for a fixed number of leaves or a fixed average leaf size, but rather for a number and size of leaves that depend on the intensity of light to which the plant is exposed. Parts B and C adapted from Sultan and Bazzaz (1993).

Natural Selection and Coat Color in the Oldfield Mouse

With these points in mind, let's now work through an example of natural selection. We will focus on an elegant set of studies by Hopi Hoekstra and her colleagues that examines the process of natural selection on coat color in populations of the oldfield mouse, *Peromyscus polionotus*. This species of small mouse, native to the American Southeast, suffers considerable mortality from visually hunting predators such as owls.

Throughout most of its range, *P. polionotus* individuals are uniformly dark in coloration. But on Santa Rosa Island off the Gulf coast of northern Florida, and along the nearby beaches and barrier islands, these mice often display a much lighter coat color. In this section, we will evaluate a number of experiments designed to test the hypothesis that natural selection favors a match between coat color and environmental background, favoring light coat color in the coastal dune populations that live on white sand, and dark coat color in inland populations that live in more vegetated environments (Figure 3.5).

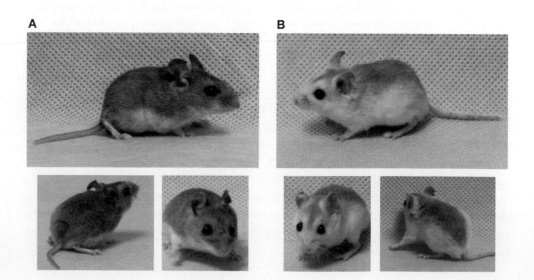

FIGURE 3.5 Coat color variation in mice. Two color variants of *Peromyscus polionotus*: (**A**) the darker inland form, and (**B**) the lighter beach-dwelling form.

Now that we have specified our trait of interest—coat color—and our populations of interest—dune and inland—we can study the process of natural selection by examining variation, heritability, and fitness in the oldfield mouse.

Variation

As we learned in the previous chapter, natural selection is a variational process, in which the properties of the members of a population change over time as a consequence of a sorting process. Thus, natural selection requires as raw material some *variation* in the trait under investigation. Without variation in a population, there is nothing for natural selection to select. If, for example, all mice had identically colored coats, natural selection with respect to coat color could not occur.

For a readily observable trait such as coat color, we can easily determine whether the first condition for natural selection—the presence of variation—is satisfied. Hoekstra and her colleagues observed considerable phenotypic variation in coat color *within* populations (Mullen et al. 2009), and they also uncovered substantial genetic variation at the *Mc1R* (melanocortin-1 receptor) locus associated with coat color. The variation in coat coloration is even more striking *between* populations, as illustrated in Figure 3.6. Although we do not presently see this wide a range of variation within any given population, the between-population variation present gives us a sense of the possible range of genetic variation in this species.

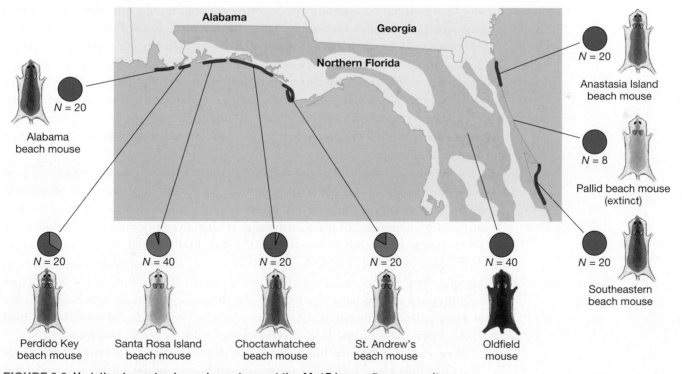

FIGURE 3.6 Variation in coat color and genotypes at the *Mc1R* locus. *Peromyscus polionotus* exhibits extensive coat color variation across localities in Florida. Red areas indicate the distribution of beach populations; gray areas denote the distribution of inland populations. Characteristic phenotypes for each population are indicated by the coat coloration sketches, but coat color varies within populations as well. The pie charts indicate that the Perdido Key, Santa Rosa Island, Choctawhatchee, and St. Andrew's beach mouse populations had more than a single variant of the *Mc1R* locus associated with coat coloration. All populations shown here are considered part of a single species—*Peromyscus polionotus*. Adapted from Hoekstra et al. (2006).

As we discussed above, phenotypes result from the interplay of genes and environment. Thus, variation in phenotype can arise through variation in genes alone, variation in environment alone, or through a combination of both. So, in principle, variation in coat color could result from genetic differences, from environmental differences such as differences in diets or in exposure to sunlight, or from some combination of these factors. Although almost any trait we might study shows both environmental and genetic variation, natural selection can operate only if there is a genetic component to variation, for reasons that we will see in the next subsection.

Heredity

As we mentioned previously, at the time that he wrote *On the Origin of Species*, Charles Darwin knew almost nothing of the mechanistic biology behind the hereditary factors that we now call genes. But any Victorian naturalist knew that offspring tend to look and act like their parents, and Darwin was a very good naturalist. So, even though Darwin didn't know about genes per se, he did know that offspring inherited something from their parents, and this something—which Darwin occasionally referred to as "gemmules"—caused offspring to resemble their parents (Darwin 1875, p. 370).

This resemblance between parents and offspring was critical for Darwin, because the process of natural selection requires *inheritance*. Without inheritance, any fitness differences among the varieties of a trait would not result in different frequencies of the trait varieties in the next generation. In the *P. polionotus* example, selection requires inheritance if it is to alter coat color in our mouse population. To see why, imagine that dark-colored mice produce five offspring on average, and light-colored mice produce ten offspring on average. If the offspring don't resemble their parents with respect to coat color, the dark parents will be no more likely to produce dark offspring than will the light parents, and vice versa. Any consequences of differing reproductive success between coat colors are lost once the parents produce new offspring.

What does it take for trait variants to be inherited? Usually inheritance in biological evolution occurs when some of the variation in the trait of interest arises from *genetic* variation. Most traits that vary do so, at least in part, because of genetic differences. Consequently, almost all traits in natural populations meet the prerequisite for inheritance (Darwin 1868; Endler 1986; Clark and Ehlinger 1987; Mousseau et al. 1999). Indeed, numerous studies from evolutionary biology, population genetics, and animal behavior suggest that many of the traits that might be relevant to natural selection—be they morphological or behavioral—are at least partially inherited from parents by their offspring (Mousseau and Roff 1987; Price and Schluter 1991; Weigensberg and Roff 1996; Hoffmann 1999).

How can evolutionary biologists show that variation in a trait is inherited? The most direct way is to identify the gene or genes responsible for this variation. In the case of the oldfield mouse, Hoekstra and her colleagues have identified several genes that are responsible for much of the coat color variation in *P. polionotus* (Hoekstra et al. 2006; Steiner et al. 2007). We will consider two of these genes here.

The first of these genes is the melanocortin-1 receptor gene (*Mc1R*), which produces a protein known to influence coat color in many species of mammals and plumage color in many species of birds. *Mc1R* functions as a critical part of a genetic

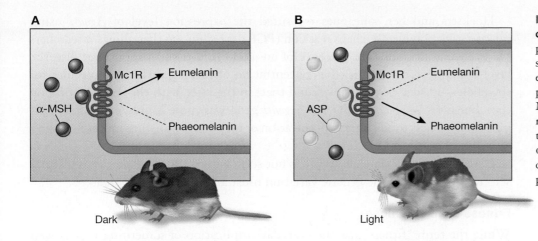

FIGURE 3.7 Genetics of coat color determination in mice. The protein Mc1R acts as a genetic switch, determining whether dark eumelanin or light phaeomelanin is produced. **(A)** α-MSH binds to the Mc1R receptor and triggers eumelanin production. **(B)** ASP binds to the Mc1R receptor, preventing α-MSH from binding and thereby causing the cell to switch over to phaeomelanin production.

switch that controls the type of pigment that is created and incorporated into hair or feathers. Depending on the environment and the interaction with other genes, this one gene switches back and forth between producing a dark pigment, known as *eumelanin*, or a light yellow pigment, known as *phaeomelanin* (Barsh 1996). When a protein called alpha melanocyte-stimulating hormone (α-MSH) is present, it binds to the Mc1R transmembrane receptor, initiating a signaling pathway that triggers the production of eumelanin. But the agouti signaling protein (ASP) can also bind to the Mc1R receptor; when it does so, it blocks the previous pathway and the cell instead produces phaeomelanin (Figure 3.7). Hoekstra and her colleagues have documented a single mutation in the *Mc1R* gene in many of the beach populations of *P. polionotus* that dwell along the Gulf coast of Florida, where oldfield mice have light coat color (Hoekstra et al. 2006) (Figure 3.8A). This mutation changes the amino acid sequence of the Mc1R protein, reducing the binding ability of that protein. This reduction in binding ability results in *reduced* eumelanin production, and thus it generates a lighter coat color.

The second major gene involved in coat color is called *Agouti*. This gene's product functions by binding to McR1 and inhibiting its further action, thereby shifting pigment production away from the darker eumelanin to the lighter phaeomelanin and generating a lighter coat color (Figure 3.8B). Hoekstra and her colleagues found that beach mice typically carry a recently evolved form of the *Agouti* allele that contributes to their lighter coat color (Hoekstra et al. 2006).

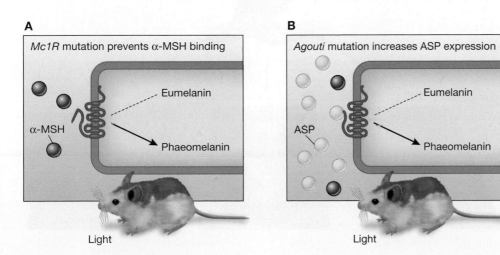

FIGURE 3.8 Two mutations contribute to light coat coloration in two different ways. (A) A mutation in the Mc1R protein reduces the ability of α-MSH to bind, and thus limits eumelanin production. **(B)** A mutation in the regulatory region of the *Agouti* gene increases ASP expression and thus further inhibits eumelanin production.

Hoekstra and her colleagues measured the expression level of *Agouti* using quantitative polymerase chain reaction (PCR), a technique that allows researchers to determine not only the presence of an allele in a tissue sample, but also the level of expression—that is, the concentration of messenger RNA molecules for the allele—in that tissue. They found that, in the mice with the *Agouti* mutation that generates light coat color, the *Agouti* gene was more highly expressed. This presumably leads to a greater concentration of the agouti signaling protein, leading to a lighter coat.

Genetic variation alone, however, is not sufficient to allow the process of natural selection to operate. The genetic variation must also have *fitness consequences*.

Fitness Consequences

While the term "fitness" has the everyday implication of something that is well matched—or *fit*—to its circumstances of life, the formal definition in evolutionary biology pertains to reproductive success. The fitness of a trait or allele is defined as the expected reproductive success of an individual who has that trait or allele *relative* to other members of the population. So, when we speak of fitness here, we are referring to the *differential effect* of the trait on the expected reproductive success of an individual relative to other individuals in its population (Fisher 1958; Williams 1966; Clutton-Brock 1988; Reeve and Sherman 1993). In many instances, it will be apparent that a trait has an effect on fitness; in the case of the mouse *P. polionotus*, we will see in a moment that coat color influences survival. The reason is straightforward. Coat color influences the visibility of mice against their background. More visible mice are more readily captured by predators; less visible mice are more likely to survive and reproduce.

To see the fitness effect of coat color, let us first examine a 1974 experiment by G. C. Kaufman in which pairs of mice, one with a dark coat and one with a light coat, were released into a large cage with an owl present (Kaufman 1974). For each environmental background—dark soil with sparse vegetation, light soil with sparse vegetation, and light soil with dense vegetation—Kaufman recorded the percentage of the time that the owl captured mice of each color. As can be seen in Figure 3.9, this experiment demonstrates a selective advantage to mice with coats

FIGURE 3.9 Early work on predation, coat color, and fitness in the oldfield mouse. In his experiments on predation, Kaufman exposed mice with light and dark coats to owl predators in three different environments: dark background with sparse vegetation (**A**), light background with sparse vegetation (**B**), and light background with dense vegetation (**C**). The percentile figures in each panel indicate the fraction of mice of each color captured by the owl. In all cases, owls captured a higher percentage of "color-mismatched" mice—namely, those with coat colors that failed to match their environments.

A	B	C
53.7% 32.8% Captured	32.1% 55.1% Captured	12.2% 23% Captured

that match the color of their background environment. Those mice are more likely to escape predators and thus to survive long enough to reproduce.

In an elegant follow-up to the Kaufman experiment, Hopi Hoekstra and her colleagues constructed silicone models that they painted to mimic either the dark or light coated oldfield mice, and they placed 125 models of each type in the natural environment of light sandy beaches or darker inland habitats (Vignieri et al. 2010). Attacks by predators could then easily be detected by looking at the plastic models for marks from teeth, talons, or beaks. They found strong evidence for a fitness advantage to mice that matched the color of their environment (Figure 3.10).

It is important to understand that small differences in fitness can translate into large changes in allele frequencies over time. For example, suppose that individual mice whose coat colors matched their environments produced just 1% more offspring per generation than those whose coats did not. Over evolutionary time, this small difference could result in a population composed completely of individuals matching their backgrounds. On average, the frequency of a gene that contributed to 1% more offspring per generation would double every 70 generations. In a population of 10,000 individuals, this gene could easily increase from a single copy to a frequency of 100% in a few thousand generations

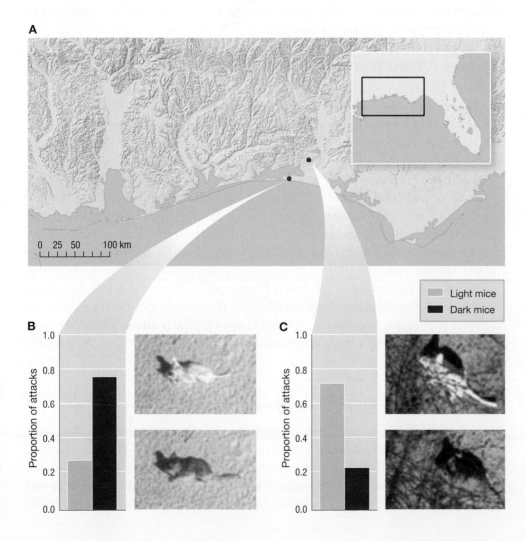

FIGURE 3.10 Predation, coat color, and fitness in the oldfield mouse, using plastic models in the field. Hoekstra and colleagues placed light and dark silicone mouse models in light and dark environments to test predation rates. **(A)** The experimental sites: a light beach environment and a dark inland environment. **(B)** Proportion of attacks against light and dark mice in the light environment. **(C)** Proportion of attacks against light and dark mice in the dark environment. Adapted from Vignieri et al. (2010).

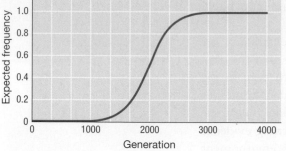

FIGURE 3.11 Small differences lead to large changes over time. Starting from a single individual in a population of 10,000—that is, starting from an initial frequency of 0.0001—a codominant gene that confers a 1% advantage in homozygous form could easily go to a frequency of 100% in only a few thousand generations.

(Figure 3.11). While that may seem like a long time, it's a very short interval on the scale of evolutionary time.

There is good reason to believe that the oldfield mouse populations that we have been discussing did indeed evolve their present coat colors fairly rapidly. Geological evidence reveals that the barrier islands on which these mouse populations live were formed only 6000 years ago. Presumably divergence in coat color has occurred since that time (Hoekstra et al. 2006).

Based on the oldfield mouse studies, natural selection appears to operate very strongly in the oldfield mouse example. Indeed, we say that coat color in the oldfield mouse example is an adaptation. Let us now examine adaptations in greater detail.

3.2 Adaptations and Fit to Environment

In Chapter 2, we discussed early theories that tried to explain the remarkable match between the structure of organisms and the environments they inhabit. Now that we understand how the process of natural selection shapes the traits of organisms, we will use the word *adaptation* to describe the results of this process. Through the production of adaptations, natural selection has shaped organisms to be well suited to the challenges they face.

Defining Adaptation

The word adaptation has been defined in many ways over the years, so we need to be specific in our own usage of this term (Mayr 1982; Sober 1987; Mitchell and Valone 1990; Reeve and Sherman 1993). An **adaptation** refers to an inherited trait that makes an organism more fit in its abiotic (nonliving) and biotic (living) environment, and that has arisen as a result of the direct action of natural selection for its primary function.

Let us unpack this definition. When we talk about an adaptation, we are referring to a trait that benefits an organism in its current environment. But not every beneficial trait is an adaptation. To be an adaptation, a trait or feature must have been shaped by natural selection. Horseshoes benefit horses by preventing cracked hooves—but the trait of having shod hooves is obviously not the direct consequence of natural selection, nor is it genetically inherited. Thus, we would not call horseshoes an adaptation, beneficial as they might be. By contrast, light coat color in the dune populations of oldfield mice is clearly an adaptation, because it is both beneficial in the current environment and it arose by natural selection for its current function.

Adaptations and Fit to Environment

Adaptations help organisms deal with both the abiotic and biotic aspects of their environment. Consider a saguaro cactus in the Sonoran Desert. The waxy coating on its surface, its shallow root system, and its low surface area-to-volume ratio are adaptations to its abiotic environment: They help it gather and retain water and survive the high temperatures and often low humidity to which it is exposed. Its

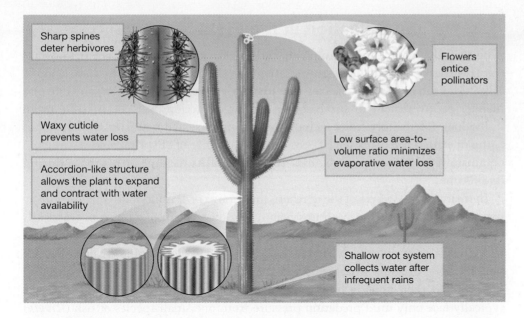

FIGURE 3.12 Adaptations of a cactus. A saguaro cactus exhibits adaptations to its abiotic environment (waxy stem coating, shallow root system, low surface area-to-volume ratio) and to its biotic environment (spines to keep away herbivores, flowers to attract pollinators).

Sharp spines deter herbivores

Waxy cuticle prevents water loss

Accordion-like structure allows the plant to expand and contract with water availability

Flowers entice pollinators

Low surface area-to-volume ratio minimizes evaporative water loss

Shallow root system collects water after infrequent rains

spines, meanwhile, are an adaptation to its biotic environment, in that they serve to protect the valuable water stored inside from herbivores that might otherwise rip open and consume the plant (Figure 3.12).

To be considered an adaptation, a trait must have been shaped by natural selection *to serve the same primary function or functions that make it beneficial today* (Williams 1966; Sober 1984). Suppose, for example, that we found that fruit flies on the Hawaiian island of Maui have thick cuticles (a hardened layer covering the body), and that thick cuticles insulate these flies against heat stress. If we could demonstrate that during the evolutionary history of this population of Hawaiian fruit flies' cuticle thickness increased as the temperature increased, then we could speak of cuticle thickness as an adaptation to heat stress in this population.

But what if we found that the frequency of thick cuticles did not increase as temperature increased over evolutionary time? What if, instead, the fossil evidence indicates that thicker cuticles only became prevalent after the introduction to the Maui population of a now-extinct predator that bored through the fruit fly's cuticle and killed it? In that case, even though thicker cuticles might protect our *present-day population* against heat stress, this trait would have evolved for another reason (protection from predators), and we would not call a thick cuticle an adaptation to heat stress. The term for a trait that serves one purpose today, but evolved under different selection conditions and served a different function in the past, is **exaptation** (Simpson 1953; Bock 1959; Gould and Vrba 1982). We will treat exaptations in detail in Section 3.6.

The term "adaptation" has a long history in the field of evolutionary biology, and it has been used in different ways by different people. If we restrict our definition of an adaptation to a trait that is shaped by natural selection for the same primary function that makes it beneficial today, then we can generate testable hypotheses about how natural selection produces adaptations. Evolutionary biologists can do just this, both in the field and in the laboratory, although at times this is a difficult and very time-consuming process. In the next section, we examine how such studies are designed, what hypotheses they test, and how the data collected have helped biologists understand the process of natural selection.

3.3 Natural Selection in the Field

A beautifully documented example of studying natural selection in the field comes from decades of work on life history strategies in the guppy *Poecilia reticulata* (Houde 1997; Magurran 2005). A species' **life history strategy** refers to the schedule and manner of investment in survivorship and reproduction over the lifetime of an individual. Life history traits include the timing of sexual maturity, the timing of aging or senescence (Chapter 20), the number and size of offspring, and whether an organism reproduces repeatedly over the course of its lifetime or just once during its lifetime.

In many of the streams of the northern mountains of Trinidad and Tobago, guppy populations can be found both upstream and downstream of a series of waterfalls (Seghers 1973; Houde 1997; Magurran 2005). Upstream and downstream sites in a stream may only be separated by a very small geographic distance (a few hundred feet in some instances), but the waterfalls act as a physical barrier to guppies and their aquatic predators alike. Upstream of such waterfalls, guppies typically face only mild predation pressure from one small species of fish (*Rivulus hartii*). Downstream of the waterfalls, however, populations of guppies are often under severe predation pressure from voracious predators such as the pike cichlid (*Crenicichla alta*).

Because upstream and downstream populations face different predation pressures, evolutionary biologists have hypothesized that natural selection should favor different suites of traits across these populations. Indeed, this turns out to be the case, and between-population comparisons in guppies have found differences in color, antipredator behavior, and numerous life history traits, including the number of offspring born in each clutch, the size of offspring at birth, the age at reproduction, and the timing of senescence (Endler 1995; Reznick 1996; Houde 1997; Magurran 2005). Let us examine some of these in more detail.

David Reznick and his colleagues found that guppies from downstream, high-predation sites mature faster than fish from upstream, low-predation sites (Reznick 1996). Females from downstream sites also produce more broods (clutches of offspring) than their counterparts in upstream sites, and broods from downstream females contain many small fry (newborn fish), while broods from upstream females tend to contain larger but fewer fry (Figure 3.13). Why? That is, why should differences in predation lead to such differences across our guppy populations?

To understand why these guppy populations have diverged, let us examine the different selective conditions at downstream and upstream sites. At upstream sites, the small fish (*Rivulus hartii*) is the only predator that guppies face. If females can produce offspring that start off relatively large and can quickly grow past a certain size threshold, such offspring will be safe from predation by *R. hartii*. So, females face a **trade-off**: Larger offspring may survive with higher probabilities, but because such offspring require more resources during their in-utero development than do smaller offspring, fewer larger offspring can be produced (Figure 3.13B).

At high-predation sites, guppy predators are much larger; they can eat a guppy no matter how large it gets. At such sites, natural selection should favor producing many smaller fry. That is, because a predator can eat a guppy fry no matter how big it is, then natural selection should now favor females that produce as many fry as possible, rather than producing larger but fewer fry, because such females will have

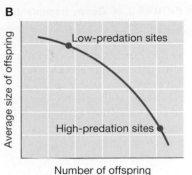

FIGURE 3.13 Natural selection and predation in guppy populations. (A) Natural selection acts differently on guppy populations from high-predation sites below waterfalls (with *Crenicichla alta*) and low-predation sites above waterfalls (with *Rivulus hartii*). At high-predation sites, selection favors guppies that produce many small young, but at low-predation sites, selection is reversed, favoring larger, but fewer, offspring. (B) Female guppies face a trade-off (red curve) between the number of offspring they can produce and the size of those offspring at birth. The optimal point along the trade-off curve illustrated depends on the predation pressures that the offspring experience.

higher reproductive success. This pattern is precisely what we see when we study reproduction in downstream females (Reznick 1996).

In the guppy system, evolutionary biologists can do more than infer adaptation by testing whether different selective conditions (high predation, low predation) correlate with the expected life history differences observed (many small fry, larger but fewer fry). In the mountain streams of Trinidad and Tobago, biologists can *experimentally manipulate* natural selection on guppy populations, make specific predictions about the changes that should occur, and test these predictions.

David Reznick, John Endler, and their colleagues experimentally manipulated predation stress in wild guppy populations by transplanting a group of 100 male and 100 female guppies from a high-predation, downstream site into a low-predation, upstream site, and they cordoned off the transplanted guppies so they could track the populations over time (Figure 3.14). *If* it is correct that producing larger but fewer offspring at upstream sites is an adaptation to predation pressure there, then given sufficient genetic variation for offspring size, we would expect that over many generations, natural selection will favor the descendants of those fish transplanted from high-predation sites who produce larger but fewer offspring than their recent ancestors (Reznick et al. 1990).

When Reznick and his team sampled the descendants of the transplanted populations 5 and 12 years after the original transplant, they found that the descendant population had evolved in the predicted direction, with females producing larger but fewer offspring than their ancestors from a high-predation site (Reznick et al. 1990). The researchers then brought guppies from the area of

FIGURE 3.14 Guppy transplant experiment. David Reznick transplanted guppies from high-predation sites below a waterfall to low-predation sites above a waterfall to test whether descendants of transplanted individuals evolved adaptations to their new selective conditions.

Guppies from high-predation site transferred to low-predation site

the transplant into the laboratory and found that the new life history strategy was inherited. Guppies from the descendant population born and raised in the laboratory adopted the same life history strategies in the lab as in the field, suggesting that the differences in life history were not solely caused by environmental differences. Thus, experimental manipulation of natural selection led to evolutionary changes in life history strategy, just as predicted.

Natural selection has also operated on various aspects of guppy behavior (Endler 1995; Reznick 1996; Houde 1997; Magurran 2005). One suite of behaviors that has been studied extensively in natural populations of guppies is their antipredator activities (Seghers 1973; Magurran et al. 1995; Magurran 2005). Depending on whether they evolved in populations with heavy or light predation stress, natural selection has produced a different suite of antipredator behaviors in guppies.

Because swimming in large, tight groups provides more protection from predators than swimming in smaller, looser aggregations, we might expect that guppies from high-predation sites would shoal in larger, tighter groups than guppies from low-predation sites (Magurran et al. 1995; Houde 1997; Magurran 2005). Data collected from natural populations confirm this prediction (Figure 3.15).

As with the work on reproductive allocation, evolutionary biologists can do more than correlate behavior with selective conditions. We can conduct manipulative experiments to see whether and on what timescale changes in selective conditions lead to changes in behavior. In the early 1990s, Anne Magurran and her colleagues

A Paria

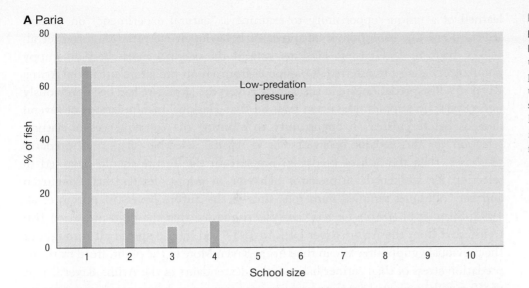

B Guanapo

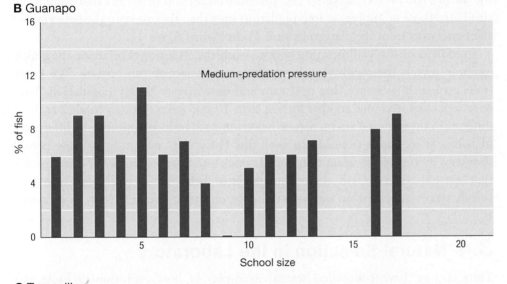

C Tranquille

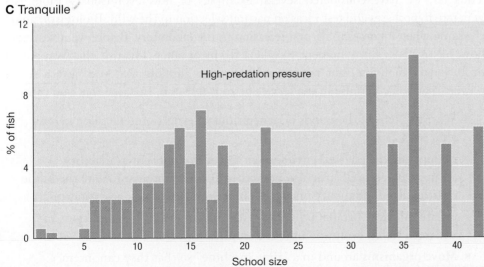

FIGURE 3.15 Natural selection, predation, and group size in guppies. Group size increases as predation pressure increases: **(A)** low-predation site, **(B)** medium-predation site, and **(C)** high-predation site. The percent of fish found in larger groups increases with increasing predation pressure. Adapted from Magurran and Seghers (1991).

learned of a unique opportunity to examine a "natural experiment" on natural selection and the evolution of antipredator behavior in guppies (Magurran et al. 1992). Back in 1957, C. P. Haskins, one of the original researchers on guppy population biology, transferred 200 guppies from a high-predation site in the Arima River to a low-predation site in the Turure River; the latter site had been previously unoccupied by guppies. Magurran realized that Haskins' manipulations of several decades before created an opportunity to examine the consequences of natural selection on antipredator behavior. For if natural selection shapes antipredator responses, then the lack of predation pressure in the Turure should have led to selection for weakened antipredator behavior in guppy descendants. Magurran and her colleagues sampled numerous sites in the Turure River (Magurran et al. 1992; Shaw et al. 1992). Genetic analysis suggested that the high-predation fish transferred from the Arima River back in 1957 had indeed spread all throughout the previously guppy-free site in the Turure River. More to the point, freed from the predation stress of their former habitat, the descendants of the Arima River fish at the Turure had evolved shoaling and predator inspection behaviors that were more similar to those of guppies at low-predation sites than they were to the behaviors of their ancestors from the dangerous sites in the Arima River.

In addition to nicely illustrating how we study the evolution of behavior, the guppy example reveals the rapidity with which natural selection can operate. We know from geological evidence that upstream and downstream guppy populations have been separated from one another for less than 10,000 years, yet as a result largely of differences in predation stress, natural selection has produced significant differences in behavior in guppy populations over this fairly brief evolutionary time period (Endler 1995). Indeed, Magurran and Reznick's transfer experiments demonstrate that natural selection can act on antipredator behavior in wild populations of vertebrates even faster than that—in this case, on the timescale of years to decades.

3.4 Natural Selection in the Laboratory

Thus far, we have considered several examples of how evolutionary biologists generate hypotheses and test ideas on natural selection in the wild. Biologists can do the same when it comes to natural selection in the laboratory. Before we investigate how, let us pause for a moment to take a flight of fancy. Imagine that you are an evolutionary biologist, but not an ordinary one. Suppose that you have a set of powers that you could use in the service of your research. Imagine that you can:

- Watch as tens of thousands of generations of evolution take place before your eyes.
- Manipulate the physical environment to control nutrient availability, temperature, spatial structure, and other features, and manipulate the biotic environment, adding or removing competitors, predators, and parasites.
- Create multiple parallel universes with the same starting conditions in which to watch evolution unfold in replicate worlds.
- Move organisms around in a "time machine" so that they can interact with—and compete against—their ancestors or their descendants.
- Go back in time to rerun evolution from any point, under the same or different environmental conditions.

■ Easily measure both allele frequencies and fitnesses to accuracies of 0.1% or smaller.

If you could do all of these things, how would you study the process and consequences of evolution? What questions would you ask, and what experiments would you do?

Lenski's Long-Term Evolution Experiment

As far-fetched a fantasy as this may seem, researchers indeed can do all of this and more when they study bacterial evolution in the laboratory. One of the most striking examples has been provided by Richard Lenski and his colleagues, who have been tracking evolutionary change for over 50,000 generations in the bacterium *Escherichia coli*. Let us examine Lenski's experimental system in some detail and see how it allows him to perform the seemingly superhuman manipulations enumerated earlier and to test fundamental ideas in evolutionary science.

Lenski's study species, *E. coli*, reproduces rapidly, dividing at rates upward of once per hour under favorable environmental conditions. As a result, Lenski and his colleagues have been able to observe evolution occurring in real time, and they have been able to monitor over 50,000 generations of bacterial evolution. To put this number into perspective, Lenski's bacterial evolution experiment now encompasses more generations than there have been in the entire history of our species, *Homo sapiens*.

Starting with a genetically homogeneous strain of *E. coli* bacteria, Lenski created 12 parallel experimental lines—the original colonists of 12 parallel "universes"—differing only by an unselected **marker gene** that allowed researchers to keep track of which experimental line was which. All 12 lines were kept in identical experimental conditions, but the 12 lines were never mixed with one another (Figure 3.16). Instead, every day, Lenski and his students transferred cells from

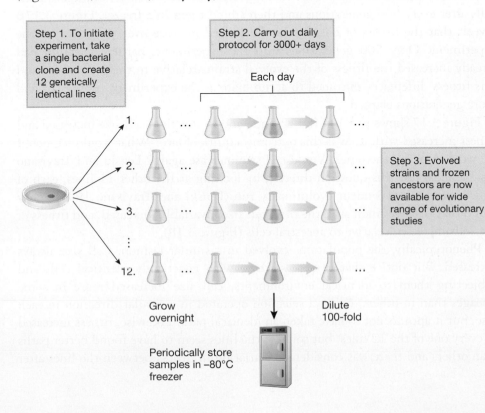

Step 1. To initiate experiment, take a single bacterial clone and create 12 genetically identical lines

Step 2. Carry out daily protocol for 3000+ days

Each day

Step 3. Evolved strains and frozen ancestors are now available for wide range of evolutionary studies

Grow overnight

Dilute 100-fold

Periodically store samples in −80°C freezer

FIGURE 3.16 Lenski's experimental evolution system. The basic protocol for the Lenski *E. coli* experiment. Each day, Lenski and his students transferred cells from the 12 lines into fresh growth medium. These cells went through 6 to 7 generations of replication overnight, and the next day the process started anew. Periodically, Lenski froze a sample of the cells from each line in a −80°C freezer. This open-ended system allows for a large number of potential experiments.

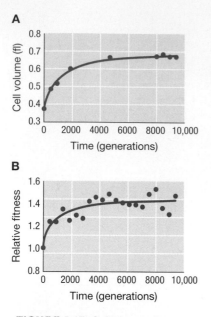

FIGURE 3.17 Cell size and fitness in one *E. coli* line.
(**A**) Change in average cell volume in one of Lenski's 12 long-term lines. (**B**) Change in fitness for the same line, relative to its ancestor. Values greater than one indicate higher fitness than the ancestor. From Lenski and Travisano (1994).

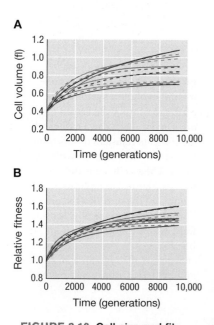

FIGURE 3.18 Cell size and fitness in 12 *E. coli* lines. Change in (**A**) cell volume and (**B**) relative fitness in each of the 12 lines. Values greater than one indicate higher fitness than the ancestor. From Lenski and Travisano (1994).

each of the 12 lines into fresh growth medium. Overnight these cells went through 6 to 7 generations of replication, and the next day the process started anew. Periodically, Lenski froze a sample of the cells from each line in a −80°C freezer. This freezer served as his "time machine": Researchers could thaw those cells at any point and could let them compete with their descendants. They could even use them to "start over" and could thus replicate the experiment from any point in time.

Evolutionary Change: Predictability and Quirks

So what can you do with an experimental system like this? We only know about one history of life: the one that actually took place on Earth and of which we are a living part. One question that has always fascinated evolutionary biologists is, what if you could "run evolution over again"? Would the same phenotypes evolve the second time around? Or would we see something completely different? And if the same phenotypes did evolve, would the same underlying genetic changes be responsible, or would natural selection find a different genetic path to a similar phenotypic outcome?

Lenski and his colleague Michael Travisano set out to address this question by comparing what happened in the 12 replicate lines—the 12 parallel runs of evolutionary history—in their experiment (Lenski and Travisano 1994). To do so, they looked at a trait that evolved rapidly early in their experiment: the physical size of the individual *E. coli* cells. These cells could be thawed at any time and allowed to compete against their descendants in order to see whether the descendants had increased in fitness or whether they had merely changed in phenotype (Box 3.1). As Figure 3.17A illustrates, the average cell volume increased substantially over the first 2000 to 3000 generations of the experiment.

In the course of their experiment, the researchers removed a sample of *E. coli* cells after every 500 generations and then stored them in a freezer. Figure 3.17B reveals that the fitness of *E. coli* cells did indeed increase over the course of the experiment. Only 500 generations into the experiment, natural selection had already increased the fitness of the evolved strains relative to their ancestors, and this fitness difference continued to accumulate as the experiment progressed and more generations elapsed.

Figure 3.17 shows just 1 of the 12 lines, and in this line, cell size increased and fitness increased with it. Was this outcome a quirk of fate? What would happen if we were to replay the tape? Would cell size increase again? Lenski and Travisano were able to test this question directly, by looking at the other 11 lines, each of which was an independent evolutionary run (Lenski and Travisano 1994). They found that in these lines, as in the first, cell size invariably increased, and fitness of the cells increased relative to ancestral cells (Figure 3.18).

Phenotypically, the populations evolved in a similar fashion. Cell size always increased. But notice that despite starting with genetically identical cells and subjecting them to identical environments, cell size increased more in some lineages than in others. Natural selection operated in a similar direction in each case, but it appears not to have taken an identical path. Likewise, fitness increased in every one of the 12 lines, but some of the lines seem to have found better paths than others and there was considerable variation in fitness between the lines after

BOX 3.1 Measuring Allele Frequencies and Fitnesses in *E. coli*

Studying natural selection in the wild can be hard, partially because of the challenges of measuring allele frequencies and fitness differences in a wild population of mobile animals such as salmon or sandpipers. When evolution is studied in the laboratory using microbial organisms, these measurements are substantially easier to perform. Researchers studying bacterial evolution in the laboratory commonly work with genetically labeled strains of bacteria. One of the most straightforward approaches to labeling is the so-called Ara$^{+/-}$ marker system. This system uses genetic markers within the *ara* operon that have no selective consequences. The strains, however, can be distinguished easily: Ara^{-} strains form red colonies and Ara^{+} strains form white colonies when grown on tetrazolium–arabinose agar. To measure the relative frequencies of two different strains, a researcher can simply spread a diluted solution containing *E. coli* cells from the population of interest, allow the cells to grow into visible colonies, and count the number of colonies of each color. Other marker systems include alternative color markers and differences in antibiotic resistance or sensitivity that a researcher can use to screen the colonies and thus distinguish the genotypes.

Measuring fitness differences is only slightly more complicated. To measure the fitness of a strain of *E. coli* relative to some other strain (for example, its ancestor), we grow each strain separately in a flask, then mix together samples from each flask, dilute, and plate as above. This allows us to measure the frequency of each strain before they begin to compete. We then grow the strains together in the same flask for some period of time, often 1 day. After this period of growth, we again dilute and plate the bacterial cells, and count colonies (Figure 3.19). From any shift in the frequencies of the two strains relative to our initial sample, we can estimate the fitness difference

between the two strains. By using the same basic approach, but with automated single-cell sorting techniques replacing the process of plating and counting colonies, researchers have been able to measure differences in fitness as small as 0.1%.

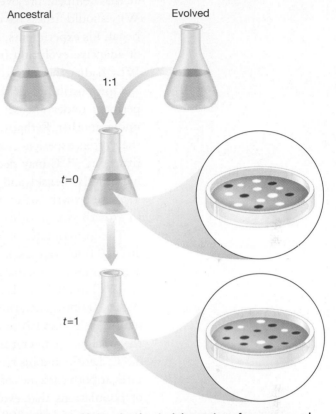

FIGURE 3.19 Measuring bacterial genotype frequency and fitness in the laboratory. Ancestral and descendant populations are competed against each other, and fitness is assayed using the neutral Ara^{+} (white) and Ara^{-} (red) markers to count colonies. Adapted from Elena and Lenski (2003).

10,000 generations. Lenski and Travisano's results highlight the fact that evolution by natural selection is in some aspects a predictable, repeatable process—and yet it is also one in which random events, such as which mutations occur, or the order in which they occur, can play a significant role in shaping the course of history.

Over the past two decades, Lenski and his colleagues have studied numerous additional traits in these 12 bacterial lines, and in doing so, they have tested a number of evolutionary hypotheses. In the next section, we will look at a thermal adaptation experiment that Lenski and colleagues used to test another important question in evolutionary biology: What are the constraints on what natural selection can achieve? Why are organisms not perfectly adapted to all environmental conditions?

Thermal Adaptation and Antagonistic Pleiotropy in *E. coli*

Let a bacterial population evolve for a few hundred generations under any particular set of laboratory conditions, and fitness under those conditions will tend to increase significantly. For example, *E. coli* is a gut bacterium that is commonly exposed to a temperature of 37°C within its hosts. Yet, Lenski and his team found that *E. coli* lines grown at a steady temperature of 37°C evolved higher fitnesses at that temperature over the course of their experiment. What is going on here? Why should fitness have increased in this experiment? After all, before Lenski ever began his experiments, *E. coli* had already undergone many billions of generations of adaptive evolution in which they might have evolved higher fitness at 37°C. Why hadn't they already done so?

One possibility is that there are trade-offs between an organism's ability to perform under one set of environmental conditions and its ability to perform under another. Perhaps *E. coli* cells are not optimized for growth at 37°C because they often experience other temperatures as well—and adaptations that increase fitness at 37°C may decrease fitness at those other temperatures. To address this hypothesis, Lenski and his colleagues asked whether evolutionary changes that increase growth rate at one specific temperature will be associated with a reduction in growth rates at other temperatures (Huey and Hertz 1984; Palaima 2007).

The growth rates of *E. coli* cells from generations 2000, 5000, 10,000, 15,000, and 20,000 were each compared to the original population of cells, and this comparison of growth rates was repeated across an array of temperatures from 20°C to 42°C in all 12 of Lenski's *E. coli* universes (Cooper et al. 2001) (Figure 3.20). After 20,000 generations in an environment where the temperature was 37°C, natural selection led to an increase in growth rate at that temperature. Moreover, the optimal temperature for growth shifted from approximately 40°C to near 37°C. Lenski and his team also found an evolutionary change toward *lower* growth rates at both extremes of the temperature range—20°C and 42°C—in the majority of populations that evolved optimal performance at 37°C (Cooper et al. 2001; Bennett and Lenski 2007).

Why did this happen? Why did evolving an optimal performance at 37°C lead to suboptimal results at the other temperatures (20°C and 42°C)? One possibility is a nonselective explanation: Perhaps after growing for 20,000 generations at 37°C, Lenski's lines had accumulated mutations that reduced their ability to grow at 20°C or 42°C. Because the bacteria were never exposed to those temperatures, natural selection would not have acted against such mutations. But Cooper and his colleagues were able to find evidence against this hypothesis in a clever way. Among their 12 lines, 3 lines evolved to become so-called *mutator* strains, with vastly higher mutation rates than those observed in the other 9 lines. If the decline in performance at 20°C and 42°C had been due to the accumulation of unselected mutations, Cooper and his team reasoned, the decline in performance should be greater in the mutator strains, because these strains accumulated far more mutations. But they found no such difference. Simple mutation accumulation seems an unlikely explanation for the fitness decline at the extreme temperatures.

Instead, the researchers suggest that their results are best explained by a phenomenon known as antagonistic pleiotropy. The **antagonistic pleiotropy** hypothesis proposes that the same gene(s) that codes for beneficial effects—here, rapid growth at 37°C—also codes for deleterious effects in other contexts—in

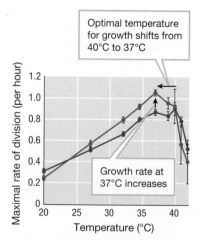

Optimal temperature for growth shifts from 40°C to 37°C

Growth rate at 37°C increases

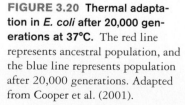

FIGURE 3.20 Thermal adaptation in *E. coli* after 20,000 generations at 37°C. The red line represents ancestral population, and the blue line represents population after 20,000 generations. Adapted from Cooper et al. (2001).

this case, poor performance at 20°C and 42°C (Figure 3.21). When genes, such as those hypothesized here, affect more than one characteristic, they are referred to as **pleiotropic genes**. And because we are testing whether such pleiotropic genes have a negative effect in one context but a positive effect in another, we refer to this as *antagonistic* pleiotropy. Thus, antagonistic pleiotropy results in a trade-off between fitness under one set of conditions and fitness under another set of conditions.

One prediction from the antagonistic pleiotropy hypothesis is that the negative components to fitness—in this case, poor performance at 20°C to 42°C—should build up quickly and early in the tested populations, because variation in response to temperature will be high at the start of the process, and hence selection for optimal performance will be most powerful. The experimental results provide support for the antagonistic pleiotropy hypothesis because suboptimal performance at extreme temperatures evolved fairly quickly in their populations, with most selection occurring in the first 5000 of the 20,000 generations of their laboratory populations of *E. coli*.

Here we have seen that because antagonistic pleiotropy involves evolutionary trade-offs, it may limit the ability of organisms to perform across a broad range of environmental conditions. In the next section, we will explore more generally the constraints on what natural selection is able to achieve.

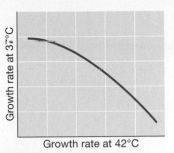

FIGURE 3.21 Antagonistic pleiotropy. The antagonistic pleiotropy hypothesis predicts a trade-off between two characters. Shown here is a hypothetical trade-off between growth rates at 37°C and 42°C.

3.5 Constraints on What Natural Selection Can Achieve

In our efforts to understand the process of natural selection, it is critical to recognize the limitations on what natural selection can achieve. In the short term, there may be limits on the genetic variation available for natural selection to operate on (Futuyma 2010). Evolutionary biologist J. B. S. Haldane captured this point in *The Causes of Evolution*:

> A selector of sufficient knowledge and power might perhaps obtain from the genes at present available in the human species a race combining an average intellect equal to that of a Shakespeare with the stature of Carnera. But he could not produce a race of angels. For the moral character or for the wings he would have to await or produce suitable mutations. (Haldane 1932/1990, p. 60)

This sort of constraint on what natural selection can achieve has been examined experimentally many times by evolutionary biologists, including in another set of *E. coli* experiments conducted by Lenski and his team. They found that, under certain conditions, the rate of adaptation in *E. coli* was proportional to the supply of new variation available (Arjan-G. et al. 1999).

Even if there is variation in a given character, selection may be unable to act on that character if the genes involved have effects on other characters that are also under selection. Another short-term constraint on natural selection is that gene flow into a local population can limit the degree of local adaptation—that is, a peripheral population may be unable to adapt to its local environmental circumstances because of continual gene flow from a larger population that faces different selective conditions.

In the long term (assuming nonextinction), these limitations may be overcome. Even in small populations, mutations that overcome some constraint may *eventually* become available; it may simply be a matter of waiting long enough. Correlated characters may become uncoupled once the appropriate mutations arise, removing the constraints associated with pleiotropy. Reproductive isolating mechanisms can reduce or eliminate gene flow into the peripheral population and thus allow local adaptation. This does not, however, mean that natural selection is free of any constraints. Rather, even in the long term, there are a number of limitations to what natural selection can achieve. First we will look at some of these limitations; then we will look at how they may be overcome.

Physical Constraints

From a spider's web, with its miniscule weight and exceptional tensile strength, to an owl's fringed feather edges that muffle any sound from its wings as they cut through the air, natural selection has fashioned countless material marvels. Nonetheless, natural selection is limited in what it can do. It operates on physical structures in the material world, and as such it is constrained by the same physical and mechanical laws that limit the realm of possibility for human engineers.

Compare the placement of the eyes in an ostrich to that in an owl (Figure 3.22). The ostrich—which must remain vigilant against predators—has eyes that are set on either side of the head, allowing a nearly 360° field of view, but affording almost no stereoscopic vision because the field of each eye scarcely overlaps with that of the other. The owl—a visual predator—has eyes that are set on the front of the head,

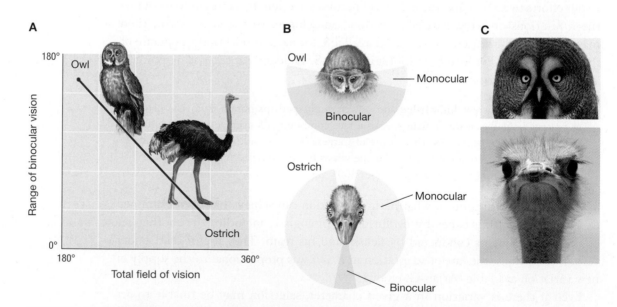

FIGURE 3.22 Trade-offs in binocular vision. **(A)** Birds face a trade-off between the total field of vision (*x*-axis) and the range of binocular vision (*y*-axis). Because of the different challenges they face, the ostrich and the owl have evolved to different points along this trade-off curve. **(B)** The position of the eyes determines where along the trade-off curve a species falls. The eyes of an owl are positioned side by side in the front of the head, limiting the field of view to about 180°, but with the benefit of binocular vision across this field. The eyes of an ostrich are set on opposite sides of its head, yielding a nearly 360° field of view. **(C)** Great gray owl and ostrich.

allowing a fully stereoscopic view of its environment, including prey species, but presenting a much more limited field of view than that enjoyed by the ostrich.

The ostrich and the owl represent two extreme manifestations of the response to the constraint that a two-eyed organism can have a 360° field of view or binocular vision across most of the visual field, but it cannot have both. For their part, owls have evolved a partial solution to this constraint: An owl can turn its neck nearly 180° over its back without shifting its perch (Figure 3.23A). The jumping spider goes one step further. It has eight eyes, allowing it to see in 360° and at the same time to enjoy a binocular (or even multiocular) forward view for visual hunting (Figure 3.23B).

Other simple physical constraints become apparent when we look at the sizes and shapes of animals (Thompson 1917; Haldane 1928; Gould 1974). Why are there no insects that are the size of wolves? Why don't single-celled swimmers have the same streamlined shape that we see in dolphins, tuna, or penguins? Why are there no elephant-sized creatures with spindly spiderlike legs?

The answer to each of these questions lies in the constraints that the laws of physics place on the form and structure of living organisms. As an example, let us consider in detail the last of these questions—why are there no elephant-sized creatures with spindly spiderlike legs? When we look at Salvador Dali's sculpture, *Space Elephant*, our intuition about the world tells us that this creature is absurd

FIGURE 3.23 Overcoming constraints. (A) A partial solution to the limited field of view: Owls can turn their heads nearly 180° to look behind themselves, as shown by this short-eared owl. **(B)** A different solution: The jumping spider has eight eyes, allowing both stereoscopic forward vision for visual hunting and a 360° field of view.

FIGURE 3.24 Art and the violation of physical constraints. In his sculpture, *Space Elephant*, Salvador Dali depicts an elephant with long, thin legs, as he did in his famous 1946 painting, *The Temptation of Saint Anthony*, which showed four elephants with long, spindly, fragile legs. Such thin legs would never support a flesh-and-blood creature of elephant-like size.

(Figure 3.24). Why? We know that, at least for elephant-sized creatures made of flesh and blood, legs like that would be too fragile to support the immense bulk of the body held high above.

Indeed, if we look at leg size (diameter) relative to body mass, we see that mammals, from the tiny pygmy shrew to the massive African elephant, conform to a tightly defined relationship between body mass and leg diameter. Figure 3.25 plots the diameter of the femur against total body mass for different species of mammals (Alexander et al. 1979). All of the mammals measured lie along a tight line across a millionfold difference in body mass. Why is this? Why has natural selection not chosen *some* solutions *somewhere* off this line? Is it an accident of history, or is there some physical constraint that shapes the relation between body mass and femur diameter?

All else being equal, organisms with longer, thinner legs will be faster and lighter. So, perhaps we should not be surprised that there are no organisms with small bodies and thick legs. But why don't we see the converse—organisms with large bodies and thin legs as illustrated by Dali? We can find the answer in the simple scaling laws of support structures, as illustrated in Figure 3.26. Looking at an ensemble of similarly shaped organisms, notice first that body mass increases with the third power of size (for example, measured as body length or height): mass $\sim$ size^3. But the strength (that is, the ability to resist compressional stress) of a supporting structure is proportional to its cross-sectional area, which scales with the second power of size: cross-sectional area $\sim$ size^2.

Because of this scaling relationship, legs must get proportionally thicker, relative to size, as an animal gets larger. Thus, it is not that we cannot have creatures with the relative proportions of Dali's elephant; it is merely impossible to have elephant-

FIGURE 3.25 Femur size and body mass. Femur diameter exhibits a tight relationship with body mass for mammals ranging in size from the 3-gram pygmy shrew to the 5000-kilogram elephant. Both the *x*- and *y*-axis are plotted on a logarithmic scale. Adapted from Alexander et al. (1979).

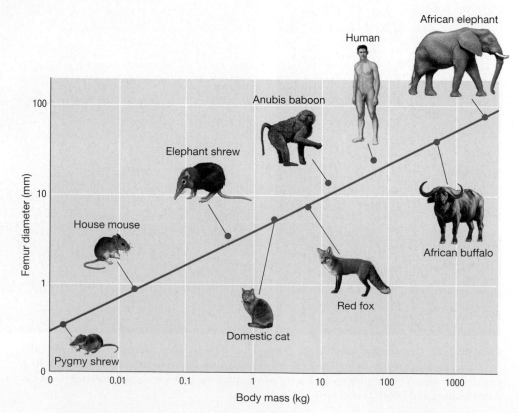

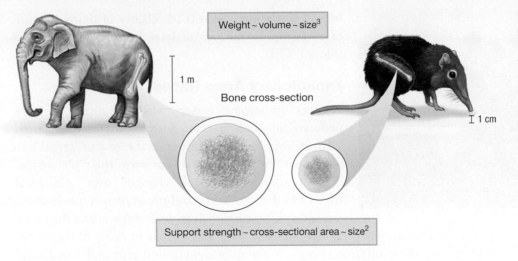

Weight ~ volume ~ size³

Bone cross-section

1 m

1 cm

Support strength ~ cross-sectional area ~ size²

FIGURE 3.26 Elephants require proportionally thicker legs. Body mass scales with the third power of size, but support strength scales with the second power of size. As a result, larger animals such as elephants require proportionally thicker legs than small animals such as elephant shrews. This physical scaling relationship underlies the pattern illustrated in Figure 3.25.

sized creatures of these proportions. The harvestman arachnids (sometimes called daddy longlegs) and *Pholcus* spiders provide examples of how, at tiny size scales, natural selection can produce creatures with a limb geometry akin to that of Dali's elephant (Figure 3.27).

Selection, no matter how strong, is hard pressed to overcome the sort of physical constraints we have discussed. We see this in striking fashion with thoroughbred racehorses, which for centuries have been bred for the extreme speed that comes from having long, thin limbs. There has been sufficient genetic variation to allow breeders to successfully change the leg geometry of these horses—but at the cost of breeding horses that do not stand up particularly well in the real world. Thoroughbred horses suffer an extraordinary rate of limb fractures and other musculoskeletal injuries, and lameness afflicts a high proportion of racehorses. Epidemiological studies from several U.S. states indicate that in a single race, a

A

B

FIGURE 3.27 Small size as an escape from constraints. Arachnids show us that the relative dimensions of Dali's elephant—a large body on long, tiny legs—are not impossible in and of themselves. The problem is having these dimensions at the size of an elephant. **(A)** The harvestman (order Opiliones) is not a true spider. **(B)** The cellar spider (*Pholcus* sp.) is a true spider, with a form that illustrates convergent evolution with the harvestman.

FIGURE 3.28 Racehorses frequently suffer catastrophic limb injuries. The racehorse Barbaro won the Kentucky Derby in 2006, but in the Preakness Stakes that year, his right hind leg shattered in 20 places, leading to his eventual death by euthanasia in January 2007.

horse has a greater than 0.1% chance of dying because of catastrophic musculoskeletal injury (Stover 2003) (Figure 3.28).

Evolutionary Arms Races

Another important reason why organisms are not perfectly adapted to their surroundings is that their surroundings do not present a stationary "target" to which natural selection can optimize their phenotype. The abiotic environment changes over geological timescales: Ice ages come and go, oxygen concentrations rise and fall, continents shift, and temperatures fluctuate. Natural selection may produce organisms with adaptations to many of these slow changes, but there are faster changes in the abiotic environment as well. Conditions vary from season to season; on a slightly longer timescale, some years are drier or wetter, hotter or colder than others. But even more important evolutionarily are the changes in the biotic environment. Much of what is significant about an organism's environment is provided by other organisms, *who themselves are evolving by natural selection as well.* It is to this topic that we now turn.

Let us look at a couple of examples in which evolutionary change in one species can affect selective conditions for a second species—a phenomenon known as **coevolution**. As a case in point, why are almost all organisms—ourselves included—vulnerable to infectious diseases? Why haven't we evolved better defenses against pathogens? We will explore this question in further detail in Chapter 20, but let us now briefly consider just one of the major reasons: We have not evolved impenetrable defenses against pathogens because our pathogens are evolving, too. As a pathogen's *hosts* evolve to deter or fight off infection more effectively, natural selection on the pathogen population intensifies, favoring variants that are able to elude the host's defenses.

The simultaneous action of natural selection on each side of the host–pathogen interaction is known as an **evolutionary arms race**, analogous to the bilateral weapons buildup that characterized the Cold War between the United States and the Soviet Union. Each side is selected to keep adding new weapons or new defenses to be able to hold its own against the other.

We see a similar evolutionary arms race in the interaction between predators and prey. Prey are selected to become increasingly effective at escaping their predators; their predators in turn are selected to become increasingly good at capturing these ever-more-elusive prey. The prey is not always able to escape, and the predator is not always able to capture its mark—because they are locked into a coevolutionary struggle. We will explore the coevolutionary process in detail in Chapter 19.

Natural Selection Lacks Foresight

A third reason why organisms are not perfectly adapted to their environments is that the process of natural selection lacks foresight. Natural selection has no way of anticipating the future beyond reacting to the past, nor can it plan ahead by multiple steps. Selection favors changes that are immediately beneficial, not changes that may be useful at some time in the future. Thus, if a new structure is to arise by natural selection alone, every step along the way must be favored.

To get a sense of just how difficult it can be to evolve major new structures by incremental changes, consider the following challenge. Suppose that we play a game in which we are given an old jalopy and a warehouse full of auto parts. Our goal is to convert the jalopy into a sleek and powerful race car—but there is a catch. Each time we swap even a single part on the car, the rules state that the car has to be in running condition. Worse yet, after each swap, we have to be able to drive the car around a racetrack in faster lap time than it could achieve prior to the swap. This certainly restricts our options for how we do the work. We cannot, for example, strip the entire car down and change the whole transmission or the whole engine in one major overhaul. Instead, we have to find a path of gradual changes, switching single bolts and single belts and single pistons one by one, always improving the lap times, and eventually producing the race car.

Natural selection has to do something similar as body plans change and new structures evolve. Those evolutionary changes that arise by natural selection tend to make the organism more fit than it was before the changes took hold. And, of course, natural selection doesn't have intentionality; it does not have a goal or target "in mind." We could even say that, in our metaphor of the race car, the player doesn't know what the parts are or what they do. The player simply tinkers with the car, making little changes, keeping those that are faster, discarding those that are not.

Despite these difficulties, this problem is not insurmountable. There may be a sequence of single part swaps that enables the car to go from jalopy to race car, always reducing the lap times. This may require that some parts of the car change functions. For example, rather than fashioning a spoiler from scratch, we might build it out of another part of the car. Perhaps we might convert the lid of the trunk into a spoiler. Why not? Race cars don't need a trunk for carrying luggage. Another possibility is that we might add new parts to the jalopy before removing old ones. We could add disc brakes before removing the current drum system. We could even add parts that we would later remove entirely; we could add structural supports to carry the car through some of the intermediate stages, and then remove them later to reduce weight.

Natural selection can take analogous paths on the way to evolving new structures. And, of course, natural selection is not the only evolutionary process operating; as we will see later in the book, mechanisms including genetic drift, founder effects, genetic hitchhiking, and many other processes also play important roles in determining the direction of evolutionary change. Thus, new structures can arise from a combination of selective and non-selective processes. In the next section, we will look at some of the ways in which new complex structures evolve by natural selection.

3.6 Origin of Complex Traits

Ever since Darwin published *On the Origin of Species*, evolutionary biologists have been fascinated by the problem of how natural selection can produce the exquisite match between organism and environment that we often observe, and how even in the absence of foresight, natural selection can create complex traits with many interdependent components.

How, for example, can we explain the exquisite complexity and detail of the human eye? How can we explain the production of milk in mammals and the associated nursing behaviors that make it such a valuable strategy for parental care? And how do we account for the coupling of wing geometry and variable wing angle that allows a dragonfly to produce the high-lift wing-tip vortices that confer its remarkable flight abilities (Thomas et al. 2004)?

In this section, we will examine two possible explanations for the evolution of such complex traits. The first explanation centers on the idea that each intermediate step on the way toward the evolution of complex traits was itself adaptive and served a function similar to the modern-day function. The second explanation—co-option of a trait to serve a new purpose—posits that intermediate stages of complex traits were functional and selected, but they did not serve the same function in the past as they do today. We will treat these in turn.

Intermediate Stages with Function Similar to Modern Function

When looking at an organ as complex as the eye, we are struck by the incredible complexity of a trait that requires so many intricate parts, all of which must work together. How could such a complex trait ever evolve in the first place? Darwin raised this issue in *On the Origin of Species*:

> To suppose that the eye, with all its inimitable contrivances for adjusting the focus to different distances, for admitting different amounts of light, and for the correction of spherical and chromatic aberration, could have been formed by natural selection, seems, I freely confess, absurd in the highest possible degree. (Darwin 1859, p. 186)

But Darwin was certain that natural selection *could* and did surmount this difficulty by small successive changes, each of which provided a benefit compared to the last version of the eye. The very next sentence of Darwin's quote reads,

> Yet reason tells me, that if numerous gradations from a perfect and complex eye to one very imperfect and simple, each grade being useful to its possessor, can be shown to exist; if further, the eye does vary ever so slightly, and the variations be inherited, which is certainly the case; and if any variation or modification in the organ be ever useful to an animal under changing conditions of life, then the difficulty of believing that a perfect and complex eye could be formed by natural selection, though insuperable by our imagination, can hardly be considered real. (Darwin 1859, pp. 186–187)

Evolutionary biologists L. V. Salvini-Plawen and Ernst Mayr have expanded on Darwin's hypothesis, laying out a series of intermediate forms that could represent one plausible sequence by which the eye evolved in gradual steps (Salvini-Plawen and Mayr 1977). Because eyes are made of soft tissue that does not fossilize well, Salvini-Plawen and Mayr used currently living species to show examples of the sorts of eye morphologies that may have been present in ancestral forms, and they found that indeed current forms can be arranged into a series of steps, each only slightly more complex than the previous, that would lead from a simple light-sensing pigment spot to a focusing eye with a lens. The aim was not to reconstruct the exact sequence by which eye evolution did occur—in fact, there is no single answer to this question, given that the lensed eye evolved in parallel in several different lineages (Figure 3.29). Rather, this work was meant to illustrate that the focusing eye, elaborate as it may seem, could have evolved in gradual steps, each of which was fully functional and each of which improved on the visual acuity of its predecessor.

But is this feasible? Is there enough time for this to have happened? Dan-Erik Nilsson and Susanne Pelger used computer simulations to explore how long it might

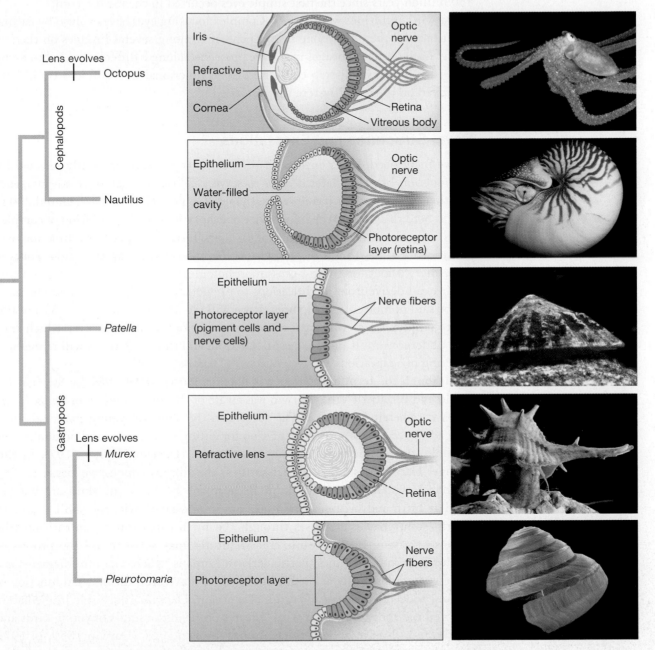

FIGURE 3.29 The evolutionary history of the eye in mollusks. Taking a phylogenetic perspective on eye morphology in the mollusks, we see that complex eyes with a lens evolved independently in the cephalopods and in the gastropods (Oakley and Pankey 2008). From top to bottom: The octopus eye uses a lens to focus light on the retina, much as does the vertebrate eye. The nautilus eye functions like a pinhole camera, casting a sharp image on the retina at the expense of a loss in brightness due to its small aperture. The limpet *Patella* has only a light-sensitive patch that can distinguish between light and dark. The predatory snail *Murex* uses a simple lens to focus incoming light. The snail *Pleurotomaria* has an indented eye cup that can detect the direction of a light source. Phylogeny is inspired by Oakley and Pankey (2008) and informed by Ponder and Lindberg (1997).

take to evolve a focusing eye from a simple light-sensitive patch (Nilsson and Pelger 1994). They assumed that individual mutations had only small phenotypic effects, and they made conservative assumptions about the rate at which natural selection would proceed under these circumstances. They found that the focusing eye could have evolved in fewer than half a million years—a very short time compared to the 550 million years since the first simple eyes occurred in the fossil record.

Darwin's intuitions were correct. Complex focusing eyes have evolved by natural selection, and they have done so independently along several lineages on the tree of life. Each of these lineages may have proceeded along a different path; but along each path, every small step could have been functional in itself, and could have improved on the visual system that preceded it.

Novel Structures and Exaptations

As we mentioned earlier in this chapter, some traits were originally selected for one function but were later *co-opted* to serve a different, selectively advantageous function. We refer to such traits as *exaptations* (Gould and Vrba 1982; Gould 2002).

It is hard to overstate the importance of exaptations in the evolution of complex traits. Any time a structure, behavior, or characteristic adopts a new function over evolutionary time, this is an exaptation. Because of the way evolution works— tinkering without foresight—gross morphological structures rarely arise de novo, but instead derive from modifications to previously existing structures. The same can be said of molecular structures, as we will see later in this section. As a result, most complex traits will have extensive evolutionary histories over which they have undergone multiple changes in function, and thus such traits will represent a "layering of adaptations and exaptations" (Thanukos 2009).

Although the term exaptation was not introduced until 1982 by Stephen Jay Gould and Elizabeth Vrba, Darwin was aware of this phenomenon in *On the Origin of Species*, wherein he wrote, "The sutures in the skulls of young mammals have been advanced as a beautiful adaptation for aiding parturition, and no doubt they facilitate, or may be indispensable for this act" (Darwin 1859, p. 197). In this passage, Darwin described cranial sutures, the fibrous connective tissue joining the bones that make up the skull. Because the bones of the skull are not yet fused at birth and because the sutures are somewhat elastic, the skull is able to deform somewhat as it passes through the birth canal during parturition (the process of giving birth). While cranial sutures may serve to aid the process of live birth in modern times (particularly in humans, where cranium diameter is a major constraint on size at birth), this need not have been the original function of sutures. Indeed, it was *not* the original function, Darwin argued. He immediately followed the above statement with "sutures occur in the skulls of young birds and reptiles, which have only to escape from a broken egg," (Darwin 1859, p. 197). Cranial sutures could not have evolved to aid the birth process in mammals, as they predated the evolution of mammalian reproduction (Figure 3.30). The original function of cranial sutures was probably to allow the rigid protective cranium to expand with a growing brain, and indeed this function is retained (Yu et al. 2004). Only subsequent to the original function, once live birth evolved, were sutures *co-opted* to facilitate passage through the birth canal (Darwin 1859). Despite Darwin's usage of the word "adaptation" in his original description, in modern terminology, these sutures are exaptations *with respect to aiding in the mammalian birth process*.

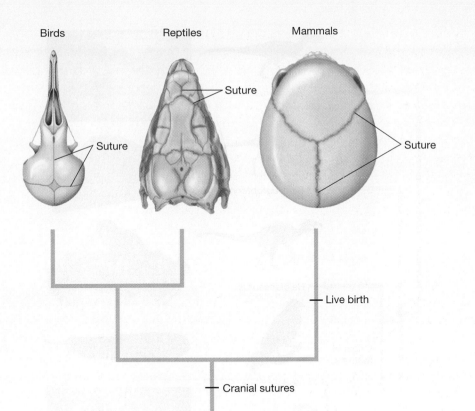

Birds Reptiles Mammals

Suture

Suture

Suture

Live birth

Cranial sutures

FIGURE 3.30 Darwin realized that cranial sutures evolved before live birth. Darwin used phylogenetic reasoning to conclude that skull sutures did not originally evolve to facilitate parturition. Because cranial sutures are present in birds, reptiles, and mammals alike, Darwin reasoned that they evolved prior to the evolutionary split between birds and reptiles and mammals, as shown. Because live birth arose after this evolutionary split, cranial sutures predated live birth and thus could not have initially evolved for the purpose of facilitating passage through the mammalian birth canal.

Let's consider another complex trait—feathers in modern-day birds—as an additional example of an exaptation. Because feathers play such a prominent role in bird flight, and because they seem so exquisitely adapted to that function, we may be tempted to assume that feathers have *always* been selected only in relation to their effect on flight.

But again, as with Darwin's example of skull sutures, phylogenetic evidence is useful for separating adaptation from exaptation (Figure 3.31). Recent paleontological discoveries from northeastern China have revealed that featherlike structures were widespread in a substantial subgroup of the bipedal *theropod dinosaurs*, which did not use these structures for flight. These dinosours ultimately gave rise to modern birds (Ji et al. 1998; Xu et al. 2001, 2009, 2010). Moreover, structural studies strongly suggest a single evolutionary origin of feathers. From this, we can deduce that the origin of feathers predates the evolution of wings and flight.

In light of the phylogenetic evidence that feathers evolved prior to flight, it would be a mistake to conclude that feathers originally evolved as an adaptation for flying. Natural selection cannot look ahead to fashion a structure that only later will become useful. Biologists Richard Prum and Alan Brush offer an appealing analogy: They say that, in light of the phylogenetic evidence, "Concluding that feathers evolved for flight is like maintaining that digits evolved for playing the piano" (Prum and Brush 2002, p. 286).

So, what might have been the original function(s) of feathers? Over the years, researchers have proposed a number of possibilities, including (1) retaining heat, (2) shielding from sunlight, (3) signaling, (4) facilitating tactile sensation, as whiskers do, (5) prey capture, (6) defense, and (7) waterproofing (Prum and Brush 2002).

Let's just look at one of these functions—thermoregulation—as an example (Evart 1921; Bock 1969; Ostrom 1974). Feathers, especially the contour feathers

FIGURE 3.31 The evolutionary origin of feathers. Phylogenetic reasoning reveals that feathers did not originally evolve for flight. Feathers arose in a lineage of theropod dinosaurs. The common ancestor of these feathered dinosaurs (including birds) is marked with a closed red circle. This species had neither wings nor the ability to fly. Therefore, feathers must have initially evolved for some other purpose. Gliding and flight subsequently evolved in the lineage leading to *microraptor*, *archaeopteryx*, and modern birds; at this stage feathers were co-opted to facilitate flight.

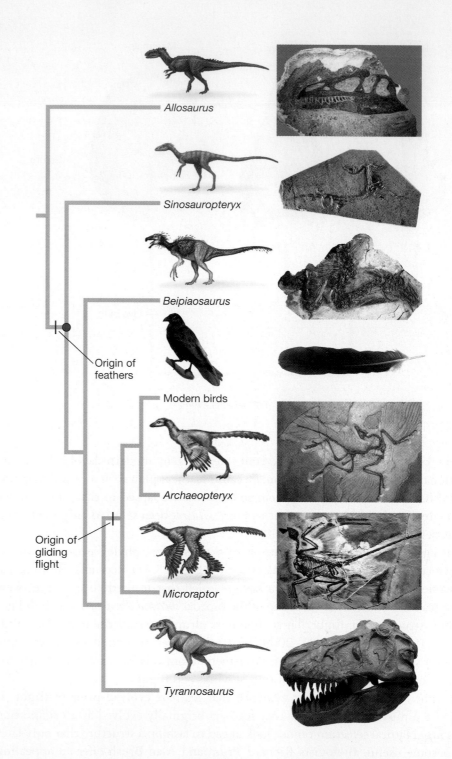

that are already seen in *Archaeopteryx*, help control thermoregulation, both because feather down is itself an insulator and because the air space between feathers acts to insulate animals against temperature change (Ostrom 1974). This early thermoregulatory function also appears to have been very important in the evolution of wings in insects (Kingsolver and Koehl 1985).

Of course, thermoregulation is not mutually exclusive with the other proposed functions. In any event, given presently available evidence, there is little prospect for distinguishing among these alternatives in identifying the original selective function or functions.

Using the arguments we developed above, we can say that the basic structure of feathers is, in part, an exaptation with respect to bird flight. That does not mean that feathers, once selected for their initial function, were not subsequently shaped by natural selection because of the fitness effects associated with flight in birds. Rather, once selected for thermoregulation or other purposes, any changes to feathers that also made them more beneficial for early flight would likely have been selected.

Notice that when a trait switches function, the organism need not lose the original function. Sometimes the trait can serve both purposes. Skull sutures facilitate brain growth and aid parturition. Feathers can serve both to insulate the bird and to facilitate flight. Next, we will consider two examples of how novelty arises at the molecular level; in each, the original function is maintained in a different way.

Novelty at the Molecular Level

Whether at the gross morphological level or at the level of individual molecules, the process of evolution is ever tinkering with extant structures. Lacking foresight, the evolutionary process rarely fashions new structures entirely from scratch. One way that new molecular functions can arise is through the process of **gene sharing**, in which a protein that serves one function in one part of the body is recruited to perform a new and different function in a second location.

There is no better illustration of the breadth and diversity of gene sharing than the lens crystallin proteins. Lens crystallins are structural proteins that form the transparent lens of the eye. While some lens crystallins are used only in the lens, many are dual-function proteins that are also used as enzymes elsewhere in the body. Table 3.1 lists a number of the lens crystallins that also function as enzymes.

The process of **gene duplication** provides another evolutionary pathway by which a protein can switch functions, without loss of the original function. In a gene duplication event, an extra copy of a working gene is formed. Once an organism has two copies of the gene, one of the two gene copies might change to a new function, while the other can remain unchanged and thus preserve the original function. We conclude this section with one such example.

TABLE 3.1

Examples of Gene Sharing: Lens Crystallins with Separate Enzymatic Functions

Crystallin	Species	Enzyme
δ	Birds and reptiles	Argininosuccinate lyase
ε	Birds and crocodiles	Lactate dehydrogenase D4
τ	Lamprey, fish, reptiles, and birds	α-Enolase
λ	Rabbit	Hydroxyacyl-CoA dehydrogenase
ζ	Guinea pig	Alcohol dehydrogenase

Adapted from Piatigorsky and Wistow (1989).

FIGURE 3.32 Lock-and-key systems. The lock-and-key mechanism of many hormone–receptor pairs.

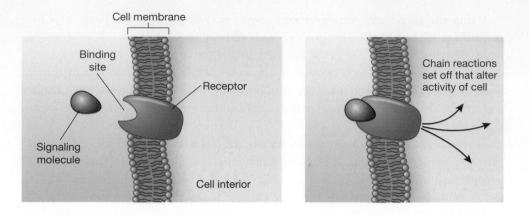

One particularly complex phenotypic suite of traits is the lock-and-key mechanism of many hormone–receptor pairs, with their exquisite specificity (Figure 3.32). These hormone–receptor pairs pose a chicken-and-egg problem: How could a signaling protein possibly evolve to match a receptor that has not yet arisen, or, conversely, how could a receptor evolve to accept a signal that does not yet exist?

Jamie Bridgham and her colleagues worked out a detailed answer to this question for one such lock-and-key pair: the mineralocorticoid receptor (let's call it the M receptor) and the steroid hormone, called aldosterone, that triggers it (Bridgham et al. 2006). The M receptor, which is involved in controlling the electrolyte balance within cells, arose in a gene duplication event from an ancestral glucocorticoid receptor.

But how did this gene duplication lead to a novel and highly specific aldosterone–M receptor pair? Again, a phylogenetic approach was the key to unraveling this mystery. By sequencing the mineralocorticoid receptor genes from a wide range of vertebrates, Bridgham's team was able to infer the genetic sequence of the ancestral receptor that was duplicated to produce both the M and modern glucocorticoid receptors.

Bridgham and her colleagues found that the ancestral receptor binds not only cortisol (a glucocorticoid hormone) but also aldosterone. This is surprising because it means that the ancestral receptor could bind a hormone that didn't exist when the ancestral receptor was in place—aldosterone evolved much later. But cortisol was already in existence at the time of the ancestral receptor. Evolutionary biologists have hypothesized that, after the gene duplication, a pair of mutations altered the shape of what is now the glucocorticoid receptor, so that it retained its ability to bind cortisol but would no longer bind aldosterone. At the time, aldosterone wasn't present yet, but over millions of years, genetic changes in biosynthetic pathways (associated with cytochrome P-450) by chance eventually led to the production of aldosterone. Because aldosterone could now trigger the M receptor without interfering with the glucocorticoid receptor, there was a new signal–receptor pair that could be used independently to regulate other cellular processes. Now we know which came first in this chicken-and-egg problem. The ability of the receptor to bind aldosterone preceded the evolution of aldosterone itself (Figure 3.33).

We have seen how the process of natural selection requires three components—variation, heritability, and fitness differentials. When a trait has been under natural selection for a specific function in a specific population, and that trait

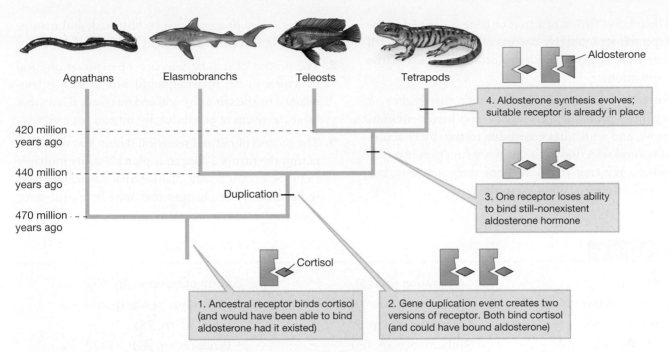

FIGURE 3.33 Gene duplication and the evolution of the aldosterone receptor. Neither the aldosterone hormone nor the aldosterone receptor were present in the vertebrate lineage 470 million years ago. (1) A single glucocorticoid receptor bound cortisol—and would have bound alderstone, had it been present. (2) Around 450 million years ago, a gene duplication created a second copy of the glucocorticoid receptor. (3) Subsequently, genetic changes to one of these receptor copies shifted its structure so that it would not be able to bind aldosterone. The other retained aldosterone binding ability. (4) In the tetrapods, when aldosterone synthesis arose, a receptor was already in place that could bind aldosterone. Because other glucocorticoid receptor had changed structure so that it could bind cortisol but not aldosterone, that pathway was not disrupted by the advent of aldosterone synthesis. Adapted from Bridgham et al. (2006).

serves the same primary function or functions today as it did in the past, we call it an adaptation. Adaptations can be studied both in the wild, as we saw with oldfield mice and guppies, as well as in the laboratory, as we discovered in our discussion of cell size and temperature sensitivity in *E. coli*. Through the use of studies that have ranged from the scale of the molecule to the whole organism, we have also explored various ways that the evolutionary process can lead to complex traits, such as the vertebrate eye and feathers in birds, through both classic step-by-step adaptation for a specific function as well as through exaptation.

We now shift our emphasis from natural selection and the adaptations it produces, to common descent and phylogeny in Chapters 4 and 5.

SUMMARY

1. Much of the research on evolution by natural selection done since Darwin relies on a solid and ever-expanding understanding of the process of genetic transmission.

2. Evolution by natural selection is the inevitable consequence of three simple conditions: variation, inheritance, and differential reproductive success.

3. Natural selection does not act directly on genotypes: It operates on phenotypic differences among the individuals in a population.

4. Evolution by natural selection is a process by which the characteristics of a population—not those of an individual—change over time.

5. The relative fitness of a trait or gene is defined as the expected reproductive success of an individual with that trait or gene, *relative* to other members of the population.

6. An adaptation is an inherited trait that makes an organism more fit in its abiotic and biotic environment, and which has arisen due to the direct action of natural selection for its primary function. An exaptation is a trait that serves one purpose today, but served a different function in the past.

7. Evolutionary processes can be observed, and manipulated, in real time in the field and in the laboratory.

8. The process of natural selection operates on physical structures in the material world, and as such is constrained by the same physical and mechanical laws that limit the realm of possibility for human engineers.

9. The process of natural selection has no way of anticipating the future\, nor can it plan ahead by multiple steps. Selection favors changes that are immediately beneficial, not changes that may be useful some time in the future.

KEY TERMS

adaptation (p. 72)

antagonistic pleiotropy (p. 82)

coevolution (p. 88)

differential reproductive
    success (p. 63)

evolutionary arms race (p. 88)

exaptation (p. 73)

gene duplication (p. 95)

gene sharing (p. 95)

inheritance (p. 63)

life history strategy (p. 74)

marker gene (p. 79)

norm of reaction (p. 64)

pleiotropic genes (p. 83)

trade-off (p. 74)

variation (p. 63)

REVIEW QUESTIONS

1. We have focused on genes as the means by which information is transferred across generations. Is that the only way that such a transfer of information can occur? Richard Dawkins and many others have suggested that cultural transmission is another. Examples given by Dawkins include musical tunes, fashions in clothing, and architectural techniques. Could some analogue of natural selection operate when culture is the means by which information is transferred from one generation to another?

2. A norm of reaction maps the way that genes are expressed in different environments. Distinguish this from the Lamarckian idea of the "inheritance of acquired characteristics" we discussed in Chapter 2.

3. Figure 3.3 shows how the heights of yarrow plants depend on genotype and environment. Redraw the data from this figure as a set of norm of reaction curves, analogous to those shown in Figure 3.4.

4. How has experimental evolution—along the lines of the *E. coli* experiment we discussed—revolutionized the sorts of questions evolutionary biologists can now test?

5. Why do unicellular swimming organisms (**A**) have a very different body shape than do swimming vertebrates (**B**)? Unicellular swimmers lack the streamlined form of large swimming vertebrates. Hint: At the size scale of unicellular organisms, the inertial forces that make up our everyday world become negligible, and instead shearing forces dominate. As a result, fluid flow is smooth and "laminar" at the unicellular scale instead of turbulent as it is at the size scale of swimming vertebrates.

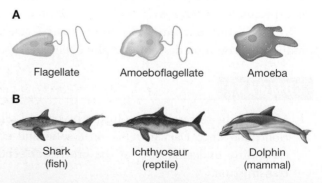

A

Flagellate   Amoeboflagellate   Amoeba

B

Shark   Ichthyosaur   Dolphin
(fish)   (reptile)   (mammal)

6. Counter the following argument: "Exaptations are common, therefore natural selection is not nearly as important as many biologists have claimed."

7. What sort of medical consequences might the antagonistic pleiotropy hypothesis have for our understanding of diseases that are often associated with old age (for example, Alzheimer's disease)?

8. Jacques Monod said that evolution operates like a "tinkerer." What do you think he meant by this?

9. Some bivalve mollusks use the same organ for feeding and breathing. How might this be thought of as a kind of historical constraint that shapes the sort of evolutionary change that might occur in this group?

10. Explain how it can be true that natural selection acts on *phenotypes*, but the result of natural selection is often measured in terms of changes to *gene* frequencies?

SUGGESTED READINGS

Hoekstra, H. E., J. M. Hoekstra, D. Berrigan, S. N. Vignieri, A. Hoang, C. E. Hill, P. Beerli, and J. G. Kingsolver. 2001. Strength and tempo of directional selection in the wild. *Proceedings of the National Academy of Sciences of the United States of America* 98: 9157–9160. A somewhat technical review of a series of studies on the strength of natural selection in different systems studied by evolutionary biologists.

Lenski, R. E. 2004. Phenotypic and genomic evolution during a 20,000-generation experiment with the bacterium *Escherichia coli*. *Plant Breeding Reviews* 24: 225–265. A very good review of Lenski's long-term experiment on evolutionary change in *E. coli*.

Orr, H. A. 2009. Testing natural selection. *Scientific American* 300: 44–50. A general overview of the process of natural selection written for a lay science audience.

Reeve, H. K., and P. W. Sherman. 1993. Adaptation and the goals of evolutionary research. *The Quarterly Review of Biology* 68: 1–32. A (somewhat lengthy) review of the concept of adaptation and its role in the evolutionary process.

Weiner, J. 1995. *The Beak of the Finch: A Story of Evolution in Our Time*. Vintage Books, New York. A wonderful book on Peter and Rosemary Grant's work on the Galápagos finches that so fascinated Darwin. The book is nonfiction, but it reads like an adventure tale.

 Visit StudySpace at wwnorton.com/studyspace.

4

Phylogeny and Evolutionary History

◀ A delicately branching whisk fern, *Psilotum nudum*, rises from recently deposited lava in Hawaii Volcanoes National Park.

The world is filled with a bewildering diversity of forms; nowhere is this more true than in the biological domain (Figure 4.1). To make sense of the world with all of its variation, we categorize the objects in it—but this is a difficult endeavor in its own right. What is the best way to break up the infinite variety out there in the world into a set of discrete categories? The Argentine writer Jorge Luis Borges describes one fanciful approach, as taken in a fictional Chinese encyclopedia known as the *Celestial Emporium of Benevolent Knowledge*:

> In its distant pages it is written that animals are divided into (a) those that belong to the emperor; (b) embalmed ones; (c) those that are trained; (d) suckling pigs; (e) mermaids; (f) fabulous ones; (g) stray dogs; (h) those that are included in this classification; (i) those that tremble as if they were mad; (j) innumerable ones; (k) those drawn with a very fine camel's-hair brush; (l) et cetera; (m) those that have just broken the flower vase; (n) those that at a distance resemble flies. (Borges 1952, in Simms 1964, p. 103)

FIGURE 4.1 An artist's view of biodiversity. Henri Rousseau's *Exotic Landscape* (1910) and a detail of *The Merry Jesters* (1906).

To most of us, this classification scheme seems strange and disorienting—and that was exactly Borges' intent. But what is the "right" way to divide up the diversity of living things?

Evolutionary biology provides an answer to this question. A bit of history shows how. The basic *Linnaean taxonomy* and resulting system of scientific names that biologists have used for nearly three centuries did not derive from evolutionary thinking. The taxonomic system was developed by Carolus Linnaeus (1707–1778), a Swedish botanist, zoologist, and physician, who wrote *Systema Naturae*. This taxonomy has proved so very useful because of Linnaeus' insight that organisms can be arranged in a hierarchical classification. Linnaeus recognized that not only can we assign species or subspecies to groups of highly similar organisms, we can also array these groups of similar species into larger groups of moderately similar organisms, and these larger groups can in turn be categorized into yet larger groups of somewhat similar organisms, and so forth, until we have accounted for all living things. Linnaeus came to this realization without having a theoretical basis for *why* these hierarchical patterns of similarity should exist. As we discussed in Chapter 2, Darwin provided the answer for why these patterns are seen. He recognized that an evolutionary process of branching descent with modification would generate nested hierarchies of similarity as the natural results of phylogenetic history. Not only did Darwin's idea of a branching pattern of descent with modification provide a theoretical foundation for the hierarchical patterns Linnaeus suggested, but Darwin's approach has also led to changes in the classification of many species, genera, and families.

German biologist Willi Hennig (1913–1976) eventually revisited the problem of taxonomy using Darwin's ideas and, in doing so, established the modern approach to classification (Zuckerkandl and Pauling 1962; Hennig 1966). The title of Hennig's classic 1966 book—*Phylogenetic Systematics*—is instructive, because it emphasizes that, in addition to documenting evolutionary history, phylogenetic trees can help us classify, or *systematize*, the world we see around us. We could classify organisms in many ways—for example, by how large they are, by where they are located, or by their morphology. But in **phylogenetic systematics**, we classify organisms according to their evolutionary histories—and phylogenetic trees are our way of representing these evolutionary relationships.

Our goal in this chapter is to introduce the central role of phylogenetic thinking within evolutionary biology. In so doing, we will address the following questions:

- How do we read and interpret a phylogenetic tree?

- How do phylogenetic trees help us make sense of—and classify—the diversity of life?

- How do phylogenetic trees help us understand the evolutionary origin of similarities among species and differences between species?

- How do we map characters onto phylogenetic trees to generate and test hypotheses about evolutionary events?

4.1 Phylogenies Reflect Evolutionary History

One of the principal aims of modern evolutionary biologists is to reconstruct and understand patterns of descent, and to use knowledge about the patterns of descent to understand the evolutionary events that have transpired throughout the history of life on Earth. This is the study of **phylogeny**—the branching relationships of populations as they give rise to multiple descendant populations over evolutionary time.

On a grand scale, the study of phylogeny allows us to reconstruct the tree of life—the historical relationships that connect all living things—and to understand the major events in evolutionary history. On a narrower scale, we may be interested in understanding the history of descent and relationships among genera with a family of organisms, species within a genus, or even among populations of a single species (Figure 4.2). Doing so requires taking a historical perspective and probing for evidence of common ancestry, as well as for information that sheds light on how various species are related to one another (Box 4.1).

The study of phylogeny rests on our observations of traits displayed by organisms. **Traits** can be any observable characteristics of organisms; for example, they may be anatomical features, developmental or embryological processes, behavioral patterns, or genetic sequences. Until the major advances in molecular genetics that occurred in the 1970s, almost all trait measurements used in the study of phylogeny were morphological or anatomical—bone length, tooth shape, and so on. With the advent of molecular genetics, actual DNA sequences are now the most common traits used to reconstruct phylogenies of extant organisms.

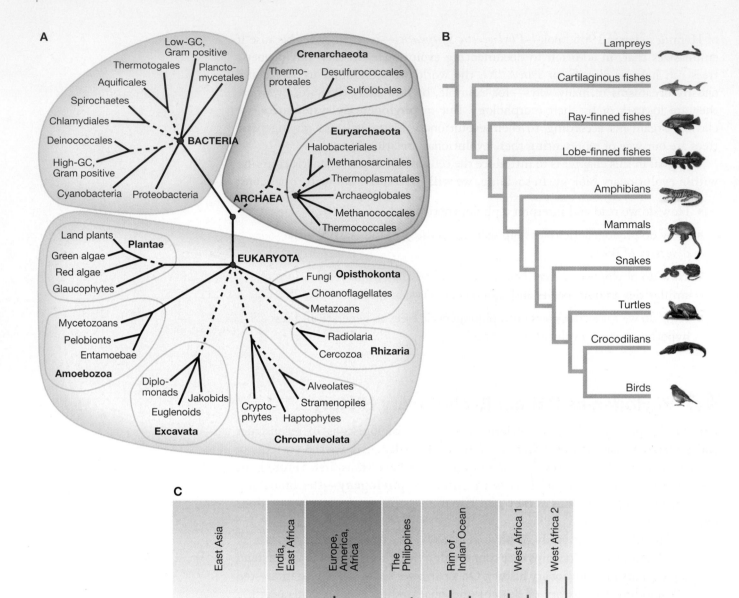

FIGURE 4.2 Phylogenies at different scales. **(A)** The tree of life represents the historical relationships among all living things. Dashed lines represent hypothetical relationships. The entire animal kingdom is contained in the tiny orange branch at far right (in the metazoans). Adapted from Delsuc et al. (2005). **(B)** A phylogeny of vertebrates. Adapted from the Center for North American Herpetology (2010). **(C)** A phylogeny of *Micobacterium tuberculosis* isolates from human patients, with geographical origins indicated. Adapted from Comas et al. (2010).

BOX 4.1 What Is the Difference between a Pedigree and a Phylogeny?

If you have ever studied your own family history, you may have come across diagrams known as *family trees* or *pedigrees*. An example is shown in Figure 4.3.

In some ways, pedigrees may seem very much like phylogenies. Both represent patterns of ancestry using treelike branching diagrams. But there are important distinctions. A pedigree tells us about the ancestry of *individuals*, whereas most phylogenies tell us the ancestry of *populations*. Thus, the *nodes* in a pedigree represent individuals, while the nodes in a phylogeny typically represent populations. Moreover, because every individual of a sexual species has two parents, each node in a pedigree has

two immediate ancestors (mother and father), and can leave any number of immediate descendants. By contrast, in a conventional phylogeny, we assume populations *split* in two, but never recombine. Thus, in a phylogeny, each node has a single direct ancestor and two direct descendants (if any). As a result, a pedigree tends to expand as one looks backward in time: two parents, four grandparents, eight great-grandparents, and so forth. By contrast, a phylogeny expands as we move forward in time. Both are often drawn in a fanlike shape, broad at the top and narrow below; by convention, time typically runs downward in a pedigree and upward in a phylogeny.

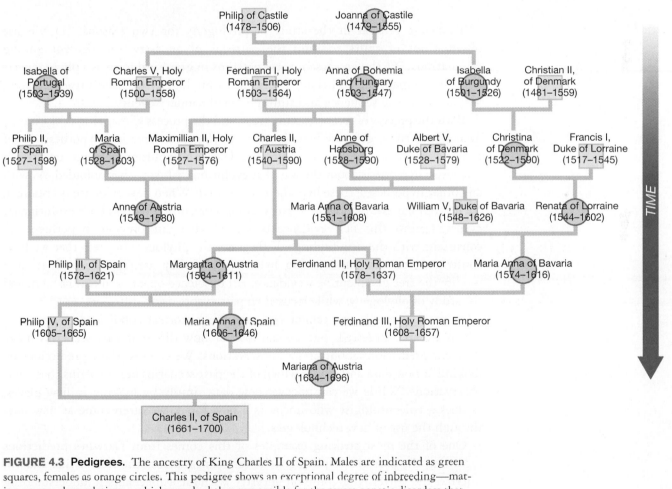

FIGURE 4.3 Pedigrees. The ancestry of King Charles II of Spain. Males are indicated as green squares, females as orange circles. This pedigree shows an exceptional degree of inbreeding—mating among close relatives—which was doubtless responsible for the severe genetic disorders that crippled Charles II, the last of the Spanish Habsburgs. Adapted from Wikimedia Commons (2006).

FIGURE 4.4 Traits and trees. We use traits both to reconstruct phylogenetic trees and to generate hypotheses about the timing of events in evolutionary history. **(A)** One set of traits—here genetic sequence data—is used to infer a phylogenetic tree for the species of interest. **(B)** A second set of traits, here flower color and morphology, are mapped onto the tree, helping us to reconstruct evolutionary events. The origin of the dark flower coloration is indicated by the filled horizontal bar. The origin of the novel flower shape is indicated by the open horizontal bar.

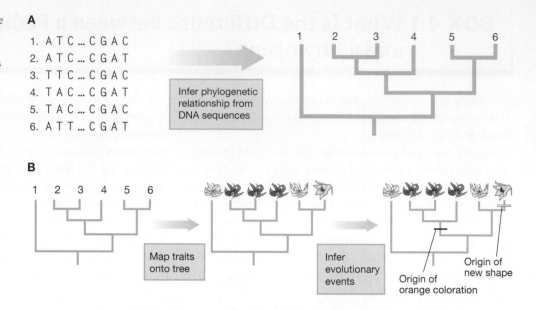

Traits are critical in the study of phylogeny for two reasons: (1) We use observations of traits to infer the patterns of ancestry and descent among populations. We then represent these patterns in graphical form as a phylogenetic tree. (2) By mapping additional traits onto a phylogeny we have already created, we can study the sequence and timing of evolutionary events (Figure 4.4).

Both the process of reconstructing trees, and the process of mapping evolutionary events onto trees, generate hypotheses. A phylogenetic tree is a hypothesis about evolutionary relationships. The location and order of evolutionary events on a tree is likewise a hypothesis about the way that evolutionary history has unfolded. As with any other hypotheses, these hypotheses are tested: When new evidence is obtained, we test our current phylogenetic trees, or our current inferences about evolutionary events, against this new evidence to see whether our previous hypotheses are consistent with the new findings. If they are, the phylogenetic trees that we have constructed remain our working hypotheses; if they aren't, we reevaluate and modify the tree given our new evidence. All of science operates in this fashion, and the study of phylogeny, while focused on past events, is no different.

In most instances, we cannot replicate the historical conditions or events in which we are interested, but we can look at how different past scenarios make different predictions about present observations. We can test these predictions by looking at new data and seeing which of the past scenarios best explains these new observations. While we can uncover new data simply by looking in new places, as does a paleontologist who uncovers a new fossil, we often come at new data through the use of new technologies.

One of the most striking examples of this comes from Darwin's predictions regarding the patterns of phylogenetic relatedness across the tree of life. Darwin inferred the patterns of common ancestry without a mechanistic understanding of genes, DNA, or heredity. His hypothesis about past events—the patterns of common ancestry of all living things—made a strong prediction that later became testable. Once DNA was identified as the carrier of hereditary genetic information and the revolution in molecular genetics allowed researchers to easily read off this information by DNA sequencing, scientists had a vast body of new data with which to test Darwin's hypotheses about ancestry. If Darwin's theory of descent with modification is correct, patterns of DNA sequence similarity should reflect the patterns of common ancestry that have been inferred from other evidence, such as morphological characters, fossil evidence, and phylogeography. We would not expect such patterns of DNA sequence similarity under hypotheses of special creation or Lamarckian spontaneous generation. It has been a major triumph for evolutionary biology that the enormously rich data about genetic sequences, although entirely unknown to Darwin, strongly support the very mechanisms that he proposed for the origin of species.

4.2 Reading Phylogenetic Trees

Before going further, let us explore how to read a phylogenetic tree. The tree in Figure 4.5 shows the pattern of evolutionary relationships among the vertebrates. In this phylogeny, each branch tip represents a group of related organisms, or a **taxon**. This phylogeny shows the relationships among such taxa (the plural of taxon) as birds, crocodilians, and mammals. Figure 4.5 shows two different

FIGURE 4.5 Two equivalent ways of drawing a phylogeny. The two phylogenies of the vertebrates shown each illustrate exactly the same information. The phylogeny on the left (**A**) is sometimes referred to as a *tree* representation, whereas that on the right (**B**) is termed a *ladder* representation. In each, time flows from left to right, so that the branch tips at the right represent current groups, whereas the *interior nodes* (nodes on the inner section of the tree) represent ancestral populations. For example, the red dot indicates the common ancestor of birds and crocodilians, whereas the blue dot indicates the common ancestor to all tetrapods. The orange line segment is the root of the tree, the ancestral lineage from which all other lineages on the tree are derived. Adapted from the Center for North American Herpetology (2010).

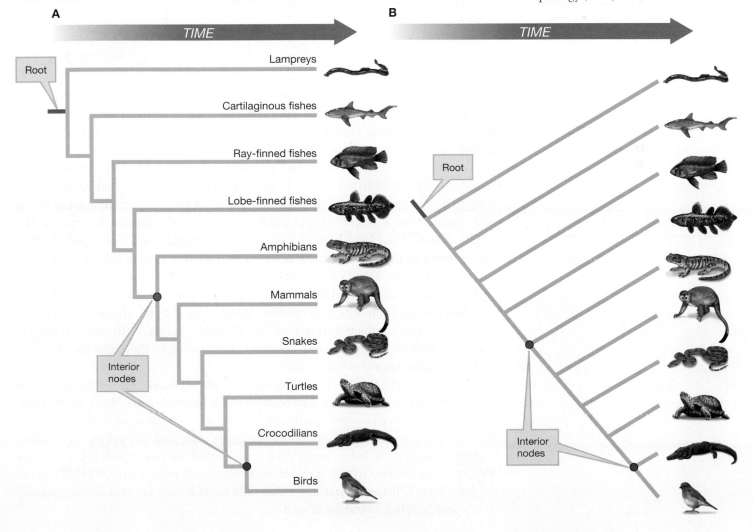

ways of conveying exactly the same information: In Figure 4.5A, the phylogeny is drawn in *tree format*, as a set of nested rectangular brackets; in Figure 4.5B, the same phylogeny is illustrated in a slanting structure known as *ladder format* (Novick and Catley 2007). These two ways of drawing a phylogeny are entirely interchangeable, and typically a phylogeny will be represented using one (but not both) of these equivalent approaches. Similarly, orientation of the tree does not matter: Phylogenetic trees can be drawn with the root at the left and the branch tips at the right as in Figure 4.5A or, equivalently, with the root at the bottom and the branch tips at the top, as in Figure 4.6. It makes no difference to the meaning of the tree. Trees can even be drawn with the root at the right and the tips at the left, or with the root at the top and the tips at the bottom, although we seldom see these orientations in practice.

FIGURE 4.6 Interior nodes represent common ancestors. Finding the common ancestor for a group involves tracing backward in time. Follow the dashed lines to see the common ancestors of different groups in this phylogeny.

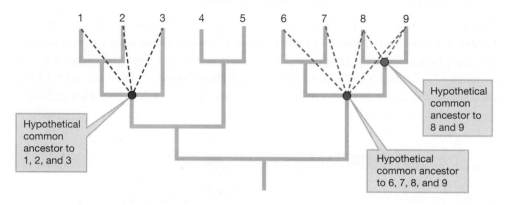

The branch points where the tree splits are called **nodes**. These represent common ancestors to the species that come after the splitting or branching point. All branch tips arising from a given node are descendants of the common ancestor at that node. For example, in Figure 4.5 a red dot highlights the node representing the common ancestor to birds and crocodilians, and the blue dot indicates the common ancestor of all tetrapods. It is important to recognize that each interior node in a phylogenetic tree represents a population that existed some time in the past, rather than a present-day population. Thus, the common ancestor to the tetrapods would not have been identical to any currently living tetrapod. Rather, evolutionary change has occurred along each and every branch leading from this ancestor to the species we observe in the world around us today.

At the base of the tree, indicated in orange in Figure 4.5, we see the **root**—the common lineage from which all species indicated on the tree are derived. To find the most recent common ancestor of two or more species, then, we can simply trace backward along the tree until the branches leading to these species converge. Figure 4.6 illustrates this idea.

One of the things that can be confusing about phylogenetic trees is that any given set of evolutionary relationships can be depicted in multiple ways. As a case in point, in Figure 4.7, notice that you can flip or "rotate" any node on a phylogenetic tree— for example, reversing the position of the green cube and the yellow pyramid— without changing the evolutionary relationships that the tree represents. If the tree indicates that any two species A and B are more closely related to each other than to a third species C before a rotation, it will indicate that they are more closely related to one another after a rotation as well.

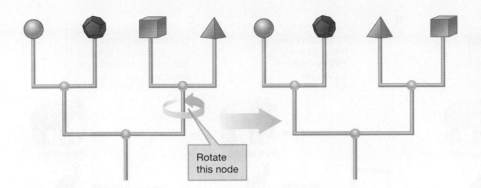

FIGURE 4.7 Rotating around any node leaves a phylogeny unchanged. Imagine that a phylogenetic tree was constructed of balls for nodes and sticks for branches. One could rotate any node 180° in space without changing the structure of the tree itself. The tree may look different, but notice that the relationships between nodes remain unaltered by the rotation.

As a result, there are a number of different ways that we can draw the very same phylogenetic tree as Figure 4.8A illustrates. In panel i, we see a phylogenetic tree for four species: 1, 2, 3, and 4. As previously described, however, we can rotate at any node—or any combination of nodes—without altering the evolutionary relationships that the tree depicts. Panels ii, iii, iv, and v show four equivalent trees generated from the rotations in panel i.

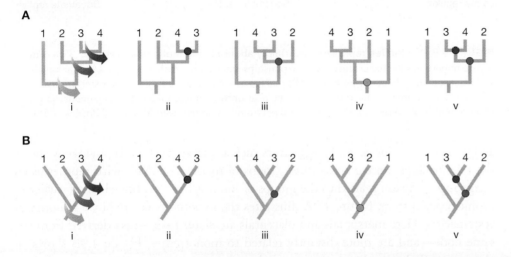

FIGURE 4.8 Rotating phylogenetic trees. Whether a phylogeny is represented as a tree (**A**) or ladder (**B**), one can rotate at any node—or any combination of nodes—without changing the structure of the tree. Thus, the leftmost tree shown in each row is identical from a phylogenetic perspective to the trees shown to the right. The colors indicate the nodes that were rotated in each case.

From this equivalence of trees, two important observations follow. First, there is nothing special about the "backbone" of the ladder representation, the apparent trunk from which the other branches arise. After all, in different representations of the same phylogeny the backbone leads to a different species. In panel i of Figure 4.8B, the backbone leads from the root to species 4, while in panel ii, it leads from the root to species 3. The second observation is that the relative positions from left to right of the branch tips do not tell us anything about how closely related two species are. In panel v, for example, species 1 is immediately adjacent to species 3, whereas species 2 is more distant, left-to-right, from species 3. Yet, as we can see by tracing back along the tree to the most recent common ancestor, species 2 is more closely related to species 3 than to species 1.

Clades and Monophyletic Groups

As we mentioned, phylogenetic trees are hypotheses. Figure 4.9 shows two competing hypotheses for the evolutionary relationships among the mammalian groups of placentals (for example, elephants), marsupials (for example, kangaroos),

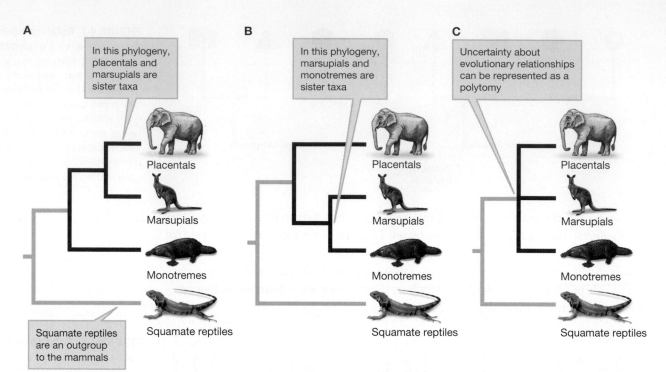

FIGURE 4.9 Polytomies represent uncertainty about phylogenetic relationships. Two competing hypotheses for the evolutionary relationships among mammalian groups: **(A)** Marsupials and placentals may be sister groups, or **(B)** marsupials and monotremes may be sister groups. **(C)** We can capture the uncertainty about the relationship among placentals, marsupials, and monotremes by representing the groups as a polytomy. Adapted from Meyer and Zardoya (2003).

FIGURE 4.10 Clades and descent from common ancestor. Clades are *nested* one within another. Different colors represent different clades, with "green" being the most encompassing of the clades here. The red, yellow, and blue clades are nested within the green clade. Thus, a given species is a member of multiple clades at multiple levels. Adapted from Understanding Evolution (2010).

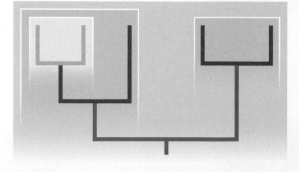

and monotremes (for example, egg layers such as platypuses). Each phylogeny shows the relationships among these three groups of mammals, along with reptiles as an **outgroup**—a taxon related to the groups of interest but that branched off earlier in evolutionary history. Figure 4.9A illustrates the hypothesis favored by a majority of systematists. Here marsupials and placentals are **sister taxa**—taxa derived from the same node—and are more distantly related to monotremes. Figure 4.9B shows an alternative phylogeny in which marsupials and monotremes are sister groups and are more distantly related to placentals. In cases where the relationships among three or more groups are unresolved, we can communicate the uncertainty as a **polytomy**—a node with more than two branches arising from it (Figure 4.9C).

A key concept in phylogenetic taxonomy is that we can use a phylogenetic tree to tell us what constitute "natural" groupings of organisms. Here the principal idea is that the natural groupings, which we call clades, are monophyletic groups (Baldauf 2003). A **monophyletic group** is defined as a taxonomic group consisting of all descendants of the group's most common ancestor and no other members. A **clade**, then, always consists of a group of species that share a single common ancestor. All species that descended from this ancestor are in the clade, and, furthermore, all species not descended from this ancestor are *not* members of that clade (Figure 4.10).

To better understand the concept of a monophyletic group, let us look at how a group can fail to be monophyletic. Figure 4.11 is a partial phylogeny of the mammals. In this figure, we see numerous monophyletic groups. For example, the group "elephants, manatees, and hyraxes" is one such monophyletic group, the group "tapirs and rhinoceroses" is another, and "tapirs, rhinoceroses, and horses" is yet a third.

But the group of organisms known as the *pachyderms*—elephants, rhinoceroses, and hippopotamuses—is not a monophyletic group because it includes neither the common ancestor of the members, shown at the root of our tree, nor all descendants of that common ancestor. A disjointed group such as pachyderms is called a **polyphyletic group**. Because polyphyletic groups do not represent proper evolutionary clades, groups such as pachyderms are no longer used in modern systematics.

There is another, perhaps more subtle way that a group can fail to be monophyletic. A **paraphyletic group** is one that contains the group's most common ancestor but not all of its descendants. We turn to yet another tree to illustrate this point. In Figure 4.12 we revisit our phylogenetic tree of the vertebrates.

Here again we see numerous monophyletic groups; for example, the tetrapod vertebrates are the monophyletic group that includes birds, crocodilians, turtles, snakes, mammals, and amphibians. The group "fish"—lampreys, cartilaginous fishes, ray-finned fishes, and lobe-finned fishes—might seem to be another natural group. Of these taxa, fish share a common ancestor that we would also classify as a fish. But not all descendants of that common ancestor are fish; after all, its descendants also include all of the tetrapod vertebrates, none of which we would call fish. Thus, fish are a paraphyletic grouping.

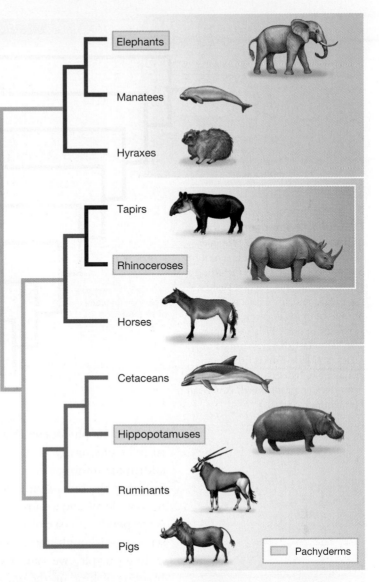

FIGURE 4.11 Monophyletic clades of mammals. A partial phylogenetic tree of the mammals shows examples of monophyletic groups. Elephants, manatees, and hyraxes form one monophyletic group, tapirs and rhinoceroses form another, and tapirs, rhinoceroses, and horses form a third. However, *pachyderms*—elephants, rhinoceroses, and hippopotamuses—are not a monophyletic group. Adapted from Murphy et al. (2001).

Rooted Trees and Unrooted Trees

Thus far, all of the trees we have looked at have been what are called **rooted trees**. On a rooted tree, the common lineage from which all the species on the tree are derived is indicated at the base of the tree. As a result, direction in a rooted tree indicates the passage of time. We see the arrow of time indicated explicitly, as in Figure 4.12. There, as we move from left to right, we are moving forward in time from the past toward the present.

In phylogenetic analysis, we often see **unrooted trees** as well. One such tree is illustrated in Figure 4.13. In contrast to rooted trees, unrooted trees do not fully indicate the direction of time. Branch tips represent more recent species than do *interior nodes* (nodes on the inner section of the tree). But given two interior nodes on an unrooted tree, we cannot say, based on the tree topology alone, that one node represents a more recent population than the other. Unrooted trees are

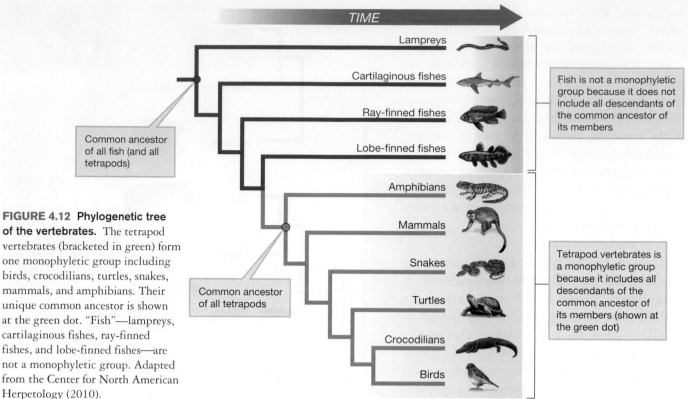

FIGURE 4.12 Phylogenetic tree of the vertebrates. The tetrapod vertebrates (bracketed in green) form one monophyletic group including birds, crocodilians, turtles, snakes, mammals, and amphibians. Their unique common ancestor is shown at the green dot. "Fish"—lampreys, cartilaginous fishes, ray-finned fishes, and lobe-finned fishes—are not a monophyletic group. Adapted from the Center for North American Herpetology (2010).

common in phylogenetic analysis because many of the algorithms that we use to infer phylogeny generate unrooted, rather than rooted, trees, and we lack the additional information needed to confidently assign a root.

Given the rooted/unrooted distinction, what exactly is the relation between an unrooted tree and a corresponding rooted tree or trees? In fact, every unrooted tree corresponds to a set of rooted trees. Figure 4.14 illustrates an unrooted tree and several—although not all—of the corresponding rooted trees.

In principle, we can "root" an unrooted tree at different points on the tree. Imagine picking up the unrooted tree at point A, and pulling this point down until it becomes the root. Doing so, we are left with the figure labeled A in the lower panel. If instead we pick up the unrooted tree at point B and pull that point down, we are left with tree B in the lower panel. Similarly if we pick up the tree at C, we arrive at the third rooted tree labeled C.

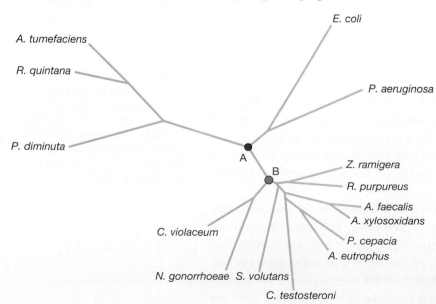

FIGURE 4.13 Unrooted tree of proteobacteria. An unrooted tree illustrates the evolutionary relationships among the proteobacteria, a large group of bacteria including human-associated species such as *E. coli* and nitrogen-fixing species such as *A. tumefaciens*. Because the tree is unrooted, it does not indicate whether, for example, interior node A represents a more recent or less recent population than does interior node B. Adapted from Shin et al. (1993).

In general, we can root an unrooted tree around any of its branches. Thus, if an unrooted tree has k branches, there will be k corresponding rooted trees. Of course, only one of these rooted trees will be correct in the sense that it accurately reflects the historical sequence of branching events. In Chapter 5, we will explore the methods that we can use to find the correct one—that is, methods used to assign the root to an unrooted tree.

It is important to realize that where we decide to root the tree influences which clades we hypothesize to be monophyletic. For example, in rooted tree A in Figure 4.14, species 1, 2, and 3 form a monophyletic group. But in trees B and C, which correspond to the same unrooted tree, species 1, 2, and 3 form a paraphyletic group.

Branch Lengths

Many trees, such as the primate phylogeny shown in Figure 4.15A, are shown with all of the branch tips aligned. Such trees are intended to convey only the pattern of relationships among the various species displayed. But sometimes we will see trees drawn with branches of different lengths, as for

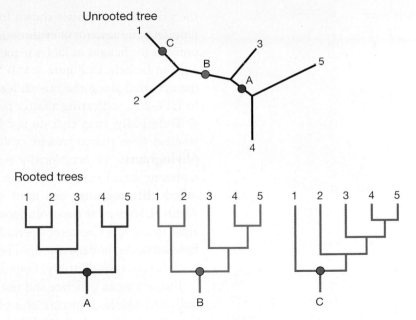

FIGURE 4.14 Rooted trees from unrooted trees. An unrooted tree and three corresponding rooted trees. Each rooted tree is rooted around the labeled point on the unrooted tree.

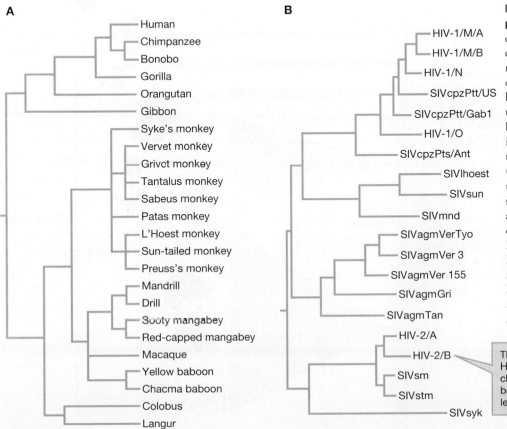

FIGURE 4.15 Cladograms and phylograms. Phylogenies can indicate evolutionary relationships only, or they can convey information regarding the amount of character change that has occurred along each branch. **(A)** A cladogram, such as this phylogeny of the primates, has the branch tips aligned and indicates only the evolutionary relationships among the species shown. **(B)** A phylogram indicates evolutionary relationships and also represents the amount of sequence change along each branch by means of differing horizontal branch lengths. Here we see a phylogram of primate lentiviruses, including human immunodeficiency viruses HIV-1 and HIV-2, and various forms of simian immunodeficiency virus (SIV). Adapted from Beer et al. (1999).

The longer branch leading to HIV-2/B indicates that more change has occurred on that branch than on the branch leading to HIV-2/A

the primate lentiviruses shown in Figure 4.15B. In this case, the branch lengths represent the amount of evolutionary change—measured as the actual or estimated number of changes in DNA sequence or other characters—that has occurred along a given branch. In Figure 4.15B, for example, we see that more sequence change has occurred along the branch leading to HIV-2/B than along the branch leading to HIV-2/A, indicating a faster rate of evolution in the HIV-2/B clade.

Technically, trees that do not have branch lengths are known as **cladograms**, whereas trees that represent evolutionary change with branch lengths are called **phylograms**. We occasionally see a third type of tree in which branch lengths represent actual time rather than the amount of evolutionary change. Such trees, called **chronograms**, are most common in paleontology. The chronogram in Figure 4.16 depicts the evolutionary history of the orchids (*Orchidaceae*). This clade arose in the late Cretaceous period. Two of its subfamilies, the *Orchidoideae* and the *Epidendroideae*, underwent rapid bouts of speciation about 60 mya, shortly after the K–T (Cretaceous–Tertiary) boundary.

Just as we can generate and test hypotheses using the evolutionary relationships indicated by the structure of a phylogenetic tree, we can also generate and test hypotheses using the branch lengths on a phylogenetic tree. Stephen Smith and

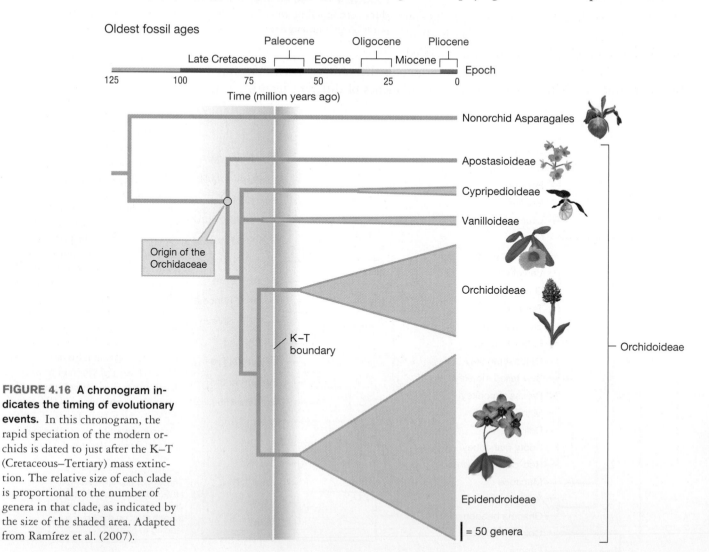

FIGURE 4.16 A chronogram indicates the timing of evolutionary events. In this chronogram, the rapid speciation of the modern orchids is dated to just after the K–T (Cretaceous–Tertiary) mass extinction. The relative size of each clade is proportional to the number of genera in that clade, as indicated by the size of the shaded area. Adapted from Ramírez et al. (2007).

Michael Donoghue did this in order to study the question of whether a plant's generation time affects its rate of evolution (Smith and Donoghue 2008). Ever since DNA sequence data became widely available, evolutionary biologists have hypothesized that species with shorter generation times experience more rapid rates of evolution as measured by changes in DNA sequence (Wu and Li 1985; Martin and Palumbi 1993). The primary reason is thought to be that in short-lived and long-lived species, germ-line cells go through roughly the same number of rounds of replication within an individual's lifetime, and thus they have roughly the same opportunity for mutational change *per lifetime*. Because the short-lived species have a shorter lifetime, they have a higher rate of mutational change in the germ line *per year*.

To test this generation time hypothesis that the rate of evolution is higher for shorter-lived species, Smith and Donoghue constructed phylogenetic trees for five large clades of plants, encompassing over 7000 species. Because precise generation time data were not available for these species, Smith and Donoghue divided the species into two categories: (1) herbaceous plants and (2) shrubs/trees. Plants in the former category tend to have a shorter generation time than do plants in the latter. Smith and Donoghue reasoned that if the generation time hypothesis is correct, there will be a slower rate of DNA sequence change along the branches of the phylogeny that represent the long-lived shrubs and trees than along the branches that represent the short-lived herbaceous plants.

Smith and Donoghue's phylogenies are shown in Figure 4.17. In these phylogenies, herbaceous species are colored in green, whereas trees and shrubs are colored in brown (the interior branches are colored as well; the authors inferred the lifestyle—herbaceous or treelike—for each ancestor using a statistical model).

FIGURE 4.17 The rate of evolution in short- and long-lived plants. A phylogeny of five major plant clades, constructed from DNA sequence data. Herbaceous species are shown in green, shrublike or treelike species in brown. For the herbaceous species, the branch lengths tend to be longer and the rates of sequence change faster. Adapted from Smith and Donoghue (2008).

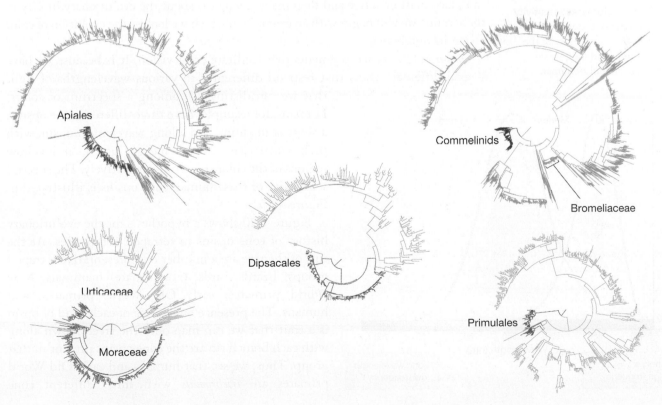

FIGURE 4.18 Different ways to depict phylogenetic relationships. Phylogenetic relationships can be represented by (**A**) a ladder phylogeny, (**B**) a tree phylogeny, or (**C**) a circular phylogeny. All three phylogenies depict the same relationships among species 1–4. Adapted from Baum (2008).

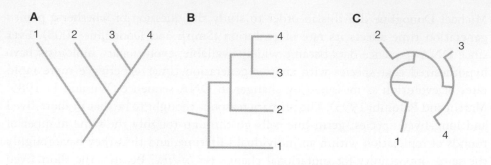

These trees, which look somewhat different from any we have seen thus far (Figure 4.18), are drawn using a method that lays out the phylogeny in an arc to make best use of the space on the page.

Even at a glance, the figures appear to support the generation time hypothesis: The brown tree-and-shrub branch lengths tend to be shorter than the green herbaceous branch lengths. Statistical analysis confirms this impression; the rates of evolution differ significantly between the herbaceous groups and the treelike groups. Indeed, the herbaceous groups had median rates of evolution 2.7 to 10 times as high as the median rates in shrub and tree species.

4.3 Traits on Trees

If a phylogenetic tree represents a hypothesis about the evolutionary history of a set of populations, then by looking at where a given trait appears on a tree, we can generate a hypothesis about when and how this trait has evolved. To get a feel for how we can place traits on a tree and then make inferences about the evolutionary history of these traits, we will begin with an example in which we look at the evolution of color vision in vertebrates.

Opsins are the visual pigments that facilitate color vision. It is because we have several different opsins that respond differently to various wavelengths of light that we can distinguish among a spectrum of colors. Humans, for example, have three different cone opsins, a short, a medium, and a long wavelength opsin, with peak sensitivities in the indigo, green, and yellow regions of the color spectrum, respectively. The spectral sensitivity of these human cone opsins is illustrated in Figure 4.19.

Figure 4.20 shows a hypothesis for the evolutionary history of cone opsins in tetrapod vertebrates. At the tips of the tree are a number of representative tetrapod groups: lizards, birds, (nonprimate) mammals, New World primates, early Old World primates, and humans. The presence or absence of each kind of opsin is a trait that we can map onto the tree. Shown along with each branch tip are the cone opsins present in that group. Thus, we see that humans and early Old World primates are *trichromats* with three different cone

FIGURE 4.19 Spectral sensitivity of the human cone opsins. Normalized spectral sensitivity of the short, medium, and long wavelength opsins found in human cones.

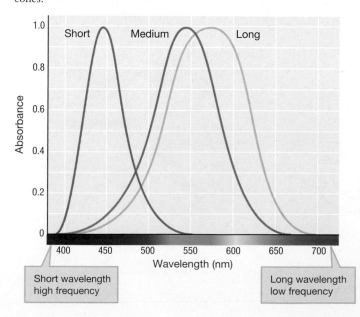

opsins; most other mammals are *dichromats* with only two different cone opsins; birds and lizards are *tetrachromats* with four different cone opsins. At the base of the tree, we see the hypothesized state of the common ancestor to these groups: The figure indicates that the common ancestor was most likely a tetrachromat like the birds and lizards.

In addition to placing traits at the tips and root of the tree, we can indicate where along the branches of the tree we think each trait has arisen or has been lost. Along the branch leading from the common ancestor to the mammalian clade, we see the loss of two intermediate wavelength opsins (the dark and light blue triangles in Figure 4.20). These evolutionary losses were perhaps associated with the nocturnal lifestyle of the early mammals, which had limited use for color vision (Goldsmith 1990). Along the branch leading from New World primates to Old World primates, we see the gain of a new medium wavelength opsin (green triangle in Figure 4.20) due to the duplication and subsequent divergence of the gene coding for the long wavelength opsin. This addition is thought to have been favored because it allowed primates to better detect and identify ripe fruit or tender young leaves, each of which may have a reddish cast (Surridge et al. 2003).

Thus, by placing traits and changes in traits on a tree that we have already constructed using other data, we represent a hypothesis about the evolutionary history of those traits and the species in which they occur.

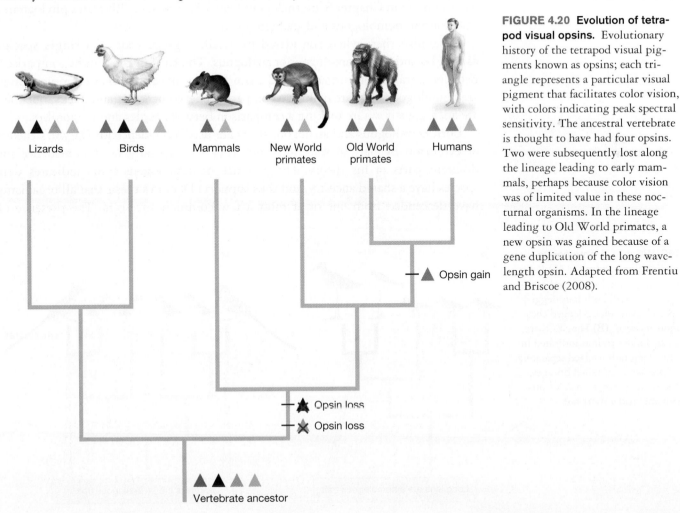

FIGURE 4.20 Evolution of tetrapod visual opsins. Evolutionary history of the tetrapod visual pigments known as opsins; each triangle represents a particular visual pigment that facilitates color vision, with colors indicating peak spectral sensitivity. The ancestral vertebrate is thought to have had four opsins. Two were subsequently lost along the lineage leading to early mammals, perhaps because color vision was of limited value in these nocturnal organisms. In the lineage leading to Old World primates, a new opsin was gained because of a gene duplication of the long wavelength opsin. Adapted from Frentiu and Briscoe (2008).

4.4 Homology and Analogy

When we look at the range of living forms that populate our planet, we notice not only the vast diversity, but also many powerful similarities that are shared across species and larger groups of organisms. Some—but not all—of these similarities are the consequence of shared ancestry. Others are the consequence of natural selection operating in similar ways on divergent groups of organisms. If we want to use similarities among organisms to deduce the historical relationships among them, we need to distinguish between these two basic sources of similarity—homology and analogy—between the traits of different species.

A **homologous trait** is a trait that is found in two or more species because those species have inherited this trait from an ancestor. All female mammals produce milk for their young, and they all possess this homologous trait because mammals share a common ancestor that produced milk. Similarly, all vertebrates have a vertebral column because the common ancestor to vertebrates had a vertebral column (or something like it).

In contrast to homologous traits, **analogous traits** are shared by two or more species, not because of a history of common descent, but instead because some other evolutionary process, usually natural selection, has independently fashioned similar traits in each species. Many of the shared adaptations for desert living that we examined in Chapter 3 are analogous traits. Figure 4.21 illustrates phylogenies that contain homologous and analogous traits.

Recognize that, when considered by itself, a given trait of a single species cannot be said to be homologous or analogous. These terms refer to the comparison between a trait of one organism and a similar trait of another. For example, wings are homologous if we are making a comparison between eagles and ducks, but they are analogous if we are making a comparison between eagles and dragonflies.

Both homologous and analogous traits are used as evidence for Darwin's theory of evolution by natural selection—but they are typically used as evidence for different parts of his theory. The presence of homologous traits indicates that species have a shared ancestry, and thus supports Darwin's thesis that all organisms have descended from one or at most a few common ancestors. The presence of

FIGURE 4.21 Homologous and analogous traits. (A) Long legs are a homologous trait as indicated in red; both long-legged species share a long-legged common ancestor. **(B)** Long tails are an analogous trait as indicated in blue; long tails evolved separately in the two long-tailed lineages, and their common ancestor presumably had a short tail.

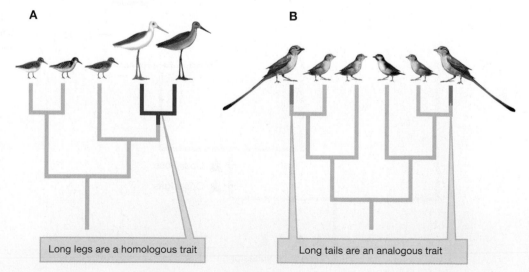

A

B

Long legs are a homologous trait

Long tails are an analogous trait

analogous traits reveals that natural selection generates structurally or functionally similar solutions to similar problems, often many times in parallel. This provides support for Darwin's thesis that the process of natural selection leads to organisms that are well adapted to their environments—and that natural selection can act as a creative force in generating these adaptations.

A discussion of homology and analogy leads us to the concepts of divergent and convergent evolution. **Divergent evolution** occurs when closely related populations or closely related species diverge from one another because natural selection operates differently on each of them. We have already seen a striking example of divergent evolution in the coat color variations of the oldfield mouse *P. polionotus* (Chapter 3). Inland, where the mice must hide against dark soils, dark coat coloration has evolved. In dune habitats along the coast and on the barrier islands, where they must hide against light soils, lighter coat colors have evolved.

Convergent evolution occurs when two or more populations or species become more similar to one another because they are exposed to similar selective conditions—that is, convergent evolution leads to analogous traits in whatever populations or species we are examining. We can again look at coloration for an example of convergent evolution. This time, however, rather than comparing the coloration of mice in one habitat to that of mice in another, we will compare the coloration of pocket mice (*Chaetodipus intermedius* and *Perognathus flavescens*) in various habitats to the coloration of fence lizards (*Sceloporus undulatus*) in those same habitats (Hoekstra 2006; Rosenblum et al. 2010). Within a span of less than 20 miles in the Tularosa Basin of New Mexico, we see three distinctly different soil types: light-colored dunes, mid-toned desert grasslands, and dark lava fields. The mouse and lizard inhabitants of these areas have evolved remarkably similar coloration patterns that render themselves hard to detect against their surroundings (Figure 4.22).

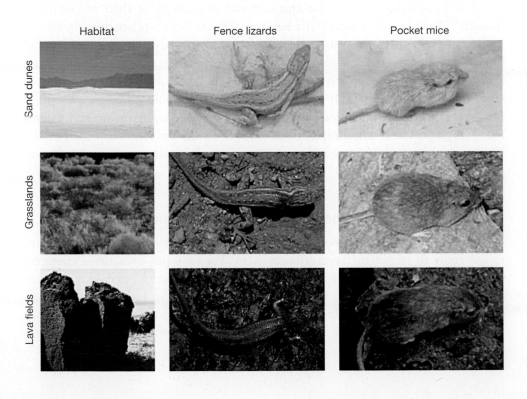

Habitat	Fence lizards	Pocket mice

Sand dunes

Grasslands

Lava fields

FIGURE 4.22 Convergent evolution for coloration. Fence lizards and pocket mice have evolved similar patterns of cryptic coloration in each of three different habitats.

FIGURE 4.23 Convergent evolution in body forms. The thunniform body design, which is well suited for open-ocean predators, represents an analogous trait when we compare tuna (left) and mako sharks (right). Adapted from Donley et al. (2004).

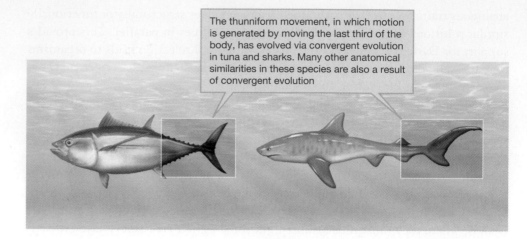

The thunniform movement, in which motion is generated by moving the last third of the body, has evolved via convergent evolution in tuna and sharks. Many other anatomical similarities in these species are also a result of convergent evolution

A second example of convergent evolution generating analogous traits comes from the body shapes of mako sharks (*Isurus oxyrinchus*) and tuna. Sharks and bony fishes diverged from each other about 400 million years ago. But lamnid sharks, such as the mako shark, and certain large predatory bony fish, such as tuna, have a similar predatory lifestyle in the open ocean. Studies show that natural selection has produced a variety of similar traits in these two groups (Lighthill 1969; Sfakiotakis et al. 1999; Graham and Dickson 2000, 2004; Bernal et al. 2001; Katz 2002).

By studying morphological traits and the kinematics—that is, the mechanics of motion—associated with swimming in mako sharks, researchers found that mako sharks and tuna both have a thunniform body design in which motion is generated by moving only the last third of the body. This sort of body design is well suited for open-ocean predators (Donley et al. 2004) (Figure 4.23).

Mako sharks and tuna display similar modes of locomotion, particularly in terms of kinematics, and similar anatomy and physiology of their red (aerobic) muscles. These are all critical traits necessary for the bursts of energy required to capture prey in the open water. Indeed, in many aspects of its morphology and physiology, the mako shark resembles the tuna more than it resembles any other shark—clear evidence for convergent evolution in these two species (Donley et al. 2004; Shadwick 2005). If we used only these analogous traits in building a phylogenetic tree that included these two species, we might incorrectly infer a closer phylogenetic relationship between lamnid sharks and tuna than truly exists. This is one reason why it is important to use multiple traits when developing phylogenetic trees.

Synapomorphies, Homoplasies, and Symplesiomorphies

The distinction between homologous and analogous traits is crucial when we aim to use traits to reconstruct evolutionary trees. For example, consider a character such as coat color. We might observe a population in which, over evolutionary time, the coat color trait changes from light to dark, as in Figure 4.24A. Here, dark color is a **derived trait**: It has been derived from an ancestor with a light color trait. So far this tells us little about phylogeny; we have only a single population and no branching structure.

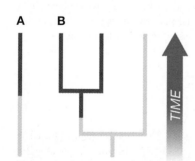

FIGURE 4.24 Derived traits.
(A) We say that the trait dark coloration is a *derived trait* when it has evolved from another trait, such as light coloration in this example.
(B) When the derived trait is shared because of a pattern of common ancestry, we call it a shared derived trait or *synapomorphy*.

But suppose that the population now splits into two descendant populations prior to the evolution of dark coloration, and it splits again after the evolution of dark coloration, as in Figure 4.24B. Now changes in coat color traits tell us something about evolutionary history. In this case, dark coloration is not only a derived trait; it is shared by two populations because it is a homologous trait in those two populations. We call a shared derived trait such as this a **synapomorphy**.

When building evolutionary trees, we are looking for synapomorphies because they help us uncover the evolutionary relationships among groups on the tree. If we could arrange to use only synapomorphies to reconstruct evolutionary trees, the entire process of constructing phylogenies would be relatively straightforward. The more traits that two species had in common, the more closely related they would be. The problem is that not all shared traits are synapomorphies. There are other ways that two species could share a common trait as well. Let us see how.

When we are using traits to reconstruct an evolutionary tree, we often see only the current state of each population. Thus, if we are trying to build an evolutionary tree for three taxa, two of which have dark coats and one of which has light coats, we would know only what we see in Figure 4.25A. In this figure, our uncertainty about the evolutionary relationships among the three populations—labeled 1, 2, and 3—is represented by the polytomy between these three populations. Our uncertainty about the history of the character states is represented by the fact that we have not colored the interior of the tree, but rather only its tips. As in our previous example, two populations have dark coats and one has light coats. Can we assume that having dark coats is a shared derived trait? No. One problem is that the trait could have changed multiple times in the tree. For example, the actual evolutionary tree could look like that seen in Figure 4.25B.

Our common dark coloration trait could be analogous, rather than homologous. We call an analogous trait like this a **homoplasy**. (*Confusion alert:* A homology is a trait that is shared by two or more species because it has been inherited from a common ancestor. A homoplasy is a trait that is similar in two or more species even though it was not present in their common ancestor. Thus, a homoplasy is an analogous trait, not a homologous one.) Homoplasies can be misleading when we try to reconstruct an evolutionary tree. In Figure 4.25, species 1 and 2 share a common trait—dark coloration—that is not shared by species 3—but species 1 and 2 are *not* more closely related to one another than they are to species 3. If we mistakenly thought that this trait was a synapomorphy, we would conclude otherwise.

Another problem is that the dark coat might not even be the derived trait at all. It could be that the lighter coat coloration is the derived trait. For example, the true evolutionary tree could be as in Figure 4.26A. In this particular case, mistakenly thinking that dark coat color is the derived trait doesn't cause us any problems; if we were to treat it as a synapomorphy we would still conclude, correctly albeit for the wrong reason, that species 1 and 2 were more closely related to one another than to species 3. We might have the wrong idea about the trait of the common ancestor of all three populations, but at least we would get the right idea about the

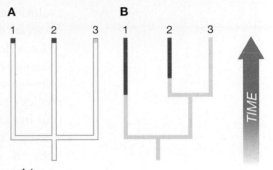

FIGURE 4.25 An example of homoplasy. (A) If we know only the current character states and not the ancestral history, we represent this as a polytomy. **(B)** Dark coloration is an analogous trait in this phylogeny.

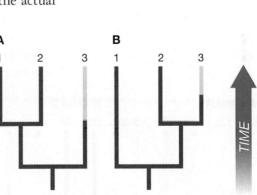

FIGURE 4.26 Derived traits and symplesiomorphy. (A) If we mistakenly believe that dark coloration is a derived trait, this doesn't necessarily lead us to misinterpret the relationships among species 1, 2, and 3. **(B).** If a derived trait has arisen recently and appears in only one of the two most closely related species, the two more distantly related species share the same trait. In this case, we call the trait a *symplesiomorphy*.

evolutionary relationships among the populations. But mistaking a derived trait for an ancestral trait might lead to the wrong conclusions.

Consider the tree in Figure 4.26B. Here we have a trait—light color—that is so recently derived that it is not shared. This leaves us with a shared trait—dark color—that is ancestral, and in fact is not shared by the two most closely related species (2 and 3). A trait of this type is called a **symplesiomorphy**. Using such a trait in reconstructing a tree would incorrectly cause us to think that species 1 and species 2 are more closely related to one another than to species 3.

So, if using traits other than synapomorphies poses such a problem for phylogenetic inference, what can we do about it? One approach is to pick traits that are likely to be synapomorphies rather than symplesiomorphies. Particularly when using phenotypic traits for building trees, we can use a thorough knowledge of the natural history of the organisms we are studying to select characters that are prone to change slowly rather than those that are prone to fluctuate rapidly over evolutionary time. This will help us avoid inadvertently choosing homoplasies and symplesiomorphies.

Another important strategy is to use a large number of characters. If we use a sufficient number of characters, we might expect the synapomorphies to outweigh any homoplasies or symplesiomorphies accidentally included in the set of characters. When we build trees based on genetic sequence data, we rely heavily on this approach.

A third approach is to use an outgroup, a group with a known evolutionary relationship to the taxon we are studying. By including multiple outgroups, we can better estimate the **polarity**—the order of appearance in evolutionary time—of the traits we are using. This can be particularly useful in helping us to avoid symplesiomorphies.

The idea of using outgroups is that, when we begin the process of phylogenetic reconstruction, we do not know the relationship among the species in the taxon we are studying, but we do know the relationship of this taxon to the outgroups. Consider two cases shown in Figure 4.27. In case 1, outgroups have the light coloration (Figure 4.27A); in case 2, outgroups have the dark coloration (Figure 4.27B).

Notice the polytomy between groups 1, 2, and 3, indicating our uncertainty about the evolutionary relationships there, but the well-resolved branch for the outgroups, indicating that we know they diverged from groups 1, 2, and 3, before 1–3 diverged from one another. With this information in place, we can infer the most likely ancestral state for each case. This likely ancestral state is the state found in the outgroups. In Figure 4.27A, we infer that the polarity of the trait is light color → dark color; whereas in Figure 4.27B, the polarity of the trait is dark color → light color.

How does this help us resolve the branching pattern among groups 1, 2, and 3? Figure 4.28 allows us to answer that question. Suppose that the common ancestor to 1, 2, and 3 was light colored. Then if species 1 and 2 are more closely related to one another than to any other species—that is, if they are sister groups—we can explain the tree by a single evolutionary event (indicated by the red arrow). But, if groups 1 and 2 are not sister groups—that is, if species 1 and 2 are not more closely

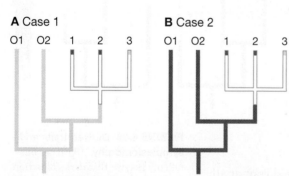

A Case 1   **B** Case 2

FIGURE 4.27 Using outgroups to infer the ancestral state. Outgroups provide information about a trait's polarity; we assume that the ancestral trait is the trait shared by the outgroups and some members of the clade of interest. O1 = outgroup 1; O2 = outgroup 2.

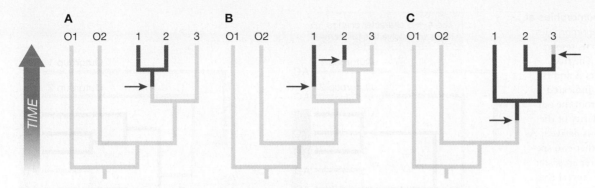

FIGURE 4.28 Case 1: The outgroups help resolve the polytomy. If species 1 and 2 are sister groups, we can explain the observed traits with a single evolutionary event (**A**). If species 2 and 3 are sister groups, we require two evolutionary events, either (**B**) two independent arisals of dark coloration, or (**C**) the evolution of dark coloration early, with a subsequent reversion to light coloration in one lineage later. Red arrows indicate evolutionary changes in the trait. O1 = outgroup 1; O2 = outgroup 2.

related to one another than to any other species we require at least two evolutionary events (Figure 4.28B, C).

In Figure 4.28, knowing that light coloration is the ancestral character supports the inference that groups 1 and 2 are likely to be sister groups. This approach of trying to explain the observed character states by a minimum number of evolutionary changes is known as parsimony. We will explore parsimony, along with other methods for inferring evolutionary trees, in Chapter 5.

If dark coloration is the ancestral character (case 2), matters are less clear-cut. In this case, we can explain the tree at hand with a single evolutionary event (red arrow), whether 1 and 2 are sister groups, as in Figure 4.29A, or not, as in Figure 4.29B.

In Figure 4.29, with dark coloration as the ancestral state, the pattern of characters that we have observed is less informative about how to resolve the branching between groups 1, 2, and 3. Under parsimony, it tells us nothing at all about how to resolve these three groups.

In the preceding examples, we have shown how synapomorphies at one level of the tree can help us resolve the branching pattern among three groups on the tree. As we try to reconstruct the evolutionary history of larger numbers of groups, we need to have synapomorphies at different levels of the tree. Figure 4.30 illustrates why.

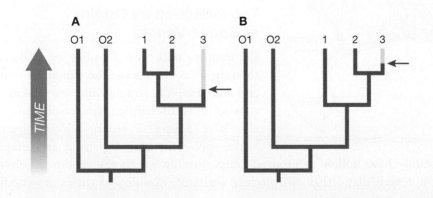

FIGURE 4.29 Case 2: The outgroups do not help resolve the polytomy. When the outgroups share a common trait with two of the members of the polytomy, we get no information from that trait to help us resolve the polytomy. The observed pattern requires only one evolutionary event, depicted at the red arrow, regardless of how the polytomy is resolved. O1 = outgroup 1; O2 = outgroup 2.

FIGURE 4.30 Synapomorphies at different levels. Synapomorphies at different levels help us resolve the entire phylogeny. The first character (with variants A and B) resolves the polytomy indicated by the upper arrow. From the outgroups, we see the polarity of the trait: A is ancestral; B is derived. As a result, we conclude that two species with the B character are sister groups. The outgroups reveal the polarity of the second character. C is the ancestral trait; D is derived. This resolves the polytomy indicated by the lower arrow. The two clades with the D character are sister groups.

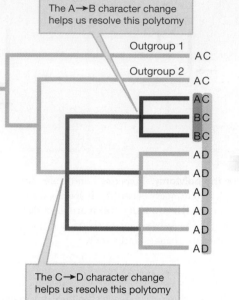
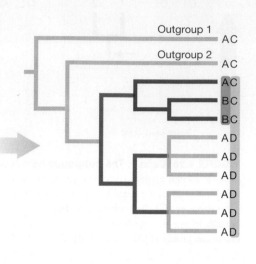

The A→B character change helps us resolve this polytomy

Outgroup 1   AC
Outgroup 2   AC
AC
BC
BC
AD
AD
AD
AD
AD
AD

The C→D character change helps us resolve this polytomy

Outgroup 1   AC
Outgroup 2   AC
AC
BC
BC
AD
AD
AD
AD
AD
AD

4.5 Using Phylogenies to Generate Evolutionary Hypotheses

Evolutionary trees or phylogenies are hypotheses about historical relationships among organisms; evolutionary biologists test these hypotheses when new sources of information about relationships and descent—for example, new fossils, new molecular data, or new phylogeographic data—become available.

When evolutionary biologists place traits on a phylogenetic tree, they are also generating hypotheses—hypotheses about when traits evolved and which traits may be shared among which groups of relatives. In this section, we will begin with a striking example in which a phylogenetic picture of snake and lizard venom led to the discovery that many supposedly nonvenomous snakes, and even nonvenomous lizards, actually produce and use venom in capturing their prey.

The Evolutionary Origins of Snake Venom

Commonly, only two families of snakes are thought of as venomous: the Viperidae (vipers) and the Elapidae (including sea snakes and cobras); a third family, Atractaspididae, may also have advanced venom-delivery systems. Snake species in both Viperidae and Elapidae commonly have hollow or grooved fangs through which the venom is delivered from a venom gland that can produce and store sizable quantities of venom, as illustrated in Figure 4.31.

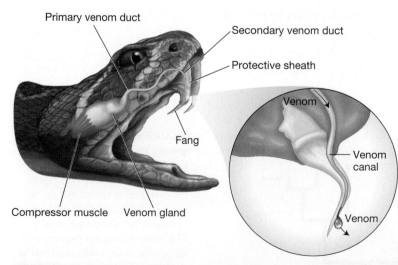

Primary venom duct

Secondary venom duct

Protective sheath

Fang

Venom

Venom canal

Venom

Compressor muscle   Venom gland

FIGURE 4.31 Snake fangs and venom. The morphology of the venom-delivery system in a venomous viperid snake.

Early phylogenetic analysis suggested that these advanced venom-delivery systems evolved independently in each family of snakes—that is, that they were analogous traits. Researchers assumed that there was no venom without a delivery system, and so they concluded that venomousness must be a highly derived trait seen in a relatively small fraction of all snake species. But more recent phylogenetic analysis, combined with careful morphological study, has forced herpetologists to reevaluate and revise this conclusion (Figure 4.32). This work suggests that numerous other families of snakes are able to produce salivary toxins in organs known as the Duvernoy's gland, even though they lack grooved/hollow fangs or advanced venom-delivery pumps (Vidal 2002; Fry 2003b).

Given the broad distribution of basic toxin production capacity, herpetologists have hypothesized that toxin production is homologous among snakes, having arisen once rather than repeatedly over the evolutionary history of this group. Evolutionary biologist and venom expert Bryan Fry reasoned that if this hypothesis was correct, many so-called nonvenomous snakes should actually be capable of producing toxic venom. Based on this phylogenetic reasoning, Fry and his colleagues decided to study the salivary secretions of a purportedly nonvenomous snake common in the pet trade, the rat snake *Coelognathus radiatus*. They obtained a number of individuals of the species, and milked the snakes to obtain their salivary secretions. Surprisingly—but in line with Fry's conjecture—they found that the most abundant peptide in the salivary secretions of this supposedly

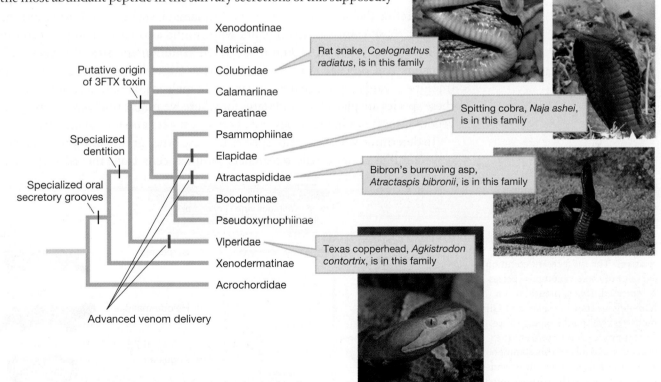

FIGURE 4.32 Phylogeny of advanced snakes (Caenophidia). A partial phylogeny of the Caenophidia indicates the distribution of (1) specialized oral secretory glands (for example, Duvernoy's glands), (2) specialized dentition, and (3) advanced venom-delivery systems. Because the three-finger toxin (3FTX) peptides are shared among the Elapidae, the Atractaspididae, and the supposedly harmless species *Coelognathus radiatus* (but not present in the vipers), Fry and his colleagues hypothesized, and subsequently demonstrated, early evolution of the 3FTX toxin family, just after the divergence of the Viperidae. Adapted from Vidal (2002) and Fry (2003b).

harmless snake is a close homologue of the three-finger toxins (3FTX) produced by the highly poisonous elapid snakes (Fry 2003a) (Figure 4.32). The supposedly harmless rat snake turned out to be producing a potent neurotoxin closely related to that in cobra venom!

Buoyed by their successes finding toxins in the saliva of purportedly nonvenomous snakes, Fry and his colleagues decided to see if they could trace the origin of venom production even further back into evolutionary history (Fry et al. 2006). In addition to venomous snakes, the helodermatid lizards (Gila monsters and beaded lizards) are known to be venomous. But venomousness in snakes and venomousness in lizards were thought to be analogous traits—that is, snakes and helodermatid lizards were thought to have independently evolved the capacity to produce and deliver venom. The venomous snakes produce their venom in specialized glands in the upper jaw and deliver it through hollow or grooved fangs on the upper jaw, whereas the helodermatid lizards produce their venom in glands on the lower jaw and deliver it through a row of grooved teeth on the lower jaw. But after discovering homologies in snake venoms, Fry hypothesized that perhaps some snake and lizard venoms are homologous as well.

Again, this hypothesis generated a strong testable prediction. If venom had evolved early, so that it was a homologous trait in snakes and these venomous lizards, other descendants of their common ancestor might share the ability to produce venom. So, Fry and an international team of herpetologists used genetic data to refine the phylogeny of the order Squamata (snakes and lizards) and thereby identify those common descendants who might also have venom. According to their phylogeny, shown in Figure 4.33, the common ancestor of snakes and Gila monsters had descendants that include the Anguidae (glass lizards), Varanidae (monitor lizards), and Iguania (iguanas, chameleons, anoles, and relatives). Thus these species are plausible candidates for where we might find venom production if venomousness is a homologous trait between snakes and Gila monsters.

To determine whether or not species in these other groups also produced venoms or venomlike proteins, the researchers sampled cells from the salivary glands or

FIGURE 4.33 Venomousness as a homologous trait between snakes and Gila monsters. Phylogeny of snakes, venomous helodermatid lizards, and their relatives. The most recent common ancestor of these venomous species is indicated. If venomousness is a homologous trait in snakes and Gila monsters, we should expect to see venom production in some of the other descendants of this common ancestor, such as the monitors and iguanas shown in the tree. Adapted from Fry et al. (2006).

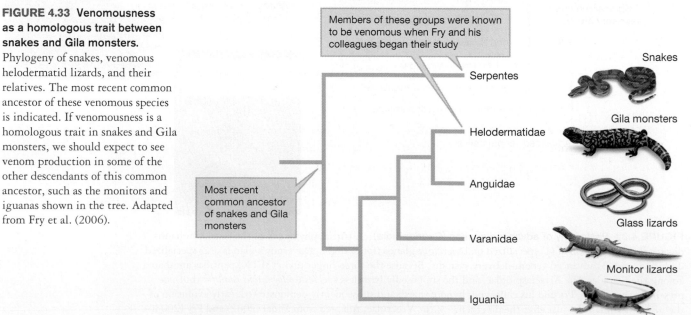

Members of these groups were known to be venomous when Fry and his colleagues began their study

Most recent common ancestor of snakes and Gila monsters

Serpentes — Snakes
Helodermatidae — Gila monsters
Anguidae — Glass lizards
Varanidae — Monitor lizards
Iguania — Iguanas

secretions of these species. They then looked at the genes that are expressed in those cells. They found nine genes coding for toxins that were shared between lizard species and snakes; seven of these were previously known only from snakes. An Australian lizard, the eastern beaded dragon (*Pogona barbata*), produces a toxin previously known only from rattlesnake venom. The lace monitor, a varanid, produces toxins that inhibit blood clotting and induce a catastrophic drop in blood pressure. While these various toxins may not be lethal or even severely debilitating when these lizards bite humans, they may be delivered at high enough doses to be extremely effective in disabling the lizards' smaller prey. The widespread presence of the toxins may also help explain the prolonged bleeding and extreme swelling that humans suffer when bitten by monitor lizards.

All in all, this study provides very strong evidence of an early emergence of venom production capability in the squamate reptiles. Here again, phylogenetic thinking was key in the discovery of the other lizards' venoms. Phylogenetic reasoning suggested to Fry and his team that lizards other than the Helodermatidae may also produce venom—and the phylogeny that these researchers constructed gave them a map of where in the lizard group to look for other venomous species. In the end, this discovery may be of more than general biological interest. Compounds derived from snake venoms are used extensively in medicine—for example, they are used as anticoagulants, in diagnosing various blood-related disorders, and to lower blood pressure (Koh et al. 2006). The diverse lizard toxins that Fry and his colleagues identified will offer a new array of potentially useful molecules for medical researchers to explore.

Deep Homology: Lipid Droplets

Some biological processes and pathways are so fundamental to life itself that we see homology in these processes across a very disparate and diverse group of species— we will refer to this as deep homology. As an example of deep homology, let us examine the way that organisms store energy as triglyceride fats (Martin 2006; Kadereit et al. 2008).

The means by which organisms store triglycerides is remarkably conserved across species, from yeast to fruit flies to humans. When we say a trait is conserved, we mean that the same set of genes controls this trait in a widely disparate group of organisms—in our case, from yeast to humans. In all these creatures, triglycerides are stored in what are called lipid droplets, also called oil bodies and adiposomes. From yeast to humans, lipid droplets are structurally similar; they are produced in the endoplasmic reticulum and then break off into the cytoplasm (Figure 4.34).

Recent work demonstrates that this similarity in structure and production is in part due to the fact that a gene labeled *FIT2* plays a role in lipid droplet formation in mammals, amphibians, birds, insects, worms, and yeast (Kadereit et al. 2008) (Figure 4.35). Moreover, the DNA sequences of the *FIT2* genes in these organisms are remarkably similar.

How can we explain all these similarities in lipid droplets from yeast to humans? Without common descent as our conceptual framework, we would have to posit that the *FIT2* gene, as well as all the similarities in lipid droplet formation, must have emerged independently in worms, fruit flies, and humans, which is virtually impossible. A much more likely explanation, based on common descent, is that the last common ancestor to mammals, amphibians, birds, insects, worms, and

FIGURE 4.34 Deep homology has been seen in lipid droplets. The formation of lipid droplets in the endoplasmic reticulum (ER) and their budding off in the cytoplasm. Adapted from Martin (2006).

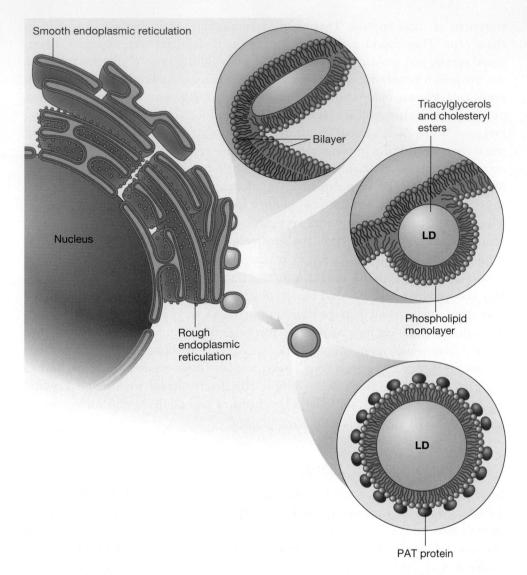

yeast had the *FIT2* gene, or something very similar to it (Figure 4.35). The genetic similarity in lipid metabolism across such an array of different groups—a similarity that arises through deep homology—allows us to use organisms like yeast as model systems to study medical issues related to lipid and fat formation in humans.

FIGURE 4.35 Phylogeny based on the *FIT2* gene. This phylogeny, which is based on DNA sequences of the *FIT2* gene associated with lipid droplet formation, exactly reflects the correct phylogenetic relationships among the species illustrated. Adapted from Kadereit et al. (2008).

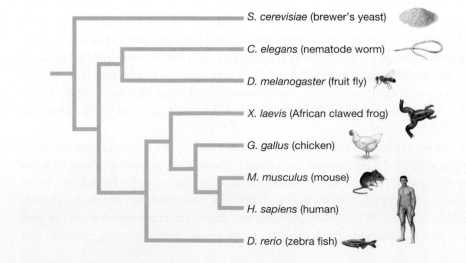

Vestigial Traits

One interesting class of homologous traits used in phylogenetic reconstruction are known as **vestigial traits**—Darwin often referred to these as "rudimentary" characteristics. Vestigial traits are those that have no known current function but appear to have been important in the evolutionary past. In *The Descent of Man and Selection in Relation to Sex*, Darwin wrote of the upper incisor teeth that never break through the gums of some ruminants as an example of a vestigial trait, because ruminant herbivores likely descended from carnivores, whose incisor teeth are very important in prey capture and consumption (Darwin 1871).

Why vestigial traits remain in place when they serve no current function will probably vary from trait to trait. There are at least three possible explanations: (1) the trait is not costly to the organism, and so natural selection does not act against it; (2) there is some natural selection against a vestigial trait—it is on its way out, and eventually it will be lost; or (3) the trait has some function that we have simply failed to identify. In this last instance, the trait would not really be vestigial, so let's confine ourselves to the former two cases.

Vestigial traits allow evolutionary biologists to trace common descent by comparing a now functionless trait in species 1 to the same trait in functional form in species 2—our assumption being that species 1 and 2 share this trait because of descent from a common ancestor who also possessed it. For example, consider the nictitating membrane—or inner eyelid—found in birds and mammals. This membrane can be drawn across the eye of birds. It can moderate incoming light, clean the eye of dust, and (in birds) prevent excessive drying of the eye during flight. Most mammalian species, including humans, also have a vestigial version of the nictitating membrane called the plica semilunaris, or semilunar fold (Figure 4.36). As far as we know, this membrane has no working function in humans and most other mammals. But it tells us something about common descent. The fact that birds and mammals share the complex trait of a nictitating membrane/semilunar fold, even though this trait has no known function in the latter group, is indicative of their common ancestry—that is, it suggests that an ancestor common to both these groups had some version of this trait. Indeed, we can say more, because reptiles also have a functioning nictitating membrane, which suggests that birds, reptiles, and mammals share a common ancestor that had such a membrane, and it was only when mammals diverged from these other groups that the nictitating membrane lost its function.

A

Nictitating membrane

Plica semilunaris

B

FIGURE 4.36 The nictitating membrane. The nictitating membrane in an eagle (**A**) is homologous to the plica semilunaris in a human (**B**). The plica semilunaris has no known function in humans, while the nictitating membrane serves many functions in birds.

Evolutionary biologists have also examined vestigial traits and phylogeny in the context of limblessness in whales and snakes. In snakes, the evidence from vestigial limbs suggests that modern snakes evolved from a limbed reptilian ancestor (Carroll 1988; Lee and Caldwell 1998). Evidence from limb structure, in both modern and extinct snake species, is most consistent with the following evolutionary history: Fossil evidence suggests that the common ancestor to all snakes had fully developed hind limbs and forelimbs and a skeleton with distinct regions. The earliest snakes had already lost forelimbs, but they had functional hind limbs. Modern snakes then went through three stages: (1) a reduced pelvic area (with hind limbs present), (2) the reduction of the hind limbs to vestigial buds, and then (3) the complete loss of hind limbs. The data on the phylogeny of snakes as it relates to vestigial traits can be summarized in the evolutionary tree shown in Figure 4.37.

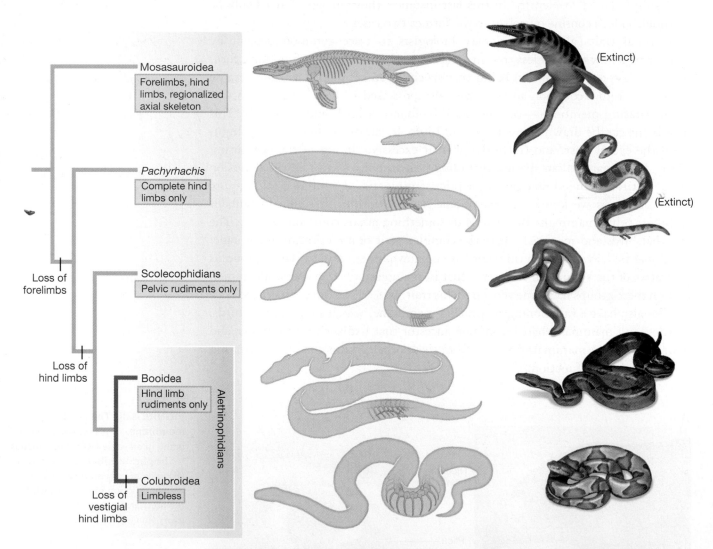

FIGURE 4.37 Vestigial limblessness in snakes. A phylogenetic history of snakes shows the gradual loss of limbs from their reptilian ancestors. Species in the superfamily Booidea (boas and pythons) retain vestigial hind limbs, whereas developmental changes in the colubrid snakes have eliminated even these vestigial hind limbs. Adapted from Cohn and Tickle (1999).

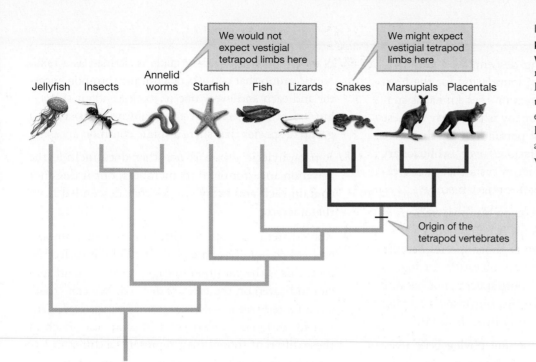

We would not expect vestigial tetrapod limbs here

We might expect vestigial tetrapod limbs here

Origin of the tetrapod vertebrates

FIGURE 4.38 Common ancestry predicts where we should find vestigial limbs. We expect that we may find vestigial tetrapod limbs in limbless clades with limbed ancestors, such as snakes. But we do not expect to find vestigial tetrapod limbs in limbless species without a tetrapod ancestor, such as earthworms.

Evolutionary biologists are beginning to gather data on the molecular genetic underpinnings of vestigial limbs in snakes. Python snakes, for example, have hundreds of vertebrae, but they lack limbs. Yet, a detailed analysis of developmental patterns in pythons shows evidence of vestigial hind limb "buds"—small stubs that typically develop into limbs in other reptiles. Researchers have found that the distribution and expression of *Hox* genes and their products may be responsible for the modern snake body shape (elongated, with many vertebrae), the loss of limbs, and the hind limb "buds" that remain in some species such as pythons (Cohn and Tickle 1999).

Vestigial traits serve as a strong test of Darwin's theory of evolution by common ancestry. If all organisms have arisen from one or a few common ancestors by a branching process of descent, we would expect to see vestigial traits shared with species that share a common ancestor subsequent to the evolution of that trait—but not among species whose most recent common ancestor predates the evolution of that trait. For example, think about where on the tree of life we might expect to find vestigial tetrapod limbs. Under the explanation provided here, we might expect to see vestigial limbs in some of the currently limbless descendants of ancestral tetrapod vertebrates. But we would not expect to find vestigial limbs in species that diverged prior to the origin of limbs. Thus, Darwin's theory predicts that we may find vestigial limbs in snakes, but that we should not find them, for example, in earthworms (Figure 4.38). Indeed, such predictions have been borne out time and again in the study of comparative morphology.

In this chapter, we have emphasized the central role that common descent and phylogenetic history play in evolutionary biology. In the next chapter, we will move on to a more detailed analysis of how phylogenetic trees are constructed in the first place.

SUMMARY

1. Darwin's idea of branching descent with modification provided a theoretical foundation for the hierarchical patterns of classification that Linnaeus suggested. The study of phylogeny is the study of these branching relationships of populations as they give rise to descendant populations over evolutionary time. Phylogenetic systematics casts that classification scheme in terms of evolutionary history.

2. The study of phylogeny rests on our observations of traits displayed by organisms. A homologous trait is a trait that is found in two or more species because those species share a common ancestor. Analogous traits are shared by two or more species, not because of a history of common descent, but instead because they have arisen independently in each species.

3. Both the process of reconstructing phylogenetic trees, and the process of mapping evolutionary events onto trees, generate hypotheses. For example, by looking at where a given trait appears on a tree, we can generate a hypothesis about when and how this trait has evolved.

4. Evolutionary biologists use synapomorphies—shared, derived traits—to infer the structure of phylogenetic trees.

5. There are many equivalent ways to draw the same phylogenetic tree.

6. The points where a phylogentic tree branches—the nodes—represent common ancestors to the species that come after the branch point. All branch tips arising from a given branching point are descendants of the common ancestor at that branching point.

7. A monophyletic group or clade is defined as a taxonomic group that consists of a unique common ancestor and each and every one of its descendant species, but no other species. A clade always consists of a group of species that share a single common ancestor.

8. A paraphyletic group is one that does include the common ancestor of all its members, but it does not contain each and every species that descended from that ancestor.

9. Rooted trees indicate the direction of time; unrooted trees do not. The base of a rooted tree is called the root; this is the common lineage from which all species indicated on the tree are derived. We can "root" an unrooted tree at different points on the tree, generating different rooted trees in each case. Each of these different rooted trees represents a different hypothesis about which nodes are most ancestral.

10. Many trees are shown with all of the branch tips aligned. Such trees, called cladograms, convey only the pattern of relationships among the various species displayed. Phylograms are drawn with branches of different lengths; in a phylogram, branch lengths represent the amount of evolutionary change—measured as the actual or estimated number of changes in DNA sequence or other characters—that has occurred along a given branch.

11. Vestigial traits are those that have no current function but appear to have been important in the evolutionary past. Such traits allow us to test evolutionary hypotheses about common origin.

KEY TERMS

analogous traits (p. 118)
chronograms (p. 114)
clade (p. 110)
cladograms (p. 114)
convergent evolution (p. 119)
derived trait (p. 120)
divergent evolution (p. 119)
homologous trait (p. 118)
homoplasy (p. 121)
monophyletic group (p. 110)

node (p. 108)
outgroup (p. 110)
paraphyletic group (p. 111)
phylogenetic systematics (p. 103)
phylogeny (p. 103)
phylograms (p. 114)
polarity (p. 122)
polyphyletic group (p. 111)
polytomy (p. 110)
root (p. 108)

rooted tree (p. 111)
sister taxa (p. 110)
symplesiomorphy (p. 122)
synapomorphy (p. 121)
taxon (p. 107)
traits (p. 103)
unrooted tree (p. 111)
vestigial traits (p. 129)

REVIEW QUESTIONS

1. Find the common ancestor of species 3, 5, and 6 on this tree. Find the common ancestor of species 1, 2, and 4.

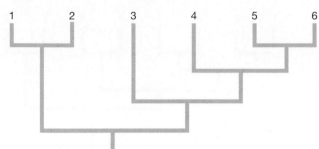

2. The tree below is an unrooted tree. Draw the three corresponding rooted trees if this tree is rooted at points A, B, and C respectively.

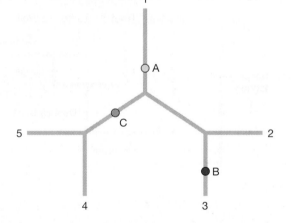

3. For the tree below, (a) draw how it would appear after rotating around node A, (b) draw it after rotating around node B, and (c) draw it after rotating around both nodes A and B.

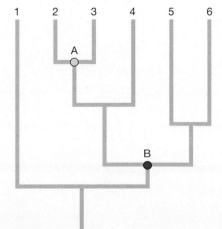

4. Depict the following tree in slanted (ladder) form:

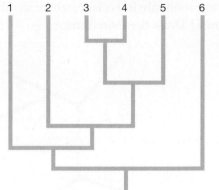

5. On the tree below, the numerals 1–7 represent seven different species. (a) Which pair of species is more closely related: 4 and 5, or 5 and 7? (b) Which pair is more closely related: 1 and 2, or 2 and 7? (c) Which pair is more closely related: 3 and 5, or 2 and 4?

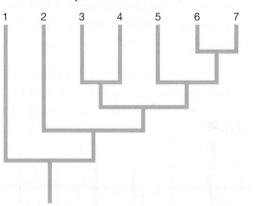

6. This unrooted tree shows the evolutionary relationships between species 1–7. If species 1, 4, 5, 6, and 7 form a monophyletic clade, and species 2 and 3 form a monophyletic clade, where should the tree be rooted? Draw the rooted tree.

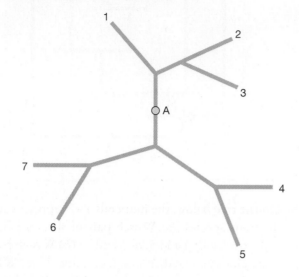

7. Suppose that the tree in question 6 is rooted around point A. What groups are monophyletic clades in this case?

8. On the tree below, what is the smallest monophyletic clade that includes species 4, 5, and 6? What node is the most recent common ancestor of the members of this clade?

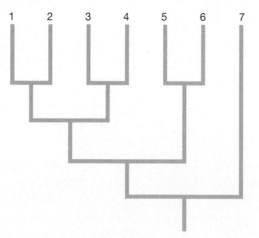

9. The tree below shows the phylogenetic relationships among eight species. How many monophyletic clades are there with exactly two members? How many with exactly three members? How many with exactly four?

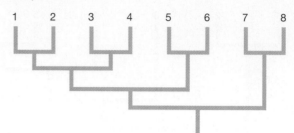

10. The origin of five traits—a rasping tongue, jaws, the dentary bone, lungs, and vivaparity—are shown on the tree below. According to the diagram, which of these five traits do sharks have?

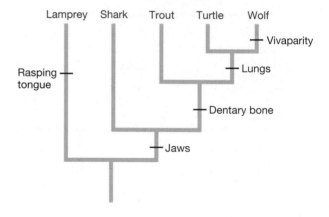

11. Bridge et al. (2005) developed a chronogram of tern species, based on a mitochondrial DNA sequence. The figure below illustrates this chronogram, with the head and beak color of each bird shown.

    a. *S. sumatrana* and *S. trudeaui* both have white head coloration. According to this phylogeny, is this a homoplasy or a homology?

    b. *Gygis alba* and *A. tenuirostris* also both have white head coloration. Is this a homoplasy or a homology?

    c. Which character is more highly conserved in this clade: beak color or head color?

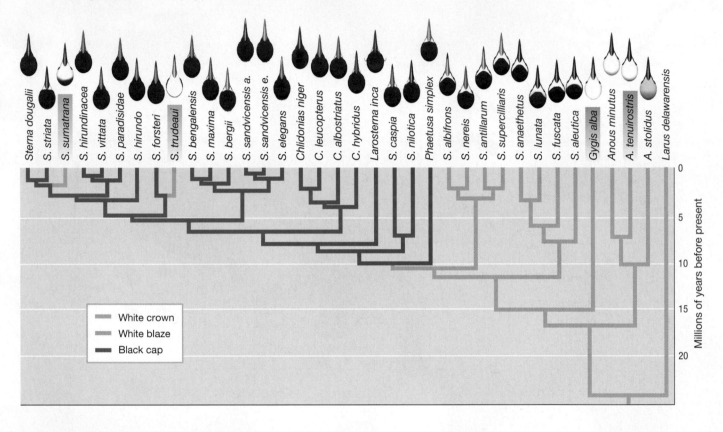

SUGGESTED READINGS

Meyer, A., and R. Zardoya. 2003. Recent advances in the (molecular) phylogeny of vertebrates. *Annual Review of Ecology, Evolution, and Systematics* 34: 311–338. A technical but important review of the role of genetics in building phylogenetic trees.

Shadwick, R. E. 2005. How tunas and lamnid sharks swim: an evolutionary convergence. *American Scientist* 93: 524–531. A short, popular, article on convergent evolution.

Surridge, A. K., D. Osorio, and N. I. Mundy. 2003. Evolution and selection of trichromatic vision in primates. *Trends in Ecology & Evolution* 18: 198–205. A more in-depth review of an example we covered in this chapter.

Vonk, F. J., J. F. Admiraal, K. Jackson, R. Reshef, M. A. G. de Bakker, K. Vanderschoot, I. van den Berge, M. van Atten, E. Burgerhout, A. Beck, et al. 2008. Evolutionary origin and development of snake fangs. *Nature* 454: 630–633. More on the evolution of venom-related traits in snakes, as per the example we discussed in the chapter.

Inferring Phylogeny

◄ Avian diversity is shown in this sample of bird eggs from the Western Foundation of Vertebrate Zoology, Los Angeles, California.

I n the spring of 1999, five Bulgarian nurses and a Palestinian medical intern working at Benghazi Hospital in Libya were accused of a horrifying crime. More than 400 children had become infected with the HIV virus at the hospital—and these six medics were alleged to have deliberately infected those children with a genetically engineered strain of HIV. Prosecutors claimed that the entire outbreak was masterminded by an unknown foreign secret service—perhaps the CIA or the Israeli Mossad—as part of a conspiracy to cause civic disruption in Libya.

But did these six medics really commit this unspeakable act? Or were they merely scapegoats for a tragedy that resulted from inexcusably poor hygienic practices in the hospital? Multiple lines of evidence suggest the latter. *If* the medics were guilty, then all of the infections should have been noted after they arrived, but the evidence shows that some of the infections were recorded as occurring before the workers came to Libya (more on this in a moment). Moreover, one child was even infected after the medics had already been imprisoned. Nonetheless, the "Benghazi six" were convicted in a Libyan court in May 2004 and sentenced to death by firing squad. And

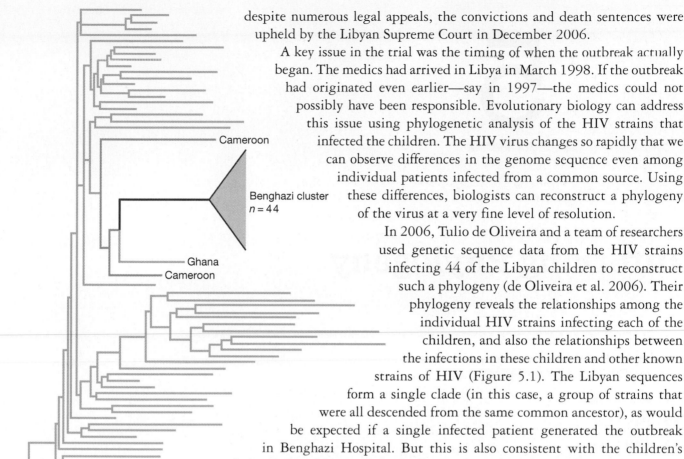

FIGURE 5.1 The Libyan HIV sequences. A phylogeny of HIV sequences that infected the Libyan children form a single clade (red), and this clade is closely related to strains from Ghana and Cameroon. This suggests that a single introduction was responsible for the outbreak in Benghazi Hospital, and that West Africa was a likely source of the strain that caused the outbreak. Adapted from de Oliveira et al. (2006).

despite numerous legal appeals, the convictions and death sentences were upheld by the Libyan Supreme Court in December 2006.

A key issue in the trial was the timing of when the outbreak actually began. The medics had arrived in Libya in March 1998. If the outbreak had originated even earlier—say in 1997—the medics could not possibly have been responsible. Evolutionary biology can address this issue using phylogenetic analysis of the HIV strains that infected the children. The HIV virus changes so rapidly that we can observe differences in the genome sequence even among individual patients infected from a common source. Using these differences, biologists can reconstruct a phylogeny of the virus at a very fine level of resolution.

In 2006, Tulio de Oliveira and a team of researchers used genetic sequence data from the HIV strains infecting 44 of the Libyan children to reconstruct such a phylogeny (de Oliveira et al. 2006). Their phylogeny reveals the relationships among the individual HIV strains infecting each of the children, and also the relationships between the infections in these children and other known strains of HIV (Figure 5.1). The Libyan sequences form a single clade (in this case, a group of strains that were all descended from the same common ancestor), as would be expected if a single infected patient generated the outbreak in Benghazi Hospital. But this is also consistent with the children's being infected by a single medic. Fortunately, other evidence allows us to distinguish between these possibilities. For example, the HIV strains in this clade are most closely related to strains observed in areas of West Africa from which numerous migrants have come to Libya seeking employment—strongly suggesting accidental introduction from the Libyan population.

It is also possible to estimate the timing of the infection from the phylogenetic information. The older a clade is, the more time it has had for phylogenetic diversification. In a very recent clade, all members would be expected to share very similar sequences, whereas in an older clade, we would see more sequence divergence among the clade members. The team of researchers measured the genetic divergence among the HIV strains in the Libyan clade. Given the rate at which the HIV sequence changes over time, they concluded that the Libyan clade was too diverse to have arisen as late as March 1998. Rather, the infections must have started early, possibly in 1997, and almost certainly prior to the medics' arrival in Libya. Comparable analysis of the hepatitis C virus strains also infecting many of the children revealed the same thing: The infections were too diverse to have begun spreading as late as March 1998.

While the Libyan courts were unwilling to heed this scientific evidence, the clear science behind the case intensified international political pressure on the Libyan government. Not the least of those campaigning on behalf of the "Benghazi six" were 114 Nobel laureates in the sciences, who, based on the scientific evidence we have detailed, published an appeal for their release in the journal *Nature* (Roberts and Nobel Laureates 2006). These pleas from the scientific community, coupled with continued diplomatic efforts, paid off. On July 16, 2007, the Libyan Supreme

Council for Judicial Authority commuted all six death sentences to sentences of life imprisonment. A week later, after 8 years in a Libyan prison, the six medics were returned to Bulgaria to serve out their terms. Back in Bulgaria, they immediately received a pardon from the Bulgarian president and were released. This is a happy ending, of a sort—but of course no such eleventh-hour reprieve was possible for the more than 400 HIV-positive children who were also victims of this tragedy.

It was, in part, due to the construction of phylogenetic trees, and the ability to make inferences from such trees, that innocent lives were spared in this case. Of course, in most instances, no lives will be spared when phylogenetic trees are constructed and interpreted, but they are still an extraordinarily powerful tool for understanding evolutionary history.

As we will explore in much more depth, evolutionary biologists use many different methods for constructing phylogenetic trees and employ various types of data when they do so. Phylogenetic trees are used both to construct hypotheses about common ancestors and how various species are related to each other, as well as to test hypotheses about such relationships.

In this chapter, we will examine the following questions:

- What is the general procedure for creating a phylogenetic tree?
- What are some of the methods used to construct phylogenetic trees, and what are their limitations?
- How do different sources of information—including information derived from molecular genetic sequences, the fossil record, and geographic patterns—enable evolutionary biologists to build phylogenetic trees?
- How do biologists control for phylogenetic history when using a technique known as the comparative method to study evolutionary patterns?

5.1 Building Trees

The task of creating a phylogenetic tree is fundamentally a problem in statistical inference—that is, we wish to make inferences about the world from a data set. In the case of phylogenetic inference, we typically have information about traits such as genetic sequences of the species we are considering, and from these data we aim to infer the historical evolutionary relationships among these species. Before we look at how this is done, take a moment and think about how powerful such techniques can be in principle. What we are aiming to do is use data we can measure *right now* to make inferences about events in the evolutionary past, often millions of years in the past.

The basic conceptual approach to phylogenetic tree building is straightforward. We select a number of species (or other taxa) for which we wish to build a tree. We collect information about the **characters** (also called traits) of these species, and we look at which species have which characters in common. The logic of tree building is that species with many characters in common are more likely to be closely related to one another than are species with fewer characters in common. For example, we presume that mammalian species—species in which females produce milk and feed their young, and in which all individuals have hair, have a middle ear with three bones, and share numerous other traits—are more closely related to one another than they are to species that lack these traits, such as lizards.

This logic assumes that shared characters are homologies—that is, characters that are shared because of shared common ancestry. Otherwise, we would not expect species with more characters in common to be more closely related phylogenetically. Although this logic seems straightforward, the devil is in the details. How do we test the possibility that common characters are analogous rather than homologous? How do we resolve conflicts in the data regarding the evolutionary relationships among the species we are studying? How—by what algorithm or procedure—do we go about actually finding the best tree corresponding to a given set of character data? Evolutionary biologists have developed a number of different *phylogenetic methods*, each of which handles these challenges in a different way. In this chapter, we will look at a number of these methods, with an aim both to understand the logic of each approach and to understand its strengths and weaknesses.

We begin by looking at what are called *parsimony methods*, in which we search for trees that have the minimum number of evolutionary changes. We touched briefly on parsimony analysis in Chapter 4 when we examined phylogenies in which the character of interest was coat coloration; here we explore the topic in more depth. Advantages of the parsimony approach include its conceptual simplicity, and the existence of straightforward algorithms for constructing parsimonious trees.

Next, we turn to *distance methods*. As we mentioned, the basic logic of phylogenetic reconstruction is that species with large numbers of common characters tend to be more closely related to one another than species with smaller numbers of common characters. One of the simplest approaches to reconstructing trees is simply to count up the number of commonalities, and to use this information directly to cluster closely related species together. This is what distance methods do.

While both parsimony methods and distance methods can be quite effective in inferring evolutionary history, both use tricks of a sort: Parsimony methods assume that the fewer changes required, the more plausible the tree; distance methods assume that more similar species are more closely related. Neither incorporates an explicit statistical model of how evolutionary change takes place. **Maximum likelihood** methods aim to remedy this by using explicit models of how characters change through the evolutionary process and by applying conventional techniques of statistical inference to find the phylogenetic tree that best explains the data. **Bayesian inference** methods do something similar. The difference between the maximum likelihood and Bayesian inference methods lies in the interpretation of what "best explains" should mean. Maximum likelihood methods and Bayesian inference methods require a modest background in probability theory, so we will defer our treatment of these topics to the appendix entitled, "Likelihood Methods and Bayesian Methods for Phylogenetic Inference," located at the end of this book.

5.2 Parsimony

The fundamental idea behind **parsimony** is that the best phylogeny is the one that both explains the observed character data and posits the fewest evolutionary changes. To find the best phylogenetic tree, one first must be able to evaluate a given tree and calculate how many character changes are necessary to explain the observed character pattern on that particular tree. An example helps. Suppose we are trying to evaluate the phylogenetic tree in Figure 5.2 as a hypothesis regarding the evolutionary relationships among species 1–4.

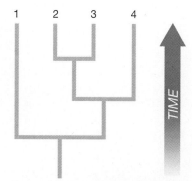

FIGURE 5.2 A phylogenetic tree represents a hypothesis for evolutionary relationships. This particular tree indicates the hypothesized relationships among species 1–4.

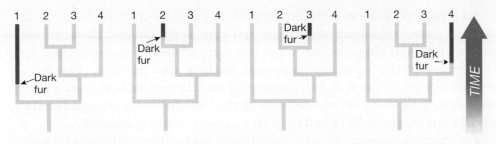

FIGURE 5.3 A single species differs from the others. If the character state of just one species differs from the others, we can always explain this by a single evolutionary change denoted by the arrows on each of these four trees. In our example, dark lines represent dark fur, and light lines, light fur.

Here we have data for a particular character—say, coat color—and we want to know how many evolutionary changes would be required to explain the current coat colors if our tree were correct. If just one of the four species has a dark coat and the others all have light coats, we can obviously explain this by hypothesizing a single evolutionary event: Dark coat color arose by a single evolutionary change occurring after our dark-colored species diverged from the other species on our tree (Figure 5.3).

But if two species have dark coats and two have light coats, matters get more interesting. If sister species 2 and 3 share a common character—say, dark coats—and species 1 and 4 share the other character (light coats), our tree can again explain the pattern with a single evolutionary event, as shown in Figure 5.4.

If species 1 and 2 instead share a common character and species 3 and 4 share a common character, our tree would require two evolutionary changes to explain the character data. Interestingly, there are a number of different ways to explain this situation with two character changes. One possibility is that dark coats arose once and were subsequently lost along one branch of the tree, or alternatively they may have arisen twice. Figure 5.5 shows two such possibilities.

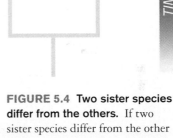

FIGURE 5.4 Two sister species differ from the others. If two sister species differ from the other species on the tree, we can explain this pattern by a single evolutionary event as well. In our example, dark lines represent dark fur, and light lines, light fur.

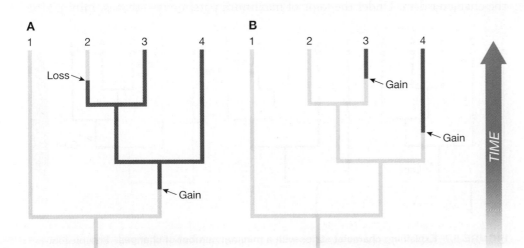

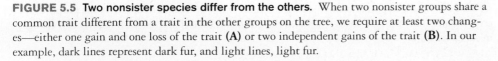

FIGURE 5.5 Two nonsister species differ from the others. When two nonsister groups share a common trait different from a trait in the other groups on the tree, we require at least two changes—either one gain and one loss of the trait (**A**) or two independent gains of the trait (**B**). In our example, dark lines represent dark fur, and light lines, light fur.

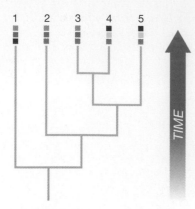

FIGURE 5.6 A phylogeny with three observed characters in five taxa. Three character states (dark/light blue, dark/light green, dark/light purple) and a hypothetical phylogenetic tree relating the species. We want to evaluate this tree using a parsimony approach.

So, given a tree and a set of character states for a particular character, we can figure out how many evolutionary changes are necessary. In the example above, we only looked at a single character, but in practice there are usually multiple characters to consider. In the parsimony framework, working with multiple characters is straightforward. We look at each character in turn, determine how many changes are necessary for that character, and sum up the total number of changes necessary for all characters in order to find the total number of changes required.

For example, suppose we have information about three different characters, as shown in Figure 5.6. To use the parsimony approach, we need to know the minimum number of changes in each character that are needed to explain our data. To do this, we tally the number of changes required, given our tree. In our case, our tree requires 1, 2, and 2 character changes, respectively, to explain the purple, green, and blue characters. In Figure 5.7, we show one way in which each of the character states could be explained by the minimum number of changes.

Notice that, while each tree in Figure 5.7 shows how a minimal number of changes can be placed on our trees to explain character changes, these patterns of change are not unique. For example, the purple tree could alternatively be explained with a single change if light purple were the ancestral state and the dark purple character arose via a change along the branch leading to species 1 alone.

Because it is inconvenient to have to draw out a separate tree for each character, we often summarize the changes in all characters with a diagram like that shown in Figure 5.8. We saw this sort of representation when we looked at the process of placing traits on trees in Chapter 4.

Once we have found a way to represent the minimum number of character state changes on a tree, we can define this number as a parsimony score *for that particular tree*. In order to use maximum parsimony to infer phylogenetic history, we look at various possible trees and select the one with the lowest parsimony score.

In Figure 5.8, for example, we saw that it takes five character changes to explain the character data on that particular tree. But we can explain the same character data with fewer changes by means of a different phylogenetic tree. Figure 5.9 illustrates this. For this tree, only three character changes are necessary to explain the character data. Under the logic of maximum parsimony—that is, minimizing

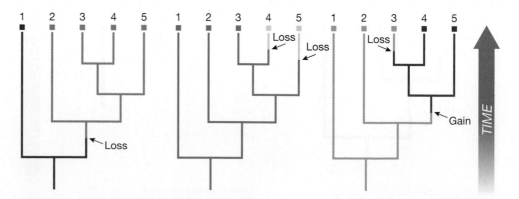

FIGURE 5.7 Explaining character states with a minimal number of changes. Possible locations of character changes—gain or loss of the trait indicated by the darker color—for three character states along the hypothetical tree shown in Figure 5.6. For this particular tree, the purple character requires only a single change, whereas the green and blue characters each require two changes in character state.

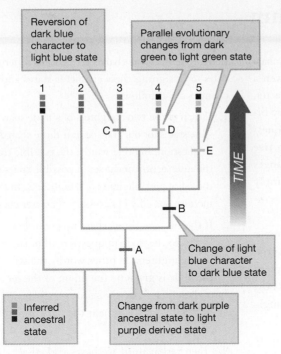

FIGURE 5.8 Showing multiple characters on a single phylogeny. We can show all of the changes on a single diagram by indicating the inferred ancestral state and then marking each change in character state.

the number of evolutionary changes required to explain our tree—we prefer this tree to the previous one because it can explain our data with fewer changes. Sometimes several different trees may be tied for the lowest parsimony score. In this case, each is said to be equally parsimonious; the parsimony approach does not give us cause to prefer any one of these most parsimonious trees over any other.

How do we know when we have found the most parsimonious tree? In Figure 5.9, it is straightforward to tell: We have only one change per variable character, so we know we cannot possibly do better. But we still need a general way to figure out how many changes a tree will require given a certain set of characters. Fortunately, there are a number of algorithms that allow us to determine the number of changes necessary to explain a given character pattern on a given tree. Box 5.1 describes one of the simplest of these, the Fitch algorithm.

Parsimony has the advantage of conceptual simplicity, but parsimony approaches are not without problems. The worst of these problems is that parsimony is not a *consistent estimator*—that is, an estimation procedure that, given enough data, will ensure that we get the right answer. Thus, if we use parsimony to reconstruct a phylogeny, it is possible for us to get the wrong tree, no matter how much

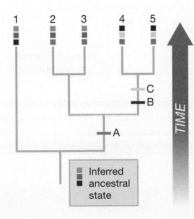

FIGURE 5.9 A more parsimonious tree for our character data. Only three character changes are necessary to explain the character data using this phylogenetic tree.

BOX 5.1 The Fitch Algorithm

Parsimony algorithms search trees to explain the observed character data with a minimum number of changes. But given a hypothetical tree and the character states for a given character, how many evolutionary changes are required? Evolutionary biologist Walter Fitch developed a method to answer this question (Fitch 1971). The *Fitch algorithm* applies to a given tree and a single character trait at a time; the number of changes required to explain multiple characters on that tree is simply the sum of the number of changes required to explain each individual character.

The Fitch algorithm does not find the best tree; it simply tells us how many character changes are required for a given tree. We then would need to repeat the process for other plausible trees in order to find the most parsimonious. In this box, we illustrate the application of the Fitch algorithm to a single character on one sample tree.

Figure 5.10 illustrates a tree in which we wish to evaluate the character values red, blue, or yellow for each of seven species on that tree. The Fitch algorithm proceeds in a series of steps (Felsenstein 2004). We begin at the branch tips, taking sister groups and working downward to the base of the tree. Beginning with zero, we keep a running count of how many character changes are necessary. As we work our way down the tree, each internal node is assigned one or more character states, and we update the tally of character changes where appropriate. The rules for assigning these character states and tallying character changes are as follows:

1. If each of the two daughters of a node share one or more possible states for our trait, assign those shared states to the node in question. In other words, the possible traits at the node are the intersection of the set of possible traits of daughter 1 and the set of possible traits of daughter 2. In this case, we do not increase our tally of necessary character changes.

2. If the two daughters share no possible states in common, assign to the node in question all of the possible states for both daughters. In other words, the set of possible traits at the node is given by the union of the set of possible traits of daughter 1 and the set of possible traits of daughter 2. In this case, we augment the tally of necessary character changes by one.

We then repeat until we have worked all the way to the root of the tree.

In the figures that follow, we carry out this process for our example tree. In Figure 5.11, we assign character states to nodes

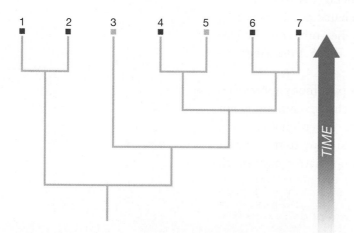

FIGURE 5.10 How many character changes are necessary for this tree? We will use the Fitch algorithm to determine the minimum number of evolutionary changes required to explain the character states (red, yellow, or blue) of the seven species on this tree.

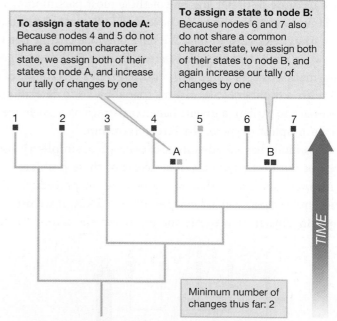

To assign a state to node A: Because nodes 4 and 5 do not share a common character state, we assign both of their states to node A, and increase our tally of changes by one

To assign a state to node B: Because nodes 6 and 7 also do not share a common character state, we assign both of their states to node B, and again increase our tally of changes by one

Minimum number of changes thus far: 2

FIGURE 5.11 Assigning possible character states to nodes A and B. Here we see how to use the Fitch algorithm to assign possible character states to nodes A and B.

A and B. In each case, the daughter nodes share no possible character states in common. We thus take the union of the daughters' character states and increase our tally of character changes by one each time. Node A has two daughters: species 4, which is blue, and species 5, which is yellow. Thus, node A is assigned both blue and yellow as possible character states. Node B has two daughters: species 6, which is red, and species 7, which is blue. Thus, node B is assigned both red and blue as possible character states. In each case, the daughters share no possible traits in common, and so we have to augment our tally of character changes each time. This gives us a total of two necessary character changes thus far.

Figure 5.12 illustrates how we continue downward along the tree. Node C has two daughters: node A with states blue and yellow, and node B with states blue and red. These share a common possible state, blue, and so we assign that state to node C. Because its daughters share a common state, we do not have to augment our tally of character changes to account for node C. We then move on to node D. Node D has two daughters: species 3 with state yellow, and node C with state blue. Because these daughters share no common character states, we assign to node D the union of their character states, blue and yellow,

and we increase our tally of character changes by one more, to a total of three.

In Figure 5.13, we assign character states to the two remaining nodes, node E and node F. Node E has two daughters: species 1, which is blue, and species 2, which is red. We thus assign the possible character states of blue and red to species E, and we augment our tally of changes again, giving us a total of four. Node F has two daughters: node E and node D. These daughters share blue as a possible character state, so we assign blue to node F, and we do not need to further increase our tally of changes.

At this point we have assigned character states to each node of the tree, and the algorithm is complete. Our tally of character changes is four. By the algorithm, this is guaranteed to be the minimum number of changes necessary to explain the character data on this particular tree.

It is important to realize that the Fitch algorithm does not tell us the most likely character states for each ancestral node. In the algorithm, the process of assigning states to interior nodes is simply a way to count the number of changes, not a reconstruction of ancestral types.

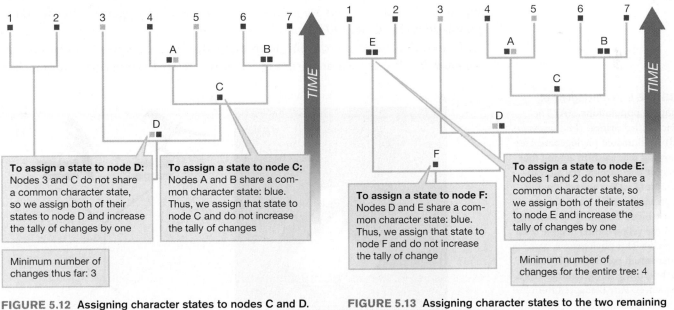

To assign a state to node D: Nodes 3 and C do not share a common character state, so we assign both of their states to node D and increase the tally of changes by one

To assign a state to node C: Nodes A and B share a common character state: blue. Thus, we assign that state to node C and do not increase the tally of changes

Minimum number of changes thus far: 3

FIGURE 5.12 Assigning character states to nodes C and D. Here, we see how to use the Fitch algorithm to assign character states to nodes C and D.

To assign a state to node F: Nodes D and E share a common character state: blue. Thus, we assign that state to node F and do not increase the tally of change

To assign a state to node E: Nodes 1 and 2 do not share a common character state, so we assign both of their states to node E and increase the tally of changes by one

Minimum number of changes for the entire tree: 4

FIGURE 5.13 Assigning character states to the two remaining nodes in the tree, nodes E and F. Here we use the Fitch algorithm to assign character states to nodes E and F.

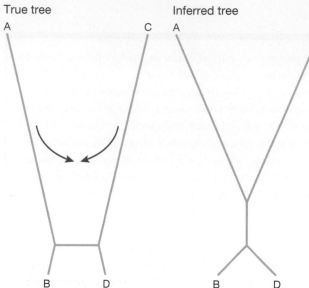

True tree

A   C

Inferred tree

A   C

B   D   B   D

FIGURE 5.14 Long-branch attraction. On the true tree shown in the left panel, A and B are sister groups and C and D are sister groups. Because evolutionary change is occurring more quickly in taxa A and C, the corresponding branches are much longer. As a consequence, parsimony methods may incorrectly infer a tree of the form shown in the right panel.

data we have available. Sequencing additional loci, or tabulating additional morphological characters, may not help us in the least. Such an undesirable outcome is particularly likely when evolutionary changes occur at different rates on different branches of the phylogeny, as illustrated in Figure 5.14. In that case, parsimony methods may incorrectly infer too close a relationship between the rapidly evolving branches. This tendency is known as **long-branch attraction**, because species on long branches of the phylogenetic tree are "pulled together" by the inference procedure used in parsimony analysis (Felsenstein 1978; Bergsten 2005).

5.3 Rooting Trees

In most of the parsimony examples that we have discussed, we illustrated our trees as if they were rooted. Strictly speaking, however, a maximum parsimony approach does not distinguish among the multiple alternative rooted trees that correspond to the same unrooted tree. Any two rooted trees corresponding to the same unrooted tree will require the same number of changes, and so there is no way to distinguish among them using parsimony criteria alone. If we want to work with rooted trees, then, it will be important to have ways of *rooting*—assigning a root to—the unrooted tree that we can get from a maximum parsimony analysis.

The most common approach to rooting a tree is to use an outgroup. Suppose we have an unrooted phylogenetic tree of several magpie populations, as shown in Figure 5.15, and from this we wish to derive a rooted phylogenetic tree for these populations (Lee et al. 2003).

To root this tree using the outgroup method, we pick another population that we know in advance to be an outgroup—that is, a related population that branched

FIGURE 5.15 Phylogeny of magpie populations. (A) The black-billed magpie (*Pica hudsonia*). **(B)** An unrooted phylogenetic tree showing relationships among four magpie populations: the Korean magpie (*Pica pica sericea*), the Eurasian magpie (*Pica pica pica*), the black-billed magpie (*Pica hudsonia*), and the yellow-billed magpie (*Pica nuttalli*). This phylogeny is based on a maximum parsimony phylogeny derived using mitochondrial DNA sequences. Part B adapted from Lee et al. (2003).

A

B

Korean magpie

Eurasian magpie

Black-billed magpie

Yellow-billed magpie

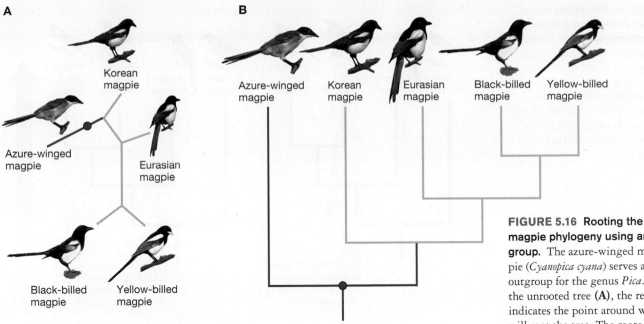

A

Korean
magpie

Azure-winged
magpie

Eurasian
magpie

Black-billed
magpie

Yellow-billed
magpie

B

Azure-winged
magpie

Korean
magpie

Eurasian
magpie

Black-billed
magpie

Yellow-billed
magpie

FIGURE 5.16 Rooting the magpie phylogeny using an outgroup. The azure-winged magpie (*Cyanopica cyana*) serves as an outgroup for the genus *Pica*. On the unrooted tree (**A**), the red dot indicates the point around which we will root the tree. The rooted tree (**B**) has the azure-winged magpie as an outgroup. Adapted from Lee et al. (2003).

off earlier in evolutionary history from the entire clade that we are considering. In this case, the azure-winged magpie (*Cyanopica cyana*) works well. The azure-winged magpie is a fairly close relative of the group we are considering, but it is less closely related to the members of the *Pica* genus than they are to one another. We can therefore construct another phylogenetic tree that includes our outgroup, as shown in Figure 5.16A.

We can form a rooted tree from an unrooted tree simply by picking a branch around which to root the tree. Using the outgroup method, we select the branch leading to the outgroup—namely, the branch connecting the magpies to the azure-winged magpie. We then draw a tree rooted around a point (the red dot in Figure 5.16A) on this branch. Figure 5.16B shows the rooted tree that we get by this process.

As we discussed in Chapter 4, rooting the tree can be useful because a rooted tree (unlike an unrooted tree) informs us about something that evolutionary biologists are keen to know—the polarity of character changes. For example, consider the light-colored beak that is unique to the yellow-billed magpie. From the unrooted tree in Figure 5.15B, we cannot tell whether having a light bill is ancestral or derived, because we do not know along which branch the root lies. If the tree were rooted along the branch between the yellow-billed magpie and the rest of the tree, having a yellow beak could have been the ancestral state, which was then lost in the branch leading to the other magpie populations. But once we find the root, we see that a yellow beak is very likely to be a derived character. Even ignoring the fact that the outgroup also has a dark beak, we see that we would require multiple character changes to explain the beak color character if yellow beaks were ancestral, whereas we can explain this character with a single character change given that yellow beaks are derived (Figure 5.17).

Knowing the root of the tree can also tell us about **phylogeography**: the story of how a group of populations or species moved across the globe over the course of their evolutionary history. The conventional explanation of magpie evolution had

FIGURE 5.17 Parsimony suggests that yellow beaks are a derived character. (**A**) If yellow beaks are ancestral, multiple character changes are required to explain the distribution of beak color on the phylogeny; one such tree is shown here. (**B**) If yellow beaks are derived, we can explain the distribution of beak color with a single change, as shown here.

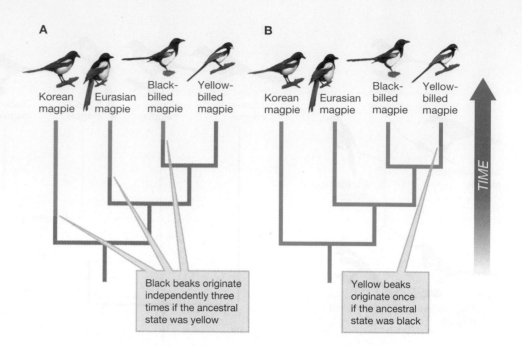

been that magpies arose in Asia and subsequently colonized North America in two separate waves, once early to found the yellow-billed magpie population, and again later as the black-billed magpie. But the form of the rooted tree suggests an alternative hypothesis (Lee et al. 2003). It suggests that magpies arose in Asia, where the Eurasian magpie diverged from the Korean magpie. Eurasian magpies subsequently invaded North America a single time and their descendant lineages branched into the black-billed and yellow-billed magpie species found there (Figure 5.18).

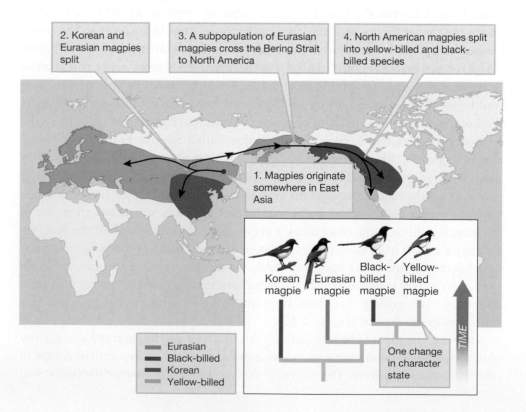

FIGURE 5.18 Magpie phylogeography as inferred from the rooted phylogeny. Magpies appear to have originated in East Asia, where they diverged into the Korean magpie lineage and the Eurasian magpie lineage. A subpopulation from the Eurasian lineage then crossed the Bering Strait to the New World, and subsequently speciated to produce the black-billed and yellow-billed species now found in North America.

Of course, we could only follow this outgroup rooting procedure because we knew that the azure-winged magpie is a suitable outgroup to the genus *Pica*. In other words, we already knew quite a bit about the patterns of evolution in the larger corvid clade that includes the genus *Pica*, and this knowledge helped us get a more detailed picture of evolution within the magpies.

5.4 Distance Methods

Phylogenetic distance methods provide a second approach to inferring phylogenetic trees. The basic idea behind distance methods is that if we can measure the pairwise "distances" between species, then we can use these distances to reconstruct a tree. A warning here: "Distance" is not being used in the literal geographic sense of feet, miles, and so on. Instead, it is a measurement of morphological or genetic differences between species. Our aim is to find a tree with branches arrayed such that the distance along the branches between any two species is approximately equal to the distance that we measured between those two species.

To do this, we need to address two questions: (1) how do we measure distance between species, and (2) once we have these distance measurements, how do we find the best tree given these distance data? We will address these in turn.

Measuring Distances between Species

There are a number of different ways we can measure the distance between any two species or, more generally, between any two populations. Prior to molecular systematics, distances were often computed from morphological measurements or by tallying the number of character differences between species. Such methods remain important when using fossil data to build phylogenies for extinct organisms. But when we study living species, it is now far more common to use DNA sequences from the two species, suitably aligned (Box 5.2). One of many ways to do this is simply to count up the number of base pair differences, and to use this tally as the distance between the two species (Figure 5.19). If we have amino acid sequence data instead of DNA sequence data, we can look at the number of amino acid substitutions between the two clades and count this fraction as the molecular distance between those clades (Figure 5.20).

In the examples above, we assumed that each population is homogeneous with respect to the trait we are measuring, or at least that we have a characteristic sequence from that population. If instead we have information about allele frequencies in each population, we can look at the differences in allele frequencies and use these differences to compute a **genetic distance** between the two populations. The idea is that populations with similar allele frequencies may be more closely related than those with more divergent allele frequencies. This approach is more commonly

Species A TAGAGCTAAACTTC
Species B TAAAGCTACACTTC

FIGURE 5.19 Measuring the distance between two species using DNA sequence data. Species A and species B differ in the two marked positions. One way to measure molecular distance between species A and species B is to count the number of differences, in which case the molecular distance between these species is 2.

Sequence A: Tyr – Pro – Tyr – Asp – Val – Pro – Asp – Tyr – Ala
Sequence B: Tyr – Pro – Tyr – Asp – Val – Pro – Asp – Val – Ala

FIGURE 5.20 Measuring the distance between two species using amino acid sequence data. Sequence A and sequence B differ in the one marked position. If we quantify the molecular distance by counting the number of amino acid differences, the molecular distance between species A and species B is 1.

BOX 5.2 Sequence Alignment

If we want to use any phylogenetic method that relies on DNA or amino acid sequence data, we face the problem of *sequence alignment*. Because of insertions, deletions, and other changes to the structure of the DNA, the sequences from species from the various groups being studied may not line up—or align—cleanly, making comparison very difficult. To see this more concretely, let's first look at a case where sequence alignment is *not* a problem, as in Figure 5.21.

Now suppose there has been a deletion in the DNA sequence of species A. Figure 5.22 illustrates the consequences. Because

of this deletion, the species A sequence doesn't align with the others directly; it would have to be adjusted, leaving a gap at this position, in order to align correctly. In general, there can be multiple deletions at different places in different species, as well as multiple insertions. Alignment becomes more difficult as the number of such instances increases. As such, evolutionary biologists have created various computer program methods for handling this "alignment problem," although we note that many sequences are frequently aligned by hand for verification (Feng and Doolittle 1987; Higgins and Sharp 1988; Baldauf 2003).

FIGURE 5.21 Sequence alignment and constructing a phylogeny. (A) A case where sequence alignment is not a problem. Here we have nucleotide sequence data for eight species, and the data align. We see differences across species at seven positions. From these we can construct a phylogenetic tree **(B)**. Adapted from University of Illinois (2011).

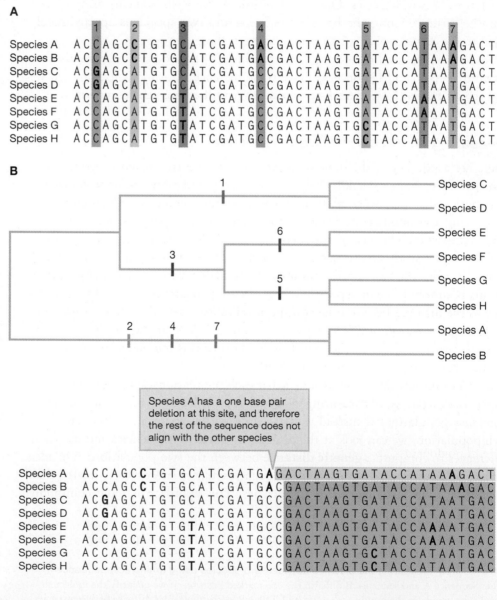

FIGURE 5.22 Deletions or insertions affect sequence alignment. Here we see the same sequences as in Figure 5.21, but with a single base pair deletion at the indicated position in species A. Notice that the subsequent base pairs in species A are now shifted relative to those in the other species. To see this, shift the orange shaded area one position to the right and observe how sequences in the blue and orange shaded areas will once again align. Adapted from University of Illinois (2011).

used when attempting to construct phylogenetic trees showing the relationships among different populations of a single species. This is a topic of great interest to evolutionary biologists—for example, those studying the process of speciation—and so population geneticists have developed a number of different ways to compute distances based on allele frequencies.

Constructing a Tree from Distance Measurements

Regardless of which type of distance measure we are using, the process of constructing a phylogenetic tree from distance information proceeds as follows: After measuring our distances between species, we have a list of the distances between each species pair in our sample. For example, if we are trying to infer the relationships among four species, A, B, C, and D, we use six pairwise measurements, as shown in Figure 5.23A.

Researchers often represent these in the form of what is called a *distance matrix*— that is, a table that lists the distance between each species pair. The distance between each species and itself is zero, so the diagonal entries of this matrix are all zero. Figure 5.23B is the distance matrix corresponding to the genetic distances shown in Figure 5.23A.

Once we have these measurements, our aim is to find a way of arranging all six segments along a single tree. One way to envision the problem is to imagine that each of the six colored line segments in Figure 5.23 is a cable made of rubber. We want to lay these out along a four-species phylogenetic tree such that the cables undergo a minimum of compression or stretching. To try to make this work, we get to choose the shape of the tree, which species go on which nodes, and how long to make each branch of the tree.

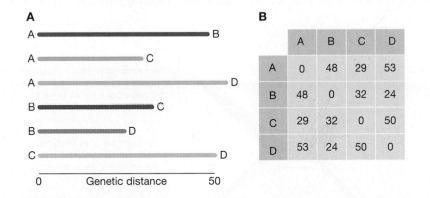

	A	B	C	D
A	0	48	29	53
B	48	0	32	24
C	29	32	0	50
D	53	24	50	0

FIGURE 5.23 Genetic distances between species A–D. (A) There are six pairwise distances among four species. Here each distance is indicated by a colored band of the appropriate length. **(B)** The distance matrix for these genetic distances.

For a phylogenetic tree relating four species, there is only one basic tree shape, as shown in Figure 5.24. Given this tree shape, there are three distinct ways to arrange the four species on the four branch tips. All other arrangements can be reached by rotating the tree around one of the interior nodes, and so they do not represent distinct trees; they are just different visual perspectives on three ways that are shown in Figure 5.25.

Our job is now to choose which of these three arrangements is best, and how long each branch should be to minimize the stretching necessary as we lay out our imaginary cables. There are a number of different algorithmic procedures for doing this, including what are called *weighted least squares*, *UPGMA* (unweighted

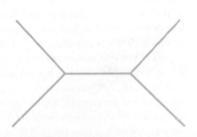

FIGURE 5.24 The only possible unrooted tree relating four species. If we had more than four species, multiple shapes of an unrooted tree would be possible.

A

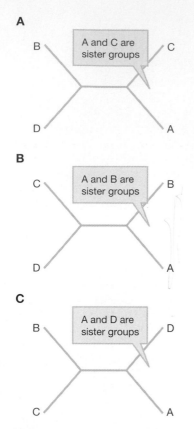

FIGURE 5.25 **Three different arrangements of four species.** Four species can be assigned to an unrooted phylogenetic tree in three different ways, as shown.

pair group method with arithmetic mean), and *neighbor-joining methods*. Each has its strengths and weaknesses; we illustrate the weighted least squares solution in three figures below. Since we are only looking at four species, we can already guess which tree shape is most appropriate without even using the weighted least squares algorithm. Looking at our distances in Figure 5.23, we see that species A is more closely related to species C than to any other species, while species B is more closely related to species D than to any other species. This means that the assignment of species to nodes on our tree will be that shown in Figure 5.25A. Now we want to lay down the six distances with a minimum of stretching. In doing so, we can adjust the lengths of the five line segments that make up the tree. Figure 5.26 illustrates the best way to do this.

Evolutionary biologists have readily available phylogenetic inference software (one of the most common is a program named PHYLIP) which can be used to construct such trees, given both the tree topology—the shape and assignment of species to branch tips—and the branch lengths. Figure 5.27 shows the weighted least squares tree for our example.

While distance methods are conceptually straightforward, and computationally among the fastest, they are not without problems. One of the biggest concerns to many researchers is a philosophical one: Distance methods lack any sort of underlying evolutionary model. Rather, they are fundamentally *phenetic* in their approach, meaning that they group species together according to similarity without attempting to reflect the underlying historical evolutionary relationships among those species. The assumption being made here is that the similarity we are measuring is a reflection of homology, rather than analogy. Sometimes this is correct, and sometimes it is not. When we use these methods, we accept the risk that some traits we employ are analogous in order to obtain the benefit of having many easily measurable characters to use when building our tree. That said, most

FIGURE 5.26 **Assigning distances to the tree.** When we use a distance-based method to infer tree topology and branch lengths, our aim is to find a tree topology in which each pairwise distance is as close as possible to that inferred from the data. For this example, with four species and six pairwise distances, our aim is to arrange the six measured distances or "cables" to best fit together in a phylogenetic tree. **(A)** If we pick the wrong tree, the fit will be very poor: Some of the cables representing each pairwise distance will be much too long, and others will be much too short. **(B)** For the best tree, the cables are too long or too short by only a small margin.

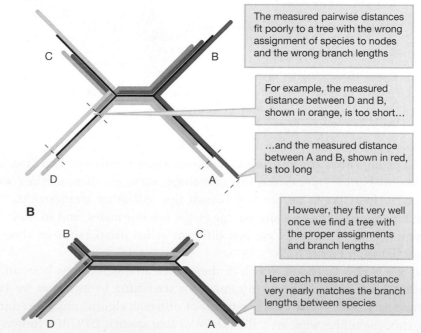

contemporary evolutionary biologists prefer *cladistic* methods that aim to explicitly reconstruct evolutionary relationships.

There is a another problem with distance methods as well. When we use genetic distances in the process of building phylogenies, we are assuming that the more that DNA sequences differ from each other, the more distantly related our species are. But what if some species in our taxa of interest are evolving faster than others (as in Figure 5.14)? In that case, it is possible that quickly evolving species cluster together because of the speed at which they evolve, rather than because of true phylogenetic similarity. Although this is beyond the scope of what we will cover in this chapter, we note that evolutionary biologists have developed a number of statistical techniques to attempt to deal with these difficulties.

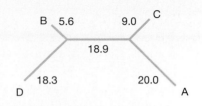

FIGURE 5.27 Weighted least squares tree for our example. Branch lengths are indicated by the measurements listed alongside each branch. These values were obtained using readily available software packages.

5.5 How Many Different Trees Are There?

We have discussed several ways of inferring phylogenies. In each of these cases, biologists can use computational algorithms to determine how strongly any particular phylogeny is supported by the data. Why, then, is phylogenetic inference a difficult and computationally intensive problem? The answer lies in the fact that there are simply too many possible phylogenetic trees to search, even with the fastest of computers. Instead, researchers must devise clever ways to search within the "space" of possible trees.

In this section, we will develop a basic intuition for the problems evolutionary biologists face regarding the number of possible trees: Just how big is the space of possible trees, and how rapidly does the space grow as we add species or other taxa (Felsenstein 2004)? We will begin by considering unrooted trees. There is only one unrooted tree relating three species A, B, and C, as shown in the center of Figure 5.28.

Now think about the different ways we could add a fourth branch to this tree to create an unrooted tree for four species. Our three-species tree has three branches, each leading from the internal node to one of the three tips. To create a four-species tree, we could add a new branch leading to a new species D to any of these three branches. Each point of attachment creates a *different* four-species tree, as illustrated in Figure 5.28. Thus, there are three different unrooted four-species trees.

Each of those four-species trees has five branches. We can create a five-species tree by adding a new branch, with a new species E, to any of those five branches. Each choice of attachment location again produces a *different* tree. Thus, from *each* of our three four-species trees, we can produce five different five-species trees. This gives us a total of $3 \times 5 = 15$ different five-species trees.

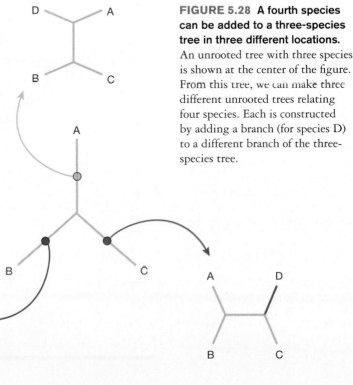

FIGURE 5.28 A fourth species can be added to a three-species tree in three different locations. An unrooted tree with three species is shown at the center of the figure. From this tree, we can make three different unrooted trees relating four species. Each is constructed by adding a branch (for species D) to a different branch of the three-species tree.

We can continue adding branches in this way and counting the resulting trees. Each time we add a new branch, we get a tree with two additional branches: one of these is the one we just added, and the other comes from splitting the branch to which our new branch is attached. This means that our five-species trees will have 7 branches and 7 potential attachment points, our six-species trees will have 9 branches, and so forth. There will be $3 \times 5 \times 7 = 105$ six-species trees, and $3 \times 5 \times 7 \times 9 = 945$ seven-species trees. As shown in Table 5.1, even a relatively small number of species can be arrayed on unrooted trees in an exceptionally large number of ways.

To give you a sense of just how rapidly these numbers increase, there are more 13-species trees than there are people on the planet (just shy of 7 billion at present). There are more 22-species trees than there are stars in the universe (approximately

TABLE 5.1

The Number of Different Unrooted Trees for 3–30 Taxa

Number of Taxa	Unrooted Trees
3	1
4	3
5	15
6	105
7	945
8	10,395
9	135,135
10	2,027,025
11	34,459,425
12	654,729,075
13	13,749,310,575
14	316,234,143,225
15	7,905,853,580,625
16	213,458,046,676,875
17	6,190,283,353,629,375
18	191,898,783,962,510,625
19	6,332,659,870,762,850,625
20	221,643,095,476,699,771,875
21	8,200,794,532,637,891,559,375
22	319,830,986,772,877,770,815,625
23	13,113,070,457,687,988,603,440,625
24	563,862,029,680,583,509,947,946,875
25	25,373,791,335,626,257,947,657,609,375
26	1,192,568,192,774,434,123,539,907,640,625
27	58,435,841,445,947,272,053,455,474,390,625
28	2,980,227,913,743,310,874,726,229,193,921,875
29	157,952,079,428,395,476,360,490,147,277,859,375
30	8,687,364,368,561,751,199,826,958,100,282,265,625

10^{23}). There are more 36-species trees than there are water molecules in all of Earth's oceans (approximately 10^{47}). There are more 53-species trees than there are atoms in the universe (approximately 10^{80}).

This is just the number of possible *unrooted* trees. As we have seen, each unrooted tree corresponds to numerous rooted trees. From an initial unrooted tree, we can form a distinct rooted tree by rooting on each of its branches. An unrooted tree with k species has $2k - 3$ branches, which means that there will be ($2k - 3$) times as many rooted trees as there are unrooted trees. So, for our 53-species tree, there are 10^{80} (the unrooted case) $\times$ 103 (that is, $2k - 3$) possible trees.

Clearly, with so many possible trees for even a few dozen species, it is not feasible to check each and every one of them to see how well it explains a given set of character data. As a result, computer programs for reconstructing phylogenies have to be very clever in the way that they search the set of possible trees, only checking a very small fraction of those trees. Researchers continue to develop increasingly good algorithms for selecting which trees to check and which can be safely ignored; this *search problem* makes up much of the challenge of phylogenetic inference.

5.6 Phylogenies and Statistical Confidence

Throughout this chapter, we have stressed that constructing a phylogeny involves sampling characters and making assumptions about homology, and that any phylogeny is a hypothesis about the true evolutionary history of a group of organisms. As a result, it is essential that we develop statistical measures of support for our phylogenetic hypotheses. Yet, thus far, we have only looked at how we find a "best estimate" of the real phylogeny, and not at another component of statistical inference: how we measure our confidence in that best guess.

Once we have used our character data to infer a tree, how certain are we that this tree—or some component of this tree—is correct? How do we know when we can reject a hypothesis of the form "the clade X is monophyletic?" or "species A and B are sister groups"? These are issues of *statistical confidence*. Typically we might aim to ascertain whether we can reject a hypothesis with 95% confidence—that on average, for every 100 instances in which we reject a hypothesis, we are doing so correctly in 95 instances.

Researchers have developed a number of techniques for quantifying how strongly our data support a given phylogeny. In this section, we explore two of these approaches. The first, known as **bootstrap resampling**, can be used with any technique for phylogenetic inference, be it parsimony, a distance method, or a model-based method such as maximum likelihood or Bayesian inference. The second, **odds ratio testing**, can only be used with the model-based frameworks of maximum likelihood or Bayesian inference.

Bootstrap Resampling

Suppose we infer a phylogenetic tree such as that in Figure 5.29 from a set of character data. How certain are we that this is the "correct" tree—that is, the actual phylogeny of the groups we are studying? As we start looking at even modest numbers of species, we will rarely be sure—our statistical confidence is low—that

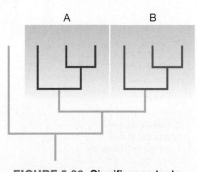

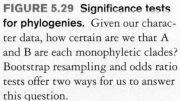

FIGURE 5.29 Significance tests for phylogenies. Given our character data, how certain are we that A and B are each monophyletic clades? Bootstrap resampling and odds ratio tests offer two ways for us to answer this question.

we have *exactly* the right tree. Because there are so many possible trees, and because many of them may be very similar, it is rare that we will have a single tree that is 95% likely given our data.

What this means is that typically we will not want to make confidence statements about the entire tree. Instead, we will make statements about *features* of the tree. In essence, we can break down our problem into more manageable bits. Because we are interested in inferring patterns of shared ancestry, one of the most important features of a tree is the set of monophyletic clades that it implies. Thus, a common aim of confidence assessment in phylogenetics is to say how strongly the data support a given monophyletic clade. That is, in Figure 5.29, how certain are we that clade A is indeed monophyletic? How certain are we that clade B is monophyletic?

Bootstrap resampling offers a powerful way to answer questions of this sort, by creating many new data sets from the observed data to get a representative distribution of results. To illustrate, suppose we have observed 10 different characters for 5 species. For this example, we will assume that these are *binary characters*—namely, characters that have two possible states, which we will call 0 and 1. Any heritable trait that can be broken into two categories can be represented by a binary trait. For example, binary traits include whether individuals in a species engage in parental care, whether they have cryptic coloration, and whether their sex determination depends on chromosomes or on environmental factors. We can represent our observations as a *character matrix*, a table that lays out the character states for each trait in each species. Such a matrix is shown at the top of Figure 5.30.

To carry out a bootstrap analysis, we *resample* from our original character matrix to create a collection of *bootstrap replicate* data sets—that is, a set of alternative character matrices. Essentially, this procedure involves picking a set of traits, *with replacement*, from the original set of traits and using these picks to form a new data set. Figure 5.30 illustrates the basic type of procedure that we might follow to generate a single replicate character matrix. In a bootstrap analysis, we create several hundred such replicate matrices.

We then apply the same tree-building methods that we used on our original data set to each replicate character matrix. This gives us a collection of *bootstrap replicate phylogenies*. Finally, we look to see how often the feature we are interested in—say, one particular set of species forming a monophyletic clade—occurs among our replicate phylogenies (Figure 5.31). If, for example, these species

1. Construct the character matrix: a list of traits and character states

2. Pick a new set of traits at random with replacement

3. Create a new character matrix from these traits only

4. This gives you **one** bootstrap replicate. Repeat the procedure many times to create additional replicates

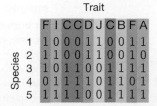

Trait

	A	B	C	D	E	F	G	H	I	J
1	1	0	0	1	1	1	1	0	0	1
2	0	0	0	1	0	1	1	0	1	1
3	1	1	1	0	0	1	1	0	0	0
4	1	1	1	1	0	1	1	1	1	0
5	1	1	1	0	1	1	0	1	1	0

Resampled traits
F, I, C, C, D, J, C, B, F, A

Trait

	F	I	C	C	D	J	C	B	F	A
1	1	0	0	0	1	1	0	0	1	1
2	1	1	0	0	1	1	0	0	1	0
3	1	0	1	1	0	0	1	1	1	1
4	0	1	1	1	1	0	1	1	0	1
5	1	1	1	1	0	0	1	1	1	1

FIGURE 5.30 Resampling character data. Here we have a character matrix made up of binary character data for 10 traits in 5 species. A single bootstrap replicate is created by resampling—by picking traits one at a time from the original data set to include in the replicate data set. Because sampling occurs with replacement, it is possible to draw the same trait more than once, and to draw other traits not at all. In the illustration here, trait C appears three times in the replicate data set and trait F appears twice. Traits E, G, and H do not appear at all. Note that for each species, the character states do not change when resampling occurs. This procedure resamples at the level of which characters are included in the analysis, but it does not cause changes in character state assignments.

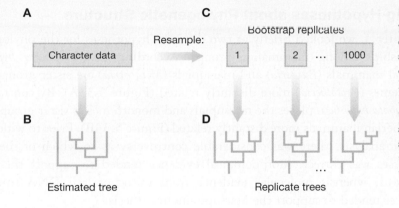

A

Character data

Resample:

C

Bootstrap replicates

1   2   ...   1000

B

Estimated tree

D

Replicate trees

...

FIGURE 5.31 An overview of a bootstrap analysis. Given our character data **(A)**, we construct our estimated phylogeny **(B)**, just as always. We also resample from the original character data to create multiple bootstrap replicate data sets **(C)**. For each replicate data set, we construct a phylogenetic tree using the same procedure that we used on the original character data. This gives us a replicate tree for each replicate data set **(D)**. To assess the support for any feature of our original tree, we count up the percentage of replicate trees that also display this feature.

form a monophyletic clade in 90% of the replicate phylogenies, we say that this clade has 90% *bootstrap support.*

Often, when presenting a phylogenetic tree, researchers will indicate the level of bootstrap support for each clade. This is done by placing a percentile number along the branch leading to that clade, as in Figure 5.32. Here the number 90 indicates that the highlighted clade, just above the number, appears as a monophyletic clade in 90% of the bootstrap replicates.

Although bootstrap support levels and statistical significance levels (statements such as "We can reject the hypothesis that A is not a monophyletic clade with 98% confidence") are both percentages used to indicate the support that our data provide for our conclusions, they are not the same thing and should not be confused for one another. Note that we sometimes see clades with bootstrap support values of 100%. This means that the clade in question appears in all bootstrap replicates—but it does not mean that we can reject the hypothesis that this is not a monophyletic clade with 100% certainty.

Odds Ratio Testing

Although bootstrap support levels are not statistical significance levels, there are other procedures by which we can construct statistical confidence tests for whether we have correctly depicted various features of our phylogenetic tree. When using likelihood or Bayesian methods for phylogenetic inference, we can do this using an approach known as odds ratio testing.

Suppose that once we reconstruct a phylogenetic tree, we want to determine how strongly our character data support a given feature of this phylogenetic tree. For example, suppose that again we want to know how strongly the data support whether clade A is monophyletic, as shown in Figure 5.32. To answer this question, we can compare the best possible tree overall against the best possible tree in which clade A is not monophyletic. We have already found the former. This is simply the tree that we constructed in the basic process of phylogenetic inference. We can find the latter by *constraining* our search of phylogenetic trees to consider only those in which clade A is not monophyletic.

We can then see *how much better* the best tree with clade A monophyletic is, relative to the best tree without clade A monophyletic. Various statistical procedures have been developed for making this comparison and determining when the difference is statistically significant.

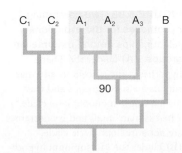

C_1 C_2 A_1 A_2 A_3 B

90

FIGURE 5.32 Numbers at a branch point indicate bootstrap support. The number 90 indicates that the highlighted clade (species A_1, A_2, and A_3) appears as a monophyletic clade in 90% of the bootstrap replicates.

Testing Hypotheses about Phylogenetic Structure

In Chapter 4, we looked briefly at two different hypotheses for the phylogenetic relationships among mammalian groups. According to the *Theria hypothesis*, placental mammals (*Eutheria*) and marsupials (*Metatheria*) are sister groups, with monotremes (*Prototheria*) more distantly related (Figure 5.33A). By contrast, the *Marsupionta hypothesis* places the marsupials and monotremes as sister groups, with the eutherian mammals more distantly related (Figure 5.33B). Prior to widespread genomic analysis, there was considerable controversy as to which of these two hypotheses was correct. Morphological evidence tended to support the Theria hypothesis, whereas molecular evidence from mitochondrial DNA (mtDNA) sequences tended to support the Marsupionta hypothesis.

In an effort to bring a new source of data to bear on the problem of distinguishing between these alternative hypotheses, Keith Killian and his colleagues obtained DNA sequences of a large nuclear gene known as *M6P/IGF2R* from 11 placental, 2 marsupial, and 2 monotreme species (Killian et al. 2001). They used this DNA sequence data to construct a phylogeny of the mammals. They reasoned that if the Theria hypothesis was correct, the placentals and marsupials would form a single monophyletic clade, whereas marsupials and monotremes would not form a monophyletic clade. If the Marsupionta hypothesis was correct, the reverse pattern would hold: Marsupials and monotremes would be a monophyletic clade, but placentals and marsupials together would not be monophyletic.

When Killian and his colleagues constructed a maximum likelihood tree, they found a pattern of relationships consistent with the Theria hypothesis. Their tree, shown in Figure 5.34, places *Eutheria* and *Metatheria* as sister groups.

But how much should we make of this result? Does the Theria hypothesis do a much better job of explaining the data from the *M6P/IGF2R* gene, or is the Marsupionta hypothesis a close second? In other words, can we quantify how strongly the data support the Theria hypothesis relative to the Marsupionta hypothesis? This is where the method of bootstrap resampling comes in. Killian and his colleagues created 100 bootstrap replicate data sets by performing the resampling procedure we have described. When they constructed phylogenetic trees for each replicate, they found that the placental mammals and marsupials

FIGURE 5.33 Two competing hypotheses for the evolutionary relationships among mammalian groups. (A) Under the Theria hypothesis, the placentals and marsupials are sister groups and thus form a single monophyletic clade, whereas marsupials and monotremes are not a monophyletic clade. **(B)** Under the Marsupionta hypothesis, the marsupials and monotremes are sister groups and form a monophyletic clade, but placentals and marsupials together are not monophyletic. Adapted from Meyer and Zardoya (2003).

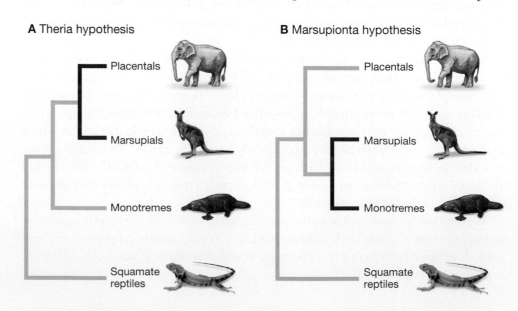

formed a monophyletic clade in every one of the 100 replicate trees (shown by the magenta 100 on the tree). This indicates that *these particular data* very strongly support the Theria hypothesis. As shown in Figure 5.34, other clades are much less well supported. For example, the bat and hedgehog formed a monophyletic clade in only half of the bootstrap replicates (shown by the magenta 50 on the tree).

Because Killian and his colleagues were using maximum likelihood to construct their phylogeny, they could also use an odds ratio test to evaluate the strength of support for the Theria hypothesis. To do so, they compared the maximum likelihood tree shown in Figure 5.34 with the maximum likelihood tree *given the constraints of the Marsupionta hypothesis*. That is, they compared their maximum likelihood tree with the highest-likelihood tree in which the marsupials and monotremes formed a monophyletic clade. A likelihood ratio test allowed them to reject (at the $p < 0.001$ level) the hypothesis that there is no difference in likelihood between the maximum likelihood tree (which happens to support the Theria hypothesis) and the best tree that is consistent with the Marsupionta hypothesis. Like the bootstrap resampling approach, the odds ratio test approach showed that Killian's data strongly supported the Theria hypothesis over the Marsupionta hypothesis.

Since the publication of Killian's paper, numerous additional mammalian phylogenies have been constructed using nuclear DNA. These have overwhelmingly tended to support the Theria hypothesis, and today the majority of researchers would agree that placental mammals and marsupials are sister groups, and that monotremes are more distantly related.

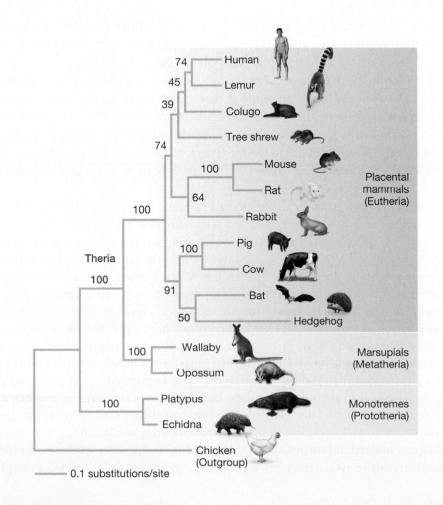

FIGURE 5.34 A maximum likelihood tree for the mammals. Killian and his colleagues inferred this maximum likelihood tree based on sequence data from the *M6P/IGF2R* gene. Numbers represent bootstrap support values for each clade. Theria—the group comprising placentals and marsupials but not monotremes—has 100% bootstrap support as a monophyletic clade. Other clades, such as that comprising bats and hedgehogs, have much lower bootstrap support. Adapted from Killian et al. (2001).

5.7 Evidence Used for Reconstructing Phylogenetic Trees

Evolutionary biologists can use many different kinds of traits to reconstruct evolutionary trees, from fossil evidence to anatomical features of modern organisms, from embryological processes to genetic sequence data, from behavioral patterns to chromosome structure. DNA sequences are the most frequently used character for phylogenetic construction today, but DNA may not always be available, as in the case of the fossil record (although recent advances in extracting DNA from some types of fossilized remains are making molecular phylogenetics possible even for extinct groups). Even when DNA sequences are available, alternative characters—be they morphological, behavioral, or otherwise—can provide additional lines of evidence with which to test the evolutionary hypotheses that our molecular trees represent. In general, we see a high degree of concordance (agreement) among phylogenies constructed using various types of traits, although often some of the smaller details can vary, depending on the choice of characters.

In this section, we will look at two additional types of evidence that evolutionary biologists can use in reconstructing phylogeny. We will begin with fossil characters, and then we will discuss what we can learn about phylogeny from biogeography—the geographic pattern of distribution of living organisms.

The Fossil Record

Especially for extinct taxa, the fossil record is a primary source of data for constructing phylogenetic trees. But the fossil record is much more than just an inert and static database from around the globe, because scientists can use the fossil record to formulate hypotheses about phylogenetic relationships. For example, Wallace, Darwin, and others recognized that extant (that is, not extinct) species from a given location tend to resemble fossils uncovered at that same spot more so than fossils found at other locations. From this and other sources of evidence, 4 years before Darwin published *On the Origin of Species*, Wallace concluded that "Every species has come into existence coincident both in space and time with a pre-existing closely allied species."

Indeed, this pattern of local resemblance among fossils has been observed so often, and at so many locations, that it is sometimes called the law of succession. Moreover, it generates a hypothesis: Common ancestry explains the similarity between extant and fossil species at location 1 and the similarity between extant and fossil species at location 2, and so on. What's more, if common ancestry explains the similarity of fossil and living forms at a given location, then by knowing enough about the geological and ecological conditions that were present at this location at various points through evolutionary time, we can generate and test hypotheses about how natural selection and other evolutionary processes may have been responsible for many of the differences between fossil and extant species. If, for example, the type of prey that was consumed in the group we are studying changed over time, that might help us to explain why the modern and fossil species were generally very similar but had differences in morphological traits associated with foraging (tooth shape, beak size, and so on).

To better understand the many ways that evolutionary biologists have employed the fossil record to reconstruct phylogenies, we will now examine two examples.

The first focuses on the use of fossil data to reconstruct the evolutionary history of horses, and the second examines how fossil evidence explains an important development in the history of animals—the transition from life in the sea to life on the land.

Phylogenetic Relationships in Equidae

The reconstruction of the phylogenetic relationships in Equidae, the family that includes the modern horse, is largely but not exclusively based on fossil evidence. Although there is some debate on the details of this phylogeny (Weinstock et al. 2005), the overall picture is clear (MacFadden 1992; Martin 2004) (Figure 5.35). The earliest horse fossils are between 50 and 60 million years old, from the Eocene. Evidence from fossilized bones and teeth indicate that these "dawn horses," or *Eohippus*, were small compared to modern-day horses. They weighed only about

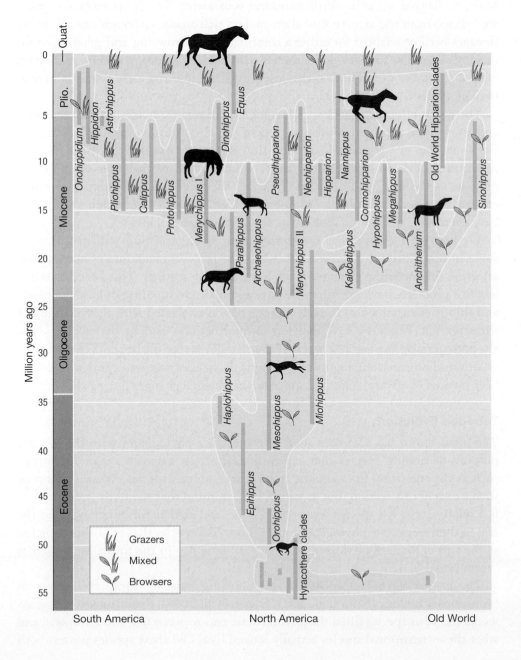

FIGURE 5.35 The evolutionary history of horses from 58 million years ago to the present. While not an explicit phylogeny, this diagram helps us understand the evolutionary origin of modern horses. Horse lineages increased in size, speed, and limb morphology, and snout shape changed as they adapted to life in emerging grasslands. Adapted from MacFadden (2005).

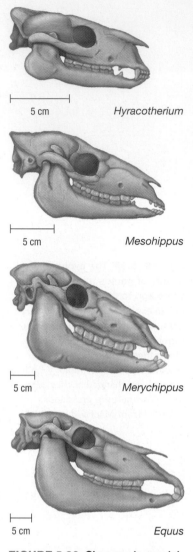

5 cm   *Hyracotherium*

5 cm   *Mesohippus*

5 cm   *Merychippus*

5 cm   *Equus*

FIGURE 5.36 Changes in cranial shape in horse lineages. *Hyracotherium* (*Eohippus*) existed about 50 million years ago, *Mesohippus* about 30 million years ago, *Merychippus* about 15–20 million years ago, and *Equus* from about 4.5 million years ago to the present. From Martin (2004).

5 kg (modern horses weigh about 500 kg), and they were primarily browsers (feeding on leaves) rather than grazers (feeding on grasslands), with teeth adapted to that mode of foraging (Figure 5.36). Most strikingly, *Eohippus* had hind limbs with three toes, and they had forelimbs with four toes, rather than the hooves of modern horses.

As we move forward in evolutionary time (toward the present) to the Oligocene, the fossil record shows a general trend in which horse lineages such as *Miohippus* and *Mesohippus* became somewhat larger in body size (approximately 10–50 kg), with a more elongated snout and larger molars than *Eohippus*. The general anatomy of these lineages also changed in a way that suggests that natural selection favored the ability to run more swiftly. During the Miocene, horses underwent a large-scale radiation, with different lineages evolving a diversity of body sizes, some larger and some smaller than those of their Oligocene ancestors. Their feeding ecology changed as well. With grassland ecosystems becoming more common, we can see from the structure of their molars that many, although not all, horse lineages became adapted for either a combination of browsing and grazing, or for grazing alone. The fossil evidence reveals that, along the lineage leading to modern horses, a number of forelimb bones fused together, and the early stages of hooves became evident.

The genus of modern horses, *Equus*, includes domestic horses, zebras, donkeys, and asses; *Equus* appeared in the fossil record about 4.5 million years ago, emerging from just one of the lineages of late Miocene horses. Around this time, natural selection appears to have favored larger animals with teeth better designed for grazing in the new environments in which they lived. These animals also had completely fused forelimbs and hind limbs, with a muscle and tendon system that gave them the "springing" motion we see in trots and gallops.

We end with a somewhat cautionary note. When working with fossils, it is sometimes tempting to use post-hoc—after the fact—explanations of how natural selection produced the changes in the lineage being studied. This becomes much less of a problem, however, when we have a good understanding of how the biotic and abiotic environments changed over the period associated with the fossils under investigation. When we have that sort of information—and we do for the case of the horse fossils—we can test whether the changes we see in the traits of the fossils we are studying are consistent with the sorts of changes that we expect would have been favored by natural selection, given environmental changes during that period.

Tetrapod Evolution

The fossil record has also been used to reconstruct phylogenies with the specific purpose of finding species that represent transitions between major life-forms, such as the transition from aquatic to terrestrial animal species. We will examine such a case in this section, but as we do, keep in mind that the term "transitional" is a relative one. All species were once extant, and at that time they were on the tips of their respective phylogenetic trees. Likewise, species that today are depicted as the tips of modern phylogenetic trees will some day in the future be viewed as "transitional."

The origin of the tetrapods has been a long-standing topic of interest in evolutionary biology (Ruta et al. 2003; Coates et al. 2008). Evolutionary biologists wondered what species filled the phylogenetic gap between fish and tetrapods, and what these transitional species actually looked like. Did these species possess both fish- and tetrapod-like features, and if so, which features, and why? In 2005, researchers took a big step toward answering these questions when paleontologist Ted Daeschler and his colleagues uncovered a set of striking fossils on Ellesmere Island, 800 miles from the North Pole in northern Canada (Daeschler et al. 2006; Shubin et al. 2006).

Daeschler was examining the evolution of tetrapods from lobe-finned fish (sarcopterygians) in the Late Devonian period (385–359 million years ago). This evolutionary transition represents not only the emergence of the group that contains our own species, but also the evolution of new forms of locomotion, respiration, and hearing. Consider Daeschler's list of remarkable changes that occurred during this transition:

> The proportions of the skull were remodeled, the series of bones connecting the shoulder and head was lost, and the region that was to become the middle ear was modified…, robust limbs with digits evolved, the shoulder girdle and pelvis were altered, the ribs expanded, and bony connections between vertebrae developed. (Daeschler et al. 2006, p. 757)

Evolutionary processes were dramatically reshaping this lineage. So, what did organisms look like when these modifications were under way? The fossil remains of three individuals from a recently discovered species called *Tiktaalik roseae* provide some answers to this question (Figure 5.37).

The fossil remains of *T. roseae*, a transitional form between lobe-finned fish and tetrapods, show the scales, gills, and fins characteristic of fish but also evidence of features that are associated with life on land. These features include ribs, a neck, and also limbs with primitive shoulders, elbows, and wrists that are the precursors of the arms and legs of land-dwelling forms.

By comparing anatomical traits such as scales, gills, fins, ribs, neck, and limbs in *T. roseae* to those species in the fossil record that came before and after, evolutionary biologists have been able to produce a more comprehensive tree depicting the transition from fish to tetrapods (Figure 5.38).

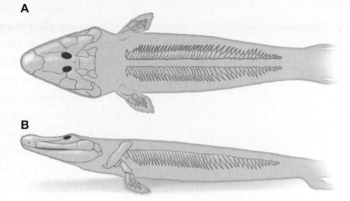

FIGURE 5.37 *Tiktaalik roseae.* (**A**) Dorsal view. (**B**) Lateral view. Drawings are based on fossil remains of three individuals, ranging in size from 4 to 9 feet. From Daeschler et al. (2006).

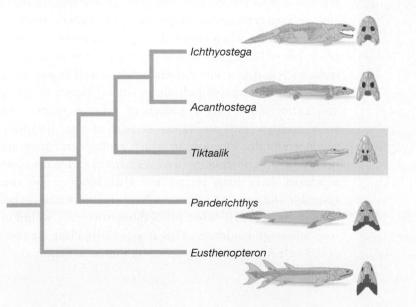

FIGURE 5.38 A bridge between fishes and tetrapods. The lineage that led to modern tetrapods includes several animals—for example, *Tiktaalik*—that are morphologically intermediate between fishes and tetrapods. Skull roofs show the loss of the gill cover (blue) and a size reduction in postparietal bones (green), as well as a reshaping of the skull. Adapted from Ahlberg and Clack (2006).

Tiktaalik roseae lived in shallow water on a floodplain in a subtropical or tropical climate (at the time these creatures lived, the land that now lies near the North Pole was located near the equator). Unlike more primitive lobe-finned fish, *T. roseae* had a flattened body that was capable of complex movements. Its ribs were modified in a way to make it capable of supporting itself on the solid substrate at the bottom of the shallow waters it inhabited, as well as on land (at least for short periods of time). The anatomy of *T. roseae* had been modified so it could move its head in a much more independent fashion than can lobe-finned fish, perhaps allowing it to feed in novel ways at the water–land interface. This species was also intermediate between lobe-finned fish and tetrapods in terms of its respiration, and anatomical analysis of the fossil evidence suggests that it was capable of breathing both in the water and in the air.

Common Descent and Phylogeography

Earlier in this chapter, we looked at how branch lengths represent the amount of evolutionary change that has occurred along sections of the phylogenetic tree. But these are not measurements of *chronological* time. To assign chronological time periods to such trees, we often need to obtain data beyond the molecular data we used in constructing the trees. That is, if we want to be able to say how far back in time two groups diverged from a common ancestor, we need to use additional data to map absolute time onto our phylogeny. There are numerous ways to do this.

The most common method for assigning absolute time to our molecular genetic phylogeny is to "anchor" our molecular genetic data to data obtained from the fossil record. For example, if we had fossil data to suggest that the common ancestor of all of the species on a tree we were studying had lived approximately 5 million years ago, we could anchor our timescale to this information and mark the start of the tree at 5 million years ago. From there, we could estimate the time span when other species in the tree existed by making some assumptions about the rate of molecular change—the rate at which "molecular clocks" tick. We will discuss the way that molecular clocks are calibrated in Chapter 8, so here we simply introduce the idea of such clocks, and note that, in conjunction with data from the fossil record, they can be used to put absolute time estimates on molecular genetic phylogenetic trees.

Phylogenies, when coupled with information about time and place, can tell us a great deal about the pattern of *microevolutionary events* such as migration and dispersal, as well as *macroevolutionary events* such as adaptive radiations—rapid bursts of speciation—and extinctions. This approach of tying together phylogeny with geology and geography is part of the discipline called phylogeography. One way in which geology and geography have been linked to phylogeny involves what is called *continental drift*. Scientists have long recognized that modern-day continents resemble the parts of a disassembled jigsaw puzzle—that is, the continents look as if, when pushed together, they would interlock into one large landmass. This is especially clear for the African and South American continents, as shown in Figure 5.39.

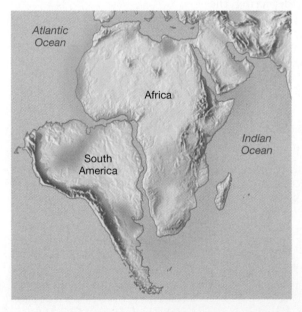

FIGURE 5.39 Once neighbors. South America and Africa look as if they are separate pieces of a larger puzzle that would interlock if moved together.

Yet, there was not a good theory to explain this interesting observation until the early twentieth century. In 1912, Alfred Wegener hypothesized that the continents are not static entities fixed in place, but rather move about Earth's surface. We now know how: The continents sit on massive crustal plates, and heat from Earth's interior causes convection currents that slowly shift these plates over vast periods of time.

According to the theory of continental drift, about 250 million years ago, all landmasses on Earth were fused together into a supercontinent called Pangaea (Figure 5.40A). Geological evidence suggests that about 175 million years ago, Pangaea split into two large continents: Laurasia in the Northern Hemisphere, and Gondwana in the Southern Hemisphere (Figure 5.40B). By 100 million years ago, Laurasia and Gondwana had broken up into the landmasses we now recognize as continents, but these continents were much closer to one another than they are currently. Slowly, the continents drifted farther apart, eventually leading to the

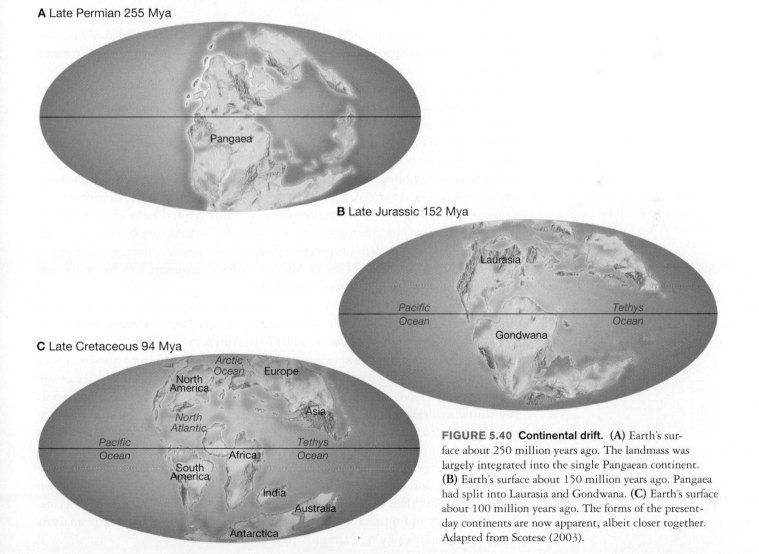

A Late Permian 255 Mya

Pangaea

B Late Jurassic 152 Mya

Laurasia

Pacific
Ocean

Tethys
Ocean

Gondwana

C Late Cretaceous 94 Mya

Arctic
Ocean

North
America

Europe

Asia

North
Atlantic

Pacific
Ocean

Africa

Tethys
Ocean

South
America

India

Australia

Antarctica

FIGURE 5.40 Continental drift. (A) Earth's surface about 250 million years ago. The landmass was largely integrated into the single Pangaean continent. **(B)** Earth's surface about 150 million years ago. Pangaea had split into Laurasia and Gondwana. **(C)** Earth's surface about 100 million years ago. The forms of the present-day continents are now apparent, albeit closer together. Adapted from Scotese (2003).

FIGURE 5.41 Modern ratites.
The distribution of extant ratite birds in the Southern Hemisphere suggests a common ancestor that originated on the ancient southern supercontinent called Gondwana. Adapted from Brown and Lomolino (1998).

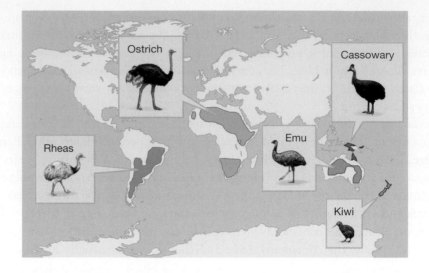

modern geography of the Earth's surface (Figure 5.40C). This continental drift has led to some testable predictions regarding the history of common descent in many taxa (Cracraft 1974a; Wiley 1988).

In the 1970s, Joel Cracraft was studying the geographic distribution of a group of flightless birds called ratites, which include ostriches, emus, kiwis, and other species (Figure 5.41). Because this group of birds is flightless, they cannot easily disperse across large bodies of water, and so their current geographic patterns may reflect very old geological roots. The geographic distribution of these birds led Cracraft to posit a phylogenetic hypothesis that all ratites shared a common ancestor that was widely distributed on Gondwana. As Gondwana broke up into the modern continents, the drifting landmasses carried the ratites along, eventually producing the geographic distribution of this group of birds (Cracraft 1973, 1974b). As they became more and more geographically isolated from one another, ratite species diverged, leading to rheas in South America, ostriches in Africa, emus and cassowaries in Australia, and kiwis and the now-extinct moas in New Zealand.

The most definitive test for the common ancestry of ratite birds comes from a series of detailed molecular genetic comparisons between species in this group (Haddrath and Baker 2001). Researchers compared the mtDNA sequences from seven ratite species: the ostrich, emu, southern cassowary, lesser rhea, great spotted kiwi, and two species of moa (because the moa species are extinct, their DNA was extracted from bone fragments). They also looked at the mitochondrial sequences of the closely related tinamous, a group of birds that are able to fly, but only weakly. When the mtDNA of those species was compared, similarities and differences at the level of DNA and at the level of proteins revealed evidence for a single common ancestor for ratites (Figure 5.42). Rheas, for example, were more closely related to flightless ratites on other continents than to the tinamous with which they lived in South America. The phylogeny in Figure 5.42 demonstrates how the molecular genetic analysis supports Cracraft's phylogeographic hypothesis.

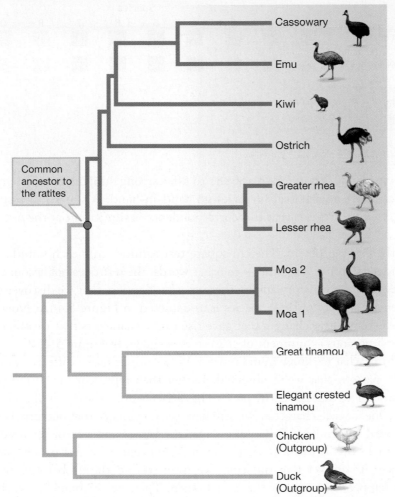

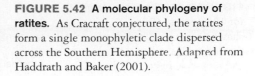

FIGURE 5.42 A molecular phylogeny of ratites. As Cracraft conjectured, the ratites form a single monophyletic clade dispersed across the Southern Hemisphere. Adapted from Haddrath and Baker (2001).

5.8 Phylogeny and the Comparative Method

One of the principal ways to understand the large-scale effects of natural selection and other evolutionary processes is by taking a *comparative* approach. By comparing traits across groups of species, we can look for trends and patterns in evolutionary events. Do ecological generalists speciate at lower rates than ecological specialists? Do species with parental care have delayed sexual maturation? Do long-lived species evolve larger brains and increased cognitive capacity? Do chromosome duplications lead to more rapid morphological differentiation? These are the types of questions that we can approach using the comparative method in evolutionary biology.

To properly apply the comparative method, it is critical to recognize that the species we study share a common evolutionary history and that historical relationships among them are represented by a phylogeny. A simple example illustrates this point (Felsenstein 2004). Suppose we are interested in understanding whether two traits, say, nocturnal activity and an arboreal (tree-based) lifestyle, tend to evolve together. We might think to simply collect information about the lifestyle of a number of species, and enumerate these in a table (see Table 5.2). Suppose we find the pattern of characters in Figure 5.43.

TABLE 5.2

An Association between Activity and Habitat[a]

	Nocturnal	Diurnal
Arboreal	4	0
Terrestrial	0	6

[a]A chi-square test reveals an association between time of activity and habitat, significant at the $p < 0.0016$ level.

FIGURE 5.43 Character states for 10 species. Characters are shown as nocturnal in dark gray, diurnal in blue, arboreal in green, and terrestrial in beige. Adapted from Felsenstein (2004).

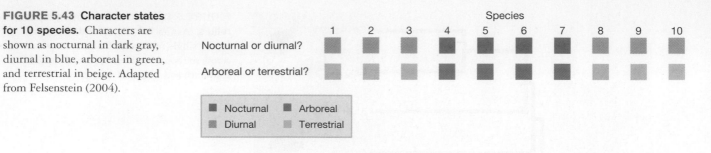

At first glance, Figure 5.43 appears to offer strong support for the hypothesis that nocturnal and arboreal lifestyles go hand-in-hand. A statistical test known as a *chi-square test* reveals that this correspondence is significant at the $p < 0.0016$ level.

But there is a problem. The chi-square test assumes that each sample evolved independently from every other—in other words, the test does not account for any shared evolutionary history among these species. Suppose that we discover that the phylogenetic history of these species is as depicted in Figure 5.44A. Now we can infer the evolutionary changes that gave rise to the characters that we observe. The most parsimonious assignment of characters is shown in Figure 5.44B.

Knowing what we know from Figure 5.44, we might take a different view of the character pattern that we've observed. Rather than representing 10 independent samples, we note that the entire pattern has arisen from a *single pair* of evolutionary changes, one for each character. We still have some evidence that nocturnal behavior and arboreal life go hand-in-hand, because the two changes both occurred on the same branch. But is this a statistically unlikely event, or could it have happened by chance? To answer that question, we need to find the probability that both changes happened on exactly the same branch. There are 18 branches on this tree, so, ignoring branch lengths, this probability is 1/18 or 5.5%, a value that is no longer significant at the 5% level (that is, with $p < 0.05$). If we fail to consider the phylogenic relationships among the species we are studying, the comparative method can give misleading estimates of the significance of the patterns that we observe.

FIGURE 5.44 Traits on a phylogeny are not independent. (A) The phylogenetic relationship among our 10 species. **(B)** The most parsimonious assignment of character changes has nocturnal activity and arboreal living each evolving a single time. Adapted from Felsenstein (2004).

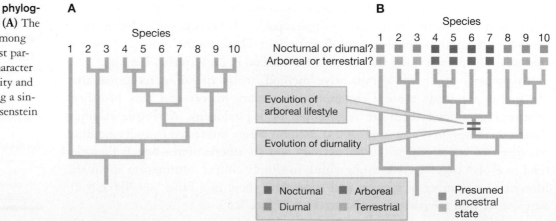

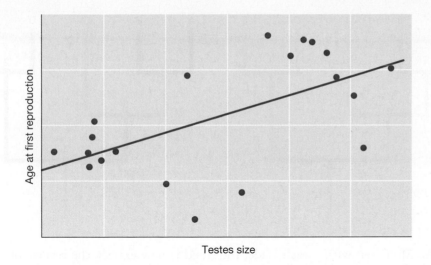

FIGURE 5.45 Testes size versus age at first reproduction. The solid line is the best fit linear regression for the 20 hypothetical species.

A similar problem arises if we try to look at comparative relationships among continuous quantitative characters without regard for the underlying phylogeny. Figure 5.45 shows a hypothetical set of measurements of testes mass and age-at-first-reproduction for 20 species. Interpreted independently from the phylogeny, it appears that there is a positive relationship between these quantities: Species with earlier age at first reproduction also have a larger testes size. One might conclude from this that these two traits are selected to change together: As one increases, the other increases as well.

But again these points are not statistically independent observations, but rather they are linked up by a shared evolutionary history. Suppose that the evolutionary history shows a single early divergence event, as in Figure 5.46A. This information radically changes our interpretation of the pattern in Figure 5.45. We now see that a single evolutionary event led to the separation between the two major

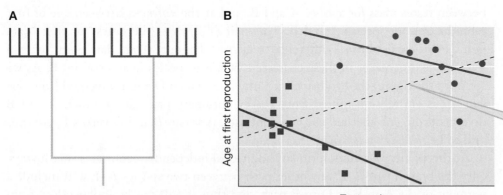

Line from Figure 5.45, which does **not** take into account shared evolutionary history

FIGURE 5.46 The phylogenetic relationship among the 20 species and evolutionary trends within each clade. (A) The partially resolved phylogeny (that is, there are polytomies) reveals that an early divergence event created two separate clades, which recently radiated to form 10 species per clade. Adapted from Felsenstein (1985). **(B)** Testes size versus age at first reproduction, with clade membership indicated by color and symbol shape. Lines indicate the best-fit linear regressions for each 10-species clade considered independently. Once each clade is considered separately, we observe a negative relationship between testes size and age at first reproduction, rather than the positive relationship (dashed line) we found in Figure 5.45, when the clades were grouped together.

FIGURE 5.47 Independent contrasts. This five-species tree features four independent contrasts: A versus B, D versus E, 1 versus C, and 2 versus 3. Here the labels 1–4 represent the inferred character states of the internal nodes.

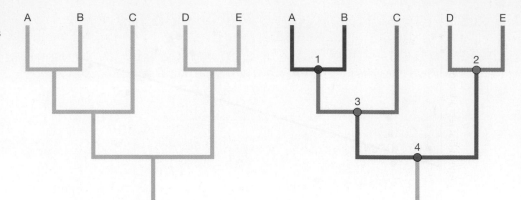

clades. Moreover, within each clade, the trend is now exactly the reverse of what we had originally thought: Testes size tends to decrease with increasing age of first reproduction. Figure 5.46B illustrates our reinterpretation of the data, coloring each species according to its clade membership and looking at the trend within each clade separately.

Thus far, we have seen the ways that we could potentially be misled by applying the comparative method without properly accounting for phylogeny. How do we cope with this problem? The method of **independent contrasts** provides a solution (Felsenstein 1985). The solution is not to look at each species as an independent data point, but rather to look at estimated changes that occur along various branches of the tree, and to pick these branches in such a way that evolution along each segment can be considered independently of every other segment.

Figure 5.47 illustrates how we can find four independent comparisons to make in a five-species tree. The key here is that we are not looking at the absolute character states, but rather at the differences in character states between each pair that we are considering in a given contrast. That is, if we are studying testes mass and age at first reproduction as our characters of interest, we look at the *difference* between testes mass for species A and B, and at the *difference* between age of first reproduction for species A and B. This pair of *differences* becomes our first "data point"; this data point is a difference or *contrast.* For our second data point, we can look at the differences in these characters between species D and species E. As we see from the figure, the evolutionary path along which D and E diverged from one another is entirely disjointed from the evolutionary path along which A and B diverged from one another; the two contrasts, A versus B and D versus E, are thus said to be *independent contrasts.*

At this point, we cannot form any additional independent contrasts that involve only the branch tips A–E; any other path between two species A–E will include a segment of the A-to-B or D-to-E path, and thus it will not be independent from the two contrasts that we have already accumulated. We are not finished, however. We can form additional independent contrasts by considering internal nodes. The comparison between internal node 1 and branch tip C follows an evolutionary path that is disjoint from those traced by the A-to-B and D-to-E paths, and it provides us with a third independent contrast. Although we do not know the character state of internal node 1 directly, we can and do infer it from the character states of nodes A and B using a model of evolutionary change. Finally, by using similar logic, we can find a fourth and final independent contrast in the internal node 2 versus the internal node 3 comparison.

Having accumulated a set of independent contrasts in this way, we can now proceed with well-established statistical analyses, such as linear regression, on the contrasts.

Independent Contrasts: A Test of the Flammability Hypothesis

Organisms are not merely the passive victims of external environmental conditions; rather, they actively impact the environment around them. The role of organisms in this process of **niche construction**—shaping their own environmental conditions—can feed back into evolutionary processes in interesting and complex ways. Fire ecology provides an excellent example. Trees, shrubs, and other plants not only suffer the effects of fire; they also provide the necessary fuel for fire, and thus it is reasonable to say that an ecosystem's flora create the conditions for their own immolation. Certain physiological characters—twig structure, needle morphology, and oil content—tend to enhance the rate and intensity of fire. Trees that retain their dead branches on the trunk make a particularly large contribution to the potential for frequent and severe fire. Dead branches are drier and burn much hotter than living branches; thus, by retaining dead branches instead of dropping them to the forest floor to decompose, branch-retaining trees greatly add to the volume of highly combustible fuel in the forest.

Dylan Schwilk and David Ackerly hypothesized that when plant species construct the fire conditions around them, this has evolutionary consequences (Schwilk and Ackerly 2001). Specifically, they conjectured that those plants that create the conditions for frequent and severe fire also induce natural selection *on themselves* for traits that allow rapid regeneration after fires have passed through (Figure 5.48).

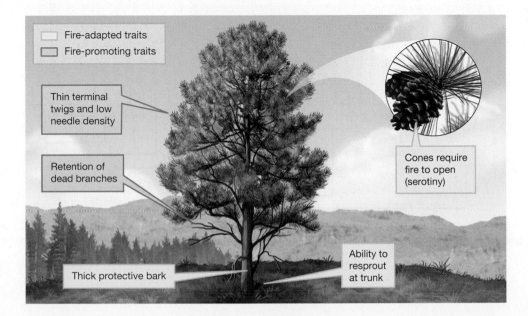

FIGURE 5.48 Fire-adapted traits and fire-promoting traits. Many pines have traits that promote fire in the environment; these species also tend to have traits that help them deal with the frequent occurrence of fire.

In order to test this hypothesis, Schwilk and Ackerly adopted a comparative approach, looking to see if pine species that create conditions for frequent and severe fire also tend to have traits that allow rapid regrowth after fire, such as the ability to resprout from surviving underground tissue, or *serotiny*, the fire-induced release of seeds from seed cones. They reasoned that if their hypothesis was correct, they would observe an association between traits that promote fire and traits that promote regeneration after fire.

For 38 pine species in the subgenus *Pinus*, the researchers collected data on a number of traits that affect the fire ecology of the landscape and on a number of traits that indicate regenerative ability after fire. Here we will focus on one particular pair: the retention of dead limbs on the tree as a fire-affecting trait, and serotiny as a regenerative trait.

Because pines are linked by evolutionary history, Schwilk and Ackerly faced a classic case of the phylogenetic nonindependence we have discussed throughout this section. To correct for this, the method of independent contrasts was necessary. They constructed a phylogenetic tree of their study species and from this phylogeny identified a set of independent contrasts between the species therein (Figure 5.49). For the characters of branch retention and serotiny, they calculated each of the contrasts for the 38 species and found a statistically significant positive correlation between serotiny and the retention of branches.

By applying the method of independent contrasts, Schwilk and Ackerly were able to demonstrate a statistically significant association between branch retention and serotiny, accounting for the shared phylogenetic histories of their study species. They found similar associations between numerous other flammability-enhancing traits and regenerative traits. These associations support their hypothesis that flammability-enhancing tree species are selected for the ability to regenerate rapidly after fire.

FIGURE 5.49 Phylogeny and the independent contrasts method. A consensus phylogeny of the 38 species of *Pinus*. This phylogeny allowed researchers to apply the method of independent contrasts to their hypothesis regarding traits that promote fire and traits that promote regeneration after fire. Adapted from Schwilk and Ackerly (2001).

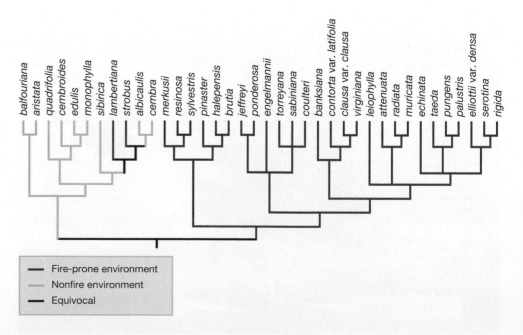

In this chapter and the previous chapter, we have learned how to read and interpret phylogenetic trees. We have seen how phylogenetic trees can be used to generate and test hypotheses, and we have explored the methods that evolutionary biologists use to infer or reconstruct phylogenies from character data. We will not be leaving phylogeny behind with the close of this chapter, however. Phylogenetic reasoning is a fundamental ingredient in almost every area of evolutionary biology, as we will see throughout the remainder of this book.

SUMMARY

1. The task of creating a phylogenetic tree is a problem in statistical inference. That is, we wish to make inferences about the historical evolutionary relationships among populations based on some data set.

2. At the most basic level, to build a phylogenetic tree, we collect information about the characters (also called traits) of some species, and we look at which species have which characters in common. We begin by assuming that species with many characters in common are more likely to be closely related to one another than are species with fewer characters in common. This logic assumes that common characters are homologies—characters that are due to shared common ancestry.

3. Evolutionary biologists have developed a number of different phylogenetic methods to test whether characters that are shared across species are analogous rather than homologous.

4. Parsimony methods search for trees that have the minimum number of evolutionary changes. The best phylogeny is assumed to be the one that both explains the observed character data and posits the fewest evolutionary changes.

5. Phylogenetic distance methods are a second approach to inferring trees. The idea behind distance methods is that, if we can measure the pairwise "distances" between species, then we can use these distances to reconstruct a tree. First, researchers have to measure these distances, and then they have to use statistical methods to find the best tree given these distance data. The goal is to find a tree with branches arrayed so that the distance along the branches between any two species is as close as possible to the distance that we measured between those two species.

6. Maximum likelihood methods and Bayesian inference methods use explicit models of how characters change through the evolutionary process. By applying techniques of statistical inference, they attempt to find the phylogenetic tree that best explains the data.

7. For any comparison involving more than a few species, there are too many possible phylogenetic trees to search exhaustively, even with the fastest computers, and so researchers have devised clever ways to search within the "space" of possible trees.

8. Evolutionary biologists have developed numerous statistical measures of support to test between different phylogenetic hypotheses. Once they have used character data to infer a tree, they can test how certain they are that a tree—or some component of a tree—is correct. Bootstrap resampling is one technique for doing this; the odds ratio test is a second technique used to address such questions.

9. A common method for assigning absolute time to our molecular genetic phylogeny is to "anchor" our molecular genetic data to data obtained from the fossil record.

10. Phylogeography links the phylogenetic history to the geographic distribution of organisms in an effort to reconstruct migrations and patterns of speciation over time and space.

11. When using the comparative method for studying how natural selection operates, we must account for any shared evolutionary history among the species we are studying. The method of independent contrasts allows evolutionary biologists to do this.

KEY TERMS

Bayesian inference (p. 140)

bootstrap resampling (p. 155)

characters (p. 139)

genetic distance (p. 149)

independent contrasts (p. 170)

long-branch attraction (p. 146)

maximum likelihood (p. 140)

niche construction (p. 171)

odds ratio testing (p. 155)

parsimony (p. 140)

phylogenetic distance
    methods (p. 149)

phylogeography (p. 147)

REVIEW QUESTIONS

1. How would the sort of analysis we discussed in the "Benghazi six" example at the start of the chapter be helpful when epidemiologists are responding to an ongoing epidemic?

2. Which of the two trees illustrated below offers a more parsimonious explanation for the observed character states?

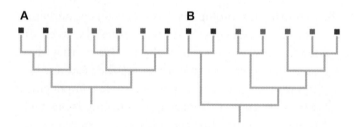

3. For the same character data in question 2, can you draw an even more parsimonious tree than either of the two shown? If so, draw it. If not, explain why it is not possible to do so.

4. Given the tree below and the character states for the three characters illustrated, assign possible locations of character changes on the tree. Be sure to indicate the presumed ancestral state.

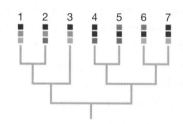

5. Are your assignments of state changes from question 4 parsimonious? How do you know?

6. Is there only one maximally parsimonious way to assign state changes to the tree in question 4? If so, why? If not, show two different ways.

7. The figure below illustrates an unrooted phylogeny (after Zhang and Ryder 1994) of several bear species: the polar bear (*Ursus maritimus*), the brown bear (*Ursus arctos*), the American black bear (*Ursus americanus*), and the spectacled bear (*Tremarctos ornatus*), with the giant panda (*Ailuropoda melanoleuca*) as an outgroup. Using the outgroup method, redraw this unrooted phylogeny as a rooted phylogeny.

Spectacled bear

Giant panda
(outgroup)

American
black bear

Brown bear

Polar bear

8. Use the Fitch algorithm to find the minimum number of character changes necessary to explain the distribution of the characters indicated on the tree below.

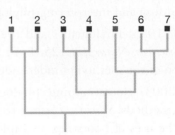

9. Show how we can obtain five different rooted trees corresponding to the single unrooted tree below.

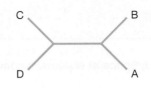

10. Indicate how six independent contrasts can be obtained from the tree below.

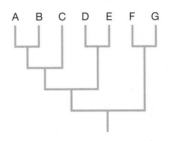

SUGGESTED READINGS

Baldauf, S. L. 2003. Phylogeny for the faint of heart: a tutorial. *Trends in Genetics* 19: 345–351. A concise review of building and interpreting phylogenetic trees.

Felsenstein, J. 1985. Phylogenies and the comparative method. *American Naturalist* 125: 1–15. The original presentation of the method of independent contrasts.

Holmes, S. 2003. Bootstrapping phylogenetic trees: theory and methods. *Statistical Science* 18: 241–255. A review of the uses of bootstrap resampling in reconstructing phylogenetic trees.

MacFadden, B. J. 2005. Fossil horses—evidence for evolution. *Science* 307: 1728–1730. An overview of how evolutionary biologists have reconstructed the evolutionary history of horses.

Shubin, N. H., E. B. Daeschler, and F. A. Jenkins. 2006. The pectoral fin of *Tiktaalik roseae* and the origin of the tetrapod limb. *Nature* 440: 764–771. An important paper on *Tiktaalik* and tetrapod evolution.

Ⓢ **Visit StudySpace at wwnorton.com/studyspace.**

PART II

Evolutionary Genetics

A collection of golden beetles, *Plusiotis optima*, from the National Institute of Biodiversity, San Jose, Costa Rica.

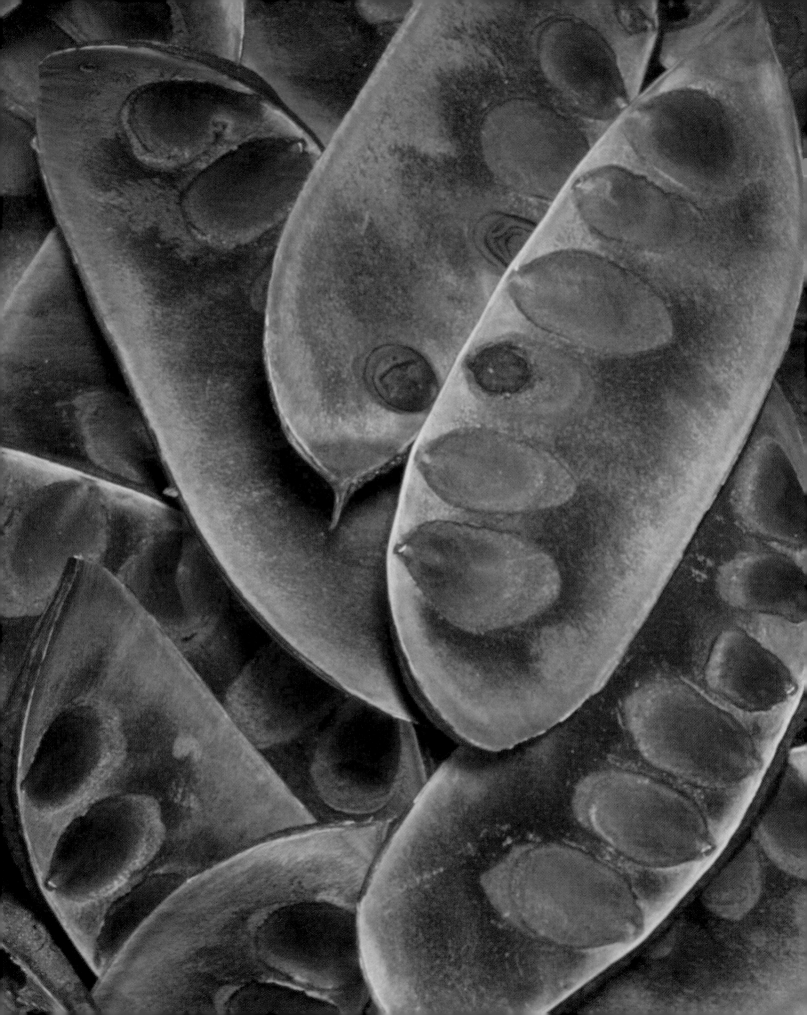

6

Transmission Genetics and the Sources of Genetic Variation

6.1 Mendel's Laws

6.2 Transmission Genetics

6.3 Variation and Mutation

6.4 Effects of Mutations on Fitness

About 10,000 years ago, people began selectively planting certain varieties of seeds to improve their crops. Those involved in these early attempts at artificial selection must have had a basic understanding that traits present in the parental stock of one generation somehow affected the traits in offspring generations. Millennia later, the Greek philosopher and physician Hippocrates suggested that offspring contained the blended "seeds" from their two parents, and that these seeds made them what they were. After Hippocrates, over the subsequent centuries, theories of heredity took some interesting twists and turns, including a hypothesis that all individuals contain within them "preformed" tiny versions of all the individuals that will ever come from their lineage. But for the next 2200 years or so, blending inheritance was the predominant mode of thought regarding what we would now call heredity.

At almost the same time that Charles Darwin was publishing his book *On the Origin of Species* in Great Britain, Gregor Mendel, an Augustinian monk and amateur plant breeder in the Austro-Hungarian Empire, was examining tens of thousands of pea plants that he had bred, and he was

◀ Seed pods from the Luangwa Valley in Zambia.

quietly undertaking some of the most important studies ever done in biology (Henig 2001). Mendel was the only child of peasant farmers, and at age 21, he entered the St. Thomas Augustinian monastery. After a short stint with pastoral duties, he became a student at the University of Vienna, where he studied mathematics and biology, hoping to teach these subjects as part of his duties as a monk. It was at the University of Vienna that Mendel became practiced in scientific research.

In his now famous experiments of the 1850s and 1860s, Mendel bred pea plants and examined the way that traits were passed down across generations. His discoveries set the foundation for the field of genetics, as we will see in Section 6.1.

In this chapter, we will review what DNA is and how it directs the synthesis of proteins. We will also include an overview of **transmission genetics**—the mechanisms by which genes are passed from parents to offspring—and a discussion of genetic variation and mutation. In the course of this brief review, we will ask the following questions:

- How does an understanding of DNA, amino acids, and proteins help us understand the evolution of life?

- What is transmission genetics, and how does our understanding of this topic affect the way that we study the process of evolution?

- How does mutation generate genetic variation, and how do mutations affect the evolutionary process?

When discussing these topics, our goal is not simply to provide a refresher on basic genetics, but rather to emphasize how knowledge of these "nuts and bolts" issues is critical for a comprehensive understanding of evolution. This chapter also sets the stage for the next four chapters, which focus on population genetics. We will include many pointers to later chapters, where the concepts we raise here are discussed in more detail.

6.1 Mendel's Laws

We begin by briefly summarizing Mendel's famous experiments on pea plants. Mendel examined seven different traits (also known as "characters"). Among the traits that he studied was flower color—specifically, he looked at whether the flowers of the pea plants were purple or white. He began 2 years of breeding experiments to determine if his pea plants always bred "true"—that is, always produced a specific type of offspring: purple-flowered offspring when a purple-flowered parent was self-fertilized, and white-flowered offspring when a white-flowered parent was self-fertilized. This assured him that his plants were what today we would call homozygotes—that is, each plant contained alleles (gene variants) for only one trait variant, in this case, a specific flower color.

Mendel's protocol was simple, but powerful. In the parental generation, he crossed a true-breeding parent plant homozygous for purple flowers with a true-breeding parent plant homozygous for white flowers. All of the offspring from these matings—known as the F_1 generation (the first generation of offspring)—produced purple flowers. Mendel then crossed F_1 plants to produce an F_2 generation (the second generation of offspring). The F_2 generation exhibited the following proportion of trait variants: three-quarters had purple flowers, while one-quarter had white flowers (Figure 6.1).

Mendel was able to derive a number of important conclusions about the genetics of diploid organisms—organisms with two copies of each chromosome—from these experiments. These conclusions have come to be known as "Mendel's laws."

The Law of Segregation

Mendel described the genetic contribution of both parents to their offspring. He deduced that, even though all F_1 plants produced purple flowers, they must have received and retained genetic information from *both* parents; otherwise, he would not have seen white flowers return in the F_2 generation. Mendel's results demonstrated that each parent plant had two copies of what he called "factors," but what we now call genes, and that the two gene copies separate with equal probability into the **gametes** (sex cells) of the pea plants. Much work has confirmed this finding, and we now speak of Mendel's first law, or the **law of segregation**, which states that each individual has two gene copies at each **locus** (the physical location of gene copies on the chromosome) and that these gene copies segregate during gamete production, so that only one gene copy goes into each gamete.

Moreover, Mendel concluded that, because all F_1 plants were purple-flowered but contained a copy of genetic information from both parents, purple color in flowers was **dominant** to white color—that is, purple flower color was revealed when both gene copies were purple or when one was purple and the other white. White flower color was **recessive**—that is, it appeared only when both gene copies were white. Hence, each gene copy retained its particulate individuality, whether or not it was expressed in external appearance.

The Law of Independent Assortment

Mendel also conducted breeding experiments in which he tracked other characters, such as seed shape (round or wrinkled). From these studies, he discovered what has since become known as Mendel's second law, or the **law of independent assortment**. This law states that which allele is passed down to the next generation at one locus (for example, the locus associated with seed shape) is independent of which allele is passed down to the next generation at another locus (for example, the locus associated with flower color). Today, we know that this holds true only for what are called unlinked loci.

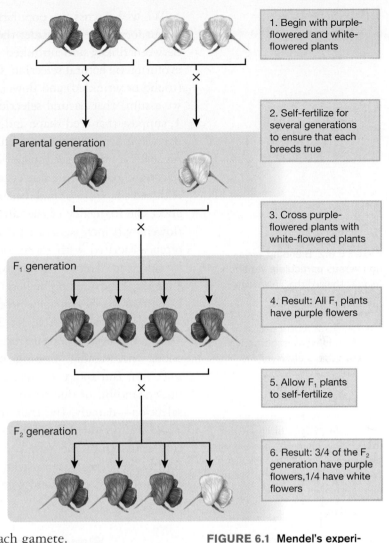

1. Begin with purple-flowered and white-flowered plants

2. Self-fertilize for several generations to ensure that each breeds true

Parental generation

3. Cross purple-flowered plants with white-flowered plants

F_1 generation

4. Result: All F_1 plants have purple flowers

5. Allow F_1 plants to self-fertilize

F_2 generation

6. Result: 3/4 of the F_2 generation have purple flowers, 1/4 have white flowers

FIGURE 6.1 Mendel's experiments. Mendel's experiments on the genetics of flower color and other traits in peas helped reveal the laws of genetic inheritance. Mendel found that when he crossed true-breeding purple-flowered plants with white-flowered plants in the parental generation, all of the F_1 offspring had purple flowers. But if he allowed the F_1 offspring to self-fertilize to produce an F_2 generation, approximately 3/4 of the F_2 plants had purple flowers, while approximately 1/4 had white flowers.

We will discuss the population genetics of linked and unlinked loci in Chapter 9, but for now let's consider the following two scenarios to see how the distinction between linked and unlinked loci has important implications for the process of evolution by natural selection. Consider two cases, both of which involve seed shape (round or wrinkled) and flower color (purple or white) in pea plants, and in which we assume that natural selection favors purple flowers over white flowers. As case 1, suppose that seed shape and flower color are unlinked. In this case, selection can operate independently on each trait. Purple flowers should increase in frequency regardless of which seed shape is favored by selection. As case 2, imagine that natural selection favors purple flowers over white flowers, but that the loci for flower color and seed shape are linked. Changes in frequency of the alleles at one locus will then affect the frequency of the alleles at the other locus. Now, to determine if purple flowers will increase in frequency, we need to know whether purple color is more often associated with round or wrinkled seeds, and which seed shape is favored by selection. This is not always a straightforward problem, as we will discuss in Chapter 9, but our point here is that whether the loci are linked or unlinked has important implications for predicting how natural selection will operate.

As we learned in Chapter 2, Mendel's work remained unnoticed until about 1900. And even when his results were rediscovered, there was an intense debate about what Mendel's findings meant for our understanding of evolution by natural selection. But today we recognize that Mendel's results provide us with a basic understanding of one of the three prerequisites for a trait to evolve by natural selection—namely, the trait must be passed down across generations. Mendel's work also provided empirical evidence disproving once and for all the early idea that traits from the two parents were permanently blended in the offspring. Rather, he clearly demonstrated that heritable "factors" are particulate—that is, they are passed down across generations even when they are not visibly expressed in offspring.

FIGURE 6.2 Blending variation versus particulate variation. Mendel showed that inheritance was particulate. The hereditary particles responsible for inherited physical characteristics behaved not like **(A)** colored dyes, but rather like **(B)** colored filters for a camera lens. Just as blue and yellow dyes can come together to make green, so can blue and yellow filters be combined to make a green one. But unlike colored dyes, filters are not irretrievably blended when they are combined. They can be separated again with ease, so that the variation in filter colors is not lost.

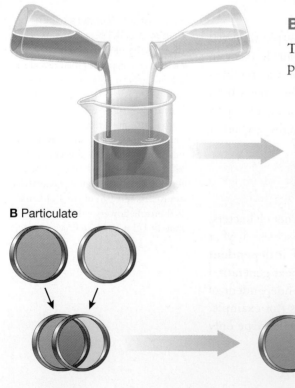

A Blending

B Particulate

Blending versus Particulate Inheritance

The demonstration that biological heredity was fundamentally particulate served to resolve one of the major challenges to Darwin's theory. As we noted in Chapter 2, one major problem for Darwin was to explain how sufficient variation could be maintained in populations to allow natural selection to continue to operate. Not only does selection itself reduce variation by favoring some forms over others, but according to Darwin's view of heredity, the very mechanism of genetic transmission would also reduce variation.

Darwin, like most of his contemporaries, envisioned heredity as a blending process analogous to that of mixing colored liquids together (Figure 6.2A). It is true that mechanisms of blending inheritance would result in the sort of resemblance between parent and offspring that is needed for heredity, and thus for evolution. After all, when mixing colors together, the resulting mixture tends to resemble the original colors: two shades of blue yield another shade of blue, not red or orange, for example. The problem is that blending of this sort also eliminates variation. If we mix yellow and blue liquids together, we get a green liquid from which we cannot reconstruct the yellow and blue precursors. Blending decreases our color variation from two colors (yellow and blue) to one color (green).

Mendel's particulate theory of inheritance suggested that colored filters (Figure 6.2B) would be a better metaphor for heredity. While the phenotypic effects of the particles carrying heritable information may blend, the particles themselves remain distinct, and they can be separated again in future reproductive events.

The theory of particulate inheritance thus resolved a major concern with Darwin's theory, which was first raised in 1867 by the engineer Fleeming Jenkin (Morris 1994). Jenkin's objection was this: Given the supposed blending nature of inheritance, how can new mutations ever have significant effects on the characteristics of a population? Under theories of blending inheritance, a favorable new mutation in a large population would, over the course of many generations, be swamped as it blended with the more prevalent character (Figure 6.3). As a result, natural selection would not be able to take a new allele to fixation, because the new allele would blend away before selection could increase its frequency enough to make a lasting difference. With Mendelian inheritance, this problem disappears. A new mutation retains its particulate nature and is not blended into obscurity. If the mutation has positive effects on fitness, its frequency can increase via natural selection.

FIGURE 6.3 Blending inheritance would swamp any rare favorable mutation. Much as a small volume of one color is diluted when added to a large volume of another color, under blending inheritance any new beneficial variant would be swamped as it blended with the most prevalent character in the population.

6.2 Transmission Genetics

For most of the last 4 billion years, DNA—deoxyribonucleic acid—has been the chemical underpinning of life on Earth. At a very basic level, it is DNA, and more critically, changes in DNA sequences, as well as the ways these sequences are expressed, that underlie the process of biological evolution. Because changes to DNA can lead to changes in which traits are expressed, they may affect fitness. We primarily will be looking at DNA as it relates to transmission genetics in this chapter. But, for now, keep in mind two things that we have already seen numerous times in this book. First, a small change to DNA that is passed down across generations can have a large effect on fitness. We saw this in Chapter 3 in our example of dark and light coat coloration in oldfield mice. The avian influenza virus offers another good example; a change to just one component of a single protein in the H5N1 virus makes this virus much more dangerous to mammalian hosts (Li et al. 2009). Second, as we saw in Chapters 4 and 5, changes in DNA sequences across populations and species are used by evolutionary biologists to reconstruct phylogenetic relationships.

DNA and Chromosomes

FIGURE 6.4 The chemical structure of DNA. DNA is a double-stranded molecule held in place with hydrogen bonds, denoted here by dotted lines. The two strands are wound together so that they are oriented in opposite directions. The nitrogenous bases (A, T, C, and G) are positioned on the interior part of each strand. This figure is magnified as you work down. Adapted from Slonczewski and Foster (2011).

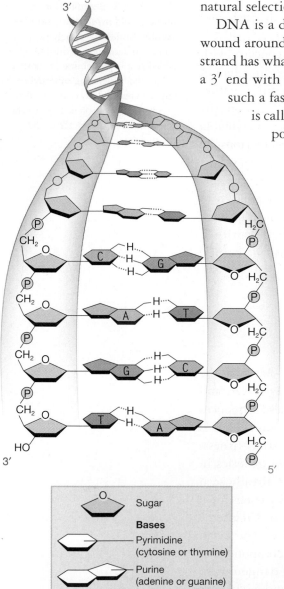

DNA is a polymer—that is, it is a macromolecule composed of repeating units linked together in a chain. The building blocks of this macromolecule are four nucleotides: adenine (A), guanine (G), cytosine (C), and thymine (T). Each nucleotide is composed of a pentose (five-carbon) sugar known as deoxyribose, a phosphate group (a phosphate atom and four oxygen atoms), and a nitrogen base. Adenine and guanine are purines, nitrogenous bases that contain a six-sided ring and a five-sided ring. Cytosine and thymine are pyrimidines, nitrogenous bases that consist of only a six-sided ring. It is a triumph of modern biology that we are capable of describing the stuff of life in such succinct terms. There remains much more that we need to learn about DNA, but we have a basic understanding of the biochemical basis of the genetic material underlying the phenotypes on which natural selection acts.

DNA is a double-stranded molecule: Two strands of connected nucleotides are wound around one another, held in place with hydrogen bonds. Chemically, each strand has what is called a 5′ (five prime) end with a terminal phosphate group and a 3′ end with a terminal hydroxyl group. The two strands are wound together in such a fashion that they are oriented in opposite directions—that is, in what is called an antiparallel fashion. The nitrogenous bases (A, T, C, and G) are positioned on the interior part of each strand. The two strands of DNA are complementary, in that adenine on one strand always pairs with thymine on the other strand, and cytosine on one strand always pairs with guanine on the other strand (Figure 6.4).

In Chapter 11, we will see that evolutionary biologists have hypothesized that the first life on Earth may have been RNA-based. RNA is similar to DNA, but single-stranded. Moreover, a nucleotide called uracil is used instead of thymine. DNA is thought to have replaced RNA as the genetic underpinning of life for a suite of reasons, including the fact that mistakes are more easily corrected in double-stranded DNA than they are in single-stranded RNA. Also, DNA is chemically more stable than RNA. RNA is very reactive with other chemicals, and so it is a good catalyst—a substance that promotes chemical reactions between other substances. But because of this reactivity, RNA is more prone to chemical reactions that potentially break down the genetic information that it carries than is DNA. The lower chemical stability of RNA may have been critical in its replacement by DNA as the chemical backbone of life.

Within cells, DNA is packed into chromosomes. Humans, for example, have 23 pairs of chromosomes, while chimpanzees have 24 pairs (see Chapter 1) (Table 6.1). Diploid organisms such as humans and chimps have two copies of each chromosome. Organisms that have a single copy of each chromosome are known as haploids. In the cells of eukaryotic organisms, most chromosomes are threadlike

TABLE 6.1

Total Chromosome Number for Various Plant and Animal Species[a]

Species	Chromosome Number	Species	Chromosome Number
Field bean	12	Ascaris roundworm	2
Garden pea	14	Mosquito	6
Onion	16	Fruit fly	8
Cabbage	18	Housefly	12
Maize	20	Frog	24
Rice	24	House bee	32
Wheat	42	Cat	38
Potato	48	Mouse	40
Cotton	52	Human	46
Sugarcane	80	Horse	64

[a]In some cases, there is variation in chromosome number among different populations in the same species.

structures composed of tightly coiled DNA that is wrapped around proteins called histones.

Eukaryotic cells contain a nucleus and organelles, which are smaller units within the cell. Bound by a lipid membrane, organelles perform specific functions, such as generating energy for the cell. Some organelles, including mitochondria and chloroplasts, have their own genomes, which are typically made up of a single chromosome with a circular structure. The fact that mitochondria and chloroplasts have their own chromosomes has a number of important evolutionary consequences, including the following: (1) We can use rates of change in nuclear genes or in mitochondria and chloroplast genes to build phylogenetic trees (Chapter 5). This allows researchers to get multiple estimates from distinct sources to use when reconstructing phylogenetic histories. (2) Sometimes the selective process that operates on genes on organellar chromosomes is different from the selective process operating on genes that reside on nuclear chromosomes. Conflicts can then occur within a genome, as we will discuss in Chapters 10 and 18.

Prokaryotic cells lack a nucleus and most lack any type of membrane-bound organelles. The prokaryotic chromosomes in these cells are not packed around histone proteins. As we will see in Chapter 10, prokaryotic cells typically have a single circular chromosome that resembles the circular chromosome of the mitochondria and chloroplasts found in eukaryotes. Some of the DNA in prokaryotic cells is located on *accessory genetic elements* such as plasmids, which have DNA that replicates independently of the cell's chromosome.

In Chapter 12, we will examine the evolutionary hypothesis that organelles such as mitochondria and chloroplasts were once independent prokaryotic life-forms that entered into a mutually beneficial relationship with other organisms and, over evolutionary time, were incorporated into these other cells, eventually becoming what we now call organelles. This basic idea is known as the endosymbiosis hypothesis, which was first examined in detail by Lynn Margulis, who hypothesized that initially free-living bacterial species capable of energy production and photosynthesis began to reside within early eukaryotic cells. These free-living bacterial species provided their hosts with energy and food and, in return, were protected from environmental dangers by residing inside another organism (Margulis 1970). Over evolutionary time, this symbiotic relationship became so strong that it developed into an obligate relationship in which the endosymbionts were no longer able to live on their own, and their hosts could not survive in the endosymbionts' absence.

From DNA to Proteins

For natural selection to operate, the genetic information encoded in DNA must produce an effect on an organism's phenotype—its observable physical, developmental, and behavioral characteristics. This is a complicated process, and we are still uncovering many of the finer details. The basic process of going from DNA to the phenotype is as follows: The double strands of DNA are "unzipped" when the hydrogen bonds that keep the strands wound around one another are broken. When the sections of DNA are unwound, portions are copied into RNA by the process of **transcription**.

Transcription occurs when a complementary and antiparallel strand of RNA is synthesized from a strand of DNA (Figure 6.5). In order to determine which portions of the DNA are to be transcribed and when, RNA polymerase binds to a **promoter**—a short DNA sequence before the transcribed part of the gene—and this serves as a signal to begin transcription.

Once RNA polymerase is bound to the promoter, the RNA polymerase unwinds the double helix, separating the two strands of DNA. One of the separated DNA strands—called the template strand—is then used to synthesize a complementary RNA strand, with DNA nucleotides binding to RNA nucleotides (T in DNA

FIGURE 6.5 The process of transcription. When RNA polymerase (not shown) binds to a promoter, double-stranded DNA is unwound, allowing the polymerase to access that segment of DNA and to synthesize a complementary RNA molecule. Adapted from Pierce (2010).

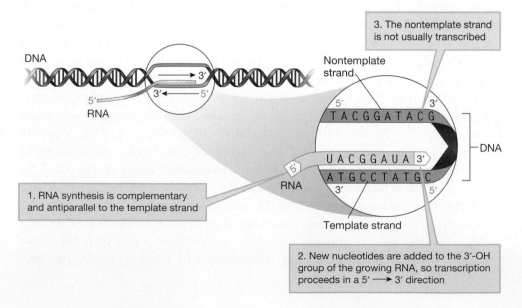

binds with A in RNA, G in DNA binds with C in RNA, C in DNA binds with G in RNA, and A in DNA binds with U in RNA). The nucleotides compose a sequence of bases that encodes genetic information. When the sequence of bases is disrupted or changed, genetic variation is created. Such variation may have effects on the synthesis of proteins, which ultimately may affect the organism's phenotype. We discuss this in more depth in a moment.

The RNA that is synthesized during transcription has numerous functions, including protein synthesis. Messenger RNA (mRNA) directs protein synthesis as part of the translation process, in which the base pair sequence of the mRNA specifies the sequence in which **amino acids** are to be linked together to form proteins (more on proteins below). Amino acids are specified by nucleotide triplets called **codons**. There are also stop codons, which terminate the process of translation and do not specify any amino acid.

Not all RNA serves as a template for protein synthesis. Some types of transcribed RNA act directly without being translated. These include ribosomal RNA (rRNA), which is a key component of the ribosomes that guide the process of protein production, making the covalent bonds that link amino acids together to form proteins; transfer RNA (tRNA), which is used to transport amino acids to ribosomes and to recognize and associate each codon triplet with the appropriate amino acid; and microRNA, which plays a number of roles in gene regulation—that is, when genes are "switched" on or off.

Proteins, which are produced by translation from mRNA instructions, are long strings of amino acids that are essential building blocks of life and serve many different functions within cells. Some proteins act as enzymes that initiate and regulate chemical reactions, while other proteins serve as chemical signals that are used in communication within and between cells. Some proteins bind to DNA and help to regulate when and how DNA is expressed; others serve structural functions, forming the cytoskeleton or elements of the extracellular matrix. Still other proteins transport materials within and between cells. All of these processes are critical for virtually every stage of development for most life-forms. If the wrong protein is produced, this may affect when a signal occurs for DNA to be expressed or turned off, or it may affect the kind of structure that is made, and hence have significant effects on fitness.

Proteins are constructed using 20 different amino acids, each of which corresponds to a nucleotide triplet. Collectively, this is known as the genetic code. As illustrated in Figure 6.6, most amino acids can be encoded by more than one nucleotide triplet; for this reason, we say that the genetic code is redundant, or *degenerate*. Given the redundancy of coding for amino acids, many nucleotide changes at the third position of a codon do not change the amino acid that is specified by the codon. In Chapter 8, we will explore the very important evolutionary consequences of redundancy in the genetic code. That discussion will center around the difference between base pair changes that affect the production of amino

FIGURE 6.6 The genetic code. The genetic code specifies the relation between codon triplets and the amino acids for which they code. To read this figure, begin at the inside of the circle and move out, reading off three codons followed by the amino acid or stop codon that they specify. For example, CCU specifies the amino acid proline, whereas UAG specifies a stop codon.

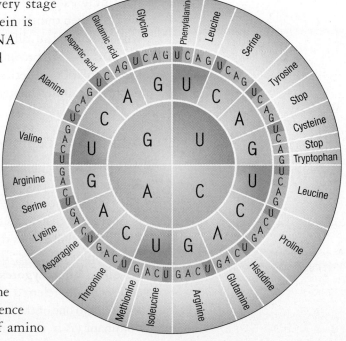

FIGURE 6.7 The processes of transcription, RNA splicing, and translation in eukaryotes. A gene is first transcribed in its entirety, including both the coding exons and the noncoding introns. The introns are subsequently excised during RNA splicing and the remaining exons are linked together to form a mature mRNA. This mRNA is in turn translated to produce a protein.

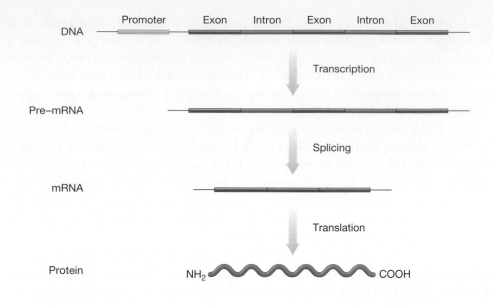

acids—such changes are likely to have important consequences for fitness—and base pair changes that do not change which amino acid is incorporated—such changes are not likely to affect fitness.

While there are many definitions of a **gene**, most reflect the notion that a gene is a sequence of DNA that specifies a functional product. This product is most often a protein, but it can also be rRNA, tRNA, or other small RNAs involved in a variety of regulatory processes. In eukaryotes, protein-coding genes are typically composed of **exons**—stretches of DNA that code for protein products—interspersed with **introns**, stretches of DNA that do not normally encode proteins (Figure 6.7). After transcription of a primary RNA, the introns are spliced—that is, they are cut out—typically by an RNA–protein complex called the spliceosome, and the remaining exons are linked together. The product of this splicing is an mRNA, which is then translated into a chain of amino acids. A single gene can be and often will be spliced in different ways: Many human genes encode multiple different proteins that are produced by this process of alternative splicing. We will explore the population genetics and evolutionary consequences of introns and exons in further detail in Chapter 10.

Alleles and Genotypes

As noted earlier in the chapter, different variants of the same gene are known as **alleles**, and the physical location of a gene on a chromosome is known as a locus. The combination of alleles that an individual has at a given locus is known as its **genotype** at that locus (sometimes the term "genotype" may instead refer to the combination of alleles that an individual has at *all* loci).

In diploid species, individuals with two copies of the same allele at a locus are called **homozygotes** (for that locus), and those with copies of different alleles at a locus are referred to as **heterozygotes**. If the heterozygote is phenotypically identical to one of the homozygotes, then the allele in that homozygote is said to be dominant (as in the dominant alleles for purple flowers in Mendel's peas), and the

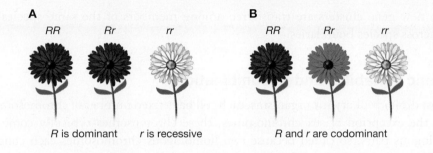

A

RR   Rr   rr

R is dominant   r is recessive

B

RR   Rr   rr

R and r are codominant

FIGURE 6.8 Dominant, recessive, and codominant alleles for floral color. **(A)** The *R* allele is dominant and the *r* allele is recessive, so the *RR* homozygote and the *Rr* heterozygote reveal the same phenotype. **(B)** The *R* and *r* alleles are codominant, so the *Rr* heterozygote manifests a phenotype that is intermediate between that of the *RR* homozygote and the *rr* homozygote.

allele in the other homozygote is said to be recessive (as in the recessive alleles for white flowers in Mendel's peas). If the heterozygote is phenotypically intermediate between the homozygotes, the alleles are said to be **codominant** (Figure 6.8).

At the turn of the twentieth century, British geneticist Reginald Punnett (1875–1967) devised the Punnett square, an elegant but simple diagram that could be used to predict the results of genetic crosses involving dominant, recessive, and codominant alleles. Figure 6.9 shows the Punnett square for a cross involving a single trait that has a recessive allele and a dominant allele. We will examine the population genetics of dominant and recessive alleles—in particular, how selection operates on alleles that are dominant or recessive—in Chapter 7.

Regulatory Elements

Stretches of DNA called **regulatory elements** control the rate at which RNA molecules are transcribed from the DNA, thereby influencing levels of gene expression and affecting the phenotype. This process is known as transcriptional regulation. Regulatory elements that increase the rate of transcription are called **enhancers**, and those that decrease the rate of transcription are known as silencers. For example, in Chapter 13, we will see how such regulatory elements affect the color of the body and wings of fruit flies and how such color patterns are critical in the context of the evolution of morphology and sexual behavior.

When regulatory elements affect genes at nearby sites on the same chromosome, they are called *cis* **regulatory elements**. By contrast, *trans* **regulatory elements** modify the expression or activity of genes on a different chromosome. *Trans* regulatory elements often do so by encoding soluble proteins that can act at remote locations on DNA.

6.3 Variation and Mutation

As we discussed in Chapter 3, natural selection requires genetic variation to operate. New genetic variation—in the form of new alleles or new allelic combinations—enters a population from one of four sources: recombination, mutation, migration, or lateral gene transfer (we discuss how sexual reproduction creates new genetic variation at the level of the *genotype* in Chapter 16). In the cases of recombination and mutation, which we will discuss here, new variation arises within a population. In the cases of migration and lateral gene transfer, new variation enters the population from outside. In Chapter 7, we will discuss migration. In Chapters 10 and 11, we will look at the process and evolutionary importance of lateral gene transfer, in

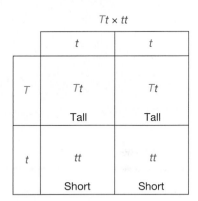

Tt × tt

	t	*t*
T	*Tt* Tall	*Tt* Tall
t	*tt* Short	*tt* Short

FIGURE 6.9 A Punnett square. In addition to flower color, Mendel also examined the genetics of pea plant traits including height. He found that the allele for Tall (*T*) was dominant and the allele for short (*t*) was recessive. A Punnett square allows us to predict the proportion of tall and short plants, given a set of parental genotypes. Here we cross a heterozygous tall individual (shown on the column) with a recessive homozygous short individual (shown on the row). As a result of the law of segregation, to predict genotype proportion in offspring, we simply fill in the four boxes with the corresponding alleles expected in possible gametes of the parents in the appropriate row and column. Our prediction in this example is a 1:1 ratio of short to tall plants. Adapted from Pierce (2010).

which new gene clusters are transferred among members of the same species or even across species boundaries.

Genetic Variability and Recombination

In most diploid eukaryotic organisms, each cell has a fixed number of chromosomes. With the exception of sex chromosomes, these chromosomes typically come in homologous pairs, so called because two homologous chromosomes each consist of the same loci (although often they will carry different alleles at some of those loci). One copy of a homologous pair of chromosomes in an individual comes from each parent as a result of meiosis, a process that leads to the production of the gametes—haploid sex cells that have one set of chromosomes. In animals, these gametes are the egg from the mother and the sperm from the father.

Meiosis begins with a single diploid cell; a single round of DNA replication followed by two rounds of division produces the four haploid gametes. Later, when fertilization occurs—that is, when two individuals mate and their gametes fuse in a process called syngamy—diploidy will be restored. The offspring produced will have a full complement of pairs of homologous chromosomes, with one chromosome in each pair coming from each parent.

Sexually reproducing organisms generate huge amounts of genetic variability among their offspring through **crossing-over**—the physical exchange of segments of DNA. Crossing-over, when areas of homologous chromosomes are exchanged, is one type of recombination that occurs during meiosis (Figure 6.10). Crossing-over occurs after the homologous chromosomes have each duplicated, when sections of one homologous chromosome may swap positions with sections on the other homologous chromosome during meiosis. Such crossing-over creates four daughter cells, each of which contains chromosomes that may differ from the chromosomes in the original cell. In Chapter 9, we will explore how recombination influences the associations among alleles at different loci. In Chapter 16, we will explore the evolution of recombination in much greater detail, including the costs and benefits of recombination and of sexual reproduction in general.

Genetic Variability and Mutation

Recombination remixes existing variation into new combinations, but where does this variation come from in the first place? Mutation, defined as a change to the DNA sequence of the organism, is the ultimate source of all genetic variation.

FIGURE 6.10 Crossing-over and recombination during meiosis. Here we have three genes (*A*, *B*, *C*), each with two alleles (*A*, *a*; *B*, *b*; *C*, *c*). Crossover occurs between one of the red *ABC chromatids* (that is, chromosome copies) and one of the blue *abc* chromatids, at a location between the *B* locus and the *C* locus. As a result, four different daughter chromatids are produced: *abc*, *abC*, *ABc*, and *ABC*. Thus, recombination generates new allele combinations not present in the original individual. Adapted from MacAndrew (2003).

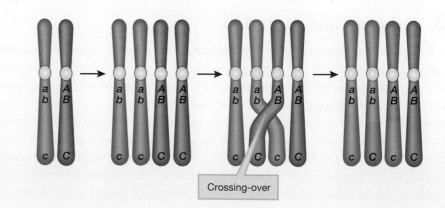

Crossing-over

In species such as humans that have a well-defined separation between germ-line cells (sex cells) and somatic (body) cells, it matters a great deal where mutations occur. When a mutation occurs in a somatic cell, it can have fitness consequences for the individual—for example, most cancers result from somatic mutations—but the mutation itself will not be transmitted to the next generation. Thus, somatic mutations do not present the type of heritable variation required for evolution by natural selection at the level of the organism. When a mutation occurs in the germ line, however, it can be transmitted to the next generation; it is these germ-line mutations that provide the underlying variation on which natural selection operates.

Mutations include many different kinds of changes to DNA. The most basic form of mutation is a base substitution (one type of point mutation). A base substitution is a change to one base—for example, from a cytosine to a thymine, or from a guanine to an adenine. When a purine (adenine or guanine) is replaced by a purine, or a pyrimidine (cytosine or thymine) is replaced by a pyrimidine, we call it a **transition**. When a purine replaces a pyrimidine, or vice versa, we call it a **transversion** (Figure 6.11).

We can also categorize base substitutions by their effects on the resulting amino acid sequence. If a base substitution does not change the amino acid that a codon normally produces, it is known as a **synonymous mutation**, also called a **silent mutation** (we will discuss synonymous mutations in much more detail in Chapter 8).

If the substitution causes the production of a different amino acid, it is known as a missense mutation: For example, a missense mutation in mice has been shown to lead to a degeneration of the neural pathways associated with locomotion (Martin et al. 2002). Sometimes, by chance, a missense mutation can prove beneficial. For example, twice a year the bar-headed goose (*Anser indicus*) migrates across the Himalaya Mountains, where the oxygen pressure is very low. In these geese, a missense mutation leading to the substitution of the amino acid proline by the amino acid leucine allows these birds to better bind oxygen during their migrations, and so this mutation has been favored over evolutionary time by natural selection. A similar scenario has been documented in the Andean goose (*Chloephaga melanoptera*), which also spends long periods of time in the low-oxygen environment of the Andes Mountains. In the case of the Andean goose, the missense mutation allowing the birds to better bind oxygen involved a change from the amino acid leucine to serine, a substitution that was subsequently favored by natural selection (Jessen et al. 1991; Weber et al. 1993; McCracken et al. 2010).

If a base substitution creates a stop codon where there was not one previously, it is known as a nonsense mutation. For example, a nonsense mutation interferes with growth rates of cattle, leading to dwarfism (Koltes et al. 2009) (Figure 6.12).

Mutations don't only involve the substitution of one nucleotide for another. An insertion mutation involves the addition of one or more nucleotides to a

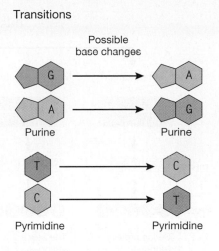

Transitions

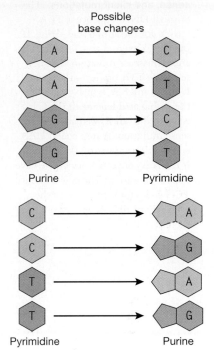

Transversions

FIGURE 6.11 Transitions and transversions. A transition occurs when a purine is replaced by a purine, or a pyrimidine is replaced by a pyrimidine. A transversion occurs when a purine replaces a pyrimidine, or a pyrimidine replaces a purine. Adapted from Pierce (2010).

FIGURE 6.12 Missense, nonsense, and silent mutations. The original DNA sequence is TCA, coding for the amino acid serine. If the C is converted to a T, this generates a missense mutation: The new sequence TTA produces the codon UUA in mRNA, which codes for the amino acid leucine. If the C is converted to an A, we have a nonsense mutation: A stop codon UAA is created, terminating the protein. If the A is converted to a G, we have a silent mutation: The new sequence TCG codes for serine, just as the original one did. Adapted from Pierce (2010).

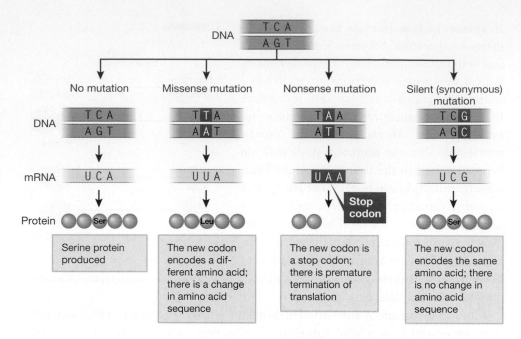

sequence, while a deletion mutation occurs when one or more nucleotides are deleted from a sequence. Because codons are made up of three nucleotides, when an insertion or deletion mutation involves a multiple of three nucleotides it does not disrupt the reading frame—the way in which adjacent base pairs are grouped into triplets and translated into amino acids. On either side of the mutation, the base pair triplets remain grouped as before. Such insertions and deletions are known as in-frame mutations. If an insertion or deletion does not occur in a multiple of three nucleotides, however, it produces a **frameshift mutation**, which affects the translation of other codons and affects the production of amino acids and proteins (Figure 6.13). For example, at least eight frameshift mutations are associated with Tay-Sachs disease in descendants of European Jewish populations (Myerowitz 1997).

Mutations can also occur at the whole gene or chromosome level. Gene duplications involve the duplication of regions of DNA that contain entire genes. For example, a gene duplication event has been linked to the ability to digest new food types in a primate species called the douc langur (Zhang et al. 2002). We will discuss the evolutionary implications of gene duplication in Chapters 9, 10, and 13. Chromosomal rearrangements are large-scale mutations at the level of the chromosome. A **chromosomal duplication** occurs when a section of a chromosome is duplicated. This results in a change in ploidy—the number of chromosomes of each type—of a cell.

FIGURE 6.13 In-frame and frameshift mutations. (A) Insertions or deletions of three nucleotides, or multiples of three nucleotides, do not shift the reading frame. **(B)** An insertion or deletion of any other length generates a frameshift mutation.

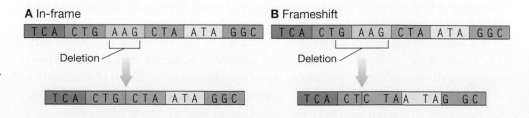

Changes in ploidy in animals are typically fatal, in that they typically disrupt the normal developmental process. But this is not always the case. For example, related species of some frogs differ primarily in the fact that some species are diploid and others are *tetraploid* with four copies of each chromosome (Holloway et al. 2006). And for reasons that we do not completely understand, changes in ploidy are *often* maintained in plant populations. For instance, many crops that humans rely on as food sources are species that have emerged from a ploidy change event in the past (Figure 6.14). These ploidy changes can have important consequences for the process of speciation, as we will discuss in more depth in Chapter 14.

A **chromosomal deletion** entails the loss of a large section of a chromosome. Another form of chromosomal rearrangement is an **inversion**, which involves a 180° flip in a section of a chromosome. A **translocation** is a mutation in which a section of one chromosome moves to another chromosome. Chromosomes can also break apart into stable new configurations (chromosomal fission) or fuse together to create new chromosomes (chromosomal fusion). Chromosomal duplications, deletions, inversions, and translocations are depicted in Figure 6.15 and will be discussed further in Chapter 10.

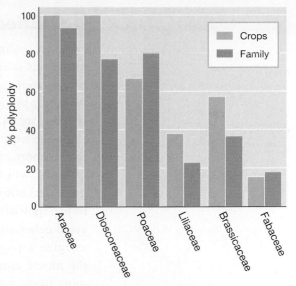

FIGURE 6.14 Polyploidy in plants. The proportion of polyploid species—species with more than two copies of each chromosome—in six families of plants, with domesticated (crop) species within these families shown separately. Adapted from Hilu (1993).

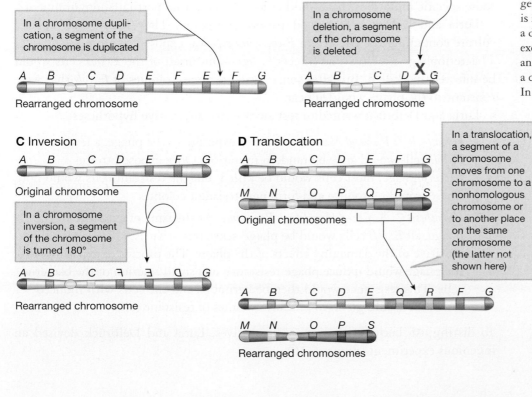

A Duplication

In a chromosome duplication, a segment of the chromosome is duplicated

B Deletion

In a chromosome deletion, a segment of the chromosome is deleted

C Inversion

In a chromosome inversion, a segment of the chromosome is turned 180°

D Translocation

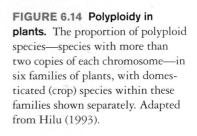

In a translocation, a segment of a chromosome moves from one chromosome to a nonhomologous chromosome or to another place on the same chromosome (the latter not shown here)

FIGURE 6.15 Chromosomal duplications, deletions, inversions, and translocations. In a duplication (**A**), a second copy of a gene region, here the *E* and *F* loci, is inserted into the chromosome. In a deletion (**B**), the *E* and *F* loci are excised from the chromosome. In an inversion (**C**), the direction of a chromosomal region is inverted. In a translocation (**D**), a section of one chromosome is moved to a different chromosome. Adapted from Pierce (2010).

6.4 Effects of Mutations on Fitness

From an evolutionary perspective, perhaps the most important way to categorize mutations is in terms of their effect on fitness. With respect to changes in relative fitness, mutations can be beneficial, deleterious, or neutral. One common sort of neutral mutation would be the synonymous mutation we discussed earlier in the chapter; we will discuss neutral mutations in greater depth in Chapter 8.

Before we discuss the frequency and distribution of different types of mutations, it is important to understand one of the most basic principles in evolutionary genetics, which is that mutations are *undirected*. In other words, mutations are generated *at random* with respect to their effects on fitness. There are no mechanisms to preferentially generate mutations that will have a positive effect on fitness or to avoid generating mutations that will have a negative effect on fitness. For example, imagine a population of dark mice introduced into a beach environment, as in the mouse example we discussed in Chapter 3. Lighter coat color would make it more likely for a mouse to survive and reproduce in its new environment. When it comes to mutations that affect coat color, however, there is no way for mice to preferentially produce mutations that result in a lighter coat color or to avoid mutations that result in a yet darker coat color. Thus, natural selection operates as a two-stage process: the *random* generation of variation, followed by the *differential* replication of certain variants.

The random nature of mutation was established through one of the most elegant experiments in the history of biology. In 1943, before geneticists knew for certain that DNA was the hereditary material, Salvador Luria and Max Delbrück wanted to understand the nature of the mutation process (Luria and Delbrück 1943). Evolutionary biologists had proposed that mutations occurred at random, independent of whether or not they would be favored by natural selection. But was this really correct? Or did the conditions in the environment somehow induce those specific mutations that would be beneficial in that particular environment?

Luria and Delbrück had good reason to wonder. They knew that when a culture containing the bacterium *Escherichia coli* was exposed to a high density of a bacteriophage—a virus that infects *E. coli*—almost all of the *E. coli* cells would be infected and killed. But after some period of time, colonies of *E. coli* that were resistant to the phage would appear.

Luria and Delbrück wanted to test among two alternative hypotheses:

1. *Hypothesis 1: Random Mutation.* Prior to exposure to the phage, a few resistant *E. coli* cells would arise by random mutation. Once exposed to the bacteriophage, most cells would be killed, but the resistant cells would not. These would reproduce and form new resistant colonies.

2. *Hypothesis 2: Acquired Hereditary Resistance.* At the time of exposure to the phage, all *E. coli* cells would be phage-sensitive—that is, all cells would be sensitive to the damaging effects of the phage. The process of exposure to the phage would induce phage resistance in a small fraction of the bacterial cells. This resistance would then be heritable, and the cells with induced resistance would go on to produce colonies of resistant cells.

To distinguish between these two alternatives, Luria and Delbrück devised an ingenious experiment (Figure 6.16).

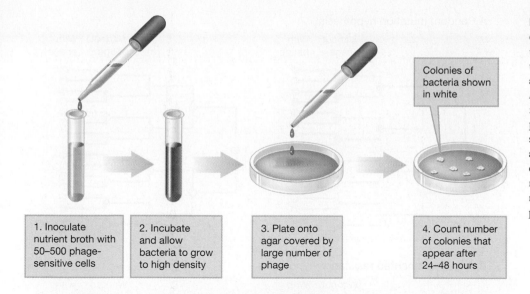

FIGURE 6.16 Luria–Delbrück experiment. To determine the distribution of phage-resistant mutants that arise from a phage-sensitive ancestor, Luria and Delbrück grew *E. coli* to high density before spreading onto an agar plate covered with phage. Only the phage-resistant strains were able to grow on the agar plate, so Luria and Delbrück could count the number of resistant mutants by simply counting the number of colonies that grew on the plate.

Colonies of bacteria shown in white

1. Inoculate nutrient broth with 50–500 phage-sensitive cells

2. Incubate and allow bacteria to grow to high density

3. Plate onto agar covered by large number of phage

4. Count number of colonies that appear after 24–48 hours

Luria and Delbrück began by inoculating multiple cultures of nutrient broth with 50 to 500 phage-sensitive bacterial cells each. Next, they incubated the cultures until the bacteria reached high density—approximately 10^8 to 5×10^9 cells/ml. They then took some samples of each culture with its high density of bacterial cells and spread the samples out on agar plates that had already been covered with a high density of phage particles. Sensitive bacteria will grow readily on agar plates, but if the phage particles are present, they are instead killed. Resistant bacteria will grow readily on agar even in the presence of phage. Finally, they incubated the agar plates for 24 to 48 hours, at which point a number of *E. coli* colonies—populated by resistant bacteria—had appeared on each plate. Each colony was composed of the descendants of a single resistant cell. The experimenters then counted the number of colonies present on each plate. From this information alone, they were able to distinguish between the two hypotheses listed above. How?

The key to understanding this experiment is to use phylogenetic reasoning. In any single culture, the large number of cells present at the time that the bacteria are transferred to the agar plate have arisen through a process of successive cell division, and they thus are related by a phylogenetic pattern, as illustrated in Figure 6.17. Once we start thinking about this phylogeny, we can see that the random mutation hypothesis and the acquired inherited resistance hypothesis make different predictions.

Under the random mutation hypothesis, resistant cells that are present once the phage is added must have had their origin in mutations that occurred earlier, during the process of bacterial growth. If one of these mutations happens to arise early in this growth process, it will give rise to a large cluster of colonies full of resistant individuals, as illustrated in Figure 6.17A, top left. If, instead, the first resistant mutation arises late in the growth process, it will generate a much smaller cluster of colonies full of resistant individuals (Figure 6.17A, top right). As a result, some cultures will have a large number of resistant cells, and others will have a small number. Thus, the random mutation hypothesis predicts that the experimenters should observe a wide variety in the number of resistant colonies on each plate.

Under the acquired inherited resistance hypothesis, resistance would never arise until the phage was added. At that point, each cell would acquire resistance,

FIGURE 6.17 Random mutation or acquired inherited resistance? The random mutation hypothesis and the acquired inherited resistance hypothesis make different predictions about the distribution of resistant mutants that will be observed on exposure to the phage. The random mutation hypothesis predicts that resistant cells arise by random mutation even before the phage is present. **(A)** In some cultures, a mutation may arise early (arrow), resulting in many resistant cells, as shown in red in the top left panel. In other cultures, a mutation may occur only late (arrow), resulting in few resistant cells, as shown in the top right panel. Thus, under the mutation hypothesis, the number of resistant cells fluctuates widely from culture to culture. **(B)** The acquired inherited resistance hypothesis predicts that resistance is only induced by the presence of the phage. Resistance arises independently with some probability in each cell once the phage is present, and clusters around the average, as shown in the bottom panels.

A Random mutation hypothesis

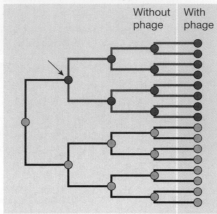 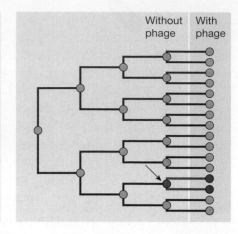

B Acquired inherited resistance hypothesis

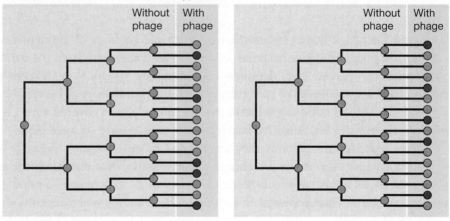

or not, independently from each other cell. Because there are a relatively large number of cells in each culture and a nontrivial fraction of these acquire resistance, then by the law of large numbers (discussed further in Chapter 8), each culture should have a similar number of resistant cells. Thus, the acquired inherited resistance hypothesis predicts that the experimenters should observe a similar number of resistant colonies on each plate (Figure 6.17B). More precisely, the acquired inherited resistance hypothesis predicts that the number of colonies on each plate should follow a *Poisson distribution*, with its variance equal to its mean. The mutation hypothesis predicts that the number of colonies on each plate should follow a different distribution—now known as the *Luria–Delbrück distribution* in honor of this experiment—with its variance much larger than its mean. Luria and Delbrück demonstrated this with a detailed mathematical model.

To distinguish between the two hypotheses, Luria and Delbrück carried out their protocol repeatedly, and they counted the number of resistant colonies that arose from each of a large number of cultures. As predicted by the random mutation hypothesis, they observed a dramatic variation from culture to culture in the number of resistant colonies. From this, they concluded that phage resistance was likely to be a product of random mutation that occurred at different times prior to the presence of the phage. At least for this trait, mutation worked as evolutionary biologists had predicted: randomly and independently of selection.

For this and other contributions, Luria and Delbrück won the 1969 Nobel Prize in Medicine or Physiology. Since their original experiment, more than half a century of subsequent developments in molecular genetics have revealed that, indeed, randomly generated mutation is the rule throughout biology. But without thinking of the phylogenetic structure of a growing population, Luria and Delbrück could never have designed their beautiful experiment and made such an important leap forward in our understanding of mutation and—consequently—the evolutionary process.

Mutation Rates

Mutation rates can be measured in many ways, and they vary considerably across species and across different tissue types in the same species, and from nucleotide to nucleotide within a single genome (Baer et al. 2007; Kondrashov and Kondrashov 2010; Lynch 2010a) (Figure 6.18 and Table 6.2). Evolutionary biologists have focused their studies of mutation primarily on deleterious mutations and on neutral mutations (Loewe and Hill 2010). The focus on deleterious and neutral mutations is due to the fact that these are more common. Because mutations are generated at random with respect to fitness, and because most traits have been under selection for long periods of time, any single arbitrary genetic change is likely to have a negative effect on fitness, or at best, no effect on fitness.

It may strike you as odd that, while beneficial mutations are what provide the fuel that drives adaptive change, most work on mutations is not focused on these sorts of mutations. And, in fact, some evolutionary biologists have begun a concerted effort to focus research on beneficial mutations (Orr 2009, 2010). In 1996, population geneticist Brian Charlesworth said that if he had a single question that he could ask a fairy godmother, it would be what the relative frequencies of deleterious, beneficial, and neutral mutations were in nature (Charlesworth 1996). Today, Charlesworth's fairy godmother is beginning to deliver, and data on mutation effects at the level of the whole genome are beginning to trickle in for a few species.

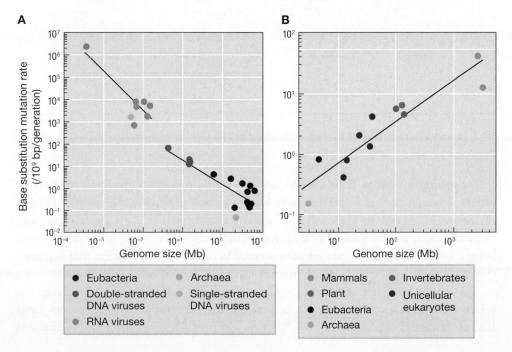

FIGURE 6.18 Base substitution mutation rate. Here we see the base substitution mutation rate per nucleotide site per generation as a function of genome size in **(A)** microbes, and **(B)** cellular organisms including cellular microbes. Among microbes, the per-site mutation rate decreases with genome size, whereas among cellular organisms, the per-site mutation rate increases with genome size. Adapted from Lynch (2010a).

TABLE 6.2

Mutation Rates per Nucleotide Site ($\times 10^{-9}$) in Different Tissues

Species	Tissue	Approximate Cell Divisions per Generation	ESTIMATED MUTATION RATES	
			per Generation	per Cell Division
Homo sapiens	Germ line	216	12	0.06
	Retina	55	54	0.99
	Intestinal epithelium	600	160	0.27
	Fibroblast (culture)			1.34
	Lymphocytes (culture)			1.47
Mus musculus	Male germ line	39	38	0.97
	Brain		77	
	Colon		83	
	Epidermis		90	
	Intestine		120	
	Liver		240	
	Lung		170	
	Spleen		130	
Rattus norvegicus	Colon		180	
	Kidney		170	
	Liver		180	
	Lung		220	
	Mammary gland		58	
	Prostate		450	
	Spleen		100	
Drosophila melanogaster	Germ line	36	4.6	0.13
	Whole body		380	
Caenorhabditis elegans	Germ line	9	5.6	0.62
Arabidopsis thaliana	Germ line	40	6.5	0.16
Saccharomyces cerevisiae		1	0.33	0.33
Escherichia coli		1	0.26	0.26

Adapted from Lynch (2010a).

For example, Sandra Trindade and her colleagues measured mutation rates in 50 different populations of *E. coli* that were known to have especially high mutation rates (Trindade et al. 2010). They followed these populations for 1150 generations and, although the researchers were not able to directly measure the frequency of deleterious, beneficial, and neutral mutations, they were able to gather information that suggests that most, although not all, of the mutations they recorded were deleterious.

Joan Peris and her colleagues took a different approach to measuring the relative frequencies of deleterious, beneficial, and neutral mutations in the virus Bacteriophage f1 (Peris et al. 2010). They experimentally produced point mutations

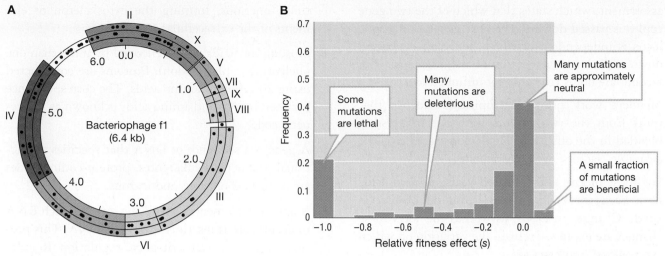

FIGURE 6.19 Deleterious, beneficial, and neutral mutations in the virus Bacteriophage f1.
(A) The circular genome of the Bacteriophage f1 virus. Each small black circle indicates where an experimental mutation occurred. Colors indicate gene functions: blue is associated with replication, green with maturation, yellow with capsid production, and red with extrusion. Rings indicate the type of mutation: synonymous mutations are shown on the outer ring, missense mutations on the middle ring, and nonsense mutations on the inner ring. **(B)** The distribution of fitness effects (s) of Bacteriophage f1 mutations. Values of s greater than zero indicate beneficial mutations. Values of s less than zero indicate deleterious mutations. Mutations with $s = -1.0$ are lethal mutations. Adapted from Peris et al. (2010).

at 100 nucleotide sites. Because they knew where mutations had occurred, they were able to measure the frequency of deleterious, beneficial, and neutral mutations in Bacteriophage f1 (Figure 6.19). They found that approximately two-thirds of the mutations caused a change in the amino acid that was incorporated. Perhaps most important, the vast majority of those mutations that had a nonneutral effect on fitness were deleterious, although a few mutations were beneficial.

Although studies such as these are beginning to inform us about the distribution of mutational effects, we are not yet in a position to make general statements about the relative frequency of deleterious versus beneficial versus neutral mutations across species (Keightley and Eyre-Walker 2010; Loewe and Hill 2010). Considerable work awaits.

Darwin was able to develop the theory of natural selection without knowing the details of genetic transmission. He simply needed to understand that traits were passed down from parents to offspring. But much of the work on evolution that has been done since Darwin's time has relied on a solid and ever-expanding understanding of the mechanisms underlying genetic inheritance. In this chapter, we have discussed some key subjects in this area, with an emphasis on their connection to key concepts in evolutionary biology. We are now ready to proceed to a series of chapters on population genetics (Chapters 7–10), moving in turn through single-locus models in large populations, to single-locus models in small populations, to multilocus models, and, ultimately, to genome evolution.

SUMMARY

1. At almost the same time that Charles Darwin was publishing *On the Origin of Species*, Augustinian monk Gregor Mendel was breeding and scoring tens of thousands of pea plants. This work gave birth to the field of genetics, including transmission genetics.

2. Mendel's laws are (a) the law of segregation, which states that each individual has two gene copies at each locus and these gene copies segregate during gamete production, so that only one gene copy goes into each gamete, and (b) the law of independent assortment, which states that which of the two gene copies is passed down to the next generation at one locus is independent of which gene copy is passed down to the next generation at the other loci. The second law holds true only for unlinked loci.

3. Mendel's work provided empirical evidence that traits from the two parents were not irreversibly blended in the offspring. He demonstrated that the heritable "factors" were particulate.

4. For most of the last 4 billion years, nucleic acids have been the chemical underpinning of life on Earth. Changes in DNA, as well as the ways these changes are expressed, create the variation on which evolutionary processes act.

5. DNA is packed into chromosomes. Diploid organisms have two copies of each chromosome. Organisms with a single copy of each chromosome are known as haploids. In cells of eukaryotic organisms, most chromosomes are threadlike structures composed of tightly coiled DNA that is wrapped around proteins called histones.

6. Proteins are long strings of amino acids that are essential building blocks of life and serve many different functions within cells. Some proteins act as enzymes that initiate and regulate chemical reactions. Other proteins are chemical signals that are used in communication within and between cells.

7. Some proteins bind to DNA and help to regulate when and how DNA is expressed; others serve struc-

tural functions, forming the cytoskeleton or elements of the extracellular matrix.

8. Most amino acids can be encoded by more than one nucleotide triplet (codon). Proteins are constructed using 20 different amino acids. The correspondence between codons and amino acids is known as the genetic code.

9. A gene is a sequence of DNA that specifies a functional product. In eukaryotes, protein-coding genes are composed of exons and introns.

10. Regulatory elements control the rate at which RNA molecules are transcribed from the DNA. This process is known as transcriptional regulation. Regulatory elements that increase the rate of transcription are called enhancers, while those that decrease the rate of transcription are known as silencers.

11. Genetic variation enters a population from one of four sources: recombination, mutation, migration, or lateral gene transfer.

12. Mutation rates can be measured in many ways and differ across species and even across different tissue types in the same species.

13. With respect to changes in relative fitness, mutations can be beneficial, deleterious, or neutral, but they are always undirected. Estimating the relative frequency of beneficial, deleterious, and neutral mutations is an active area of research within the science of evolutionary biology.

KEY TERMS

alleles (p. 188)

amino acids (p. 187)

chromosomal deletion (p. 193)

chromosomal duplication (p. 192)

cis regulatory elements (p. 189)

codominant (p. 189)

codons (p. 187)

crossing-over (p. 190)

dominant (p. 181)

enhancers (p. 189)

exons (p. 188)

frameshift mutation (p. 192)

gametes (p. 181)

gene (p. 188)

genotype (p. 188)

heterozygotes (p. 188)

homozygotes (p. 188)

introns (p. 188)

inversion (p. 193)

law of independent assortment (p. 181)

law of segregation (p. 181)

locus (p. 181)

promoter (p. 186)

recessive (p. 181)

regulatory elements (p. 189)

silent mutation (p. 191)

synonymous mutation (p. 191)

trans regulatory elements (p. 189)

transcription (p. 186)

transition (p. 191)

translocation (p. 193)

transmission genetics (p. 180)

transversion (p. 191)

REVIEW QUESTIONS

1. We emphasized that mutation is undirected. Why is this such a critical concept? How can a misunderstanding about this lead to a complete failure to grasp how the evolutionary process operates?

2. Why is it fortunate that Mendel picked traits in the pea plant that were, for the most part, unlinked? Why would it have been much more difficult for Mendel to come up with his law of independent assortment if he had chosen linked traits?

3. For many traits, across many species, the heterozygote phenotype is intermediate between the homozygote phenotypes. How is this different from the idea of blending inheritance?

4. In Table 6.1, a substantial majority of the species listed have an even number of unique chromosomes (that is, the total number of chromosomes in these diploid species is a multiple of 4). Propose an explanation for why even numbers of different chromosomes might be more common than odd numbers.

5. How might the fact that virtually all life on Earth relies on the transcription and translation of DNA help inform us in our search for life on other planets?

6. How does the presence (versus the absence) of regulatory elements exponentially increase the amount of genetic variation possible in a population?

7. How do mutations in somatic cells tie in to Lamarck's interesting, but incorrect, ideas on the inheritance of acquired characteristics?

8. Even though Andean geese and bar-headed geese are both species of geese, why should we view their increased ability to bind oxygen as a case of convergent evolution?

9. How will an increased interest in beneficial mutation, in conjunction with continued work on deleterious and neutral mutations, broaden our understanding of both adaptation and phylogenetic history?

10. Why is knowledge about the redundant nature of the genetic code fundamentally important to understanding evolutionary change?

SUGGESTED READINGS

Baer, C. F., M. M. Miyamoto, and D. R. Denver. 2007. Mutation rate variation in multicellular eukaryotes: causes and consequences. *Nature Reviews Genetics* 8: 619–631. A thorough overview of variation in mutation rates.

Henig, R. 2001. *The Monk in the Garden: The Lost and Found Genius of Gregor Mendel, the Father of Genetics.* Mariner Books, Boston. A beautiful book on Mendel's work and its place in the history of science.

Luria, S. E., and M. Delbrück. 1943. Mutations of bacteria from virus sensitivity to virus resistance. *Genetics* 28: 491–511. A classic paper using phylogenetic thinking to demonstrate that conditions in the environment did not induce specific mutations that would be beneficial in that particular environment.

McCracken, K. G., C. P. Barger, and M. D. Sorenson. 2010. Phylogenetic and structural analysis of the HbA (α^A/β^A) and HbD (α^D/β^A) hemoglobin genes in two high-altitude waterfowl from the Himalayas and the Andes: bar-headed goose (*Anser indicus*) and Andean goose (*Chloephaga melanoptera*). *Molecular Phylogenetics and Evolution* 56: 649–658. A technical article that ties together transmission genetics, natural selection, and phylogenetics to help understand adaptation to low oxygen at high altitudes.

Otto, S. P., and J. Whitton. 2000. Polyploid incidence and evolution. *Annual Review of Genetics* 34: 401–437. An overview of changes in ploidy and their effects on the evolutionary process.

7

The Genetics of Populations

◀ Greater flamingos (*Phoenicopterus roseus*) migrate in huge numbers to breed in the Makgadikgadi salt pans of Botswana.

I n the previous chapter, we provided an overview of Gregor Mendel's work on the nature of genetic inheritance, and mentioned that it is famous as one of the great "lost discoveries" in the history of science. What we did not explore in that chapter was the intense controversy that arose on the rediscovery of this work. This controversy itself makes a fascinating story.

When, after 34 years of obscurity, Mendel's work was finally rediscovered in 1900, his ideas were met with great excitement but not with broad and immediate acceptance. Instead, the renewed attention around Mendel's paper triggered a vigorous debate about the nature of heredity. Were the peculiar rules of inheritance that Mendel described simply a strange quirk of a few traits in one particular species, the garden pea? Or were they more fundamental to biology, telling us about the process of inheritance throughout the living world?

Critics attacked Mendel's conclusions on multiple grounds. First, Mendel's examples did not seem to accord with most biological observations: The traits Mendel studied were discrete characters that take on one of a fixed

A Spiral direction is a discrete trait

FIGURE 7.1 Discrete versus continuous traits. (A) The succulent plant *Aloe polyphylla* spirals either clockwise (left) or counterclockwise (right). The direction of the spiral is a discrete trait. **(B)** Human skin color is a continuous trait.

A Spiral direction is a discrete trait

Clockwise   Counterclockwise

B Human skin color is a continuous trait

BB and *Bb*

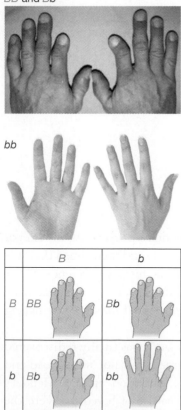

bb

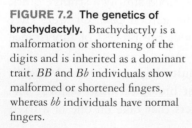

FIGURE 7.2 The genetics of brachydactyly. Brachydactyly is a malformation or shortening of the digits and is inherited as a dominant trait. *BB* and *Bb* individuals show malformed or shortened fingers, whereas *bb* individuals have normal fingers.

set of possible values, whereas most biological variation appeared to be continuous (Figure 7.1). Second, at the time it was unclear whether a Mendelian system of inheritance could be consistent with Darwin's theory of evolution by natural selection. We will defer discussion of these issues—and their ultimate resolution— until the beginning of Chapter 9.

For now, we will focus on a third critique, levied by leading biologists of the time against the best examples of so-called *Mendelian traits*—discrete traits passed on to offspring in the expected Mendelian ratios. These critics thought that trait frequencies as observed in nature were not consistent with the frequencies expected under Mendelian inheritance. A satisfactory resolution to this problem required a mathematical way of linking the rules of individual inheritance to their population consequences. It drew in one of the leading mathematicians of the twentieth century. And it led to the development of the initial foundation of the field known as *population genetics*.

A concise version of the story centers around a 1908 paper presented by Reginald Punnett to the Royal Society of Medicine. Punnett, who is also known for introducing the Punnett square (Chapter 6), was a leading advocate of Mendel's ideas. In his paper, Punnett laid out a series of examples of human traits subject to Mendel's laws of inheritance. Among these was *brachydactyly*, a genetically inherited condition leading to shortened or malformed fingers and toes. Based on an analysis of human pedigrees, Punnett noted that heredity of this trait was consistent with Mendel's model of inheritance. We now know that Punnett was correct: Brachydactyly is controlled by a single autosomal locus with the allele conferring the brachydactylous state dominant to the normal, or so-called *wild type* (Figure 7.2).

Punnett's paper was followed by spirited discussion. G. Udny Yule (1871–1951), a British statistician who also wrote important papers on Mendelian genetics, is reported to have attacked the brachydactyly case, as he believed that it was an invalid example of a Mendelian trait. Supposedly, Yule expected that any dominant Mendelian trait should occur in a 3:1 ratio, reflecting the 3:1 ratio Mendel had found with his peas.

Across the fog of a century, it is hard to reconstruct exactly who believed precisely what, but Yule appears to have reasoned along the following lines: Mendel's rules predict that heterozygote crosses yield a 3:1 ratio of dominant to recessive phenotypes among offspring. Therefore, if Mendel's rules are correct, a heterozygous trait should be observed in a 3:1 ratio *in a population* (this inference turns out to be false, as we will see later in this chapter). But brachydactyly—one of the favorite examples used to support Mendelian arguments—does not occur in a 3:1 ratio in human populations. Rather, as simple observation reveals, brachydactyly remains rare in human populations. From this observation, Yule erroneously concluded that the brachydactylous trait must not be strictly Mendelian in nature. Yule reportedly took this empirical observation as evidence against the Mendelian hypothesis.

Unable to counter Yule's critique on his own, Punnett turned for help to his friend G. H. Hardy (1877–1947), a renowned British mathematician. Hardy developed a straightforward mathematical model to predict the *population-level* consequences of Mendelian inheritance. This model allowed Hardy to mathematically test—and refute—Yule's presumption that Mendel's rules necessarily produce a 3:1 ratio of dominant to recessive phenotypes at the population level. The model undercut Yule's criticism, and it showed that Punnett's examples of rare Mendelian traits, including brachydactyly, could be valid even though nothing close to a 3:1 ratio was observed.

Hardy's model also cleared up a second misperception surrounding the population-level implications of Mendel's laws. Many biologists believed that under Mendelian inheritance, dominant alleles would replace recessive alleles over time, simply by the nature of heredity. Hardy showed otherwise. According to Hardy's model, the frequency of an allele neither increases nor decreases simply because its effects are dominant or recessive. In other words, an allele's dominant or recessive mode of *expression* has nothing to do with the mechanics of its *transmission*. Other factors, such as selection or mutation, may lead to changes in allele frequencies. But in the absence of such factors, dominant alleles do not increase in frequency simply because they are dominant, nor do recessive alleles decrease in frequency simply because they are recessive.

Over the next three chapters, we will learn how to construct some simple population genetic models. In this chapter, we will limit ourselves to considering how allele frequencies change *at a single locus* in a large population. Our goal will be to understand how genotype frequencies in the offspring population relate to genotype frequencies in the parental population. We will begin our quantitative treatment of the subject by exploring the model that Hardy developed at Punnett's request. In doing so, we will see how this model serves as a null model against which we can compare observations of genotype frequencies and the way that they change over time. We will then examine natural selection, mutation, and migration to see how each can produce changes in gene frequencies and thus affect the evolution of traits. We will look at the following questions:

- How do allele frequencies change over time in the absence of natural selection and other evolutionary processes?
- How do we build a mathematical model of natural selection?
- How do mutation, nonrandom mating, and migration affect genotype and allele frequencies in a population?

7.1 Individual-Level versus Population-Level Thinking

The field of transmission genetics, which we reviewed in Chapter 6, characterizes the way in which the genotype of an *individual* offspring is related to the genotypes of its parents. The field of **population genetics** then investigates how the genotype frequencies in an offspring *population* are related to the genotype frequencies in a parental population. This shift from the individual-level thinking that is so prevalent in the study of genetics, to the sort of population-level thinking that we tend to associate with ecology and evolution, is critical in understanding the process of evolutionary change, because individuals live but one lifetime, whereas evolution results in changes in populations across generations. We illustrate the difference between these individual-level and population-level approaches in Figure 7.3.

Individual-level thinking: What gametes and offspring are produced, in what frequencies, from a given pair of parents?

$A_1A_2 \times A_1A_2$

	A_1	A_2
A_1	A_1A_1	A_1A_2
A_2	A_2A_1	A_2A_2

Population-level thinking: How do the characteristics of the population change over time as the result of evolutionary processes?

A_1A_1 A_1A_2 A_1A_2 A_2A_2 A_1A_2 A_1A_1

A_1A_2 A_1A_1 A_2A_2 A_1A_2 A_1A_1 A_1A_1

FIGURE 7.3 Individual-level thinking versus population-level thinking. Individual and population-level approaches ask different questions.

Quantitative versus Qualitative Predictions

In the previous chapters, we examined the evolutionary process and its consequences in *qualitative* terms. For example, we learned that in order for natural selection to operate on a trait such as the coat color of oldfield mice, there must be variation in coat color, fitness differences associated with the different coat colors, and heritability of coat color. From this, we can then predict whether coat color in a given population is likely to change over evolutionary time. If lighter-colored mice are less likely to be eaten by predators, we expect to see the allele variants that contribute to lighter coloration become more common over evolutionary time. In essence, if any measurable trait has a genetic basis, we can make predictions about whether the alleles for that trait will increase or decrease in frequency.

Evolutionary biologists are not limited to making qualitative predictions about the course of evolutionary change. We can also make *quantitative*, or numerical, predictions about evolutionary dynamics. Evolutionary change occurs because certain alleles or genotypes become more common and others become less common. At its most basic level, *biological evolution occurs when genotype frequencies change over time.* The field of population genetics provides a formal structure with which to look at this process. Using population genetics, we can develop a mathematical description of how these frequencies change over time—and thus a mathematical description of the evolutionary process itself. This greatly facilitates the testing of evolutionary hypotheses.

It is not only change that we are interested in. We also want to understand stasis; we want to understand when genotype frequencies or allele frequencies will stay the same. Are there "steady-state" frequencies for which no further change will occur? Such frequencies are known as the *equilibria* of our models. In general, we say that a physical or mathematical system is at equilibrium if the system has reached a state where it does not change in the absence of outside forces or processes acting on it. In population genetics, we typically track the genotype frequencies in a *population*. An equilibrium is then a state of the population such that genotype frequencies do not change from generation to generation. Box 7.1 illustrates several types of equilibria.

BOX 7.1 Types of Equilibria

Typically, when we think about an equilibrium, we think about a *stable equilibrium*, for which two conditions hold:

1. When at this point, the system does not change.

2. If perturbed or displaced by some small amount, the system will return to its original position at rest.

The first condition ensures that we have an equilibrium; the second ensures that our equilibrium is stable.

Perhaps the simplest way to envision a stable equilibrium is by thinking about a marble in a rounded cup (Figure 7.4). The bottom of the cup is a stable equilibrium for the marble, because a marble at rest at this point does not move further, and if perturbed with a small push, the marble will return to the equilibrium point at the bottom of the cup.

**FIGURE 7.4
Stable equilibrium.**
A marble at the bottom of a rounded cup represents a stable equilibrium.

But stable equilibria are not the only kind of equilibria. There are also *unstable equilibria*. At an unstable equilibrium, two conditions hold:

1. When at this point, the system does not change.

2. If perturbed or displaced by some small amount, the system will move away even further from its initial position at rest.

Corresponding to our marble in a cup, we can think of an unstable equilibrium as a marble perfectly balanced on the top of a hill (Figure 7.5). In the absence of external forces, it is not going anywhere. But give

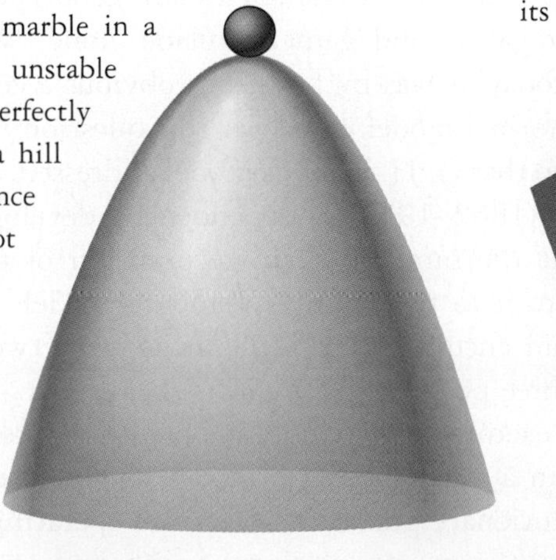

**FIGURE 7.5
Unstable equilibrium.**
A marble balanced on top of a hill represents an unstable equilibrium.

FIGURE 7.6 Neutral equilibrium. A marble at rest on a tabletop represents a neutral equilibrium.

it the slightest push in any direction, and it will tumble off the hill rather than return to its starting position.

In addition to stable equilibria and unstable equilibria, there are also *neutral equilibria*. A neutral equilibrium is a state of the system such that these conditions hold:

1. When at this point, the system does not change.
2. If perturbed or displaced by some small amount, the system will stay in its displaced position, rather than returning to the original position as it would in a stable equilibrium, or moving further away as it would in an unstable equilibrium.

Here we can think about a marble on a flat tabletop (Figure 7.6). If we move it slightly to the left or right, front or back, it neither returns to its original position nor falls off the table. It will simply sit at rest in its new position.

An equilibrium can also be stable with respect to perturbations in one direction, but neutral with respect to perturbations in another. We will call this a *mixed equilibrium*. One example of such an equilibrium is the position of a marble in a half-pipe (Figure 7.7).

When displaced leftward or rightward, up the sides of the half-pipe, the ball will return to its position in the center, as with a stable equilibrium. But when displaced forward or backward along the bottom of the half-pipe, the ball will remain in its newly displaced position.

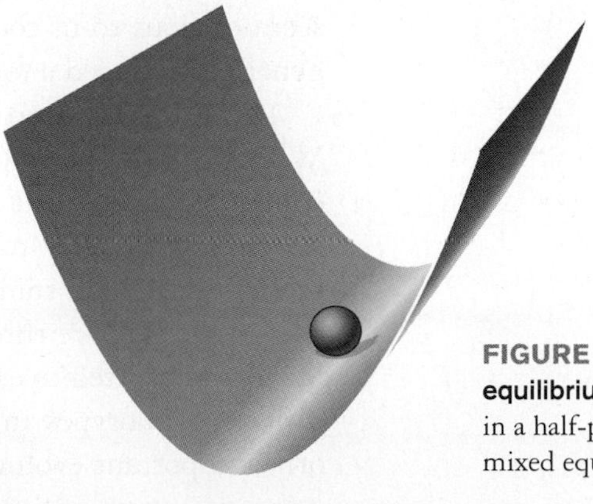

FIGURE 7.7 Mixed equilibrium. A marble in a half-pipe represents a mixed equilibrium.

7.2 The Hardy–Weinberg Model: A Null Model for Population Genetics

Population-level thinking sets the stage for the construction of a mathematical model of evolutionary change. But, in order to understand the effects of any natural process, we need a baseline model for comparison.

The Role of Null Models in Science

The role of a null model in science is to provide such a baseline. In physics, Newton's first law provides a baseline to help us understand the effects of forces acting on objects. The first law states that if no force is acting, an object in motion continues that motion and an object at rest stays at rest. With this baseline in place, we can see that objects in motion speed up or slow down only when they are acted on by forces. If we want to understand the effects of biological processes such as natural selection or mutation on the frequencies of genotypes in a population, we also need a null model. The **Hardy–Weinberg model** provides such a null model. It tells us what happens to genotype frequencies when natural selection and other important drivers of evolutionary change are not operating. Then when we observe change in genotype frequencies relative to Hardy–Weinberg predictions, we will be able to make inferences about the sorts of evolutionary processes necessary to explain our observations.

It only became possible to construct such a null model once biologists had a rudimentary understanding of the mechanistic basis of heredity. With this basic understanding in place, evolutionary biologists could scale up their thinking about how genes are transmitted, using the rules of heredity at the individual level in order to model the rules of heredity at the level of populations. In other words, they could now model how the frequency of traits might change in populations.

The Hardy–Weinberg Model

Taking the most basic case, suppose that a single trait at a single genetic locus is encoded by a single pair of alternative alleles. What will happen over time to the frequencies of these alleles, as well as to the genotypes in which they are found, in the absence of any significant evolutionary processes? In other words, what will happen to the frequencies of these alleles and genotypes due to the dynamics of chromosomal segregation and gametic fusion alone? While the answer may seem obvious to us today, it was by no means obvious a century ago. Population geneticists needed a formal model to answer this question definitively.

This is the question that G. H. Hardy's model addressed. The German physician Wilhelm Weinberg (1862–1937) independently developed and published a comparable model at the same time; in recognition of this parallel discovery, we commonly refer to it as the Hardy–Weinberg model. The Hardy–Weinberg model examines a trait encoded by a single locus, with two alleles A_1 and A_2. In this case, there are three possible genotypes—A_1A_1, A_1A_2, and A_2A_2. Hardy and Weinberg wanted to examine what would happen to the frequencies of these three different genotypes in a simple genetic model in which natural selection—and other important evolutionary processes—were *not* operating. Their solution, now called the **Hardy–Weinberg equilibrium**, serves as a null model for studies of allele frequencies and genotype frequencies in populations. The model provides three important conclusions:

1. The frequencies of the A_1 and A_2 alleles do not change over time in the absence of evolutionary processes acting on them.

2. Given allele frequencies (the frequencies of A_1 and A_2) and random mating, we can predict the equilibrium genotype frequencies (the frequencies of A_1A_1, A_1A_2, and A_2A_2) in a population in which evolutionary processes are not acting. Today, these are referred to as *Hardy–Weinberg equilibrium frequencies*.

3. If no evolutionary processes are operating, a locus that is initially not at Hardy–Weinberg equilibrium will reach Hardy–Weinberg equilibrium in a single generation.

The first conclusion tells us how allele frequencies change in the absence of evolutionary processes. The second conclusion tells us how genotype frequencies relate to allele frequencies in the absence of evolutionary processes. The third conclusion tells us how long it takes to reach these genotype frequencies.

The Hardy–Weinberg Assumptions

Every mathematical model begins with a list of assumptions, and the Hardy–Weinberg model is no exception. When modelers list their assumptions, they are, in essence, laying out for the reader what will and will not be included in a model. This process of enumerating the assumptions is one of the most important aspects of any model, because it allows the reader to understand the scope, as well as the limitations, of the mathematics to follow.

The Hardy–Weinberg model begins by making a number of basic assumptions about the individuals and population under study, as well as the evolutionary processes in operation. In addition to assuming that we are studying a sexually reproducing diploid organism that reproduces in discrete generations (all parents reproduce synchronously and then die), the Hardy–Weinberg assumptions state that none of five important evolutionary processes are operating:

1. Natural selection is *not* operating on the trait or traits affected by the locus in question.

2. Individuals have no preference for others with similar (or dissimilar) genotypes. Thus, mating in the population is random with respect to the locus in question.

3. No mutation is occurring.

4. There is no migration into or out of the population.

5. The population is effectively infinite in size, so that chance fluctuations in allele frequencies (known as "genetic drift," which we will discuss in Chapter 8) are negligible.

We begin by developing the model using these assumptions. Later in this and the subsequent chapters, we will explore what happens when each of the assumptions listed above is removed. By comparing what happens when we remove assumptions with what happens in the basic Hardy–Weinberg model when all of the assumptions are operating, we can get a sense of how processes such as natural selection, nonrandom mating, mutation, migration, and genetic drift influence genotype frequencies.

Deriving the Hardy–Weinberg Model

Every organism in our population must have one of the three possible genotypes A_1A_1, A_1A_2, or A_2A_2. Let us call the frequencies of these three genotypes $f[A_1A_1]$, $f[A_1A_2]$, and $f[A_2A_2]$. Because each individual has one of these three genotypes, the sum of genotype frequencies is one: $f[A_1A_1] + f[A_1A_2] + f[A_2A_2] = 1$. (Box 7.2 summarizes the rules of probability used in this chapter.)

From these genotype frequencies, we can compute the allele frequencies directly. Allele A_1 is found only in individuals with the A_1A_1 or A_1A_2 genotypes. Because each A_1A_1 individual possesses two A_1 alleles, and each A_1A_2 individual possesses a single A_1 allele, we can devise a simple mathematical relationship between genotype frequencies and allele frequencies. This allows us to calculate p, the frequency of the A_1 allele from the genotype frequencies:

$$p = f[A_1A_1] + \frac{f[A_1A_2]}{2}$$

We are counting the A_1A_2 genotypes only half as much as the A_1A_1 genotypes because, in the former, only half of the alleles at the A locus are A_1 alleles, whereas in the latter, both of the alleles at the A locus are A_1 alleles. Similarly, because half of the A alleles in an A_1A_2 heterozygote are A_2 alleles, whereas all of the A alleles in an A_2A_2 individual are A_2 alleles, the frequency of the A_2 allele, which we denote as q, is given by

$$q = f[A_2A_2] + \frac{f[A_1A_2]}{2}$$

Finally, because we have only two alleles in our system, it must be true that $p + q = 1$ because every A allele is either an A_1 or an A_2.

We want to see how genotype frequencies change over time, so we need to calculate the new genotype frequencies after individuals in our population mate with one another and produce offspring. One way to do this is to go through all possible mating pairs that can occur in our population, compute how commonly such mating pairs occur, and determine what type of offspring are produced from such matings. But doing the calculations in that way would involve a large amount of tedious algebra even in this simple one-locus, two-allele case. Fortunately, if the Hardy–Weinberg assumptions are met, we can bypass all of that algebra. We can take advantage of the very convenient fact that in this model, gametes assort at random—that is, they pair up at random to produce offspring—just as if they were all mixed together in one great gamete pool and then drawn out randomly in pairs (Figure 7.8). The composition of this hypothetical gamete pool is simply proportional to the frequency of the alleles in the parental generation.

FIGURE 7.8 A gamete pool approach. When individuals mate at random with respect to the genotype we are studying, we can take a gamete pool approach. Using this approach, the frequencies of the offspring produced are equal to those expected if the parental generation were to simply combine their gametes into one large gamete pool, from which pairs of gametes are drawn at random to form new offspring.

Parents

A_1A_2
A_1A_1
A_2A_2
A_1A_2
A_1A_1
A_1A_2

We imagine that parents combine gametes into one large gamete pool

Gamete pool

A_1 A_2
A_2 A_1
A_2 A_1 A_2
A_2 A_1
A_1 A_1 A_1

Then pairs of gametes are drawn at random to form new offspring

Offspring

A_1A_1
A_1A_2
A_2A_2
A_1A_1
A_1A_1
A_1A_2
A_1A_2

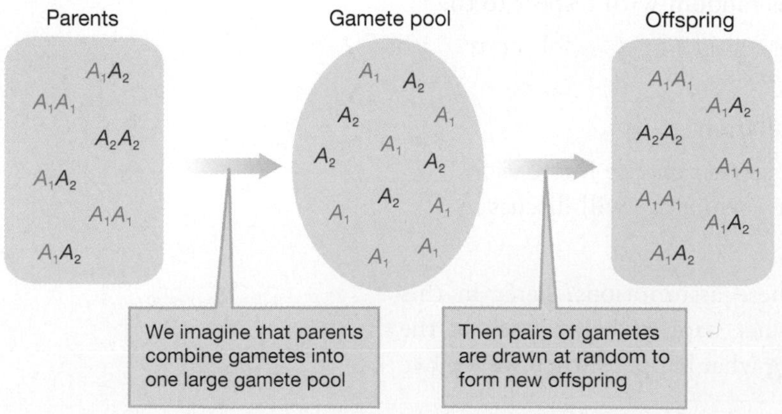

BOX 7.2 Basic Probability Calculations

In probability, we study the chance that certain outcomes—which we will call *events*—are observed. Suppose P_1 is the probability that a given outcome—call it event E_1—occurs, and suppose that P_2 is the probability that another event E_2 occurs.

Probability of a Sure Event and Probability of an Impossible Event

1. If the event E_1 is certain to occur, we say that its probability is 1.

2. If the event E_2 is certain not to occur, we say that its probability is 0.

Probability That an Event *Does Not* Occur

If E_1 occurs with probability P_1, the probability that E_1 does not occur is given by $1 - P_1$.

Events Can Be Assembled from Other Events

We can create new events using other events as building blocks. For example, we could define E_3 as the event that both E_1 and E_2 occur; we could define E_4 as the event that neither E_1 nor E_2 occurs.

Probability of Event 1 *and* Event 2

If event E_1 and event E_2 are *independent* events—that is, the chance of E_2 happening does not depend on whether E_1 happened, and vice versa—then the probability that both E_1 and E_2 occur is given by the product of their probabilities:

$$\Pr(E_1 \text{ and } E_2) = P_1 \times P_2$$

For example, let E_1 be the event that you roll a 1 on a fair die, and E_2 be the event that you get heads on the flip of a fair coin. The probabilities of these events are $P_1 = 1/6$ and $P_2 = 1/2$, respectively. These are independent events; the result of the coin flip does not depend on the result of the die roll and vice versa. Therefore, the probability that you both roll a 1 on the die and get heads on the coin flip is $\Pr(E_1 \text{ and } E_2) = P_1 \times P_2 = 1/12$.

Probability of Event 1 *or* Event 2

If the events E_1 and E_2 are *mutually exclusive* events—that is, it is impossible for E_1 and E_2 to both occur—then the probability that either E_1 or E_2 occurs is given by the sum of their probabilities: $\Pr(E_1 \text{ or } E_2) = P_1 + P_2$. The probability that they both occur is, of course, 0.

For example, let E_1 be the event that you get a 1 when you roll a die, and E_2 be the event that you get an even number on the same roll. These are mutually exclusive events. If they have probabilities $P_1 = 1/6$ and $P_2 = 1/2$, respectively, the probability that E_1 and E_2 both occur is $\Pr(E_1 \text{ and } E_2) = 0$, and the probability that P_1 or P_2 occurs is $\Pr(E_1 \text{ or } E_2) = P_1 + P_2 = 2/3$.

More generally, for *any* two events E_1 and E_2, independent or not, mutually exclusive or not, the probability that one or the other occurs is given by

$$\Pr(E_1 \text{ or } E_2) = P_1 + P_2 - \Pr(E_1 \text{ and } E_2)$$

We can rewrite this as a general expression for the probability that E_1 and E_2 both occur as:

$$\Pr(E_1 \text{ and } E_2) = P_1 + P_2 - \Pr(E_1 \text{ or } E_2)$$

Frequencies and Probabilities

In population genetics, we often speak of the frequencies or expected frequencies of different genotypes or alleles—that is, of the fraction of the population that we expect to be composed of each genotype or allele. If we assume that each offspring is produced independently by the same random process that leads to the production of every other offspring, the *frequencies* in a very large population will be equal to the *probabilities* of producing each type of offspring in a single reproduction event. In the Hardy–Weinberg model, and many (but not all) other population genetic models, we indeed make this assumption. Therefore, we can and will use the laws of probability laid out above in order to compute the frequencies of genotypes and alleles.

The offspring, produced by random draws from this gamete pool, occur with frequencies that we can calculate using the rules of probability detailed in Box 7.2:

Genotype	Hardy–Weinberg Equilibrium Frequency
A_1A_1	p^2
A_1A_2	$2pq$
A_2A_2	q^2

The frequency of the A_1A_1 genotype among the offspring is just the frequency of the A_1 allele, squared. The Hardy–Weinberg model, then, predicts that, in the absence of evolutionary processes, the expected frequency of the A_1A_1 genotype is equal to the fraction of the time that we would expect a random draw from

BOX 7.3 Hardy–Weinberg Equilibrium Is a Mixed Equilibrium

The key to a deep understanding of the Hardy–Weinberg equilibrium is to recognize that it is a mixed equilibrium. How is this so?

Recall that for a single locus A with alleles A_1 and A_2 at frequencies p and q, the Hardy–Weinberg model predicts that:

1. A population not at Hardy–Weinberg equilibrium will return to Hardy–Weinberg genotype frequencies after a single generation of random mating.

2. In the absence of evolutionary processes acting on the population, allele frequencies remain constant.

This first condition indicates that Hardy–Weinberg genotype frequencies $f[A_1A_1] = p^2$, $f[A_1A_2] = 2pq$, and $f[A_2A_2] = q^2$ represent a stable equilibrium, given the allele frequencies p and q.

The second condition indicates that the allele frequencies p and q are themselves a neutral equilibrium. In the absence of external processes (for example, natural selection, drift, migration, mutation), they don't change. But once displaced from their initial values to new values p' and q', the allele frequencies do not return to the initial values, but rather they remain at the new values until further influenced by external processes.

We can represent this graphically by plotting the frequency p of the A_1 allele on the x-axis and the frequency $f[A_1A_2]$ of the heterozygote on the y-axis. (These two quantities are sufficient to determine all three genotype frequencies and thus the entire state of the system.) The curve in Figure 7.9 indicates the Hardy–Weinberg heterozygote genotype frequency as a function of the frequency of the A allele.

Returning to the metaphor of marbles on surfaces that we developed in Box 7.1, the Hardy–Weinberg mixed equilibrium is like a marble on a curved half-pipe as shown in Figure 7.10A. The marble can be shifted left to right along the bottom of the half-pipe, and it simply stays in its new position; allele frequency p is a neutral equilibrium. But if the marble is pushed forward or backward up the side of the half-pipe, it will return once again to the corresponding rest position at the bottom of the pipe; genotype frequency $f[A_1A_2]$ is a stable equilibrium.

Returning to the story at the opening of this chapter, at last we can see where Yule and his colleagues went wrong in their

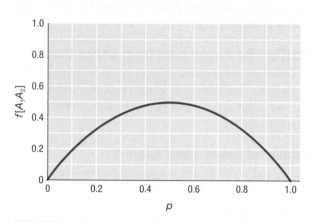

FIGURE 7.9 Heterozygote frequency at Hardy–Weinberg equilibrium. The Hardy–Weinberg equilibrium frequency $f[A_1A_2]$ of the heterozygote is a function of the allele frequency p of the A_1 allele: $f[A_1A_2] = 2pq = 2p(1 - p)$.

a gamete pool with A_1 at frequency p to yield two A_1 alleles: p^2. Similarly, the frequency of the A_1A_2 genotype is equal to the fraction of the time that a random draw would select one A_1 allele and one A_2 allele. This is $2pq$ rather than pq because there are two ways to draw an A_1A_2 individual: by drawing an A_1 first and an A_2 second, or by drawing an A_2 first and an A_1 second. The frequency of the A_2A_2 genotype is equal to the fraction of the time that two A_2 alleles would be drawn: q^2. This is a general result; the frequencies at Hardy–Weinberg equilibrium are always those that we would find if the gametes were paired randomly.

What is remarkable is that Hardy–Weinberg genotype frequencies represent a type of equilibrium, such that in the absence of outside processes, the genotype frequencies will remain constant at the Hardy–Weinberg frequencies over time. Box 7.3 expands on this point.

Thus, we see that the Hardy–Weinberg model settles down to equilibrium genotype frequencies of p^2, $2pq$, and q^2 after a single generation. And, provided that the assumptions of the model are met—that is, the organism is diploid, it reproduces sexually, natural selection is not operating, mating is random, there

intuitions about what Mendel's rules predicted for population-wide genotype frequencies. Yule and his colleagues expected that Mendel's rules predicted a stable equilibrium for both genotype frequencies *and* allele frequencies, with allele frequencies returning to an even 1:1 ratio (Figure 7.10B). But instead, as Hardy and Weinberg each showed, Mendel's rules predicted a mixed equilibrium. Genotype frequencies are stable for *given* allele frequencies, but allele frequencies themselves are at a neutral equilibrium.

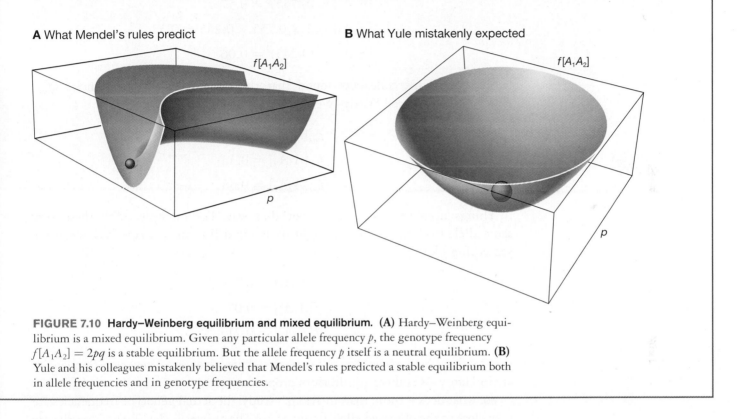

A What Mendel's rules predict

$f[A_1A_2]$

p

B What Yule mistakenly expected

$f[A_1A_2]$

p

FIGURE 7.10 Hardy–Weinberg equilibrium and mixed equilibrium. **(A)** Hardy–Weinberg equilibrium is a mixed equilibrium. Given any particular allele frequency p, the genotype frequency $f[A_1A_2] = 2pq$ is a stable equilibrium. But the allele frequency p itself is a neutral equilibrium. **(B)** Yule and his colleagues mistakenly believed that Mendel's rules predicted a stable equilibrium both in allele frequencies and in genotype frequencies.

is no mutation or migration, and the population is extremely large—genotype frequencies remain at these values indefinitely.

An Example of Hardy–Weinberg Genotype Frequencies: The Myoglobin Protein

To see an example of the Hardy–Weinberg genotype frequencies in a human population, we will look at a polymorphism of the gene coding for the myoglobin protein. Myoglobin is a muscle protein that supplies oxygen to the muscles when needed. Molecular genetic analysis reveals that human myoglobin alleles are typically one of two forms—let's call them A_1 and A_2—that differ by only two bases (Takata et al. 2002).

To study the distribution of these two alleles in a Japanese population, Tomoyo Takata and his colleagues collected blood samples from 100 Japanese volunteers, and they used a molecular genetic technique known as PCR–SSCP (polymerase chain reaction–single strand comformation polymorphism) to determine the genotype of each individual at the myoglobin locus. They found that, among their subjects, the frequency of the A_1 allele was $p = 0.755$, while that of the A_2 allele was $q = 0.245$.

If no other evolutionary processes are in operation, the equilibrium frequencies of genotypes in this example can be predicted from allele frequencies by using the Hardy–Weinberg model. As we have seen, when the allele frequencies are p and q, we expect the genotype frequencies to be p^2, $2pq$, and q^2 at Hardy–Weinberg equilibrium. In this case, that means that the expected Hardy–Weinberg genotype frequencies (f_{exp}) will be

$$f_{exp}[A_1A_1] = (0.755)^2 = 0.57$$
$$f_{exp}[A_1A_2] = 2 \times 0.755 \times 0.245 = 0.37$$
$$f_{exp}[A_2A_2] = (0.245)^2 = 0.06$$

Takata and his colleagues found that the *actual* genotype frequencies (f_{act}) matched the expected Hardy–Weinberg frequencies very closely:

$$f_{act}[A_1A_1] = 0.59$$
$$f_{act}[A_1A_2] = 0.33$$
$$f_{act}[A_2A_2] = 0.08$$

Things need not have turned out that way. For example, with these very same allele frequencies Takata might have found that the *genotype frequencies* were something like

$$f[A_1A_1] = 0.72$$
$$f[A_1A_2] = 0.07$$
$$f[A_2A_2] = 0.21$$

Box 7.4 describes how one can test to see whether a given set of genotype frequencies are in Hardy–Weinberg equilibrium proportions.

So, what does it mean that, in Takata's study, the actual genotype frequencies are very close to the observed allele frequencies? The answer is that Takata's results are consistent with the hypothesis that there are no evolutionary processes operating on the A_1 and A_2 alleles. There is good reason to think that this may be the case. It turns out that neither of the base pair differences that distinguish the A_1 and A_2 alleles changes the amino acid sequence, and therefore the proteins produced by each allele are identical. Because this genetic difference should have no phenotypic effect, we would expect to find no fitness differences between A_1 and A_2. Similarly, in the absence of any phenotypic differences between the two alleles, we do not expect assortative mating with respect to this locus. Because it takes a pair of perfectly placed point mutations to convert A_1 to A_2 or vice versa, mutation rates between these two loci are low enough to be negligible. Migration is also negligible: Migration into the Japanese population has traditionally been low, presumably low enough that the frequencies at these alleles have been unaffected. Finally, the population studied is very large. There are more than 125 million people in Japan. Thus, we would expect that the large population assumption of the Hardy–Weinberg model has been satisfied as well.

BOX 7.4 Testing for Hardy–Weinberg Equilibrium

At Hardy–Weinberg equilibrium, both allele frequencies and genotype frequencies remain unchanged from generation to generation. Thus, if we observe a change in the allele or genotype frequencies in a population, we can safely infer that either (1) the population was not initially at Hardy–Weinberg equilibrium, or (2) at least one of the five Hardy–Weinberg assumptions has been violated.

We also know that the Hardy–Weinberg model predicts that if a population is initially away from Hardy–Weinberg equilibrium, the equilibrium genotype frequencies will be reached in one generation—without any change in allele frequencies. Therefore, if we observe allele frequencies changing at all, or if we observe genotype frequencies continuing to change over multiple generations, we can again conclude that at least one of the five Hardy–Weinberg assumptions has been violated.

But what if we observe a population in which neither the allele frequencies nor the genotype frequencies are changing? This still does not necessarily mean that the population is in Hardy–Weinberg equilibrium.

How can we tell? Using the model we have developed thus far, we can use the known genotype frequencies of a population to test whether that population is at or near Hardy–Weinberg equilibrium. For example, suppose that we have a population with the following genotype frequencies:

$$f[A_1A_1] = 0.59 \qquad f[A_1A_2] = 0.16 \qquad f[A_2A_2] = 0.25$$

We use these genotype frequencies to calculate the allele frequencies:

$$p = f[A_1A_1] + \frac{f[A_1A_2]}{2} = 0.67$$

$$q = \frac{f[A_1A_2]}{2} + f[A_2A_2] = 0.33$$

These are the *actual* allele frequencies in our population. Next, we calculate the *expected* Hardy–Weinberg genotype frequencies for a population with these allele frequencies. Call these expected Hardy–Weinberg frequencies $f_{exp}[A_1A_1]$, $f_{exp}[A_1A_2]$, and $f_{exp}[A_2A_2]$.

$$f_{exp}[A_1A_1] = p \times p = 0.45$$
$$f_{exp}[A_1A_2] = 2pq = 0.44$$
$$f_{exp}[A_2A_2] = q \times q = 0.11$$

These expected genotype frequencies are considerably different from our observed genotype frequencies, and thus we can conclude that our population is not in Hardy–Weinberg equilibrium. We can also definitively conclude that at least one of the evolutionary processes in question is operating to shift genotype frequencies away from Hardy–Weinberg equilibrium. Researchers will often use a statistical test such as a χ^2 test to determine whether the observed frequencies differ enough from the expected frequencies for us to be confident that the discrepancy is not due to sampling error alone.

Suppose that the population had been at or near Hardy–Weinberg equilibrium. What could we have concluded then? Could we have concluded that none of the evolutionary processes described in our assumptions are operating? The answer is no. In this case, we could not directly rule out the possibility that all of the Hardy–Weinberg assumptions are met, nor could we rule out the possibility that some of the assumptions are violated. For example, ongoing mutation could occur without shifting the genotype frequencies away from Hardy–Weinberg proportions.

While both our knowledge of the biology of these two alleles and the results of the Takata study are consistent with the Hardy–Weinberg model, the study does not definitively demonstrate that the Hardy–Weinberg assumptions are met for this locus. For one thing, while we know that the genotype frequencies are currently in Hardy–Weinberg proportions, we do not yet know that they will remain there. Furthermore, even if we could show that the Hardy–Weinberg assumptions were met at one locus, this would not mean that they would be met at *all* loci in the human genome. Indeed, we know that they are not; we have evidence that the evolutionary processes of natural selection, assortative mating, mutation, and drift all operate on human populations. Later in this chapter, we will look at an example in which two of these processes, mutation and selection, oppose one another in human populations.

7.3 Natural Selection

In the previous section, we examined what happens to genotype frequencies when the Hardy–Weinberg assumptions are met. In this section, we will extend our model to include the action of natural selection. Before doing so, let us begin by sketching out an example of natural selection that occurs in the wild. From there, we will use the data to make predictions about allele frequency change.

Selection for Coat Color in Pocket Mice

As an example of natural selection in the wild, we will return to the trait of coat color in mice, which we discussed in depth in Chapter 3. Here we will consider not the oldfield mouse, but instead a related species, the rock pocket mouse (*Chaetodipus intermedius*), that Hopi Hoekstra studied with Michael Nachman and Susan D'Agostino (Nachman et al. 2003; Hoekstra et al. 2004; Nachman 2005). Pocket mice live in rocky areas at low elevations in the Sonoran and Chihuahuan deserts and are well adapted to desert life. Within the confines of the desert, *C. intermedius* lives in one of two very different types of habitat—either on light-colored rocks or on much darker rocks associated with lava flows. Mice that live on light-colored rocks tend to have a sandy, gray coat color, while mice that inhabit lava fields are darker (Figure 7.11) (Benson 1933; Dice and Blossom 1937). Just as with the oldfield mice in Chapter 3, coat color influences predation risk for pocket mice. Pocket mice whose coat colors match their environment are much less susceptible to predation than mice that stand out against the rocks they inhabit (Dice 1947). We would expect, then, that natural selection would favor individuals with coat colors that offer camouflage in their natural environment.

FIGURE 7.11 Pocket mice live in light and dark rock habitats. (A) Light-colored rock habitat, and light- and dark-coated mice on light rock, (B) dark lava field habitat of the rock pocket mouse, and light- and dark-coated pocket mice on dark rock.

The genetic control of coat color in rock pocket mice is also very similar to the genetic control seen in oldfield mice. In pocket mice, coat coloration is influenced by the same melanocortin-1 receptor (Mc1R) that we described for oldfield mice in Chapter 3. In these mice the *Mc1R* locus has two alleles that we will call *D* and *d*. The *D* allele is associated with dark coloration, whereas the *d* allele is associated with light coloration (Nachman et al. 2003). *D* is dominant to *d*, so that *DD* and *Dd* individuals both display dark coloration, and only individuals with the *dd* genotype display light coloration.

Here we have a system in which an important trait—coloration—is associated with a single locus and clearly tied to survival. But just how beneficial is it for an individual to have the allele coding for a coat coloration that matches the background environment—that is, how advantageous is it for mice on the dark lava fields to be *DD* or *Dd*, and for mice along the light-colored rocks to be *dd*?

To address this question, Nachman and his colleagues collected individuals at both lava sites and light-colored rock sites in an area along the border between Arizona and Mexico (Nachman et al. 2003; Hoekstra et al. 2004). Most individuals at the lava sites were dark-colored, and most individuals at the light-colored rock sites were light-colored. Each population, however, had a number of individuals that were "mismatched"—that is, individuals whose coats did not match their environment. From their data on survival and migration, the researchers were able to demonstrate that light-colored pocket mice living in the dark lava fields suffered higher rates of mortality. Their chances of survival ranged from 60 to 98% of the chances of survival of dark-colored mice on the dark lava fields. With these data in hand, we can now start to make specific predictions about how the frequencies of the *D* and *d* alleles should change as a result of natural selection. To do so, we must build a mathematical model of natural selection that we can then use to examine the pocket mouse example.

A Simple Model of Natural Selection

We begin with the Hardy–Weinberg model, but we will relax Hardy–Weinberg assumption number 1: We will now allow natural selection to operate on our population. In order to use the terminology we developed earlier in discussing the Hardy–Weinberg model, but also to allow us to link back to the pocket mouse example, let us again consider two alleles—allele A_1 (at frequency p) and allele A_2 (at frequency q). Think of A_1 as the *D* allele for dark coloration in our mouse example, and let A_2 represent the *d* allele for light coloration. Because A_1 is dominant to A_2, both the A_1A_1 and A_1A_2 genotypes display dark coloration. But against a dark lava field, only the A_2A_2 individuals stand out and suffer a reduced survival probability. On the lava fields, natural selection is thus acting against the A_2 allele.

To quantify the strength of natural selection against allele A_2, we use a parameter called the **selection coefficient**, labeled s, to describe the fitness reduction of the light phenotype relative to the dark phenotype. By convention, the fitness of one type—here the dark phenotype—is set to 1. The fitness of the other phenotype—here the light phenotype—is set to $1 - s$. The value $s = 0$ indicates no selection against an allele; $s = 0.25$ indicates a 25% reduction in fitness, $s = 0.50$ indicates

TABLE 7.1

Fitnesses for a Dominant Locus

A_1 DOMINANT TO A_2

Genotype	Fitness
A_1A_1	1
A_1A_2	1
A_2A_2	$1 - s$

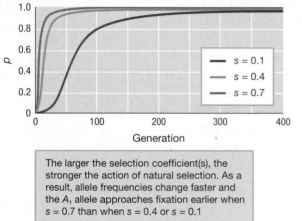

The larger the selection coefficient(s), the stronger the action of natural selection. As a result, allele frequencies change faster and the A_1 allele approaches fixation earlier when $s = 0.7$ than when $s = 0.4$ or $s = 0.1$

FIGURE 7.12 The consequences of natural selection favoring a dominant allele. Here we plot the *trajectory*—the path over time—of the frequency p of the dominant A_1 allele for three different selection intensities. The horizontal axis indicates time in generations, and the vertical axis, ranging from 0 to 1, indicates the frequency of the A_1 allele. The initial frequency of the A_1 allele is 0.005, and this allele increases to near-fixation in all three cases albeit at different rates for our three values of s.

a 50% reduction in fitness, and so forth. For light-colored mice in dark lava environments, Nachman and his team measured survival probabilities ranging from 98 to 60% of that experienced by the dark-colored mice, depending on the population examined. As a result, they estimated selection coefficients against light coloration ranging from 0.02 to 0.40. In our mathematical example, we will use a selection coefficient $s = 0.1$.

Our goal now is to look at the change in allele and genotype frequencies over time as the result of natural selection, with intensity quantified by the selection coefficient s. We begin by constructing a simple table of genotypes and their corresponding fitness values (Table 7.1). In this table, fitness is a measure of the relative lifetime reproductive success of our three genotypes.

For example, imagine that, before natural selection operates, we have 100 A_1A_1, 100 A_1A_2, and 100 A_2A_2 individuals in our population, but after selection, the numbers are reduced to 60 A_1A_1, 60 A_1A_2, and 54 A_2A_2. If we denote the fitness of A_1A_1 and A_1A_2 as 1, the relative fitness of A_2A_2 is $(54/100)/(60/100) = 54/60 = 0.9$. As such, $s = 0.1$. Box 7.5 demonstrates how we can make detailed predictions regarding allele frequency change when natural selection is operating in the case of the pocket mouse.

For example, Box 7.5 demonstrates that, when A_1 is dominant, the frequency of the A_1 allele should increase by $pq^2s/(1 - q^2s)$ in every generation. Figure 7.12 maps how the allele frequency of A_1 would change over evolutionary time. In our rock pocket mouse example where $s = 0.1$, if the frequency of the dominant dark allele started at a frequency of just 0.01, we would expect that, within only 500 generations, it would increase to a frequency near 1.

Modes of Frequency-Independent Selection

In the example we just considered, the genotypes producing the dark phenotype are favored over the genotypes producing the light phenotype, irrespective of the frequency of each type. Whether dark mice are rare or common, we expect them to have the same fitness. Our mouse example is an instance of **frequency-independent selection**, where the fitness associated with a trait is not directly dependent on the frequency of the trait in a population.

In general, there are a number of ways in which frequency-independent selection can operate. These differ in how the relative fitnesses of the A_1A_1, A_1A_2, and A_2A_2 genotypes vary in relation to one another.

Directional Selection

The most straightforward type of frequency-independent selection is known as **directional selection**. Under directional selection, one allele is consistently favored over the other allele. As a result, selection drives allele frequencies in a single direction, toward an increasing frequency of the favored allele. Eventually, the favored allele will become *fixed* in the population: It will replace all other alternative alleles at the same locus. When an allele becomes fixed, we say that it has reached **fixation**. (Strictly speaking, an allele will never reach complete fixation under the

BOX 7.5 Natural Selection Favoring a Dominant Allele

Here we build a model in which natural selection acts on a trait controlled by the A locus, where the A_1 allele is dominant to the A_2 allele. As in Table 7.1, fitnesses are:

Genotype	Fitness
A_1A_1	1
A_1A_2	1
A_2A_2	$1-s$

Genotype frequencies before and after selection are therefore as follows:

Genotype	A_1A_1	A_1A_2	A_2A_2
Frequency before selection	p^2	$2pq$	q^2
Frequency after selection	p^2	$2pq$	$q^2(1-s)$

We know that $p^2 + 2pq + q^2 = 1$, so $p^2 + 2pq + q^2(1-s)$ must be less than 1. Yet, the A_1A_1, A_1A_2, and A_2A_2 genotypes make up our entire population after selection, so their frequencies must sum to 1. To arrive at a sum of 1, we take the frequency of each genotype after selection and divide it by the sum of these frequencies:

A_1A_1	A_1A_2	A_2A_2
$\dfrac{p^2}{\left[p^2+2pq+q^2(1-s)\right]}$	$\dfrac{2pq}{\left[p^2+2pq+q^2(1-s)\right]}$	$\dfrac{q^2(1-s)}{\left[p^2+2pq+q^2(1-s)\right]}$

Expanding the denominator, we get $p^2 + 2pq + q^2 - q^2s$. If we replace $p^2 + 2pq + q^2$ with 1, our denominator equals $1 - q^2s$.

A_1A_1	A_1A_2	A_2A_2
$\dfrac{p^2}{1-q^2s}$	$\dfrac{2pq}{1-q^2s}$	$\dfrac{q^2}{1-q^2s}$

Because $p = f[A_1A_1] + f[A_1A_2]/2$, it follows that p', the frequency of allele A_1 in the offspring, will be

$$p' = \frac{p^2 + pq}{1-q^2s} = \frac{p}{1-q^2s}$$

To see this, factor the numerator, $p^2 + pq$ to $p(p + q)$ and replace $p + q$ with 1.

We can use the fact that $p + q = 1$ to write our last equation slightly differently, as:

$$p' = \frac{p}{1 - s(1-p)^2}$$

This form of the equation for p' is called a recursion equation, because it shows us what p' is in direct relation to p.

We can also write an expression for the change in allele frequency $p' - p$:

$$p' - p = \frac{p}{1 - q^2s} - p$$

To give both terms a common denominator, multiply the second term by:

$$\frac{1 - q^2s}{1 - q^2s}$$

This gives us:

$$p' - p = \frac{p}{1-q^2s} - \frac{p(1 - q^2s)}{1 - q^2s}$$

$$p' - p = \frac{p}{1 - q^2s} - \frac{p - pq^2s}{1 - q^2s}$$

$$p' - p = \frac{p - p + pq^2s}{1 - q^2s}$$

$$p' - p = \frac{pq^2s}{1 - q^2s}$$

---

infinite population size assumption of the Hardy–Weinberg model, but it will get arbitrarily close. For simplicity of language, we will speak of this as fixation.) Figure 7.13 illustrates various ways in which directional selection could operate in favor of the A_1 allele. When A_1 is dominant to A_2, the genotypes A_1A_1 and A_1A_2 produce the same phenotype and have the same fitness, but A_2A_2 has a lower fitness (see also Box 7.5). When A_1 and A_2 are codominant, all three genotypes produce different phenotypes, with the A_1A_2 heterozygote presenting a phenotype intermediate to the two homozygote phenotypes. In this case, A_1A_1 has the highest fitness, A_1A_2

FIGURE 7.13 Directional selection at one locus with two alleles. (A) In directional selection, one allele A_1 is favored over another, A_2. This can occur in different ways: A_1 can be dominant (red), A_1 and A_2 can be codominant (blue), or A_1 can be recessive (orange). (B) The trajectories of p, the frequency of the A_1 allele, are illustrated from a starting value of $p = 0.005$.

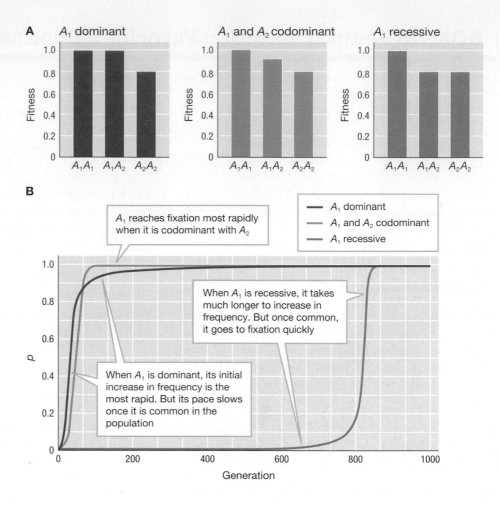

has an intermediate fitness, and A_2A_2 has the lowest fitness. When A_1 is recessive to A_2, the A_1A_1 genotype has the highest fitness, while A_1A_2 and A_2A_2 produce the same phenotype and share the same lower fitness value.

Rates of Fixation under Directional Selection

The plot of allele frequency trajectories in Figure 7.13 not only tells us that the A_1 allele eventually goes to fixation in all three cases, but it also informs us about the rate at which this process occurs. Given the same fitness difference between the A_1A_1 and A_2A_2 homozygotes, the A_1 allele approaches fixation most rapidly in the codominant case, somewhat less rapidly in the dominant case, and much more slowly in the recessive case.

Looking at these trajectories, there are two qualitative features we would like to explain. First, why does a rare A_1 allele quickly increase in frequency in the dominant and codominant cases, but not in the recessive case? Second, once A_1 is common, why does it take a long time to go to fixation in the dominant case but not in the codominant or recessive case? We address these questions in turn.

We can understand why a rare A_1 allele quickly increases in frequency in the dominant and codominant cases, but not in the recessive case by looking at the genotypes in which A_1 and A_2 typically occur, and the average selective differences that result. Suppose that A_1 is initially rare, as shown in Figure 7.13. Then, initially, most copies of the A_1 allele appear in A_1A_2 heterozygotes. When A_1 is dominant to or codominant with A_2, these heterozygotes enjoy a selective advantage, and thus the frequency of the A_1 allele responds immediately with a sizeable increase. But when A_1 is recessive, the heterozygotes have the same fitness as the A_2A_2 homozygotes that make up the majority of the population. Selection increases the frequency of the allele A_1 in the rare events in which A_1A_1 homozygotes—which see the fitness benefits—are produced.

Once the A_1 allele becomes more common in the population, it starts to occur in A_1A_1 homozygotes an appreciable fraction of the time, but the A_2 allele now typically appears in heterozygotes. When A_1 is dominant to A_2, this means that most A_2 alleles now appear in individuals with the same phenotype as A_1A_1 homozygotes. Because this is an advantageous phenotype, there is no longer strong selection against the A_2 allele. Selection slows down, and it takes a very long time to entirely eliminate the A_2 allele from the population. Rare recessive alleles mostly reside in heterozygotes where they suffer no fitness disadvantage. When A_1 and A_2 are codominant, however, there is no way for A_2 alleles to hide from the effects of selection. Thus, in the codominant case, selection against A_2 continues to be strong even once A_1 becomes very common. As a result, the A_2 allele is more quickly removed from the population.

Overdominance and Underdominance

There are two additional ways that frequency-independent selection can act on one locus with two alleles. In the case of **overdominance**, also known as **heterozygote advantage**, the A_1A_2 heterozygote has a higher fitness than either the A_1A_1 or the A_2A_2 homozygotes (Figure 7.14). In this case, the direction of natural selection depends on the current allele frequencies in the population. When A_1 is rare, it will usually occur in heterozygotes. As a result, the average fitness of individuals carrying the A_1 allele will be higher than the average fitness of all individuals in the population. But when it is common, the A_1 allele will usually occur in A_1A_1 homozygotes that have a lower fitness than the population average. As a result, A_1 increases in frequency when rare and decreases in frequency when common. Natural selection leads to a **balanced polymorphism**—a stable equilibrium that is *polymorphic*—that is, in which both alleles are present. Because this is a stable equilibrium, allele frequencies will return to their equilibrium values after a perturbation away from the equilibrium. We refer to selection that leads to a balanced polymorphism as **balancing selection**.

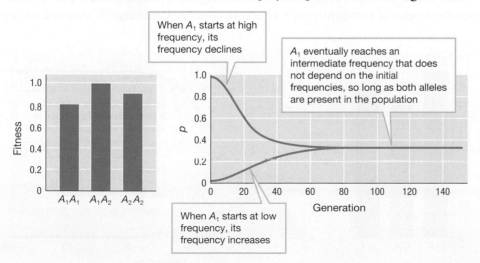

When A_1 starts at high frequency, its frequency declines

A_1 eventually reaches an intermediate frequency that does not depend on the initial frequencies, so long as both alleles are present in the population

When A_1 starts at low frequency, its frequency increases

FIGURE 7.14 Overdominance. In the case of overdominance, the heterozygote has a higher fitness than either homozygote. Irrespective of the initial frequencies, so long as both alleles are present in the population, the resulting fitness trajectory leads to an intermediate frequency of the A_1 and A_2 alleles. Here we show trajectories with random mating and initial frequencies $p = 0.025$ and $p = 0.975$.

A

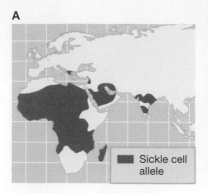

Sickle cell allele

FIGURE 7.15 The sickle cell allele is common where malaria is endemic and rare elsewhere. (A) Geographic range of the sickle cell allele in human populations. (B) Historical geographic distribution of endemic malaria. Adapted from Piel et al. (2010).

B

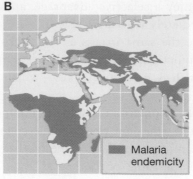

Malaria endemicity

The sickle cell mutation in the human hemoglobin gene is a classic example of overdominance. In its homozygous form, the sickle cell allele induces a change in the shape of red blood cells, from a round disk to a "sickle" shape, hence the name. These sickle-shaped cells clump together, preventing blood from flowing smoothly through the circulatory system. As a result, affected individuals suffer numerous health problems, including anemia, chronic pain, bacterial infection, organ damage, and ultimately reduced life expectancy. But in its heterozygous form, the sickle cell mutation does not cause pathology. Rather, it is partially protective against infection by the malaria parasite. As a result, sickle cell heterozygotes actually have a fitness advantage in areas where malaria is endemic. Consequently, the sickle cell allele has reached relatively high frequencies in populations that originated in these areas but is rare in other populations (Figure 7.15).

Nonetheless, we should not let the elegance of the sickle cell example generate a misleading impression about the importance of overdominance as a mechanism of preserving genetic polymorphism. While a few such cases are well known—the HLA loci involved in immune recognition and the ABO locus that determines blood type are good examples—overdominance is probably quite rare in general. When we see a balanced polymorphism, we should not rush to conclude that it is the result of overdominance (Bubb et al. 2006).

Underdominance is the reverse of overdominance. In underdominance, the A_1A_2 heterozygote has a lower fitness than either the A_1A_1 or A_2A_2 genotype. In this case, natural selection will favor one allele over the other—but which allele becomes fixed in the population will depend on where the population starts (Figure 7.16). If A_1 is very rare, it will typically appear in A_1A_2 heterozygotes that have lower-than-average fitness. When A_1 is very common, the A_1 allele will typically appear in A_1A_1 homozygotes that have higher-than-average fitness. The same holds true for the A_2 allele. As a result, in the case of underdominance, there is a threshold frequency of the A_1 allele, above which A_1 will be fixed and below which A_1 will be lost.

Real-world examples of underdominance are perhaps even scarcer than are real-world examples of overdominance. One case of underdominance is seen in a mouse

FIGURE 7.16 Underdominance. In the case of underdominance, the heterozygote has lower fitness than either homozygote. The resulting fitness trajectory leads to fixation of one allele or the other, depending on the starting allele frequencies. Here we show trajectories for the frequency p of the A_1 allele, with random mating and initial frequencies $p = 0.30$ and $p = 0.36$.

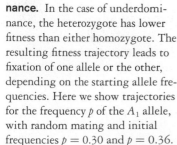

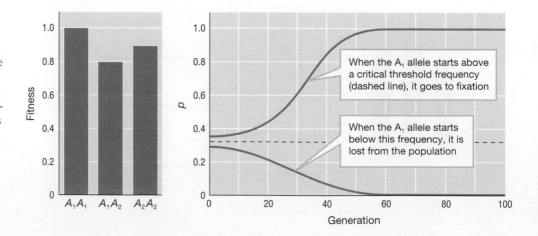

When the A_1 allele starts above a critical threshold frequency (dashed line), it goes to fixation

When the A_1 allele starts below this frequency, it is lost from the population

BOX 7.6 Equilibrium Allele Frequencies in Overdominance and Underdominance

Both the balanced polymorphism due to overdominance and the critical threshold due to underdominance are equilibria—and in both cases, allele frequencies can be computed in the same way. The key is to recognize that at an equilibrium, the average fitness of the A_1 allele is precisely equal to the average fitness of the A_2 allele. If we can find the frequency p of the A_1 allele such that the fitnesses of these two alleles are the same, we have found the equilibrium.

Let w_{11} be the fitness of the A_1A_1 homozygote, w_{12} be the fitness of the A_1A_2 heterozygote, and w_{22} be the fitness of the A_2A_2 homozygote. Assuming random mating, no mutation, no migration, and large population size, the frequencies of each genotype before selection are simply Hardy–Weinberg frequencies. Thus, before selection, a fraction p of the A_1 alleles are in homozygotes and a fraction $1 - p$ are in heterozygotes. As a result, the average fitness of the A_1 allele is $pw_{11} + (1 - p)w_{12}$. By similar logic, a fraction $1 - p$ of the A_2 alleles are in homozygotes and a fraction p are in heterozygotes, so the average fitness of the A_2 allele is $pw_{12} + (1 - p)w_{22}$. At equilibrium, these average fitnesses are precisely equal, giving us the equation

$$pw_{11} + (1 - p)w_{12} = pw_{12} + (1-p)w_{22}$$

Solving for p, we get

$$p = \frac{w_{22} - w_{12}}{w_{11} - 2w_{12} + w_{22}}$$

This is the equilibrium frequency of A_1: the frequency at the balanced polymorphism in the overdominance case, and the frequency of the critical threshold in the underdominance case.

Applying this equation to the overdominance example in Figure 7.14, we get

$$p = \frac{0.9 - 1}{0.8 - 2 + 0.9} = 1/3$$

Indeed, p = 1/3 is the frequency of A_1 at the balanced polymorphism that is reached in that example. Applying this equation to the underdominance example in Figure 7.16, we get

$$p = \frac{0.9 - 0.8}{1 - 2(0.8) + 0.9} = 1/3$$

And, indeed, $p = 1/3$ is the frequency of A_1 at the critical threshold in that example.

population that contains hybrid New Zealand Black (NZB)/New Zealand White (NZW) mice (Helyer and Howie 1963a,b; Theofilopoulos and Dixon 1985). NZW homozygotes are phenotypically normal, while NZB homozygotes exhibit a number of autoimmune defects. The NZB/NZW heterozygotes experience even more severe autoimmune disease, and they are used as a medical model of the human autoimmune disease lupus (Figure 7.17).

In both overdominance and underdominance, there is a critical equilibrium frequency of the A_1 allele at which the average fitnesses of the A_1 and A_2 alleles are the same, assuming random mating. In the case of overdominance, this equilibrium frequency is a *stable* equilibrium, and it is the allele frequency that we observe at the balanced polymorphism. In the case of underdominance, this equilibrium is an *unstable* equilibrium, and it represents the threshold allele frequency above which A_1 goes to fixation and below which A_2 goes to fixation. These equilibrium frequencies can be calculated as shown in Box 7.6.

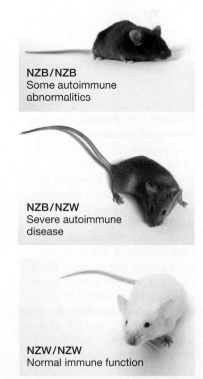

NZB/NZB
Some autoimmune abnormalities

NZB/NZW
Severe autoimmune disease

NZW/NZW
Normal immune function

FIGURE 7.17 Heterozygotes and autoimmune defects. The NZB/NZW hybrid mouse develops severe autoimmune disease and is used as a model system for the study of lupus. Because the heterozygote has a lower fitness than either homozygote, this is an example of underdominance.

In the cases of overdominance and underdominance, the direction of selection depends on the frequencies of the A_1 and A_2 alleles. So why do we classify overdominance and underdominance as forms of frequency-*independent* selection? The answer is that when we talk about frequency-independent selection or frequency-dependent selection, we are referring to the way that the fitnesses of *phenotypes* (or the genotypes that produce them) depend on the frequencies of *phenotypes*—not on the way that the average fitnesses of individuals carrying a given allele depend on the frequencies of those alleles. In both overdominance and underdominance, the fitness of each genotype, and its corresponding phenotype, is constant and independent of the frequencies of the genotypes in the population. The fitnesses of individuals carrying the A_1 and A_2 alleles vary according to the frequencies of those alleles only because these frequencies determine the chance that any given A_1 allele or any given A_2 allele ends up in a heterozygote instead of a homozygote.

Modes of Frequency-Dependent Selection

Frequency-dependent selection occurs when the costs and benefits associated with a trait depend on its frequency in the population. Frequency-dependent selection can be *positive* or *negative*; we will treat these two cases in turn. With positive frequency-dependent selection, the fitness associated with a trait *increases* as the frequency of the trait increases in a population (Figure 7.18). Thus, under positive frequency-dependent selection, each phenotype is favored once it becomes sufficiently common in the population. If the phenotypes are controlled by two alternative alleles at a single locus, one of the two alleles will eventually be fixed and the other will be lost—although which is fixed and which is lost depends on the initial allele frequencies.

For example, land snails have shells that either coil to the right or coil to the left. In the so-called "flat" snail species, individuals mate in a face-to-face position, and because of physical constraints, mating in these species can only take place between individuals whose shells coil in the same direction (Figure 7.19) (Asami et al. 1998). In such a situation, positive frequency-dependent selection operates on

FIGURE 7.18 Positive frequency-dependent selection. The more frequent a phenotype, the higher its fitness. **(A)** The fitnesses of phenotypes P_1 and P_2 depend on the frequency of the P_1 phenotype. **(B)** Suppose that the phenotype in question is determined by a single locus A, with A_1 dominant to A_2. That is, the A_1A_1 and A_1A_2 genotypes exhibit phenotype P_1, while the A_2A_2 genotype exhibits phenotype P_2. The frequency p of the A_1 allele over time then depends on the starting frequency of A_1. If A_1 is sufficiently common, it will be fixed; otherwise it will be lost.

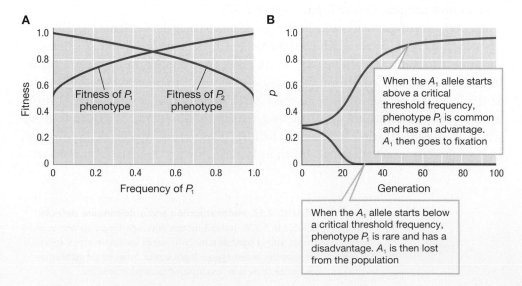

When the A_1 allele starts above a critical threshold frequency, phenotype P_1 is common and has an advantage. A_1 then goes to fixation

When the A_1 allele starts below a critical threshold frequency, phenotype P_1 is rare and has a disadvantage. A_1 is then lost from the population

the direction of the shell's coil. The higher the frequency of either type of shell—right coil or left coil—the greater the mating success of that type, because more potential mates exist and thus mates are easier to find. Interestingly, in so-called "tall" species of land snails, males mount females from above, and therefore snails with shells that coil in opposite directions can still mate. In this situation, we see much weaker frequency dependence than in flat snails.

With negative frequency-dependent selection, the fitness associated with a trait *decreases* as the frequency of the trait *increases* in a population. Thus, under negative frequency-dependent selection, each phenotype is favored when it is rare. If the phenotypes are controlled by two alternative alleles at a single locus, both alleles will be maintained in a balanced polymorphism (Figure 7.20). Thus, negative frequency-dependent selection, like overdominance, is a form of balancing selection.

A fascinating case of negative frequency-dependent selection has been documented in the scale-eating cichlid fish *Perissodus microleptis*. Two genetic morphs of this species exist in the population of Lake Tanganyika, a large lake in central Africa. One morph has a "left-handed" (sinistral) mouth opening, and the other has a "right-handed" (dextral) mouth opening (Figure 7.21). Because *P. microleptis* attacks its prey from behind, this morphological difference translates into individuals with left-handed mouths grabbing scales from the right side of their prey, and predators with right-handed mouths snatching scales from the left side of their prey.

FIGURE 7.19 Mating between two *Euhadra cingenita*, a "flat" species of snail. Mating in this species can only take place between individuals whose shells coil in the same direction.

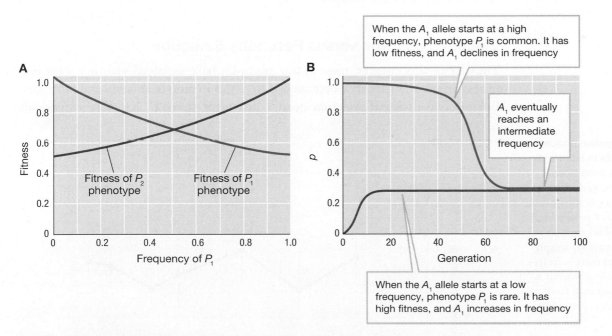

When the A_1 allele starts at a high frequency, phenotype P_1 is common. It has low fitness, and A_1 declines in frequency

A_1 eventually reaches an intermediate frequency

When the A_1 allele starts at a low frequency, phenotype P_1 is rare. It has high fitness, and A_1 increases in frequency

FIGURE 7.20 Negative frequency-dependent selection. The rarer a phenotype, the higher its fitness. **(A)** The fitnesses of phenotypes P_1 and P_2 depend on the frequency of the P_1 phenotype. **(B)** Suppose that the phenotype in question is determined by a single locus A, with A_1 dominant to A_2. That is, the A_1A_1 and A_1A_2 genotypes exhibit P_1, while the A_2A_2 genotype exhibits phenotype P_2. So long as both alleles are present from the start, the frequency p of the A_1 allele reaches an intermediate value regardless of the starting allele frequencies.

FIGURE 7.21 Two morphs of a cichlid fish. The "right-handed" (dextral) morph of *Perissodus microlepis* is shown from both sides on the top, while the "left-handed" (sinistral) is shown from both sides on the bottom. Adapted from Hori (1993).

In 1993, Michio Hori found that these mouth morphologies were inherited—via a two-allele model, in which the right-handed mouth was dominant—and that across many years, the frequency of each of these two types stayed at roughly 50% (Figure 7.22). Evidence suggests that negative frequency dependence explains the relative frequencies of the two mouth genotypes in Lake Tanganyika cichlids—the higher the frequency of either type, the lower its fitness. But why?

The answer lies in the behavioral response of prey. Using scar marks on prey to determine where the cichlids grabbed scales from the prey to eat, Hori found that when one predator mouth morphology became dominant, prey grew increasingly adept at watching for predator attacks from that particular side. This translated into greater hunting success by the rarer predator morph, and hence a higher fitness (Hori 1993). For example, imagine that the right-handed mouth (dextral) types are rare and the left-handed mouth (sinistral) types are common. Prey learn to avoid the more common sinistral morphs. As a result, the dextral types, which have greater fitness, increase in frequency because of negative frequency-dependent selection. Then the fitness advantage of the dextral morphs *decreases* as they become more common. Once the dextral morphs become more common, sinistral morphs have a fitness advantage, and they increase in frequency. The result should be along the lines of what Hori saw in nature—approximately even numbers of dextral and sinistral morphs in most years. When the frequency of one type becomes significantly greater than the other, frequency-dependent selection brings the population back to a 50–50 split of dextral and sinistral morphs.

Viability Selection versus Fecundity Selection

Thus far in this chapter, most of our examples have involved viability selection: fitness differences that arise because of differences in rates of survival and mortality. But of course natural selection doesn't just favor survival, it favors individuals

FIGURE 7.22 Negative frequency dependence results in intermediate frequencies of each phenotype. Over a 10-year period, Hori tracked the frequency of the sinistral morph of the cichlid *Perissodus microleptis.* Note the oscillations in frequency. When the sinistral morph is the more common morph, its frequency tends to decline; when it is less common, its frequency tends to increase. Adapted from Hori (1993).

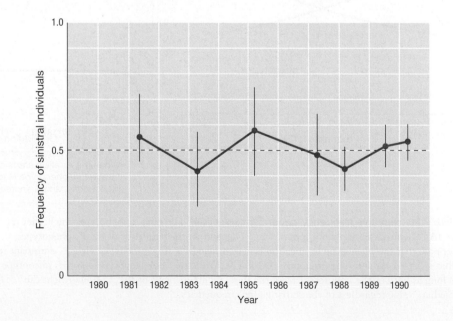

who leave the most surviving offspring. Thus, natural selection can operate on more than just survival probabilities; it can also act on the number of offspring produced, known as **fecundity**.

Recall that in Chapter 3, we defined fitness as expected reproductive success relative to other individuals in a population. In simple models (for example, those in which organisms are *semelparous*—that is, they reproduce only once at the end of their lives) it is straightforward to see how viability and fecundity differences combine to influence fitness. The expected reproductive success of an individual is equal to the probability that the individual survives to reproduce, multiplied by the number of offspring that are produced if the individual does survive. As a consequence, viability and fecundity act equivalently in basic population genetic models. Halving the number of offspring produced or halving the probability of survival each halves an organism's fitness, and thus each has the same effect in these basic models.

A dramatic example of fecundity differences without viability differences comes from a study of sunflower hybrids formed when wild sunflowers (*Helianthus annuus*) are crossed with their domesticated relatives, the crop sunflower (also called *Helianthus annuus*) (Figure 7.23). From an applied perspective, crop scientists are highly concerned with what happens to these hybrids, because it is through such hybrids that novel genes from genetically modified crops could make their way into wild populations with potentially severe ecological consequences. Charity Cummings and her colleagues studied what happens to the genes from genetically modified crops after they were introduced into a wild sunflower population by crossing wild sunflowers with domesticated crop sunflowers to produce hybrids (Cummings et al. 2002).

To find out, the researchers set up three replicate study populations, each with 100 wild plants and 100 hybrid plants. They measured both the viability and fecundity components of each plant's fitness: survival rate and lifetime seed production. In terms of viability selection, there was little difference between hybrids and wild plants. But there were large fecundity differences: The hybrids suffered a striking reduction in lifetime seed production (Table 7.2). These data provided Cummings and her colleagues with fitness estimates for their experimental populations. From these fitness estimates, they were able to use a simple model of natural selection, similar to the one that we developed for mouse coat coloration, to predict the

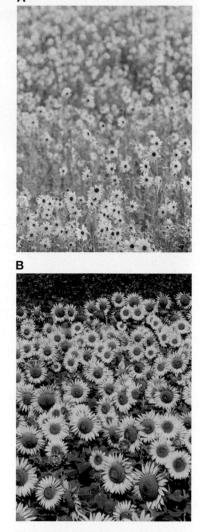

FIGURE 7.23 Wild and domestic sunflowers. (A) Wild sunflowers, and **(B)** domestic sunflowers.

TABLE 7.2

Survival and Lifetime Seed Production for Hybrid and Wild Sunflowers[a]

Site	PLANT SURVIVAL (%)		TOTAL NUMBER OF VIABLE SEEDS	
	Hybrid ($N = 100$)	Wild ($N = 100$)	Hybrid	Wild
1	100	98	15,428	635,000
2	100	100	3026	274,453
3	99	100	11,960	671,200

[a]Hybrid plants do not suffer reduced viability, but their lifetime seed production is dramatically decreased.
Adapted from Cummings et al. (2002).

change in the frequency of crop sunflower alleles over the subsequent generation. In accord with their predictions, when they went back to measure strain frequencies after a single generation, they found that the frequency of crop sunflower alleles dropped from 25% (half of their initial plants were hybrids, each with 50% crop sunflower alleles) in their initial plots to a mere 3% after one generation. Selection on fecundity, rather than on viability, rapidly eliminated crop sunflower alleles from wild populations.

7.4 Mutation

As we saw in Chapter 6, genetic mutation is the ultimate source of the variation on which natural selection acts. In that chapter, we also saw that mutation is undirected. By this, we mean that mutations are generated randomly with respect to their effects on the organism's fitness. Organisms may be able to alter the *rate* of mutation when under stress or in other circumstances (Bjedov et al. 2003), but they cannot affect the probability that a mutation will turn out to be favorable.

Mutation Can Change Allele Frequencies in a Population

Mutation, like natural selection, can influence allele frequencies in a population. As with natural selection, we can model this process mathematically. To understand how mutation rate alone can affect allele frequency, let us consider two alleles—A_1 and A_2. In the simplest case, imagine that allele A_1 mutates to allele A_2 with probability μ, and that allele A_2 mutates to A_1 with probability ν (Figure 7.24).

Suppose that the frequency of the A_1 allele in the parental generation is p and the frequency of the A_2 allele is q. In Box 7.7, we show that the equilibrium frequency of the A_1 allele, which we label p^*, is equal to $\nu/(\mu + \nu)$. Correspondingly, the equilibrium frequency of the A_2 allele is given by $q^* = \mu/(\mu + \nu)$. These are the equilibrium frequencies of the A_1 and A_2 alleles if mutation is the only process operating to change allele frequencies. Figure 7.25 illustrates the change in allele frequencies in a population with mutation rates $\mu = 0.00010$ and $\nu = 0.00005$. As expected, this population eventually approaches equilibrium when the A_1 allele is at frequency $p^* = 0.00005/(0.00005 + 0.00010) = 1/3$. Notice that this process

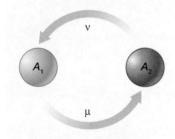

FIGURE 7.24 A model of mutation. The A_1 allele mutates to A_2 at rate μ, and A_2 mutates to A_1 at rate ν.

BOX 7.7 A Population-Genetic Model of Mutation

Suppose that the frequency of the A_1 allele in the parental generation is p and the frequency of the A_2 allele is q. Then over the course of one generation, some A_1 alleles will convert to A_2 alleles by mutation (this happens with frequency $p\mu$), and some A_2 alleles will convert to A_1 alleles (with frequency $q\nu$). Thus, after one generation, the frequency p' of A_1 alleles in the population will be

$$p' = p(1 - \mu) + q\nu = p(1 - \mu) + (1 - p)\nu$$

Correspondingly, the frequency q' of A_2 alleles in the population will be

$$q' = q(1 - \nu) + p\mu = (1 - p)(1 - \nu) + p\mu$$

If mutation is the only process changing allele frequencies in such a population, the allele frequencies of the A_1 allele will eventually reach an equilibrium value, called p^*. At that equilibrium value, the allele frequency p' in one generation is unchanged from p in the previous generation: $p = p' = p^*$. Substituting p^* for both p and p' in the first equation above, we get $p^* = p^*(1 - \mu) + (1 - p^*)\nu$. With a little bit of algebra we can solve this equation for p^*, and when we do, we get $p^* = \nu/(\mu + \nu)$. Correspondingly, $q^* = \mu/(\mu + \nu)$.

typically operates far more slowly than does natural selection. Even with the exceptionally high mutation rate of $\mu = 0.00010$ that we have used in Figure 7.25, it takes tens of thousands of generations for the allele frequencies to approach equilibrium.

We have now calculated what happens to allele frequencies as a result of mutation. But how does mutation affect *genotype* frequencies? Provided that the other Hardy–Weinberg assumptions are met (no selection, random mating, no migration, large population size), the genotype frequencies will always be in the Hardy–Weinberg proportions. Given frequencies p and q of alleles A_1 and A_2, the genotype frequencies for $A_1A_1 : A_1A_2 : A_2A_2$ will be $p^2 : 2pq : q^2$.

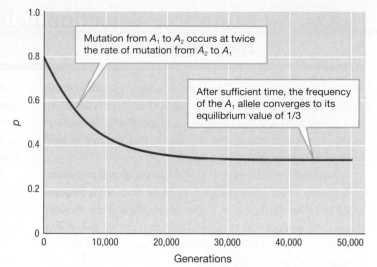

Mutation from A_1 to A_2 occurs at twice the rate of mutation from A_2 to A_1

After sufficient time, the frequency of the A_1 allele converges to its equilibrium value of 1/3

FIGURE 7.25 **Mutation can cause changes in allele frequencies.** Here mutation from A_1 to A_2 occurs at twice the rate of mutation from A_2 to A_1. The frequency p of the A_1 allele eventually converges to a value of 1/3.

Mutation–Selection Balance

Thus far, we have looked at the consequences of selection without mutation, and mutation without selection. In practice, both processes operate at the same time—and their interaction is critically important to understanding why and at what frequency deleterious mutations remain present in populations. Using the same mathematical approach we have taken throughout this chapter, we can put both evolutionary processes together in a single model to get a picture of how natural selection can act to oppose deleterious mutations.

Suppose that we have two alleles, A_1 and A_2, at the A locus in a diploid population. Let A_1 be the normal or wild-type allele, and let A_2 be a deleterious recessive allele, such that the fitnesses for the three types A_1A_1, A_1A_2, and A_2A_2 are 1, 1, and $1 - s$, respectively. Further, let's assume that the A_1 allele mutates to the A_2 allele at rate μ. To keep the algebra simple, we assume that the rate of back mutation—mutation from A_2 back to A_1—is negligible. This is a reasonable assumption if we think of A_1 as coding for a functional form of a protein, while A_2 codes for a nonfunctional version. In such a case, there will be many ways to "break" the functional protein determined by A_1, so the mutation rate from functional to nonfunctional will be relatively high. By contrast, there will typically be only one way to fix any particular nonfunctional protein coded by A_2—namely, by reversing whatever specific change made it nonfunctional in the first place. As a result, the mutation rate from nonfunctional to functional will be relatively low.

In Box 7.8, we build our model in which selection and mutation operate in turn. When mutation and selection are the only two processes operating in our model, even though the A_1 allele is favored by selection, it will never be fixed in the population because A_2 alleles are continually being regenerated by the process of mutation, as illustrated in Figure 7.26. Eventually the population will reach a steady-state or equilibrium

FIGURE 7.26 **Mutation–selection balance.** Deleterious mutation transforms A_1 alleles (blue) into selectively disfavored A_2 alleles (pink). As indicated by the red Xs, natural selection eliminates some A_2 alleles from the population. At equilibrium, these two processes exactly balance one another.

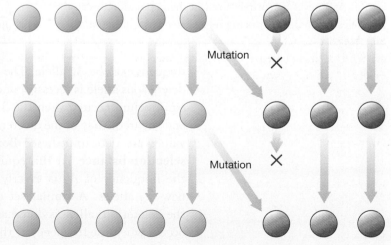

Mutation

Mutation

BOX 7.8 Mutation–Selection Balance for a Deleterious Recessive Allele

We have two alleles A_1 and A_2 at the A locus. A_2 is a deleterious recessive allele, such that the fitnesses for the three types A_1A_1, A_1A_2, and A_2A_2 are 1, 1, and $1 - s$, respectively. A_1 mutates to the A_2 allele at rate μ, and we assume that the rate of back mutation from A_2 to A_1 is negligible.

To build our model, we will assume that selection and mutation operate in turn. Suppose that we begin with allele frequencies p and q for the A_1 and A_2 alleles, respectively. First we write down the consequences of natural selection, as we did in Box 7.5. After natural selection operates, but before mutation, the frequency of the A_1 allele will be

$$p_{\text{after selection}} = \frac{p^2 + pq}{p^2 + 2pq + q^2(1 - s)}$$

$$= \frac{p(p + q)}{1 - q^2 s}$$

$$= \frac{p}{1 - q^2 s}$$

Then we allow mutation to operate

$$p_{\text{after mutation and selection}} = (1 - \mu)p_{\text{after selection}}$$

Mutation and selection are the only two processes operating in our model, so now we can write the expression for p', the frequency of the A_1 allele after one generation. We do so by simply applying the two formulae above:

$$p' = p_{\text{after mutation and selection}}$$

$$= (1 - \mu)p_{\text{after selection}}$$

$$= (1 - \mu)\frac{p}{1 - q^2 s}$$

In this model, even though the A_1 allele is favored by selection, it will never be entirely fixed in the population, because A_2 alleles are continually being regenerated through mutation.

But eventually the population will reach an equilibrium frequency of the A_1 allele. At this equilibrium, which we call the *mutation–selection balance*, the action of natural selection to decrease the frequency of A_2 is exactly balanced by the action of mutation to produce new A_2 alleles by mutation from A_1.

We can find the frequency p of the A_1 allele at the mutation–selection balance by recognizing that, at this equilibrium, the frequency p of the A_1 allele does not change from generation to generation. Thus, if we find the solution to the equation

$$p = (1 - \mu)\frac{p}{1 - q^2 s}$$

we will obtain the mutation–selection equilibrium.

Recognizing that $q = 1 - p$, our task is to solve the equation

$$p = (1 - \mu)\frac{p}{1 - (1 - p)^2 s}$$

To do so, we cancel the p terms on each side:

$$1 = (1 - \mu)\frac{1}{1 - (1 - p)^2 s}$$

We then multiply through by the denominator:

$$1 - (1 - p)^2 s = (1 - \mu)$$

We next subtract 1 from each side and then multiply each side by $-1/s$:

$$(1 - p)^2 = \frac{\mu}{s}$$

Now solving for p, we get the pair of solutions $p = 1 \pm \sqrt{\mu/s}$. Of these, only $p = 1 - \sqrt{\mu/s}$ is in the range [0,1] that is necessary for an allele frequency. This is the mutation–selection balance; the equilibrium frequency of the favored allele A_1 is $1 - \sqrt{\mu/s}$ and the frequency of the deleterious recessive allele A_2 is $\sqrt{\mu/s}$.

frequency of the A_1 allele. The value of that steady state depends on whether the deleterious allele is recessive or dominant. If the deleterious allele is recessive, that equilibrium occurs when the frequency of the wild-type allele is $p = 1 - \sqrt{\mu/s}$ and the frequency of the deleterious recessive allele is $q = \sqrt{\mu/s}$. To see how these values are determined, see Box 7.8. We call these frequencies the **mutation–selection balance**. At this equilibrium, the action of natural selection to increase the frequency of A_1 is exactly balanced by the action of mutation to produce new A_2 alleles. A similar set of calculations, not included here, reveals that if the deleterious allele is dominant instead of recessive, mutation-selection balance occurs when the frequency of the wild-type allele is approximately $p = (s - \mu)/s$ and the frequency of the deleterious dominant allele is approximately $q = \mu/s$.

Familial Adenomatous Polyposis Is Maintained by Mutation–Selection Balance

Many genetic diseases of humans have negative fitness consequences. Why haven't the alleles responsible for such diseases been eliminated from the population by selection? One answer is that some genetic diseases persist in mutation–selection balance. Familial adenomatous polyposis (FAP) is a common genetically inherited disorder, affecting approximately 1 in 8000 individuals (Bisgaard et al. 1994). Patients with FAP have large numbers of polyps that form in the colon. Left untreated, some of these initially benign polyps progress to malignant cancerous states when the patient is 35–40 years old, leading to cancer of the colon and other organs (Figure 7.27). As a result, untreated patients suffer a shortened life expectancy.

One of the major causes of FAP is the occurrence of mutations to the *APC* tumor suppressor gene (Bodmer 1999). This locus, which is approximately 9000 base pairs in length, offers numerous mutational targets, any of which are sufficient to induce the FAP condition. These are dominant mutations; individuals with a single copy of a disease allele progress to disease. To remain consistent with our notation throughout this chapter, we will refer to the normal (nondisease) *APC* allele as A_1, and to any of the disease-causing alleles as A_2.

Population genetic analysis can inform our understanding of the causes and consequences of FAP. To learn more about the mutation process that generates FAP and the fitness consequences of the disease, Marie Bisgaard and her colleagues studied 154 individuals listed in an exhaustive Danish case registry (Bisgaard et al. 1994). In order to estimate the penetrance of the disease—the probability that an individual with a disease-causing allele develops the disease—they looked at the disease status of the offspring of the registered cases, and they found that the penetrance of FAP is near 100% at age 40.

To estimate the mutation rate μ from A_1 to A_2, the researchers used pedigree data to estimate that 39 of the 154 FAP cases in the registry were due to new, rather than inherited, mutations. During the same period, there were approximately 2,000,000 total surviving births in Denmark—and thus $2 \times 2,000,000$ new *APC* gene copies in the population. This allowed them to estimate the mutation rate from A_1 to A_2 to be 39/4,000,000—that is, approximately $\mu = 10^{-5}$.

By looking at the number of surviving offspring of parents with FAP, Bisgaard and her colleagues were also able to estimate the fitness consequences and thus the selection coefficient s. The 154 affected individuals in the registry had a total

FIGURE 7.27 The development and progression of familial adenomatous polyposis. (A) A nonsense mutation occurs in the germline *APC* gene. In this particular example, we show a section of the *APC* gene. A single base deletion at position 41 creates a premature stop codon (TGA) and thus codes for a defective protein. **(B)** Polyps develop in the large intestine around the time of adolescence. These polyps are initially benign rather than cancerous. **(C)** The polyps progress to a cancerous stage, usually when the patient is 35–40 years of age. Colon cancer develops as a result. **(D)** Cancer cells metastasize to the liver and other organs.

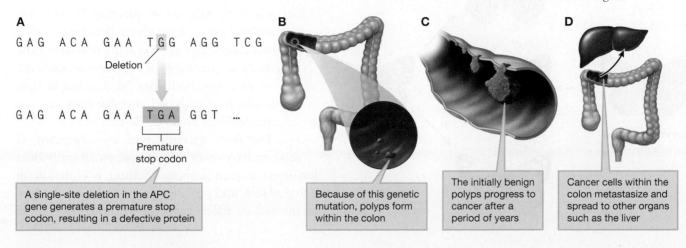

of 297 children by the end of their lives or reproductive years; a normal cohort of 154 in Denmark at the same time would be expected to have 340 children. This leads to an estimated fitness of $297/340 = 0.87$. In other words, FAP imposed a selective cost of $s = 0.13$.

Now that we have estimates of the mutation rate μ and the selection coefficient s, we can check these by computing the expected equilibrium frequencies of the normal and disease-causing alleles A_1 and A_2 under mutation–selection balance. The *APC* variant causing FAP disease is a dominant allele, so its equilibrium frequency will be approximately $q = \mu/s$. Thus, we expect the frequency of the A_2 allele at mutation–selection balance to be approximately $q = 10^{-5}/0.13 = 1/13,000$. The frequency of A_1A_2 heterozygotes in the population would then be approximately $2pq = 1/6500$. This is a close match to the observed frequency of disease in the population.

7.5 Nonrandom Mating

One of the Hardy–Weinberg assumptions is that individuals choose their mates randomly with respect to their own genotypes. All of the mathematical models we have developed in this chapter thus far have made this assumption as well. But it can easily be violated, as we saw in the case of the flat snails that could only mate if the shells of both partners coil in the same direction. If individuals tend to mate with those of the same genotype or phenotype, we call this **assortative mating** (Figure 7.28). When individuals tend to mate with those of different genotypes or phenotypes, we call this **disassortative mating**. Here we consider examples of each in turn.

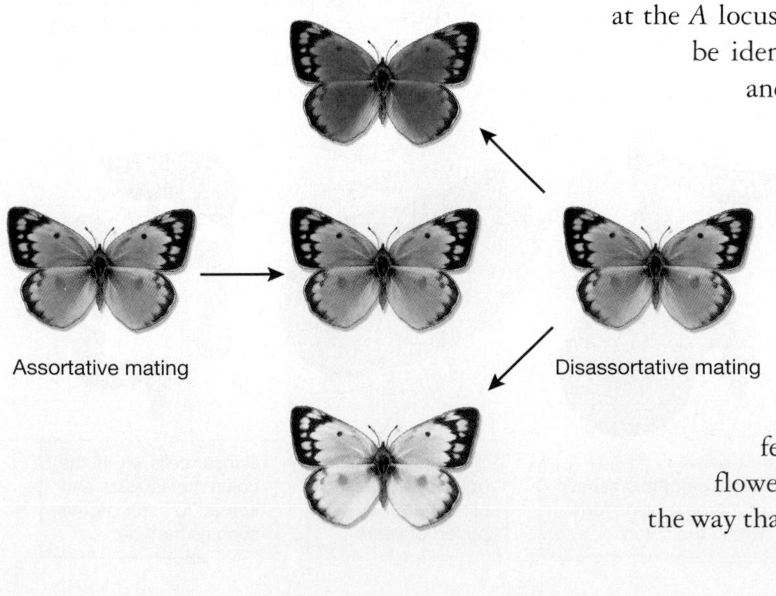

FIGURE 7.28 Assortative and disassortative mating. In assortative mating, like mates with like. In disassortative mating, individuals mate with phenotypes different from their own.

Assortative mating

Disassortative mating

Inbreeding

Inbreeding, in which individuals mate with genetic relatives, is one very common type of assortative mating. Inbreeding is assortative because in an inbred population, gametes are not paired at random, but instead they are preferentially paired with gametes from close relatives. Thus, in an inbred population, a pair of gene copies at the A locus may be **identical by descent**—that is, they may be identical because of shared descent through a recent ancestor (Figure 7.29). It is important to recognize that when we talk about identity by descent, we are making a claim about two gene copies' history, not their genetic sequence. Two gene copies can be the same in terms of sequence (for example, both may be A_1), but if they do not have a shared ancestor they are not considered identical by descent.

The most extreme type of inbreeding is self-fertilization or selfing—when an individual fertilizes its own gametes. Selfing is common in flowering plants, and provides a convenient example of the way that inbreeding affects genotype frequencies. Suppose that we have a population with allele frequencies $A_1 = 0.8$ and $A_2 = 0.2$. If the population is initially in Hardy–Weinberg equilibrium at the A locus, the genotype frequencies for the A_1A_1, A_1A_2, and A_2A_2 genotypes will be 0.64, 0.32, and 0.04, respectively. Suppose that the population now begins to reproduce exclusively by selfing; what offspring genotypes will be produced? A_1A_1 individuals produce only A_1 gametes, and thus they will produce only A_1A_1 offspring. Similarly, A_2A_2 individuals will produce only A_2A_2 offspring. The heterozygotes A_1A_2, however, do not produce only heterozygote offspring. Because they produce both A_1 and A_2 gametes, which then pair at random, these heterozygotes will produce A_1A_1, A_1A_2, and A_2A_2 offspring in a 1:2:1 ratio. As a result, after one generation of selfing, we will have more A_1A_1 and A_2A_2 homozygotes in the population and fewer A_1A_2 heterozygotes (Figure 7.30).

Note that although genotype frequencies change, the allele frequencies remain constant over time. This is generally true of inbreeding in the absence of other evolutionary processes. Because inbreeding does not add or remove alleles from the population and does not differentially pass one allele or another into the next generation, inbreeding acting alone does not cause changes in allele frequencies.

The selfing example we just explored is an extreme form of inbreeding; in general, there is a broad continuum between selfing and purely random mating. In Box 7.9, we look at how population geneticists quantify this continuum using an approach known as F-statistics. The larger the value of the inbreeding coefficient F, the closer the population lies to the selfing end of the continuum.

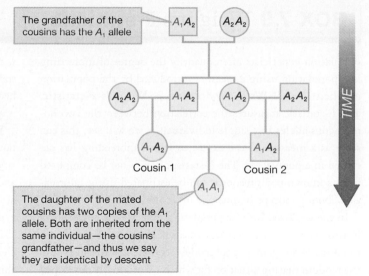

The grandfather of the cousins has the A_1 allele

The daughter of the mated cousins has two copies of the A_1 allele. Both are inherited from the same individual—the cousins' grandfather—and thus we say they are identical by descent

FIGURE 7.29 Identity by descent. Identity by descent is illustrated in a pedigree representing a mating between two full cousins. Each of the two full cousins has received an identical copy of the A_1 allele from their grandfather. Each passes this allele on to the daughter at the bottom of the figure. She therefore has two copies of the A_1 allele that are identical by descent.

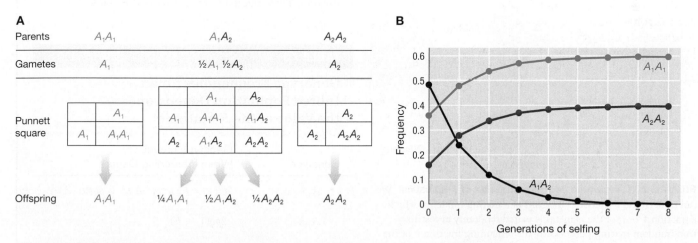

FIGURE 7.30 Reproduction by selfing. (A) A_1A_1 and A_2A_2 parents produce only A_1A_1 and A_2A_2 offspring, respectively, whereas A_1A_2 parents produce all three types of offspring. As a result, the number of homozygotes among the offspring increases and the number of heterozygotes decreases. **(B)** Selfing rapidly eliminates heterozygotes. Selfing over eight consecutive generations results in a population with very few heterozygotes. We begin with a Hardy–Weinberg population and track the frequencies of each genotype. Initially, allele frequencies of A_1 and A_2 alleles are $p = 0.6$ and $q = 0.4$, respectively. After eight generations of selfing, $f[A_1A_1]$ is nearly 0.6, $f[A_1A_2]$ is very close to 0, and $f[A_2A_2] = 0.4$. Genotype frequencies have changed dramatically, but allele frequencies have not changed; p remains 0.6 and q remains 0.4.

BOX 7.9 Wright's *F*-statistic

Population geneticists often quantify the degree of inbreeding in a population using a measure introduced by the population geneticist Sewall Wright and known as **Wright's *F*-statistic**. The *F*-statistic measures the correlation between the two homologous alleles in a single individual; as we will see, this can serve as a measure of the extent to which inbreeding has occurred in a population. The *F*-statistic can either be computed from a known pedigree, or it can be estimated using information about genotype frequencies in a population.

In the idealized infinite populations that we are considering in this chapter, the easiest way to think about the *F*-statistic is to return to our gamete pool model (Section 7.2). Recall that in the random mating situation for which we originally developed the model, it is as if all of the gametes in the population were mixed together in one large gamete pool and then drawn out in pairs at random. In an inbred population, each gamete instead has some chance of being paired with a gamete that is identical by descent and some chance of being matched at random. This suggests a modification to our gamete pool model. Imagine that instead of forming one big gamete pool, the gametes are divided into two separate gamete pools. A fraction $(1 - F)$ goes into a *random mating pool* in which they are paired at random and a fraction F goes into an *inbred pool* in which they are paired

with another gamete that is identical by descent. Thus, the off-spring from the random mating pool may be homozygotes or heterozygotes, but the offspring from the inbred pool are always homozygotes (Figure 7.31).

We can view the results of inbreeding as if they were obtained from this model of two gamete pools. The *F*-statistic simply measures the fraction that goes into the inbred pool, and thus populations that are more inbred have higher values of *F*. The larger the value of *F*, the higher the fraction of homozygotes in the population.

Using the *F*-statistic, the expected genotype frequencies under inbreeding follow in straightforward fashion from this model of two gamete pools. Suppose the frequencies of the A_1 and A_2 alleles are p and q, respectively. A fraction F of the offspring are drawn from the inbred pool; all of these will be homozygotes with a fraction p of them being A_1A_1 and a fraction q being A_2A_2. In addition, some homozygotes are produced out of the random mating pool as well, when one A_1 happens to be paired with another A_1 by chance, or when one A_2 is paired with another A_2 by chance. Of the fraction $(1 - F)$ of the offspring derived from the random mating pool, p^2 of them will be A_1A_1 homozygotes, q^2 will be A_2A_2 homozygotes, and the remaining $2pq$ will be A_1A_2 heterozygotes. Thus, the entire offspring population will be composed of a fraction $p^2(1 - F) + pF$ of A_1A_1 homozygotes, a fraction $q^2(1 - F) + qF$ of A_2A_2 homozygotes, and a fraction $2pq(1 - F)$ of A_1A_2 heterozygotes. These fractions are summarized in Table 7.3.

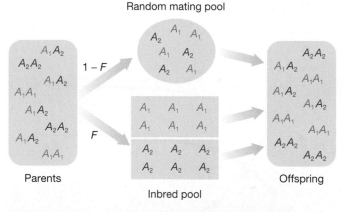

Random mating pool

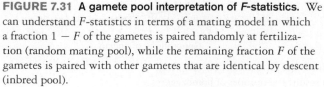

Parents / Inbred pool / Offspring

FIGURE 7.31 A gamete pool interpretation of *F*-statistics. We can understand *F*-statistics in terms of a mating model in which a fraction $1 - F$ of the gametes is paired randomly at fertilization (random mating pool), while the remaining fraction F of the gametes is paired with other gametes that are identical by descent (inbred pool).

TABLE 7.3

Genotype Frequencies for an Inbred Population Depend on the Value of the *F*-statistic

Genotype	Genotype Frequency When Inbreeding Occurs
A_1A_1	$p^2(1 - F) + pF = p^2 + pqF$
A_1A_2	$2pq(1 - F)$
A_2A_2	$q^2(1 - F) + qF = q^2 + pqF$

Inbreeding Depression

Inbreeding depression occurs when the offspring from matings between genetic relatives have reduced fitnesses. One of the most common types of inbreeding depression occurs when recessive alleles have deleterious consequences on fitness (Carr and Dudash 2003). Because inbreeding increases the frequency of homozygotes in an inbred population, recessive alleles are more likely to appear in homozygotes and thus to be expressed phenotypically. For example, Igor Rudan and his colleagues studied the extent to which inbreeding depression contributed to human hypertension. They examined the effects of inbreeding on blood pressure in 2760 individuals from 25 villages located on various Croatian islands in the Adriatic Sea. These researchers uncovered a strong positive relationship between the inbreeding coefficient F and blood pressure—as F increases across populations, so too does the average blood pressure of individuals within those populations (Rudan et al. 2003). Indeed, Rudan and his team estimated that 36% of all cases of high blood pressure (hypertension) in their sample population were caused by the effects of inbreeding (Figure 7.32). In Western societies, we tend not to worry much about such effects because of customs and laws that make inbreeding uncommon. On a global scale, however, it is estimated that 20% of all marriages involve a bride and groom who are genetic relatives (Bittles 1988; Bittles et al. 2001).

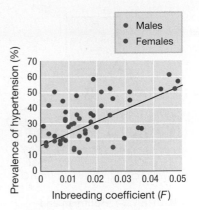

FIGURE 7.32 Inbreeding and high blood pressure. Rudan and his colleagues measured the relationship between the inbreeding coefficient F and high blood pressure (hypertension) from 2760 individuals on Croatian islands in the Adriatic Sea. Individuals with higher inbreeding coefficients were more likely to suffer from hypertension. Adapted from Rudan et al. (2003).

Disassortative Mating

Disassortative mating occurs when individuals tend to mate with partners that differ from themselves with respect to a given locus or trait. As a result, disassortative mating tends to generate an excess of heterozygotes relative to Hardy–Weinberg equilibrium.

One straightforward cause of disassortative mating is disassortative preference: a preference for individuals that differ from oneself. For example, evidence suggests that many mammals prefer mates that differ from themselves at the MHC loci—a highly polymorphic set of loci involved in the immune response. While the mechanisms responsible are only partially understood, studies on mice indicate that olfactory cues are used to discriminate among potential mates by MHC type.

Disassortative mating need not be driven by preferences alone, nor does it require that all individuals prefer mates that are different from themselves. Anne Houtman and Bruce Falls uncovered an interesting case in white-throated sparrows (*Zonotrichia albicollis*) (Houtman and Falls 1994) (Figure 7.33). White-throated sparrows have two differently colored morphs: a smaller, tan-striped form, and a larger, more aggressive, white-striped form.

These two morphs are controlled by a polymorphism involving a *chromosomal inversion*—a region of DNA that has reversed direction and often has a low recombination rate. White-striped birds are heterozygous for the inversion and tan-striped birds are homozygous for the normal chromosome.

In the wild, more than 90% of white-throated sparrows mate with individuals of the opposite morph. Houtman and Falls wanted to determine the mechanism responsible for this pattern of disassortative mating. To do so, they designed a series of mate preference experiments using captive birds (Figure 7.34). From these experiments, the researchers found that females of both morphs have a strong preference for tan-striped males, whereas males do not have a strong preference

FIGURE 7.33 Two color morphs of the white-throated sparrow. The white-throated sparrow has two distinct color morphs that differ in the brow coloration: white-striped (upper bird) and tan-striped (lower bird). More than 90% of matings occur between individuals of opposite morphs.

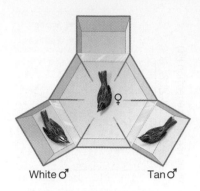

FIGURE 7.34 Houtman and Falls's experiment. Rivals of opposite morphs (males in this diagram) occupy two of the three chambers adjacent to the hexagonal arena that holds the subject (a female in this diagram). Dividers prevent the rivals from seeing one another, and prevent the subject from interacting with more than one of the rivals at a time. Adapted from Houtman and Falls (1994).

White ♂   Tan ♂

between the two female morphs. Thus, mate preferences by themselves seem unlikely to generate the observed patterns of disassortative mating.

If preferences alone are not driving the pattern of disassortative mating in the wild, what could be the source of this disassortative mating? To answer this, Houtman and Falls conducted competition experiments in which individuals were paired in a single cage with same-sex rivals of the opposite morph. They found that in males and females alike, white-striped individuals aggressively dominated tan-striped rivals.

Taken together with the preference observations, these results now suggest an explanation for the trend toward disassortative mating recorded in nature. White-striped females are able to dominate their tan-striped rivals, and thus they are able to mate with the desirable tan-striped males. Tan-striped females are unable to always obtain access to the tan-striped males, and they typically mate with white-striped mates. Thus, the majority of matings involve birds of opposite morphs.

7.6 Migration

While an idealized population may be isolated from the rest of the world, in practice populations are often linked to other populations by a flow of migrants between them. As a result, allele frequencies change within populations. When individuals immigrate into a population, they may bring new or previously uncommon alleles with them. When individuals emigrate from a population, allele frequencies may change as well, if the emigrants are more likely than members of the population at large to carry a particular allele. We can extend our models of gene frequency change to include the possibility of migration. Figure 7.35 shows a mainland–island model of migration, in which we have different allele frequencies on the mainland and on a small island. Migrants occasionally reach the island from the mainland. Migrants from the island may reach the mainland as well, but we will ignore this on the grounds that they will have a negligible effect on allele frequencies in the vastly larger mainland population.

We can develop a simple mathematical model to show what will happen to genotype and allele frequencies on the island as a result of migration. Suppose that initially the frequencies of the A_1 and A_2 alleles on the island are given by p_i and q_i, and the allele frequencies on the mainland are given by p_m and q_m. We assume that migration is the only violation of the Hardy–Weinberg assumptions—that is, we have no natural selection at the A locus, we have random mating, we have no mutation, and we have a very large population that is not subject to chance fluctuations (genetic drift); this latter assumption may be questionable for an island, but we maintain it here for simplicity. In the next chapter, we will look at what happens when genetic drift comes into play in a small island population.

Let k be the fraction of the island population made up by new migrants from the mainland. Then, after migration occurs, genotype frequencies on the island will be

$$f[A_1A_1] = (1 - k)p_i^2 + kp_m^2$$
$$f[A_1A_2] = 2(1 - k)p_iq_i + 2\,kp_mq_m$$
$$f[A_2A_2] = (1 - k)q_i^2 + kq_m^2$$

Unless allele frequencies were initially the same on the island and the mainland ($p_i = p_m$), the new frequencies on the island will not be Hardy–Weinberg proportions.

A

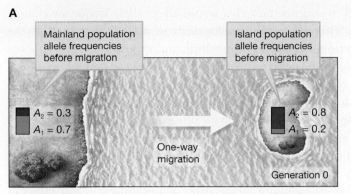

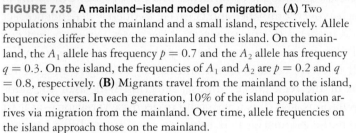

B

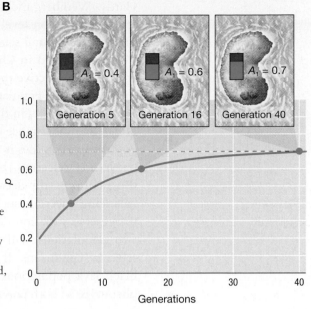

FIGURE 7.35 A mainland–island model of migration. (A) Two populations inhabit the mainland and a small island, respectively. Allele frequencies differ between the mainland and the island. On the mainland, the A_1 allele has frequency $p = 0.7$ and the A_2 allele has frequency $q = 0.3$. On the island, the frequencies of A_1 and A_2 are $p = 0.2$ and $q = 0.8$, respectively. **(B)** Migrants travel from the mainland to the island, but not vice versa. In each generation, 10% of the island population arrives via migration from the mainland. Over time, allele frequencies on the island approach those on the mainland.

What happens to allele frequencies as a consequence of migration? We can see the answer by calculating the *change* in allele frequencies on the island as a result of migration. Initially, before migration, the frequency of the A_1 allele on the island was p_i. The frequency of the A_1 after migration, call it p_i', will be

$$p_i' = (1 - k)p_i + kp_m$$

Thus, the net change in allele frequencies will be

$$\Delta p_i = p_i' - p_i = k(p_m - p_i)$$

From this expression, we can also calculate the equilibrium allele frequencies on the island if migration continues. By definition, at equilibrium, the allele frequencies no longer change, so we set $\Delta p_i = 0$ in our expression above. For nonzero migration ($k > 0$), this gives us the solution $p_i = p_m$. This means that the system reaches equilibrium only once the allele frequencies on the island and mainland are the same.

Figure 7.35B provides an illustrative example of how migration changes allele frequencies over time. Initially, the frequency of the A_1 allele on the mainland is $p_m = 0.7$ and on the island it is $p_i = 0.2$. Over the course of 40 generations, with 10% of the island population arriving by migration each generation, the frequency of the A_1 allele on the island has nearly reached its equilibrium value of 0.7, the allele frequency on the mainland.

7.7 Consequences on Variation within and between Populations

In this chapter, we used the Hardy–Weinberg equilibrium to illustrate evolutionary models of how allele frequencies change at the population level in the absence of natural selection. We then relaxed a number of assumptions of the

Hardy–Weinberg model to examine how this would affect allele frequency change at the population level. This led to the development of models that include the processes of natural selection, mutation, inbreeding, and migration.

As we learned in Chapter 3, genetic variation is the fuel for natural selection. It is thus instructive to consider what effect each of these processes has on genetic variation within populations and between populations. The Hardy–Weinberg model demonstrates that, in the absence of evolutionary processes, the ongoing process of Mendelian inheritance does not decrease (or increase) the amount of variation in a population, nor does it alter the amount of variation between populations.

Natural selection, by contrast, will tend to sort on standing variation, and thus will typically decrease the amount of variation in a population. But forms of balancing selection, such as overdominance and negative frequency dependence, act to preserve variation. Whether selection increases or decreases variation between two populations will depend on whether the populations experience similar selective conditions. If selective conditions are similar, as we might imagine for black lava populations of the rock pocket mouse, selection will favor the same phenotype in both populations, and thus it will tend to decrease variation—at least in phenotype—between the populations. If selective conditions differ between the environments, as we see when comparing rock pocket mice in light rock and dark lava populations, selection will tend to favor different phenotypes in each population, and it will tend to increase variation between the populations.

Mutation, as a source of new variation, will tend to increase the amount of variation within a population. Because different mutations may arise in different populations, mutation will also tend to increase variation between populations. In the absence of sexual selection (higher mating success for one genotype than another), nonrandom mating does not change allele frequencies on its own. Thus, it has comparably little effect on variation in this regard, although it is worthwhile to note that assortative mating and inbreeding tend to decrease the frequency of heterozygotes in a population. Yet, in doing so they may increase the average differences between individuals within a single population. Disassortative mating increases the fraction of heterozygotes within a population, and it may help stabilize polymorphism, as we saw in the example of the white-throated sparrows.

Migration will tend to bring new alleles into a population and, in this way, it will typically serve to increase the variation within a population. But, as we saw in the mainland–island model, the long-term effect of migration between populations is to equilibrate their allele frequencies—and thus migration decreases the variation between populations. Table 7.4 summarizes these conclusions about the effects of natural selection, mutation, nonrandom mating, and migration on variation within and between populations.

In all of the models we have considered in this chapter, we have assumed that evolution is occurring in very large populations, such that chance fluctuations have a negligible effect on allele or genotype frequencies. But evolution often operates in small populations that are subject to chance fluctuations, and these populations have their own evolutionary dynamics. We will explore these dynamics in the next chapter.

TABLE 7.4

Effects of Population-Genetic Processes

Evolutionary Process	Variation within Population	Variation between Populations
Natural selection	Decreases (except in cases of balancing selection)	Increases if selective conditions differ; decreases if conditions are the same
Mutation	Increases	Increases
Nonrandom mating	No effect on allele frequencies (in the absence of sexual selection)	No effect on allele frequencies (in the absence of sexual selection)
Migration	Increases	Decreases

SUMMARY

1. The field of population genetics provides a quantitative way of describing, modeling, and predicting how allele and genotype frequencies in populations change over time.

2. The Hardy–Weinberg model serves as a null model in population genetics, telling us what happens to allele frequencies and genotype frequencies when no evolutionary processes—natural selection, mutation, nonrandom mating, migration, and genetic drift—are operating.

3. When none of these five evolutionary processes are operating, the Hardy–Weinberg model makes three predictions: (a) allele frequencies will not change over time, (b) genotype frequencies will be the so-called Hardy–Weinberg equilibrium frequencies, and (c) a population with genotype frequencies away from Hardy–Weinberg equilibrium frequencies will return to these frequencies in a single generation.

4. For a locus with two alleles, A_1 and A_2, at frequencies p and q respectively, the Hardy–Weinberg genotype frequencies are as follows: $f[A_1A_1] = p^2, f[A_1A_2] = 2pq, f[A_2A_2] = q^2$.

5. Natural selection, mutation, nonrandom mating, and migration can each drive changes in genotype frequencies in a population.

6. Natural selection can take on various forms. Directional selection, overdominance, and underdominance are types of frequency-independent selection in which the fitness of a genotype is independent of its frequency in the population; these contrast with positive and negative frequency-dependent selection, in which the fitness of a genotype depends on the genotype frequencies in the population.

7. Mutation–selection balance can maintain deleterious alleles at low frequency in a population.

8. Assortative mating, in which individuals tend to mate with similar individuals, increases the frequency of homozygotes in a population; disassortative mating, in which individuals mate with dissimilar individuals, increases the frequency of heterozygotes.

9. Migration between populations brings their allele frequencies closer to one another.

10. The evolutionary processes considered in this chapter have diverse but predictable effects on variation within and between populations.

REVIEW QUESTIONS

1. Thomas keeps a very large number of socks unpaired in his sock drawer.

   a. A fraction p of them are black, and a fraction $q = 1 - p$ are blue. If Thomas were to pair up all of his socks in the dark so he could not distinguish them by color, what fraction of the pairs do you expect would be mismatches, with one blue and one black sock?

   b. Suppose that Thomas were to properly match each pair, wear them, wash them all (as singletons), and then rematch them again in the dark. Assuming that black and blue socks wear out or are lost at the same rate, assuming no new socks are added, and assuming that socks don't change color in the laundry, what fraction of the pairs will be mismatches this second time around? Explain how your answer relates to the time it takes to reach Hardy–Weinberg equilibrium.

2. Red-green color blindness is a recessive trait on the X chromosome: males with a single copy of the responsible allele and females with two copies display the trait. The frequency of red-green color blindness in males in the United States is approximately 7%. Knowing this, estimate the frequency of the trait in U.S. females. What assumptions did you have to make in order to make that estimate?

3. A biologist studies a genetic locus A, with alleles A_1 and A_2, in two adjacent populations of blue jays. She samples from each population, and finds the following genotype frequencies:

	A_1A_1	A_1A_2	A_2A_2
Population 1	0.09	0.42	0.49
Population 2	0.64	0.32	0.04

   a. Is the A locus at Hardy–Weinberg equilibrium frequencies in population 1? How about in population 2?

   b. She then combines all of her data from two populations to get the following genotype frequencies:

	A_1A_1	A_1A_2	A_2A_2
Pooled Data	0.365	0.37	0.265

This population is not at Hardy–Weinberg equilibrium. Why not? Explain which of the assumptions needed for Hardy–Weinberg equilibrium have been violated by combining the population data.

4. You observe that allele frequencies in a population are $f[A_1A_1] = 0.3$, $f[A_1A_2] = 0.2$, $f[A_2A_2] = 0.5$. How many different explanations can you think of for why this population may not be in Hardy–Weinberg equilibrium?

5. If all of the Hardy–Weinberg assumptions are met, allele frequencies stay constant and genotype frequencies are in Hardy–Weinberg proportions. Which of the Hardy–Weinberg assumptions, when violated, allows allele frequencies to change but leaves the genotype frequencies in Hardy–Weinberg proportions? Which of the assumptions, when violated, does not change allele frequencies but causes a deviation from Hardy–Weinberg proportions?

6. Suppose two alleles, A_1 and A_2, exhibit overdominance, with fitnesses $w_{11} = 0.8$, $w_{12} = 1.0$, and $w_{22} = 0.9$. Under random mating, we would expect to observe a balanced polymorphism between A_1 and A_2. What should we expect to observe if instead the population is strictly selfing? Explain why.

7. Two alleles at the A locus, A_1 and A_2, are under directional selection, with A_1 favored and A_2 disfavored. Each is currently at frequency 0.5 in the population. In which situation will A_1 be fixed more quickly: A_1 is dominant, or A_1 is recessive? Explain.

8. You hypothesize that two different varieties of a weedy plant species—call them A and B—are under positive frequency-dependent selection. Explain how you would design an experiment to test your hypothesis, and how you would interpret the results of this experiment.

9. In Box 7.5, we derived a model for how allele frequencies change because of natural selection when the favored allele A_1 is dominant to the alternative allele A_2. In particular, we found the change in the frequency of the A_1 allele from one generation to the next is equal to

$$\frac{pq^2s}{1 - q^2s}$$

Derive an analogous model for the case in which the favored allele A_1 is recessive. In this case, what is the expression for the change in frequency of the A_1 allele from one generation to the next?

10. In a given population, the wild-type A_1 allele is dominant to the recessive A_2 and A_3 alleles. Fitnesses of the A_1A_1, A_2A_2, and A_3A_3 genotypes are 1.0, 0.9, and 0.8, respectively. Mutation from A_1 to A_2 occurs at rate μ and mutation from A_1 to A_3 occurs at rate 4μ. Ignoring back mutation from A_2 and A_3 to A_1, which allele will be more common at mutation–selection balance, A_2 or A_3? Explain.

SUGGESTED READINGS

Bisgaard, M., K. Fenger, S. Bulow, E. Niebuhr, and J. Mohr. 1994. Familial adenomatous polyposis (FAP): frequency penetrance and mutation rate. *Human Mutation* 3: 121–125. A review of the FAP case we discussed.

Hardy, G. H. 1908. Mendelian proportions in a mixed population. *Science* 28: 49–50. A classic paper on what came to be known as Hardy–Weinberg equilibrium.

Houtman, A., and J. Falls. 1994. Negative assortative mating in the white–throated sparrow *Zonotrichia albicollis*: the role of mate choice and intrasexual competition. *Animal Behaviour* 48: 378–383. A behavioral overview of disassortative mating in white-throated sparrows.

Nachman, M. W. 2005. The genetic basis of adaptation: lessons from concealing coloration in pocket mice. *Genetica* 123: 125–136. A nice summary of the genetics of coat coloration.

Pauling L., H. A. Itano, S. J. Singer, and I. C. Wells. 1949. Sickle cell anemia, a molecular disease. *Science* 110: 543–548. A classic paper on the population genetics of sickle cell anemia.

(S) **Visit StudySpace at wwnorton.com/studyspace.**

8

Evolution in Finite Populations

I n the middle of the Irish Sea, between Britain and Ireland, lies a small island known as the Isle of Man. The island is home to an unusual breed of cat, the Manx, easily recognized by its shortened, or missing, tail (Figure 8.1). Manx cats have reportedly been found on this island for several hundred years. One local legend has it that they arrived in 1588 aboard a ship from the Spanish Armada that was wrecked on the sea cliffs at Spanish Head at the southwestern tip of the island.

An even more curious story for the origin of these cats appears in Joseph Train's 1845 history of the Isle of Man:

> My observations on the structure and habits of the specimen in my possession, leave little doubt on my mind of its being a...cross, between the female cat and the buck rabbit. In August, 1837, I procured a female [Manx] kitten, direct from the Island. Both in its appearance and habits it differs much from the common house cat: the head is smaller in proportion, and the body is short; a fud or brush like that of a rabbit, about an inch in length, extending from the lower vertebra, is the only indication it has of a tail. The hind legs are considerably longer than those of the common cat, and, in comparison with the

◀ A small group of chinstrap penguins (*Pygoscelis antarcticus*) climbs atop an Antarctic iceberg.

243

FIGURE 8.1 The Isle of Man, home to the Manx cat. (A) Map showing the location of the Isle of Man. **(B)** In this 1902 photograph of a Manx cat, we can see the long hind legs and the absence of a tail. **(C)** Joseph Train's illustration of a Manx cat in his 1845 *A Historical and Statistical Account of the Isle of Man.* Train incorrectly speculated that the breed resulted from a cross between a cat and a rabbit.

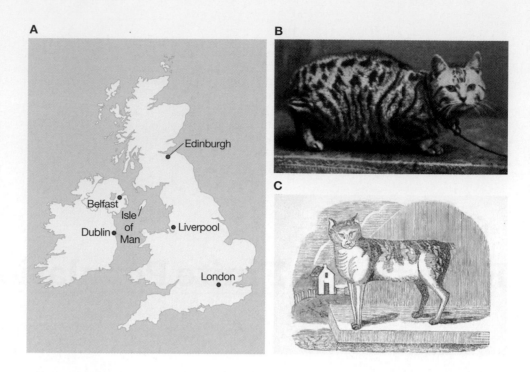

fore legs, bear a marked similarity in proportion to those of the rabbit. Like this animal too, when about to fight, it springs from the ground and strikes with its fore and hind feet at the same time. The common cat strikes only with its fore paws, standing on its hind legs. The [Manx] discharges its urine in a standing posture, like a rabbit, and can be carried by the ears apparently without pain. (Train 1845, p. 2)

But in actuality, the Manx cat is not a cat–rabbit hybrid, but rather an ordinary domestic cat carrying an unusual genetic mutation. The primary genetic determinant of the *Manx phenotype*, which includes both the reduced or absent tail and longer hind legs than forelegs, is a single autosomal locus M (for Manx). The M allele is dominant, conferring the Manx phenotype in Mm heterozygotes. In the MM homozygous form it is lethal, with most MM individuals aborted prenatally (Robinson 1993). Because homozygote lethality generates strong natural selection against the M allele, one might be surprised that the M allele, virtually unknown elsewhere, should have become common in the Isle of Man cat population. Indeed, this would be a surprising outcome in a very large population. But on a small island, natural selection is not the only process that influences allele frequencies. Allele frequencies can change because of random effects associated with low population size. Moreover, chance variation in the initial allele frequencies in a founding population may lead to dramatically different allele frequencies on an isolated island compared to on a mainland. This phenomenon, known as the *founder effect*, is most likely responsible for the prevalence of the M allele in the Manx cats on the Isle of Man. We will see how it works and consider additional examples of the founder effect in Section 8.3.

In this chapter we address the following questions:

- How does the process of evolution in small populations differ from what is seen in large populations?

- How does genetic drift work, and what are its consequences?

- How do gene copies spread through populations, and how do coalescent trees help us to understand this process?

- How do demographic processes such as population bottlenecks and the founder effect contribute to evolutionary change?

- What happens when genetic drift interacts with mutation and selection?

- What does the neutral theory of molecular evolution predict about the nature of genetic variation, and to what degree is the neutral theory supported by contemporary evidence?

8.1 Random Change and Genetic Drift

In the previous chapter, we developed simple mathematical models of how gene frequencies change with and without the action of natural selection. In those simple models of evolution, we assumed that populations were large—so large, in fact, that in every generation the law of large numbers applied to changes in gene frequencies. The law of large numbers states that, as the size of a random sample increases, the realized frequencies—those frequencies that we actually observe— usually will be very close to the expected frequencies. But when sample sizes are small, the realized frequencies will not always be close to the expected frequencies.

By way of illustration, suppose that you tossed a fair coin 1000 times. At the end of the coin tosses, the odds are that you would observe something very close to a 1:1 ratio of heads to tails. If instead you only tossed your coin 10 times, you might get 5 heads and 5 tails for another 1:1 ratio. But over 75% of the time, you'd get some other combination: 4 heads and 6 tails, or 6 heads and 4 tails, or 3 heads and 7 tails, and so on. *In experiments with small sample sizes, realized frequencies are not always very close to the expected frequencies.* The same thing happens in populations. In very large populations, the realized genotype frequencies will be very close to the expected genotype frequencies. For this reason, in the previous chapter, we assumed that the genotype frequencies of the offspring were always exactly those expected, given the genotype frequencies and the relative fitnesses of the parents.

In a small population, the realized genotype frequencies often may deviate substantially from the expected genotype frequencies for any number of reasons. By chance, in any given generation, some mating pairs may form more or less often than expected; certain genotypes may produce more or fewer offspring than expected; other genotypes may survive more or less often than expected. All of these factors will make it less likely that the actual genotype frequencies in our population will match the expected frequencies.

If we want to think about evolution in small populations in a quantitative way, we need a model of evolution in such populations. The **Wright–Fisher model**, named after its creators Sewall Wright (1889–1988) and Ronald Fisher (1890–1962), is one of the simplest such models, and it is used widely in population genetics. Loosely speaking, the basic Wright–Fisher model is a small-population version of the Hardy–Weinberg model that we developed in the previous chapter. Because we will make reference to Wright–Fisher populations throughout this chapter, we present the model in further detail in Box 8.1.

BOX 8.1 The Wright–Fisher Model

In Chapter 7, we examined the Hardy–Weinberg model, which provides an idealized picture of how genotype frequencies change over time in a very large population. Using the Hardy–Weinberg model, we were able to see what happens in a large population in the absence of such evolutionary processes as selection, migration, mutation, and nonrandom mating. We also saw how to relax some of the assumptions of the Hardy–Weinberg model in order to study the evolutionary consequences of selection and other processes.

The Wright–Fisher model can be seen as a counterpart to the Hardy–Weinberg model, *for small populations* (Figure 8.2). Again, it provides us with a baseline for how genotype frequencies are expected to change over time in the absence of selection, migration, mutation, and nonrandom mating. As in the Hardy–Weinberg model, the Wright–Fisher model assumes a population of diploid sexual organisms that reproduce in discrete nonoverlapping generations. As in the Hardy–Weinberg model, the most basic form of the Wright–Fisher model assumes:

1. Natural selection is *not* operating on the trait or traits affected by the locus in question.

2. Mating in the population is random with respect to the locus in question.

3. No mutation is occurring.

4. There is no migration into the population from other populations.

And just as with the Hardy–Weinberg model, with the Wright–Fisher model we can relax each of these assumptions to see how various evolutionary processes affect genotype frequencies over time. But unlike in the Hardy–Weinberg model, in the Wright–Fisher model we assume that the population size is small instead of very large. In doing so, we take account of chance events that influence allele frequencies in a small population.

The basic idea behind the Wright–Fisher model is to consider a population of N diploid organisms, each of which produces a large number of gametes that go into a common pool. Because the gamete pool is very large, allele frequencies in the gamete pool exactly reflect those in the parental generation. But then we draw $2N$ gametes at random from this pool. As a result of random chance, allele frequencies in this small sample of $2N$ gametes may not be exactly the same as the frequencies in the large gamete pool. This is where the model differs from the Hardy–Weinberg model. These gametes are then paired up at random to produce N new diploid offspring for the next generation. Figure 8.2 shows an example with $N = 10$. There, the frequency of the A_1 allele is 0.5 in the parental generation, but 15 of the 20 gametes drawn from the gamete pool happen to carry the A_1 allele, so the frequency of the A_1 allele in the offspring generation is now 0.75.

Because we have only one gamete pool, instead of having separate pools for gametes from male parents and gametes from female parents, this version of the Wright–Fisher model is sometimes described as modeling a *hermaphroditic* or *monoecious* species, such as many flowering plants, in which each parent produces both male and female gametes. A model with two separate sexes in each generation and two separate gamete pools for eggs and sperm, although somewhat more complicated, is conceptually similar and has similar mathematical properties.

Most of the theoretical results presented in this chapter will be based on the Wright–Fisher model of population genetics. This will allow us to explore how drift interacts with other evolutionary processes.

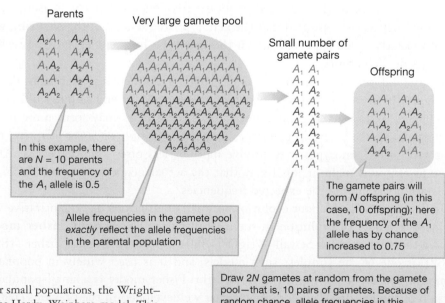

FIGURE 8.2 The Wright–Fisher model. For small populations, the Wright–Fisher model can be seen as a counterpart to the Hardy–Weinberg model. This model assumes a population of diploid sexual organisms that reproduce in discrete nonoverlapping generations.

Genetic drift is the process of random fluctuation in allele frequencies due to sampling effects in finite populations. There are three general consequences of genetic drift:

1. In a finite population, allele frequencies fluctuate over time, even in the absence of natural selection.

2. Some alleles are fixed, others are lost, and the fraction of heterozygotes in the population decreases over time.

3. Separate populations diverge in their allele frequencies and in terms of which alleles are present.

We will consider these three points in turn.

Genetic Drift Causes Allele Frequencies to Fluctuate over Time

The fundamental effect of genetic drift is to cause fluctuations in allele frequencies in a population, even in the absence of natural selection or other evolutionary processes. The *rate* at which allele frequencies fluctuate because of drift depends on the size of the population. Drift acts more powerfully in small populations than in large populations, and thus drift causes larger allele frequency fluctuations in small populations.

Figure 8.3 illustrates the result of genetic drift in populations of size 10, 100, and 1000 individuals. All three populations start with two alleles, A_1 and A_2, each at a frequency of 50%. These alleles are **selectively neutral**—that is, there is no fitness difference between them. As a result, natural selection does not act on the frequencies of these alleles. But because of genetic drift, allele frequencies change nevertheless. Over time, random fluctuations lead to rapid changes in the allele frequencies in the smallest population, modest changes in allele frequency in the intermediate population, and small changes in allele frequency in the large population.

As a result of genetic drift, one particular allele may reach a frequency of 100% in a given population, while the other alleles at that locus are lost. Recall from Chapter 7 that, when this happens, we say that the remaining allele has been fixed, or reached fixation, in the population. We see this happening in Figure 8.3 for all of the populations of size 10, but not for any of the populations of size 100 or size 1000. But if we were to run the experiment longer and longer, in the absence of mutation and migration, each and every finite-sized population, no matter how large, would *eventually* become fixed for one or the other of the two alleles.

Genetic drift is a random process. Therefore, while it is certain that some allele will eventually be fixed in each population in this model, it is not certain *which* allele will become fixed in which population. In some of the populations plotted in Figure 8.3, the A_1 allele is fixed; in others, the A_2 allele is fixed. It turns out that,

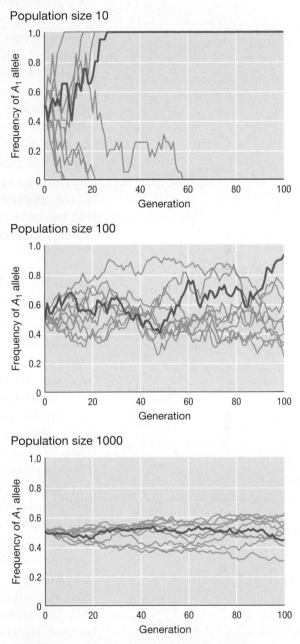

FIGURE 8.3 Genetic drift is stronger in smaller populations. The three graphs show simulations of genetic drift in diploid populations of size 10, 100, and 1000, respectively, each starting with the A_1 and A_2 alleles at equal frequency. Each graph shows 10 different runs of the simulation, with one highlighted in blue for visibility. In each case, drift causes allele frequencies to fluctuate over time, but the fluctuations are far more dramatic in the smaller populations. In each population of size 10, one allele or the other goes to fixation—a frequency of 1.0—within 100 generations.

at a given time, the probability that an allele at a neutral locus will eventually be fixed is equal to the frequency of that allele in the population at that time.

The easiest way to see this is to recognize that in a finite population, sooner or later every allele is either fixed or lost because of drift. Thus, in a population of N diploid individuals, there are $2N$ gene copies at any given locus. If the locus is neutral, each of these $2N$ gene copies is equally likely to be fixed, and so an allele that is present in only a single copy has a $1/2N$ chance of being fixed. If instead there are k copies of a given allele, each of these copies has a $1/2N$ chance of being fixed, for a total probability $k/2N$ that this particular allele is fixed.

Why are the random fluctuations that result from genetic drift important in the evolutionary process? For one thing, they cause allele frequencies to shift—and thus cause evolutionary change—in the absence of natural selection. As we saw in Chapter 3, natural selection lacks foresight, and so evolution might get stuck at a locally optimum but globally suboptimal phenotype if natural selection were the only process operating. But drift can also cause shifts in allele frequencies, even in the opposite direction of what would be favored by natural selection. Drift also has important effects on the amount of variation present in populations, and on divergence between populations. We will consider these consequences below.

Genetic Drift Causes Heterozygosity to Decrease within a Population over Time

Another important consequence of drift is that it tends to reduce variation within populations. There are at least two different ways to see this intuitively. First, we could simply notice that, in the absence of natural selection, genetic drift causes alleles to go to fixation in a finite population over evolutionary time. When alleles are fixed, variation is lost. Second, we could think about finite population size as a sort of inbreeding because in a finite population, there is a nonzero chance that individuals mate with genetic relatives. As we learned in Chapter 7, inbreeding leads to the loss of genetic variation. And as we illustrate below, we can measure this loss of variation in the wild.

Population geneticists often use quantities known as *observed heterozygosity* and *expected heterozygosity* to measure the amount of variation in a population. The **observed heterozygosity**, H_o, at a given locus is defined as the fraction of individuals in the population that are heterozygous at the given locus. For example, suppose we have three alleles, A_1, A_2, and A_3, at the A locus. If the genotype frequencies of the three homozygotes are $f[A_1A_1] = 0.2$, $f[A_2A_2] = 0.2$, and $f[A_3A_3] = 0.1$, the remaining fraction (0.5) of the individuals in the population will be heterozygotes, and the observed heterozygosity will be $H_o = 0.5$. In general, the observed heterozygosity is 1 minus the frequency of homozygotes in the population, expressed as

$$H_o = 1 - \sum_{i=1}^{n} f[A_iA_i]$$

The **expected heterozygosity** (H_e) is the fraction of heterozygotes expected under the Hardy–Weinberg model, given the allele frequencies in the population. According to the Hardy–Weinberg model (Chapter 7), if the frequency of the ith allele is p_i, the fraction of homozygotes for allele i will be p_i^2. Thus, the expected frequency of the heterozygotes will be

$$H_e = 1 - \sum_{i=1}^{n} p_i^2$$

Expected heterozygosity is often easier to measure than observed heterozygosity, especially if there are many alleles at the locus in question, because one does not need to know the frequencies of all genotypes, only the frequencies of all alleles.

In a Wright–Fisher population, expected heterozygosity decreases by an average factor of $1/2N$ in each generation (Box 8.2). When N is very large, $1/2N$ is very small, and we see little decrease in heterozygosity due to drift. When N is small, however, $1/2N$ is relatively large, and we see substantial loss of heterozygosity due to drift. Looking back at our initial simulations of genetic drift in Figure 8.3, we can see this happening. In the small populations, allele frequencies rapidly diverge from 0.5 (where heterozygosity is maximal) and eventually reach fixation or loss (where heterozygosity is zero).

We see the same thing in natural populations. Where human activities such as overfishing reduce population size, they may have evolutionary consequences as well as ecological ones—that is, they may contribute to genetic drift. To see whether this had occurred in a heavily exploited New Zealand snapper fishery in Tasman Bay, Lorenz Hauser and his colleagues looked at DNA sequences from snapper scales collected at this fishery over the period 1950–1986 and from fresh samples from 1998 (Hauser et al. 2002). Heavy commercial fishing began in this area in 1950, so the earliest samples reflect levels of heterozygosity prior to fishing, whereas the later samples reveal heterozygosity levels after extensive commercial fishing. Hauser and his colleagues reasoned that if commercial fishing were causing genetic drift in this snapper population, they should see a decline in heterozygosity over time, as one consequence of drift is to reduce population heterozygosity. Figure 8.4 shows their findings. At the set of genetic loci that they sequenced, the expected heterozygosity H_e in this fishery showed a statistically significant decline over the period 1950–1998.

From their results, the authors concluded that genetic drift was operating strongly in the population. This result might be somewhat surprising, given that this fishery is estimated to contain at least 3 million individuals—a population so large that drift might be expected to have only minimal effects. The reason that drift could nonetheless have had such a big impact is because, as is common in pelagic fish—that is, fish that live in open water areas—relatively few individuals in each generation produce most of the offspring in the next generation.

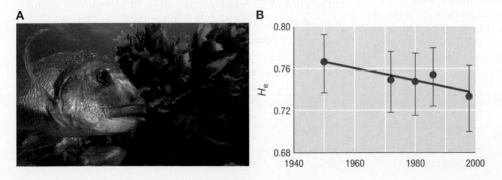

A

B

FIGURE 8.4 Loss of heterozygosity. **(A)** Genetic drift is a likely explanation for the loss of expected heterozygosity over time in an overfished population of New Zealand snapper (*Pagrus auratus*) in Tasman Bay, New Zealand. **(B)** The graph plots the expected heterozygosity of the New Zealand snapper over time. Part B from Hauser et al. (2002).

BOX 8.2 Quantifying the Effects of Genetic Drift on Variation

Wright's *F*-statistic, which we introduced in Box 7.9, provides an alternative to H_o and H_e for measuring the effects of drift on variation in a population. Recall that *F* quantifies the correlation between the two gene copies at a locus. We can think of *F* as the probability that the two gene copies at a locus in a single individual are identical by descent.

The conceptual difference between the heterozygosity approach and the *F*-statistic approach is that the former quantifies allelic similarity, whereas the latter focuses on the probability of identity by descent and thus on history irrespective of allelic state. Recall that two gene copies can be the same in terms of genetic sequence (for example, both may be the A_1 allele), but if they did not come from a shared ancestor they are not considered identical by descent. The *F*-statistic approach provides an elegant mathematical formulation of how drift reduces variation over time.

In an idealized population of infinite size with random mating, all parents will be unrelated, and therefore the two gene copies at a locus in any individual will never be identical by descent. But in a finite population, things work differently.

To see how this process increases the probability of identity by descent and thus the value of *F*, a thought experiment is helpful. Imagine that, at some arbitrary time in the past, we define all gene copies in the population as distinct—that

is, not identical by descent—irrespective of their genetic sequence (Figure 8.5). At this time, the probability that any two gene copies in a newly formed offspring are identical by descent is zero. By definition, $F = 0$ at this point. As time proceeds forward, however, some of the gene copies in the population will be lost by drift. Of those that are not lost, many will be present in multiple copies. Some of the gene copies present in multiple copies will end up paired in offspring of the next generation. In those individuals, the two alleles at our given locus will be identical by descent. The value of *F* in the population will now be greater than zero.

Using the gamete pool approach that we first presented in Chapter 7, we can derive a mathematical expression for how *F* changes over time in a finite-sized Wright–Fisher population. (Here, as in Box 8.1, we consider the case where a single individual can produce both types of gametes necessary for fertilization; this greatly simplifies the derivation and closely approximates what happens with two sexes.) Imagine that each parent contributes a large number of gametes to the common gamete pool, as shown in Figure 8.6. Offspring are then formed by drawing pairs of gametes at random from the gamete pool.

Suppose that the value of *F* in a parental population of size *N* is F_{parental}. There are 2*N* different sources of gametes—namely, each of the 2*N* gene copies in the parental generation. Therefore, with probability 1/2*N*, the two gene copies in an offspring will come from the same gene copy in the parental generation. In this case, the probability of identity by descent is 1. With probability $1 - 1/2N$, the two gene copies in an offspring will correspond to two different gene copies in the parental generation. In this case, the probability of identity by descent is F_{parental}. Putting these two cases together, the overall probability of identity by descent is

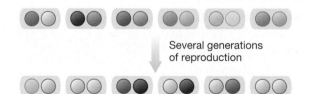

Several generations of reproduction

FIGURE 8.5 Genetic drift increases the probability of identity by descent over time. Diploid individuals (shown by blue shaded boxes) in a population each have two gene copies (indicated by colored circles) at a given locus. Initially, we label all gene copies as distinct—here indicated by color in the top row—irrespective of their allelic state. After many generations, some of the gene copies have left no descendants, while others have left multiple descent copies. Thus, some of the individuals pictured in the bottom row—the second and the sixth from left in this illustration—have gene copies that are identical by descent.

$$F_{\text{offspring}} = \frac{1}{2N} + \left(1 - \frac{1}{2N}\right)F_{\text{parental}} \qquad (8.1)$$

The value of *F* is always in the range [0,1] inclusive, so this equation ensures that $F_{\text{offspring}}$ will be greater than or equal to F_{parental} (with equality only in the case when $F_{\text{parental}} = 1$). This derivation shows that *F* will increase over time in a finite population (Hartl and Clark 2007). If *F* is equal to zero at time 0

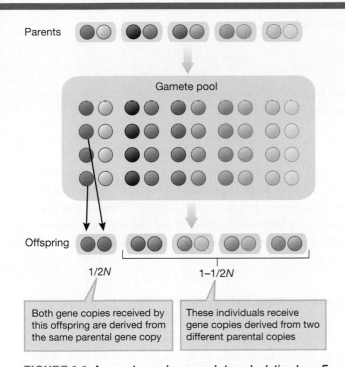

Parents

Gamete pool

Offspring

$1/2N$ $1-1/2N$

Both gene copies received by this offspring are derived from the same parental gene copy

These individuals receive gene copies derived from two different parental copies

FIGURE 8.6 A gamete-pool approach to calculating how F changes over time in a population. With probability $1/2N$, both gene copies in an offspring derive from the same gene copy in a parent, and thus they are identical by descent with probability 1. With probability $1 - 1/2N$, the gene copies in an offspring derive from two different gene copies in the parent, and thus they are identical by descent with the same probability as were gene copies in the parental generation.

(that is, $F_0 = 0$), by applying Equation 8.1 repeatedly, we find that

$$F_t = 1 - \left(1 - \frac{1}{2N}\right)^t \tag{8.2}$$

Using this expression, together with the relationship between F and H_e that we learned in the previous chapter, we can quantify the effect of drift on expected heterozygosity in a finite population.

In a Hardy–Weinberg population (that is, infinite size with no inbreeding) Wright's F-statistic is $F = 0$. There the expected fraction of heterozygotes in the population will be $2pq$ while

the expected fraction of homozygotes will be $1 - 2pq$. If instead $F > 0$, the expected fraction of heterozygotes will be

$$H_e = 2pq(1 - F) \tag{8.3}$$

We can now compare H_e values for a parental generation (call it $H_{parental}$) with H_e values for the offspring generation (call it $H_{offspring}$). Using Equation (8.1), we find that

$$\frac{H_{offspring}}{H_{parental}} = \frac{2p_{offspring}q_{offspring}(1 - F_{offspring})}{2p_{parental}q_{parental}(1 - F_{parental})}$$

Yet, we know that the expected values of p and q do not change from the parental generation to the offspring because of drift alone, so these cancel in the expression above and we can write

$$\frac{H_{offspring}}{H_{parental}} = \frac{1 - F_{offspring}}{1 - F_{parental}}$$

Rearranging Equation 8.1 for how F changes over time, we get

$$1 - F_{offspring} = \left(1 - \frac{1}{2N}\right)(1 - F_{parental})$$

and therefore

$$\frac{H_{offspring}}{H_{parental}} = \left(1 - \frac{1}{2N}\right)$$

or equivalently

$$H_{offspring} = \left(1 - \frac{1}{2N}\right)H_{parental}$$

The expected heterozygosity decreases by a factor of $1/2N$ each generation because of drift in a finite population.

It is important to recognize that, although drift causes heterozygosity to decrease *on average*, heterozygosity can increase in *particular* instances. Sometimes, drift may increase the frequency of a rare allele in a population and thus increase heterozygosity, at least for a while. But if we were to look at the effects of drift on 1000 independent populations, for example, we would see that drift reduces heterozygosity more often than drift increases it.

BOX 8.3 Effective Population Size

In populations in the real world, genetic drift does not proceed exactly as we would expect in an idealized Wright–Fisher population. The actual or "census" population size—the number of individuals we can count—will vary from generation to generation, and this influences the rate of drift (Wright 1931, 1938, 1969). In addition, individuals in real populations contribute unequally to future generations, due to differential reproductive success, differential mortality, or other factors. To account for these differences in the rate of drift, population geneticists commonly use the concept of *effective population size* as a tool with which to understand how key population parameters, such as expected heterozygosity (H_e) or Wright's F-statistic, change over time. Here we will concentrate on the most commonly used of these statistics, the inbreeding effective population size (N_e), which we use to quantify change in the value of Wright's F-statistic.

In a Wright–Fisher population, the rate at which F changes due to drift is given by Equation 8.1 in Box 8.2, which is

$$F_{offspring} = \frac{1}{2N} + \left(1 - \frac{1}{2N}\right)F_{parental}$$

In an actual population, drift may operate differently for a number of reasons, and thus F may change at a different rate. Using the statistic for inbreeding effective population size, we can quantify how drift causes F to change in a non-Wright–Fisher population. *The inbreeding effective population size N_e is defined as the size of a Wright–Fisher population that would undergo an equivalent change in the value of F.* The value of N_e is defined by the equation

$$F_{offspring} = \frac{1}{2N_e} + \left(1 - \frac{1}{2N_e}\right)F_{parental} \qquad (8.4)$$

This is simply Equation 8.1, with N replaced by N_e. Using a bit of algebra, we can rearrange Equation 8.4 into a direct expression for N_e, as shown by the equation

$$N_e = \frac{1 - F_{parental}}{2(F_{offspring} - F_{parental})}$$

When we start with an outbred population ($F_{parental} = 0$), this expression further simplifies to

$$N_e = \frac{1}{2F_{offspring}}$$

In order to understand how drift operates in populations that do not meet all of the assumptions of the Wright–Fisher model, population geneticists have a set of formulas that can be used to approximate the effective population size of various non-Wright–Fisher populations. Below we consider two such examples.

Fluctuating Population Size

Suppose we have a population that fluctuates in size from generation to generation, with N_1 individuals in the first generation, N_2 in the second, N_3 in the third, and so on, and N_m in the mth generation. What is its effective population size over these m generations? It turns out that the effective population size is closely approximated by what is known as the *harmonic mean* of the population sizes in each generation:

$$N_e \approx \frac{m}{\dfrac{1}{N_1} + \dfrac{1}{N_2} + \cdots + \dfrac{1}{N_m}} \qquad (8.5)$$

The harmonic mean heavily weights the smallest values, so that the harmonic mean of a set of numbers is typically much closer to the smallest value than to the arithmetic mean or average of those numbers. As a result, effective population size will

---

Thus, despite the large absolute population size, the population experienced rates of drift that might be expected in a population of fewer than 200 individuals. We therefore say that, although the census population is large, on the order of 3 million fish, the **effective population size** of the fishery is very small, probably fewer than 200 fish. In Box 8.3, we explore the concept of effective population size further, and we consider the sorts of demographic considerations that can cause the effective population size to be substantially less than the census population size.

be greatly diminished when a population spends even small amounts of time at low population numbers.

For example, suppose that over a 100-year period an annual population spends 95 years at size 100,000 and 5 years at size 50. Its effective population size is then given by the equation

$$N_e \approx \frac{100}{\dfrac{1}{100,000} + \dfrac{1}{100,000} + \cdots + \dfrac{1}{50} + \dfrac{1}{50}}$$

$$= \frac{100}{\dfrac{95}{100,000} + \dfrac{5}{50}} \approx 991$$

The effective population size over the 100 years, 991, is much closer to the smallest population sizes experienced than to the largest ones, even though the years with small population size are relatively rare. We will explore this effect further in Section 8.3, when we discuss population bottlenecks.

Uneven Sex Ratio

Fluctuating population size is not the only factor that influences effective population size. If the members of a population contribute unequally to future generations (and hence to the subsequent genetic variability in those future generations), effective population size is reduced. This happens whenever a population features an uneven sex ratio. In a sexually reproducing species, if we let N_m equal the number of reproductive males in a population, and let N_f equal the number of reproductive females, the effective population size is approximately

$$N_e \approx \frac{4N_m N_f}{N_m + N_f}$$

For example, loggerhead turtles (*Caretta caretta*) commonly exhibit a strong sex ratio skew, with many more females than males present in adult populations (Freedberg and Wade 2001) (Figure 8.7). Suppose that we have a breeding population of 10,000 loggerhead turtles, of which 8000 are female and 2000 are male. While the total population size is 10,000, the effective population size is

$$N_e \approx \frac{4 \times 2000 \times 8000}{2000 + 8000} - 6400$$

Thus, the skewed sex ratio reduces the effective population size of these turtles to less than two-thirds of the actual population size. This means that drift will act more strongly, and heterozygosity will be lost more quickly, in this population than it would be in a population of the same size with an even sex ratio.

Overall, effective population sizes tend to be substantially smaller than census population sizes. In a wide-ranging meta-analysis of nearly 200 studies of effective population size, Richard Frankham found that, across a range of taxa, effective population size N_e averages only one-tenth of the census population size N, and that it can drop considerably lower in some species such as marine invertebrates (Frankham 1995).

FIGURE 8.7 Skewed sex ratio. Loggerhead turtles (*Caretta caretta*) exhibit skewed sex ratios. Females make up more than 80% of many loggerhead populations and this reduces the effective population size.

Genetic Drift Causes Divergence between Populations over Time

To get a better understanding of how drift affects populations, it can be useful to look at more than one population at a time. Let us begin with a thought experiment, and then we can move to an empirical example.

A Thought Experiment

Imagine that we have an archipelago of small islands, each able to maintain a constant-size population of 10 diploid individuals (Christiansen 2008). Moreover, suppose that each island is spaced far enough from the others that there is no migration between islands. Also assume that there is no natural selection, mutation, or assortative mating. Thus, drift is the only evolutionary process in operation (Figure 8.8).

Suppose that we seed each island with 10 A_1A_2 heterozygotes, so that each island receives 10 copies of the A_1 gene and 10 copies of the A_2 gene. Because

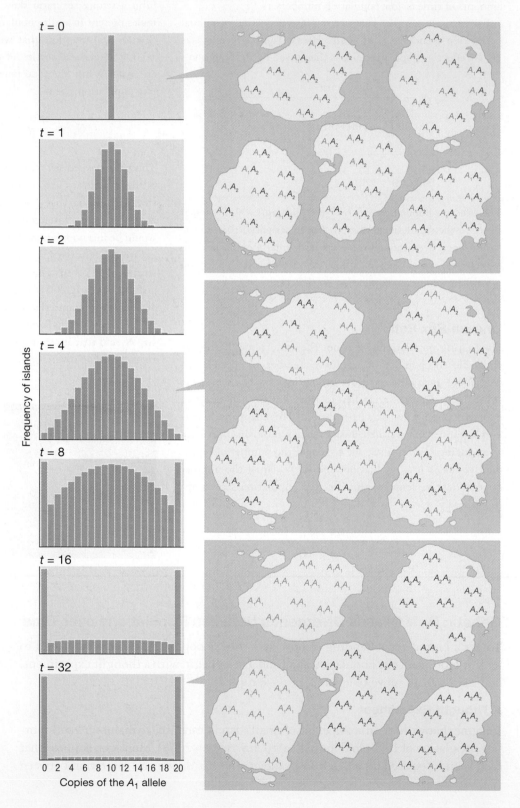

FIGURE 8.8 Genetic drift in island populations. A thought experiment illustrates how drift leads to divergence between populations. We envision a large number of islands, each with 10 diploid inhabitants. At time 0, the islands are founded by A_1A_2 heterozygotes at the neutral A locus. The inhabitants then mate randomly, and there is no mutation or migration. The bar graphs at the left show the frequency of *islands* with 0, 1, 2, and so on, copies of the A_1 allele at times $t = 0$ through $t = 32$. Over time, most islands become fixed either for the A_1 allele or for the A_2 allele. At the right are shown a group of five islands from the $t = 0$, $t = 4$, and $t = 32$ distributions shown at the left.

genetic drift is a random process, we know that different things will happen on different islands. On some islands, the A_1 allele will eventually become fixed; on others, the A_2 allele will eventually become fixed. On some islands, fixation will occur quickly; on others, it will take a long time to reach fixation.

Instead of looking at the frequencies of *different types of individuals* within a population, here we are focusing on the frequencies of *different types of populations*. The bar graphs in Figure 8.8 show the frequency of islands that have populations with 0, 1, 2, and so on, copies of the A_1 allele, in the original founding population ($t = 0$), and then the expected frequencies at subsequent times ($t = 1$, $t = 2$, $t = 4$, $t = 8$, $t = 16$, $t = 32$) under the Wright–Fisher model we outlined in Box 8.1. For example, at $t = 1$, about 17% of the islands have populations with unchanged allele frequencies—10 copies of the A_1 allele and 10 copies of the A_2 allele—but on all of the other islands, the frequency of the A_1 allele has already drifted away from 0.5. As time goes on, drift continues. By $t = 8$, an appreciable number of islands have populations that have already fixed either the A_1 allele or the A_2 allele. By $t = 32$, few of the islands have populations that remain polymorphic.

From this example, we see that genetic drift leads to divergence—differences in allele frequencies and ultimately the fixation of different alleles—among the populations on the islands in our hypothetical archipelago. In the next subsection, we will see that something similar happens in real archipelagos.

Drift and Divergence in the Galápagos Archipelago

Galápagos lava lizards, *Microlophus albemarlensis*, are moderately sized (17–25 cm in length) insectivorous lizards that inhabit dry rocky areas of numerous Galápagos islands (Figure 8.9). They are thought to disperse between islands only rarely, and they form a set of independent populations on the large island of Santa Cruz and its surrounding islets.

FIGURE 8.9 Galápagos lava lizard. Genetic drift and divergence have been studied in the Galápagos lava lizards, *Microlophus albemarlensis*.

These lizard populations have not always been separate, however (Figure 8.10). During much of the Late Pleistocene—as recently as 12,000 years ago—large volumes of water were trapped in kilometer-thick glacial ice sheets covering northern North America and Eurasia. As a result, sea levels around the world were substantially lower than at present. During this period, Isla Santa Cruz was connected to many surrounding islands and islets by land, and overseas distances to the other islands were considerably smaller. At that time, populations of lava lizards presumably were able to mix more readily. Once the glaciers receded and sea levels rose to present levels, the populations were separated, and migration between populations was eliminated or severely curtailed.

This leaves us with a situation very similar to that of the hypothetical archipelago that we studied in our previous thought experiment. To explore the consequences of genetic drift on these recently separated populations, Mark Jordan and Howard Snell assessed the genetic diversity of 17 populations by sequencing 11 different *microsatellite* markers in a sample of individuals from these populations (Jordan et al. 2002; Jordan and Snell 2008). Microsatellites are short stretches of DNA sequence in which a brief sequence—for example, CAG—is repeated several times. Microsatellites tend to make very good genetic markers for studying relatively short periods of evolutionary time. First, they are typically selectively neutral. Second, they tend to be highly variable in length

FIGURE 8.10 Geographic changes on Isla Santa Cruz. Around 12,000 years ago, sea levels around the Galápagos islands were 60 meters lower than at present and around 17,000 years ago, sea levels were approximately 130 meters lower than at present. At that time, Isla Santa Cruz was connected by land to many of the islets that now surround it. The 17 lava lizard populations that Jordan and Snell sampled are labeled on the map. Adapted from Jordan and Snell (2008).

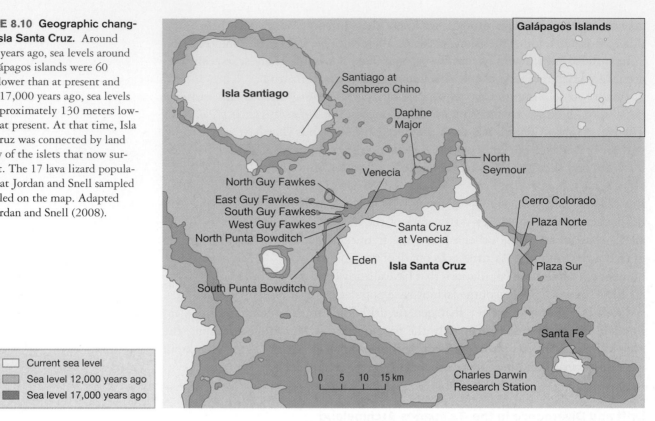

because copy number changes readily by something known as *slippage-induced mutation* (Figure 8.11).

Jordon and Snell reasoned that in the absence of gene flow between populations, genetic drift should strongly influence the patterns of diversity at these microsatellite loci. This allowed them to make a number of predictions. First, drift is expected to operate more strongly—and cause the loss of more variation—in smaller populations. Thus, the smaller lizard populations on smaller islands would be expected to have fewer microsatellite alleles than would larger populations on larger islands. As illustrated in Figure 8.12, this is exactly what Jordan and Snell found.

Jordon and Snell also found strong evidence of genetic drift in the patterns of genetic divergence *between* lizard populations on the various islands, with different islands revealing very different alleles and allele frequencies. Here we see *population subdivision*, in which there is limited or no gene flow between subpopulations of a larger population, along with genetic drift leading to divergence among subpopulations of the lava lizards on the Galápagos.

In both our thought experiment and our example from the Galápagos, we looked at genetic drift and differentiation on the islands of archipelagos. Island populations of terrestrial species make convenient systems for studying drift, because gene flow between populations on different islands is kept to a minimum. It is important to stress that genetic drift occurs not only on islands, but in every population. Moreover, population subdivision can occur without physical barriers as obvious as those imposed by the stretches of open ocean between islands. More subtle geographic barriers, or even behavioral differences, can likewise restrict gene flow and thus create population subdivision, leading to accelerated genetic drift and possible divergence among subpopulations.

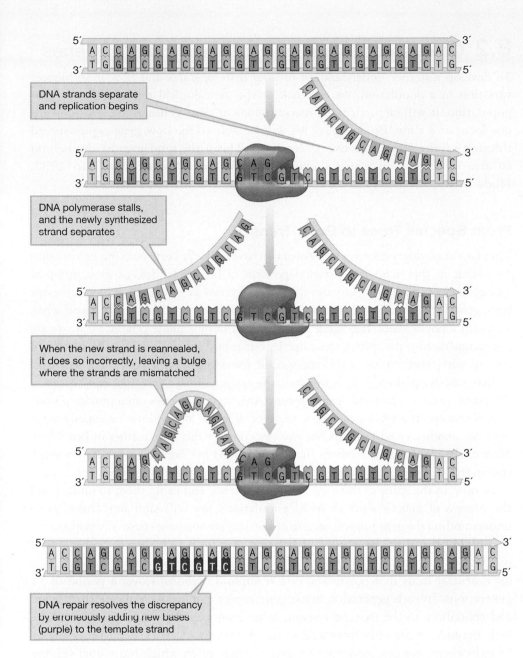

DNA strands separate and replication begins

DNA polymerase stalls, and the newly synthesized strand separates

When the new strand is reannealed, it does so incorrectly, leaving a bulge where the strands are mismatched

DNA repair resolves the discrepancy by erroneously adding new bases (purple) to the template strand

FIGURE 8.11 Slippage-induced mutation increases repeat copy number. During DNA replication, the DNA polymerase stalls. The newly synthesized strand slips and is incorrectly reannealed to the template strand. DNA repair mechanisms mistakenly add new additional bases to the template strand to fix this discrepancy. As a result, the number of repeats increases.

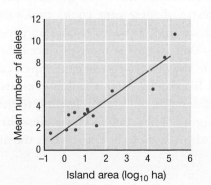

FIGURE 8.12 Lizard populations on smaller islands have lower diversity in microsatellite alleles than do populations on larger islands. Here we plot the area of the island for each sample population (horizontal axis) against the mean number of alleles per microsatellite locus (vertical axis). The former serves as a measure of population size; the latter as a measure of diversity. The statistically significant relationship between island size and genetic diversity, indicated by the solid line, suggests that genetic drift has been operating more strongly in smaller populations, as predicted. From Jordan and Snell (2008).

8.2 Coalescent Theory and the Genealogy of Genes

To develop a deeper understanding of how drift operates and how it influences variation in a population, we can look at the genealogical relationships in that population. It will be particularly useful to look at these genealogical relationships one locus at a time. By doing so, we will be able to see how gene copies spread through a finite population over generations. This is the fundamental idea behind an area of population genetics known as **coalescent theory** (Kingman 1982; Hudson 1990, Wakeley 2008).

From Species Trees to Gene Trees

Thus far, the phylogenetic trees we have drawn have typically been *species trees* or *population trees*—that is, they represent historical patterns of branching descent for a group of species or populations. We can also draw trees known as gene trees, which represent these genealogical relationships for a single locus. This is not such a new concept; when we build a phylogenetic tree using sequence data from a single genetic locus, we are not reconstructing the species tree directly, but rather we are inferring the pattern of descent with modification at this one specific locus. Such a phylogeny *is* a gene tree, in that strictly speaking it tells us about the history of that gene, not the history of the populations in which that gene appears. Although gene trees often provide a good approximation for a species tree, gene trees for different loci will not necessarily agree with one another, or with the species tree (we consider this issue further in Box 14.3). Most of the phylogenetic methods that we examined in Chapter 5 work by finding a species tree that is most consistent with the various gene trees for multiple loci.

So now, in the spirit of thinking about gene trees, and using them to understand the process of genetic drift in small populations, we will shift our attention to understanding the genealogical pattern of ancestry among gene copies in a population of diploid organisms. By way of illustration, Figure 8.13A shows a genealogical diagram—a depiction of which gene copy derived from which ancestral copy—for a neutral locus in a population of five diploid organisms over a period of 11 generations. In each generation, some gene copies manage to replicate themselves and contribute to the next generation; other gene copies fail to replicate and are lost. Because we are only interested in the genealogy of genes, not the genealogy of individuals, we can ignore which gene copies are in which individual (Figure 8.13B) and then "untangle" the genealogical graph to provide a clean picture, with no crossing lines, as in Figure 8.13C.

Suppose we are interested in the genealogical relationships among some set of gene copies in the present population. If we know the genealogical graph for the population, we can trace the ancestry of these gene copies backward in time, as illustrated in Figure 8.14A. What we find, as we trace back in time from the present, is that gene copies coalesce—that is, two or more distinct gene copies at some time point are all descended from the same ancestral gene copy. For example, in Figure 8.14A, gene copies ii and iii coalesce after a single generation. Three generations later, their lineage coalesces with the lineage leading to gene copy i, as indicated by the red circle in the figure. This circle is the **coalescent point** for gene copies i, ii, and iii—in other words, it is the gene copy that is the most recent common ancestor of i, ii, and iii.

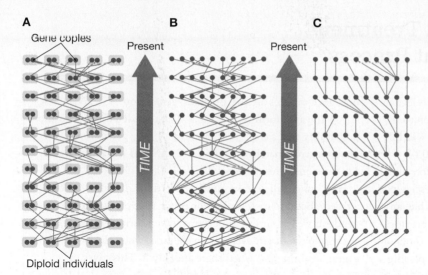

Gene copies

A   B   Present   C   Present

TIME   TIME

Diploid individuals

FIGURE 8.13 Gene genealogies for a diploid population. (A) This figure shows a simulated genealogy of gene copies (blue circles) at a neutral locus, in a population of five diploid individuals (five shaded boxes) over 11 generations. Orange lines indicate ancestry. Yet, even for this small a population and this short a time period, the graph is complex and difficult to interpret, with many crossing lines. **(B)** Because we are only interested in the gene genealogy and not in the diploid individuals, we can ignore the identity of the individuals in which each gene copy resides. **(C)** If we do not require that gene copies in the same diploid individual be placed adjacent to one another in the diagram, we can "unscramble" the graph, generating a genealogical diagram with no crossing lines. This form, which is much easier to interpret at a glance, summarizes the genealogical relationships among the gene copies present in the population. Adapted from Felsenstein (2004).

We can also look at the coalescent process for the entire population. Figure 8.14B shows what happens as we trace back in time from *all* of the gene copies in the population at the present. We have to go back further, but eventually we reach a coalescent point, indicated by the green circle in the figure, for these as well. This coalescent point is the gene copy that is the common ancestor to all gene copies in the population at the present time.

Furthermore, notice that by tracing the genealogy backward, we have created a tree structure; this *coalescent tree* shows the branching pattern of relatedness among the gene copies in the population.

Dynamics of the Coalescent Process

One of the major advantages of taking a coalescent approach is that this way of thinking is particularly amenable to mathematical treatment. The basic idea in mathematically modeling the coalescent process is to think of a genealogy as a stochastic process running *backward in time*. Suppose that we sample k gene copies from a population of N diploid individuals. At the present, which we will call time t, these k gene copies are all distinct. Now imagine that we take a step backward to time $t - 1$, and look at the previous generation. With some probability, any two or more of our k gene copies may come from the same gene copy at $t - 1$. If that occurs, we call it a coalescent event. It turns out that, for a neutral locus, we can write down an elegant mathematical model of this process. This model tells us the distribution of times until coalescence and also the distribution of gene tree topologies that arise at a neutral locus. We explore this model in Box 8.4.

For a neutral locus in a diploid Wright–Fisher population of size N, the average time to coalescence for any randomly chosen *pair* of gene copies turns out to be $2N$ generations (Hudson 1990). For a larger group of gene copies, the average time to coalescence of all of these copies is approximately $4N$ generations.

In the coalescent process for a neutral locus, much of the action happens "early"— that is, only shortly before the present. Thus, most of the coalescent events between pairs of gene copies are expected to occur early on. We see this in Figure 8.15,

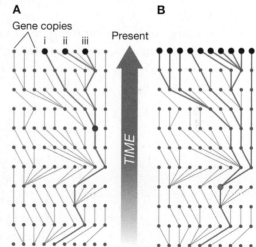

Gene copies

i   ii   iii

A   Present   B

TIME

FIGURE 8.14 Tracing back the ancestry of specific gene copies. (A) The genealogical history of the three highlighted gene copies are indicated. The three gene copies are all derived from the single gene copy four generations back (red circle). We say that these gene copies *coalesce* at the red circle. **(B)** The genealogical history of all gene copies in the population at the present time is traced back. In this case, all of the gene copies in the population are derived from a single gene copy seven generations back (green circle).

BOX 8.4 A Mathematical Treatment of the Coalescent Process

Following Kingman (1982), we can write down an elegant mathematical model that provides a close approximation to the neutral coalescent process we have discussed here. We will follow Felsenstein's simplified derivation (Felsenstein 2004).

Consider k gene copies in a much larger population of N diploid individuals. Each of the k gene copies is descended from a random ancestral gene copy, so the chance that any particular pair share a common ancestor in the previous generation is $1/2N$. But our k gene copies form a total of $k(k-1)/2$ different pairs, ignoring the order of the pairing. If we assume that N is large and that $k \ll N$, the chance that more than two gene copies come from the same copy in the previous generation is very small, as is the probability that more than one pair will coalesce at the same time. Thus, the probability that a coalescent event occurs in a single generation is approximately $k(k-1)/4N$. The waiting time until the first coalescent event is then approximately geometrically distributed with rate $k(k-1)/4N$ and average waiting time $4N/[k(k-1)]$.

After the first coalescent event occurs, there are now $k-1$ distinct lineages. Again the probability that any pair of these lineages coalesces in the previous generation is $1/2N$. These $k-1$ lineages form $(k-1)(k-2)/2$ unordered pairs, so the probability that a coalescent event in a single generation occurs is now approximately $(k-1)(k-2)/4N$ and the average waiting time until this occurs is $4N/(k-1)(k-2)$.

How long will it take until all of the k lineages have coalesced? Because each coalescent event is approximately independent, we can simply sum the average waiting times for each successive coalescent event, from the first, when there are k lineages, until the last, when there are only 2. This gives us

$$\sum_{i=2}^{k} \frac{4N}{i(i-1)} = 4N\left(1 - \frac{1}{k}\right)$$

This equation provides us with the results described in the text. When k is relatively large, this quantity is closely approximated by $4N$; hence, the average coalescent time for k gene copies in a large population is approximately $4N$. The final coalescent event occurs between two lineages that can be paired in only one way. In each generation, there is $1/2N$ probability that they will coalesce. Thus, the expected time for the last event to occur is $2N$, fully half of the total coalescent time for all k lineages.

which shows five different simulated coalescent trees for 20 gene copies at a neutral locus. In each of the five trees, the large majority of coalescent events occur very early, fewer than N generations into the process. In fact, the expected time for the population to coalesce down to just two parental lineages is only $2N$ generations. But the final coalescent event typically takes a very long time. Even once we are down to two lineages, it takes on average another $2N$ generations for the final two lineages to coalesce.

It is important to recognize that these results about coalescent times refer to expected times or averages; there is substantial variation around the mean. As a

FIGURE 8.15 Coalescent trees vary in shape. Here are five simulated coalescent trees for a sample of $k = 20$ gene copies in a population of 100 diploid individuals. Adapted from Wolfram Demonstrations Project (2011).

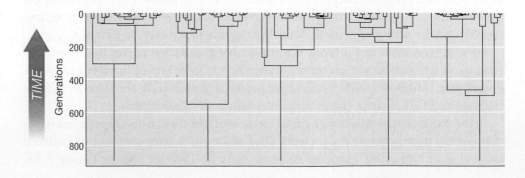

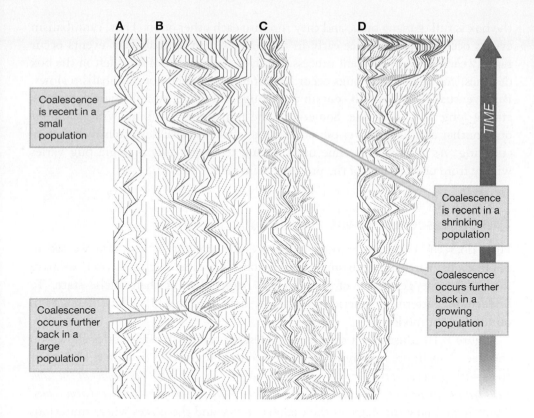

Coalescence is recent in a small population

Coalescence is recent in a shrinking population

Coalescence occurs further back in a growing population

Coalescence occurs further back in a large population

TIME

FIGURE 8.16 The effect of demography on coalescence. Gene genealogies, with the coalescent tree highlighted, in **(A)** a small population of constant size, **(B)** a large population of constant size, **(C)** a declining population, and **(D)** an expanding population.

result, different loci in the same population may have very different coalescent times. We see this in Figure 8.15 as well. Although all five trees result from simulating the same random process, the time until coalescence varies from less than 200 generations to more than 500 generations.

Coalescent times depend strongly on the demography of a population. In populations of constant size, we have seen that the coalescent time of any pair of alleles is $2N$ and the average coalescent time of a sample of k alleles is approximately $4N$. Therefore, in a small population with small N, coalescence will take less time to occur than it will in a large population with large N (Figure 8.16).

Bugs in a Box

How can we develop an intuitive understanding of these results? Coalescent trees can be hard to think about because it is not easy to envision a process running backward in time. To get around this difficulty, population geneticist Joe Felsenstein has proposed a delightful metaphor for thinking about the coalescent as a stochastic process that runs forward in time. Felsenstein envisions a box full of voracious and cannibalistic bugs (Figure 8.17). The bugs wander around the box at random; any time two bugs encounter one other, one eats the other. The process continues until the box contains only a single surviving bug. Mathematically, Felsenstein's bugs-in-a-box metaphor is identical to the coalescent process for a neutral locus, but with time running forward instead of backward. In Felsenstein's metaphor, the bugs represent gene copies. When one bug eats another, this represents a coalescent event. When only one bug is left in the box, the entire population has coalesced.

Thinking about what would happen in a box of bugs like this, we can get an intuitive feel for many of the results we have observed for the coalescent. Early on,

FIGURE 8.17 Bugs in a box. The coalescent process is mathematically analogous to a process in which hungry bugs run around inside a box and one eats another any time two meet.

the box is full of many bugs, and they run into each other often. Thus, cannibalism events occur at a rapid pace early in the process, just as coalescent events occur rapidly early in the coalescent process. Later, as the number of bugs left in the box declines, contacts among bugs occur less often, and the rate of cannibalism slows. But eventually the box will contain three bugs, then two, and ultimately, perhaps after a long wait, only one. Sometimes the remaining two bugs will encounter one another after a short period; other times they will wander extensively before colliding. As a result, the time until we are left with only a single bug varies widely from one instance of the process to the next.

The Coalescent Process and Genetic Variation

How does the coalescent process influence the amount of variation we see in populations, particularly in small populations? Thus far in this section, we have focused on the genealogy of gene copies, irrespective of their allelic state. To understand patterns of genetic variation, we now need to add allelic differences to our coalescent model.

Figure 8.18 illustrates a simulated coalescent tree with allele states shown on the tree. This figure highlights the fundamental observation that links coalescent trees with genetic variation: *Any allelic differences among a set of gene copies at the same locus must have arisen by mutation subsequent to the coalescent point for this set of gene copies.* Thus, if we know the shape of the coalescent tree and the places where mutations arose after the coalescent point, we know everything about the variation in the present population.

The structure of coalescent trees in a population tells us a great deal about the amount of variation we should expect to see. If all of the gene copies in a population coalesce only 7 generations back, then any variation present in the population must have arisen by mutation some time in the past 7 generations. If instead the population does not coalesce until 70 generations back, there will have been much more time for variation to arise by mutation. With all else equal, the deeper the coalescent point, the more variation we expect to see in the population. We will illustrate this by exploring what the coalescent tree at a neutral locus tells us about the process of genetic drift.

FIGURE 8.18 A coalescent tree with allelic states shown. Mutations generate new alleles, shown in orange and red. Notice that all of the variation at this locus has arisen subsequent to the coalescent point.

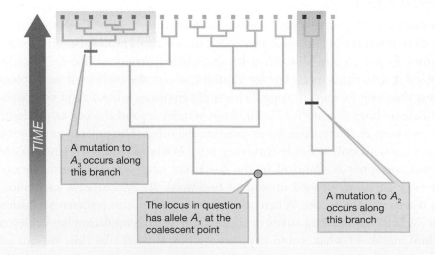

TIME

A mutation to A_3 occurs along this branch

The locus in question has allele A_1 at the coalescent point

A mutation to A_2 occurs along this branch

A Randomness in the shape of the coalescent tree

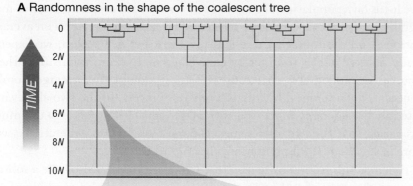

B Randomness in the location of mutation on a given tree

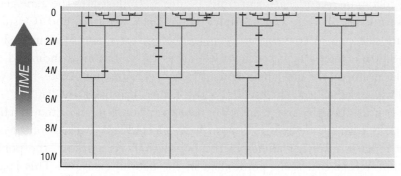

FIGURE 8.19 Separating genealogy and mutation. The distribution of variation at a neutral locus depends on two separate processes: **(A)** the random process by which the shape of the coalescent tree is determined, and **(B)** the random process of mutation events (shown by the red bars) along the branches of this coalescent tree. Notice that in a neutral model the locations of the mutations have no effect on the shape of the tree, which is determined simply by the demographic history of the population. Adapted from Wolfram Demonstrations Project (2011).

The coalescent process is particularly elegant for a neutral locus. For such loci, we can separate the genealogical history of the locus from the mutational process that takes place at that locus (Hudson 1990; Nordborg 2007). Thus, we can think of the process by which variation arises at the locus as the result of two separate processes: (1) the genealogical process by which a coalescent tree is formed, and (2) the mutation process by which variation arises along the coalescent tree (Figure 8.19). We can separate these processes because, at a neutral locus, all gene copies are equally likely to leave descendants, irrespective of their allelic state. Thus, the mutation process and the allelic states of gene copies have no effect on the genealogical process and the resulting shape of the coalescent tree.

In this case, the coalescent process tells us about the strength of genetic drift to eliminate genetic variation. In Figure 8.16A, we showed a simulated coalescent tree for a small population; in Figure 8.16B we showed a simulated coalescent tree for a larger population. As we noted, the small population has a much more recent coalescent time; thus, we expect that less variation will have been generated since coalescence in the small population. This is consistent with the finding we discussed in Section 8.1: that drift will act more strongly to reduce heterozygosity in a small population than in a large one.

The pattern of variation that we see at a neutral locus is therefore the result of two sources of randomness superimposed on one another: (1) the randomness associated with which particular genealogical history happens to occur—that is, the coalescent tree of the present population, and (2) the randomness associated with where mutations arise along this coalescent tree (Nordborg 2007; Felsenstein 2004).

If we focus on a population of constant size with no selection, assortative mating, or migration, then if two randomly selected alleles are separated by on average $4N$ generations, and the mutation rate is μ per locus per generation, we expect two randomly selected alleles to differ by an average of $4N\mu$ mutations. But there are two sources of randomness that cause variation around this average number of differences: (1) genealogical history is a random process, so the two alleles may be separated by considerably more or less than $4N$ generations, and (2) the mutation process varies, so if the two alleles are separated by say 1000 generations, we may see more or less than 1000μ mutations distinguishing them.

We conclude this section by noting that selective processes also have a substantial influence on the shape of coalescent trees. Selection drives alleles quickly to fixation, leading to a more recent coalescent time. Figure 8.20 illustrates the gene genealogy for new mutants that are (A) neutral, (B) positively selected, and (C) subject to balancing selection. A conventional gene geneology for a neutral locus is shown in Figure 8.20A. Here, a new neutral allele arises by mutation as indicated. In this particular example, the new allele drifts, by chance, to fixation. Note, however, that most newly arisen neutral alleles will be lost, rather than fixed, by drift.

Alleles under positive selection do not have to rely on drift alone to reach fixation. In Figure 8.20B, the new allele is positively selected and, because of selection, it quickly replaces all other alleles in the population. As a result, the population has a more recent coalescent point than in the neutral example. This is a useful observation. Because a recent selective event results in a more recent coalescent point, we expect to find less neutral variation—that is, fewer silent substitutions—at the locus under selection. In Chapter 10, we will see how this observation can be used to find regions of the genome that have been under natural selection in the recent past.

As we learned in Chapter 7, forms of balancing selection such as overdominance or negative frequency dependence can maintain balanced polymorphisms of two or more alleles. In Figure 8.20C, a new allele arises by mutation that is under balancing selection with the ancestral allele. Because balancing selection favors the new allele when it is rare, but favors the ancestral allele when the new allele is common, neither allele is easily able to go to fixation. As a result, both remain in the population for an extended period of time, and the coalescent point for this locus occurs further in the past than it did for the neutral and positively selected cases. Because the population is finite, we expect one allele will eventually replace the other by chance despite balancing selection. But this may take a very long time to occur, and in the meantime we observe a balanced polymorphism with a coalescent point far from the present.

FIGURE 8.20 Gene genealogies and selection. (A) Gene genealogy for a new allele subject to neutral drift. In this particular case, the gene shown drifts to fixation. **(B)** Gene genealogy for a new allele subject to positive selection. Here, natural selection quickly drives the new allele to fixation. **(C)** Gene genealogy for an allele under balancing selection. Here, the two alleles both persist indefinitely in a balanced polymorphism. Adapted from Bamshad and Wooding (2003).

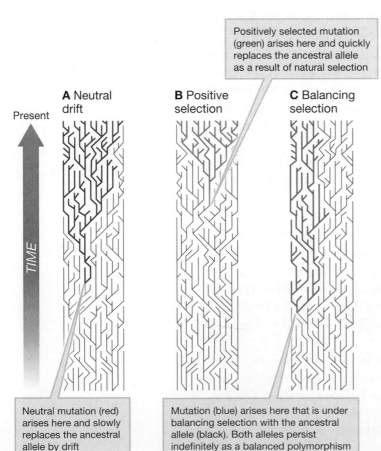

Present

TIME

A Neutral drift

B Positive selection

C Balancing selection

Positively selected mutation (green) arises here and quickly replaces the ancestral allele as a result of natural selection

Neutral mutation (red) arises here and slowly replaces the ancestral allele by drift

Mutation (blue) arises here that is under balancing selection with the ancestral allele (black). Both alleles persist indefinitely as a balanced polymorphism is maintained by selection

8.3 Demography, Biogeography, and Drift

We have seen that genetic drift operates most powerfully when population sizes are very small. Although many natural populations tend to be large most of the time, certain demographic and biogeographic processes can reduce their size considerably. Even a brief reduction in population size can cause drift to operate strongly. In the language of effective population size (Box 8.3), even a short period of having a small census population size can massively reduce the effective population size. In this section, we will consider two particularly important processes of this type: (1) population bottlenecks, and (2) the founder effect.

Original population   Bottleneck event   Surviving population

FIGURE 8.21 The population bottleneck concept. In the original population, there are three different alleles, represented here by blue, black, and yellow balls. A bottleneck cuts population size dramatically, leading to shifts in allele frequency simply by chance. Compare the frequency of black and blue balls before and after the bottleneck. Bottlenecks can even result in the loss of certain alleles. The yellow allele is lost in this example.

Population Bottlenecks

We have learned that genetic drift can be an important evolutionary process in small populations. But what happens in large populations, especially those without significant subdivision? Are they protected from the operation of drift? Not entirely, because natural populations inevitably fluctuate in size over time. Even very large populations can go through rough periods where population size becomes small. And when populations become very small, even for a short time, allele frequencies can change dramatically. This is because of the sampling that occurs during the reduction of population size and because of the accelerated pace of genetic drift in the small population. This process is so important in natural populations that population geneticists have a specific name for it: A brief period of small population size is called a **population bottleneck** (Figure 8.21).

A Simulation of the Effects of a Bottleneck

In Figure 8.22, we show the results from a simulation of 10 replicate populations of size 1000 going through a brief population bottleneck. Notice that the biggest changes in allele frequency come during the bottleneck. Even though the population consists of 1000 diploid individuals for most of the period shown, the bottleneck has a considerable effect on allele frequencies, and alleles even go to fixation in two of the replicate populations.

We can infer the effects of a bottleneck on the rate of genetic drift from the equation for effective population size that we developed in Box 8.3. There, we saw that the effective population size of a population that varies in size from generation to generation is given by the harmonic mean of the population sizes in each generation. In the case of a bottleneck, the population size is large for much of the time, only briefly becomes small, and again grows back to its usual large size. How does this affect the effective population size? The harmonic mean of these population sizes will tend to be close to the smallest population size—that is, to the size of the population during the tightest part of the bottleneck. Because the effective population size is small, we expect the rate of drift to be high—exactly as we have seen is the case when a bottleneck occurs.

FIGURE 8.22 A bottleneck causes a drastic shift in allele frequency. Here we see the results from simulations of 10 replicate populations, each with 1000 diploid individuals, going through a brief population bottleneck. Each population starts with the A_1 and A_2 alleles each at frequency 50%. One sample trajectory is highlighted for emphasis. Allele frequencies drift gradually until the population bottleneck, at which point the drift accelerates dramatically, causing large changes in allele frequency. As the populations are restored to their original sizes, the rate of allele frequency fluctuation slows.

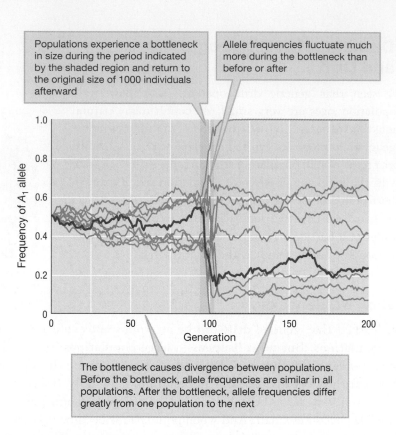

Populations experience a bottleneck in size during the period indicated by the shaded region and return to the original size of 1000 individuals afterward

Allele frequencies fluctuate much more during the bottleneck than before or after

The bottleneck causes divergence between populations. Before the bottleneck, allele frequencies are similar in all populations. After the bottleneck, allele frequencies differ greatly from one population to the next

FIGURE 8.23 Bottlenecks in a natural population. Male northern elephant seals at San Simeon, California.

A Strong Bottleneck Reduced the Heterozygosity of Elephant Seals

The effects of a population bottleneck are illustrated by one of the most remarkable recoveries from near-extinction yet observed: that of the northern elephant seal (*Mirounga angustirostris*) (Figure 8.23). This species, which breeds on the beaches of California and Baja California, was hunted to the very edge of extinction in the nineteenth century. Although the commercial harvest ceased as the seal population declined, museum collectors killed many of the remaining animals. In 1892, eight individuals, thought to be the last of the northern elephant seals, were discovered on Guadalupe Island off the west coast of Mexico (Hoelzel 1999). These were promptly killed for museum specimens!

Fortunately, these were not the last members of the species. Roughly 10–20 individuals had been missed by hunters, and from these few individuals the population began its recovery. After vigorous protection efforts, the northern elephant seal population rebounded, and it now numbers well over 100,000 individuals.

As we have seen, population genetics predicts that a bottleneck should cause a dramatic reduction in heterozygosity in the northern elephant seal population. To test this prediction, Michael Bonnell and Robert Selander took blood samples from 159 individuals at five different breeding locations (Bonnell and Selander 1974). They used a technique known as *enzyme electrophoresis* to look for molecular variation in the structure of 21 different proteins—and by this assay found no variation whatsoever. In a 1993 follow-up study, A. Rus Hoelzel and his colleagues surveyed 41 additional proteins using similar methods and again found zero variation (Hoelzel et al. 1993). As summarized in Figure 8.24, this lack of variation was in marked contrast to observations of considerable molecular variation in the proteins of the southern elephant seal (*Mirounga leonina*). This species, which is the northern elephant seal's closest relative, did not experience a comparable population bottleneck. These findings strongly support the theoretical prediction that a tight bottleneck should greatly reduce the genetic variation within a population.

Enzyme electrophoresis—while the best approach available in 1974 when Bonnell and Selander conducted their study—is a relatively coarse-grained tool for surveying the extent of molecular variation. Some protein structure variants will not be detected by this method. Moreover, because enzyme electrophoresis operates at the level of protein product rather than DNA, it is unable to detect silent substitutions in the DNA sequence. More recent studies have used DNA sequencing to take a finer-grained look at the extent of molecular variation in northern elephant seal populations. These studies have looked at the DNA sequences in several highly variable regions of DNA such as the control loop region of the mitochondrial DNA, the M2b microsatellite locus, and several major histocompatibility complex (MHC) loci (Hoelzel et al. 1993, 1999a,b; Weber et al. 2004). In each case, variation is extremely limited in the northern elephant seal and more abundant in its southern relative (Figure 8.25).

Yet, none of this work decisively shows that the population bottleneck caused the low level of heterozygosity among northern elephant seals. Perhaps this population had unusually low levels of variation even before the bottleneck. There are other reasons why we might expect low heterozygosity in elephant seals, including the highly skewed distribution of reproductive success in this species, where a dominant male mates with many different females. To demonstrate definitively that the reduction in heterozygosity occurred coincident with the bottleneck, researchers would have to take genetic samples from individual seals that lived before the bottleneck.

Fortunately, museum samples make this possible. Diana Weber and her colleagues did exactly this in a study published in 2000 (Weber et al. 2000). They extracted mitochondrial DNA from bone and dried skin of animal samples taken before, during, and after the tightest part of the bottleneck. In 149 samples taken from post–bottleneck specimens by these and other investigators, only two genotypes were found. Samples from the late nineteenth century reveal that both extant genotypes date at least as far back as the tightest portion of the bottleneck. By contrast, in the five bone samples from before the bottleneck, the researchers found four distinct genotypes. This strongly indicates greater diversity prior to the bottleneck and establishes that the bottleneck was coincident with, and presumably the cause of, the severe reduction in heterozygosity that we observe in the current elephant seal population.

Founder Effect

We began this chapter with the story of the *Manx* (*M*) mutation in the cats on the Isle of Man. The high prevalence of this mutation there, and its comparative rarity elsewhere, is probably a result of a phenomenon known as the **founder effect**. The founder effect

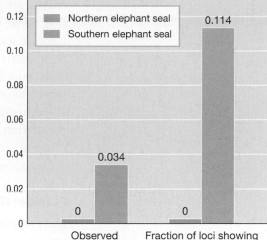

FIGURE 8.24 Low variation in northern elephant seals: enzyme electrophoresis data. No molecular variation is observed in any of the 62 northern elephant seal proteins surveyed by enzyme electrophoresis. By contrast, enzyme electrophoretic studies on southern elephant seals reveal significant molecular variation. Adapted from Hoelzel (1999).

FIGURE 8.25 Low variation in northern elephant seals: DNA sequence data. DNA sequence studies at three different highly variable loci reveal much greater variation in southern elephant seal populations than in northern elephant seal populations. Adapted from Hoelzel et al. (1999b).

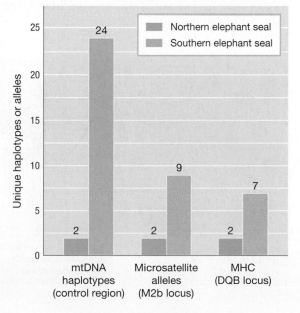

refers to the change in allele frequencies that results from the sampling effects that occur when a small number of individuals from a large population initially colonize a new area and *found* a new population. For example, islands often draw their initial inhabitants, or founders, from large mainland populations nearby. This sampling process introduces random change. Genes in founders usually represent only a subset of the genes present in the mainland population, and so the allele frequencies in the founders may deviate by chance from those in the large population. Moreover, alleles that are extremely rare on the mainland, such as the *Manx* allele, may become common on the island if carried by one of the founders of the island population.

Founder Effect in an Island Population

Darwin pointed out that many plants "migrate" to small islands by drifting on water currents, or by having their seeds transported in the mud stuck to a bird's foot. Such a scenario offers ample opportunity for founder effects to influence allele frequencies in island populations. By way of example, some plant species are polymorphic for the direction that their flowers tilt relative to the floral axis. In the plant *Heteranthera multiflora*, this tilting trait is controlled by a single locus with two alleles, labeled R for right leaning, and r for left leaning (Jesson and Barrett 2002). The R allele is dominant, so that RR and Rr individuals have right-leaning flowers, and rr individuals have left-leaning flowers. Imagine that the frequency of R is 0.3 on the mainland so that, at Hardy–Weinberg equilibrium, the frequencies of each phenotype are approximately the same—right-leaning flowers at 51% and left-leaning flowers at 49%. Five migrants move from the mainland to the island (Figure 8.26A). These five migrants, being diploid, carry with them 10 gene copies at the R locus. There is only about a 27% chance that our founding island population will have the same allele frequencies as our mainland population (Figure 8.26B)—that is, random fluctuations create a 73% chance that the founders of our island population will have different allele frequencies for the tilting trait than were found on the mainland.

Genetic drift not only affects the gene frequencies in the founding population on the island, but it also affects the *long-term frequencies* of genes in future generations

FIGURE 8.26 The founder effect. (A) Five plants from a larger mainland colonize a smaller island. **(B)** If the founders are sampled randomly from the mainland population, they may carry anywhere from 0 to 10 copies of the R allele, although it is unlikely that they would carry more than 7. The probability that the five founders of the island population carry exactly 3 copies of the R allele—and thus have the same allele frequencies on the island as on the mainland—is less than 0.27.

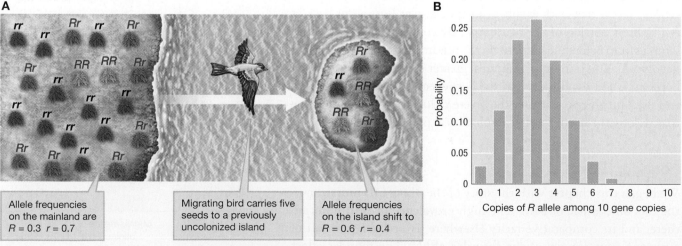

of offspring. If natural selection is not acting on alleles R and r, then over the long run our island population will become fixed for one of the two alleles—sooner or later a string of chance events will cause the loss of one of the alleles, and hence the fixation of the other. Moreover, if the island population is smaller than the mainland population, this process of genetic drift will proceed more quickly, as we saw in Figure 8.3.

As we showed earlier, the probability that a particular allele will become fixed over the long run is equal to its initial frequency on the island. This makes intuitive sense, keeping in mind that when genetic drift alone is in operation, alleles increase or decrease in frequency strictly as a result of chance. If an allele is initially at a low frequency, the odds are high that a series of chance events—ordinary events, such as the accidental failure of certain parents to reproduce—will cause that allele to disappear from the founder population. The reason that an allele that is initially at a low frequency will likely be lost from a population is simple: there were very few of these alleles to begin with. If an allele is represented at a higher frequency in our founder population, a much larger series of chance events must act against it to make that allele go extinct in our population. As in Figure 8.26, if, by chance, we ended up with six copies of the R allele and four copies of the r allele in our founder population, then in the absence of selection the probability that our island population would become fixed for R is 0.6 and the probability that it would become fixed for r is 0.4.

To better understand how founder effects operate in nature, let us consider work on founder effects in the black spruce tree.

Founder Effects, Mitochondrial DNA, and Black Spruce

Consider what happens when glaciers recede after an ice age, and a species moves back into the once-glaciated areas. Those individuals that colonize the newly uncovered land are not randomly sampled from the species, but rather they tend to come from the so-called *leading edge subpopulations* near the previous limit of the species range during the ice age. This process of colonization from the populations nearest the previous range limits is known as a leading edge expansion (Figure 8.27). Like the founder effects associated with island colonization, leading edge expansions result in reduced genetic diversity in the newly colonized region.

The genetic consequences of leading edge expansions after the recent ice ages can be observed widely throughout the Northern Hemisphere in plant and animal species alike (Hewitt 1996, 2000). Isabelle Gamache and her colleagues studied such founder effects in the subarctic black spruce (*Picea mariana*) growing in the forest tundra of the eastern coast of Canada's Hudson Bay (Figure 8.28) (Gamache et al. 2003). Glaciers disappeared from this area

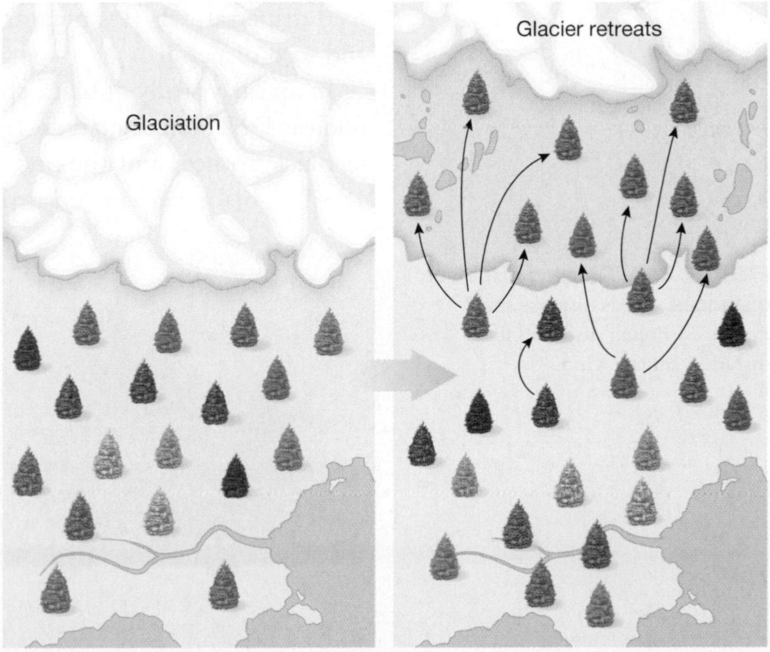

FIGURE 8.27 Leading edge expansion. A land mass is half covered by glaciation during an ice age. South of the ice sheet, the uncovered land provides a refuge for a number of populations (genetic diversity is indicated by different colors). When the ice sheet recedes at the end of the ice age, the uncovered terrain is colonized by individuals from the leading edge subpopulations—here, the populations adjacent to the former glacier. The populations farther from the leading edge contribute relatively little to the colonization. The consequence is a sort of founder effect in which we observe reduced genetic diversity in the recently colonized area.

FIGURE 8.28 Leading edge expansion of black spruce. (A) A forest of black spruce in Canada, (B) the current distribution of black spruce in North America is shown in dark green. Part B from Viereck and Johnston (1990).

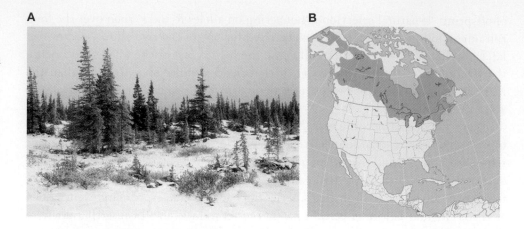

A

B

about 6000 years ago, and it was recolonized by tree species such as the black spruce, which eventually reached its northernmost latitude about 1500 years ago.

The genetics of dispersal and recolonization are particularly interesting in plants. While seeds have to be dispersed into a new area for the initial colonization to take place, seed dispersal is not the only source of genetic variation for an established population: Pollen from other populations can blow in on the wind and fertilize the plants that have become established there (Figure 8.29).

Thus, in the process of colonization, genetic material is carried by two different sources that differ dramatically in their mobility. While some seeds travel by wind, many are dispersed by animals or other range-limited processes. In contrast, pollen is much lighter and can travel much farther by wind, covering greater distances in much greater volume. We can tease apart the patterns of pollen dispersal and the patterns of seed dispersal because not *all* genetic material travels in pollen. Mitochondrial DNA is maternally inherited, and thus it is passed on only through seeds; it is absent from pollen (Figure 8.30). We might therefore expect the geographic distribution of mitochondrial DNA variants to reflect only patterns of seed dispersal, whereas nuclear DNA variants will reflect patterns of both pollen and seed dispersal.

In an early study of black spruce in the Hudson Bay area, researchers studying nuclear DNA found no reduction of genetic diversity in post–ice-age populations of black spruce, and thus no evidence of founder effects (Desponts and Simon 1987). This is perhaps unsurprising given that wind-dispersed pollen need not

FIGURE 8.29 Differing dispersal distances. Seeds disperse short distances. Pollen disperses long distances on the wind.

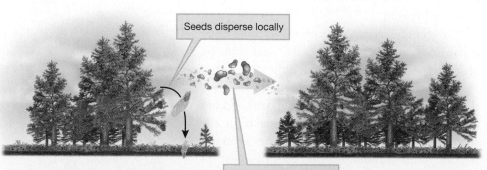

Seeds disperse locally

Pollen disperses across long distances (several hundred miles)

travel only from the leading edge populations during a recolonization event; vast quantities of such pollen can move long distances, minimizing any possible founder effects.

In addition to using nuclear DNA to study the movement of pollen, Gamache and her colleagues also examined the effect of migration via wind-dispersed seeds by using the DNA found in the mitochondria, the energy-producing organelles of cells (Gamache et al. 2003). Wind-dispersed seeds occur in much smaller numbers than wind-dispersed pollen, and hence we might expect to find genetic drift affecting mitochondrial gene frequencies. To compare nuclear and mitochondrial DNA, these researchers took foliage samples from about 30 trees in each of nine populations along a 1000-kilometer transect, or section, of forest. This transect included populations at the northernmost distribution of black spruce, as well as much larger populations to the south, and the diversity of both nuclear and mitochondrial DNA was calculated for each population.

Gamache and her team found that the migration of mitochondrial DNA via wind-dispersed seeds was much more restricted and localized than the migration of nuclear DNA via pollen dispersion. There were two lines of evidence for this. First, all the different types of nuclear DNA found in large parent populations were represented in northern subpopulations. When it came to mitochondrial DNA, however, although the southern populations contained four different types of mtDNA, every one of the northern subpopulations had one and only one type of mtDNA, called mitotype I (Figure 8.31). This suggests that, by chance, either mitotype I was able to move north into a single subpopulation and then spread even farther north through time, or that a single long-distance migration event involving mitotype I occurred. Both are consistent with the idea of founder effects.

A second line of evidence for founder effects in black spruce was that, when both southern and northern populations were compared, *between-population* variability in mitochondrial DNA were 10 times greater than between-population measures of nuclear DNA variability. In other words, northern and southern populations were very *similar* with respect to nuclear DNA, but very *different* with respect to mitochondrial DNA. Fixation for a single genetic type within a population, combined with high between-population variation, is a hallmark of genetic drift. Indeed, Gamache and her colleagues were able to use their estimates of genetic diversity to calculate the effective number of mitochondrial DNA seed "migrants" and nuclear DNA pollen "migrants" entering populations in each generation. As expected, the average number of mitochondrial DNA migrants per generation was almost 10 times lower than the average number of nuclear DNA migrants (Gamache et al. 2003).

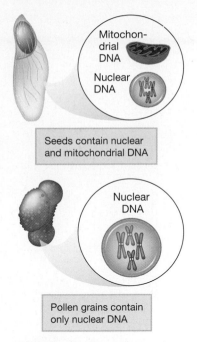

FIGURE 8.30 Seeds carry additional genetic material. Seeds contain both nuclear DNA and mitochondrial DNA, whereas pollen grains contain only nuclear DNA.

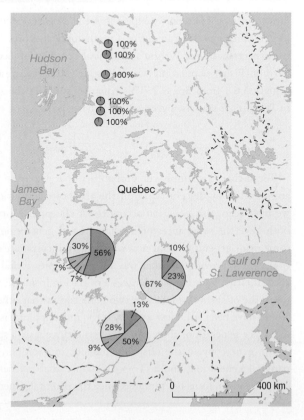

FIGURE 8.31 Limited mitochondrial diversity in leading edge expansion. Nine black spruce populations in Quebec, Canada, were sampled to determine the frequencies of mitotypes within each population. The pie charts indicate the mitotypes in each population. The six northern subpopulations of black spruce were all fixed for a single type of mtDNA, called mitotype I (shown in blue). The southern populations contained three or four mitotypes (blue = mitotype I, orange = mitotype II, green = mitotype III, yellow = mitotype IV). Adapted from Gamache et al. (2003).

8.4 The Interplay of Drift, Mutation, and Natural Selection

As we have seen, genetic drift increases the homozygosity of a population. Indeed, if drift were the only evolutionary process operating, any finite population would eventually become entirely homozygous. In practice, however, populations do not become entirely homozygous, because mutation provides a continual supply of new genetic variation. This leads to a balance or steady state in which the loss of heterozygosity due to drift is balanced by the gain in heterozygosity due to mutation. In Box 8.5, we develop a simple model that predicts the amount of variation that we expect to find at a neutral locus in a Wright–Fisher population at steady state.

The Mathematics of Selection and Drift

In our discussion of selection in Chapter 7, we looked at large populations in which drift was not operating. In our treatment of drift thus far in this chapter, we have primarily looked at neutral loci in which selection is not operating. But selection and drift are not mutually exclusive modes of evolutionary change. Both can, and usually do, operate simultaneously in natural populations. Having seen how each

BOX 8.5 Wright's *F*-statistic at a Neutral Locus with Mutation

How much variation do we expect to see at a neutral locus subject to mutation? We can derive a mathematical expression for the expected value of Wright's *F*-statistic at a neutral locus in a Wright–Fisher population at steady state. To do so, we revisit Equation 8.1 from Box 8.2. This equation specifies the change in Wright's *F*-statistic over a single generation:

$$F_{\text{offspring}} = \frac{1}{2N} + \left(1 - \frac{1}{2N}\right)F_{\text{parental}}$$

Recall that *F* is simply the probability that two gene copies are identical by descent (IBD) in the absence of mutation. As we have seen, this equation accounts for the probability that both have the same ancestor in the parental generation or in some prior generation. But now we want to incorporate mutation, which provides another way for gene copies to fail to be IBD. If either gene copy undergoes a mutation from the parental to the offspring generation, two gene copies that otherwise would have been IBD now are not. If the mutation rate is μ per locus per generation, there is a $(1 - \mu)$ chance that a specific single gene copy in the offspring has not mutated since the parental generation, and a $(1 - \mu)^2$ chance that neither gene copy at a given locus has mutated since the parental generation.

Thus, the chance that two gene copies are IBD in the presence of mutation is obtained by multiplying the right-hand side of Equation 8.1 by $(1 - \mu)^2$, so that we get

$$F_{\text{offspring}} = \left[\frac{1}{2N} + \left(1 - \frac{1}{2N}\right)F_{\text{parental}}\right](1 - \mu)^2 \quad (8.6)$$

We can find the equilibrium or steady-state level value of *F* by setting $F_{\text{offspring}} = F_{\text{parental}}$ in Equation 8.6 and solving the resulting equation to get

$$F_{\text{equilibrium}} = \frac{(1 - \mu)^2}{2N - (1 - \mu)^2(2N - 1)}$$

Because the mutation rate μ is typically small, both μ and μ^2 will be small and can be ignored in an approximation of the equilibrium value of *F*:

$$F_{\text{equilibrium}} = \frac{1}{4N\mu + 1}$$

But we cannot ignore the $N\mu$ term, because *N* can be large.

In Box 8.2, we saw that heterozygosity tends to decrease with increasing values of *F*. This means that, as intuition would suggest, heterozygosity will tend to be lower when (1) population size is small, and (2) mutation rate is low.

acts alone, we are now in a position to think about how these processes interact with one another.

Even alleles that are favored by natural selection are not guaranteed to become fixed in a population. The early population geneticist J. B. S. Haldane (1892–1964) looked at a simple model in which a new, slightly beneficial allele with a fitness of $1 + s$ arises in a large population and competes with the wild type that has a fitness of 1 (Haldane 1927). Even though the population size is large, the new mutation is surprisingly unlikely to be fixed. Haldane found that the fixation probability is approximately $2s$. This means that a new beneficial mutation that confers a 1% fitness advantage has only a 1 in 50 chance of being fixed in a large population!

The reason that drift matters here even in a large population is that we are now looking at what happens to the *initial* mutant allele. In large populations, allele frequencies fluctuate less because of drift, but a new allele begins at a lower frequency. Think about a new allele arising in a haploid population of size 100 or size 1,000,000. In a population of 100, drift can cause substantial fluctuations in allele frequencies, but the new allele will begin at a frequency of 1 in 100; relatively speaking, it doesn't have all that far to go to reach fixation. In a population of 1,000,000, drift will have less effect on allele frequencies overall, but the new allele will begin at a frequency of only 1 in 1,000,000; it will have a really long way to go if it is to reach fixation. In Haldane's model, these effects cancel out, and the probability of fixation is independent of population size.

Motoo Kimura (1924–1994), best known as the architect of the neutral theory of molecular evolution (Section 8.5), looked at more complicated models and extended Haldane's analysis. For example, when the effective population size N_e is less than the actual population size N, the probability that a small beneficial mutation is fixed is approximately $2sN_e/N$ instead of $2s$ as under Haldane's model (Otto and Whitlock 2005). Population bottlenecks, sex ratio biases, and other factors that reduce effective population size also reduce the chance that a beneficial mutation will be fixed.

While the population size term dropped out of Haldane's expression for the fixation probability of the *initial* mutation, if we look at an allele present at some intermediate frequency—say, 1% or 10% or 50%—the population size matters as well. Broadly, the interplay between selection and drift depends on the strength of the selection and the population size. When selection is strong and population size is large, selection largely determines the change in allele frequencies. When selection is weak and population size is small, drift largely determines allele frequency change. To quantify this, Kimura proposed a rule of thumb for when selection is effective and when drift dominates (Kimura 1983). In a diploid population, selection dominates when the selective advantage $s > 1/2N_e$; drift dominates otherwise. Thus, selection can operate effectively on an allele with a fitness advantage of $s = 0.001$ in a population of 10,000 individuals, but not in a population of 100 individuals.

Figure 8.32 illustrates the effectiveness of natural selection. In this graph, we show the approximate probability that a rare but selectively favored allele, A_2, initially present at frequency 1%, goes to fixation in a Wright–Fisher population. In this example,

FIGURE 8.32 Selection versus drift. Here we plot the approximate probability that a selectively favored allele, initially present at a frequency of 1%, goes to fixation. The horizontal axis indicates the population size; the vertical axis indicates the probability of fixation. The effectiveness of natural selection at fixing a favored allele depends on population size and the strength of selection.

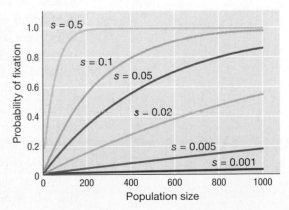

the fitness effects of the A_2 allele are multiplicative: fitnesses are 1 for the A_1A_1 genotype, $1 + s$ for the A_1A_2 genotype, and $(1 + s)^2$ for the A_2A_2 genotype. We see that when selection is very strong (for example, $s = 0.5$), the favored allele A_2 goes to fixation with high probability even in a relatively small population. By contrast, when selection is weaker (for example, $s = 0.005$), the favored A_2 allele is more likely than not to be lost even in a population of size 1000.

An Empirical Study of Selection and Drift

To examine the relationship between drift and selection in further detail, we turn to a classic 1957 study by Theodosius Dobzhansky and Olga Pavlovsky. Dobzhansky and Pavlovsky wanted to see whether drift could play an important role in populations under natural selection, or whether the effects of selection would swamp any influence of drift.

To find out, these researchers conducted a multigeneration study of the fruit fly *Drosophila pseudoobscura* in population cages in the laboratory. As a focal trait, they studied naturally occurring genetic chromosomal inversions (Dobzhansky and Pavlovsky 1957). A chromosomal inversion results from the breaking off of a section of chromosome, which then flips around and is reinserted back into the chromosome in the opposite direction. Distinguishing one type of inversion from another was straightforward: Different inversions can be detected by examining stained chromosomes under a light microscope. In their experiment, Dobzhansky and Pavlovsky worked with two inversions of the third chromosome named Pike's Peak (*P*) and Arrowhead (*A*). These inversions exhibit overdominance, such that *PA* heterozygotes have a higher fitness than either homozygote.

Dobzhansky and Pavlovsky's first step was to create 20 replicate lines of fruit flies by crossing large numbers of individuals from two wild populations—one from California and one from Texas. Fruit flies from the California population were *PP* homozygotes, and those from the Texas population were *AA* homozygotes, so that all individuals in each of the 20 replicate lines were *PA* heterozygotes at the third chromosome. But such matings also introduced a tremendous amount of genetic variation at all loci besides those associated with the chromosomal inversions under study. Dobzhansky and Pavlovsky then examined how natural selection and drift interacted to change the frequency of *P* and *A* over the course of a 17-month experiment, which represented 19 fruit fly generations.

To distinguish the effects of drift and selection, Dobzhansky and Pavlovsky looked at how allele frequency change in replicate populations depended on population size. They divided their 20 lines into two different treatments. Ten lines were assigned to the "large-founding-population treatment"; these were initiated with 4000 individuals per line. The other 10 lines were assigned to the "small-founding-population treatment"; these were initiated with 20 individuals per line. Other than the differences in initial population size, the two treatments were identical. In addition, populations were allowed to increase at normal rates in both treatments, and each line quickly—often within a single generation—stabilized at about 1000 to 4000 individuals. Any effect of drift therefore was due to *initial* population size differences across treatments (Figure 8.33).

Notice that by creating and following a set of replicate populations in the laboratory, Dobzhansky and Pavlovsky could directly observe one of the

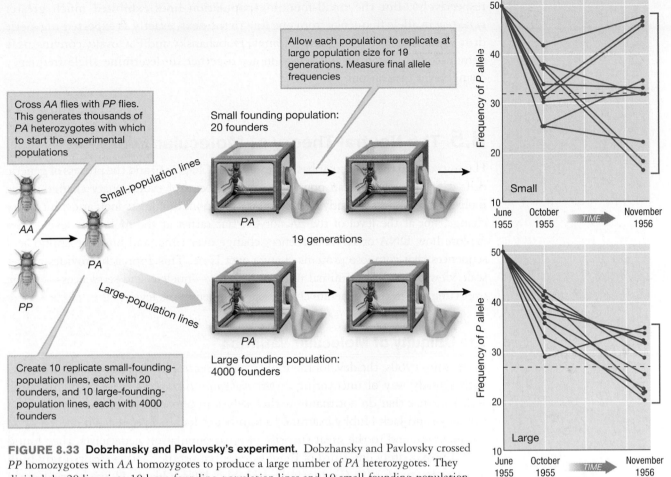

FIGURE 8.33 Dobzhansky and Pavlovsky's experiment. Dobzhansky and Pavlovsky crossed *PP* homozygotes with *AA* homozygotes to produce a large number of *PA* heterozygotes. They divided the 20 lines into 10 large-founding-population lines and 10 small-founding-population lines. Each of the 20 lines was then allowed to reproduce for 19 generations, and the frequencies of the *A* and *P* alleles in each line were measured. By the end of the experiment, the frequency of *P* had dropped in all lines; the *mean* frequency of *P* was not significantly different across large-population and small-population lines (27% and 32%, respectively). But much greater variation in final allele frequencies was observed across the small-population lines, indicative of the effects of genetic drift. Graphs adapted from Dobzhansky and Pavlovsky (1957).

consequences of drift that we discussed in Section 8.1: the way that drift leads to divergence between populations. Dobzhansky and Pavlovsky knew from theoretical considerations that drift operates more strongly in small populations. Therefore, they reasoned that, if drift played an important role in their experimental populations, they would see a greater degree of divergence between populations in their small-founding-population lines than in their large-founding-population lines.

Indeed, both natural selection and drift were operating in Dobzhansky and Pavlovsky's populations. Because the chromosomal inversions exhibit overdominance, both the large-founding-population lines and the small-founding-population lines were under balancing selection. In each treatment, balancing selection resulted in the same average frequency of the *P* allele—that is, the average frequency of the chromosomal inversion was not statistically different across the large- and small-founding-population lines (27% and 32%, respectively). But the small-founding-population lines exhibited much greater *variation* in allele frequency from one line to the next, exactly as expected if genetic drift was operating. With these findings, Dobzhansky and Pavlovsky convincingly demonstrated that drift and selection act together to determine allele frequency change over generations.

8.5 The Neutral Theory of Molecular Evolution

Having covered the molecular basis of mutation (Chapter 6) and the process of genetic drift (this chapter), we can now explore the process of evolutionary change at the molecular scale. In the study of molecular evolution, biologists look at evolutionary change, not at the level of the phenotype, but rather at the molecular level. They explore how DNA or RNA sequences change over time, and how the amino acid sequences that compose proteins change over time. This approach provides a fine-scale view of how the minimal units of heredity—nucleic acid sequences—change over time and in turn generate changes at the phenotypic level.

The Ubiquity of Molecular Variation

In the mid-1960s, the development of enzyme electrophoresis provided researchers with a ready way of uncovering *cryptic molecular variation*—differences in amino acid sequence that do not manifest themselves in phenotypic differences. Richard Lewontin and Jack Hubby examined a number of loci in a population of *Drosophila pseudoobscura*, and to the great surprise of most population geneticists, they found that approximately one-third of these were polymorphic, with a surprisingly high heterozygosity of 12% (Lewontin and Hubby 1966). Harry Harris carried out a similar study on humans; he found that 3 of the 10 loci were polymorphic with a heterozygosity of 6% (Harris 1966).

From these and other studies that followed, population geneticists were forced to conclude that molecular variation is far more common in populations than they had previously imagined. That conclusion posed a major problem. At the time, most explanations for the presence and/or maintenance of variation in a population required strong natural selection. Concurrently, it was thought that, when variation was observed at a locus, it either was maintained by balancing selection or natural selection was in the process of replacing one allele with another. But with so much variation present, researchers worried that natural selection could not be the whole story. Selection is costly in that it requires either differential survival or differential reproductive success, and researchers had found ways to quantify the "cost of natural selection" and relate it to the amount of variation in a population (Haldane 1957; Kimura 1961). There was simply not enough natural selection going on to account for this much variation. There had to be some other explanation.

The Neutral Theory Proposes That Most Substitutions Are Selectively Neutral

Perhaps the most straightforward explanation is that selection may not be acting on this variation at all. Although most heritable phenotypic differences result in fitness differences and thus are subject to natural selection, the same might not be true of molecular differences. To account for the extensive molecular variation observed in populations, Kimura proposed the **neutral theory** of molecular evolution in 1968 (Kimura 1968, 1977, 1983, 1993; Jukes and Kimura 1984; Dietrich 1994). The neutral theory proposes that at the molecular level of DNA sequence or amino acid sequence:

1. Most of the variation present within a population is selectively neutral.

2. Most of the changes in DNA or amino acid sequence over time—and thus many of the molecular differences between related species—are selectively neutral.

According to the neutral theory, most of the genetic variation within a population is neutral and thus not subject to natural selection. Therefore, when a DNA sequence does change over time, some process other than selection is usually responsible. The neutral theory argues that the critical process is genetic drift.

When studying molecular evolution, we will often be concerned with allelic **substitutions**. A substitution occurs when a new allele arises by mutation and is subsequently fixed in the population. The substitution rate, usually measured in terms of substitutions per generation, is defined as the rate at which new alleles become fixed in the population.

It is important to understand that the neutral theory proposes that most *substitutions* are neutral, not that most *mutations* are neutral. Proponents of the neutral theory universally agree that most mutations are deleterious and will be purged from the population by natural selection. But of the remaining mutations that are not purged, the neutral theory proposes that many may be neutral. Similarly, the neutral theory does not propose that most loci are selectively irrelevant in the sense that fitness doesn't depend on the DNA sequence at that locus. It only proposes that, when there are alternative alleles present at appreciable frequency, these alternative alleles are often neutral with respect to one another. The so-called *neutralist–selectionist debate* is not a dispute about the effects of typical mutations; it is a dispute about whether drift or selection is the primary driver of evolutionary change in that subset of mutations that reach a high frequency in populations.

Reasons for Selective Neutrality

The neutral theory suggests that many alternative alleles may be selectively neutral, but why should this be? There are a number of biological reasons why allelic differences might have no fitness consequences; we will explore them here.

Synonymous Substitutions

One of the predominant reasons that molecular variation may be neutral is that many molecular changes do not cause changes in phenotype. First and foremost, the degeneracy of the genetic code means that many changes in protein-coding DNA sequences do not cause changes in the amino acid sequence of the corresponding protein. Because 64 possible nucleotide triplets (codons) are used to code for only 20 amino acids (plus three stop codons), there is redundancy and most amino acids are coded for by several different codons. Typically, codons that code for the same amino acid differ in the third position (Figure 8.34). Thus, many nucleotide changes—particularly those in the third position—do not change the amino acid specified. Mutations that do not result in a changed amino acid are known as

Second

First	U		C		A		G		Third
U	UUU	Phe	UCU	Ser	UAU	Tyr	UGU	Cys	U
	UUC	Phe	UCC	Ser	UAC	Tyr	UGC	Cys	C
	UUA	Leu	UCA	Ser	UAA	Stop	UGA	Stop	A
	UUG	Leu	UCG	Ser	UAG	Stop	UGG	Trp	G
C	CUU	Leu	CCU	Pro	CAU	His	CGU	Arg	U
	CUC	Leu	CCC	Pro	CAC	His	CGC	Arg	C
	CUA	Leu	CCA	Pro	CAA	Gln	CGA	Arg	A
	CUG	Leu	CCG	Pro	CAG	Gln	CGG	Arg	G
A	AUU	Ile	ACU	Thr	AAU	Asn	AGU	Ser	U
	AUC	Ile	ACC	Thr	AAC	Asn	AGC	Ser	C
	AUA	Ile	ACA	Thr	AAA	Lys	AGA	Arg	A
	AUG	Met	ACG	Thr	AAG	Lys	AGG	Arg	G
G	GUU	Val	GCU	Ala	GAU	Asp	GGU	Gly	U
	GUC	Val	GCC	Ala	GAC	Asp	GGC	Gly	C
	GUA	Val	GCA	Ala	GAA	Glu	GGA	Gly	A
	GUG	Val	GCG	Ala	GAG	Glu	GGG	Gly	G

Coded by 6 codons
Coded by 4 codons
Coded by 3 codons
Coded by 2 codons

FIGURE 8.34 Degeneracy in the genetic code. The genetic code exhibits degeneracy, such that DNA base changes—especially in the third codon—do not always change the amino acid specified. Many amino acids are coded by six (blue), four (orange), three (green), or two (purple) different codons. Adapted from Agris (2008).

synonymous or silent mutations. Because such changes do not alter the sequence of the protein that they encode, they will typically be neutral or very close to neutral.

In a 1977 *Nature* paper, Kimura compared the sequences of messenger RNA (mRNA) across species to test his idea that many of the genetic differences that we see when comparing the same gene across two different species are in fact neutral substitutions (Kimura 1977). Using data on the sequence of mRNA from both human and rabbit hemoglobin (Salser et al. 1976), Kimura noted that, of 53 nucleotide positions that can be compared across humans and rabbits, there were differences in six base pairs. Only one of these changes, however, led to a difference in amino acid coding; the other five (83%) were synonymous mutations. By contrast, Kimura calculated that, if mutations occurred and accumulated at random, we would expect only 24% to be found at synonymous sites.

Kimura found support in similar results from Michael Grunstein's work on the rate of molecular evolution of the histone H4 protein found in two species of sea urchins, *Strongylocentrotus purpuratus* and *Lytechinus pictus* (Grunstein et al. 1976) (Figure 8.35). Grunstein and his coworkers had found that, of the 84 nucleotides in the mRNA segment that they compared across these two sea urchin species, 9 of the 10 base pair differences found were synonymous.

More recent work shows that this pattern is very common. When we compare genetic sequences in two or more related species, we see an excess of synonymous substitution over nonsynonymous substitutions in many, though not all, protein-coding genes. Figure 8.36 shows, for 835 genes compared between mice and rats, the relative rates of synonymous versus nonsynonymous substitution. The vast majority of these genes show a great excess of synonymous substitution, indicating that substitutions have been more common at silent sites than at nonsilent sites.

Nonsynonymous Substitutions with Little Effect on Function

In contrast to synonymous mutations, **nonsynonymous mutations** are mutations that do change the amino acid sequence. Many nonsynonymous mutations are not neutral because they change the way that a protein functions, and such changes

A

B

Codon site	24	25	26	27	28	29	30	31	32	33	34
L. pictus messenger RNA	GAU	AAC	AUC	CAA	GGA	AUA	ACU	AAA	CCG	GCA	AUC
S. purpuratus messenger RNA	GAC	AAC	AUC	CAA	GGU	AUC	ACG	?	?	GCU	AUC
Histone IV amino acid sequence in both species	Asp	Asn	Ile	Gln	Gly	Ile	Thr	Lys	Pro	Ala	Ile

FIGURE 8.35 Molecular differences between sea urchin species. **(A)** The sea urchins *Strongylocentrotus purpuratus* (left) and *Lytechinus pictus* (right). **(B)** In a comparison of the histone H4 protein sequences of these two species at codon sites 24 to 34, Grunstein's team found five changes at the third base pair of codons. All five are silent, nonfunctional changes. Part B adapted from Grunstein et al. (1976) and Kimura (1979).

have fitness consequences. While many nonsynonymous sites may be under selection, some nonsynonymous mutations may have minimal fitness effects. For example, changes away from the binding site of a protein often have weaker consequences on protein function than do changes at the binding site of a protein.

As an example, birds and mammals sense temperature using proteins called transient receptor potential vanilloid (TRPV) channels. One domain of these proteins binds adenosine triphosphate (ATP), which in turn modulates the receptor's response to temperature. To better understand the ATP-binding function of these channels, Christopher Phelps and his colleagues compared the DNA sequence of three closely related TRPV channels, $TRPV_1$, $TRPV_3$, and $TRPV_4$ across three species: humans, rats, and chickens (Phelps et al. 2010). They found that while the structure of the ATP-binding site was highly conserved, other regions of the protein were far more variable (Figure 8.37). This indicates that changes away from the binding site may have smaller functional consequences than changes to the binding site; some of these changes may have no effect on function, and thus they may be selectively neutral.

Noncoding Regions

In most eukaryotes, only a small fraction of the genome encodes the sequence of proteins. The rest of the genome is *untranslated*. This is not to say that it necessarily lacks any function; as we will see in Chapter 10, untranslated sections of DNA may, for example, have important regulatory functions. But it is likely that many mutations in noncoding regions of the genome will have very minor effects, or even no effect, on function and fitness. **Pseudogenes**—nonfunctional and typically untranslated segments of DNA that arise from previously functional genes—are often particularly informative about evolutionary history, as they are derived from known homologous genes and subject to neutral drift.

Because pseudogenes do not affect function, mutations in pseudogenes tend to be neutral and they accumulate rapidly over evolutionary time. Pseudogenes can arise through a number of processes. In the process of *gene duplication*, a second copy of the gene is inserted into the genome during DNA replication. As such a copy is a duplicate of another functional gene, mutations that prevent expression may not be selected against. In the process of *retroposition*, mRNA from a functional gene is reverse-transcribed by a *retrotransposon* (see Chapter 10) and inserted into the genome. Because it lacks the appropriate promoter structure, it will tend not to be expressed and thus forms a pseudogene. More rarely, through a process of *deactivation*, genes become pseudogenes without leaving behind a functional copy. In this process, mutation disables an active gene; if the gene is not strongly selected, the deactivated form can be lost as a result of drift. We humans appear to owe our susceptibility to scurvy to such a deactivation event. The primate lineage, of which we are members, arose as fructivores—fruit eaters. Because fruit is rich in vitamin C, early primates would have initially faced minimal selection costs from the loss of the L-gulono-γ-lactone oxidase gene used to synthesize that vitamin. But it is because of the loss of this gene that humans suffer from scurvy if they lack a dietary source of vitamin C.

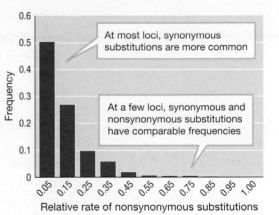

FIGURE 8.36 Most genes show higher rates of synonymous substitution. The substitution rate at nonsynonymous sites relative to the substitution rate at synonymous sites for 835 mouse–rat gene pairs. Nonsynonymous sites tend to evolve much less rapidly than synonymous sites. Adapted from Hurst (2002).

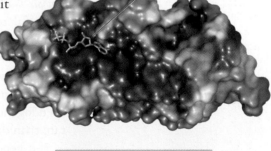

FIGURE 8.37 Conserved and divergent sites in a channel protein. In this representation of the ATP-binding domain of the $TRPV_1$ protein (bound to an ATP molecule), highly conserved amino acids are indicated in red and divergent ones in blue. The binding site of this molecule is the most highly conserved region. This suggests that amino acid sequence changes that alter this binding site will have more dramatic consequences than those that alter other parts of the molecule. At least some amino acid sequence changes in the most divergent regions may be selectively neutral or nearly so.

We will investigate the structure of the genome in further detail in Chapter 10 and look at other reasons why genes may be untranslated or nonfunctional, but for now it will be sufficient to note that mutations in noncoding regions do not change the sequence of proteins, and thus they may be neutral, at least if they do not disrupt gene regulation.

Effective Neutrality

As we saw in Section 8.4, in finite populations, natural selection cannot operate effectively on mutations that have extremely small fitness consequences. The random change in allele frequencies due to drift overwhelms any effects due to natural selection. Thus, even when alternative alleles do have an effect on function and fitness, they can be *effectively neutral* if these effects are sufficiently small. As a rule of thumb, an allele will be effectively neutral if twice the effective population size times the selection coefficient is much smaller than 1—that is, if $2N_e s$ is much less than 1.

Genomics and the Neutral Theory

Of course much has happened since Kimura first championed the neutral theory. Not only have evolutionary biologists made important empirical and theoretical advances, but the molecular genetics tools available to test the predictions of the neutral theory have also improved dramatically. Indeed, recent work in evolutionary genomics now provides researchers with the ability to undertake *genome-wide* assessments of mutation rates in some species (Lynch et al. 2008). Such powerful techniques, when fully employed, will allow biologists to better test many questions regarding mutation rates and the neutral theory.

Some of the basic insights of the neutral theory have withstood the test of time. As we have seen across a wide range of organisms, sites that are expected to have a minimal effect on phenotype—synonymous sites, as well as sites within pseudogenes, introns, and untranslated regions—evolve at a substantially higher rate than do nonsynonymous sites within coding regions.

But genome-scale analysis is beginning to reveal that positive selection has also been extremely important in driving molecular evolutionary divergence among species. For example, a series of genomic studies on *Drosophila* species has estimated that positive selection is responsible for 40%–70% of the nonsynonymous substitutions that have occurred in these species (Welch 2006). Even in noncoding regions, a large fraction of substitutions appear to have been driven by positive selection (Andolfatto 2005). Similar results have been obtained for numerous bacterial and viral taxa as well. Curiously, when comparable methods are applied to the genomes of humans and great apes, the fraction of adaptive substitutions within this clade appears to be dramatically lower (Eyre-Walker 2006).

While much early work by researchers had aimed to demonstrate the plausibility of the neutral theory and the importance of genetic drift as an evolutionary process, one of the most important contemporary functions of the neutral theory is that it serves as a *null model* against which we can test for the operation of selection or other evolutionary processes. The basic idea is straightforward: The neutral theory makes predictions about the amount of variation expected in a population, the relative rates of synonymous and nonsynonymous substitution, and other population-genetic quantities. If we wish to determine whether selection is acting on a locus, we can look at whether these quantities are consistent with what we would expect under a neutral model. If they are not, we might expect that some other process, possibly natural selection, is operating. We will explore the role of the neutral theory as a null model in Chapter 10 when we consider various population-genetic tests for natural selection.

Fixation Probability and Substitution Rate for Neutral Alleles

The neutral theory of molecular evolution makes strong mathematical predictions about rates of evolutionary change. For example, Kimura showed that we can find simple expressions for the probability that a neutral allele is fixed in the population and for the rate at which novel substitutions occur. As we saw in Section 8.1, the probability that a neutral allele is fixed is simply its frequency in the population.

Once we know that the probability of fixation of a neutral allele is equal to its frequency in the population, we are ready to calculate the rate of substitution of neutral alleles in a population. Surprisingly, this turns out to be independent of the population size. Suppose that, in a diploid population of size N, there are k neutral loci in the genome and that the mutation rate at each of these loci is ν. Then in each generation, we expect $2Nk\nu$ neutral mutations to arise in the population. Each new mutation will be at frequency $1/2N$ at the time that it arises, and thus each will have a fixation probability of $1/2N$. The rate at which neutral substitutions occurs is simply the rate at which neutral mutations arise times the probability that each is fixed, as shown by the equation

$$\text{Substitution rate} = 2Nk\nu \times 1/2N = k\nu$$

The population size terms N cancel out; thus, the substitution rate of neutral alleles in the population is simply the rate at which neutral mutations occur within a single (haploid) genome, irrespective of the population size. This is an astonishing result: Neutral substitutions occur in the *population* at the rate that neutral mutations arise in an *individual*.

Not only is this result surprising, it is also very powerful in that it contributes to the foundational logic of the concept of a so-called **molecular clock**. Because substitution rates at neutral loci do not depend on population size or other demographic parameters, proponents of the neutral theory suggest that selectively neutral mutations arise at similar rates in different taxa, and they should also be fixed at similar rates. If this is indeed the case, the substitution rate gives us a way to measure time using genetic data. We explore this in the next subsection.

The Molecular Clock Concept

In the 1960s, biochemists studying the amino acid sequences of various proteins noticed an interesting pattern in the way that these sequences differed between species. Emil Zuckerkandl and Linus Pauling observed that, for any two species, the number of amino acid differences in their hemoglobin molecules was approximately proportional to the time since they diverged on the phylogenetic tree (Zuckerkandl and Pauling 1962). Thus, closely related species have few differences, whereas more distantly related species have a larger number of differences. To account for these

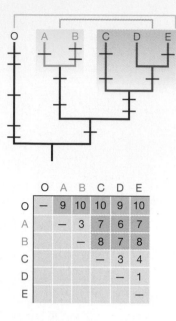

FIGURE 8.38 The genetic equidistance principle. The genetic equidistance principle suggests that, if molecular changes occur at a constant rate across lineages, the members of any given clade should be equidistant from the members of an outgroup. In the phylogenetic tree shown, bars indicate substitution events. The table shows the genetic distance between pairs of species. All species A–E are approximately equidistant from the outgroup O. Moreover, species in the clade C–E are approximately equidistant from species in the clade A–B.

observations, they hypothesized that molecular evolution proceeds in a clocklike manner, with amino acid sequences changing at a constant rate over time, and at the same rate in different lineages.

Emanuel Margoliash found a similar pattern when looking at the differences between species in the amino acid sequence of the cytochrome *c* molecule (Margoliash 1963). These findings led Margoliash to propose the principle of genetic equidistance—if molecular evolution proceeds at the same constant rate over time in different lineages, all members of a clade should be genetically equidistant from an outgroup to the clade (Figure 8.38). Margoliash gave an example: Because fish are an outgroup to the tetrapod vertebrates, we can expect the cytochrome *c* molecules in bird, mammal, and reptile species to all be about the same distance from the cytochrome *c* molecule in a fish species, where distance is measured as the number of DNA or amino acid sequence differences.

It is this principle of genetic equidistance that makes it possible to infer phylogeny from DNA or amino acid sequence data. When the principle breaks down, and evolution proceeds at different rates along different branches of the phylogenetic tree, phylogenetic inference methods run into problems such as the long branch attraction problem described in Chapter 5.

Where we have a reasonable approximation to a molecular clock, we can use molecular data to estimate not only the phylogenetic relationships among species, but also the dates of evolutionary events. If the rate of mutation is known and is approximately the same across lineages, we can use such data to predict the point in time when groups diverged from one another. This prediction can be checked against other estimates of divergence, such as those that might be obtained through the fossil record (Donoghue and Benton 2007). The larger the number of selectively neutral alleles that differ between two groups, the further back in history we must go to find the point in time when the groups diverged (Kumar 2006).

In a dramatic early application of this approach, in 1967 Allan Wilson and Vincent Sarich used the molecular clock to date the divergence time of humans and chimpanzees (Sarich and Wilson 1967). To assess divergence, they looked at the serum albumin molecule—a very common protein in blood plasma. DNA sequencing technology had not yet been developed, so they needed a way of assessing the degree of similarity between versions of proteins in different species. They looked at *immunological cross-reactivity*—the strength of an immune reaction, specific to one protein, when confronted with another—as a measure of distance. The principle is that, if molecular evolution operates in a clocklike fashion, then as species diverge, molecular changes in the structure of albumin should reduce the degree of cross-reactivity at an approximately constant rate. Using this approach, Sarich and Wilson estimated that humans and chimpanzees had diverged only 5 million years ago, far more recently than the 30-million-year estimate that other researchers had derived from paleontological data. This estimate was extremely controversial when first published. But, as we will see in Chapter 14, it is now widely accepted and is closely in accord with more recent data based on genome-scale analysis (Hobolth et al. 2007).

In the late 1970s and early 1980s, there was considerable hope that most molecular evolution would turn out to be clocklike, with nearly constant rates of change across sites and along different evolutionary lineages. Wilson and his colleagues, for example, found that rates of amino acid sequence changes in a number of proteins were approximately constant across the mammalian clade (Wilson et al. 1977) (Figure 8.39).

By the late 1980s, however, this hope began to dim. One early indication of problems came from Vawter and Brown's comparison of substitution rates across lineages and genome regions (Vawter and Brown 1986). While the earliest studies of the molecular clock had relied on protein structure assayed via electrophoresis, immunological cross-reactivity, or other techniques, Vawter and Brown used a technique known as restriction typing to look directly at changes in DNA. They compared rates of substitution in mitochondrial DNA (mtDNA) to rates of substitution in nuclear DNA, for pairs of primate species and pairs of sea urchin species. They found that, in primates, mitochondrial DNA evolved 5 to 10 times faster than nuclear DNA (Vawter and Brown 1986). Thus, different regions of the genome evolved at different rates. Worse still for the molecular clock hypothesis, they found that, in sea urchins, mitochondrial and nuclear DNA evolved at approximately the same rate. From this, they could conclude that the rates of molecular evolution of one or both of these types of DNA were different in sea urchins than they were in primates. They found that the clock turns at different rates in different lineages.

Researchers turned to the possibility that molecular clock approaches would work better when applied to a single genomic region within a single clade. If this worked, molecular clocks could still be useful for dating evolutionary events. In order to test this hypothesis, Takashi Gojobori and his colleagues, and Thomas Leitner and Jan Albert examined molecular evolution in the influenza A virus and the human immunodeficiency virus (HIV), respectively (Gojobori et al. 1990; Leitner and Albert 1999). Viruses are particularly useful organisms for testing this hypothesis, because for known strains of a given virus, we do not have to estimate the dates of evolutionary events from fossil data or other sources of information. Rather, the viruses evolve so rapidly and are sampled so intensively by medical researchers that we can often very closely determine divergence dates from epidemiological information. For viruses such as influenza and HIV, parts of the viral genome have been mapped in numerous strains at many points over time and in many populations, allowing evolutionary biologists to construct phylogenies and test ideas about the neutral theory.

As expected, Gojobori found that substitutions were more common at synonymous sites than at nonsynonymous sites (Figure 8.40). Moreover, as predicted by the neutral theory, the substitution rate was constant across different strains of influenza A over the 20-year time interval examined by the researchers (Gojobori et al. 1990). Leitner and Albert found similar results at one region, known as p17 of the HIV-1 genome, based on a study of well-characterized strains that have been tracked for 25 years in Sweden. But at a different region, known as V3, nonsynonymous changes were more frequent. In both regions, however, we see an approximately linear increase in the number of differences as a function of divergence time. Their results are illustrated in Figure 8.41 (Leitner and Albert 1999).

Another inherent limitation of molecular clock methods is that for any particular gene, the number of substitutional differences between two lineages will not

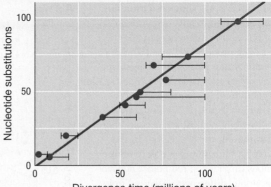

FIGURE 8.39 Nucleotide substitution rate appears to be approximately constant in mammals. Each point reflects a pair of mammalian species; the horizontal axis indicates their divergence time as estimated from fossil data, while the vertical axis indicates the estimated number of nucleotide differences in seven proteins compared across each species pair. Adapted from Wilson et al. (1977).

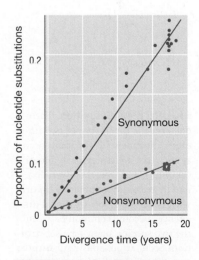

FIGURE 8.40 Clocklike molecular evolution in influenza A. The proportion of nucleotide differences as a function of divergence time for pairs of influenza A strains. Adapted from Gojobori et al. (1990).

FIGURE 8.41 Clocklike molecular evolution in HIV. Proportion of nucleotide differences in (**A**) the p17 region of HIV-1 and (**B**) the V3 region strain pairs, plotted against the divergence times of each pair. Synonymous differences are more common in the p17 region, but nonsynonymous differences are more common in the V3 region. In the graphs, blue points denote synonymous differences; red points nonsynonymous differences. Adapted from Leitner and Albert (1999).

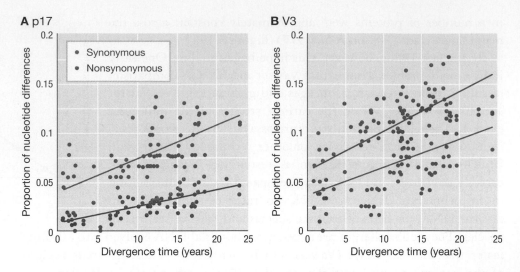

increase indefinitely with time. After the initial divergence of two lineages, most substitutions will occur at sites that were previously identical in the two species. During this period, differences will tend to accumulate at an approximately constant pace, and it is during this period that divergence will accumulate in a clocklike manner. But after two lineages have diverged substantially, further substitutions may occur at sites that already differ. Such substitutions do not contribute to increased divergence between the two lineages, and as a result the rate at which divergence increases with time begins to slow down. Once this happens, differences cease to accumulate in a clocklike fashion (Figure 8.42). This phenomenon is known as *saturation*, because the sequence has become saturated with substitutions and further substitutions will not be detected.

Statistical methods can be used to correct some of the effects of saturation, but eventually the number of sequence differences between two lineages reaches a steady state and provides no further information about the divergence time. Thus,

FIGURE 8.42 Saturation. Early on, after the divergence of two lineages, most new substitutions occur at sites that were previously identical (pink), and thus the divergence rate increases approximately linearly with time. Once substitutional differences become common between the two species, many new substitutions occur at previously substituted sites (blue), and the divergence rate slows. This phenomenon is known as saturation.

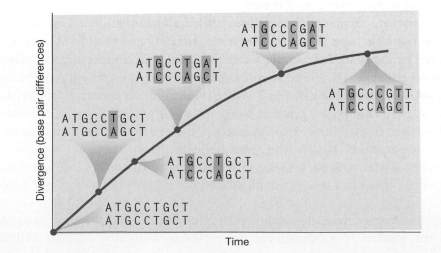

there is a natural timescale for molecular clocks. Clocks based on sites that change rapidly, such as the silent third-codon positions, are useful for looking at short time periods. They accumulate changes quickly, and so they can be used to estimate recent evolutionary events, but they also saturate relatively quickly, and thus they are useless for inferring ancient events. As an extreme example, Jaume Jorba and colleagues looked at the divergent rate of third codon sites in the rapidly evolving polio virus. While divergence initially accumulates in clocklike fashion, after only a decade the effects of saturation become important and the rate of further divergence slows (Jorba et al. 2008) (Figure 8.43). Clocks based on sites that change very slowly, such as nonsynonymous sites in highly conserved genes, do not accumulate enough differences to be useful in dating recent events, but they also are slow to saturate, and thus they can be used to date ancient events. For example, 16S ribosomal RNA sequence is useful for dating very old evolutionary events; it takes approximately 50 million years to accumulate 1% sequence divergence at this locus.

Over the past two decades, a great volume of work has used DNA sequence data to quantify rates of molecular evolution. These studies have collectively affirmed that different parts of the genome evolve at different rates. Synonymous sites—which tend to be neutral or very nearly neutral—accumulate substitutions more rapidly than do nonsynonymous sites, which tend to be under stabilizing selection. Noncoding regions tend to change more rapidly than coding regions, although some noncoding regions also appear to be under stabilizing selection, presumably because of their functional roles in gene regulation. "Housekeeping" genes that perform essential core functions tend to change less rapidly than do genes with more limited or specialized function. These differences are unsurprising; in all cases, the general pattern is that the stronger the action of stabilizing selection, the slower the substitution rate. These differences are also useful. The fact that different loci change at different molecular clock rates allows researchers to pick loci that change at a rate appropriate for answering the questions of interest. To look at a recent evolutionary divergence, one might choose to look at rapidly changing sites; to study ancient evolutionary events, more highly conserved sites would be more useful.

These studies have also revealed that, as Vawter and Brown suspected, evolutionary rates differ along different lineages. This creates further problems for the use of molecular clocks. But this does not mean that molecular information is useless for dating evolutionary events. Population geneticists have developed an ensemble of statistical methods, collectively known as *relaxed clock methods*, to partially compensate for the difficulties introduced by differing evolutionary rates (Welch and Bromham 2005). Dating based on clocklike methods remains an important tool in evolutionary biology, and how to best estimate such divergence dates from genomic information remains an active area of research.

Generation Time and the Rate of Neutral Substitution

We conclude this chapter with a puzzle, and a likely solution. The puzzle is this: For species with similar mutation rates, the neutral theory predicts a constant rate of synonymous substitution *per generation.* But empirical data suggest that the rate

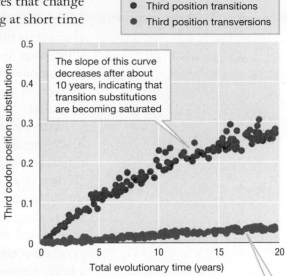

The slope of this curve decreases after about 10 years, indicating that transition substitutions are becoming saturated

The slope of this curve has not diminished appreciably. The less frequent transversions have yet to saturate after 20 years

FIGURE 8.43 Rapid saturation of sequence divergence. Changes in the third codon position accumulate rapidly in poliovirus. Within a decade, frequently occurring transitions at the third position (red) have started to saturate. By contrast, transversions (blue) occur at a lower rate, and thus even after 20 years transversion substitutions continue to accumulate in a clocklike manner. Adapted from Jorba et al. (2008).

of synonymous substitution is approximately constant *per year*, despite the fact that generation times across taxa differ dramatically. The generation time for a rat, for example, is much shorter than the generation time for an elephant. Over absolute time, then, organisms with faster generation times would produce many more generations of offspring than their slower counterparts. How can we explain why, for pairs of species such as rats and elephants, the rates of molecular evolution are so similar?

In the 1970s, using an approach known as the *relative rates test*, Allan Wilson and his colleagues found that the annual rate of molecular change in short-generation time and long-generation time mammalian lineages was approximately equal (Wilson et al. 1977). This suggests that generation time—at least in mammals—should not strongly bias a molecular evolutionary clock. More recent work supports this finding. Using data on 17,208 genetic sequences from more than 300 species of placental mammals that varied from short generation times (rodents) to long generation times (primates), Sudhir Kumar and Sankar Subramanian estimated a fairly constant mutation rate of approximately 2×10^{-9} substitutions per base pair *per year* and found that neutral mutations accumulate at the same rate in short- and long-lived mammals (Kumar and Subramanian 2002). They found this to be the case both when the divergence time between mammalian species was estimated from fossil data and when it was estimated from molecular genetic data (Figure 8.44). Thus, researchers can test the estimates of divergence time using independent estimates for when mammalian species diverged from one another.

A similar sort of debate has been going on with respect to generation time and the rate of molecular evolution of neutral traits in plants. Many researchers argue that such change should occur more quickly in annuals, which live only 1 year, than in perennials, which live more than a year. This change should also occur more quickly in taxa in which "time to flowering" is rapid compared to taxa where time to flowering is slower (Gaut et al. 1996; Laroche et al. 1997; Charlesworth and Wright 2001). In the most comprehensive work done to date on plants, however, there was little evidence for a generation time effect (Whittle and Johnston 2003). Using 24 paired comparisons between annual and perennial plant species, and 9 paired comparisons between trees with short and long generation times, no evidence was found for different rates of gene substitution in any of them.

FIGURE 8.44 Accumulation of neutral substitutions in mammals. Each point represents a pair of mammalian species. The vertical axis—evolutionary distance—is a measure of the number of genetic differences between species. The horizontal axis—divergence time millions of years ago (mya)—is determined from fossil evidence **(A)** and from molecular genetic data **(B)**. Adapted from Kumar and Subramanian (2002).

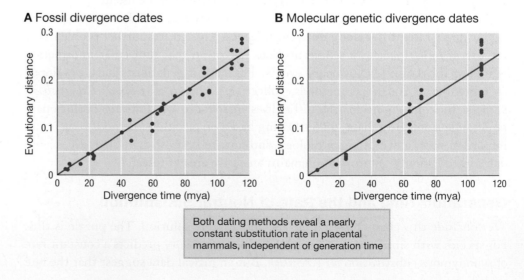

A Fossil divergence dates

B Molecular genetic divergence dates

Both dating methods reveal a nearly constant substitution rate in placental mammals, independent of generation time

The finding that mutation rate per base pair per year is similar in long- and short-generation species does make molecular clocks more useful for estimating divergence times in that we do not automatically expect different rates of evolution along branches with different generation times. But *why* should it be the case that mutation rate/base pair/year is similar in long- and short-generation species, given that more generations, and hence more opportunities for mutation, occur in the latter?

Tomoko Ohta provided an answer to this puzzle by modifying the neutral theory to account for the prevalence of mildly deleterious mutations. The **nearly neutral theory** of molecular evolution posits that most substitutions are, if not exactly neutral, only mildly deleterious (Ohta 1992). Their fate is consequently determined by the interplay between selection and drift as discussed in Section 8.4. Population size then plays a critical role in determining the balance between drift and selection. Whereas the neutral theory predicts that the substitution rate is independent of population size, the nearly neutral theory predicts that the substitution rate is higher in smaller populations, where mildly deleterious alleles can drift to fixation. This provides a possible resolution to the puzzle described above. Under the neutral theory, we would expect species with longer generation times to have lower annual substitution rates. But generation time is highly correlated with population size, so species with larger populations tend to have shorter generation times (Figure 8.45). These factors at least partly cancel one another's effects, leading to an approximately constant annual rate at which nearly neutral mutations are fixed across a wide range of generation times.

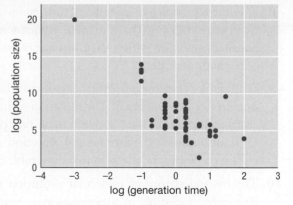

FIGURE 8.45 Population size and generation time. Species with larger populations tend to have shorter generation times. Based on data from 77 species. From Chao and Carr (1993)

Over the past two chapters, we have explored the processes by which allele frequencies change in large and small populations. In the next chapter, we will explore what happens when allele frequencies are changing simultaneously at more than one locus. It is there—in the interplay between alleles at different loci—that we will find much of the action that makes evolutionary biology so interesting.

SUMMARY

1. In finite populations, allele frequencies will fluctuate as a result of random sampling effects. This process is known as genetic drift.

2. Genetic drift operates more strongly in smaller populations than in large populations.

3. Genetic drift reduces the heterozygosity—the fraction of individuals who are heterozygous at a given locus—within a population by causing alleles to be fixed or lost even in the absence of natural selection.

4. Because genetic drift is a random process, it causes divergence between populations over evolutionary time.

5. We can trace the genealogy of individual gene copies through a population. For any sample of gene copies at a single locus, somewhere in the past there is an ancestral gene copy from which all copies in our sample are descended.

6. Tracing this genealogy of gene copies back in time, we derive the coalescent tree. In a sexual population, every locus has a different coalescent tree.

7. Population bottlenecks, in which populations are temporarily reduced to a small number of individuals, accelerate genetic drift and can cause substantial changes in allele frequencies.

8. Allele frequencies in peripheral and island populations can differ greatly from allele frequencies in the populations from which they were derived because of the founder effect.

9. Drift reduces heterozygosity in a population, but mutation creates new variation. The mutation–drift balance represents a steady state between these two processes. Drift also interacts with natural selection and can reduce the ability of selection to fix favorable alleles.

10. The neutral theory of molecular evolution proposes that most variation in a population is neutral and most substitutions that occur over evolutionary time are neutral substitutions. If so, it follows that genetic drift plays a major role in the evolutionary process.

11. Under the neutral model, the fixation rate in a population is equal to the mutation rate in an individual in that population.

12. At many loci, molecular changes occur at an approximately constant rate over time. The behavior of this molecular clock makes it possible to assign dates to the branch points on a phylogeny using DNA sequences.

KEY TERMS

coalescent point (p. 258)

coalescent theory (p. 258)

effective population
   size (p. 252)

expected heterozygosity (p. 248)

founder effect (p. 267)

genetic drift (p. 247)

molecular clock (p. 281)

nearly neutral theory (p. 287)

neutral theory (p. 277)

nonsynonymous mutations (p. 278)

observed heterozygosity (p. 248)

population bottleneck (p. 265)

pseudogene (p. 279)

selectively neutral (p. 247)

substitutions (p. 277)

Wright–Fisher model (p. 245)

REVIEW QUESTIONS

1. In Box 8.2, we showed that in a Wright–Fisher population the expected heterozygosity decreases by a factor of $1/2N$ each generation as a result of drift in a finite population. Does this mean that over time, we would expect to see fewer heterozygotes than predicted by Hardy–Weinberg proportions? Why or why not?

2. In natural populations, the effective population size N_e is typically less than the census population size N. Is there any way that N_e could exceed N? Why or why not?

3. A researcher sets up 100 replicate population cages. Each is founded with 20 *Drosophila melanogaster* individuals, drawn from a population that is polymorphic for the L_1 and L_2 alleles at the larval cuticle pseudogene locus. After many months, the L_1 allele is fixed in 11 of the 100 cages, and the L_2 allele is fixed in 89 of the cages. Estimate the frequencies of the L_1 and L_2 alleles in the original population from which these cages were founded.

4. In many polygynous songbird species, such as wrens or red-winged blackbirds, a single male holds a territory and mates with several females on that territory. In monogamous species, such as cardinals and blue jays, mated pairs typically hold a territory and males mate with only one female. In comparably sized populations, do you expect drift to have a stronger effect in a polygynous species or in a monogamous species? Explain.

5. Consider a neutral locus, in a constant-size population of 500 diploid individuals. Which is expected to take longer: coalescence of the 1000 gene copies at this locus down to 10 ancestral copies, or coales-

cence of those 10 ancestral copies down to a single ancestral copy?

6. Mutations at the *A* locus occur in approximately 1 individual of 2000, but comparisons with closely related species suggest that substitutions at the *A* locus occur approximately once every 20,000 generations. Based on this information, has the *A* locus been neutral or under selection? Explain.

7. In Section 6.6, we saw that the majority of new mutations in *Drosophila* are deleterious. Is this observation inconsistent with the neutral theory of molecular evolution? Why or why not?

8. Researchers measure genotype frequencies in a wild population of mice, and find that the observed heterozygosity is significantly lower than the expected heterozygosity for this population. Propose a hypothesis for what evolutionary process may have been responsible for this observation.

9. The following figure plots the frequency of the A_1 allele over time at a neutral locus in an isolated population. After 400 generations, this allele has become fixed in the population. Based on the graph, do you think this population has been growing, declining, or staying at a constant size?

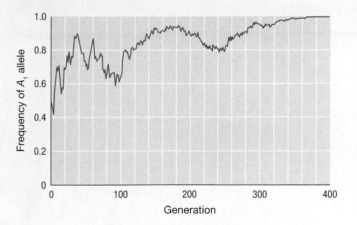

10. The gene genealogy below has not yet been untangled. Find the coalescent point for the 10 gene copies shown at the top of the diagram.

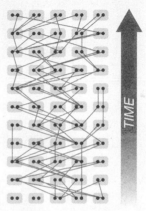

SUGGESTED READINGS

Gamache, I., J. P. Jaramillo-Correa, S. Payette, and J. Bousquet. 2003. Diverging patterns of mitochondrial and nuclear DNA diversity in subarctic black spruce: imprint of a founder effect associated with postglacial colonization. *Molecular Ecology* 12: 891–901. A case study of a leading-edge expansion of the black spruce into mid- and northern Canada.

Hoelzel, A. R. 1999. Impact of population bottlenecks on genetic variation and the importance of life history: a case study of the northern elephant seal. *Biological Journal of the Linnean Society* 68: 23–39. A fascinating history of the near-extinction of the northern elephant seal and the population-genetic consequences of this demographic event.

Hudson, R. R. 1990. Gene genealogies and the coalescent process. *Oxford Surveys in Evolutionary Biology* 7: 1–44. A clear early exposition of coalescent theory.

Jorba, J., R. Campagnoli, L. De, and O. Kew. 2008. Calibration of multiple poliovirus molecular clocks covering an extended evolutionary range. *Journal of Virology* 82: 4429–4440. A recent example in which molecular clocks have been calibrated for the polio virus.

Ohta, T. 1992. The nearly neutral theory of molecular evolution. *Annual Review of Evolution and Systematics* 23: 263–286. An overview of the nearly neutral theory and how it helps to explain clocklike molecular evolution.

Evolution at Multiple Loci

I n the first chapter of this book, we described how the use of antibiotics selects for the evolution of antibiotic-resistant strains of bacteria. Such strains represent a major public health threat and annually are responsible for tens of thousands of deaths in the United States alone. Many of these strains have resistance mutations that modify the chemical composition of an antibiotic's target: a membrane protein, a ribosomal RNA, a component of the cell wall, or some other cellular structure. For example, methicillin-resistant *Staphylococcus aureus* (MRSA) features a modified membrane protein that no longer binds methicillin. Such modifications are highly beneficial in the presence of antibiotics, but they can have substantial fitness costs to the bacterial cell in the absence of antibiotics because they hinder or eliminate the function of highly adapted elements of the cellular machinery.

Because bacteria suffer fitness costs to resistance in the absence of antibiotic use, we might expect that resistance evolution would be "reversible": In the presence of antibiotics, resistant strains would increase in frequency due to the benefits conferred under those conditions, while in the absence of antibiotics,

◀ The rugged limestone formations of Madagascar's Bemaraha National Park.

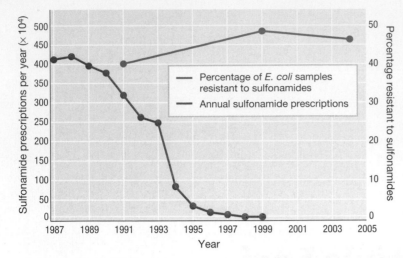

FIGURE 9.1 Sulfonamide use and sulfonamide resistance. This graph shows the estimated number of prescriptions for sulfonamide-containing drugs in the U.K. in red, and the fraction of *E. coli* samples that exhibited sulfonamide resistance in blue. Curtailing sulfonamide use did not reduce the frequency of sulfonamide resistance. Adapted from Enne et al. (2001) and Bean et al. (2005).

resistant strains would decrease in frequency and eventually be lost due to the costs of resistance when no antibiotics are present. Sometimes this is exactly what happens. In the early 1990s, Finland reduced its use of the antibiotic erythromycin almost by half. In response, the frequency of erythromycin resistance in streptococcal bacteria dropped sharply over the next 5 years (Seppälä et al. 1997).

Unfortunately, such drops do not always occur. Terminating antibiotic use is not necessarily sufficient to reverse the evolution of resistance, either within an individual patient (Sjölund et al. 2003, 2005) or at the level of the community at large (Sundqvist et al. 2010). One of the most striking examples involves a class of antibiotics known as the sulfonamides, commonly used to treat urinary tract infections caused by *Escherichia coli* and other bacteria. In 1995, concerns over possible side effects led the British government to severely curtail the use of these drugs. Use of sulfonamides in Britain promptly dropped to less than 3% of pre-1995 totals. This set up a natural experiment: What would happen to the frequency of sulfonamide resistance in the *E. coli* population once the use of the drugs was reduced more than 40-fold on a nationwide scale? The results have been striking—and discouraging (Enne et al. 2001; Bean et al. 2005). Over 9 years, from 1995 to 2004, sulfonamide resistance did not decline at all in Britain despite the cessation of drug use (Figure 9.1).

How do we explain this? Why weren't the alleles for sulfonamide resistance lost after sulfonamide use was halted? If we were to look at evolution one locus at a time—in this case, looking at the locus conferring antibiotic resistance—this would be hard to explain. But if we recognize that evolution occurs simultaneously at multiple loci, and that the loci interact to determine the course of evolution, then the long-term persistence of antibiotic resistance alleles becomes easier to understand. There are a number of reasons why antibiotic resistance does not always disappear as quickly as one might expect given its initial fitness consequences. Among them, the most straightforward is that *compensatory mutations* arise at other loci in resistant bacterial strains. These compensatory mutations do not reduce the degree of resistance, but do reduce or eliminate the fitness costs associated with the resistant phenotype. Compensatory mutations have been documented in a wide range of bacterial species for resistance to a wide range of antibiotics (Andersson and Hughes 2010).

To understand how natural selection operates on resistance and compensatory mutations, it helps to think in terms of the underlying genetics. Recall that bacteria are haploid, with only one gene copy at each locus. In the simplest cases, a mutation at one locus confers antibiotic resistance, and a compensatory mutation at a second locus reduces the fitness cost of resistance (Schrag and Perrot 1996). Call the resistance locus *R*, with alleles *r* (sensitive)—that is, susceptible to antibiotics—and *R* (resistant). Call the compensatory locus *C*, with alleles *c* (uncompensated) and *C* (compensated). The antibiotic-sensitive wild-type strain is drug sensitive and uncompensated—that is, it has the genotype *rc*. In the absence of antibiotics, the antibiotic-sensitive wild type has high fitness, but when antibiotics are in use, the fitness of the wild type is very low. Resistance arises by mutation at the *R* locus, from *r* to *R*. This gives rise to resistant uncompensated individuals with genotype *Rc*, which can now grow even in the presence of antibiotics, albeit with appreciable fitness costs. Next, suppose that a compensatory mutation arises at the *C* locus, from *c* to *C*. This results in resistant-compensated individuals with the *RC* genotype (Figure 9.2A). Not only can they grow in the presence of antibiotics, but also the cost of resistance has been reduced dramatically by the compensatory mutation, and the *RC* individuals now have almost as high a fitness as the wild type even in the absence of antibiotics (Figure 9.2B).

In the presence of antibiotics, both the initial *R* mutation and the subsequent *C* mutation lead to substantial fitness increases, and so each of the mutations is likely to increase rapidly in frequency during a period of antibiotic use. When antibiotic use is halted, the *RC* individuals have only a very slight fitness disadvantage relative to the *rc* wild type, and so resistance will not decline at anything like the rate at which it arose. Once we start thinking about both the resistance locus and the compensatory locus at the same time, we can understand why antibiotic-resistant bacteria strains increase rapidly in frequency when drugs are used but disappear slowly when drugs are withdrawn.

We will return to this example to see some of the additional complexities of evolution at multiple loci that are illustrated by resistance evolution. For now, however, this example simply shows that, to understand the evolution of a phenotype, we often need to understand how evolutionary processes operate on multiple interacting loci. We have seen how this holds for haploid organisms such as bacteria; further complexities arise when we start to consider diploid organisms such as ourselves. In Chapters 7 and 8, we developed the basic mathematical machinery to describe, one locus at a time, how allele frequencies change in large and small populations. But we also need to be able to think about how allele frequencies change at *multiple loci*—and how the process of change at one locus influences what happens at another. And so, in this chapter, we will address the following questions:

- How do multiple genes interact to determine phenotypes?
- What is linkage disequilibrium, how does it arise, and how does it change over evolutionary time?
- How does the physical arrangement of genes within the genome influence the evolutionary process?
- What are adaptive landscapes and how can they help us understand the course of evolution?
- How can we use quantitative genetic models to predict evolutionary change in traits even when we do not know the specific genetic basis for these traits?

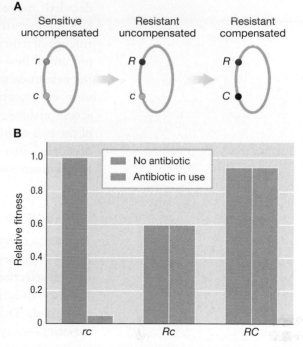

FIGURE 9.2 Evolution and fitness effects of resistance and compensation. (A) The resistance allele *R* replaces the sensitive allele *r*, and then the compensatory allele *C* replaces the wild type *c*. **(B)** In the absence of antibiotics, the wild type *rc* is the most fit. The resistant allele *R* imposes a significant fitness cost that is largely ameliorated by the compensatory allele *C*. In the presence of antibiotics, the wild type has very low fitness. Under these conditions, the resistant allele *R* provides a large fitness benefit and the compensatory allele *C* then provides an additional fitness advantage.

9.1 Polygenic Traits and the Nature of Heredity

As we discussed in Chapter 2, Darwin crafted his theory of evolution by natural selection without having even a rudimentary understanding of the mechanisms of inheritance. Without that understanding, it was not possible to construct detailed, quantitative models of the process of natural selection. Then, in the late nineteenth and early twentieth centuries, researchers began to uncover the rules of heredity. They rediscovered Mendel's work, and they developed a detailed picture of how genetic inheritance operates. Initially, it was not at all clear that the new understanding of how genes are transmitted was consistent with Darwin's view of evolution by natural selection. One of the most important scientific accomplishments of the first half of the twentieth century was the reconciliation of the two major components of biology—Darwin's theory of natural selection and Mendel's discoveries about genetic transmission—leading to the founding of the field known today as population genetics. We now examine how this occurred.

Continuous versus Discontinuous Variation

The rediscovery of Mendel's work eventually solved a number of problems for Darwin's theory but, in the short term, it posed new challenges to his ideas of natural selection. Even before researchers revisited Mendel's work on the nature of heredity, an active debate raged over the nature of the changes by which evolution proceeded. Darwin argued that evolution primarily occurred by continuous variation—very small gradual changes in form, such as the elongation of a bone, or a gradual shift in the color of an animal's fur—but even many of Darwin's closest allies disagreed.

Mendel's findings only exacerbated the debate. Over the first decade of the twentieth century, researchers amassed an impressive body of experimental evidence in support of Mendelian inheritance, but this view of inheritance was seen as incompatible with Darwinian gradualism by most biologists of the time. The problem was that the Mendelian traits studied by geneticists involved large discrete variations rather than the finely graded continuous characters that Darwin took to be essential for his theory of natural selection. This left researchers with a number of questions: Could Darwin's mechanism operate on Mendelian characters? Is Mendelian inheritance compatible with the sort of small graded variation often observed for traits such as size or height? To what degree are small graded variations heritable at all? These issues were resolved through the joint efforts of theoreticians and experimental researchers.

Polygenic Traits Can Exhibit Nearly Continuous Variation

On the theoretical side, the recognition that many traits are **polygenic**—that is, affected by many genes simultaneously—was a first step in reconciling Darwinian natural selection with Mendelian inheritance. George Yule and Ronald Fisher independently developed mathematical models demonstrating that Mendelian inheritance was compatible with small graded variations provided that multiple genes, each of relatively small effect, were involved (Yule 1902; Fisher 1918). This was a suggestion that Mendel himself had made based on observations of flower color in the bean plant *Phaseolus vulgaris*, but the idea was largely overlooked after the rediscovery of his work. The patterns of multifactorial inheritance—that is, inheritance of a polygenic trait—are more complicated than simple Mendelian inheritance for a trait controlled by a single locus, but they are still predictable in a Mendelian framework.

On the experimental side, geneticist Herman Nilsson-Ehle observed a nearly continuous gradation of kernel colors in crosses between red-kernel and white-kernel variants of winter wheat when kernel color was polygenic—in this case, involving three genes (Figure 9.3) (Nilsson-Ehle 1908). The genes in Nilsson-Ehle's kernel color system interacted in a particularly straightforward way: They had **additive genetic effects**, meaning that the phenotype of any individual could be worked out simply by summing the effects of each allele that it carried. The more alleles for dark red color that an individual carried, the darker the phenotype.

Edward East found comparable results on polygenic traits and continuous variation in his cross-breeding experiments with maize (Emerson and East 1913). Thomas Hunt Morgan, working with the fruit fly *Drosophila*, likewise was able to isolate a number of Mendelian factors of very small effect (Morgan et al. 1925). Collectively, these theoretical and empirical observations established that the small graded variations that were so important to Darwin's view of evolution were compatible with the Mendelian picture of inheritance.

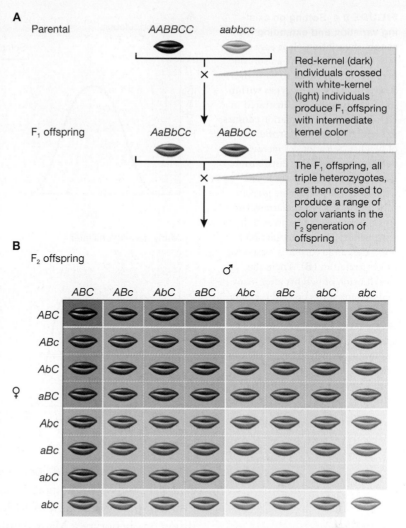

The Importance of Latent Variation

Empirical work in the early twentieth century demonstrated that polygenic traits produced the variation necessary for natural selection to operate, but it also showed that new types, not seen in a parent population, could appear in the offspring produced by that population. Where could these new types have come from? Could they all have been the result of new mutations? Or could there have been other possible "variance-generating" engines that had been overlooked?

Consider this problem: It is relatively easy to see how a variational sorting process like the one we described in Chapter 2 could act on a population with heights ranging from 5'0 to 5'10" and generate offspring with heights in the range of 5'5" to 5'10" (Figure 9.4A). But natural selection doesn't create variation; it reduces variation—by favoring some forms over others. So why should new variants—for example, offspring in the range of 5'10" to 6'4" (Figure 9.4B)—arise in a population selected for greater height?

The answer emerged through the synthesis of Mendelian heredity with Darwinian evolution. As we have seen, when multiple genetic factors are involved in determining a phenotype, variation at a relatively small number of genetic loci can potentially generate an enormous number of possible phenotypes. But these sorts of polygenic characters can also shed light on how we can see variants in

FIGURE 9.3 Multifactorial inheritance generates near-continuous variation. Grain color in winter wheat is controlled by three loci, here labeled *A*, *B*, and *C*. **(A)** Nilsson-Ehle crossed red-kernel parents with white-kernel parents, to produce F_1 progeny of intermediate grain color. He then crossed the F_1 progeny to produce offspring with a range of grain colors in the ratios illustrated in **(B)**. Here we use the background shading in each box to indicate the color of the wheat kernels.

FIGURE 9.4 Sorting on existing variation and extending the range of variation. It is easy to see how natural selection can sort on preexisting variation to shift the distribution of phenotypes within its current range, as illustrated in (**A**). Initially the population consists of broad bell-shaped distribution of heights. Selection for increased height sorts upon this distribution, narrowing the distribution and increasing the average height of individuals many generations later. But natural selection can also shift the phenotypes in a population beyond the range currently observed, as illustrated in (**B**). There the distribution of heights after many generations shifts beyond the range observed initially. But where has the new variation come from? This question was resolved in the early twentieth century through the synthesis of Mendelian inheritance with Darwin's theory of natural selection.

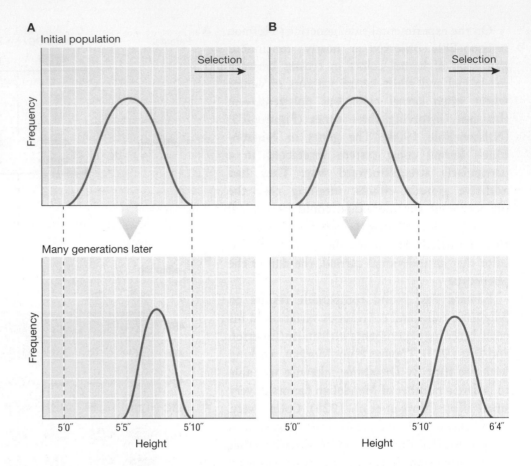

generation 2 that we did not observe in generation 1, as in our height example. For example, Nilsson-Ehle noted that, under certain conditions, the presence of two different variants at each of 10 loci would be sufficient to generate nearly 60,000 different phenotypes. Many natural populations might be too small to manifest all of these possible phenotypes. When a new phenotype is observed in subsequent generations, it need not be the result of a new mutation; it could simply be a new assortment of previously occurring Mendelian variation.

Herein lay the answer to how natural selection could drive a population beyond its original range of variation without having new mutations arise. Because natural selection changes the allele frequencies in a population, it also changes the probabilities that various allele combinations will be realized. Over time, allele combinations that might have been highly unlikely to occur in a modest-size population under the initial allele frequencies might have been much more likely to occur under the shifted allele frequencies resulting from the operation of natural selection. Box 9.1 provides a concrete illustration of such a case.

Thus, population geneticists came to recognize that, under Mendelian inheritance, populations contained latent variation—that is, there was so much Mendelian variation within populations that not all possible genotypes could be represented. As a result, selection could shift allele frequencies, and genetic reassortment could then draw out new phenotypes from the preexisting variation, even in the absence of further mutation.

BOX 9.1 A Numerical Example of How Selection and Reassortment Can Generate New Phenotypes

Suppose we are studying a diploid population in which a trait such as cell volume is controlled by 10 unlinked loci labeled A–J, each of which contributes additively to the phenotype in question. At each locus, there are two alleles that we label A_1 and A_2, B_1 and B_2, . . . , J_1 and J_2. Suppose that for each allele X_1 that an individual possesses, the trait value is increased by 1 unit, and for each copy of X_2, the trait value increases by 0 units. The phenotype is then determined simply by the number of X_1 alleles and the number of X_2 alleles that an individual carries. If an individual has only X_2 alleles, that individual has a phenotype of 0. If an individual has only X_1 alleles, that individual has a phenotype of 20. Now suppose we start with a population in which each allele is at a frequency of 50% at each locus, as illustrated in the top panel of Figure 9.5A. A population of 100 individuals might have a distribution of phenotypes, as shown in the top of Figure 9.5B. In this particular population, the range of phenotypes goes from 4 to 15. Now suppose that natural selection operates on this population, so that only individuals with phenotypes of 11 or higher survive. The distribution of survivors is then as shown in Figure 9.5B, middle.

Among the surviving members of the population, the allele frequencies have now shifted as a result of selection (and sampling effects), as shown in the bottom panel of Figure 9.5A. When these individuals produce new offspring, the offspring will now have the shifted set of phenotypes shown in Figure 9.5B, bottom.

In this example, in the course of a single generation of selection, three new phenotypes (16, 17, and 18) have arisen—not through mutation, but through reassortment of the latent variation that was present all along in the population. Prior to selection, when X_1 allele frequencies were 50%, the chances of producing an offspring with 16 or more X_1 alleles were very low. After selection, the X_1 allele frequencies increased substantially—and the chances of producing offspring with 16, 17, or even 18 X_1 alleles became sufficiently high that such offspring were observed in the next generation.

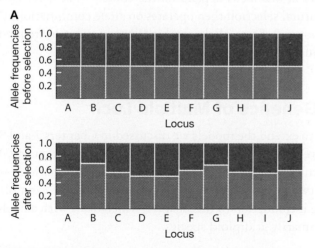

A

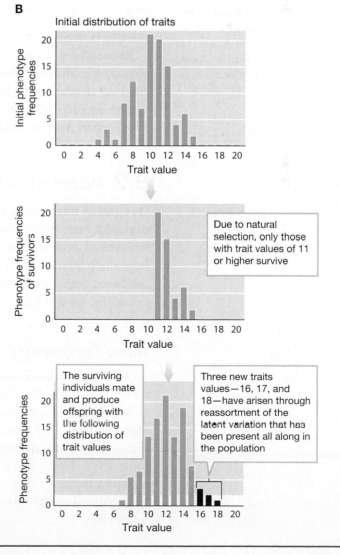

B

FIGURE 9.5 Selection reveals latent variation. (A) In the thought experiment shown here, at each of 10 loci, there are two alleles that we label A_1 and A_2, B_1 and B_2, . . . , J_1 and J_2. For each allele X_1 possessed by an individual, the phenotypic trait value is increased by 1 unit, and for each copy of X_2, the trait value increases by 0 units. Initially, each allele is at a frequency of 50% at each locus, as shown (blue: X_1 alleles, red: X_2 alleles). Selection then operates so that only individuals with phenotype trait values of 11 or higher survive. **(B)** Among the surviving members of the population, the allele frequencies have now shifted due to selection and sampling effects. The offspring of these survivors have a new distribution of trait values; in this new distribution, three new trait values—16, 17, and 18—have arisen not through mutation, but through reassortment of the latent variation that was present all along in the population.

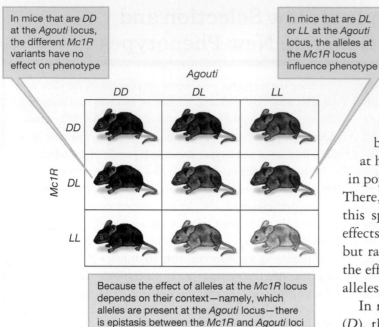

In mice that are *DD* at the *Agouti* locus, the different *Mc1R* variants have no effect on phenotype

In mice that are *DL* or *LL* at the *Agouti* locus, the alleles at the *Mc1R* locus influence phenotype

Agouti

Because the effect of alleles at the *Mc1R* locus depends on their context—namely, which alleles are present at the *Agouti* locus—there is epistasis between the *Mc1R* and *Agouti* loci

FIGURE 9.6 Epistasis between the *Mc1R* and *Agouti* loci. When both of the alleles at the *Agouti* locus are *D* (dark), different alleles at the *Mc1R* locus have no effect. In mice with at least one *L* (light) allele at the *Agouti* locus, the alleles at the *Mc1R* locus influence coat color. Adapted from Steiner et al. (2007).

Gene Interactions

Not all genes interact to produce the straightforward additive genetic effects that Nilsson-Ehle observed in his wheat kernel color system. When the alleles at two or more loci interact in *nonadditive* ways to determine phenotype, we refer to this as **epistasis**. Let us begin by considering an example. In Chapter 3, we looked at how natural selection operates on coat color variation in populations of the oldfield mouse, *Peromyscus polionotus*. There, we described two loci that influence coat color in this species: the *Mc1R* locus and the *Agouti* locus. The effects of alleles at the two loci do not combine additively, but rather the loci exhibit epistatic interactions—that is, the effect of an allele at the *Mc1R* locus depends on which alleles are present at the *Agouti* locus (Steiner et al. 2007).

In mice that are homozygous for the dark *Agouti* allele (*D*), the effects of the *Mc1R* locus are entirely masked—irrespective of the genotype at *Mc1R*, the mice have fully dark coloration. But when at least one copy of the light *Agouti* allele (*L*) is expressed, the effects of the *Mc1R* locus are revealed (Figure 9.6).

Because of epistasis, the phenotypic effects of these loci are *context dependent*; the phenotypic effects of alleles at one locus depend on the context that is set by the alleles at another locus. Natural selection then operates on allele combinations that determine particular phenotypes. Some allele combinations increase in frequency, while others may be eliminated from the population.

9.2 Population Genetics of Multiple Loci

Our goal in this section is to extend the models we discussed in Chapters 7 and 8 to deal with cases in which we are concerned with more than one locus at a time. We will aim to work out the rules—and write down the mathematical equations—for how allele frequencies at two or more loci jointly change. Unlike the example of antibiotic resistance in the introduction to this chapter, which dealt with a haploid organism, we will look primarily at diploid species.

Allele Frequencies and Haplotype Frequencies

To treat the population genetics of multiple loci, it is not enough to simply track the frequencies of the alleles at these loci. Rather, we need to track the frequencies of **haplotypes**. A haplotype is defined as a set of alleles, one at each locus under consideration. If, for example, we are interested in the *A*, *B*, and *C* loci of a diploid organism, *ABc* or *aBC* would be haplotypes, whereas *Aa BB Cc* would be a genotype. Often when population geneticists talk about a haplotype, they are referring to the set of gene copies along one particular chromosome. In organisms with haploid gametes (such as ourselves), we can also talk about the haplotype of a gamete.

By way of an example, suppose we are interested in tracking or modeling evolutionary dynamics at two loci—call them A and B—each of which has two alleles: A and a, B and b. (In this chapter we will minimize the proliferation of subscripts in our notation by using A and a instead of A_1 and A_2; this does not imply that A is dominant and a recessive.)

To understand the population genetics of both loci together, we have to keep track not only of how many A and a alleles there are and how many B and b alleles there are, but also of which alleles at the A locus are associated with which alleles at the B locus. Why do we need to do this? The answer is that the allele frequencies do not uniquely determine the haplotype frequencies. For example, imagine a population with haplotype frequencies 33% AB, 17% Ab, 0% aB, and 50% ab. The allele frequencies in this population are 50% A, 50% a, 33% B, and 67% b. But if we know only these allele frequencies and not the haplotype frequencies, we would be unable to tell that the a allele never co-occurs with the B allele. For example, given the same allele frequencies, the haplotype frequencies might instead have been 16.5% AB, 33.5% Ab, 16.5% aB, and 33.5% ab.

Similarly, to predict what sorts of offspring will be produced in a population, we need to consider not only the allele frequencies of the parents, but also the haplotype frequencies. In doing so, it will be helpful to recall how loci are physically positioned within the genome. In diploids, two loci A and B may either be located on separate chromosomes or on the same chromosomes. In the former case, the alleles at these loci will segregate independently according to Mendel's laws. An $AaBb$ parent will produce four types of gametes, AB, Ab, aB, and ab, with equal frequency (Figure 9.7). In the latter case, where the two loci are on the same chromosome, we say that there is **physical linkage** between the two loci. In the absence of recombination, physically linked loci segregate together. A parent with one AB chromosome and one ab chromosome—which we denote $AB|ab$—will produce only AB and ab gametes in the absence of recombination (Figure 9.7). Similarly, a parent with one Ab chromosome and one aB chromosome—denoted $Ab|aB$—will produce only Ab and aB gametes in the absence of recombination.

Through the presence of recombination, alleles at loci on the same chromosome can be reassorted to form new combinations. $AB|ab$ parents can produce Ab and aB gametes, and $Ab|aB$ parents can produce AB and ab gametes. The rate of recombination, and thus the proportion of gametes of each type that are produced, depends on the physical distance between the A and B loci on the chromosome. If the two loci are located very close to one another, crossover between the two loci will occur only rarely and the recombination rate will be low. If, instead, the two loci are far apart on the chromosome, there will be a high probability of crossing over between the two loci and a higher recombination rate. With this in mind, we can now extend the Hardy–Weinberg

FIGURE 9.7 Location of two loci. Two loci can be located on different chromosomes or on the same chromosome (and hence physically linked).

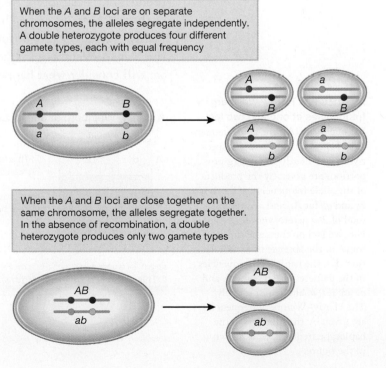

When the A and B loci are on separate chromosomes, the alleles segregate independently. A double heterozygote produces four different gamete types, each with equal frequency

When the A and B loci are close together on the same chromosome, the alleles segregate together. In the absence of recombination, a double heterozygote produces only two gamete types

model to two loci. To keep the algebra simple, we will consider the case in which the two loci are on the same chromosome and there is no recombination between them.

Hardy–Weinberg Proportions for Two Loci

Recall from Chapter 7 how to compute Hardy–Weinberg proportions for a single locus. If we have one locus A with alleles A and a at frequencies p and q in the population, the Hardy–Weinberg proportions reflect the probability that each genotype is formed by random mating. These probabilities are illustrated graphically in Figure 9.8A. In this figure, the height of each box represents the allele frequencies in the population—that is, the height p corresponds to the frequency of the A allele, and the height q corresponds to the frequency of the a allele within the population. The *area* of each block—that is, the product of the allele frequencies—is then proportional to the Hardy–Weinberg frequency of the corresponding genotype. This is because the Hardy–Weinberg frequencies of each genotype are equal to the chance that two gametes that are drawn randomly from the population compose that genotype. The chance is simply the product of the allele frequencies of the two alleles that make up that genotype. Ignoring the order of the two alleles, Aa and aA are indistinguishable; each arises with frequency pq, for a total heterozygote frequency of $2pq$.

We can take the same approach when dealing with two loci, but we have to consider haplotypes rather than alleles. This is because we need to distinguish between two different kinds of double heterozygote ($Aa\ Bb$) parents if we are to correctly predict the genotypes of the offspring. In the absence of recombination, $AB|ab$ parents will produce only AB and ab gametes, whereas $Ab|aB$ parents will produce only Ab and aB gametes.

The Hardy–Weinberg frequencies of a particular genotype are given by the product of the haplotype frequencies that make up that genotype. If the frequencies of the AB, Ab, aB, and ab haplotypes are s, t, u, and v, respectively, the two-locus Hardy–Weinberg proportions are as illustrated in Figure 9.8B. Remember that the Hardy–Weinberg frequencies shown here assume no recombination. Shortly, we will consider what happens when recombination occurs.

FIGURE 9.8 Hardy–Weinberg frequencies at one and two loci. For one locus (**A**), the allele frequencies of A and a are p and q, respectively. The Hardy–Weinberg proportions are given by the products of the allele frequencies: p^2 for AA; pq and qp for Aa and aA, giving a total of $2pq$ heterozygotes; q^2 for aa. For two loci on the same chromosome, in the absence of recombination (**B**), the haplotype frequencies in the gametes of AB, Ab, aB, and ab are s, t, u, and v, respectively. The Hardy–Weinberg frequencies are given by the products of the haplotype frequencies, as shown in the figure.

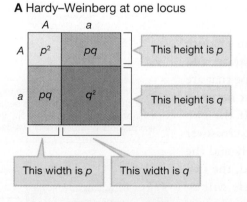

A Hardy–Weinberg at one locus

B Hardy–Weinberg at two loci

Statistical Associations between Loci

In the one-locus Hardy–Weinberg model, there are no statistical associations between one gene copy at a locus and the other gene copy at the same locus. At the Hardy–Weinberg equilibrium, knowing that an individual received the *A* allele from its mother provides us with no new information about whether it received the *A* allele or the *a* allele from its father. This would not be true if we violated the random mating assumption of the Hardy–Weinberg equilibrium under assortative mating, for example. If individuals preferentially chose similar mates, an offspring who received an *A* from its mother would be more likely than the population at large to have received an *A* from its father as well. But when all Hardy–Weinberg assumptions are met, the genotype frequencies are simply equal to the products of the allele frequencies, as illustrated in Figure 9.8A.

In the two-locus Hardy–Weinberg model, the same is true *of gene copies at each single locus considered separately.* Knowing that an individual received the *A* allele from its mother tells us nothing about whether it received the *A* or *a* allele from its father, and knowing that an individual received the *B* allele from its mother tells us nothing about whether it received the *B* or *b* allele from its father. This is why the genotype frequencies in Figure 9.8B are simply equal to the products of the haplotype frequencies.

Even though there are no statistical associations between the two gene copies at one locus in the Hardy–Weinberg model, *there can be associations between gene copies at two different loci.* How can this be? An illustration is useful. Suppose there are no *ab* haplotypes in the population and therefore all *b* alleles are in *Ab* haplotypes. In this case, if we know that an individual has the *b* allele, we also know for certain that this individual has at least one *A* allele.

When statistical associations are present between the alleles at the *A* locus and the *B* locus, we say that there is **linkage disequilibrium** in the population. To model or track the changes in haplotype frequencies in a population over time, we have to account for these associations. In the next section, we will see how to do so, and in the process we will describe a simple way to conceptualize and quantify the amount of linkage disequilibrium in a population.

Quantifying Linkage Disequilibrium

To measure the associations between allele frequencies at two loci *A* and *B*, we look at the haplotype frequencies at these loci. Let f_A, f_a, f_B, and f_b be the frequencies of the *A*, *a*, *B*, and *b* alleles, respectively, and let h_{AB}, h_{Ab}, h_{aB}, and h_{ab} be the frequencies of the *AB*, *Ab*, *aB*, and *ab* haplotypes, respectively. If the allele at the *A* locus occurs independently of the allele at the *B* locus, the haplotype frequencies will be given as (Maynard Smith 1989):

$$h_{AB} = f_A f_B \tag{9.1a}$$
$$h_{Ab} = f_A f_b \tag{9.1b}$$
$$h_{aB} = f_a f_B \tag{9.1c}$$
$$h_{ab} = f_a f_b \tag{9.1d}$$

When the alleles at the *A* and *B* loci occur independently

We define the **coefficient of linkage disequilibrium (*D*)** as the difference between the actual frequency of the *AB* haplotype (h_{AB}) and the expected frequency ($f_A f_B$) of the same haplotype if the loci are independent—that is, if there is no association between the allele at one locus and the allele at the other. Mathematically, this is written as follows:

$$D = h_{AB} - f_A f_B \qquad (9.2)$$

Here, h_{AB} is the actual frequency of the AB haplotype and $f_A f_B$ is the expected frequency of the AB haplotype if the loci were independent. When the alleles at each locus occur independently, these terms will be equal and the coefficient of linkage disequilibrium D will be zero.

When the alleles at each locus occur nonindependently, the linkage disequilibrium is nonzero. Suppose that the A allele is more likely to occur in combination with the B allele, and that the a allele is more likely to occur in combination with the b allele. Then by our mathematical definition (Equation 9.2), the coefficient of linkage disequilibrium D will be a positive value. Conversely, if A is more likely to occur with b, and a is more likely to occur with B, the coefficient of linkage disequilibrium D will be a negative value. In the classical population genetics literature, researchers often used the terms **coupling** and **repulsion** to refer to these associations. We will use these terms here because they provide a convenient language for talking about these types of associations between alleles. When A tends to occur with B and a tends to occur with b, we call this coupling. This is because the "like" alleles represented by the capital letters tend to be coupled in the haplotypes of the population, as are the like alleles represented by the lowercase letters. In contrast, when A tends to occur with b and a tends to occur with B, we call this repulsion. This is because upper- and lowercase alleles generally do *not* tend to occur together. The A allele seems to repel B in favor of b, and the a allele seems to repel b in favor of B (Figure 9.9). Thus, we can view linkage disequilibrium as a measure of whether we have excess coupling haplotypes, in which case D is positive, or excess repulsion haplotypes, in which case D is negative.

Notice that the terms *coupling* and *repulsion* reflect nothing more than our choice of nomenclature for the loci in question. Suppose we had named the alleles differently—for example, suppose that we had called B by the name c, and b by the name C. Then the coupling pair AB would be written as Ac and would be considered a repulsion pair, while the repulsion pair Ab would be written as AC and would be considered a coupling pair. Thus, it is arbitrary whether we call a given haplotype a coupling haplotype or a repulsion haplotype, but this kind of arbitrariness is an inevitable if unfortunate consequence of how the coefficient of linkage disequilibrium D is defined. The sign of D depends on the notation we choose for our alleles in the first place.

Using the mathematical definition of D (Equation 9.2), we can express the frequency of each haplotype as a function of allele frequencies and linkage disequilibrium as follows:

$$\left.\begin{array}{l} h_{AB} = f_A f_B + D \\ h_{Ab} = f_A f_b - D \\ h_{aB} = f_a f_B - D \\ h_{ab} = f_a f_b + D \end{array}\right\} \text{General case} \qquad \begin{array}{l}(9.3a)\\(9.3b)\\(9.3c)\\(9.3d)\end{array}$$

The value of the coefficient of linkage disequilibrium D depends not only on how the alleles at each locus are associated with one another, but also on the frequencies

Coupling   Repulsion

$A\ B \qquad a\ b \qquad\qquad A\ b \qquad a\ B$

FIGURE 9.9 Coupling and repulsion haplotypes. When "like" alleles (A and B, or a and b) appear together, we call this coupling. We call the converse case (A with b, or a with B) repulsion.

of each allele. In our two-locus, two-allele example, D takes on its maximum value when the frequency of each allele is 0.5 (and, as described above, reaching this maximum requires that A always co-occurs with B and a always co-occurs with b). In this case, applying Equation 9.2, $D = (0.5 \times 0.5) - (0 \times 0) = 0.25$. Similarly, D can achieve a minimum of $D = (0 \times 0) - (0.5 \times 0.5) = -0.25$. Thus, the linkage disequilibrium ranges from -0.25 to 0.25. When the associations among loci are not absolute or the allele frequencies deviate from 0.5, D will take on smaller absolute values.

Evolutionary Processes Create Linkage Disequilibrium

We have seen that linkage disequilibrium measures the statistical association between alleles at different loci. But how does this association arise, and what happens to it over time? We turn now to these questions.

Linkage Disequilibrium via Mutation

Linkage disequilibrium can arise from many of the evolutionary processes we have studied, including mutation, selection, drift, and migration. One of the simplest sources of linkage disequilibrium is the spread of a new mutation. Suppose a population is initially polymorphic at the A locus, with both A and a alleles present, but is monomorphic at an adjacent locus B on the same chromosome, with only B alleles present. Because only one of these two loci is polymorphic, there is no linkage disequilibrium. (We can also see this from Equation 9.2: if $f_B = 1$, then $f_A = h_{AB}$, and $D = 0$.)

Now suppose that, on an individual chromosome, a new allele b is formed by mutation at the B locus. This b allele will be adjacent to some allele at the A locus—for the purpose of this example, suppose that b arises adjacent to an a allele, as shown in Figure 9.10.

Prior to the mutation, only the AB and aB haplotypes were present in the population. Subsequent to the mutation, an additional haplotype, ab, has been formed. In this population, there is now a statistical association between alleles at the A and B loci. Most notably, the presence of the b allele at the B locus guarantees the presence of the a allele at the A locus. This means that the A and B loci are now in linkage disequilibrium in this population. Over time, this linkage disequilibrium may break down because of recombination, but we will discuss that process later. For now, the important point is that simple evolutionary history—the mutations that occur and the genetic background on which the mutations happen to arise—generates linkage disequilibrium among loci.

Linkage Disequilibrium via Natural Selection

Natural selection is another very important source of linkage disequilibrium. We will first provide an example of how natural selection can generate linkage disequilibrium, and then we will discuss how selection only generates linkage disequilibrium between alleles with epistatic interactions.

Consider the example in Figure 9.11. Either of two biosynthetic pathways is sufficient to produce an essential molecular product from precursor raw materials.

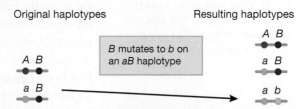

Original haplotypes   Resulting haplotypes

FIGURE 9.10 Mutation can create linkage disequilibrium. In this example, the b allele arises by mutation on a chromosome that carries the a allele at the A locus. As a result, a new coupling haplotype ab is created, but the corresponding repulsion haplotype Ab is not yet present in the population. The result is a positive coefficient of linkage disequilibrium D.

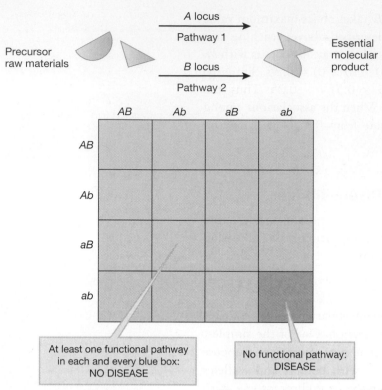

At least one functional pathway in each and every blue box: **NO DISEASE**

No functional pathway: **DISEASE**

FIGURE 9.11 Selection can generate linkage disequilibrium. In this example, an *A* allele or a *B* allele—but not both—is needed to produce an essential molecular product from precursor raw materials. Natural selection disfavors only the *ab* haplotype, and even this haplotype is disfavored only in the case that it is paired with another *ab* haplotype, resulting in *aabb* individuals who are unable to produce the essential molecules, so that disease occurs. Thus, only the *ab* haplotype will be less common than expected among surviving adults given the allele frequencies in the population.

Each pathway is controlled by a single locus, and the functional wild-type alleles (*A*, *B*) are dominant to the nonfunctional disease alleles (*a*, *b*). In this case, only *aabb* individuals are unable to produce this molecule, and thus only these *aabb* individuals manifest the disease phenotype. As a result, selection operates against the *a* allele, but only when it is part of an *ab* haplotype—and even then only if that *ab* haplotype is paired with a second *ab* haplotype. Similarly, selection operates against the *b* allele only when it is part of an *ab* haplotype. All other haplotypes can produce the needed molecular product from precursor raw materials, so no other haplotype is selected against. In this example, there will be a dearth of *ab* haplotypes relative to what would be expected given the frequencies of the *a* and *b* alleles. Natural selection has created a statistical association between the alleles at the two loci—that is, selection has generated linkage disequilibrium between the loci.

Notice that in the example above, the selective consequence of carrying the *a* allele is contingent on what allele was present at the *B* locus. In other words, this example featured epistasis between the *A* and *B* loci. But when there is no epistasis between the loci, selection *will not* generate linkage disequilibrium (Felsenstein 1965). The reason is that the selective consequences of carrying each allele will be independent of what alleles are carried at other loci, so in such a case there will be no statistical association between alleles when we examine the consequences of selection.

Linkage Disequilibrium via Migration

Migration is another source of linkage disequilibrium. For example, suppose the *a* and *b* alleles are fixed in a mainland population of lizards, and the *A* and *B* alleles are initially fixed in an island population of the same species. All the haplotypes on the mainland are the coupling haplotypes *ab*, and all the haplotypes on the island are the coupling haplotypes *AB*. Considered separately, there is no linkage disequilibrium in either population, because each population is monomorphic. If a few individuals then migrate from the mainland to the island, however, *ab* haplotypes will be introduced to the island, and on the island there will now be a statistical association between the *A* and *B* loci (Figure 9.12). In particular, even though both loci are polymorphic, all haplotypes on the island will be coupling haplotypes, so the coefficient of linkage disequilibrium *D* will be positive.

Linkage Disequilibrium via Genetic Drift

Drift can also generate linkage disequilibrium. We have already seen that drift can lead to the loss of alleles; it can do the same thing to haplotypes. For example, imagine a small population with four haplotypes: *AB*, *Ab*, *aB*, and *ab*, and a very low recombination rate between the *A* and *B* loci. Just as drift can lead to the loss of an allele, say *B*, it can also (and in fact, more easily) lead to the loss of

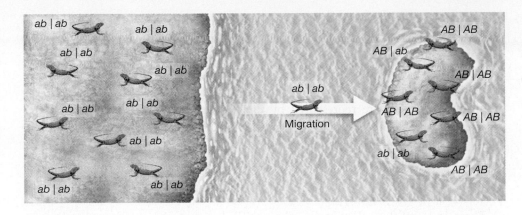

FIGURE 9.12 Migration creates linkage disequilibrium. In this example, the *a* and *b* loci are fixed on the mainland, while the *A* and *B* were previously fixed on the island. When *ab* haplotype migrants reach the island by migration, there will be a statistical association between alleles on the island—that is, there will be linkage disequilibrium.

a haplotype, say *AB*. If this were to happen, the population would be left with only three haplotypes—*Ab*, *aB*, and *ab*—and thus with a nonrandom association between the *A* and *B* loci.

More generally, drift need not entirely eliminate any haplotype in order to generate linkage disequilibrium. Simply by causing random fluctuations in haplotype frequencies, drift can generate statistical associations between alleles at different loci and thus create linkage disequilibrium. Imagine a small population without any linkage disequilibrium. If by chance fewer coupling haplotypes are passed on to the next generation than are repulsion haplotypes, negative linkage disequilibrium will result.

Recombination Dissipates Linkage Disequilibrium

Once linkage disequilibrium is present in a population, we want to understand what happens to it. The short answer is that, in the absence of other evolutionary processes, it is broken down by the process of recombination, and eventually it disappears. In this section, we will look at how this takes place.

Recombination occurs between haplotype pairs. Thus, to understand the effects of recombination, it will be useful for us to track the diploid genotypes in a population (Figure 9.13A). Returning to our two-locus model of a Hardy–Weinberg population,

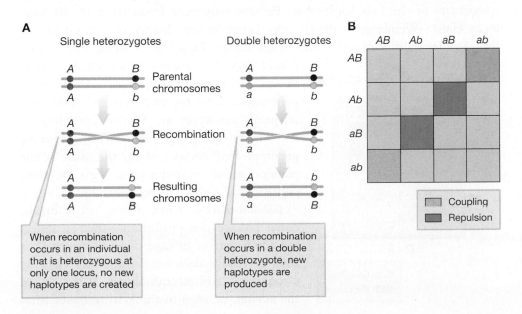

FIGURE 9.13 Recombination creates new haplotypes only in double heterozygotes.
(A) When recombination between the *A* and *B* loci occurs in single heterozygotes (for example, *AB* and *Ab*) no new haplotypes are produced. When recombination occurs in double heterozygotes, new haplotypes are produced. **(B)** For the double homozygote and single heterozygote genotypes indicated by the blue squares, recombination does not alter the haplotypes produced. For the double heterozygotes, shown along the highlighted diagonal, recombination generates new haplotypes.

recall that there are 16 different ordered genotypes as shown by the 16 sections of the box in Figure 9.13B. In 12 of these, recombination will have no effect on the haplotypes produced. For example, recombination between an *AB* haplotype and an *Ab* haplotype will produce *AB* and *Ab* gametes—just as if recombination had not occurred at all. The genotypes for which recombination has no effect on the resulting haplotypes are represented by the 12 blue boxes in Figure 9.13B. But recombination does change the haplotypes produced by the four *double heterozygote* genotypes. These genotypes are represented by the green and gold boxes in Figure 9.13B. If recombination occurs between haplotypes *AB* and *ab*, the new haplotypes *Ab* and *aB* are produced. Similarly, recombination between haplotypes *Ab* and *aB* produces *AB* and *ab* haplotypes.

There is a useful mathematical relationship between the coefficient of linkage disequilibrium *D* and the genotype frequencies in the two-locus Hardy–Weinberg model. From Equations 9.3a–d, we can derive an alternative expression for the coefficient of linkage disequilibrium *D* as follows:

$$D = h_{AB}\,h_{ab} - h_{aB}\,h_{Ab} \tag{9.4}$$

The first term of this expression, $h_{AB}h_{ab}$, is one-half of the frequency of coupling double heterozygotes in a Hardy–Weinberg population. The second term, $h_{aB}h_{Ab}$, is one-half of the frequency of repulsion double heterozygotes in a Hardy–Weinberg population. Recall that the size of each region in our geometric picture of two-locus Hardy–Weinberg frequencies (Figures 9.8B and 9.13B) is equal to the frequency of that genotype in the population. Thus, the coefficient of linkage disequilibrium *D* is simply one-half the difference between the size of the gold coupling regions and the size of the green repulsion regions. When the coupling regions are larger than the repulsion regions, the coefficient of linkage disequilibrium is positive; when the repulsion regions are larger, the coefficient of linkage disequilibrium is negative (Figure 9.14).

In our two-locus Hardy–Weinberg model, linkage disequilibrium does not change in the absence of recombination. Mathematically, we can see this from Equation 9.4, which states that linkage disequilibrium depends only on the frequencies of the four haplotypes. Because haplotype frequencies in the two-locus Hardy–Weinberg model do not change in the absence of recombination, linkage disequilibrium does not change either. Figure 9.15 illustrates this by considering the simple case in which linkage disequilibrium is at a maximum in the parental generation. Gametes are produced, and then offspring genotypes are formed according to the Hardy–Weinberg model. The offspring genotype frequencies end up the same as the parental genotype frequencies, with the same coefficient of linkage disequilibrium.

Recombination, when it does occur, breaks down linkage disequilibrium—that is, unless the coefficient of linkage disequilibrium *D* is zero, *D* always decreases in absolute value as a consequence of recombination. Why? To see the answer, it helps to think in terms of what

FIGURE 9.14 Another interpretation of the coefficient of linkage disequilibrium *D*. Based on Equation 9.4, we can view the coefficient of linkage disequilibrium as a measure of the difference between the frequency of coupling double heterozygotes and the frequency of repulsion double heterozygotes in a two-locus Hardy–Weinberg model. **(A)** A case with more coupling than repulsion; **(B)** a case with more repulsion than coupling.

A More coupling than repulsion

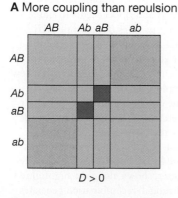

B More repulsion than coupling

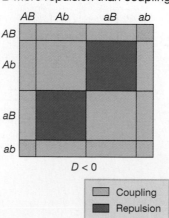

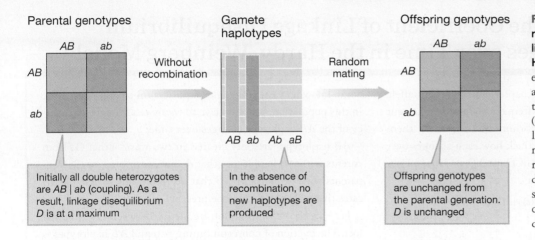

FIGURE 9.15 In the absence of recombination, linkage disequilibrium does not change in the Hardy–Weinberg model. In this example, only coupling haplotypes are present in the parental generation, and allele frequencies are (0.5, 0.5) at each locus. As a result, linkage disequilibrium is at its maximum of $D = 0.25$. Without recombination, the process of producing gametes and forming offspring from these gametes does not change the magnitude of the linkage disequilibrium.

recombination does to the various genotypes. Suppose some of the haplotype pairs undergo recombination between the A and B loci. For the double homozygote and single heterozygote genotypes, these recombination events don't change the haplotypes that are produced. But for the double heterozygotes, coupling pairs that recombine will produce pairs of repulsion gametes, and repulsion pairs that recombine will produce pairs of coupling gametes. Thus, when D is positive and there are excess coupling double heterozygotes, there will be more new repulsion gametes produced by recombination among coupling pairs than there will be new coupling gametes produced by recombination among repulsion pairs. The frequency of coupling haplotypes relative to repulsion haplotypes therefore drops, causing a decline in the absolute value of D. An analogous argument holds in the case in which D is negative.

Figure 9.16 illustrates this process. The population starts at a maximum level of linkage disequilibrium, just as in Figure 9.15. But in this case, recombination does occur, and gametes with new repulsion haplotypes are produced as a result. When

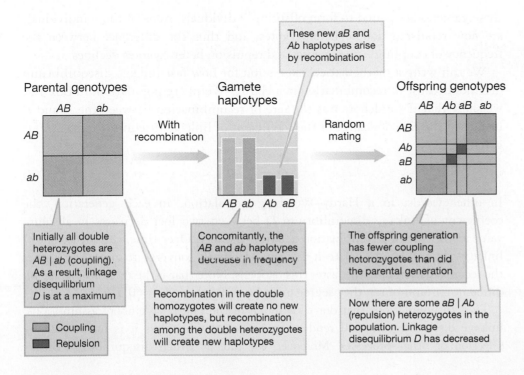

FIGURE 9.16 Recombination breaks down linkage disequilibrium. As in Figure 9.15, only coupling haplotypes are present in the parental generation, and allele frequencies are (0.5, 0.5) at each locus. As a result, linkage disequilibrium is at its maximum of $D = 0.25$. Recombination among coupling double heterozygotes forms new repulsion haplotypes. Some of these then pair together to form repulsion double heterozygotes in the offspring generation; the value of D has declined.

BOX 9.2 How the Coefficient of Linkage Disequilibrium Changes over Time in the Hardy–Weinberg Model

As we have done throughout this chapter, let f represent allele frequencies, h represent haplotype frequencies, and D represent the coefficient of linkage disequilibrium, but now make them functions of time t so that we can track how each changes over time. We will denote the change in D as $\Delta D = D(t + 1) - D(t)$. As we noted in Equation 9.3a:

$$h_{AB}(t) = f_A(t)f_B(t) + D(t)$$

and

$$h_{AB}(t + 1) = f_A(t + 1)f_B(t + 1) + D(t + 1)$$

We can rearrange these two equations to solve for $D(t)$ and $D(t + 1)$ as follows:

$$D(t) = h_{AB}(t) - f_A(t)f_B(t)$$

and

$$D(t + 1) = h_{AB}(t + 1) - f_A(t + 1)f_B(t + 1)$$

We can now calculate ΔD as follows:

$$\begin{aligned} \Delta D = f_A(t)f_B(t) - f_A(t + 1)f_B(t + 1) \\ + h_{AB}(t + 1) - h_{AB}(t) \end{aligned} \quad (9.5)$$

But, as we learned in Chapter 7, allele frequencies do not change in a Hardy–Weinberg population. Therefore, $f_A(t) = f_A(t + 1)$ and $f_B(t) = f_B(t + 1)$. This means that we can simplify Equation 9.5 as follows:

$$\Delta D = h_{AB}(t + 1) - h_{AB}(t) \quad (9.6)$$

So, to know how the linkage disequilibrium changes over time in this population, we only have to figure out how the frequency of the AB haplotype changes over time.

AB haplotypes are only created in two ways: either (1) from parents with AB haplotypes that do not recombine, or (2) from parents with genotype $A*|*B$ that do recombine (where $*$ indicates that either allele may be present).

Let r be the recombination frequency between the A and B loci. The fraction of nonrecombining parental AB haplotypes is $(1 - r)h_{AB}(t)$. The fraction of parents with one $A*$ haplotype and one $*B$ haplotype is simply $2f_A(t)f_B(t)$. Of these, a fraction r do recombine, and half of the gametes thus produced contain the A and B alleles from the $A*|*B$ pair. Thus, the total fraction of new h_{AB} haplotypes at time $t + 1$ is as follows:

$$h_{AB}(t + 1) = (1 - r)h_{AB}(t) + r f_A(t)f_B(t)$$

We can substitute this expression into Equation 9.6, and with a little bit of algebra, we find that:

$$\Delta D = -rD \quad (9.7)$$

This means that, in each generation, the coefficient of linkage disequilibrium (D) decreases in absolute value by a rate equal to the recombination frequency.

these gametes are paired to form offspring individuals, some of those individuals are now repulsion double heterozygotes, and thus the difference between the frequency of coupling heterozygotes and repulsion heterozygotes declines.

We can write a mathematical expression for how fast linkage disequilibrium declines because of recombination in a Hardy–Weinberg population. In Box 9.2, we show that if we denote r as the rate of recombination between the A and B loci, then the rate of change of the coefficient of linkage disequilibrium (ΔD) is given by

$$\Delta D = -rD \quad (9.7)$$

In other words, in a Hardy–Weinberg population, in each generation, the coefficient of linkage disequilibrium D between two loci decreases in absolute value at the rate of recombination between the loci. Over time, the coefficient of linkage disequilibrium between these two loci will converge to zero. The higher the recombination rate, the faster this happens. Note that if the A and B loci are on different chromosomes, they segregate independently and $r = 0.5$.

Even in populations that don't satisfy all of the Hardy–Weinberg assumptions, linkage disequilibrium will tend to be broken down unless it is maintained by selection or other processes. Moreover, the rate at which disequilibrium breaks down between two loci is proportional to the distance between them along the chromosome. This is the fundamental principle underlying the process of association mapping, a technique by which loci responsible for disease or other traits are located. Researchers measure the statistical association between disease state and the alleles present at a set of variable *marker loci* on a given chromosome or across the entire genome. The idea is that when the disease-related mutation arose in the population, it did so in one particular haplotype, and thus linkage disequilibrium was created between the disease allele and other polymorphic alleles in the genome. Over time, linkage disequilibrium will break down throughout the genome, but it will break down more slowly for the marker loci closest to the disease locus. In an association study, we aim to find statistical associations between marker loci and the disease state. The strengths of these associations, as we move from marker locus to marker locus along the chromosome, can potentially provide us with the information we need to pinpoint the location of the disease gene. In Chapter 10, we will look at other ways in which patterns of linkage disequilibrium can help us to understand the structure of genomes and to find alleles that have been under recent natural selection.

Consequences of Genetic Linkage

We have discussed how to think about and quantify linkage disequilibrium in populations, and we have seen that the rate at which linkage disequilibrium breaks down depends on physical linkage. Therefore, physical linkage can have important evolutionary consequences. For reasons we will describe shortly, these consequences are particularly strong in bacterial populations. Because of this, we will draw many of our examples of the evolutionary consequences of physical linkage from bacterial systems.

Genetic Hitchhiking and Background Selection

When a beneficial allele arises in a population and is subject to selection, it increases in frequency; we explored the dynamics of this process in Chapter 7. But there is a side consequence of this as well: Allele frequencies at loci that are physically linked to the locus under selection may also change. This process is known as **genetic hitchhiking** because an unselected or even disadvantageous allele is able to "ride along" with a nearby favorable allele and thus increase in frequency (Figure 9.17) (Kojima and Schaffer 1967; Maynard Smith and Haigh 1974).

Over time, recombination will break down the association between the favored allele and those around it. The association will break down quickly for alleles far from the selected locus, but slowly for nearby loci. As a result, certain alleles at loci near a selected locus—namely, those in the "genetic background" on which the favored allele arose—may increase in frequency in the population by the process of hitchhiking. A second consequence is that genetic

FIGURE 9.17 Genetic hitchhiking. A beneficial allele, *B*, arises on a genetic background with the *A*, *C*, *D*, and *E*, alleles—that is, on a chromosome that has the *A*, *C*, *D*, and *E* alleles. These alleles hitchhike along with the *B* allele, increasing in frequency. Eventually, recombination breaks up the association between the *B* allele and the *A*, *C*, *D*, and *E* alleles. Adapted from Understanding Evolution (2008).

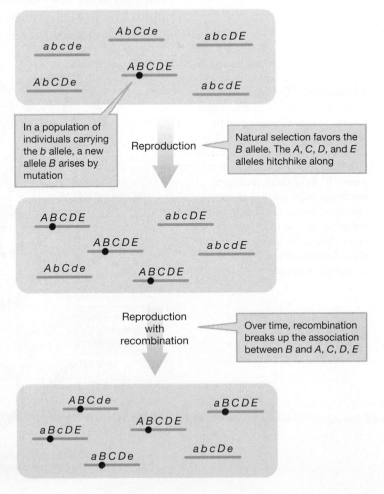

diversity at loci near the selected locus will be reduced relative to what we would expect in a neutral model.

Deleterious mutations can also cause a reduction in genetic diversity. When deleterious alleles arise in a population, they tend to be eliminated by selection. Much as beneficial mutations carry nearby alleles to fixation via hitchhiking, deleterious mutations carry nearby alleles to extinction via a process known as **background selection**. As a result of this process, genetic variation is reduced relative to what we would expect in a neutral model.

The important general points about the evolutionary consequences of selection and physical linkage are as follows: (1) Alleles can increase in frequency due to selection either because they directly code for beneficial traits on their bearers, or because they are physically linked to other beneficial alleles at other loci. (2) Natural selection, be it positive or negative, tends to cause a decrease in genetic variation at loci near the selected allele.

Periodic Selection

The phenomenon of genetic hitchhiking can have particularly dramatic consequences in bacteria, which are haploid and typically have only a single chromosome. In bacteria, any two alleles in the genome tend to be tightly linked because bacteria have neither of the key processes that break up linkage disequilibrium in diploid eukaryotes: Bacteria don't have recombination between homologous chromosomes, and they don't have independent segregation of multiple chromosomes. This is illustrated in Figure 9.18. Indeed, for this reason, the bacterial chromosome is sometimes described as one single, albeit very large, locus.

FIGURE 9.18 Physical linkage is particularly important in haploid prokaryotes. (A) Most pairs of loci in eukaryotes are not tightly linked because they can be separated by segregation if they occur on different chromosomes and by recombination if they occur on the same chromosome. (B) By contrast, linkage is much tighter in many bacterial species. Most bacteria have only a single chromosome, and there is no process of recombination between homologous chromosomes analogous to that in diploid eukaryotes.

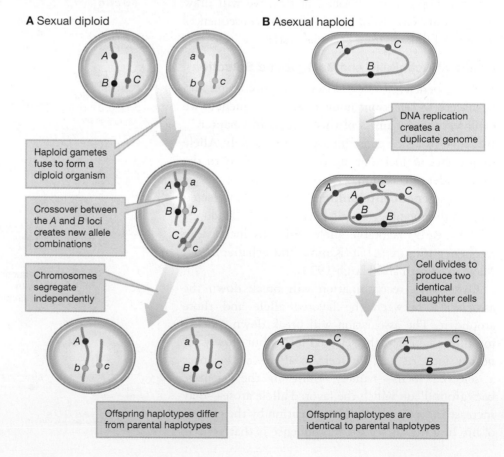

The phenomenon of **periodic selection** is one dramatic result of tight linkage across the entire bacterial genome. Periodic selection refers to the following process:

1. A new beneficial mutation arises in a bacterial population.

2. The mutation goes to fixation as a result of strong natural selection. Because the other loci on the bacterial chromosome are tightly linked, alleles at most other loci are fixed as well. This is known as a **selective sweep**.

3. After the previous beneficial allele goes to fixation, a new beneficial allele arises and the process repeats (Figure 9.19).

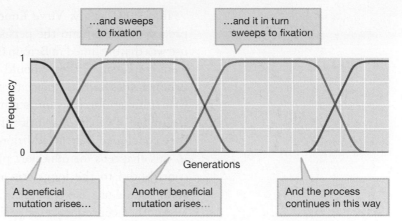

FIGURE 9.19 Periodic selection in a bacterial population. Each time a new beneficial mutation arises, it sweeps to fixation. In the absence of recombination, it carries with it the fixation of the particular haplotype on which it arose.

Periodic selection has probably contributed to the phenomenon we described at the beginning of this chapter: the long-term persistence of bacterial resistance to antibiotics that have been withdrawn from general use. The likely process is illustrated in Figure 9.20. In an environment where antibiotics are used frequently, such as a hospital or nursing home, resistance is strongly selected and often arises either by mutation or by acquisition of genes on small DNA molecules called *plasmids* or genetic elements called *transposons* (both of which we will examine in Chapter 10). When a new antibiotic is introduced into this environment, selection favors resistance to the new antibiotic. Alleles that confer resistance to the new drug will tend to arise by mutation or be incorporated by gene transfer onto a genetic background that includes resistance genes for previously encountered antibiotics. Even if the use of one particular antibiotic stops, selection on the new antibiotics continues. If alleles for resistance to the new antibiotics are linked to alleles for resistance to now-discontinued drugs, the old resistance alleles will be maintained (because of physical linkage) by selection for the new resistance alleles.

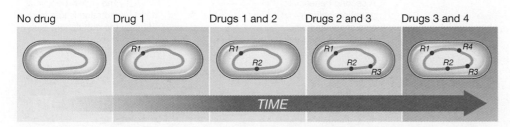

FIGURE 9.20 Periodic selection and the persistence of antibiotic resistance. We begin with a drug-sensitive strain. In response to the use of drug 1, a resistance gene to this drug (*R1*) is favored. When drug 2 comes into use, resistance to drug 2, labeled *R2*, arises and is favored in the drug 1–resistant strain. When drug 1 is phased out and replaced with drug 3, resistance (*R3*) to drug 3 evolves—on the common local strain, which is already resistant to drugs 1 and 2. When drug 2 is phased out and replaced with drug 4, resistance (*R4*) to drug 4 evolves—again on the common local strain, now already resistant to drugs 1, 2, and 3. As a result, we observe a multi-drug-resistant strain that is resistant even to drug 1, which has not been used for a long period of time.

In a 2001 study, Virve Enne and her colleagues wanted to test whether this process could explain the persistence of sulfonamide resistance after sulfonamide use was discontinued in Britain (Enne et al. 2001). They reasoned that if this process were responsible, there should be a statistical association between resistance to sulfonamides and resistance to more recently used antibiotics. Indeed, they found that sulfonamide-resistant strains were significantly more likely to be resistant to a number of other antibiotics than were sulfonamide-sensitive strains. From this, they concluded that, in addition to the compensatory mutations that we discussed in this chapter's introduction, physical linkage to other resistance genes had likely contributed to the long-term maintenance of sulfonamide resistance even after sulfonamide use had ceased.

Another consequence of periodic selection is that genetic diversity is greatly reduced in a population. Before heterogeneity has a chance to build up by mutation, it is wiped out in a selective sweep. This process happens over and over again and, when combined with frequent bottlenecks (as described in the previous chapter), it is responsible for many bacterial populations having relatively small effective population sizes despite being composed of huge numbers of individuals (Levin and Bergstrom 2000).

Clonal Interference

When we discussed periodic selection, we assumed that beneficial mutations arise one at a time, and that in the time between these events, the previous beneficial mutations have a chance to go to fixation. What happens if beneficial mutations instead arise close together in time but in different individuals? If recombination is nonexistent or limited, both alleles cannot go to fixation at the same time. Moreover, the beneficial allele that goes to fixation does so slowly, because it has to outcompete not only the lower-fitness wild type, but also the other beneficial mutation. In this way, the beneficial alleles "interfere" with one another, and the consequence is an overall reduction in the rate at which beneficial alleles are fixed. This slowing down of selection is known as **clonal interference** (Figure 9.21) (Gerrish and Lenski 1998).

FIGURE 9.21 Clonal interference. In a bacterial population with limited recombination, when two beneficial alleles arise, selection for one beneficial allele (blue) can interfere with the increase in allele frequency in the other beneficial allele (yellow).

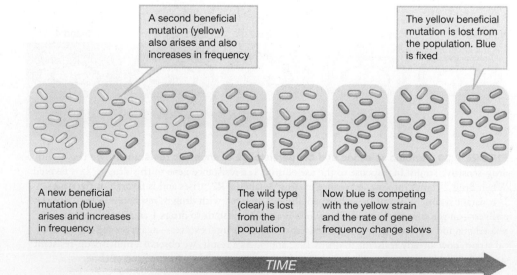

A second beneficial mutation (yellow) also arises and also increases in frequency

The yellow beneficial mutation is lost from the population. Blue is fixed

A new beneficial mutation (blue) arises and increases in frequency

The wild type (clear) is lost from the population

Now blue is competing with the yellow strain and the rate of gene frequency change slows

TIME

9.3 Adaptive Landscapes

Thus far, we've seen how statistical associations can build up between genes at two or more loci, and how we can mathematically model the changes of haplotype frequencies in populations. But the mathematics gets complicated quickly. It would be useful to have a conceptual way—even if it is only metaphorical— to think about evolution at multiple loci. The population geneticist Sewall Wright developed the **adaptive landscape** (or fitness landscape) metaphor for this purpose.

Wright was trained as a *physiological geneticist*—he studied the way that genes determine phenotype. From his doctoral research on heredity in guinea pigs and other species, Wright recognized that the relationship between genes and phenotype is seldom a simple and straightforward one (Provine 1986). Rather, interactions *among* genes are extremely important, generating a *genotype-to-phenotype map* that is complicated for many reasons that we have already discussed. To recap, these include:

- *Pleiotropy*. A single gene can have effects on multiple aspects of phenotype.

- *Epistasis*. A given phenotypic trait is often determined by complex interactions among multiple genes.

- *Norms of reaction*. A single genotype produces different phenotypes in different environments.

- *Dominance*. One allele may cover up the effects of another allele at the same locus.

- *Multiple pathways*. A common phenotype may have a different genetic basis in different individuals.

In short, natural selection acts on the phenotype, the next generation inherits only the genotype, and the relation between genotype and phenotype is complex. In Wright's mind, this picture created considerable difficulties for simpler models in which natural selection brought about change through a series of mutations, each with small additive effects. Wright developed an extensive mathematical theory to deal with these challenges. To make these results from this theory accessible to those without a mathematical background, Wright developed the adaptive landscape metaphor (Wright 1932; Provine 1986).

Phenotype Space

The idea behind the adaptive landscape approach is that we can think about different phenotypic or genotypic combinations as points on a map. This may be a map of **phenotype space**, in which the *x*- and *y*-axes represent the values of phenotypic traits, such as the size of a tree's leaves and the size of its flowers, as pictured in Figure 9.22. Within this phenotype space, each point corresponds to a pair of trait values. Thus, each individual in a population can be assigned a point in the phenotype space. Similarly, the average phenotype in a population can also be assigned a single point.

FIGURE 9.22 A two-dimensional phenotype space. The *x*-axis indicates the length of a leaf plus its petiole (leaf stem). The *y*-axis indicates the length of an inflorescence (flower cluster) plus its peduncle (flower stem). Each point represents the average phenotype for a different species of maple (*Acer* spp.). Adapted from Ackerly and Donoghue (1998).

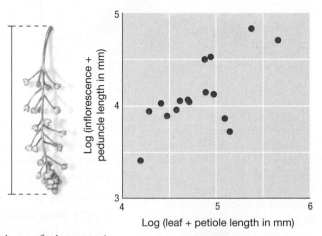

Note that Figure 9.22 specifically represents a *two-dimensional* phenotype space. In practice, we can consider three, four, or even more traits. A phenotype space that takes all of these traits into account necessarily has a high number of dimensions, and while it would be more realistic, it would also be very difficult to visualize. Fortunately, we can use even low-dimensional phenotype spaces to think about how evolutionary change occurs.

Adaptive Landscapes in Phenotype Space

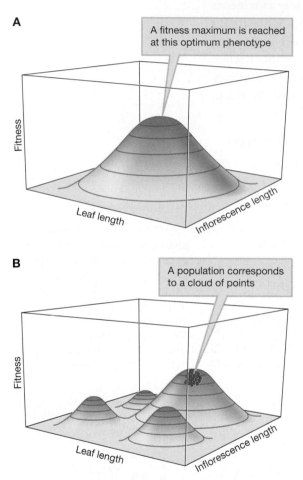

A

A fitness maximum is reached at this optimum phenotype

B

A population corresponds to a cloud of points

FIGURE 9.23 Fitness landscapes in phenotype space. In these illustrations, hypothetical fitnesses are associated with the values of two phenotypic traits, leaf length and inflorescence (flower cluster) length. The landscape in **(A)** is single peaked, with a unique local fitness optimum that is also a global optimum. The landscape in **(B)** is multipeaked, with several local fitness optima—that is, several hilltops on the landscape. A population corresponds to a cloud of points on a fitness landscape, as illustrated.

Now suppose that we want to draw an adaptive landscape. The *x*- and *y*-axes of our map contain our phenotype space: the aspects of phenotype under consideration (for example, leaf length and inflorescence length). We can then introduce the *z*-axis (elevation) to plot the corresponding fitness of an organism with that phenotype combination. This transforms our flat map of phenotype space into a landscape of "hills," where the highest points represent **fitness peaks**—combinations of traits associated with the greatest fitness values.

Figure 9.23A shows a *single-peaked* fitness landscape: A single optimal phenotype lies at the peak of the "hill," and any movement that takes a population closer to the (x, y) coordinates of the peak necessarily moves the population to a higher fitness as well. Figure 9.23B shows a *multipeaked* fitness landscape. Here, instead of a single hill, we have a range of multiple fitness peaks with **fitness valleys**—regions of lower fitness—between them.

In the adaptive landscape metaphor, each point in phenotype space corresponds to a different phenotype. Thus, a population of individuals with different phenotypes corresponds to a set or "cloud" of points, as illustrated in Figure 9.23B.

By looking at a fitness landscape, we can get a qualitative sense about how phenotypic evolution might proceed. In a large population with sufficient genetic variation for the traits in question, we would expect the population to move uphill on the adaptive landscape. Thus, selection will favor phenotypic values that result in increased fitness—that is, evolution might follow a hill-climbing trajectory on a fitness landscape. If each genetic change has a small phenotypic effect, we would expect natural selection to follow a path that moves directly uphill until a local fitness maximum is attained, rather than crossing whatever fitness valleys are necessary to reach the global fitness maximum. This illustrates the shortsightedness of natural selection that we discussed in Chapter 3. Natural selection cannot plan ahead and aim for the highest peaks; rather, it simply sorts on the existing variation, causing the population to move across the adaptive landscape like a myopic mountain climber who takes small, incremental steps, without being able to see a final goal (the red curve in Figure 9.24).

As we saw in Chapter 8, genetic drift will also be an important source of evolutionary change in smaller populations, and drift can even drive selectively disadvantageous changes in phenotype. This corresponds to downhill movement on the adaptive landscape. Drift—and processes that facilitate drift such as bottlenecks and founder

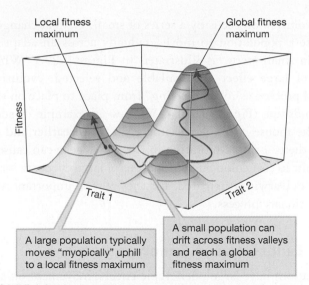

FIGURE 9.24 **Movement on an adaptive landscape.** In a large population (red curve), a population with sufficient genetic variation will climb myopically uphill to the nearest local fitness maximum. In a small population (blue curve), drift plays a significant role and the population can drift down, into, and across fitness valleys.

Local fitness maximum

Global fitness maximum

A large population typically moves "myopically" uphill to a local fitness maximum

A small population can drift across fitness valleys and reach a global fitness maximum

FIGURE 9.25 **The northwestern garter snake (*Thamnophis ordinoides*).** This species exhibits dramatic variation in coloration and behavior.

effects—can help a population move across a fitness valley so that it can subsequently climb an adaptive peak on the other side (the blue curve in Figure 9.24).

Let's examine a set of studies by evolutionary biologist Edmond Brodie II that map a simple fitness landscape in phenotype space. Brodie studied the fitness consequences of antipredator behavior and coloration pattern in garter snakes (*Thamnophis ordinoides*) (Brodie 1992). Garter snakes vary significantly in coloration, ranging from nearly unpatterned to mottled to dramatically striped (Figure 9.25). These snakes also differ in the escape behaviors that they use when threatened by a predator. Some individuals flee in a direct course, while others make a few or many "reversals"—evasive changes in direction that may confuse a pursuing predator.

Brodie measured both coloration and reversal frequency in 646 newborn garter snakes, marked each individual, and released the lot into the wild. By looking at the frequency with which the marked snakes were later recaptured over the course of a 3-year experiment, Brodie was able to estimate the fitness consequences of particular *combinations* of color pattern and reversal behavior. He observed strong nonadditive interactions between coloration and behavior: Spotted snakes that reverse and striped snakes that do not reverse have high fitnesses, while striped snakes that reverse and spotted snakes that do not reverse have low fitnesses. Figure 9.26 shows a fitness landscape inferred from these fitness estimates. This fitness landscape is in phenotype space: The *x*- and *y*-axes reflect the phenotypes of reversal rate and degree of striping, respectively.

FIGURE 9.26 **An adaptive landscape in phenotype space.** The fitness landscape for the garter snake *T. ordinoides* in the phenotype space defined by body coloration and reversal behavior during escape. Two bivariate phenotypes have high fitness; the fitness of the other two combinations is low. Adapted from Brodie (1992).

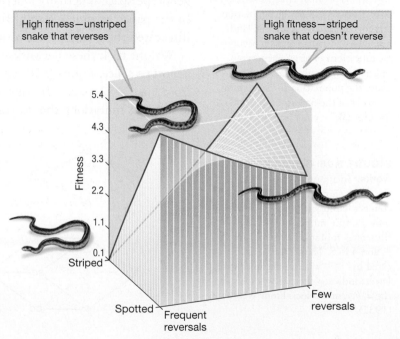

High fitness—unstriped snake that reverses

High fitness—striped snake that doesn't reverse

A

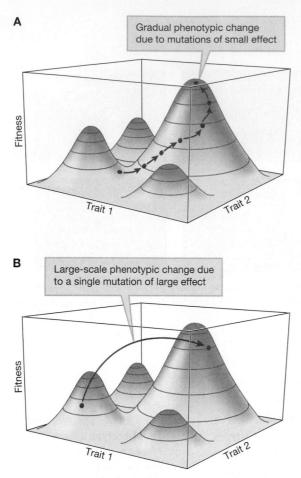

Gradual phenotypic change due to mutations of small effect

B

Large-scale phenotypic change due to a single mutation of large effect

FIGURE 9.27 Evolutionary change can be represented as movement across the adaptive landscape. When evolution proceeds gradually by a sequence of mutations of small effect, a series of substitutions causes the population to move gradually across the landscape **(A)**. When evolution occurs by mutations of large phenotypic effect, a single substitution can cause the population to "leap" from one part of the fitness landscape to another **(B)**.

When evolution proceeds by a series of small, gradual changes, it will cause a population to move gradually across an adaptive landscape in small steps, as illustrated in Figure 9.27A. When mutations of large effect are available and selected, evolution may instead proceed as if "teleporting" from place to place on the adaptive landscape (Figure 9.27B). We saw an example of such a leap in the mouse coat color example discussed earlier and in Chapter 3; there, a single mutation in the *Mc1R* gene can cause a dramatic shift in coat color. Such mutations of large effect are more common than Darwin anticipated, and they play an important role in the evolutionary process.

Adaptive Landscapes in Genotype Space

When mutations have large phenotypic effects, movement on the adaptive landscape is not smooth and gradual. Worse still, the distance between two points on the fitness landscape may be a poor indicator of how many genetic changes are needed to shift from one to the other. To get around these problems, population geneticists often conceptualize adaptive landscapes as occupying **genotype space** rather than phenotype space. This is the way that Wright initially presented his adaptive landscape metaphor. The idea here is that genotypes that are mutational neighbors—namely, those separated by a single mutation—appear close together in the genotype space, whereas those that are separated by many mutations appear far apart. When adaptive landscapes are represented in this way, nearby points are genetically very similar, even if their phenotypes vary dramatically. By the same logic, distant points are very different genetically even if they happen to correspond to very similar phenotypes.

Because mutations are discrete rather than continuous changes, the true genotype space is actually a network of genotypes rather than a continuous space. In the paper in which Wright first proposed the adaptive landscape metaphor, he illustrated the genotype network concept as shown in Figure 9.28 (Wright 1932).

Wright took these networks of loci and redrew them as adaptive landscapes on genotype space. Figure 9.29 recreates Wright's original sketch of such a landscape from his 1932 paper. In this figure, similar genotypes are close together, so individual mutations should correspond to small movements in this space, and

FIGURE 9.28 Genotypic networks. In a mutation network, the edges shown here as blue lines connect haplotypes that differ by only a single mutation. This figure illustrates mutation networks for 2, 3, and 4 loci. The wild type is indicated by "+", whereas the lowercase letters indicate alternative alleles at the *a*, *b*, *c*, and *d* loci. From Wright (1932).

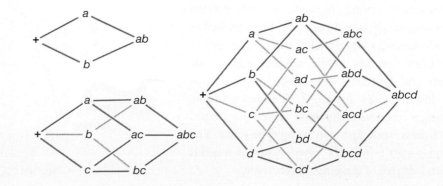

A

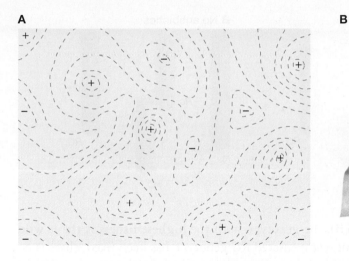

B

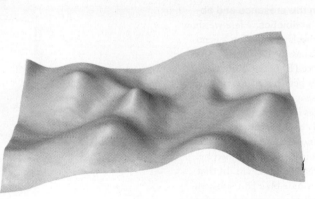

FIGURE 9.29 Wright's original sketch of a fitness landscape. (**A**) Here the landscape is drawn as a topographic map. From Wright (1932). (**B**) A three-dimensional version of the same landscape.

evolution might be expected to trace out a nearly continuous path within this space as a sequence of mutations each become fixed in turn. But, as Wright was well aware from his studies of heredity, single mutations can cause large phenotypic changes, and they can thus have dramatic fitness consequences. As a result, the fitness landscape in genotype space is likely to be *rugged*. Rather than being a smooth, gradual, and single-peaked surface, the landscape might consist of many sharp jagged peaks and ridges. In Wright's view, this metaphorical space might look more like the saw-toothed limestone Tsingy of Madagascar than the smooth volcanic slopes of Mt. Fuji (Figure 9.30).

Returning to the compensatory mutation story from the introduction to this chapter, we can see the value of thinking in terms of genotypic networks. We have already observed that compensatory mutations reduce the fitness cost of antibiotic resistance and thus make it hard to reverse the evolution of such resistance. The story is even more complex than this, however. Once resistance has evolved and been compensated, neither the loss of resistance nor the loss of compensation is immediately beneficial, even when no antibiotics are being used.

The genotypic network concept can help us develop a deeper understanding of how compensatory resistance works. Suppose that the R locus controls resistance, that the C locus is the compensatory locus, and that the wild-type rc is neither resistant nor compensated. When antibiotics are present (Figure 9.31A), the resistance allele R confers a large selective advantage, and the compensatory allele C further increases fitness. Resistance evolution proceeds along the trajectory indicated by the red arrows in the figure. When antibiotics are not present in the

A

B

FIGURE 9.30 Smooth and rugged physical landscapes. R. A. Fisher envisioned selection moving on a smooth and single-peaked adaptive landscape analogous to the physical landscape of Mt. Fuji (**A**), whereas Sewall Wright imagined that selection operated on a rugged adaptive landscape analogous to the Tsingy of Madagascar (**B**).

FIGURE 9.31 Adaptive landscapes in the presence and absence of antibiotics. Here we show the adaptive landscapes on a genotypic network in the presence (**A**) and absence (**B**) of antibiotics. The blue edges between *rc*, *Rc*, *rC*, and *RC* represent a genotypic network; these edges link genotypes that differ by only a single mutation. This figure is modeled on streptomycin resistance in *E. coli*, as studied by Schrag, Perrot, and Levin. In that system, a single base pair substitution in a ribosomal protein confers resistance, and a second mutation at a separate locus compensates for much of the fitness cost induced by the resistance mutation. Adapted from Schrag et al. (1997).

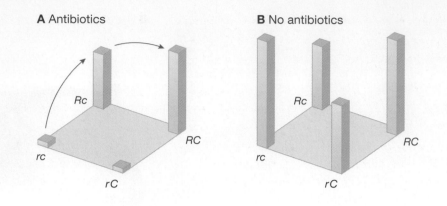

A Antibiotics

B No antibiotics

environment (Figure 9.31B), the genotype with the highest fitness is the *rc* wild type. But once the resistant, compensated *RC* genotype has been fixed, there is no direct path to return to the *rc* wild type by a series of beneficial mutations. This is because either reversion, $R \rightarrow r$ or $C \rightarrow c$, imposes a fitness cost if it occurs first. Loss of the compensatory mutation in a resistant individual obviously causes a fitness decrease. But loss of the resistance mutation in a compensated individual also causes a fitness decrease. As shown in Figure 9.31B, the compensatory allele *C* provides a fitness advantage in the presence of the resistance allele *R* but imposes a fitness cost when paired with the wild-type allele *r*. This type of fitness interaction is common for compensatory mutations (Andersson and Hughes 2010).

Antibiotic resistance evolution is akin to an evolutionary lobster trap: One can get in easily, but once in, it's not easy to get out. In the presence of the antibiotic, natural selection can readily drive the population from *rc* to *RC* by means of two sequential substitutions, each of which increases fitness. But in the absence of the antibiotic, there is no comparable sequence of fitness-increasing mutations that leads from *RC* to *rc*. Returning to the sensitive uncompensated state requires that the population cross a fitness valley, and this can take a long time.

FIGURE 9.32 Continuously varying traits. Variation in fruit size across various tomato (*Lycopersicon*) species.

9.4 Quantitative Genetics

In Section 9.1, we saw how a nearly continuous range of phenotypes could arise via Mendelian inheritance when multiple genes influence the phenotype. In this section, we will revisit continuously varying traits and, in doing so, we will explore the field of **quantitative genetics**.

Consider a continuously varying trait such as the fruit size for a tomato plant (Frary et al. 2000) (Figure 9.32). Going from plant to plant, we see a continuous range of fruit sizes. But why is this so? The general question, applicable to almost any varying trait, is "Why does one individual differ from another?"

In the preceding chapters, we have seen that a number of different factors contribute to a phenotype. First, *genes* obviously influence phenotype. As we saw in Section 9.1, when gene effects combine additively, multifactorial inheritance can generate nearly continuous variation. Patterns of inheritance and variation can become even more complicated as a result of gene interactions or epistasis.

Second, the *environment* influences a phenotype. Tomato plants may generally tend to grow larger in sunnier environments, for example, or they may grow poorly if rainfall falls short of some critical threshold.

Finally, we expect some differences in a phenotype, even for genetically identical individuals raised under the same environmental conditions. Random chance events during the process of development can give rise to considerable phenotypic differences. This source of variation is known as *developmental noise*, and it contributes significantly to phenotypic variation in some populations (Babbitt 2008).

Given that continuously varying traits are shaped by these numerous influences, often with complex interactions among them, how can we make predictions about how continuous traits will change as a result of natural selection? The field of quantitative genetics provides a way of doing this, and supplies additional tools for understanding the evolution and genetics of complex continuous traits.

The Phenotypic Value of Continuous Traits

The first step in constructing a theory of quantitative genetics is to develop a basic model of how phenotypes of the individuals in a population are determined (Christiansen 2008). *For a given individual*, we define P as the phenotypic value of the continuous trait that we are studying. In the case of tomato fruit size, P might be quantified as weight in grams. We then decompose P into two parts: the part due to the genotype (G), and the remainder, which we ascribe to environmental influences (E):

$$P = G + E \qquad (9.8)$$

Here, the genotypic value (G), is defined as the expected phenotypic value of individuals of that particular genotype. Any deviation between P and G is attributed to environmental effects or developmental noise and is quantified as the environmental deviation (E). The average or expected value of E is zero, because the environmental deviation is equally likely to be positive or negative.

The key to the quantitative genetics approach is that it enables us to track not only the phenotypes of the individuals in the population, but also the *variation* that is present in the population, and whether or not this variation has a genetic basis. This provides us with a way to make predictions about how natural selection—which requires genetic variation to proceed—will drive evolutionary change in the observed phenotypes over time.

We will measure variation by using a quantity known as the variance, which is a statistical measure of the variation in a sample. Different members of a population typically have different trait values, and the variance tells us *how different* from one another these trait values are. Let $x_1, x_2, x_3, \ldots, x_{n-1}, x_n$ be a set of observations (for example, the heights of the students in a class, measured in meters). The mean of these observations is

$$\bar{x} = \frac{1}{n} \sum_{i=1}^{n} x_i$$

The sample variance of the observations is given by the expression:

$$\mathrm{Var}\,[x] = \frac{1}{n-1} \sum_{i=1}^{n} (x_i - \bar{x})^2$$

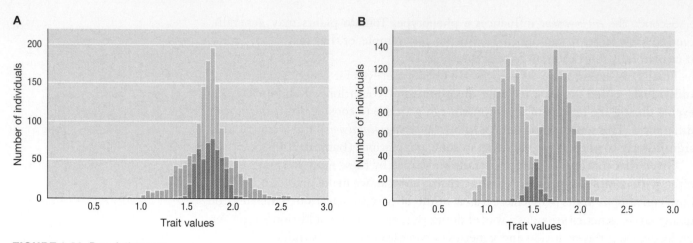

FIGURE 9.33 Population samples. Each histogram shows a sample of 1000 individuals from a population. **(A)** Both the blue and tan populations have the same mean, but the variance of the tan population is three times higher than that of the blue population. **(B)** Both the blue and tan populations have the same variance, but the mean of the tan population is higher than the mean of the blue population.

The larger the variance, the more that individuals differ from one another and from the mean. Figure 9.33A illustrates samples from two populations with the same mean but different variances. Figure 9.33B illustrates samples from two populations with different means but the same variance.

Breaking up the phenotype P into genetic and environmental influences—G and E—is helpful not so much in that we have a model of what determines the phenotype of one particular individual. Rather, the point is that this decomposition also allows us to model the contributions to the *variance* in phenotypes observed in the population, and to distinguish between the heritable and nonheritable factors involved. This means that we need to derive a mathematical equation for the phenotypic variance, denoted V_P. To do so, we use the basic fact from statistics that the variance of a sum of independent variables is equal to the sum of the variances. Because $P = G + E$, it follows that:

$$V_P = V_G + V_E \tag{9.9}$$

where V_G is the variance of the genotypic value G, and V_E is the variance of the environmental deviation E. (Here we have assumed that the genotypic value and the environmental deviation are uncorrelated.)

The contribution to phenotypic variance that derives from genotypic variance is, in principle, transmitted genetically from parents to offspring. The contribution that derives from environmental variance is not. We define **broad-sense heritability** **(H^2)** as the fraction of the variance that is potentially due to genetic causes:

$$H^2 = \frac{V_G}{V_G + V_E} \tag{9.10}$$

Broad-sense heritability quantifies the total fraction of the variation of a trait in a population that can be attributed to genetic causes. But it is not a particularly useful predictor of evolutionary change because, as we will see, selection is not able to operate effectively on all genetic variation. Therefore, evolutionary biologists more commonly work with a different quantity known as narrow-sense heritability, which is the fraction of the total variance due to additive genetic variation.

Before going into narrow-sense heritability, how can we estimate broad-sense heritability? To do so, we will need to consider how biologists can measure the terms in

Equation 9.10. All we need to know is the relative magnitude of the phenotypic variance that is due to genetic and environmental contributions. These we can find by comparing the amount of variation among genetically identical (or nearly identical) individuals with the amount of variation among unrelated individuals. For many model organisms, we can easily obtain or construct large numbers of nearly genetically identical individuals in the form of *inbred lines*. An inbred line is produced by multiple generations of repeated inbreeding (for example, between siblings) until the remaining genetic variation in the line is minimal.

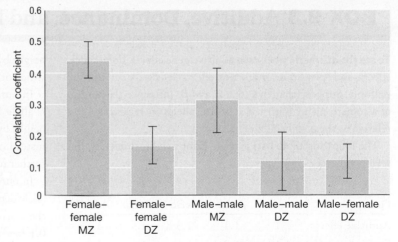

FIGURE 9.34 A study of the genetics of susceptibility to depression. Researchers compared monozygotic (MZ) twins to dizygotic (DZ) twins in order to estimate the influence of genetics on depression. The figure illustrates correlations by twin type (MZ or DZ) and sex. The error bars indicate the 95% confidence intervals for each of the correlation coefficients. These data reveal a higher correlation in depression among monozygotic twin pairs than among dizygotic twin pairs. This finding implies a genetic component to the susceptibility to depression. Adapted from Kendler et al. (2006).

Estimating the components of variation is then straightforward. Within an inbred line, there is negligible genetic variation. Therefore, all phenotype variation within an inbred line is due to environmental variation. Thus, we can estimate V_E as the average phenotypic variance among individuals *within* a single inbred line. We can then estimate the total phenotypic variance V_P by the phenotypic variance among individuals taken from different inbred lines. Subtracting V_E from both sides of Equation 9.9, we see that the genotypic variance V_G is the difference between these two quantities: $V_G = V_P - V_E$.

When studying human genetics, it is obviously not feasible to create inbred lines. In studies of humans, researchers can instead look at pairs of monozygotic ("identical") twins as a way of measuring variation among genetically identical individuals. But there is a problem: Twins typically experience very similar environmental conditions as well. To control for this, researchers typically use one of two approaches. One approach is to study monozygotic twins who were adopted at an early age and reared in separate families. In this case, the genetics are the same, but the environments are different. Such twin pairs can be hard to find, so alternatively researchers can compare monozygotic twins reared together with dizygotic (fraternal) twins reared together. Monozygotic and dizygotic twin pairs alike experience similar environmental conditions, but the genetics of the two kinds of twin pairs differ. Monozygotic (MZ) twins should be essentially genetically identical, whereas dizygotic (DZ) twins should not be more closely related than are ordinary pairs of full siblings. Figure 9.34 illustrates the use of monozygotic–dizygotic twin comparisons to study the genetic basis of depression.

Decomposing Genotypic Effects

The genotype is composed of many different genes. If all gene effects were to combine additively, an individual's genotypic value G could be represented as a simple sum of gene effects. But, of course, gene effects generally do not combine in so simple a fashion. Rather, we have to consider both the interactions between two alleles at the same locus, which we call **dominance effects**, and interactions between alleles at different loci, which we have already discussed in Section 9.1 as epistasis. Box 9.3 provides an example of each.

To account for dominance and epistasis, we can break down our equations for P and V_P. We can think of an individual's genotypic value G as the sum of three

BOX 9.3 Additive, Dominance, and Epistatic Effects

To see the distinction between additive gene effects, dominance effects, and epistasis or interaction effects, let's walk through an example. Suppose that two loci, A and B, influence the height of a tomato plant, with the A and B alleles increasing height relative to the a and b alleles.

The following table lists average plant height, in meters, for several different genotypes.

Genotype	aabb	Aabb	aaBb	AaBb	AABB
Additive effects	1	1.1	1.3	1.4	1.8
Dominance effects	1	1.2	1.6	1.8	1.8
Epistatic effects	1	1.1	1.3	1.8	1.8

In the simplest case of additive effects, each A allele increases plant height by 0.1 meters, and each B allele increases plant height by 0.3 meters. The effect of each allele is independent of which other alleles are present.

In our example of dominance effects, one copy of the A or B allele is sufficient to have the full effect of increasing plant height by 0.2 or 0.6 meters, respectively. Having two A alleles instead of one, or two B alleles instead of one, adds nothing further to the plant height.

In our example of epistatic effects, the effect of having an A allele depends on whether or not the plant also has a B allele. In the absence of the B allele, the A allele increases plant height by 0.1 meters; in the absence of the A allele, the B allele increases plant height by 0.3 meters. But when both the A allele and the B allele are present, they together increase plant height not by the additive amount $0.1 + 0.3 = 0.4$ meters, but rather by a total of 0.8 meters. (In this particular example, A and B also have dominance effects; a single copy of each is sufficient for the full effect.)

---

contributing components: an additive component A, a dominance component D, and an epistasis or interaction component I. Here A is the sum of the expected individual effects of each allele, D is the sum of the effects of dominance interactions between allele pairs at each locus, and I is the sum of the effects of epistatic interactions across loci (Barton et al. 2007). We can then write:

$$P = G + E = A + D + I + E \qquad (9.11)$$

If we assume that all components are independent of each other, we can write the variance of this sum as a sum of the variances, much as we derived Equation 9.9:

$$V_P = V_A + V_D + V_I + V_E \qquad (9.12)$$

where V_A is the variance of the additive component A, V_D is the variance of the dominance component D, V_I is the variance of the interaction component I, and V_E is the variance of the environmental component E.

The power of breaking down the variances in this way is that we can now write down a very simple and very general expression for how a phenotypic trait changes over time in a population in response to natural selection. The way we do this is by looking carefully at which components of variation can be selected on directly by natural selection.

While all of the genetic contributors to a phenotype contribute to the genetic variance, the dominance component and interaction component are highly context dependent—that is, their effects depend strongly on the genetic background in which they occur. The additive component, by contrast, is independent of context. Irrespective of genetic background, the effects of this component are the same—and, as a result, the additive contributions are more accessible to natural selection. In other words, it is primarily the additive genetic variation on which natural selection operates.

In Equation 9.10, we defined broad-sense heritability H^2 to be the fraction of the variance due to any form of genetic variation: $H^2 = V_G/(V_G + V_E)$. But natural selection cannot easily act on all of this variation. To study the response of phenotype to selection, we need to look at the fraction of the total variation that is due to the additive genetic variation. We call this fraction the **narrow-sense heritability (h^2)**. Mathematically, we define narrow-sense heritability as:

$$h^2 = \frac{V_A}{V_A + V_D + V_I + V_E}$$

(If the broad or narrow sense is not specified, the term "heritability" typically refers to the narrow sense.)

Narrow-sense heritability, as a measure of what fraction of the variation is accessible to natural selection, plays a very important role in predicting how phenotypes change over time as a result of natural selection. Before seeing how this works, let's first consider a basic interpretation of narrow-sense heritability as a population-level measure of resemblance between parents and offspring: *narrow-sense heritability reflects the degree to which offspring resemble their parents in a population.* Specifically, narrow-sense heritability is the slope of a linear regression between the average phenotype of the two parents and the phenotype of the offspring. A subtle point: Because narrow-sense heritability is calculated as the *slope* of this regression, it is not always the case that the closer the resemblance between parents and offspring, the higher the heritability. As an extreme example, if all parents are identical and all offspring are identical to them, the heritability is undefined because there is no variability in the population.

Peter Berthold and Francisco Pulido used this expression for the narrow-sense heritability to estimate the heritability of migratory behavior in a European species of blackcap warblers, *Sylvia atricapilla* (Figure 9.35A) (Berthold and Pulido 1994; Pulido

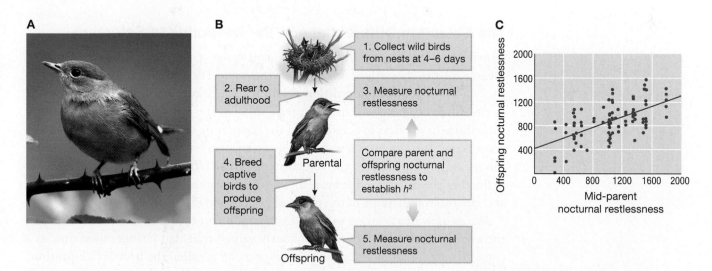

FIGURE 9.35 Heritability of migratory timing. (A) Berthold and Pulido estimated the narrow-sense heritability of migratory behavior in the blackcap warbler *Sylvia atricapilla*. **(B)** To do so, they ran an experiment in which they captured and reared wild birds to adulthood and measured their nocturnal restlessness, then bred the birds to produce offspring and measured the offspring's nocturnal restlessness. Berthold and Pulido estimated the heritability of migratory behavior by looking at the relationship of offspring migratory activity as measured by nocturnal restlessness to mid-parent (average value for both parents) migratory activity as measured by nocturnal restlessness. **(C)** The axes on the graph indicate the number of half-hour periods of nocturnal restlessness in the offspring and the parents. Part C adapted from Berthold and Pulido (1994).

et al. 2001). These researchers were interested in understanding whether migratory behavior—known to be under genetic control—could change by natural selection in response to climatic shifts such as global warming. If natural selection is going to shift migratory behavior, they reasoned, there must be additive genetic variation in the population for this behavior. For the blackcap warbler, this was at least plausible, as behavioral variation in migration in this species is well known—while many birds overwinter in the Mediterranean, some migrate as far as sub-Saharan Africa.

To estimate the heritability of migratory behavior, Berthold and Pulido designed a *parent–offspring regression study* (Figure 9.35B). They collected 186 newborn birds from wild nests, hand-reared these birds, and then measured their propensity for migratory behavior using a well-established assay: nocturnal restlessness in their cages during the autumn migratory period. The researchers then bred these captive birds (the parental generation) assortatively by nocturnal restlessness to produce a set of offspring derived from parents with known migratory behavior. They reared these offspring to adulthood and measured their migratory behavior in the same way. To estimate the heritability of this behavioral trait, the researchers plotted the nocturnal restlessness of the offspring birds as a function of the average nocturnal restlessness of their parents (Figure 9.35C). The heritability estimate is simply the *regression coefficient*—that is, the slope of the best-fit line—between the offspring and the parental average.

Using the parent–offspring regression technique, Berthold and Pulido estimated the narrow-sense heritability to be 0.453 ± 0.080. Because this value is significantly higher than zero, Burthold and Pulido were able to conclude that migratory behavior in the blackcap warbler is heritable, and indeed strongly so. These findings suggest that migratory patterns could change rapidly by natural selection, even over the course of a few generations. This is consistent with recent observations that the migratory patterns of this species have already begun to shift, perhaps in response to warmer winter temperatures in Europe.

The Selection Differential and the Response to Selection

When studying the evolution of a quantitative character, we need a way of measuring the strength of selection on the trait. The simplest approach is to use the concept of the **selection differential (S)**. The selection differential S is defined as the difference between the mean trait value of the individuals who successfully contribute to the next generation and the mean trait value of all individuals in the population. We also want a way to measure the consequences of selection. Here we can measure what is called the **selection response (R)**. The selection response R is defined as the difference between the mean trait value of the offspring population and the mean trait value of the parental population (Figure 9.36).

We are now in a position to write an expression for the mean trait value—that is, the average phenotype—of a continuously valued trait that changes over time as a consequence of natural selection. This expression is called the **breeder's equation**, which relates the narrow-sense heritability, the strength of selection measured as S, and the consequences of selection measured as R:

$$R = h^2 S \tag{9.13}$$

This simple equation predicts evolutionary change for quantitative traits. For example, suppose we select on fruit size in tomatoes, and the narrow-sense

$\overline{P}_0 = 0.60$ $\overline{P}_1 = 0.80$ $\overline{O}_0 = 0.72$

Parental generation
before selection

Parental generation
after selection

Offspring generation
before selection

FIGURE 9.36 Calculating the selection differential and selection response. Here a population of turtles differs in a quantitative trait, shell color, which can take on any value from light (0.0) to dark (1.0). In this example, selection favors darker-colored turtles, and most of the light-colored individuals in the parental generation die before reproducing. We calculate the selection differential S as the difference between the mean trait value of individuals in the parental generation who survive to reproduce, and that of all individuals in the parental generation, whether they survive or not. In this example, $S = \overline{P}_1 - \overline{P}_0 = 0.20$. We calculate the selection response R as the difference between the mean trait value of individuals in the offspring generation prior to selection, and the mean trait value of individuals in the parental generation prior to selection. In this example, $R = \overline{O}_0 - \overline{P}_0 = 0.12$.

heritability of this trait is $h^2 = 0.5$. If the fruit from plants that we allow to reproduce are, on average, 2 grams heavier than the population mean, we can use the breeder's equation $R = h^2 S$ to predict the fruit size in the offspring population. In our case, $h^2 = 0.5$ and $S = 2$ grams, so $R = (0.5)(2 \text{ grams}) = 1$ gram. We would therefore expect the offspring generation to have a mean fruit size that is 1 gram heavier than that of the (preselection) parental generation.

Quantitative Genetic Analysis of an Artificial Selection Study

To see how quantitative genetics can be applied to understand the process and consequences of selection on a quantitative trait, we will look at the longest-running selection experiment in crop plants, the Illinois Long-Term Selection Experiment on Corn (*Zea mays*) (Moose et al. 2004). This study, parts of which are still running today, was initiated in 1896 by C. G. Hopkins and has operated continuously since that time, except for a 3-year interruption during World War II.

The Illinois study is a long-running truncation selection experiment, so named because it involves truncating, or limiting, a population in terms of which individuals breed and which do not. The aim of this study was to look at the genetic basis of kernel oil. To do so, the investigators initially set up two different breeding lines from the same starting stock. One line was selected for high oil concentration, and one for low oil concentration. The truncation selection regime was relatively severe; only the top 20% of the high oil concentration line was used to seed the next generation, and the bottom 20% was used to seed the low oil concentration line. The response of each line, over the subsequent century, is shown in Figure 9.37.

The degree to which phenotypes continued to shift under artificial selection is remarkable. In Section 9.1, we discussed

FIGURE 9.37 Long-term phenotypic response in the Illinois Long-Term Selection Experiment on Corn. Truncation in this experiment was relatively severe; only the top 20% of the high oil concentration line was used to seed the next generation of the high oil concentration line, and only the bottom 20% of the low oil concentration line was used to seed the next generation of the low oil concentration line. The response of each line, over the subsequent century, is shown. Adapted from IDEALS (2011).

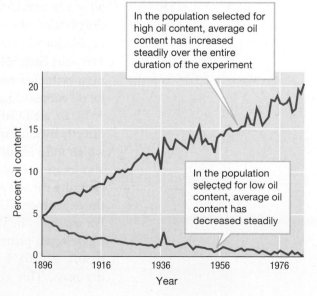

In the population selected for high oil content, average oil content has increased steadily over the entire duration of the experiment

In the population selected for low oil content, average oil content has decreased steadily

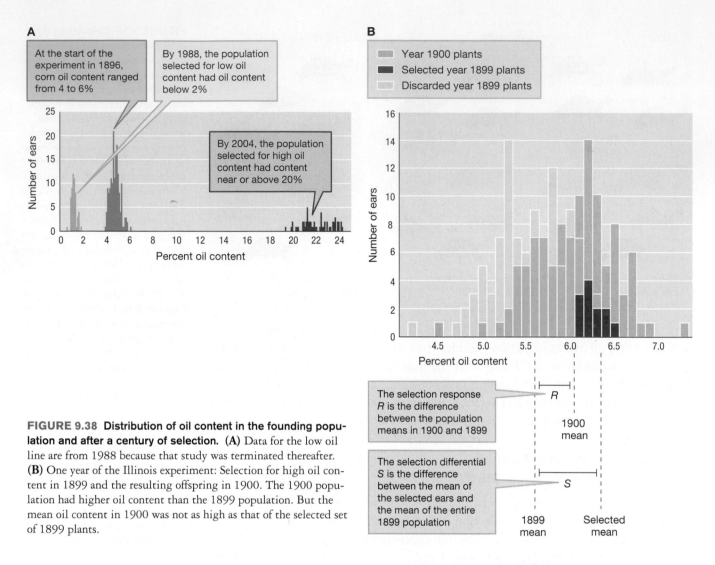

FIGURE 9.38 Distribution of oil content in the founding population and after a century of selection. (A) Data for the low oil line are from 1988 because that study was terminated thereafter. **(B)** One year of the Illinois experiment: Selection for high oil content in 1899 and the resulting offspring in 1900. The 1900 population had higher oil content than the 1899 population. But the mean oil content in 1900 was not as high as that of the selected set of 1899 plants.

the concept of latent variation and showed how selection could generate phenotypes beyond the range of those observed in the initial parental generation; this kernel oil study provides a striking example of latent variation. Figure 9.38 shows the distribution of oil content, measured as percent dry mass, in the initial founding population of 1896 (green), and in the high oil (red) and low oil (blue) lines nearly 100 years later. Selection has driven the production of phenotypes far beyond even the most extreme forms present in the founding population. In addition, novel mutations for oil content likely arose and were fixed over the course of the experiment.

In an artificial selection experiment such as this one, the experimenters can directly measure quantitative traits for each individual. Moreover, they know exactly which individuals reproduce and which do not (in Box 9.4 we discuss how we can also locate the loci involved when studying quantitative traits). Therefore, we can compute the selection differential S and the selection response R from the results of the experiment. Using these values, we can then estimate narrow-sense heritability. For example, in 1899, the mean oil content was 5.65% dry mass, but the mean of the selected plants was 6.30% dry mass. This gives a selection differential of $S = 6.30\% - 5.65\% = 0.65\%$ dry mass. The mean of all plants in 1900 was 6.10% dry mass. This gives a selection response of $R = 6.10\% - 5.65\% = 0.45\%$ dry

BOX 9.4 Mapping Quantitative Trait Loci

It can be difficult to identify the precise loci that are responsible for quantitative traits (so-called **quantitative trait loci**, or **QTLs**), but **QTL mapping** is a powerful way of finding at least the general region of the genome in which quantitative trait loci reside. The idea is that we can use *marker loci* that are easily assayed, but causally unrelated to the trait in question, in order to identify the approximate locations of the unknown alleles that affect the trait of interest. Figure 9.39 illustrates the basic concept behind the QTL mapping procedure.

Step 1. We typically begin the process by selecting two parental strains that (1) differ considerably in their values of the quantitative trait and (2) differ at a set of marker alleles. Parental strain 1 has a lower distribution of trait values than does strain 2; strain 1 is homozygous for the A, B, and C alleles, while strain 2 is homozygous for the a, b, and c alleles.

Step 2. The next step is to cross these two strains to produce a set of F_1 progeny. If the parents are homozygous at the marker loci, these F_1 progeny will be heterozygous at each marker locus, and typically they will manifest intermediate values of the quantitative trait.

Step 3. The F_1 individuals are then mated to produce an F_2 generation. For the F_2 individuals, we measure (1) the genotypes at the marker loci, and (2) the value of the quantitative trait. From this information, we can infer which marker loci are most closely associated with QTLs for the trait in question. The F_2 generation in Figure 9.39 illustrates the basic logic behind this inference. In each frame, the quantitative trait values are plotted with the genotypes sorted according to one of the marker loci. At left, we see a large difference in the quantitative trait values associated with the AA, Aa, and aa genotypes. This does not mean that the A marker locus is itself influencing the quantitative trait value, but it does imply that this locus is linked to an important quantitative trait locus. In particular, there appears to be a large positive QTL associated with the a allele. At center, we see essentially no difference in the quantitative trait values associated with the different genotypes at the B locus. Apparently there are no important QTLs near this B marker locus. At right, we see a modest association

between the value of the quantitative trait and the genotype at the C locus, again suggesting the presence of a quantitative trait locus in the vicinity of the C marker locus.

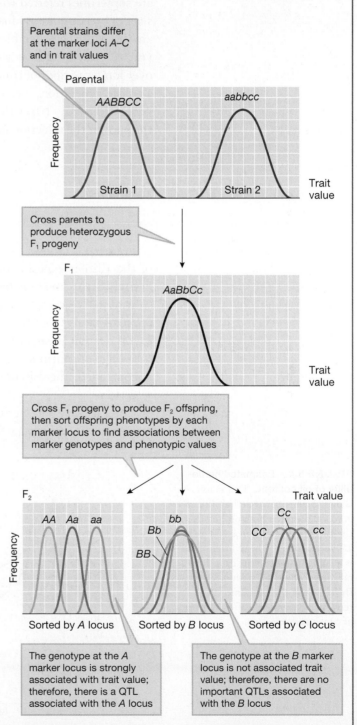

FIGURE 9.39 A schematic diagram of the principle behind the QTL mapping process. Note that in the F_2 generation, the heterozygotes will be twice as common as either homozygote, so the frequency distributions in the figure have been scaled to make each distribution directly comparable.

mass. From the response and selection differential, we can compute the narrow-sense heritability h^2 directly using the breeder's equation (9.13):

$$h^2 = \frac{R}{S} = \frac{0.45}{0.65} = 0.69$$

Heritabilities estimated from the selection differential and the selection response are sometimes referred to as **realized heritabilities**. In any given year, our sample sizes of breeding individuals are relatively small, and thus the heritability estimates are subject to considerable stochastic variation—that is, random fluctuation—from year to year. To deal with this problem, we might want to look at heritabilities over longer periods of time. Strictly speaking, the breeder's equation only holds for a single generation, but we can estimate the heritability over a modest number of generations (say, 10) as the sum of the selection responses in each year, divided by the sum of the selection differentials in each year:

$$h^2 \approx \frac{\sum\limits_{i=1}^{n} R_i}{\sum\limits_{i=1}^{n} S_i} \qquad (9.14)$$

Figure 9.40 shows the estimated heritability for oil content in the high oil line of the Illinois experiment. Here we have used a 10-year window—that is, each heritability estimate is based on the sum of 10 years' selection responses and the sum of 10 years' selection differentials. Heritabilities are initially quite high—approximately 0.4. Over time, heritability declines as some of the genetic variation for oil content is exhausted. Nonetheless, heritability remains above zero even after 100 years of continued directional selection. This suggests that, if the experiment is continued, the oil content will continue to increase in response to continued selection.

The fact that the heritability of oil content changes over time highlights an important concept regarding heritability: *Heritability is a statistical property of a*

FIGURE 9.40 Estimated heritability of oil content. This graph shows the estimated heritability of oil content in the high oil line from the Illinois experiment, based on a 10-year window. The red trend line is a cubic best-fit to the data. Heritability declines substantially over the course of the experiment, but it remains nonzero even after 100 years of continued selection.

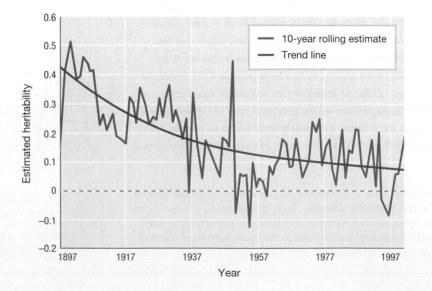

population, not a general fact about the genetic basis of a phenotypic trait (Barton et al. 2007). It is meaningless to say something like "the heritability of seed weight is 0.4" without specifying a population and the associated environmental conditions. This is because one population may have considerable additive genetic variance for seed weight while another has little or none. One population may experience dramatic environmental variation in seed weight (leading to reduced heritability), while another experiences highly uniform environmental conditions, and thus minimal environmental variation. For the same reason, a heritability estimate obtained from one population does not tell us about the heritability of the same trait in another population unless we have other reasons to believe that environmental and genetic variance in the two populations are similar. Finally, it is important to realize that heritability estimates are *within-population measures*, not between-population measures—that is, heritability tells us about the sources of differences *within* a population, but not about the sources of differences between populations. Just because we observe a high heritability for differences within one population, we cannot conclude that differences between this population and another population are also due to genetic factors.

Quantitative Genetic Analysis of Natural Selection in the Wild

Quantitative genetic tools are also useful for studying natural selection in the wild. By way of illustration, we return to an example from the start of Chapter 3: the drought-induced shift in flowering time of the annual plant *Brassica rapa* in southern California (Franks et al. 2007). In Chapter 3, we described one of the basic qualitative findings of that study: A *Brassica rapa* population sampled from a relatively wet habitat indeed evolved a more rapid flowering time over the period 1997–2004, presumably as a response to the drought of 2000–2004.

Researchers can go beyond simple qualitative assessments of this sort, however, as did Steve Franks and his colleagues. They wanted to determine whether the magnitude of the observed changes in flowering time are consistent with the operation of natural selection, given what can be determined about the genetic variation in the population and the strength of selection on this particular trait. Questions such as this one are not of purely academic interest. Our planet is currently going through a period of rapid climate change, and the ability of plant species to track these ongoing changes in temperature and precipitation will depend on whether there is sufficient genetic variation for traits such as reproductive timing.

To address this question, the researchers used a quantitative genetics framework. As we have seen, the breeder's equation allows us to predict the magnitude of evolutionary change in a trait, given (1) the narrow-sense heritability of that trait, and (2) the selection differential associated with the trait. Franks and his colleagues were able to measure both of these quantities in a straightforward fashion.

To determine the heritability of flowering time in the initial 1997 population, the researchers raised parent individuals in the greenhouse and recorded their flowering times. Using artificial pollination, they crossed known pairs of parents to produce F_1 offspring. They raised these F_1 offspring from seed in the same greenhouse, and they measured their flowering times. As we have already seen, narrow-sense heritability can be estimated directly from such data: It is the *regression coefficient* between the offspring flowering time and the average parental flowering time. Franks and his colleagues found that the heritability of flowering time was high in this population: $h^2 = 0.46$.

To determine the selection differential, the researchers first recorded the flowering time of each plant, and subsequently, once the seeds had set, they counted the number of seeds produced. From these data, they could estimate the selection differential. They found that, in 2003, the selection differential in this population was -7.67 days—that is, plants that reproduced successfully had flowered, on average, 7.67 days earlier than did those that failed to reproduce.

The breeder's equation predicts that in one generation, the change in mean flowering time should be $R = h^2 S = (0.46) \times (-7.67 \text{ days}) = -3.53$ days. But in their study Franks and his colleagues were not comparing flowering times of 2003 plants to flowering times of 2004 plants. They were comparing the flowering times of 1997 plants to 2004 plants—that is, they looked at the changes over seven generations. If we assume that, in each year the selection response was the same, the model then predicts that the total change in flowering time over the 7-year period should have been -3.53 days $\times 7 = -24.7$ days. The plants should have flowered 24.7 days earlier.

In practice, average flowering time shifted by only 8.5 days in this population—still a large amount, but not as large as predicted. What can we make of this? One conclusion we can draw is that, given the heritability of flowering time, selection was more than strong enough to shift flowering times by the 8.5 days observed. Not only did the researchers observe rapid change in flowering time in response to a multiyear drought, but also they were able to show that a response of this magnitude is easily consistent with the operation of natural selection. But why might the observed selection response have been less than predicted? Franks and his colleagues suggest that a number of factors may have contributed. Selection for early flowering may have been stronger—and thus the selection differential greater—in the 2003 population that they measured than in the other years between 1997 and 2004. This seems likely, given that the drought began only in 2000. Moreover, there may not have been seven full generations of selection between 1997 and 2004. If some fraction of the seeds remained dormant for one or more years before germinating, this would mean fewer generations of selection. Finally, the heritability estimate $h^2 = 0.46$ was based on studies in the greenhouse, not in the wild. Because environmental variance may be reduced under homogeneous greenhouse conditions, this heritability value may be an overestimate of heritability in the wild.

In this chapter, we have seen how multiple loci interact with one another in the evolutionary process. In the next chapter, we will further expand our view, to look at evolution on a genome-wide scale.

SUMMARY

1. Interactions between alleles at different genetic loci play an important role in the evolutionary process.

2. When traits are polygenic—that is, influenced by multiple loci—Mendelian inheritance can give rise to a near-continuous range of variation.

3. To create population genetic models of evolutionary change at multiple loci, we need to track haplotype frequencies rather than merely tracking allele frequencies.

4. When there are statistical associations between alleles at different loci, we say there is linkage disequilibrium in a population. The magnitude of these associations is quantified by the coefficient of linkage disequilibrium D.

5. Linkage disequilibrium can be created by evolutionary processes, including mutation, selection, migration, and drift.

6. Genetic recombination breaks down linkage disequilibrium over time.

7. Physical linkage on the chromosome facilitates the processes of genetic hitchhiking, background selection, periodic selection, and clonal interference.

8. The adaptive landscape metaphor provides a way to think about how phenotypes or genotypes change over evolutionary time as a consequence of natural selection.

9. Quantitative genetic approaches allow us to model and predict how continuous or quantitative characters change as a result of natural selection.

10. Narrow-sense heritability measures the fraction of phenotypic variation in a population due to the additive genetic variation on which selection can efficiently operate. If we know the narrow-sense heritability of a trait in a population, we can use the breeder's equation to predict how that trait will change in response to natural or artificial selection.

KEY TERMS

adaptive landscape (p. 313)

additive genetic effects (p. 295)

background selection (p. 310)

breeder's equation (p. 324)

broad-sense heritability (H^2) (p. 320)

clonal interference (p. 312)

coefficient of linkage disequilibrium (D) (p. 301)

coupling (p. 302)

dominance effects (p. 321)

epistasis (p. 298)

fitness peaks (p. 314)

fitness valleys (p. 314)

genetic hitchhiking (p. 309)

genotype space (p. 316)

haplotype (p. 298)

linkage disequilibrium (p. 301)

narrow-sense heritability (h^2) (p. 323)

periodic selection (p. 311)

phenotype space (p. 313)

physical linkage (p. 299)

polygenic (p. 294)

QTL mapping (p. 327)

quantitative genetics (p. 318)

quantitative trait loci (QTLs) (p. 327)

realized heritabilities (p. 328)

repulsion (p. 302)

selection differential (S) (p. 324)

selection response (R) (p. 324)

selective sweep (p. 311)

1. Imagine a species in which the A locus controls temperament, such that A individuals are aggressive and a individuals are timid. Further suppose that the B locus, on a different chromosome from the A locus, controls tooth morphology: B individuals have big sharp teeth, whereas b individuals have small delicate teeth. Assume that A and B are dominant to a and b. In this species, aggressive behavior is good if one has big teeth, but not if one has small teeth; timid behavior is good if one has small teeth but deleterious if one has big teeth. As a result, fitnesses are as indicated below:

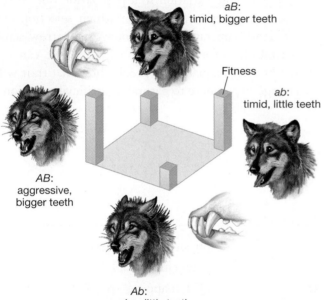

aB:
timid, bigger teeth

Fitness

ab:
timid, little teeth

AB:
aggressive,
bigger teeth

Ab:
aggressive, little teeth

Do you expect to observe linkage disequilibrium between the A and B loci? If so, will the value of D be positive or negative? Explain.

2. A population geneticist once quipped that "linkage disequilibrium can be neither." He meant that linkage disequilibrium can exist without physical linkage, and there can be linkage disequilibrium between loci in an equilibrium population. Explain.

3. Do you think that increasing the amount of epistasis would increase or decrease the ruggedness of an adaptive landscape in genotype space?

4. At the A locus, the frequency of the A allele is 0.25 and the frequency of the a allele is 0.75. At the B locus, the frequency of the B allele is f_B and the frequency of the b allele is $f_b = 1 - f_B$. Plot the largest possible value of D as a function of f_B. What is the largest possible value of the coefficient of linkage disequilibrium D in this population? At what value of f_B does it occur?

5. To estimate the broad-sense heritability of chirping rate in crickets, researchers create a set of inbred lines. They find that, at 68°C, the variance in chirping rate within inbred lines is 2 (chirps/minute)^2, and the variance between inbred lines is 10 (chirps/minute)^2. Estimate the broad-sense heritability of chirping rate from these data.

6. Design an experiment in which you are able to measure the narrow-sense heritability of flower size in an annual plant, without selecting on flower size or any other trait.

7. Design an experiment in which you can estimate the narrow-sense heritability of bristle number in the fruit fly *Drosophila melanogaster*, in which you do not need to know which offspring come from which parent.

8. Which will have a higher value, the broad-sense heritability or the narrow-sense heritability? Which is more useful for predicting evolutionary change? Explain your answers.

9. In a very large random-mating population of mice, haplotype frequencies for the AB, Ab, aB, and ab genotypes are 0.1, 0.4, 0.3, and 0.2, respectively. Compute the coefficient of linkage disequilibrium D in this population. If Hardy–Weinberg assumptions are met and the recombination rate between these two loci is $R = 0.2$ per generation, what will the coefficient of linkage disequilibrium be one generation later? Five generations later?

10. In a very large random mating population of zebrafish, the A and B loci are on separate chromosomes, and thus they are physically unlinked. Under what conditions would you expect to see linkage disequilibrium between these two loci?

SUGGESTED READINGS

Andersson, D. I., and D. Hughes. 2010. Antibiotic resistance and its cost: is it possible to reverse resistance? *Nature Reviews Microbiology* 8: 260–271. This review looks at the reasons why antibiotic resistance does not readily disappear after antibiotic use is halted. In doing so, it provides a number of good examples of how evolution operates on multiple interacting loci.

Berthold, P., and F. Pulido. 1994. Heritability of migration activity in a natural bird population. *Proceedings of the Royal Society B: Biological Sciences* 257: 311–315. This paper presents the study of the heritability of migration behavior that we described in this chapter.

Franks, S. J., S. Sim, and A. E. Weis. 2007. Rapid evolution of flowering time by an annual plant in response to a climate fluctuation. *Proceedings of the National Academy of Sciences of the United States of America* 104: 1278–1282. This paper describes the clever study on *Brassica rapa* flowering time that we introduced in Chapter 3 and treated in further detail here.

Maynard Smith, J., N. H. Smith, M. O'Rourke, and B. G. Spratt. 1993. How clonal are bacteria? *Proceedings of the National Academy of Sciences of the United States of America* 90: 4384–4388. A classic paper that explores the relationship between linkage disequilibrium in the genome and the population structure of bacterial species.

Ⓢ **Visit StudySpace at wwnorton.com/studyspace.**

10

Genome Evolution

◀ The Kau silversword, *Argyroxiphium kauense*, on Mauna Kea, Hawaii.

L ong before biologists had cause even to dream of whole-genome sequencing, researchers were already asking questions about the evolution of genomes and making comparisons of genomic data. Perhaps most notably, they were comparing the absolute sizes of genomes from species across the tree of life. Beginning in the 1950s, researchers compared genome size by measuring the amount of DNA per cell, originally called the C-value, across numerous species (Mirsky and Ris 1951). The results of such comparisons were surprising, and decidedly counterintuitive. Researchers found that genome sizes vary by more than 100,000-fold across living organisms, and by more than 10,000-fold even among the eukaryotes (Figure 10.1).

Yet, more perplexingly, as new genetic techniques were developed to estimate the number of genes in a genome, researchers discovered that differences in genome size do not correlate in any straightforward way with the number of protein-coding genes that an organism has, nor with its phenotypic complexity. This observation, known as the **C-value paradox**, or C-value enigma, was profoundly puzzling. If an organism's genes are encoded in its DNA, why is there scant correlation between the number of genes and

FIGURE 10.1 Genome size varies widely across the tree of life.
Genome sizes are measured in millions of base pairs, called *megabases* (Mb). The data are displayed here on a logarithmic scale. Thus, the largest genomes (lungfishes, some flowering plants) are more than 100,000 times the size of the smallest genomes (archaea). Adapted from Gregory (2011).

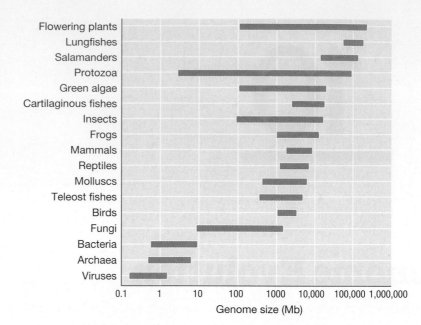

the amount of DNA in a genome? Why should lungfish require 40 times as much genetic information as humans? Why would a single-celled amoeba have a genome that is 1000 times the size of the genome of a complex multicellular puffer fish?

We will address those questions and numerous others in this chapter. In Chapters 7–9, we explored the field of population genetics, which concentrates on evolutionary processes operating at a single locus or at sets of loci. With the advent of whole-genome sequencing, evolutionary biologists now have a rich set of tools for studying population genetics at the genome-wide scale, and for exploring how entire genomes evolve over time. This is the field of **evolutionary genomics**.

A major theme of this chapter is that genome structure arises from a combination of selective and nonselective processes. Whereas many facets of genome organization may be explained by natural selection operating at the organismal level, others involve natural selection operating on "selfish" genetic elements within the genome, sometimes at a fitness detriment to the organism itself. Neutral or nearly neutral processes also play major roles in structuring genomes.

In this chapter, we present an overview of evolutionary genomics. We will:

- Begin with a look at the development of whole-genome sequencing technology.

- Explore the evolution of genome size as a way to build up our intuition for thinking about how genomes evolve.

- Examine the genome structure and composition of viruses, prokaryotes, and eukaryotes and how they are fashioned by a combination of selective and nonselective processes.

- Conclude by considering how biologists scan for the signatures of natural selection in genome sequence data.

Because the study of genome evolution is so young, open questions and unsolved problems still outnumber the resolved issues. As a result, this chapter is more descriptive than many of the others in this book. Rather than providing the last word on genome evolution, however, our aim here is to provide an overview of what patterns are to be found in the genomes of organisms from viruses to vertebrates, and to discuss the various genome-scale processes that may be involved in generating the patterns. In many cases, the relative importance of these processes remains to be seen, and some processes may still be undiscovered. It is an exciting time to be working on evolutionary genomics.

10.1 Whole-Genome Sequencing

Genetic sequencing is unquestionably one of the most important technological achievements of the past half-century; it has had an enormous impact on nearly every area of the life sciences. In 1976, only a decade after the full genetic code was deciphered, Robert Fiers and his colleagues reported the entire genome sequence for the bacteriophage MS2, an RNA virus. This was the first whole-genome sequence for any microbe (Fiers et al. 1976). In 1977, Fred Sanger and his colleagues sequenced all 5386 base pairs of the genome of the bacteriophage φX174 (Sanger et al. 1978). This was the first entire DNA genome ever to be sequenced, a feat possible at that time only because of the exceptionally small size of the genome of this phage. Even very small bacteria have genomes several orders of magnitude larger than the genomes of bacteriophage. Thus, it was another 18 years before researchers at Johns Hopkins University sequenced the entire genome of a bacterium, *Haemophilus influenzae*. In a 1995 paper, they reported this accomplishment, which was the first whole-genome sequence obtained for an independently living organism (Figure 10.2) (Fleischmann et al. 1995). A year later, the first genome sequence of a eukaryote was released: Researchers had sequenced the 12-million-base-pair (that is, 12-*megabase*) genome of the yeast *Saccharomyces cerevisiae* (Goffeau et al. 1996). In 1998, the first genome of a multicellular organism, *Caenorhabditis elegans*, was published (*C. elegans* Sequencing Consortium 1998), and in 2001, the initial draft of the human genome was released (International Human Genome Sequencing Consortium 2001). The final draft was completed in 2003.

Today, a large number of additional genomes have been sequenced. As of summer 2010, over 100 eukaryotic genomes have been completed and published, and researchers are in the process of sequencing hundreds of other genomes. Over 1000 prokaryotic genomes have been sequenced as well. Biologists now have the ability to generate enormous quantities of genomic sequence data; many of the greatest remaining challenges in genomics involve finding ways to best organize and make use of the data.

FIGURE 10.2 Some landmark genome sequencing projects. The numbers below each organism indicate the size of its genome in megabases (Mbs).

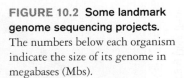

0.0036 Mb	0.0054 Mb		1.8 Mb	12.5 Mb		100 Mb	123 Mb	115 Mb	3100 Mb
1976	1977	//	1995	1996	1997	1998	1999	2000	2001
Bacteriophage MS2	Bacteriophage φX174		*Haemophilus influenzae*	*Saccharomyces cerevisiae*		*Caenorhabditis elegans*	*Drosophila melanogaster*	*Arabidopsis thaliana*	*Homo sapiens* (draft; completed 2003)

10.2 Resolving the Paradoxes of Genome Size

In the introduction to this chapter, we posed the C-value paradox: Why is there such enormous variation in genome size, and why does organismal complexity not correlate well with genome size? This paradox was largely resolved once researchers discovered the prevalence of **noncoding DNA**. In large genomes, such as those of animals and land plants, only a small fraction of the total genome is devoted to coding sequence. The remainder is made up of noncoding DNA of various types, which we will consider later in this chapter. Moreover, most of the differences in genome sizes among complex multicellular organisms result from differences in the quantity of noncoding DNA—that is, the organisms all have roughly similar amounts of coding DNA (Figure 10.3). With this observation in hand, our previous question—why a lungfish would require 40 times as much genetic information as a human—becomes easier to resolve. The answer is that the lungfish does not require 40 times as much information; it just happens to have a genome that is 40 times bigger than the human genome, with most of that difference due to extra noncoding DNA.

The mysteries of genome size are not entirely solved by the simple observation that genome size differences result largely from differences in the quantity of noncoding DNA. We would also like to explain the causes of the variation that we observe. Why do some species have vastly larger genomes than others, despite similar degrees of apparent phenotypic complexity? A number of hypotheses have been proposed regarding possible mechanisms that determine genome size. One possibility is that changes in genome size are favored because of their structural effects on the size of the nuclear envelope, the volume of the cell, and other aspects of cell physiology (Cavalier-Smith 1978). Although these factors are certainly influenced by genome size, it is unclear that there has been sufficient individual-level selection on these factors to account for the enormous genome size differences.

An alternative hypothesis was proposed in 1980 in a pair of back-to-back papers published in the journal *Nature* (Doolittle and Sapienza 1980; Orgel and Crick 1980). This view holds that genome size is the result of a balance between two types of processes. On the one hand, proliferation of self-replicating genetic units, such as **transposable elements** (or transposons)—small genetic elements capable either of catalyzing their own movement within the genome or of moving with the assistance of other transposable elements—may drive an increase in genome size over time. On the other hand, selection for replication speed, small cell size, and energetic efficiency may favor reductions in genome size. Different species face different ecological challenges—some may need cells that can divide very quickly, while others may not. Similarly, different species bring different evolutionary histories with them. Some, such as those in the genus *Drosophila*, have large numbers of active

FIGURE 10.3 Coding DNA versus total genome size. Whereas smaller genomes consist largely of coding DNA, many larger genomes are made up mostly of noncoding DNA. Here the number of coding base pairs is plotted against the total number of base pairs in the genome for organisms ranging from viruses to animals, with both axes on a logarithmic scale. The dashed lines indicate the fraction of the genome composed of coding DNA. In viruses, nearly 100% of the genome is coding sequence. In many prokaryotes and unicellular eukaryotes, the coding fraction drops below 50%. In land plants and in animals, the fraction drops further, to below 1% in some organisms. Adapted from Lynch (2007).

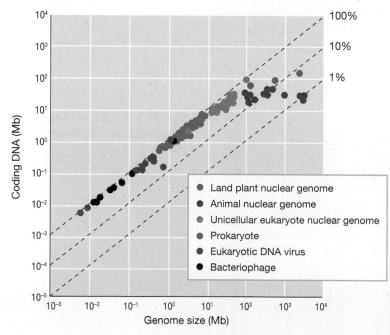

transposons in their genomes; others, such as humans, may have a past history of transposon accumulation but relatively few active transposons in their current genome. As a result, the processes above will balance one another in different ways in different species, leading to the broad variation in C-values that we observe across taxa.

One inescapable consequence of large genome size is that larger genomes require larger cell nuclei, and thus larger cells. Figure 10.4 shows the relationship between cell volume and genome size. Cell volume influences a number of aspects of phenotype that are relevant to fitness, including rate of cell division, metabolic efficiency, rates of protein and ion exchange, and in many taxa, overall body size. This association between C-value and cell size may be one of the important drivers of selection on genome size (Gregory 2001).

Evolutionary biologist Michael Lynch has proposed that an additional process may be important in the evolution of genome size. Recall from Chapter 8 that the strength of natural selection to eliminate deleterious mutations and to fix beneficial ones depends on the population size. As the population size grows larger, natural selection can operate effectively on smaller and smaller fitness differences. This may contribute to a general trend for prokaryotes to have smaller genome sizes than unicellular eukaryotes, which in turn tend to have smaller genome sizes than large multicellular eukaryotes. Larger organisms, because they tend to have smaller population sizes, will be less able to eliminate mildly deleterious variants that result from a minor increase in noncoding DNA. According to this view, the expanded

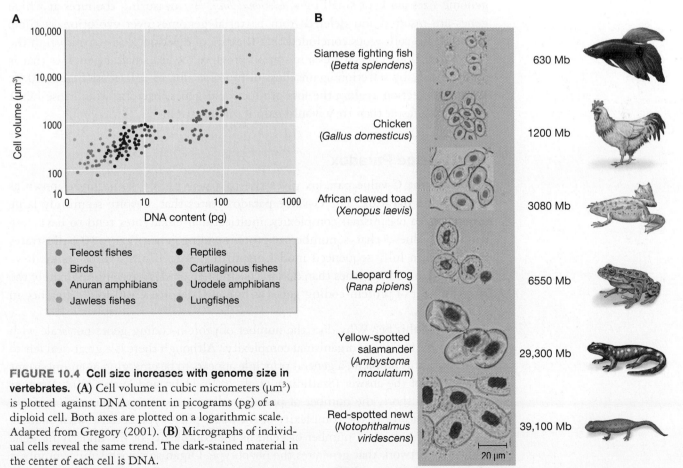

FIGURE 10.4 Cell size increases with genome size in vertebrates. (A) Cell volume in cubic micrometers (μm^3) is plotted against DNA content in picograms (pg) of a diploid cell. Both axes are plotted on a logarithmic scale. Adapted from Gregory (2001). (B) Micrographs of individual cells reveal the same trend. The dark-stained material in the center of each cell is DNA.

FIGURE 10.5 A model of genome size evolution in prokaryotes. Prokaryotic genomes increase in size principally by gene duplication events and by acquiring nonhomologous genes from external sources. They decrease in size by deletion of active genes or by deletion of inactivated pseudogenes. The latter processes operate at a higher rate than the former, and genome size remains at a steady state only because of natural selection against deletions that cause loss of function. Adapted from Mira et al. (2001).

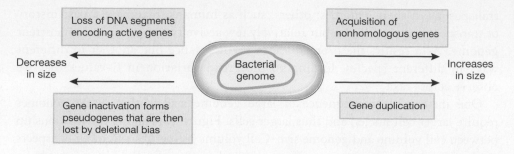

genome sizes of eukaryotes result from nonadaptive processes at the organismal level. Nonetheless, the additional genetic material may be co-opted in any number of ways, allowing the subsequent evolution by natural selection of complex genome organization in eukaryotes (Lynch and Conery 2003; Lynch 2007).

In addition to population size, other factors may be involved in keeping prokaryotic genome sizes small. One hypothesis is that prokaryotes have small genomes because they are under strong selection for rapid growth by cell division. Given that prokaryotic genomes have a single origin of replication (ORI) on the chromosome from which DNA synthesis is initiated, larger genomes take proportionally longer to copy. The larger the genome, the slower the minimal doubling time. But Alex Mira and his colleagues have argued against that view (Mira et al. 2001). They note that there is no association between genome size and maximum replication rate in bacteria, and propose that, instead, prokaryotic genome sizes are kept small by a *deletional bias.* By measuring the rates at which genes are inserted and deleted from bacterial genomes over evolutionary time, Mira and his colleagues concluded that the rate of deletion tends to outweigh the rate of insertion. This creates a mutation pressure toward smaller genomes that is opposed only by selection against loss of function due to deletions (Figure 10.5). Without selection against the loss of functional genes, bacteria would lose DNA via deletion faster than they would gain it via insertion.

The G-Value Paradox

Resolving the C-value paradox gives rise to a new puzzle, sometimes known as the **G-value paradox**. The G-value paradox states that, despite seemingly large differences in organismal complexity, multicellular eukaryotes tend to have very similar G-values—that is, numbers of protein-coding genes. Figure 10.6 illustrates this for seven fully sequenced model organisms. Surprisingly, slime molds have more protein-coding genes than do insects, and nematodes have approximately the same number of protein-coding genes as humans, despite a 300-fold difference in genome sizes.

How can this be? Why does the number of protein-coding genes not scale with our intuitive notions of organismal complexity? Although there is a great deal left to learn about this problem, a general principle appears to be emerging as an important component of the answer (Szathmary et al. 2001; Hahn and Wray 2002; Wray et al. 2003). In short, the number of protein-coding genes in an organism's genome is a poor indicator of the complexity of the adult phenotype because what matters more than the absolute number of genes is the complexity of the underlying gene regulatory network that generates the phenotype. Organisms with similar numbers of genes can have very different gene regulatory network structures.

For example, **transcription factors** play a very important role in gene regulation. Transcription factors are proteins that bind to specific regions of DNA in order to regulate when, where, and to what degree specific genes are expressed. Despite similar numbers of protein-coding genes, the nematode, fruit fly, and human genomes have very different numbers of transcription factors: approximately 500, 700, and 2000, respectively (Szathmary et al. 2001; Tupler et al. 2001). Because transcription factors often act on one another, these differences can translate into even bigger differences in regulatory complexity. If transcription factors were to operate in pairs, then humans, who have 4 times as many such proteins as nematodes do, would have approximately 4^2 or 16 times as many possible combinations of transcription factor proteins. If transcription factors were to operate in trios, that ratio would be 4^3 or 64 times as many. Although the networks of interactions among transcription factors are obviously more complicated than simple pairwise or three-way interactions, the same general principles of scaling probably apply.

Another important element affecting regulatory complexity may be the degree of regulatory control exercised by noncoding regions of the genome. In particular, organisms with larger noncoding genomic regions may have a larger number of regulatory elements that are involved in specifying complex patterns of expression. A third contributing factor is that one protein-coding gene does not necessarily correspond to one protein. Through the process of alternative splicing, a single gene with multiple exons can be spliced in a number of different ways to produce a variety of protein products. For example, humans have more alternatively spliced genes, and more introns per gene (allowing a greater range of alternative splicing products), than do nematodes. In addition to alternative splicing, various forms of posttranscriptional modification—alterations made to newly transcribed RNA—can potentially increase the diversity of an organism's functional protein products (Hahn and Wray 2002).

Together, these observations lead us to at least a partial resolution of the G-value paradox. Complexity depends less on how many genes a species has than on how those genes are connected. The total number of coding genes in a genome matters less than the complexity of the regulatory network through which those coding genes interact.

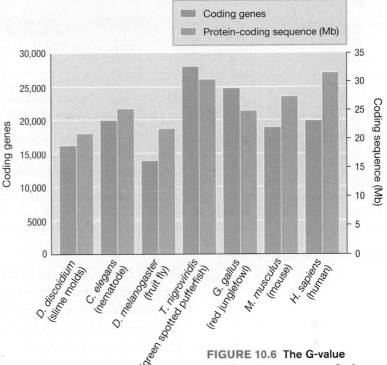

FIGURE 10.6 The G-value paradox. Across a range of eukaryotes that appear to vary enormously in their structural complexity, there is relatively little variation in the number of coding genes or in the total amount of protein-coding sequence. Adapted from Taft et al. (2007).

10.3 Content and Structure of Viral Genomes

As we have just noted, genome sizes differ in part because the genomes of different taxonomic groups are made up of different types of genomic elements. In this and the two subsequent sections, we look at the components of viral, prokaryotic,

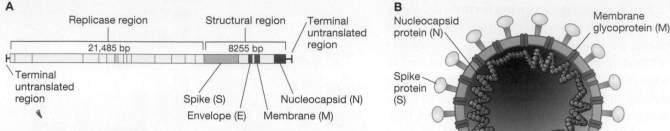

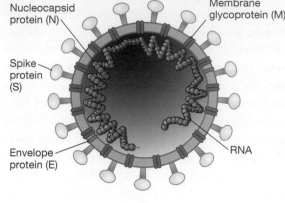

FIGURE 10.7 The SARS coronavirus. The SARS coronavirus, responsible for the 2002–2003 SARS epidemic, is a single-stranded positive-sense RNA virus with a genome composed of a single linear chromosome. **(A)** This diagram is a simplified map of the SARS genome that shows the gene segments for the replicase region with the genes involved in genome replication and the gene segments for the structural region with the genes responsible for the structural proteins that the virus requires. **(B)** This virus diagram shows the structural components corresponding to the gene segments in the map. Adapted from Stadler et al. (2003).

and eukaryotic genomes in turn. We will look at what sorts of genetic elements are present, in what quantities, in genomes of different types. We will look at how these elements are arranged spatially within the genome. And we will look at the processes, selective or otherwise, by which the content and organization of genomes have been shaped.

The genome structures of viruses are extremely diverse. Even the genetic material itself varies. Whereas many viruses have DNA-based genomes as do prokaryotes and eukaryotes, many others have RNA-based genomes. Some viruses, known as retroviruses, are capable of reverse transcription, in which DNA is synthesized from an RNA template. Many retroviruses, including the human immunodeficiency virus (HIV), use that capability to integrate their own genomes into the host's chromosome. Both DNA-based and RNA-based genomes may be either double- or single-stranded. Single-stranded RNA viruses may be either *positive-sense* viruses, in which case the genome is effectively the same as the viral mRNA, or *negative-sense* viruses, in which case the genome is complementary to the viral mRNA. Viral genomes may consist of a single linear chromosome, a set of linear chromosomes (in which case we refer to the genome as being *segmented*), or a circular chromosome. Figures 10.7, 10.8, and 10.9 illustrate a number of different viral genome structures.

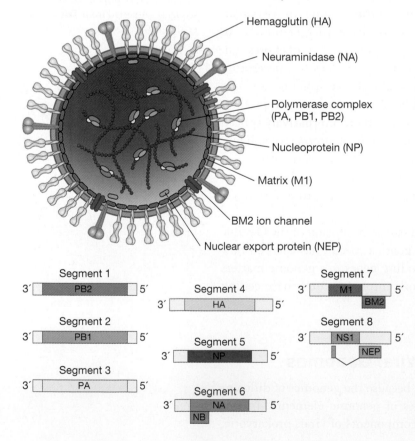

FIGURE 10.8 The influenza virus. The influenza B virus is a single-stranded negative-sense RNA virus of approximately 13,600 nucleotides in length. Its genome is segmented into 8 linear chromosomes encoding its 11 protein-coding genes. Color coding indicates the correspondence between structural proteins in the virus and the corresponding protein-coding genes in the genome.

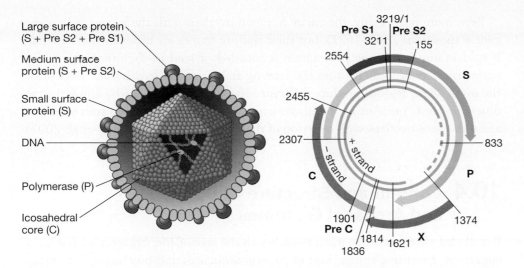

FIGURE 10.9 The hepatitis B virus. The hepatitis B virus (HBV) is a very small circular double-stranded DNA virus of approximately 3.3 kb. The inner rings represent the DNA chromosome. As shown by the dashed line, the positive strand is incomplete in the virus capsule but is completed by synthesis from the negative strand after infection. The colored arrows indicate the location and direction of each gene in the genome. Note the substantial overlap of different genes. Adapted from Park et al. (2006).

Viral genomes tend to be extremely compact (Carter and Sanders 2007). One reason is that many—although not all—viruses have RNA-based rather than DNA-based genomes. RNA is both structurally more fragile than DNA, and it is subject to higher mutation rates; those factors severely limit the maximum possible size of RNA-based genomes. The SARS coronavirus shown in Figure 10.7 is at the upper end of the RNA virus size range, with a 30 kilobase (kb) genome. As a result, RNA viruses typically encode only a few proteins, as illustrated in Figure 10.8.

DNA-based viruses can be much larger, up to a megabase in length, but even this is relatively small in comparison with all eukaryotes and the vast majority of prokaryotes. One reason that even DNA viruses are relatively small is that most viruses undergo strong natural selection for rapid replication. The shorter the genome, the faster it can be copied. Another reason may involve natural selection on physical size. The very small physical size of virus particles constrains the amount of genetic material that can be packaged within them (Cann 2005).

Because viral genomes are under such strong selection for reduced size, most of a viral genome consists of a protein-coding sequence, with terminal untranslated regions in the linear genomes. One of the most remarkable aspects of viral genomes is the tremendous degree of compression achieved. Multiple genes may be packed into a single region in two different ways: (1) in the same reading frame but only partially overlapping, or (2) with different reading frames (Figure 10.10). The hepatitis B virus (HBV) in Figure 10.9 uses both methods. In HBV, the three different surface antigen proteins, Pre S1, Pre S2, and S, are all derived from a single gene (bases 2554–833) with different ATG start codons but a shared stop codon. In this case, all three proteins are produced by reading in the same reading frame; they just start in different places.

FIGURE 10.10 Two kinds of overlapping code. Here a string of three-letter words represents a set of nucleotide triplets. From the basic string "gnu are too new hot ads awe tom any day," we can get multiple messages in two different ways. We can read in the same frame, but start and stop in different places. Alternatively, we can read in a different frame. Viruses employ both methods of coding for multiple proteins using a single region of the genome.

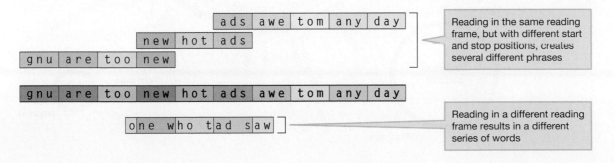

Reading in the same reading frame, but with different start and stop positions, creates several different phrases

Reading in a different reading frame results in a different series of words

Even more remarkably, the entire S region overlaps with the longer polymerase gene P (positions 2307–1621), but their reading frames are offset by one base pair. If read in one frame, the polymerase is encoded; if read in the other, the surface antigens are produced! Because the reading frames are shifted by one nucleotide, the overlapping regions produce different sequences of amino acids. Similarly, two different core C proteins are produced from another gene with two start sites, and again the gene overlaps with a section of the polymerase gene (Zaaijer et al. 2007).

10.4 Content and Structure of Bacterial and Archaeal Genomes

Recall that there are three main branches to the tree of life: bacteria, archaea, and eukaryota. Extensive comparison of genetic sequences and biochemical pathways has revealed that the archaea are phylogenetically closer to the eukaryota than to bacteria. Despite this phylogenetic divide, archaeal and bacterial genomes have evolved to have qualitatively similar structures and organization (for a discussion of the differences, see Karlin et al. 2005). Thus, we will consider prokaryotic genomes—bacterial and archaeal—jointly in this chapter. Prokaryotic genomes tend to be structured as a single circular chromosome, present in a single copy per cell. Figure 10.11 illustrates this type of organization, using as an example the

FIGURE 10.11 The *E. coli* O157:H7 genome. Like most bacteria, *E. coli* O157:H7 has a single circular chromosome. In addition, the strain carries two additional small DNA molecules: a virulence plasmid and a second miniplasmid. Here the virulence genes are shown in red. Adapted from Genome Center of Wisconsin (2011).

Selected features of the O157:H7 genome

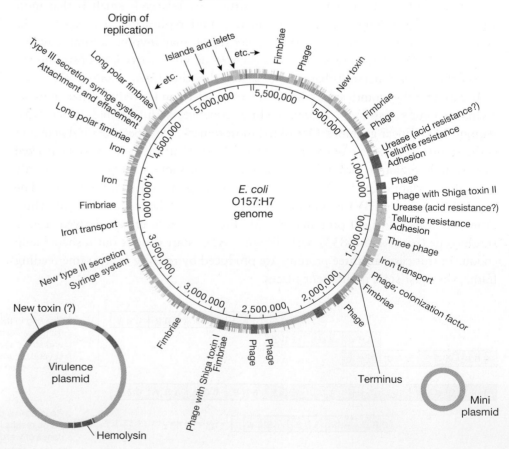

genome of the pathogenic *E. coli* strain O157:H7. There are exceptions, however. Some bacteria have more than one circular chromosome. Others, such as species of *Borrelia* and *Streptomyces*, have a linear chromosome.

Although the genomes of bacteria and archaea tend to be larger than those of all but the very largest viruses, they remain relatively compact and indeed are smaller than those of all but the smallest eukaryotes. Typically, a prokaryotic genome consists of 85–95% protein-coding sequence. Intergenic regions—the stretches of DNA between genes—are minimal. In *E. coli*, for example, adjacent protein-coding genes are commonly separated by roughly 100 base pairs. Many of the prominent elements of eukaryotic genomes are absent or rare: Bacteria do not have the spliceosomal introns that are so prevalent in eukaryotes (see the next subsection). The introns that are found in prokaryotic genomes tend to be located in ribosomal RNA (rRNA) or transfer RNA (tRNA), and are self-splicing. Prokaryotic genomes do have pseudogenes—recall that these are nonfunctional and typically untranslated sequences of DNA—although they tend to be less common in prokaryotes than in eukaryotes. Finally, although transposable elements are found in prokaryotes, they typically make up only a small fraction of the genome. In a few cases, however, that fraction can exceed 10% (Gregory and DeSalle 2005).

Bacterial genomes often include DNA from **prophages**, viral genomes that insert themselves into bacterial chromosomes. Prophage DNA can make up 10% or more of the bacterial genome. Most of these prophages are no longer functional, but new ones are readily incorporated into the genome. Over time, this process contributes to genomic differences not only between species, but even between strains of the same bacterial species. Figure 10.12 shows how prophage DNA occurs in different places in three different strains of the human pathogen *Streptococcus pyogenes*.

Prophages often encode **virulence factors**, specialized genes that assist bacteria in exploiting eukaryotic hosts by aiding colonization, producing toxins, entering host cells, and evading immune responses (Wagner and Waldor 2002). For example, the shiga toxins of the pathogen *E. coli* O157:H7 are encoded on prophages (O'Brien et al. 1984). The majority of prophages have lost the ability to form viral particles because of the accumulation of mutations that have eliminated their ability to replicate independently. However, some prophages remain active, and thus they can directly transmit virulence factors to new strains of bacteria. Evolutionarily, it remains unclear why virulence factors are so commonly encoded on prophages. One possibility is that they provide a stable long-term reservoir for the associated virulence genes (Muniesa et al. 1999).

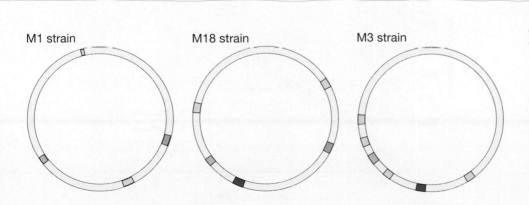

M1 strain   M18 strain   M3 strain

FIGURE 10.12 Prophage DNA in three strains of *Streptococcus pyogenes*. Three different strains of *S. pyogenes* have prophages incorporated in different positions around the chromosome. Genetically similar prophages are shown in the same color, revealing that these three strains share many of the same prophages. However, these prophages have entered the genomes of each strain in separate insertion events at different genomic positions. Adapted from Canchaya et al. (2003).

A

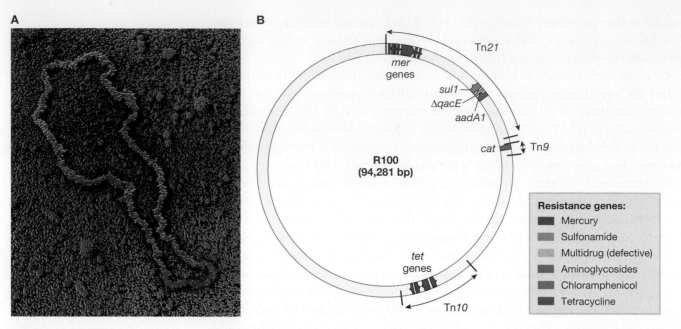

B

FIGURE 10.13 Bacterial plasmids. (A) An electron micrograph of a bacterial plasmid reveals that it forms a closed loop. **(B)** A genetic map of the antibiotic resistance plasmid R100. Note that the genes conferring resistance are themselves located within transposons on the plasmid. These transposons are indicated as Tn*9*, Tn*10*, and Tn*21* on the diagram. Other regions of the plasmid genome predominantly include genes involved in plasmid replication and conjugation. Part B from Nikaido (2009).

In addition to their main chromosomes, many prokaryotes carry one or more **plasmids** (Figure 10.13A). Plasmids are small, nonessential circular DNA molecules that often code for additional functions beyond the ability to move by the process of conjugation, which we will shortly discuss in detail. Plasmids often encode genes for resistance to one or more antibiotics. For example, the R100 resistance plasmid is shown in Figure 10.13B.

In the introduction to this chapter, we noted that when we look across the entire tree of life, there is little correlation between the number of genes in an organism and the size of its genome. Within the prokaryotes, however, these quantities are tightly correlated, as illustrated in Figure 10.14. This should not be all that surprising. As we have noted, prokaryotic genomes are largely composed of protein-coding sequences and thus we would expect such a relationship. Free-living prokaryotes have larger genomes than do those that live in obligate associations with eukaryotes, probably because they tend to need more genes to perform a wider range of functions than do obligately parasitic or symbiotic species (Figure 10.15).

FIGURE 10.14 Genome size and number of genes. In bacterial (blue) and archaeal genomes (gold), genome size is tightly correlated with the number of genes. Adapted from Gregory and DeSalle (2005).

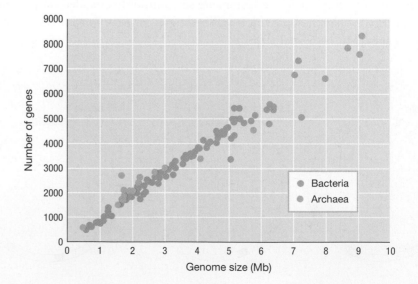

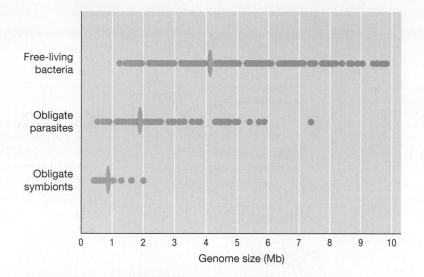

FIGURE 10.15 Genome sizes of symbionts, parasites, and free-living bacteria. Obligate symbionts tend to have smaller genomes than obligate parasites, which in turn have smaller genomes than free-living bacteria. Mean genome sizes for each class of bacteria are indicated in gold. Adapted from Gregory and DeSalle (2005).

Horizontal Gene Transfer and Prokaryote Genomes

Horizontal gene transfer (HGT), also known as **lateral gene transfer**, is an important source of genetic variation for microbes. Both phages and plasmids are common vehicles of horizontal gene transfer, which involves the transfer of genetic material from one organism to another by one of three processes: transduction, transformation, or conjugation (Figure 10.16).

Transduction occurs when a phage packages bacterial DNA instead of its own within its capsule. When such a phage infects a new host, it injects the bacterial DNA into that host, where it can be incorporated into the genome by homologous or nonhomologous recombination.

In the process of transformation, a cell takes up free-standing double-stranded DNA—such as that released when other cells die—from the environment (Dubnau 1999). This DNA can subsequently be incorporated into the genome by recombination. Some species, including the human pathogens *Streptococcus pneumoniae, Haemophilus influenzae,* and *Neisseria gonorrhoeae,* are naturally *competent—* that is, they have active mechanisms for acquiring DNA by transformation. Microbiologists have proposed several possible functions for competence, including (1) the acquisition of nucleotides as "food," (2) use of acquired DNA in the process of DNA repair, and (3) the generation of variability. At present, the relative importance of each explanation remains unresolved.

In the process of conjugation, a plasmid is passed from a donor cell to a recipient cell. A donor bacterium creates hairlike *conjugative pili* that pull a recipient bacterium close, and then it opens up a *conjugative junction* between the two cells through which a copy of the plasmid is transferred. Both structures can be seen in Figure 10.17; the dark pili surround the cell on the left, while the conjugative junction joins the two cells. *Conjugative plasmids* encode all of the genes necessary for carrying out the conjugation process; other *nonconjugative plasmids* do not encode this machinery and can only undergo horizontal gene transfer when they are facilitated by the presence of a conjugative plasmid in the same cell.

Although horizontal gene transfer is sometimes referred to as "bacterial sex," the process is very different from sexual reproduction. First, it is decoupled from the process of reproduction and occurs far less frequently than the once-per-generation rate that we observe for sexual reproduction in most multicellular

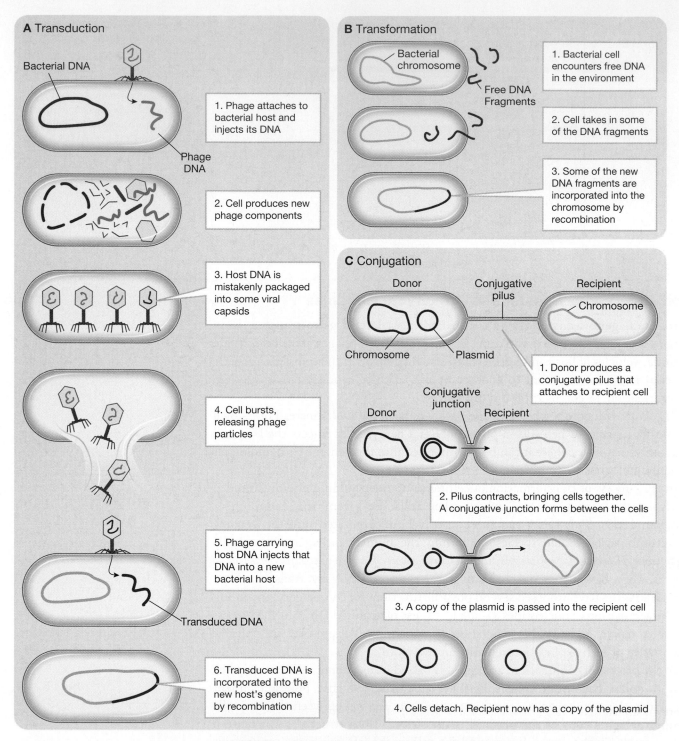

A Transduction

Bacterial DNA

Phage DNA

1. Phage attaches to bacterial host and injects its DNA

2. Cell produces new phage components

3. Host DNA is mistakenly packaged into some viral capsids

4. Cell bursts, releasing phage particles

5. Phage carrying host DNA injects that DNA into a new bacterial host

Transduced DNA

6. Transduced DNA is incorporated into the new host's genome by recombination

B Transformation

Bacterial chromosome

Free DNA Fragments

1. Bacterial cell encounters free DNA in the environment

2. Cell takes in some of the DNA fragments

3. Some of the new DNA fragments are incorporated into the chromosome by recombination

C Conjugation

Donor Conjugative pilus Recipient

Chromosome

Chromosome Plasmid

1. Donor produces a conjugative pilus that attaches to recipient cell

Conjugative junction

Donor Recipient

2. Pilus contracts, bringing cells together. A conjugative junction forms between the cells

3. A copy of the plasmid is passed into the recipient cell

4. Cells detach. Recipient now has a copy of the plasmid

FIGURE 10.16 Three modes of horizontal gene transfer. Horizontal gene transfer occurs by the processes of **(A)** transduction, **(B)** transformation, and **(C)** conjugation.

eukaryotes. Second, although horizontal gene transfer is most common between individuals of the same or closely related species, bacteria are by no means limited by species boundaries when exchanging genes horizontally. In fact, genes have been transferred horizontally not only between different species, but even between members of completely different domains of life, such as the transfer from prokaryotes to eukaryotes (Woese et al. 2000; Koonin et al. 2001; Thomas and Nielsen 2005). Furthermore, unlike sexual reproduction, gene exchange is not reciprocal: In conjugation, the donor does not receive DNA from the recipient.

Horizontal gene transfer can have important health implications. For example, *E. coli* K-12 is a harmless enteric strain that resides in the human gut. But the closely related strain of *E. coli* known as O157:H7 is a pathogen, and one that is potentially life threatening in humans. Often acquired by consuming undercooked beef, *E. coli* O157:H7 causes bloody diarrhea and, in some cases, hemolytic uremic syndrome, which leads to kidney failure. A comparison of the two strains suggests that many of the virulence genes (shown in red in Figure 10.11) that make *E. coli* O157:H7 a human pathogen were obtained via horizontal gene transfer (Perna et al. 2001).

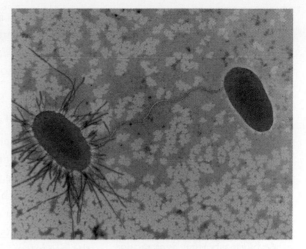

FIGURE 10.17 *E. coli* during conjugation. In the process of conjugation, a plasmid is passed from a donor cell to a recipient cell. Here we see *E. coli* conjugation. A donor cell (left) creates hairlike pili that pull a recipient bacterium close, and then opens up a conjugative junction between the two cells through which a copy of the plasmid is transferred.

Horizontal gene transfer is intriguing not only for its consequences—delivering important new genes and clusters of genes—but also as a selected trait in its own right. Why would natural selection favor the ability to engage in horizontal gene transfer? There are potential costs of "accepting" genetic material from other cells—particularly cells of a species that is only distantly related to the recipient cell. The genes of the donor species would have evolved in a different genetic background and would have been selected to work in a different cellular environment. Moreover, the donor species itself may have been exposed to very different selective conditions over evolutionary time. Such genetic material, even if beneficial to the donor species in its environment, may reduce the fitness of recipient cells that live in a different environment. Indeed, evidence suggests that, over time, natural selection has favored cells that more finely control the circumstances under which HGT occurs (Pal et al. 2005; Thomas and Nielsen 2005).

Despite the costs, HGT can also have beneficial consequences. Sometimes genes obtained via HGT will increase the fitness of recipient cells, and hence be favored by natural selection—which means that HGT is a contributor to important evolutionary and developmental change (Yanai et al. 2002; Koonin 2003). When organisms can receive new genes—sometimes whole groups of genes that are unrelated to the genes they already possess—then, all at once, new evolutionary pathways can emerge. This can lead to increasingly complex cellular life-forms that are better able to survive and reproduce in the environments in which they live. This is especially true when the genes transferred are associated with one or more modular functions (Woese 2000, 2002), by which we mean some function that is not extensively integrated with other functions in a cell. Many fundamental cell functions are tightly integrated within the cell and are not likely to be replaced by horizontal gene transfer (Woese 2002). For example, genes associated with glycolysis—the process in which sugars are broken down and converted into energy—are tightly integrated with other genes and seldom appear to be taken up by cells via HGT (Pal et al. 2005).

It is difficult to overstate the importance of horizontal gene transfer in bacterial evolution. Recall that natural selection requires a supply of variation on which to act; this is what HGT supplies at a large scale for bacteria. If useful genetic variants arise anywhere in the bacterial world, the variants can be and often are transferred into other species by HGT. As a result, the supply of genetic variation available to a species such as the human gut microbe *Enterococcus faecalis* is not limited to the variation currently present in *E. faecalis*, but rather includes much of the variation in the entire bacterial domain. When humans developed the antibiotic vancomycin and thus imposed positive selection for vancomycin resistance on *E. faecalis*, the evolution of vancomycin resistance did not occur from scratch by de novo mutation. Instead, *E. faecalis* acquired genes for vancomycin resistance by horizontal gene transfer from soil microbes that already carried such resistance genes.

Gene Order in Prokaryotes

The arrangement of loci within bacterial genomes shifts rapidly on an evolutionary timescale. To get a picture of these changes, syntenic dot plots are useful tools for comparing the gene order of two different strains or species (Figure 10.18). They provide us with a picture of the genomic reorganization that has occurred. In doing so, they allow us to deduce the translocations, inversions, deletions, and other genomic events that have occurred over evolutionary time since the divergence of the organisms in question. To create a syntenic dot plot comparing two species' genomes, researchers choose one of the two organisms as a reference; this organism's genes are then represented from left to right along the *x*-axis. The position of each homologous gene in the second organism is then plotted on the *y*-axis. If no genetic rearrangement has occurred, the gene positions will form an unbroken band along the 45° (*x* = *y*) line. Other events have other characteristic patterns, as shown in Figure 10.18. Because most bacteria have circular chromosomes, gene position is typically plotted in the clockwise direction beginning at the origin of DNA replication.

FIGURE 10.18 Syntenic dot plots. The arrangement of loci within bacterial genomes shifts rapidly over evolutionary time. Syntenic dot plots compare the gene order of two different strains or species, providing a picture of the genomic reorganization that has occurred. Here we have two syntenic dot plots, one comparing genome 1 with genome 2, and one comparing genome 1 with genome 3. Each dot in the plot shows the relative position of a gene within the genome.

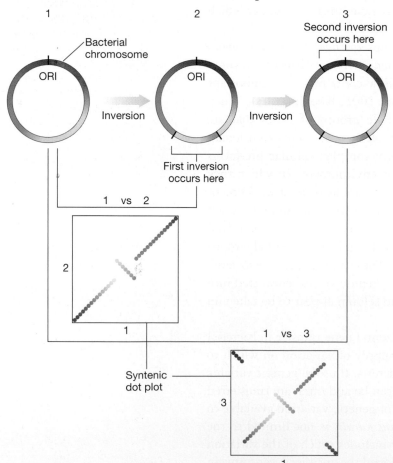

Syntenic dot plots for bacterial genomes reveal extremely rapid change in genome structure in prokaryotes. Figure 10.19A compares two closely related substrains of *E. coli* K-12. There we see a single major inversion (red), but otherwise highly similar gene order (green). However, similarities in gene order can break down entirely in even closely related species (Figure 10.19B).

Codon Usage Bias

The genetic code is degenerate; as we saw in Chapter 6, most amino acids are coded by several different codon triplets. But in prokaryotes and eukaryotes alike, the various triplets encoding a given amino acid tend not to be equally common in a given organism's genome. Rather, when we look at the protein-coding regions within a genome, we observe **codon usage bias**, in which some codons occur more frequently than others that specify the same amino acid. Furthermore, different species have different codon usage patterns.

Why do we see codon usage bias? One process that contributes to this bias is mutation itself. Mutation rates from one base pair to another are

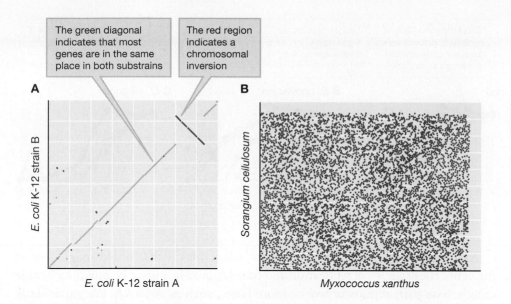

The green diagonal indicates that most genes are in the same place in both substrains

The red region indicates a chromosomal inversion

A

E. coli K-12 strain B

E. coli K-12 strain A

B

Sorangium cellulosum

Myxococcus xanthus

FIGURE 10.19 Comparing gene order across strains. A syntenic dot plot indicates the relative positions along the chromosome of homologous genes in two genomes. **(A)** Two closely related substrains of the K-12 strain of *E. coli*. The green diagonal shows that most genes are in the same place in both substrains; the red line with the opposite slope indicates a chromosomal inversion along one of the two lineages. Adapted from CoGePedia (2009a). **(B)** Even among closely related species pairs, similarity in gene order can be lost because of continual genomic reorganization. Here we see a dot plot for two myxobacterium species: *Myxococcus xanthus* and *Sorangium cellulosum*. Essentially all similarity in gene order has disappeared. Adapted from Schneiker et al. (2007).

not equal, and as a result, we would not expect all codons to be equally frequent, even in the absence of selection.

In addition to mutation, natural selection also appears to play an important role. If codon usage is advantageous, we would expect a greater advantage in genes that are more highly expressed. In a study of codon bias and gene expression in *E. coli*, Mario dos Reis and his colleagues found exactly this pattern: The genes that are the most highly expressed exhibit the greatest bias in codon usage rates (dos Reis et al. 2003) (Figure 10.20).

But why would selection favor one codon over another synonymous codon, given that each specifies the same amino acid? There are a number of possible reasons, and evolutionary geneticists have amassed evidence for several of them. We consider two here.

One reason that a codon might be preferred over a synonymous alternative is that the frequencies of transfer RNAs (tRNAs) are not equal. A codon for which complementary tRNAs are common can be translated more quickly, with lower probability of error, than can a codon for which complementary tRNAs are rare. As a result, we might expect a match between the frequencies of tRNAs and the frequencies of codon usage. This is what Toshimichi Ikemura observed for bacteria and yeast in a now classic series of papers written in the early 1980s (Ikemura 1981a,b, 1982). The same pattern has now been established throughout the tree of life. Less clear is what aspect of this pattern is cause, and what is effect. Does natural selection set tRNA frequencies to match codon usage bias patterns that arise for other reasons, or does selection favor codon usage bias patterns that track tRNA frequencies? The jury is still out on this.

FIGURE 10.20 Codon usage bias and gene expression. Codon usage bias is higher for more highly expressed genes in *E. coli*. This figure shows the relative expression level on the vertical axis as a function of the codon usage bias on the horizontal axis. Note the general upward trend indicated by the trend line. Adapted from dos Reis et al. (2003).

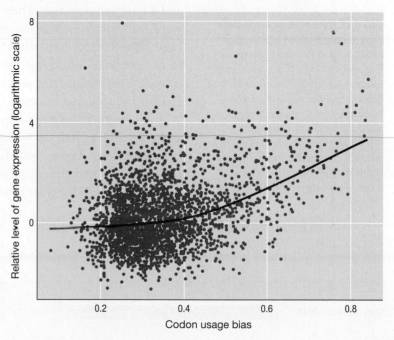

Relative level of gene expression (logarithmic scale)

Codon usage bias

FIGURE 10.21 Long mono-nucleotide repeats are rare. Ackermann and Chao found that while short (four to five base pairs) mononucleotide repeats are about as common as would be expected at random, longer repeats are significantly underrepresented. This finding provides evidence of selection against inaccuracy in replication, transcription, or translation. The lengths of nucleotide repeats are shown in **(A)** bacteria (*E. coli*), **(B)** yeast (*S. cerevisiae*), and **(C)** nematode worms (*C. elegans*). Adapted from Ackermann and Chao (2006).

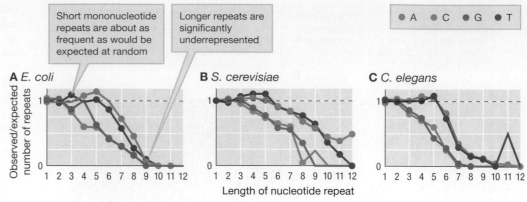

FIGURE 10.22 Mononucleotide repeat spanning an intron. Ackermann and Chao looked at the frequency of mononucleotide repeats spanning introns. The extended repeat structure of these intron-spanning repeats does not appear until the intron has been excised after transcription. Thus, these repeats do not reduce the accuracy of replication or transcription; their main effects are on the accuracy of translation.

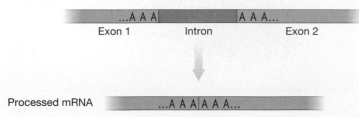

Another possible explanation for codon usage bias is that codon usage choices influence the accuracy of replication and translation. Most notably, mononucleotide (single base pair) repeats of five or more bases, such as AAAAA, are particularly prone to replication slippage—that is, the DNA polymerase may slip forward or backward during replication. This results in frameshift mutations. Mononucleotide repeats also reduce the fidelity of transcription and translation. Thus, it is plausible that selection would favor either an increase in the frequency of mononucleotide repeats if a higher mutation rate is advantageous, or a decrease in the frequency of such repeats if a lower mutation rate is advantageous.

To assess that hypothesis, Martin Ackermann and Lin Chao looked at the prevalence of mononucleotide repeats in the genomes of the bacterium *E. coli*, the yeast *S. cerevisiae*, and the nematode *C. elegans* (Ackermann and Chao 2006). They reasoned that, if selection favors an increased mutation rate, they should see more mononucleotide repeats than expected at random (holding the amino acid sequence constant), whereas if selection favors a decreased mutation rate, they should see fewer mononucleotide repeats than expected at random. Figure 10.21 shows their results for the entire protein-coding regions of the *E. coli*, *S. cerevisiae*, and *C. elegans* genomes. They found that short repeats of four to five base pairs were just as common as one would expect if the codon for each amino acid had been chosen at random. But long repeats of more than five base pairs were scarce in the genome, indicating selection against long repeats due to selection against increased mutation rate, selection against transcriptional inaccuracy, or selection against translational inaccuracy.

To distinguish between selection against mutation or *transcriptional* inaccuracy, and selection against *translational* inaccuracy, Ackermann and Chao devised an ingenious test. As we will discuss in detail in Section 10.5, eukaryotic genes often include untranslated regions known as introns that are spliced out of the transcribed mRNA before translation. Ackermann and Chao reasoned that repeats that span introns (Figure 10.22) have an effect on the process of translation, but because they are formed only after transcription occurs, they have no effect on replication or transcription. Such repeats are therefore ideal for distinguishing selection on replication and transcription from selection on translation. Ackermann and Chao found that mononucleotide repeats within an exon are much rarer than expected. By contrast, mononucleotide repeats that span introns are not uncommon in the genome. This indicates that translational accuracy has less of an impact on fitness than does the accuracy of replication or transcription.

GC Content

Because of base pairing, the fraction of A nucleotides in a genome will always be the same as the fraction of T nucleotides in that genome. Similarly, the fraction of G nucleotides will always be the same as the fraction of C nucleotides. But the fraction of G and C nucleotides need not be the same as the fraction of A and T nucleotides. Indeed, organisms vary widely in their **GC content**—that is, the fraction of G and C nucleotides. Some organisms, such as the soil microbe *Streptomyces coelicolor*, are GC-rich; others, such as the malaria parasite *Plasmodium falciparum*, are extremely GC-poor. Figure 10.23 depicts GC content values for a number of fully sequenced genomes. As illustrated, GC content varies widely both in prokaryotes and in eukaryotes, so in this section we will consider eukaryotes as well.

As with codon usage bias, there are both nonselective and selective explanations for differences in GC content. One important nonselective consideration is a bias in mutation rates. For biochemical reasons, mutations from G to A and C to T are more common than mutations from A to G and T to C. As a result, mutation tends to drive genomes toward decreased GC content.

GC content in eukaryotes does not appear to be set by mutation rates alone, however. There are two lines of evidence for this. First, there is an overall excess of mutations from G to A and C to T. In a model at mutational equilibrium, we would expect to see the same number of mutations from G to A and C to T as from A to G and T to C. Second, when we look at GC composition in eukaryotes, it is not as low as we would expect given the excess rate of mutation from GC to AT relative to that from AT to GC. From both of these observations, we can infer that something other than mutation must be elevating GC content (Lynch 2010b).

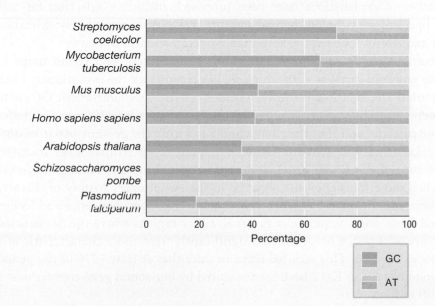

FIGURE 10.23 Genomic GC content varies widely across organisms. Here we show genomic GC and AT content for seven representative species. Adapted from Borodina et al. (2005); Cole et al. (1998); Gardner et al. (2002); Ruvinsky and Marshall Graves (2005); Wood et al. (2002).

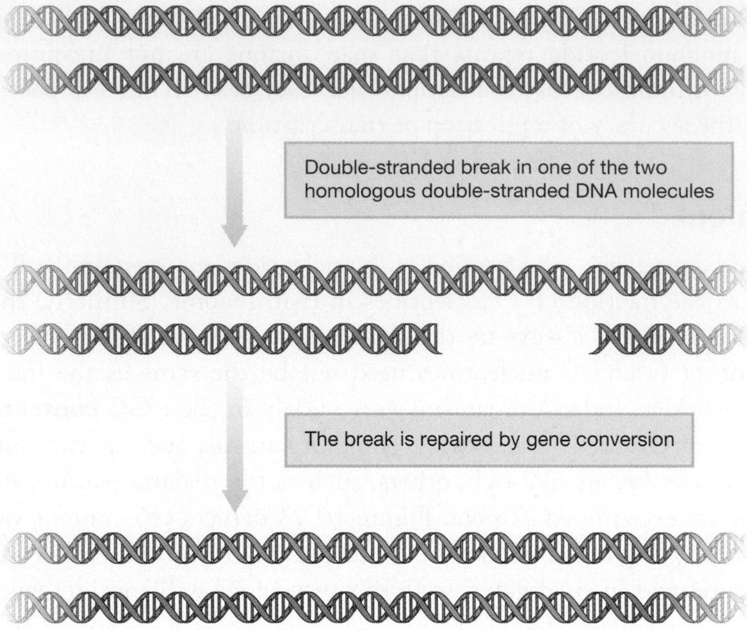

FIGURE 10.24 Gene conversion.
In this figure, a double-stranded break occurs in the red DNA molecule. A homologous stretch of DNA from the blue molecule is used as a template for repairing the double-stranded break. As a result, gene conversion occurs. The sequence from the blue molecule is incorporated into the red molecule. The reverse does not occur, so this is a nonreciprocal recombination event.

Double-stranded break in one of the two homologous double-stranded DNA molecules

The break is repaired by gene conversion

One candidate for the increase in GC content is the process of gene conversion, a common process of homologous, but nonreciprocal, recombination that is often associated with the repair of double-stranded breaks (Figure 10.24). A site initially has a G or C on one strand and an A or T on the other. When gene conversion takes place, a strand with G or C is more likely to replace a strand that has A or T than the converse. This process tends to increase the frequency of G and C nucleotides at the expense of A and T nucleotides, thereby compensating for GC loss due to mutation (Galtier et al. 2001).

The interactions of mutation and gene conversion are not by themselves sufficient to explain the GC content levels observed across the tree of life, however. In the absence of other obvious nonselective processes influencing GC content, evolutionary biologists have hypothesized that natural selection also plays a role. A number of explanations have been proposed, including selection for codon usage bias and selection on thermal stability of DNA or, more likely, selection of functional RNAs. The relative importance of each remains unknown.

Returning our focus to prokaryotes now, GC content and codon usage bias provide powerful markers with which we can reconstruct the evolutionary history of genomes. The basic idea is that each species has its own characteristic GC content and codon usage bias; genes acquired by horizontal gene transfer may not conform to those patterns, and thus they may stand out within the genome. As an example, Jeffrey Lawrence and Howard Ochman wanted to determine what fraction of the genome of the *E. coli* K-12 strain was acquired by horizontal gene transfer, and when those transfer events occurred in the evolutionary history of this strain (Lawrence and Ochman 1998). To answer those questions, Lawrence and Ochman scanned the genome sequence of *E. coli* K-12 for regions where the frequencies of base pairs or of codon usage differed significantly from those characteristic of the genome as a whole. This scan led them to infer that at least 17% of the genes in the genome of *E. coli* K-12 have been acquired by horizontal gene transfer over the last 100 million years.

Lawrence and Ochman were also able to estimate when these various gene transfer events occurred by using a clever technique (Lawrence and Ochman 1998): When a gene is first acquired by horizontal transfer, it will have a GC content and codon usage pattern characteristic of the species from which it was received. But over evolutionary time, processes of mutation, gene conversion, and natural selection will act to drive GC content and codon usage toward patterns characteristic of the recipient species. If we knew the source of each horizontally acquired gene, we could simply see how much the GC content had changed and we could use this information to estimate the time since acquisition. But the sources of the acquired genes are rarely known. Fortunately, there is another way to proceed. The first, second, and third positions of each codon have different probabilities of generating a synonymous versus nonsynonymous change, and thus each codon position changes at a different rate toward the characteristic GC content and codon usage patterns of the recipient. This provides the information necessary to infer the time since acquisition by horizontal gene transfer. Figure 10.25 shows Lawrence and Ochman's estimates of times since transfer for the horizontally acquired genes in the *E. coli* K-12 genome.

Studies of the genome teach us a great deal about the processes of divergence and speciation in bacteria. Of the genes that are present in either *E. coli* or its sister species *Salmonella enterica*, but not in both, the vast majority have been acquired by horizontal gene transfer subsequent to the divergence of the two species. As Lawrence and Ochman note, this suggests that speciation and diversification in bacteria proceed very differently than in eukaryotes. If *E. coli* and *S. enterica* are representative of bacteria more broadly, it appears that the ecological specializations responsible for evolutionary divergence are more often a result of wholesale acquisition of novel genes than a result of gradual accumulation of mutational differences.

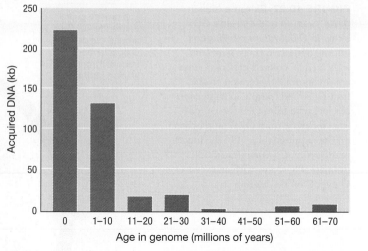

FIGURE 10.25 Age of horizontally acquired genes in the *E. coli* K-12 genome. Overall, at least 755 genes—more than one-sixth of the *E. coli* K-12 genome—have been acquired by HGT. Of these, the majority have been transferred quite recently. Adapted from Lawrence and Ochman (1998).

FIGURE 10.26 Leading and lagging strands of DNA. Here a DNA strand is shown in the process of replication. The replication fork moves in the $3' \rightarrow 5'$ direction of the leading strand. DNA synthesis proceeds in the $5' \rightarrow 3'$ direction on the newly synthesized strand. The leading strand is replicated in a single continuous fragment; the lagging strand is replicated in a sequence of shorter fragments, known as *Okazaki fragments,* that are ligated together. RNA polymerase also moves along the template DNA strand in the $3' \rightarrow 5'$ direction, so all else being equal, collisions between the RNA polymerase and the DNA polymerase will occur less often when the RNA polymerase moves away from the replication fork along the leading strand than when it moves toward the replication fork along the lagging strand.

GC Skew and Leading/Lagging Strand Gene Position in Prokaryotes

Prokaryotes typically have a single origin of replication. DNA replication is initiated at this AT-rich noncoding region, and it proceeds bidirectionally around the chromosome until a single replication terminus is reached at the other side. In prokaryotes, important genes tend to be located on the leading strand—that is, the strand that is synthesized continuously in the direction of the moving replication fork, rather than on the complementary lagging strand (Figure 10.26).

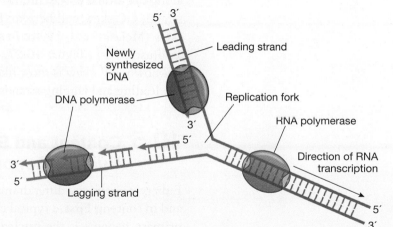

FIGURE 10.27 Extreme bias in gene location and extreme GC skew in *Clostridium perfringens*. On the outer two rings, genes are indicated by colored regions. Genes are predominantly located on the leading strand (purple moving counterclockwise from ORI; green moving clockwise from ORI) rather than on the lagging strand. On the inner orange ring, *C. perfringens* exhibits dramatic GC skew, with an excess of G over C on the leading strand.

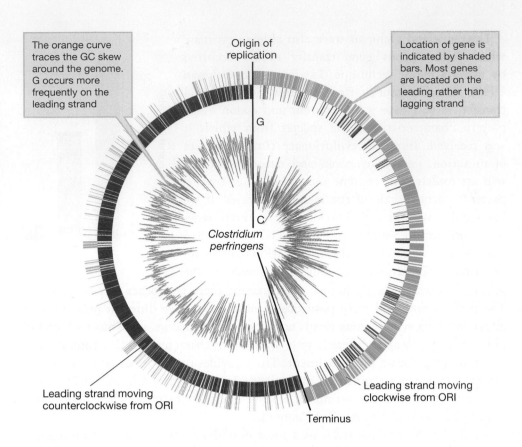

The orange curve traces the GC skew around the genome. G occurs more frequently on the leading strand

Origin of replication

Location of gene is indicated by shaded bars. Most genes are located on the leading rather than lagging strand

G

C

Clostridium perfringens

Leading strand moving counterclockwise from ORI

Terminus

Leading strand moving clockwise from ORI

This may function to reduce head-on collisions between the DNA polymerase involved in replication and the RNA polymerase involved in transcription; the processes of transcription and of DNA replication often occur concurrently in prokaryotes. Figure 10.27 illustrates an unusually strong excess of genes on the leading strand in the bacterium *Clostridium perfringens*.

Pairing constraints ensure that G = C and A = T in the genome at large, but on an individual leading or lagging strand, no such constraint is necessary. In principle, G might occur more often on the leading strand, while C might occur more often on the lagging strand. Because of mutational differences between the leading and lagging strands, this turns out to be exactly what we observe (a similar pattern holds for T on the leading strand and for A on the lagging strand). The difference is often measured as **GC skew**, the ratio $(G - C)/(G + C)$ in a sliding window moving along one strand of the chromosome. If G and C occur with equal frequency on each strand, GC skew will be zero. However, many prokaryotes exhibit substantial GC skew (McLean et al. 1998). In some of these, GC skew can be extremely dramatic, as illustrated in Figure 10.27. While the precise mechanisms responsible remain unknown, GC skew is most likely a consequence of different mutation patterns on the leading and lagging strands (Eppinger et al. 2004).

10.5 Content and Structure of Eukaryotic Nuclear Genomes

Eukaryote genomes differ dramatically from prokaryotic genomes both in structure and in content. First, a typical eukaryote can be said to have multiple genomes. The primary genome is the **nuclear genome**, which comprises a set of chromosomes contained in the nucleus. In addition, certain eukaryotic organelles—including mitochondria and chloroplasts—have their own separate genomes, presumably a relic of the ancient endosymbiosis events in which the formerly free-living life-forms were incorporated into the eukaryotic cell. We will defer our treatment of endosymbiosis and organellar genomes until Chapter 12; here we will consider the nuclear genome of eukaryotes.

Most eukaryotes have nuclear genomes that are made up of multiple linear chromosomes. Unlike in prokaryotic genomes, only a relatively small fraction of the total DNA sequence in eukaryotic genomes codes for proteins. Another small fraction of the genome codes for functional RNAs such as tRNAs, rRNAs, and microRNAs. The rest of the genome is composed of noncoding regions, including transposons, introns, and structural elements such as centromeres and telomeres (Hellmann and Nielson 2008). Figure 10.28 illustrates the proportions of these elements in the composition of the human genome. In this section, we will look at each of these components and consider the evolutionary processes by which they came to be.

Transposable Elements

Transposable elements, or transposons, represent a major fraction of the genomes of many multicellular eukaryotes; they make up approximately half of the human genome (Figure 10.28). Most unicellular eukaryotic genomes also feature transposable elements, although at substantially lower frequencies (typically 1–5% of the genome). Transposons move around within the genome in a variety of ways. **Conservative transposons** simply excise the original DNA element and reinsert it at another site. In this way, the transposon jumps to a new location, but the old copy is lost. **Nonconservative transposons** leave the original copy intact and create a new copy elsewhere. DNA transposons use a DNA intermediate, whereas retrotransposons copy the original element first to RNA and then back to DNA via a reverse transcriptase.

A number of classes of transposons are present in the human genome. The most common transposons in the human genome, by total sequence length, are retrotransposons known as **LINE-1** elements (or L1 elements), where LINE is an abbreviation for "long interspersed elements." The human genome includes over 500,000 of these elements; each is about 6000 base pairs in length, and together they make up roughly 17% of the human genome. L1 elements are called *autonomous transposons* because they encode the enzymes necessary to catalyze their own movement within the genome. But because of breakdowns that result from new mutations, the vast majority of these elements in the human genome have decayed and are no longer capable of transpositional activity. It is estimated that, in the human genome, only 100 or so L1 elements retain the ability to transpose (Cordaux and Batzer 2009). Still, this 100 is a sufficiently large number to make L1 transposition events responsible for occasional instances of genetic disease in humans (Callinan and Batzer 2006).

SINE elements (**SINEs**)—short interspersed elements—represent another common class of transposable elements in the human genome. SINEs are *non-autonomous transposons* because they lack the capacity for independent replication.

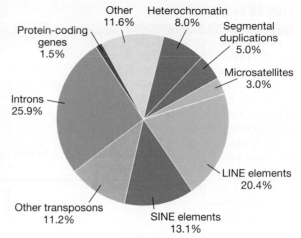

FIGURE 10.28 Composition of the human genome. In most eukaryotic genomes, only a relatively small fraction of the total genome is composed of protein-coding sequence, while a little over a quarter of the genome is made up of introns. Transposable elements make up almost half of the genome; these include LINE elements and SINE elements. Other categories include heterochromatin, segmental duplications produced by gene duplication events, and short nucleotide repeats known as microsatellites. Adapted from Gregory (2005) based on data from the International Human Genome Sequencing Consortium (2001).

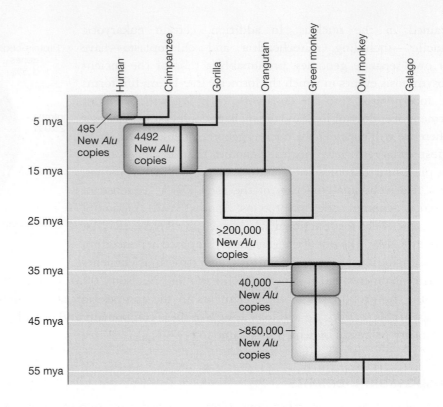

Like nonconjugative plasmids that rely on conjugative plasmids to move among
bacterial cells, nonautonomous transposons rely on the machinery provided by
autonomous transposons to move around the genome. In humans, SINEs rely on
the protein products encoded by active L1 elements for their ability to move. A
class of SINEs known as *Alu elements* outnumbers L1 elements by a substantial
margin—there are over a million *Alu* elements in the human genome. But *Alu*
elements are much smaller in size than L1 elements—approximately 300 base
pairs in length—and thus they represent a somewhat smaller total fraction of the
genome (Cordaux and Batzer 2009). As with L1 elements, most *Alu* copies in the
human genome are not currently active because the *Alu* promoter region is not by
itself sufficient to initiate transcription. If it is to be active, an *Alu* copy has to be
inserted by chance adjacent to the right types of flanking sequences (Batzer and
Deininger 2002).

Alu elements appear to have arisen and proliferated at an extraordinary rate early
in the evolution and radiation of the primate clade (Figure 10.29). During this
initial phase, new *Alu* copies were substituted into the genome at a rate of one per
generation. The process of expansion has continued throughout primate evolution,
with ongoing amplification of various *Alu* families along different branches of the
primate phylogeny. But the rate of insertion has dropped approximately 200-fold,
such that the rate of new insertions is now substantially reduced relative to that of
55 million years ago (Batzer and Deininger 2002).

Transposons are classic examples of **selfish genetic elements**. Selfish genetic
elements are stretches of DNA that do not normally perform a useful function at
the whole-organism level, but instead act to ensure their own survival and even
replication within the genome. Transposons do this by copying themselves within
genomes. The ability of transposons to do so allows them to increase in frequency in

at least three ways. First, consider a transposon in a haploid asexual organism. The transposon has no immediate way to move beyond the lineage in which it arises, but its ability to copy itself within the genome can reduce the chance that it is lost from its lineage. A single copy of any genetic element is always at risk of being lost, either by a segmental deletion or by mutational decay. But if that element can make multiple copies of itself within the genome, it is able to "hedge its bets" across those multiple copies. If one copy is lost by some mutational process, others will still remain and allow the transposon to persist within the genome (Figure 10.30A).

A reduced rate of loss is not the only benefit that transposition confers on transposons in asexual haploids. Many bacterial species have plasmids, which provide a further advantage to transposons. In such species, transposons also spread to new lineages, shuttled from one to the next on plasmids or other accessory genetic elements (Figure 10.30B).

In sexual diploid species, transposons can copy themselves onto new chromosomes. This increases their chances of being passed on to offspring of the next generation. If a transposon in the germ line jumps from one chromosome to a homologous chromosome that lacks that transposon, all subsequent meiotic products will include a copy of the transposon, and thus the transposon can spread through the genomes in the population (Figure 10.30C). This additional benefit of transposition may be one of the reasons that transposons are particularly numerous in sexual species.

In each of these explanations, the transposon does not benefit the organism in which it resides—that is, it does not confer any selectively advantageous trait on that organism. Rather, it benefits only itself, acting "selfishly" to minimize its own rate of loss from the genomes in which it resides and/or to maximize its own rate of spread into other genomes in the population. In this way, transposons are much like parasites. They are not capable of independent replication, but instead they rely on the replicative machinery of their "hosts"—in this case, the genomes of the organisms in which they reside. They persist over evolutionary time, not because of any benefit that they confer to their hosts, but rather because the genes that they do encode operate to facilitate

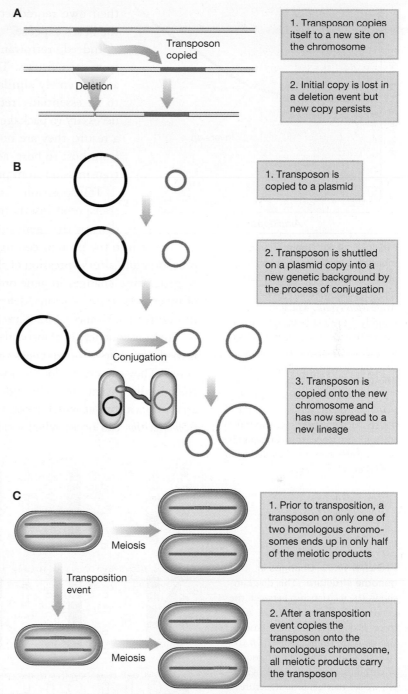

1. Transposon copies itself to a new site on the chromosome

2. Initial copy is lost in a deletion event but new copy persists

1. Transposon is copied to a plasmid

2. Transposon is shuttled on a plasmid copy into a new genetic background by the process of conjugation

3. Transposon is copied onto the new chromosome and has now spread to a new lineage

1. Prior to transposition, a transposon on only one of two homologous chromosomes ends up in only half of the meiotic products

2. After a transposition event copies the transposon onto the homologous chromosome, all meiotic products carry the transposon

FIGURE 10.30 Three processes that favor transposition. Transposition confers a selective advantage at the level of the transposable element, but not at the level of the whole organism, in each of these cases. **(A)** Transposition creates additional transposon copies within the genome. **(B)** Transposition onto an accessory genetic element such as a plasmid facilitates the movement of a transposable element into a new genome. **(C)** In a sexual diploid species, transposition copies an element from one chromosome to its homologue, and thus it ensures that the transposon will be present in all meiotic products.

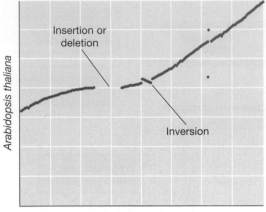

FIGURE 10.31 Changes in chromosome structure. A syntenic dot plot for two closely related plant species, *Arabidopsis thaliana* and *Arabidopsis lyrata*, which diverged roughly 5 million years ago. At left, we see that a set of continuous loci in *A. lyrata* is absent in *A. thaliana*; this indicates an insertion on the lineage leading to *A. lyrata* or a deletion on the lineage leading to *A. thaliana*. Toward the center of the figure, we see a short segment with the opposite slope; this corresponds to an inversion event. Adapted from CoGePedia (2009b).

FIGURE 10.32 Changes in genome structure. This diagram shows the relationship between the genome structures of humans and mice. Each human chromosome is shown at the center, flanked by the corresponding mouse chromosome or chromosomes at each side. Overall, we see that genome structure has been shuffled considerably, with segments moving within or among chromosomes subsequent to the divergence of the lineages leading to mice and to humans approximately 80 million years ago. Nevertheless, we see that, within segments, the basic arrangement and order of genes is conserved, and that in some cases—notably the X chromosome—rearrangement has been minimal. From Lewis et al. (2002).

their own reproduction and spread into other genomes, possibly at the host's expense.

Indeed, retrotransposons are thought to have evolved directly from retroviruses. The LTR (long terminal repeat) retrotransposons are extremely similar in structure to retroviruses, and they appear to be essentially retroviruses that have lost the genetic machinery necessary to package themselves as independent replicating units. As a result, they are no longer capable of horizontal gene transmission from host to host, and instead they rely exclusively on vertical gene transmission from parent to offspring (Lynch 2007).

Transposition events can have a number of consequences. If a transposon inserts into the middle of a protein-coding gene, it will disrupt that gene and cause the loss of that protein. Even if it does not insert into the protein-coding region itself, it might interfere with the gene's promoter and alter expression of the gene. Transposons also play an important role in generating changes in gene order and chromosome structure, driving the sorts of inversions, translocations, deletions, and rearrangements that we see illustrated in Figures 10.31 and 10.32 (Curcio and Derbyshire 2003).

The consequences of transposition tend to be deleterious. Transposons can insert into the middle of other genes or can delete segments from the middle of other genes. They can create double-stranded breaks that generate mutation. Because of their high copy number throughout the genome, transposons also can set up an array of locations at which recombination errors can arise. As a consequence, *ectopic recombination* can occur when a transposon in one location is accidentally aligned

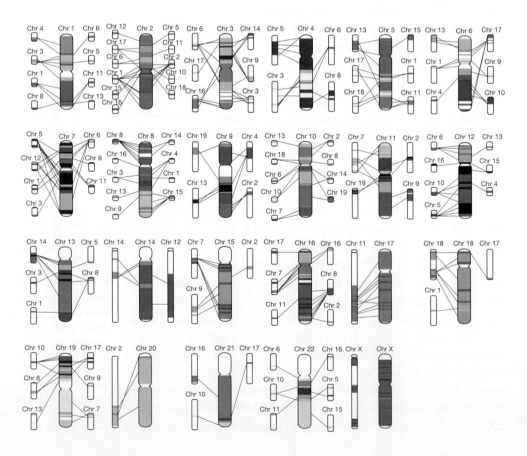

with an identical transposon in another location and crossover occurs within the two misaligned transposons. Finally, a transposon may accidently copy some of the adjacent DNA as well, moving it to a new location in the genome when the transposon is inserted.

Each transposition event induces mutations of one form or another, and these mutations are typically deleterious. As a result, organisms have evolved a number of mechanisms that suppress the activity of transposons. A mechanism of *posttranscriptional gene silencing* known as *RNA interference* appears to reduce transposition activity by eliminating transposon messenger RNA. The same pathway may also be involved in *pretranscriptional silencing*, with the RNA products from the RNA interference pathway serving as guides to prevent the transcription of transposon DNA. Several other systems have been proposed as additional mechanisms to limit transposon activity (Lynch 2007).

Occasionally, however, the mutations caused by transposition will turn out to have beneficial effects. As such, transposition can potentially have advantages as well. As sources of mutation and particularly of genomic rearrangement, transposons almost certainly accelerate adaptive evolution of the host organism—even though it is unlikely that the selective advantage from doing so can explain their widespread evolutionary success. Transposons likely persist in huge numbers despite the costs they impose on their hosts, not because of the benefits they confer.

Origins of Replication, Centromeres, and Telomeres

Whereas prokaryotic chromosomes typically have only a single origin of replication, eukaryotic chromosomes have multiple origins of replication. There is good reason for this: Because eukaryotic genomes are so much larger than prokaryotic genomes, and because eukaryotic DNA synthesis is considerably slower, replication would take a prohibitively long time if eukaryotic chromosomes had only a single origin of replication. These origins of replication are thought to make up a larger fraction of the genome than do coding sequences.

Again in contrast to prokaryotes, eukaryotic chromosomes contain centromeres—that is, regions of DNA that form the attachment points for the kinetochore proteins to which the spindle binds in order to pull apart the chromosomes during cell division. Centromeres appear to be marked for this purpose not by specific DNA sequences, but rather by the presence of a particular type of DNA packaging protein, the centromeric histone CenH3. The centromeres are typically, although not always, a discrete region somewhere in the middle of a chromosome, and they are usually composed of satellite repeats extending for hundreds of kilobases, interspersed with frequent insertions of transposons.

Centromeres present a fascinating puzzle in genome evolution. While their function is critical to successful replication and their presence is a highly conserved trait, the actual sequences of the centromeres are evolving rapidly (Henikoff et al. 2001). In fact, the DNA sequence of the centromeric regions is among the most rapidly evolving of any region of the chromosome. At the same time, the CenH3 histones and other proteins involved in structuring the centromere are also rapidly evolving, in marked contrast to other noncentromeric histones, which are highly constrained evolutionarily—that is, the noncentromeric histones have not changed much over time (Malik and Henikoff 2001).

FIGURE 10.33 Meiosis in females and males. The process of meiosis in females produces one egg cell and three inviable polar bodies; meiosis in males produces four sperm.

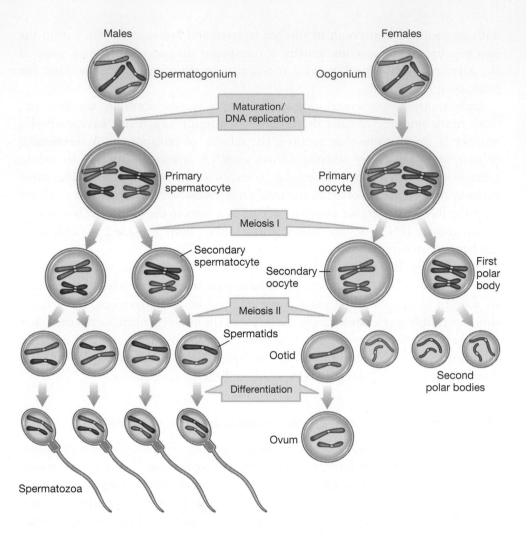

To explain these surprising patterns, Steve Henikoff and his colleagues proposed the **centromere drive** hypothesis (Henikoff et al. 2001; Malik and Henikoff 2001, 2002). When a female produces gametes by meiosis, only one of the four meiotic products forms a viable egg (oocyte); the other three form *polar bodies* that are discarded (Figure 10.33). As a result, selection at the level of the chromosome will favor any mutation to the centromere that increases its chance of segregating to the oocyte instead of to the polar bodies—for example, a change that allows the centromere to recruit more microtubules. Thus, the centromere might increase its number of repeat sections, providing a larger target area to which microtubules could bind. A chromosome with such a centromere would end up in a disproportionate number of oocytes and would rapidly sweep through the population because of its advantage during the process of meiotic segregation. (We will treat this phenomenon, known as meiotic drive, in detail in Chapter 18.)

Yet, a centromere that increases its chance of segregating to the oocyte instead of to the polar bodies might cause meiotic problems such as nondisjunction—that is, the failure of homologous chromosomes to separate during meiosis I. In that case, natural selection would favor modifications at the protein level that counter the effects of the deleterious centromeric mutations. Such modifications are particularly likely to occur in the CenH3 histone, as illustrated in Figure 10.34. If this process played out repeatedly along different lineages, it would generate the observed patterns of genomic variation—that is, rapid evolutionary divergence between species both in the centromeric sequence and in the sequence of CenH3. Henikoff and his colleagues speculate that, by rapidly generating genetic differences in the meiotic machinery of closely related populations, this process could even contribute to reproductive isolation and eventual speciation.

Compared to prokaryote genomes, another major difference in the genome structure of eukaryotes is that they have telomeres. Telomeres, the extended regions of short repeats at the ends of eukaryotic chromosomes, are thought to be a solution to a problem that arises from having linear, instead of circular, chromosomes. Recall that DNA polymerase can operate only in the 3′ to 5′ direction along the template strand. At the 5′ end of the template strand, this is unproblematic: The DNA polymerase can simply begin at an origin of replication and continue until it runs off the end of the strand, with the 5′ end successfully replicated. But there is no way to replicate the far 3′ end of a linear chromosome. Along that strand, replication proceeds by ligating (joining together) short fragments known as *Okazaki fragments*; at some point, there is no longer sufficient room to add another such fragment, and the 3′ end will remain unreplicated (Figure 10.35A). As a result, the ends of the chromosome would shorten by approximately 100 base pairs with each replication (as indeed they do during ordinary mitotic cell division of somatic cells).

The solution to this problem is that eukaryotic chromosomes end with telomeres, which can be replaced by the action of a protein–RNA complex known as telomerase. Telomerase extends the 3′ end of a chromosome, adding a specific repeat sequence, such as TTAGGG in vertebrates (Figure 10.35B). This compensates for the loss of base pairs due to incomplete replication.

> Centromeric region expands, allowing it to recruit more microtubules

> A mutation to histone CenH3 compensates for changes to centromere structure and goes to fixation

FIGURE 10.34 The centromere drive model. Prior to mutation, centromere strength is balanced. After mutation (here, an expansion of the centromere), the mutant form recruits more microtubules. This goes to fixation, but it causes other problems in meiosis. Modifications to the CenH3 histone resolve the problem. Adapted from Henikoff et al. (2001).

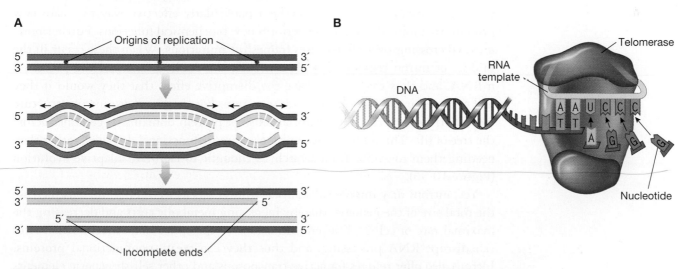

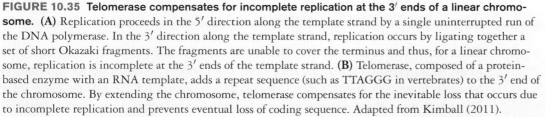

FIGURE 10.35 Telomerase compensates for incomplete replication at the 3′ ends of a linear chromosome. **(A)** Replication proceeds in the 5′ direction along the template strand by a single uninterrupted run of the DNA polymerase. In the 3′ direction along the template strand, replication occurs by ligating together a set of short Okazaki fragments. The fragments are unable to cover the terminus and thus, for a linear chromosome, replication is incomplete at the 3′ ends of the template strand. **(B)** Telomerase, composed of a protein-based enzyme with an RNA template, adds a repeat sequence (such as TTAGGG in vertebrates) to the 3′ end of the chromosome. By extending the chromosome, telomerase compensates for the inevitable loss that occurs due to incomplete replication and prevents eventual loss of coding sequence. Adapted from Kimball (2011).

Much of the DNA in centromeres and telomeres is tightly packed in what is known as *heterochromatin*. Because it is so densely packed, it is largely inaccessible for transcription; therefore, gene expression from these regions is limited. Recombination is also greatly reduced in these regions.

Introns

Most protein-coding genes in prokaryotes comprise a single contiguous run of nucleotide bases, but this is not the case for most eukaryotic genes. Recall that in eukaryotes, protein-coding genes are typically composed of exon regions that code for protein products, interspersed with intron regions that are spliced out before translation. If they are not translated, why are introns there at all? According to the *exon theory of genes*, the organization of eukaryotic genes into intron and exon regions is evolutionarily ancient, and many current genes arose by rearrangement of exons into new combinations. The idea is that individual exons often code for modular units of a protein, such as functional protein domains. When homologous recombination occurs within the introns between the exon-encoded domains, different allelic forms of each domain can form new combinations. When nonhomologous recombination occurs at locations within introns, the result is a new protein made up of a combination of functional domains—each coded by an exon (Gilbert 1987).

By increasing the length of protein-coding genes, introns increase the probability that recombination events can occur within individual genes. Moreover, they have a strong effect on *where* these events can occur. In the absence of introns, unequal recombination within the gene is likely to disrupt functional protein domains. But in the presence of introns, unequal recombination is now likely to occur between the exons, creating new combinations of protein domains without disrupting the structure of the individual domains themselves. Creating new proteins out of well-established modular subunits may be a particularly effective way to create new proteins that fold effectively and perform new biochemical functions. Furthermore, unequal crossing-over often causes frameshift mutations. When these occur in the middle of intron regions, they do not shift the reading frame of the processed mRNA, and thus they do not have the disruptive effect that they would if they had occurred in the middle of a coding region. For these reasons, intron structure may contribute to the combinatorial reuse of protein domains in genomes across the tree of life. This ability to recombine and reuse functional domains, rather than needing them to evolve from scratch, is thought to facilitate adaptive evolution (Figure 10.36).

Yet, introns may impose substantial fitness costs as well. First, introns increase the total size of the genome, thereby increasing metabolic costs and decreasing the maximal rate of cell replication. Mutations to the spliceosomal recognition sites can disrupt RNA processing, and thus they can create nonfunctional proteins. Introns also offer refuges for active transposons and other selfish genetic elements that can subsequently cause deleterious mutations.

There has been a major debate surrounding the evolutionary origins of introns (Rodriguez-Trelles et al. 2006). The *introns-early model* proposes that introns arose in ancestral prokaryotes. If so, they probably evolved to facilitate recombination between protein domains. One of the major challenges for the introns-early view is to explain the absence of spliceosomal introns in bacteria and archaea. Although

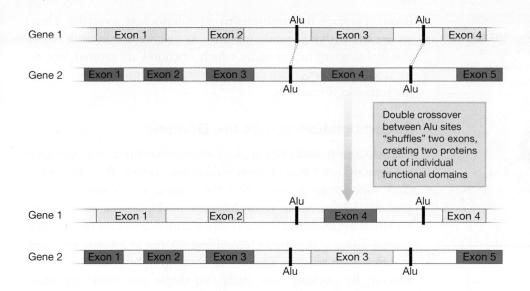

FIGURE 10.36 Exon shuffling. In the absence of introns, unequal recombination within the gene is likely to disrupt functional protein domains. In the presence of introns, the much greater length of the gene increases the probability of recombination within the gene. Moreover, with introns present and accumulating transposons, ectopic recombination is now likely to occur between the exons, creating new combinations of protein domains without disrupting the structure of the individual domains themselves. Adapted from Studentreader.com (2011).

those organisms do have introns known as *class I* and *class II* introns, these are simpler, self-splicing introns. Thus, the introns-early model suggests that selection on genome size led to an elimination of spliceosomal introns and the subsequent loss of the spliceosome in these lineages.

In contrast, the *introns-late model* proposes that introns arose within eukaryotes subsequent to the archaea–eukaryota split, possibly through the action of transposable elements (Cavalier-Smith 1978). According to the introns-late model, the present distribution of introns is due to their movement within the genome subsequent to that point rather than the result of phylogenetically conserved positions within the genome. With the additional evidence made possible by the genomics revolution, it is now clear that the common ancestor to modern eukaryotes had spliceosomal introns. But we still do not know precisely when these first evolved.

Isochores

We have already seen that genomes vary considerably in their GC content and that within a single genome, GC content may vary by region because of a history of horizontal gene transfer. In the tetrapod vertebrates, GC content also varies region by region within the genome of a single organism. Within the lineage, the chromosomes are structured in extended blocks with similar GC content. These blocks, which span 100 kb or more, are called **isochores** (Figure 10.37). Horizontal gene transfer is not a plausible explanation for the existence of isochores as it is not

FIGURE 10.37 Isochores. Chromosomes are sometimes structured in extended blocks with similar GC content, called isochores. Isochore structures of human chromosomes 1 and 2 are shown here. Colors and bar height indicate GC content. Adapted from Constantini et al. (2006).

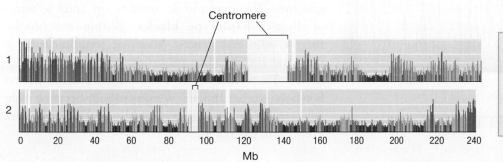

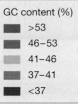

GC content (%)
- >53
- 46–53
- 41–46
- 37–41
- <37

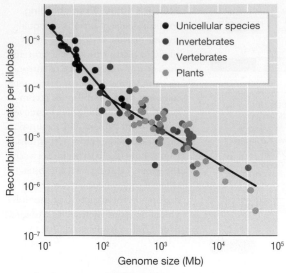

FIGURE 10.38 Recombination rate decreases with genome size. Recombination rates vary across species. Species with larger genomes tend to have lower recombination rates per base pair. Adapted from Lynch (2007).

FIGURE 10.39 Recombination rate along human chromosome 12. The histogram shows the local recombination rate (in centimorgans per megabase) along human chromosome 12. Note the frequent recombination hotspots represented by spikes in the recombination rate, and the particularly low recombination rate around the centromere. Adapted from Myers et al. (2005).

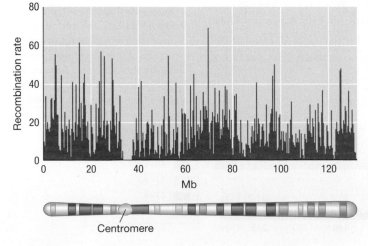

frequent enough in vertebrates to explain isochore structure. Thus, we are left with a genomic mystery. As yet, we do not understand the evolutionary processes by which isochores are created, nor do we understand why they are largely restricted phylogenetically to the tetrapod vertebrates.

Recombination across the Genome

Homologous recombination plays an important role in structuring the genomic contents of most eukaryotic species. Recombination rates vary across species, with the general trend toward larger genomes having lower recombination rates per base pair (Figure 10.38). Rates also differ dramatically within the genome of any given species. To assess patterns of recombination, researchers use a number of different techniques. One of the most straightforward is *pedigree analysis*. By tracking how often two single-gene traits segregate together within a large pedigree, we can estimate the probability of recombination between those two genes. But the resolution of the method is low. We can obtain a much finer degree of resolution by *sperm-typing*, in which large numbers of sperm are genotyped. The sperm-typing method allows sample sizes that are vastly larger than those that can be obtained from pedigrees, but it can only provide estimates of recombination rates in males (Li et al. 1988). To obtain a recombination rate map of comparable resolution that is not male specific, but rather that is averaged over the whole population, geneticists have developed a number of statistical tests that allow the use of population-wide patterns of linkage disequilibrium along the genome to estimate local recombination rates (Stumpf and McVean 2003).

Using the various techniques to map recombination rates across the genome, researchers have found that, in many organisms, recombination occurs largely at **recombination hotspots**—that is, small regions of the genome that are particularly prone to serving as locations of crossover. Basing their analyses of linkage disequilibrium in human populations, Simon Myers and his colleagues estimated that 80% of recombination events in humans occur at sites located in only 10–20% of the genome (Myers et al. 2005). Figure 10.39 shows a fine-scale recombination map for human chromosome 12. We see dramatic variation in recombination rates along the chromosome, with numerous hotspots at which recombination occurs at high rates and other regions where the local recombination rate approaches zero. Hotspots tend to occur near, but not within, coding genes.

The dramatic variation in recombination rate along each chromosome has important consequences for patterns of linkage disequilibrium in the human genome. The genome is broken up into a series of discrete **haplotype blocks**. Within the blocks, there is minimal genetic diversity, recombination is rare, and linkage disequilibrium is high (Daly et al. 2001). These blocks are bounded by recombination hotspots, so that linkage disequilibrium between even adjacent haplotype blocks is rapidly broken down over evolutionary time.

We do not yet have a detailed understanding of the factors that determine local recombination rate, but we do know that recombination hotspots appear to shift around the genome at an evolutionarily rapid pace. Highlighting this, evolutionary biologists have found that hotspots are not conserved between closely related species such as humans and chimpanzees (Ptak et al. 2005).

10.6 Tests for Selection

Much as geological strata carry within their layers a fossilized record of biotic diversity over time, the genomes of extant organisms carry within them the statistical traces of past evolutionary events. By examining genomic variation within and between species, and comparing these patterns to the patterns we would expect under a neutral model of evolution (as discussed in Chapter 8), we can search this record for evidence of past selection events. Genome-wide scans for the signatures of natural selection can help evolutionary biologists develop a better understanding of what traits and alleles have been favored over evolutionary time (Nielsen 2005). In particular, researchers often aim to distinguish between **purifying selection**, in which the major consequence of selection has been to remove deleterious mutations at a locus, and **positive selection**, in which the major consequence of selection has been to favor new beneficial mutations at that locus. In this section, we consider a few of the methods that biologists can use to detect evidence of natural selection from genetic data.

Ratio of Nonsynonymous to Synonymous Changes

In Chapters 6 and 8, we discussed the degeneracy of the genetic code and the fact that, as a result of this degeneracy, not all nucleotide substitutions in a protein-coding region change the amino acid sequence of the protein specified. This fact proves useful in determining the nature of selection on that locus.

The basic approach is to compare the pattern of substitutions actually observed with the pattern that would be expected if the variation at a particular gene were selectively neutral—that is, with a neutral model of evolution in which the variation would not be under selection. Under a neutral model, nonsynonymous mutations that change the amino acid sequence of a protein would be just as likely to go to fixation by genetic drift as would synonymous mutations that do not change the amino acid sequence. Thus, if the variation at a protein-coding gene is selectively neutral, we would expect to see as many nonsynonymous substitutions as synonymous substitutions. When making this comparison, we need to correct for the fraction of mutations that give rise to nonsynonymous and synonymous substitutions; the former are about three times as frequent as the latter, although this can vary given biases in mutation rates, GC content, and codon usage.

One major advantage of this approach is that—unlike many other tests of selection—comparisons of nonsynonymous and synonymous changes tend not to be affected by demographic events such as population bottlenecks or expansions. To compare nonsynonymous and synonymous substitutions, researchers align protein-coding sequences for two or more species and then look at the ratio K_a/K_s, where K_a is defined as the number of nonsynonymous substitutions per nonsynonymous site, and K_s is defined as the number of synonymous

TABLE 10.1

Interpreting the K_a/K_s Ratio

Nature of Selection	K_a/K_s Ratio
Purifying selection	$K_a/K_s < 1$
Near neutrality	$K_a/K_s \approx 1$
Positive selection	$K_a/K_s > 1$

Val - Thr - Pro - Glu - Glu - Lys - Ser

CTG ACT CCT GAG GAG AAG TCT
TTG ACA CCG GTG GAG AAA AGT

Val - Thr - Pro - Val - Glu - Lys - Ser

▨ Nonsynonymous
▨ Synonymous

FIGURE 10.40 Comparing K_a and K_s. The two hypothetical sequences differ by six synonymous substitutions and one nonsynonymous substitution. Because the rate of synonymous substitution has been higher than that of nonsynonymous substitution, this suggests that purifying selection has been operating on the gene.

FIGURE 10.41 Adaptive radiation and adaptations. Because of their evolutionary radiation about 5 million years ago, species in the silversword plant alliance grow in a range of habitats and forms, including **(A)** cactuslike rosettes (*Argyroxiphium sandwicense*, ssp. *macrocephalum*), **(B)** "cushion plants" (*Dubautia waialealae*), **(C)** climbing vines (*Dubautia latifolia*), and **(D)** treelike shrubs (*Dubautia reticulata*).

substitutions per synonymous site. If the variation at a protein-coding gene is selectively neutral, we expect the same substitution rate at nonsynonymous sites as at synonymous sites, and thus we expect that the K_a/K_s ratio will be approximately 1. If a gene is under directional selection, we expect that nonsynonymous substitutions will occur as a result of selection more rapidly than synonymous substitutions will occur as a result of drift, and thus the K_a/K_s ratio will exceed 1. If a gene is under purifying selection, we expect that nonsynonymous substitutions will be rare relative to synonymous substitutions, which are most likely caused by drift, and thus the K_a/K_s ratio will be less than 1 (Table 10.1) (Nei and Gojobori 1986). Figure 10.40 illustrates a hypothetical situation in which K_a is substantially less than K_s.

In some papers, the quantities K_a and K_s are replaced with the similar quantities dN and dS instead, where dN is the *rate* of nonsynonymous substitutions at nonsynonymous sites, and dS is the *rate* of synonymous substitutions at synonymous sites.

Positive Selection Associated with a Recent Adaptive Radiation

In an elegant study, Marianne Barrier and her colleagues used a K_a/K_s comparison to test the hypothesis that changes in regulatory genes play an important role during evolution in novel environments (Barrier et al. 2001). The Hawaiian silversword alliance is a group of plants that underwent a recent evolutionary radiation—a rapid burst of speciation—on the Hawaiian Islands about 5 million years ago. These species evolved numerous adaptations for living and reproducing in a wide range of ecosystems, from bogs to forests to the harsh, high-altitude barrens of several Hawaiian volcanoes (Robichaux et al. 1990) (Figure 10.41).

The silversword alliance provided Barrier and her colleagues with a way of testing the hypothesis that regulatory gene evolution has been important in this adaptive radiation. If changes in regulatory genes have influenced the radiation of the silversword alliance on the Hawaiian Islands, we would expect species in the alliance to exhibit more evidence of positive selection on regulatory genes than closely related mainland species. To test this, Barrier and her colleagues looked at two regulatory genes involved in floral development. They computed K_a/K_s ratios for the two regulatory genes in the silversword alliance species, and then they compared them to the K_a/K_s ratios for regulatory genes in a set of closely related mainland

A   B   C   D

A

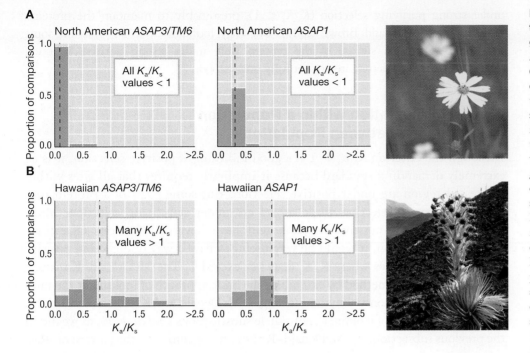

B

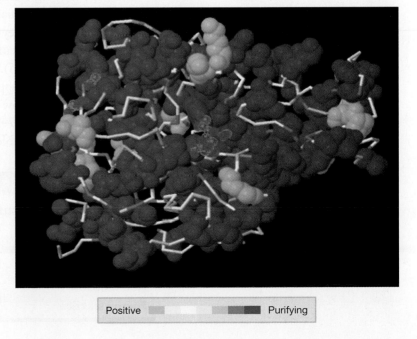

FIGURE 10.42 Positive selection in the Hawaiian silversword group. The bar graphs indicate the K_a/K_s ratios for two regulatory loci (*ASAP3/TM6* and *ASAP1*) in pairs of species in (**A**) North American tarweeds and (**B**) Hawaiian silverswords. The dashed lines are the mean K_a/K_s ratios. North American tarweeds have K_a/K_s ratios that are well below 1.0, indicating that the genes in this group have been under strong purifying selection. We see a very different pattern in the Hawaiian silverswords, where K_a/K_s ratios commonly exceed 1.0, indicating that the genes in this group have been under positive selection, presumably associated with the adaptive radiation and physiological divergence that the Hawaiian silverswords have undergone over the past 5 million years. Adapted from Barrier et al. (2001).

species known as North American tarweeds. Their results are shown in Figure 10.42. The North American tarweed species, which have not undergone a recent adaptive radiation, have low K_a/K_s ratios at both regulatory loci, indicating that the loci have been under purifying selection—that is, that selection has opposed changes at these loci. The silversword alliance species reveal a different pattern. The K_a/K_s ratios for these genes exceed 1.0 in many of the silverswords, indicating a history of positive selection. These results suggest that the adaptive radiation of the silversword alliance species was facilitated by natural selection favoring changes in the regulatory loci, which, in turn, caused the changes in phenotype that have allowed these species to diversify into the many niches that they now inhabit.

Mapping Selection within a Single Protein

By looking at the rates of nonsynonymous and synonymous substitutions, evolutionary biologists can resolve the effects of selection down to the scale of individual amino acids within a protein. To understand how selection operates on sialidase, a key protein product of the avian pathogens *Mycoplasma synoviae* and *Mycoplasma gallisepticum*, microbiologists Meghan May and Daniel Brown computed K_a/K_s ratios for each codon of the sialidase gene across 20 strains of the two pathogen species. By mapping the K_a/K_s ratios onto a physical model of the protein as bound to its substrate (Figure 10.43), they were able to reveal regions under positive selection (gold in the figure) and regions under purifying selection (magenta in the figure). They found that the binding site and the regions involved in catalytic activity were

FIGURE 10.43 Positive selection within a single protein. In the sialidase protein of two *Mycoplasma* species, some of the surface regions have undergone positive selection (gold), whereas much of the internal structure, especially around the binding site (gray and red substrate shown bound) is under strong purifying selection (magenta).

Positive ▭▭▭ ▭▭▭ Purifying

under strong purifying selection ($K_a/K_s < 1$), presumably to maintain the protein's basic function (May and Brown 2009). But they found that, in *M. synoviae*, some sites on the external surface of the protein were under strong positive selection ($K_a/K_s > 1$), for reasons that are not yet fully understood.

Comparing Variation within a Population to Divergence between Populations

Although the criterion $K_a/K_s > 1$ is a good indicator of positive selection, it is an extremely demanding standard because it implicitly requires that all sites within the tested region are under positive selection. But some parts of a protein may be under positive selection while others are under stabilizing selection. Many legitimate cases of positive selection will have lower K_a/K_s ratios (Kreitman 2000). For this reason, it would be very useful to have a way of detecting positive selection even in cases where K_a/K_s does not exceed 1. The McDonald–Kreitman test provides such a method by extending the basic approach of looking at the rates of synonymous and nonsynonymous changes (McDonald and Kreitman 1991; Egea et al. 2008). Instead of looking at only allele substitutions *between species*, as we did in the previous subsection, the McDonald–Kreitman test compares the pattern of allele substitutions *between species* to the pattern of allelic polymorphisms *within species*.

The McDonald–Kreitman test compares ratios of nonsynonymous to synonymous change across two different timescales: a short timescale represented by polymorphism within a species, and a long timescale represented by sequence divergence between species. This provides a powerful tool for detecting positive selection. The basic logic is as follows: Under a neutral model of evolution, selection neither acts on variation at the nonsynonymous sites nor on variation at the synonymous sites. Thus, under the neutral model of evolution, the ratio of nonsynonymous to synonymous polymorphism within a population (sometimes called *pN/pS*) should be the same as the ratio of nonsynonymous to synonymous substitutions between populations (in the McDonald–Kreitman test, this ratio is typically called *dN/dS* instead of K_a/K_s). If a locus is under purifying selection, deleterious mutations will create some level of polymorphism within a population; these deleterious variants are unlikely to be fixed, however, and therefore they will contribute very little to differences between populations. Thus, under purifying selection, we expect the *pN/pS* ratio to exceed the *dN/dS* ratio (*pN/pS* > *dN/dS*). If a locus is instead under positive selection for different traits in the different populations, beneficial mutations will go to fixation relatively quickly, leading to low levels of polymorphism within populations but high levels of divergence between populations; the *dN/dS* ratio will exceed the *pN/pS* ratio (*pN/pS* < *dN/dS*) (Table 10.2).

TABLE 10.2

Interpreting the Results of the McDonald–Kreitman Test

Nature of Selection	Comparing Ratios
Purifying selection	*pN/pS* > *dN/dS*
Near neutrality	*pN/pS* ≈ *dN/dS*
Positive selection	*pN/pS* < *dN/dS*

The McDonald–Kreitman test can be used at a genome-wide scale to characterize the types of selection operating on populations. In one notable example, Carlos Bustamante and his colleagues had access to an extraordinary data set: the genomic sequences of more than 20,000 loci from 39 different humans, obtained from Celera Genomics. What could the data tell them about human evolution? They realized that when they augmented the data by including the genomic sequence from a chimpanzee, they could use the McDonald–Kreitman approach to explore what types of selection have been operating on the human genome, and even to identify genes that have been under particularly strong positive or purifying selection (Bustamante et al. 2005).

The researchers narrowed down the data to 11,000 protein-coding loci that could be properly aligned with the chimpanzee sequence, and they compared the diversity among the 39 human samples to the divergence between humans and chimpanzees (Figure 10.44). Within humans, they found that 0.169% of the nonsynonymous sites and 0.470% of the synonymous sites were polymorphic, for a *pN/pS* ratio of 0.360. Comparing human and chimpanzee sequences, they found that the two species differed at 0.242% of the nonsynonymous sites and 1.02% of the synonymous sites, for a *dN/dS* ratio of 0.237. Because the *pN/pS* ratio was significantly larger than the *dN/dS* ratio, they were able to conclude that much of the nonsynonymous variation seen in humans is under purifying selection and thus due to mildly deleterious mutations.

This genome-wide approach also helps us to understand the nature of the evolutionary differences between humans and chimpanzees by homing in on the adaptive significance of the genetic substitutions that have occurred since these species diverged from their common ancestor. Using a conceptually similar approach, Bustamante and his co-workers singled out the genes that have been under positive or purifying selection. They found a relationship between the functional roles of these genes and the types of selection that they had experienced. Their findings hint at the important selection processes that have driven the divergence between humans and chimpanzees. For example, the team found that many transcription factors have been under positive selection; therefore, positive selection on gene regulation appears to have been important in the recent evolutionary history of these species. Similarly, they found evidence for positive selection at a number of loci associated with immune function, the formation of gametes, and sensory function. By contrast, selection on basic structural and metabolic functions does not appear to have been particularly important in driving the divergence between humans and chimpanzees. Bustamante and his colleagues found that most of the genes involved in cellular structure and biosynthetic function show evidence of purifying selection and thus, relative to transcription factors, such genes tend to be highly conserved between humans and chimpanzees (Bustamante et al. 2005).

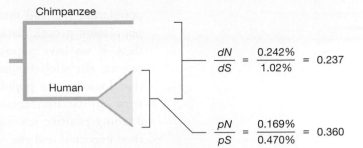

$$\frac{dN}{dS} = \frac{0.242\%}{1.02\%} = 0.237$$

$$\frac{pN}{pS} = \frac{0.169\%}{0.470\%} = 0.360$$

FIGURE 10.44 Using the McDonald–Kreitman test to find human loci under selection. The McDonald–Kreitman test identifies loci under selection by comparing the ratio of nonsynonymous to synonymous variation within a species (*pN/pS*) to the ratio of nonsynonymous to synonymous substitutions between species (*dN/dS*). Bustamante and his colleagues compared variation within humans to divergence between humans and chimpanzees.

The Distribution of Allele Frequencies Reveals Past Selective Events

If the variation at a locus is selectively neutral, we expect to see a few common alleles and a larger number of less common alleles. It turns out that the allele frequency distribution in a neutral model depends only on the product of effective population size and mutation rate, a quantity that can be easily estimated from population genetic data. Selection causes deviations from this distribution, and thus, if we have population data on the distribution of allele frequencies, we can use the allele frequency distribution to tell us about the selective history of that locus in the population. If a population has a number of common alleles with similar frequencies, this indicates the presence of balancing selection or ongoing positive selection. If, instead, the most common allele is more common than expected and the less common alleles are rarer than expected, this suggests purifying selection against deleterious variants or, alternatively, that a recent selective sweep has occurred. A number of statistical approaches test for natural selection by examining the deviations from the distribution of allele frequencies that would be expected under a neutral model (Tajima 1989). The disadvantage of many of these approaches is that they do not readily distinguish between selection and other demographic events such as gene flow or population expansions.

Extended Haplotype Blocks Indicate Recent Positive Selection

When an allele increases rapidly in frequency in a population as a result of natural selection, it carries along with it the alleles at nearby locations on the chromosome. The phenomenon of genetic hitchhiking, which we treated in the previous chapter, occurs because the nearby alleles are physically linked to the selected allele; recombination may take a long time to break down the resulting linkage disequilibrium. Population geneticists have developed numerous statistical tests that take advantage of this phenomenon to screen for selected sites. Here, we describe one such test, which Benjamin Voight and his colleagues developed to screen the human genome for loci that currently are or recently have been under strong positive selection (Voight et al. 2006).

As a consequence of hitchhiking, strongly selected alleles will be surrounded by relatively long blocks of a single haplotype. Voight and his colleagues had an important insight about how to use this fact not only to locate loci under positive selection, but also to figure out which allele at that locus is favored by selection. Suppose we have a polymorphic locus D, with 70% D alleles and 30% d alleles. How can we tell which allele (if any) has been under positive selection—that is, which allele has been favored by natural selection? If we know only the frequencies of the D and d alleles, there is no way to tell. We don't know whether the D allele is at high frequency because it is neutral and drifting, because it is favored and on its way to fixation, or because it is disfavored but the d allele arose only recently. But, if we can look at the other nearby loci, we may be able to resolve this puzzle.

Suppose that the D allele is ancestral, and that the d allele has recently increased in frequency as a result of strong natural selection. On haplotypes containing the D allele, recombination will have had plenty of time to mix up the alleles at surrounding loci. By contrast, because the d allele arose recently and reached its present frequency as a result of strong selection, recombination will not have had time to mix up the alleles at nearby loci. At the loci nearest to the D locus, we will tend to see the same alleles that were present on the haplotype where the d mutation first arose. Figure 10.45 illustrates this idea.

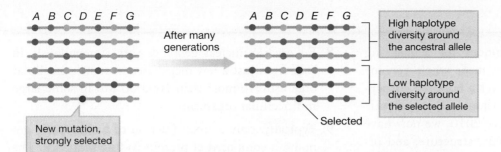

A B C D E F G

After many generations

A B C D E F G

High haplotype diversity around the ancestral allele

Low haplotype diversity around the selected allele

Selected

New mutation, strongly selected

FIGURE 10.45 Voight's test for selection. Bars indicate the section of the chromosome on which the loci *A–G* reside; colors indicate allelic variants. Left side: A new mutation arises at the previously monomorphic *D* locus. Right side: After generations of positive selection for the new mutation, it has become common in the population. Haplotypes that include the original *D* allele (shown in gold) are highly variable at the other loci, just as they were in the original population. Haplotypes that include the new variant *d* (shown in red) are much less variable, as recombination has not yet had time to separate the favorable *d* mutation from the background of alleles at surrounding loci on which it arose.

Using genetic data from 209 individuals of African, Asian, and European descent, Voight and his colleagues applied this approach to explore how evolutionary processes have shaped the human genome. They identified a number of polymorphic loci for which one allele was surrounded by long haplotype blocks (indicating recent selection), while the other was not (indicating that it was the ancestral form). Figure 10.46 shows a plot of the haplotypes around the *Rs1561277* allele. This gene is responsible for the persistence of lactose tolerance into adulthood, and thus it has been strongly favored in populations that raise dairy cattle. The red allele, *Rs1561277*, exhibits long haplotype blocks, whereas the blue variant has quite short ones. This indicates that, as expected, the allele conferring lactase persistence has been the subject of recent strong natural selection.

In this chapter, we have explored the rapidly growing field of evolutionary genomics, and we have looked at how genome-wide sequencing contributes to our understanding of the evolutionary process. This concludes Part II of the book. In the next chapter, we will turn to the origin and history of life.

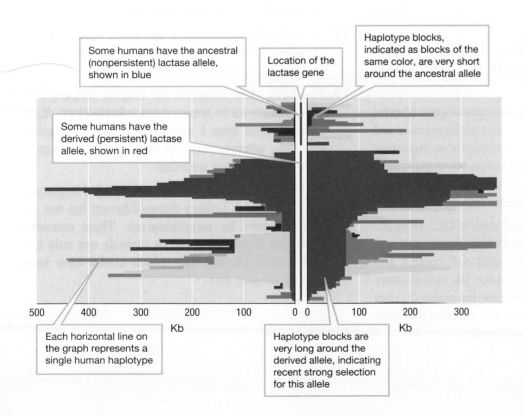

Some humans have the ancestral (nonpersistent) lactase allele, shown in blue

Location of the lactase gene

Haplotype blocks, indicated as blocks of the same color, are very short around the ancestral allele

Some humans have the derived (persistent) lactase allele, shown in red

500 400 300 200 100 0   0   100 200 300
Kb   Kb

Each horizontal line on the graph represents a single human haplotype

Haplotype blocks are very long around the derived allele, indicating recent strong selection for this allele

FIGURE 10.46 Recent strong selection in the human genome. Voight and his colleagues used haplotype variation to find alleles that have been under recent strong selection. Here, we see the haplotypes for the *Rs1561277* (lactase) allele. Lactase persistence has been under strong positive selection in cattle-growing populations, as we see from the strong conservation of haplotype structure around the selected lactase allele (red) relative to the haplotype structure around the other ancestral allele (blue). Adapted from J. K. Pritchard (personal communication).

SUMMARY

1. The development of techniques for sequencing large amounts of genetic data made whole-genome sequencing possible. With over a hundred eukaryotic genomes and over a thousand prokaryotic genomes fully sequenced as of 2010, we now have the data to study the content, structure, and organization of entire genomes and to consider how genomes themselves evolve.

2. Genome sizes vary dramatically across organisms. Viruses tend to have the smallest genomes, followed by prokaryotes, unicellular eukaryotes, and then multicellular eukaryotes.

3. Within multicellular eukaryotes, genome size does not correlate closely with organismal complexity. Much of the variation in eukaryotic genome size results from variation in the amount of noncoding DNA.

4. Viral genomes may be a single chromosome or a series of chromosomal segments. They are extremely compact, ranging in size from about 2 kb to just over 1 Mb, and they often achieve additional compression by means of overlapping coding regions. Some viral genomes, particularly those of RNA viruses, encode fewer than a dozen proteins.

5. Prokaryotic genomes are often organized as a single circular chromosome, supplemented by accessory genetic elements such as plasmids. They tend to be relatively compact, ranging in size from roughly 0.6 Mb to over 10 Mb.

6. Bacteria engage in frequent horizontal gene transfer by the processes of transduction, transformation, and conjugation. Horizontal gene transfer is an important source of genetic variation in prokaryotic populations, and appreciable fractions of some bacterial genomes have been acquired by horizontal transfer.

7. In most organisms, the frequencies of GC versus AT base pairs and the frequencies of alternative synonymous codon triplets are not equal. GC content and codon usage bias can tell us about the evolutionary history of genes within genomes; for example, they allow us to identify regions of the genome that have been acquired by horizontal transfer.

8. Eukaryotic nuclear genomes vary tremendously in size, from just a few megabases in some unicellular organisms to more than 100,000 Mb in some large multicellular organisms.

9. Typically, only a small fraction of a eukaryotic genome is composed of protein-coding sequence. The remainder is made up of introns, transposons, and other genetic elements; their distribution across the genome is the result of both selective and nonselective processes.

10. Transposons are selfish genetic elements that facilitate their own replication and movement within the genome of their eukaryotic "hosts." By moving and replicating within genomes, transposons increase their chance of being represented in the next generation. The action of transposons is an important driver of mutation, including changes in chromosome structure.

11. In addition to protein-coding regions and transposons, eukaryotic chromosomes include important structural regions such as centromeres and telomeres. Again, the structure of these components of the genome is fashioned by a combination of selective and nonselective evolutionary processes.

12. Natural selection leaves a statistical signal on the genome; genes that have been under selection can be identified by genome-wide scans.

13. The K_a/K_s ratio measures the relative frequency of nonsynonymous to synonymous substitutions. K_a/K_s values greater than 1 indicate a history of positive selection, whereas very low K_a/K_s values indicate a history of purifying selection.

14. Recent natural selection creates extended haplotype blocks in which linkage disequilibrium has not yet been removed by recombination. These extended haplotype blocks can help us identify not only loci under selection, but also which alleles have been favored at these loci.

KEY TERMS

centromere drive (p. 362)

codon usage bias (p. 350)

conservative transposons (p. 357)

C-value paradox (p. 335)

evolutionary genomics (p. 336)

GC content (p. 353)

GC skew (p. 356)

G-value paradox (p. 340)

haplotype blocks (p. 366)

horizontal gene transfer (HGT) (p. 347)

isochores (p. 365)

lateral gene transfer (p. 347)

LINE-1 (p. 357)

noncoding DNA (p. 338)

nonconservative transposons (p. 357)

nuclear genome (p. 356)

plasmids (p. 346)

positive selection (p. 367)

prophages (p. 345)

purifying selection (p. 367)

recombination hotspots (p. 366)

selfish genetic elements (p. 358)

SINEs (p. 357)

transcription factors (p. 341)

transduction (p. 347)

transposable elements (p. 338)

virulence factors (p. 345)

REVIEW QUESTIONS

1. Figure 10.10 illustrates a series of three-letter words that can be read in two different reading frames. Come up with your own example of this phenomenon.

2. Given the genome structures pictured in Figures 10.8 and 10.9, would you expect to see greater linkage disequilibrium between the HA and NA loci of the influenza virus, or between the S and C loci of the hepatitis B virus? Explain.

3. In which organisms do you expect a transposon could more easily spread despite imposing a small fitness cost on its host: in an asexual haploid species, or in a sexual diploid species? Explain.

4. Along a lineage from ancestor A_1 to modern species S_1, chromosome translocation occurs as follows:

   ABCDEFGHIJKLMNOP → EFGHIJKLMNOP**DCBA**

   Along the lineage from ancestor A_1 to modern species S_2, gene order along the ancestral chromosome remains unchanged. Sketch a syntenic dot plot comparing this chromosome in species S_1 and S_2.

5. Suppose that, when comparing the DNA sequences at a given locus across various *Drosophila* species, you observed that the nonsynonymous-to-synonymous substitution ratio (K_a/K_s) was approximately 0.5. What would you tentatively conclude about the history of selection at the locus within the *Drosophila* clade?

6. If you then observed that within a single species, *Drosophila melanogaster*, the ratio of nonsynonymous to synonymous polymorphism (pN/pS) was 0.2, how would you revise your previous conclusion?

7. Bacteria engage in horizontal gene transfer via the process of transduction, transformation, and conjugation. Which of the three processes is least likely to have evolved by natural selection for its present purpose—that is, which of these processes is least likely to be an adaptation?

8. If LINE-1 transposons are viewed as selfish genetic elements, explain why one might view *Alu* elements as hyperselfish genetic elements.

9. Describe three processes that are nonadaptive at the organismal level but that play an important role in the evolution of the structure and content of eukaryotic genomes.

10. The human rhinovirus is responsible for many cases of the common cold. Its surface is covered with a number of protein-based pentamer subunits, as illustrated here. To create this image, Amy Kistler and her colleagues sequenced a number of rhinovirus genomes and computed *dN/dS* ratios for each amino acid that makes up the pentamer (Kistler et al. 2007). The *dN/dS* values are indicated by color in the figure. Where on the capsid is purifying selection the strongest? Where is positive selection the strongest? Propose a hypothesis for why this pattern is observed.

A — External capsid surface

B — Capsid cross-section

C — Internal capsid surface

dN/dS   0.05   0.13   0.27

SUGGESTED READINGS

Lawrence, J. G., and H. Ochman. 1998. Molecular archaeology of the *Escherichia coli* genome. *Proceedings of the National Academy of Sciences of the United States of America* 95: 9413–9417. A beautiful study in which genome sequence is used to reconstruct evolutionary history.

Lynch, M., and J. S. Conery. 2003. The origins of genome complexity. *Science* 302: 1401–1404. An overview of how genome structure arises from the interplay of adaptive and nonadaptive processes.

Nielsen, R. 2005. Molecular signatures of natural selection. *Annual Review of Genetics* 39: 197–218. A technical review of how we can use the statistical traces that natural selection leaves behind in the genome to find genes that have been under selection.

Robichaux, R. H., G. D. Carr, M. Liebman, and R. W. Pearcy. 1990. Adaptive radiation of the Hawaiian silversword alliance (Compositae Madiinae): ecological, morphological and physiological diversity. *Annals of the Missouri Botanical Garden* 77: 64–72. The study of positive selection in the silversword adaptive radiation that we examined in this chapter.

 Visit StudySpace at wwnorton.com/studyspace.

PART III

The History of Life

Prior to the Permian–Triassic mass extinction
251 million years ago, crinoids were abundant in
the Earth's oceans. Although the stalked crinoids
here look somewhat like plants—hence, the name
"sea lilies"—these creatures are echinoderm
animals, related to starfish and sea urchins.

11

The Origin and Evolution of Early Life

◀ Stromatolites—layered mats of sediment trapped within bacterial biofilms—provide some of the oldest fossil evidence of life. In a few locations around the world, including Shark Bay, Australia, as pictured here, bacteria continue to form stromatolites today.

Emmanuelle Javaux and her team of researchers found something more precious than gold in the Agnes gold mine of South Africa. Using sediment samples that they obtained from five drill holes stretching 600 meters below the surface of the gold mine, they discovered remarkable evidence of life that had existed billions of years ago: tiny fossils, called *microfossils*, that were approximately 3.2 billion years old. At the time of their discovery, these microfossils were 1.4 billion years older than the oldest microfossil samples known (Buick 2010; Javaux et al. 2010). Javaux and her team studied the microfossils using state-of-the-art equipment, including a scanning electron microscope. In 22 of their 55 samples, they found fossils that were 30–300 micrometers (μm) in diameter (300 μm = 0.01 inch) (Figure 11.1). Detailed analysis suggested that these fossilized creatures would have had organic walls around them, possessed other cell-like qualities, and would have been part of populations of other such entities, probably in an environment on the edge of a marinelike coast. Although questions remain about exactly what kind of organisms were captured in these microfossils, Javaux's analysis suggests that more than 3 billion years ago, microorganisms possessed something similar to cell walls.

FIGURE 11.1 Microfossils of organisms from 3.2 billion years ago. (A, B) Examples of microfossils from the approximately 3.2-billion-year-old sample from South Africa. Scale: 100 micrometers (µm) = 0.004 inches. **(C, D)** Wall structure in the microfossils. Scale: 200 nanometers (nm) = 0.000008 inches.

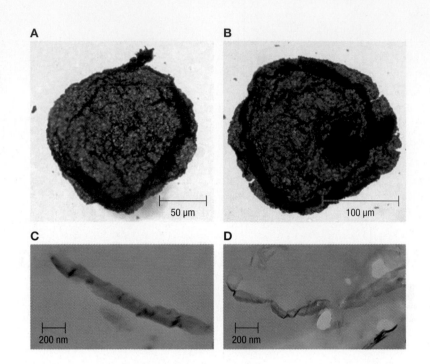

A

B

50 µm

100 µm

C

D

200 nm

200 nm

Understanding that life on Earth could have existed 3.2 billion years ago is difficult for most people, as this is an almost incomprehensibly long time (Figure 11.2). Until the late eighteenth century, most people thought that Earth was only a few thousand years old, and they could not imagine that it could actually be billions of years old. As more information was gathered from across the sciences, however, estimates of Earth's age began to increase from thousands to millions to billions of years, with current estimates being about 4.5 billion years. But these were largely estimates about the age of the planet, not estimates as to how long life has been present on Earth.

In this chapter, we will examine the following questions:

- What is life?
- What are some hypotheses about the origins of life on Earth?
- How might early evolution and the diversification of life have occurred?

Conceptually, evolutionary biology can readily adapt well-established ideas and theories to address the early evolution and diversification of life. Addressing the *origins* of life on Earth is also possible, but more difficult, for reasons we will discuss shortly.

11.1 Origin and Diversification of Life on Earth

Throughout this book, we have seen how evolutionary processes explain the diversity of life on Earth. If variation, fitness differences, and heritability are present, evolution by natural selection will occur. Selection will weed out some life-forms and favor others, resulting in organisms that are well suited to their environments. This process applied to early life-forms that existed billions of years ago, just as it does to life-forms in the world we see around us today. Of course, understanding how natural selection and other evolutionary processes acted on organisms that were present more than 3 billion years ago and then using that information to generate testable predictions is not easy—but we will examine numerous ways that modern evolutionary biologists do just this. In addition, we can reconstruct the evolutionary relationships between early organisms by employing the phylogenetic techniques that we discussed in Chapters 4 and 5, and we can use our phylogenies to make concrete predictions about many questions of interest to biologists. Again, this is not always easy, and we will see that there are debates about how the early branches on the tree of life were structured. But our point is that evolutionary biologists can use well-established theories and techniques to understand both the process of natural selection and the phylogenetic relationships among early organisms. So, we have a conceptual and theoretical foundation from which to work when it comes to the goal of understanding the evolution and diversity of early life.

But understanding the origin of life lies somewhat beyond the purview of evolutionary biology. The theory of evolution, both as conceived by Darwin and Wallace and as developed subsequently by thousands of scientists over the past century and a half, neither offers nor aims to offer an explanation of how life arose on Earth. Rather, the theory of evolution explains how life diversified subsequent to its origin. Understanding the origin of life itself is an inherently interdisciplinary project, requiring evolutionary biologists to collaborate with chemists, geologists, atmospheric scientists, and researchers from other disciplines (Gould 1987; Rice et al. 2010). Chemists can help us to understand what the initial chemical building blocks of life might have been, and geologists and atmospheric scientists can shed light on the possible physical characteristics of the environment in which life originated. Work from these areas has provided information on the conditions on Earth 3.5 billion years ago, including the composition of the atmosphere, the geology and chemistry of Earth's surface, and Earth's temperature patterns (Knauth and Lowe 2003; Robert and Chaussidon 2006).

For example, independent work from geology and chemistry on the chemical signatures from the remains of ancient oceans, as well as studies involving the phylogenetic reconstruction of the way that proteins react to temperature in organisms that lived billions of years ago (Gaucher et al. 2007), all suggest that ancient Earth was much hotter than today's planet. Ocean temperatures, for example, have cooled 30°C from 3.5 to 0.5 billion years ago (Figure 11.3).

When addressing questions about the *origin* of life, evolutionary biologists are deprived of one of their strongest tools: phylogenetic reconstruction. To understand why, think about life at the base of the tree of life—what is often referred to as the **last universal common ancestor (LUCA)**. LUCA is not meant to be thought of as a single organism, but rather as a population of organisms.

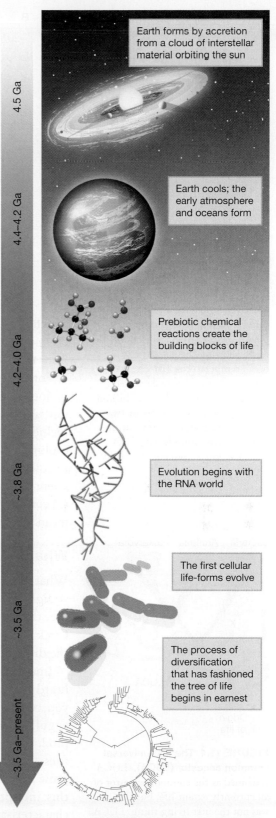

4.5 Ga

Earth forms by accretion from a cloud of interstellar material orbiting the sun

4.4–4.2 Ga

Earth cools; the early atmosphere and oceans form

4.2–4.0 Ga

Prebiotic chemical reactions create the building blocks of life

~3.8 Ga

Evolution begins with the RNA world

~3.5 Ga

The first cellular life-forms evolve

~3.5 Ga–present

The process of diversification that has fashioned the tree of life begins in earnest

FIGURE 11.2 Early events in the history of life on Earth. Dates are in billions of years or *gigaannum* (Ga) before the present time. Adapted from Joyce (2002).

A

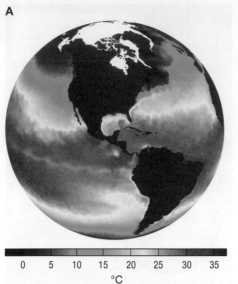

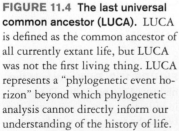

FIGURE 11.3 Ocean temperatures now and over the past 3.5 billion years. (**A**) A snapshot of Earth showing the present-day ocean temperatures. (**B**) The average ocean temperature decreased dramatically from 3.5 to 0.5 billion years ago. The lines represent two different estimates of temperature based on maximum levels of ^{18}O. Part B adapted from Gaucher et al. (2007).

B

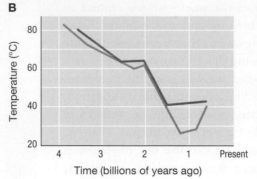

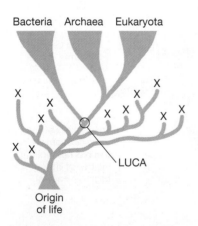

FIGURE 11.4 The last universal common ancestor (LUCA). LUCA is defined as the common ancestor of all currently extant life, but LUCA was not the first living thing. LUCA represents a "phylogenetic event horizon" beyond which phylogenetic analysis cannot directly inform our understanding of the history of life.

And LUCA itself was not the first life-form, nor was it the only life-form present at the time. LUCA was presumably just one of many life-forms on Earth at the time, but it is—by definition—the only one that left any descendant lineages that remain to this day. And because whatever else that was present left no descendants to the present day, we cannot reconstruct a phylogeny of its descendants, and so we cannot see back beyond LUCA. That means that if we use phylogenetic analysis based on extant (that is, currently living) species, LUCA is the common ancestor to any group of living species that we might choose to analyze. As such, we might say that LUCA represents a **phylogenetic event horizon**: a point in the history of life beyond which phylogenetic analysis cannot possibly see. As a result, it is impossible to use the tools of phylogenetic analysis to infer what happened during the period of time prior to LUCA when life on Earth first originated (Figure 11.4).

Fortunately, there are other tools besides phylogenetic analysis that evolutionary biologists can use when studying the origin of life on Earth. For example, we can use population thinking and our understanding of evolutionary processes, including drift and natural selection, to analyze questions relating to the origin and evolution of early life. As we will see, evolutionary biologists have used these tools, in collaboration with chemists, molecular biologists, geologists, and atmospheric scientists, to make progress in understanding the origin of life on Earth. Before we consider the *origin* of life, however, we will step back and think about what we mean when we talk about life in the first place.

What Is Life?

What does it mean for something to be alive? While this may appear to be a straightforward question, the harder we try to pinpoint the defining features of life, the more difficulties come to mind (Schrödinger 1944; Fox and Dose 1977; Crick 1981; Dyson 1985). The following thought experiment illustrates some of the difficulties of defining exactly what we mean by life.

Imagine that you are the lone person on a remote island and that you have never heard of, or seen, fire. You then observe a fire and watch what it does. Should you conclude that the fire is alive? The fire grows and it appears to move. It engulfs living material in its path, and smoke and ash appear to be waste products produced by fire. The fire even seems to reproduce by splitting off new, smaller fires. Based on your everyday intuition that living things acquire nutrients to grow and reproduce, you might well conclude that the fire is alive. But, of course, this interpretation would be wrong. Fire isn't alive; it just shares some of the characteristics of living things.

Rather than trying to construct a definition of life, perhaps the best we can do is to identify a set of properties that are typically, if not always, associated with living things. These properties include

- homeostasis: the ability to adjust the internal environment to maintain a stable equilibrium,
- structural organization: the ability to maintain distinct parts and the connections between them,
- metabolism: the control of chemical reactions,
- growth and reproduction,
- response to environmental conditions or stimuli.

In addition to these characteristics, there is a very important property shared by all the living things that we have observed on this planet: All life is subject to, and appears to have evolved by, the process of natural selection. This is a critical observation in that it shapes what we have to explain if we are to understand how life originated on Earth. The origin of life on Earth was more than just the origin of self-replicating entities; the origin of life that we are interested in as biologists is the origin of life that is subject to natural selection—we need replication, but we also need heritable variation in traits that cause fitness differences (see Chapter 3). To highlight the difference between self-replicating entities and self-replicating entities subject to natural selection, Figure 11.5 presents a thought experiment devised by evolutionary biologists John Maynard Smith and Eörs Szathmary (Maynard Smith and Szathmary 1997, 1999). This thought experiment involves a self-reproducing machine that uses energy from some energy source (for example, the sun) and materials from the environment to make copies of itself.

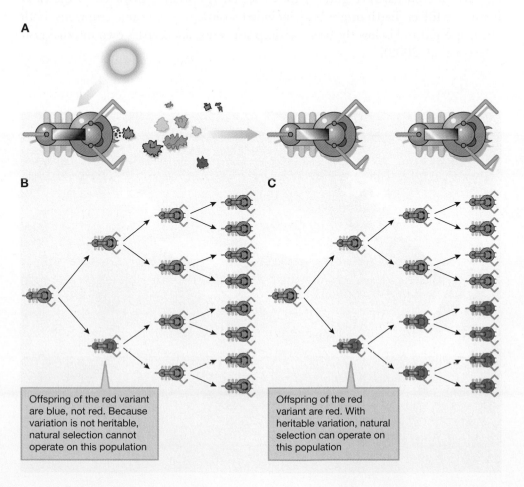

A

B

Offspring of the red variant are blue, not red. Because variation is not heritable, natural selection cannot operate on this population

C

Offspring of the red variant are red. With heritable variation, natural selection can operate on this population

FIGURE 11.5 Self-replication and the origin of natural selection.
(**A**) A self-replicating machine, here using materials in the environment and sunlight as a source of energy to replicate itself. (**B**) A lineage of such machines, capable of self-replication, but not capable of passing on any variations that might occur. (**C**) A lineage of machines that self-reproduce with heritable variation. Because the variation is heritable, natural selection can drive phenotypic change provided that different types leave different numbers of offspring. (For simplicity, differential reproduction is not shown in the figure.) The point is that the origin of life, as we observe it on this planet, involved not only the origin of self-replication, but also the origin of heritable variation and thus of natural selection. Adapted from Maynard Smith and Szathmary (1999).

In Section 11.2, we will look at how both self-replication and heritable variation tied to fitness differences could arise in a world in which RNA was the basis of life: what is known as the **RNA world**. But first, we will consider the chemical processes that may have given rise to the necessary molecular building blocks for the earliest life-forms.

Prebiotic Chemical Reactions and the Origin of Life

To understand the origin of life, we begin by asking a question: What were the prebiotic chemical reactions that created the complex molecules needed for life to arise? Darwin himself envisioned this process as taking place in a "warm little pond, with all sorts of ammonia and phosphoric salts, lights, heat, electricity, etc., present, so that a protein compound was chemically formed ready to undergo still more complex changes" (letter from Darwin to Hooker, February 1, 1871). Many hypotheses have been put forth since then. Some of these have stood the tests of time and experimentation, and others have not. Here we will focus on what is known as the **prebiotic soup hypothesis** for the origin of life.

In the 1920s, both Aleksandr Oparin and J. B. S. Haldane elaborated on Darwin's "warm little pond" idea and proposed the following hypothesis for the origin of life on Earth: In an atmosphere that lacked oxygen—as primitive Earth's atmosphere did—ultraviolet light and lightning might serve as sources of energy that converted atmospheric gases into a range of molecules that served as the basis for early life on Earth (Figure 11.6). Other energy sources involved in the early history of life on Earth might have included cosmic rays, volcanic eruptions, both above ground and below the ocean at deep sea vents, and Earth's own internal heat (Martin et al. 2008).

FIGURE 11.6 Energy sources for prebiotic chemical reactions. Lightning and ultraviolet light might both have served as sources of energy for the formation of a wide range of molecules from atmospheric gases. Other energy sources might also have included cosmic rays (denoted by white arrows), volcanic eruptions, and Earth's own internal heat. The dark area in the lower portion of the figure represents underwater processes. Adapted from Lazcano (2006).

This prebiotic soup—meaning the pool of molecules that existed in liquid form before life arose—would have developed over time, growing richer and richer in living and nonliving matter. This pool of molecules might have grown more diverse as a result of matter arriving on extraterrestrial objects such as carbon-rich meteorites and comets, which are known to carry amino acids, purines, and pyrimidines (Oro 1961; Kvenvolden et al. 1970; Cooper et al. 2001) (Figure 11.7). Oparin and Haldane hypothesized that the earliest life-forms may have emerged from such a prebiotic soup, and that they may even have used various parts of this soup as a source of energy and nutrients.

In the early 1950s, Stanley Miller and Harold Urey tried to test the plausibility of the prebiotic soup hypothesis for the origin of life. Their approach was to simulate in their lab—on a very small scale—the conditions outlined by Oparin and Haldane to see if some of the chemical precursors to life would emerge. To simulate lightning in the ancient atmosphere, they set an electric current between two electrodes. This current, in turn, interacted with a mixture of gases. The gases used were methane (CH_4), hydrogen (H_2), ammonia (NH_3), and water (H_2O) because, based on evidence from chemistry and physics, these were thought at the time to best represent the characteristics of the atmosphere of the Earth about 4 billion years ago (Miller 1953) (Figure 11.8).

Depending on which combination of gases was used, numerous common amino acids, such as glycine, alanine, and valine, were produced in the experimental apparatus. From their simple experimental protocol, Miller and Urey produced some of the building blocks of life. While debate continues over whether the combination of gases used by Miller and Urey are an accurate representation of what was present in the ancient atmosphere—it is possible that the atmosphere contained carbon dioxide (CO_2) not used by Miller and Urey (Kasting 1993)—the fact that complex chemicals such as amino acids formed in this experiment hints at how the building blocks of life may have come into existence.

As with almost all groundbreaking experiments in science, Miller and Urey's experiment, and the related experiments that followed shortly thereafter, raised as many questions as they answered. For example, in these early experiments, the critically important sugar ribose was found in very low supply, and a mechanism for joining the amino acids together to make proteins was completely absent. Some of the key parts of complex proteins could be produced in an environment that was meant to simulate that of early Earth, but how did those parts come to be joined together? This problem was addressed in 1977, when Sidney Fox mixed a number of different amino acids together at a high temperature (120°C) in an environment lacking water (Fox and Dose 1977).

When Fox mixed large amounts of the amino acids aspartic acid and glutamic acid, and subsequently placed the mixture into water, the amino acids present were strung together in a peptidelike structure. The bonds between the amino acids, however, were weak and unstable, and they were different in structure from the peptide bonds that join amino acids in most organisms. Subsequent work by Claudia Huber and her colleagues found that amino acids do link together via stable peptide bonds if a compound such as carbon monoxide (CO)—which is thought to have been present in early Earth's atmosphere—is used in the laboratory experiments (Huber and Wachtershauser 1997; Huber et al. 2003).

Many questions remain regarding the origin of life. The prebiotic soup model, the most well studied of the models for the origin of life, provides one explanation

FIGURE 11.7 The Murchison meteorite. A sample of the Murchison meteorite that fell on Murchison, Australia, on September 28, 1989, is shown here. This meteorite is rich in carbon, purines, and pyrimidines—substances that may have enriched the "prebiotic soup" that was present early in Earth's history by adding organic substances.

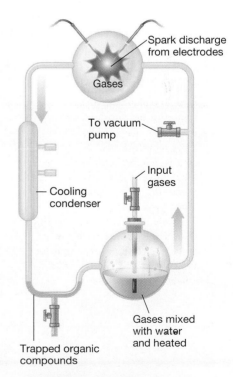

FIGURE 11.8 The Miller–Urey experiment. The experimental device used in the 1953 experiments by Stanley Miller and Harold Urey. Adapted from Lazcano (2006).

for how the necessary complex molecules could be formed (Powner et al. 2009), but it does not explain how the first organisms capable of variation, multiplication, *and* heritability came into existence. For that we must turn to something called the RNA world.

11.2 The RNA World

Because RNA has the capacity for heredity and metabolism, evolutionary biologists have been fascinated by its role in early life. Gerald Joyce describes this fascination as an archaeological mystery:

> It is as if a primitive civilization had existed prior to the start of recorded history, leaving its mark in the foundation for a modern civilization that followed. Although there may never be evidence for an RNA-based organism, because the RNA world has likely been extinct for almost four billion years, molecular archaeologists have uncovered artifacts of the ancestral era. . . . (Joyce 2002, p. 214)

In 1986, Walter Gilbert coined the phrase "RNA world" to capture the idea that early evolution—from about 4 billion to 3.5 billion years ago—may have been RNA based (Gilbert 1986). There is some evidence from laboratory work mimicking conditions on early Earth that ribose, phosphate, purines, and pyrimidines—all the essential parts of RNA—likely existed in the prebiotic environment (Robertson and Miller 1995a). Evolutionary biologists hypothesize that these molecules may have emerged from conditions similar to those simulated in the prebiotic-soup experiment, or they may have arrived on meteorites, which often contain high amounts of carbon. In either case, if these compounds had bonded together, RNA-based life-forms might have resulted (Benner et al. 1989; Joyce 2002).

Experimental Evidence on the Origins of Natural Selection

A critical step in understanding how basic chemical reactions could lead to entities capable of variation, multiplication, and heritability came in a fascinating experiment by Sol Spiegelman and his colleagues (Mills et al. 1967; Spiegelman 1970). In an early experiment on the origins of natural selection, Spiegelman and his colleagues placed a "primer" strand of RNA, made up of about 4000 RNA nucleotides—adenine (A), guanine (G), cytosine (C), and uracil (U)—into a small test tube. To this mixture, they added more A, G, C, and U nucleotides, and a replicase enzyme, which functions to make copies of an RNA molecule. The researchers heated and incubated the mixture, and then they transferred a small drop to a new test tube.

The new test tube contained the replicase enzyme and nucleotides A, G, C, and U, but it did *not* contain primer RNA. This new test tube was heated and incubated, followed by yet another transfer to a new test tube, and so on, for 75 serial transfers (Figure 11.9). Spiegelman and his colleagues found that RNA made copies of itself in these test tubes. The interesting thing was not that the RNA was copied—Spiegelman had added replicase enzyme to ensure that this would happen—but rather that natural selection took place on these copies, leading to a change in their characteristics. Let us examine how.

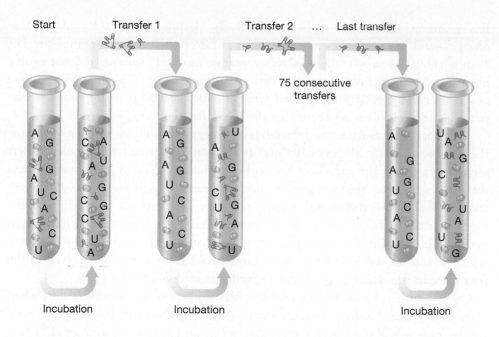

Start   Transfer 1   Transfer 2   ...   Last transfer

75 consecutive transfers

Incubation   Incubation   Incubation

FIGURE 11.9 Spiegelman's experiment on the origins of life. A "primer" strand of RNA made up of about 4000 nucleotides, along with other nucleotides of A, G, C, and U, and a replicase enzyme were placed into a small test tube. The mixture was incubated, and then a small drop was transferred to a new test tube. The new test tube contained replicase enzyme, and units of A, G, C, and U, but no additional primer RNA was added. Incubation then occurred, followed by transfer to a new test tube, and so on for 75 serial transfers. RNA is shown in green; replicase enzyme in blue. Adapted from Maynard Smith and Szathmary (1997).

The process of RNA replication carried out by the replicase enzyme involved errors; these errors produced new mutant forms of RNA that differed both in their length and their nucleotide sequence. This generated variation in RNA types on which natural selection might act. Because the replicase enzyme copies whatever strands are present, these changes should be heritable as well.

We already have mentioned that two of the three ingredients for natural selection were present in the experiment: variation and heritability (which was built into the design of Spiegelman's experiment). The third ingredient for natural selection is differential survival or replication, and we would expect this to be present in Spiegelman's experiment as well. We know that shorter RNA sequences will replicate more quickly, and so, in general, selection should favor sequences shorter than the original 4000-nucleotide primer strand. But, if RNA is very short—fewer than 50–100 nucleotides or so—the error rate in replication is so very high that these strands can no longer be copied reliably by the replicase enzyme. As a result, we would expect strands shorter than 50–100 base pairs to be selected against.

At the end of the serial transfer experiment, if natural selection was operating in Spiegelman's test tubes, we would expect to see a strand that is less than 4000 nucleotides long and greater than 50–100 nucleotides in length. And that is what Spiegelman found—a strand that was a little over 200 nucleotides long. Apparently, selection for a moderately short strand was very strong, as the end product RNA was much closer to the 50–100 nucleotide minimal length we discussed than the 4000-nucleotide-long original primer. Shorter RNA sequences, which took a shorter time to replicate, were favored by selection.

Using Spiegelman's protocol, Manfred Sumper and his colleagues further examined natural selection and the early stages of RNA-based life (Kuppers and Sumper 1975; Sumper and Luce 1975; Eigen et al. 1981). They ran an experiment similar to the one described earlier; in addition, they added a chemical called acridine orange. Acridine orange is a dye used in fluorescence microscopy that binds to the RNA and typically inhibits replication by the replicase enzyme. Researchers found that, while replication was initially inhibited, within a few hours there arose

FIGURE 11.10 A chemical structure of the first ribozyme discovered. Colored ribbons show the path of what are called RNA "backbones," and the red star indicates the active site of the ribozyme. This particular ribozyme is a self-splicing intron: It catalyzes its own excision from a precursor RNA molecule.

RNA variants that could replicate effectively, despite the presence of acridine orange—and in fact these variants replicated faster with acridine orange present than without. Moreover, these "acridine orange adapted" variants did not evolve in other experiments where acridine orange was not present. Taken together, these observations suggest strong natural selection for the ability to replicate in the particular chemical environment that the RNA molecules experienced.

In essence, the results from the studies by Spiegelman and by Sumper show how the genetic building blocks of life may have evolved. If small variations in RNA sequences exist, if the sequences can replicate themselves, and if there are fitness differences between these sequences, natural selection will lead to evolutionary change. But as we mentioned earlier, these experiments all required that the replicase enzyme be added by the experimenters at the start of each incubation period. RNA plus replicase is sufficient to get natural selection going—but from where would the replicase have come? How could RNA ever have evolved in the first place in the absence of enzymes to facilitate its replication?

One possible solution to this problem was suggested in the early 1980s when Thomas Cech and Sidney Altman independently discovered that enzymes need not be proteins. Rather, RNA itself can act as an enzyme (Kruger et al. 1982; Guerrier-Takada et al. 1983). Such RNA enzymes are called **ribozymes** (Figure 11.10). Although ribozymes are much less stable and much less efficient than protein enzymes, a number of different ribozymes have been documented (Lilley 2003; Orgel 2004).

Scientists have hypothesized that, in the RNA world, RNA molecules would have replicated by using ribozymes. But because ribozymes do not replicate themselves in the *modern* world, nor have they done so for billions of years, evolutionary biologists have created laboratory conditions under which reproduction could occur in order to examine whether RNA molecules are capable of using ribozymes for independent reproduction (Issac and Chmielewski 2002; Voytek and Joyce 2007).

In an ingenious experiment, Natasha Paul and Gerald Joyce demonstrated that RNA can catalyze reactions involved in its own assembly, using what is called the R3C ribozyme (Paul and Joyce 2002). R3C ligates—that is, joins together—two RNA molecules, let's call them RNA A and RNA B, by forming a bond between them (Figure 11.11). Using genetic engineering techniques, Paul and Joyce redesigned R3C so that the product formed when RNA A is joined to RNA B by R3C is identical to the R3C itself. In ligating RNA A and RNA B, R3C operates both as a template—positioning A and B in relation to one another—and as an enzyme, catalyzing the chemical reaction that joins them. From simpler precursors RNA A and RNA B, R3C creates a copy of itself; these copies could in turn create copies of *themselves*. But Paul and Joyce found that in practice their system could only undergo about two rounds of replication before further copying reactions failed. This is because the RNA A and RNA B were producing AB compounds instead of binding to the R3C ribozyme. But this is in no way a fundamental limitation of these types of self-replicating systems; in a follow-up experiment, Lincoln and Joyce (2009) were able to design a similar system that could replicate indefinitely so long as substrate was present.

Whereas the Paul and Joyce experiment demonstrated that RNA species can self-replicate, that system was not capable of evolution by natural selection because it lacked variation for natural selection to act on. To construct a system capable of

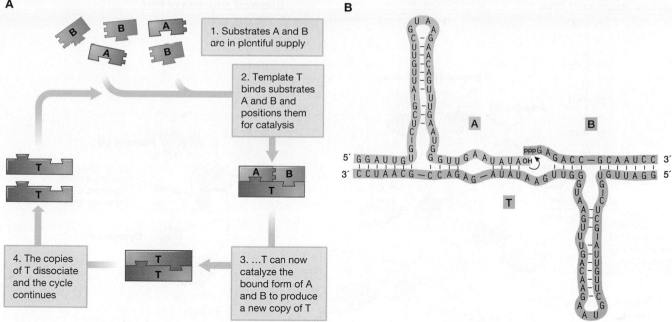

A

1. Substrates A and B are in plentiful supply

2. Template T binds substrates A and B and positions them for catalysis

3. ...T can now catalyze the bound form of A and B to produce a new copy of T

4. The copies of T dissociate and the cycle continues

FIGURE 11.11 A self-replicating RNA system. **(A)** A schematic diagram of a simple self-replicating system in which a template molecule T positions the substrates A and B, and catalyzes the chemical reaction, joining them together. **(B)** In Paul and Joyce's experiment, the substrates A and B, and the template molecule T, were constructed out of RNA as shown. We can verify that, once they are bound together, A and B form a new copy of T. Adapted from Paul and Joyce (2002, 2004).

evolution by natural selection, researchers needed to design a system that featured variation that affects something about successful replication in the RNA molecules present. Moreover, these self-replicating molecules needed to have the property illustrated in Figure 11.5C: a variant template T′ needs to catalyze synthesis of other variant T′ molecules, rather than copies of the original template T.

In 2009, Lincoln and Joyce found a way to do this. They constructed a system much like the one from the 2002 experiment, albeit now with two different templates and four different substrates. After establishing that the template RNAs in this system could self-replicate, they conducted an experiment in which several variant template molecules were supplied with variant substrate molecules. In this system, new variant forms of the templates arose via various mutations. This variation was heritable, and different templates replicated at different rates. The conditions for natural selection were now met in their system. Lincoln and Joyce found that those self-replicating ribozymes that had more efficient catalytic activity and the ability to grow quickly soon began to dominate their populations of self-replicating ribozymes (Lincoln and Joyce 2009).

From RNA to DNA

How do we move from an ancient RNA world to one in which DNA is the primary means by which genetic transmission occurs? To understand this, we need to answer two questions: (1) What sort of biochemical changes produced DNA in the RNA world, and (2) Why would DNA be favored once it was present? Let us address the latter question first.

In the RNA world, natural selection would have favored any transmission system that was more efficient than that of RNA and ribozymes (Paul et al. 2006). For a number of reasons, evolutionary biologists have hypothesized that DNA-based transmission may be just such a system. DNA is chemically more stable than RNA, primarily because DNA's deoxyribose sugar is less reactive than is RNA's ribose sugar (Figure 11.12).

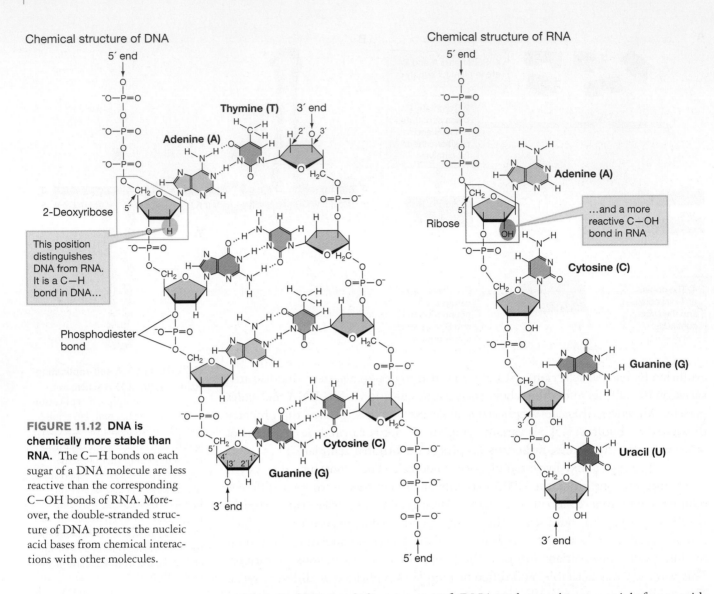

Chemical structure of DNA

Chemical structure of RNA

FIGURE 11.12 DNA is chemically more stable than RNA. The C—H bonds on each sugar of a DNA molecule are less reactive than the corresponding C—OH bonds of RNA. Moreover, the double-stranded structure of DNA protects the nucleic acid bases from chemical interactions with other molecules.

The double-stranded structure of DNA reduces the potential for outside molecules to interact with and disrupt the nitrogenous bases that encode sequence information. DNA replication systems also have "proofreading" capabilities that are not present in RNA replication. During DNA synthesis, an exonuclease checks each newly added base to make sure it is complementary to that on the parent strand. Finally, as a result of its double-stranded structure, DNA has repair mechanisms that are not available to RNA (Figure 11.13). For example, if only one strand of the double helix is damaged, cells can use the complementary strand of a repair template to correct errors.

The higher fidelity associated with DNA proofreading and repair is evolutionarily important because it dramatically lowers mutation rates. Lowered mutation rates allow longer genes, and thus more information to be stored in the genome. When mutation rates are as high as they are in the RNA genomes of RNA viruses, there is a relatively low upper limit—approximately 10,000 base pairs—to the size of a genome that can reliably create error-free or nearly error-free copies of itself. By lowering mutation rates, DNA proofreading and repair allow DNA genomes to increase in size by many orders of magnitude, thus allowing organisms to store and transmit far more genetic information than would be possible with an RNA genome. Another reason that DNA-based transmission may have been favored by natural selection is because it allowed for specialization within cells—DNA could act as a genetic storage system, while RNA could be involved in other cell functions (for example, it could serve as a cell messenger system), and proteins could perform most enzymatic functions.

Given the selective benefits that are associated with DNA-based over RNA-based transmission, evolutionary biologists are building mathematical models to decipher what sort of molecular genetic changes might have occurred to produce DNA in an RNA world, and to construct experiments to simulate the conditions of an RNA world to find the molecular bridge from RNA to DNA (Robertson and Miller 1995b; Alberti 1997; Saladino et al. 2004).

Some laboratory work has been able to use experimental evolution to select for a DNA-like version of a ribozyme, which would be one possible bridge out of the RNA world (Paul et al. 2006). Other work has been based on the hypothesis that formaldehyde (CH_2O)—which is believed to have been produced on early Earth—played a role in producing DNA. To test this hypothesis, Michael Robertson and Stanley Miller mixed formaldehyde with the RNA nucleotide uracil (Pinto et al. 1980; Robertson and Miller 1995b). From the resulting chemical reaction, they found that formaldehyde added something called 5-hydroxymethyluracil to uracil. This is important because 5-hydroxymethyluracil has structures that are similar to the side chains of most of the 20 amino acids in proteins, providing an indirect link between the RNA world and the proteins that are so critical in DNA-based genetic transmission. This is an example of the type of research that is being conducted to address this problem. Yet, much work remains to be done to fully understand the move from an RNA world to a world dominated by DNA and proteins.

11.3 The Evolution of Single-Celled Organisms

The cell is such a fundamental entity in modern biology that it is hard to imagine a world without it. But how did cells arise in the first place? We have little, if any, phylogenetic evidence with which to answer this question. All organisms on the tree of life have a cellular structure, so presumably LUCA did as well. Viruses are acellular, and that is one of the reasons that they are not integrated into the tree of life at present (Box 11.1). It would be nice if we could use phylogenetic inference to determine the structure of the first life-forms, but again we run into the phylogenetic event horizon that we discussed earlier. Fortunately, we can use an understanding of natural selection to generate hypotheses about how and why the first cells may have arisen. In this section, we will explore one such hypothesis about how natural selection could have favored the early evolution of single-celled life-forms from acellular precursors.

While biologists distinguish between a variety of different types of cells, for our purposes, we can define two basic cell forms: prokaryotic and eukaryotic. Eukaryotic cells have membrane-bound organelles and a distinct nucleus containing DNA. Prokaryotic cells are structurally simpler and evolved much earlier than eukaryotic cells; they typically lack membrane-bound organelles and their DNA is not contained in a nucleus (Figure 11.14). We will begin with the evolution of prokaryotic cells here, and we will return to the evolution of eukaryotic cells in the next chapter.

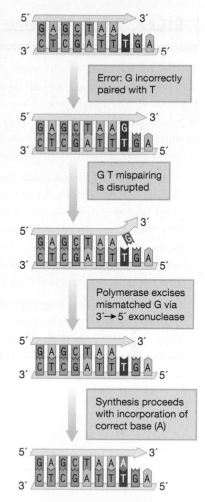

FIGURE 11.13 DNA checking and repair mechanisms. During DNA synthesis, an exonuclease associated with the DNA polymerase removes inappropriate pairings, allowing the polymerase another try to make the correct pairing. Adapted from Cooper and Hausman (2009).

BOX 11.1 Where Did Viruses Come From?

The tree of life contains all known cellular life-forms, but it does not include viruses. We still do not know the exact phylogenetic relationship between viruses and cellular organisms. What can we say about where viruses came from? This remains an unresolved question, but there are three leading hypotheses (Forterre 2006; Koonin 2006; Domingo et al. 2008):

1. The *escaped genes hypothesis* posits that viruses have their origins as selfish genetic elements that replicate within a "host" genome. At some point in their evolutionary history, these rogue stretches of parasitic DNA or RNA somehow evolved or assimilated the necessary protein capsules and packaging mechanisms to allow themselves an independent existence outside of their cellular hosts.

2. The *reduction hypothesis* proposes that viruses have their origins in parasitic cellular organisms. Over evolutionary time, the genomes of these cellular parasites were greatly reduced as the parasites came to rely more and more on the functions of their hosts. Eventually, they abandoned cellular structure, metabolism, and independent replication entirely, taking on protein capsules for existence outside of their hosts.

3. The *relics of the RNA world hypothesis* suggests that viruses are remnants of the original RNA world. According to this hypothesis, they have existed alongside cellular life for its entire history, and as such represent a link back to the first precellular life that existed long before LUCA.

It is possible that more than one of these answers are correct. Viruses almost certainly do not have a single origin, but rather they arose multiple times in the history of life. Different origins could easily have occurred along different pathways. For example, RNA viruses may have arisen as escaped genes, while DNA viruses may have come from reduced organisms and viroids (small, circular noncoding RNAs that infect plants), and may be relics from the RNA world.

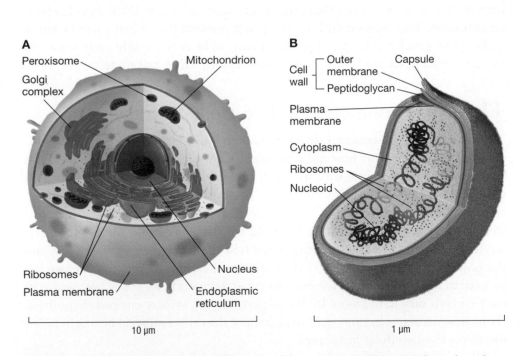

FIGURE 11.14 Eukaryotic and prokaryotic cells. (A) A eukaryotic cell has membrane-bound organelles and a distinct nucleus containing DNA. **(B)** Prokaryotic cells lack membrane-bound organelles, and their DNA is not contained in a nucleus. (These cells are not to scale.)

Hypercycles and Encapsulation in Cell Membranes

To understand how and why cells may have evolved from the simpler RNA-based life-forms that we have been discussing so far, it is helpful to look at an analogy from modern forms of life. When members of two species interact in ways that benefit both, we say that the individuals are engaged in a mutualistic interaction, or **mutualism**. Mutualistic relationships—in which each species provides something to the other—are win–win scenarios in that they increase the fitnesses of all parties involved. Mutualistic relationships can also occur at the molecular level. If two or more molecular substrates each contribute in a positive way to the replication of the others, we would call this a **molecular mutualism**. Such molecular mutualisms may have been important among *replicators* (entities that can replicate themselves) in the RNA world. The **hypercycle model** proposed in 1977 by Manfred Eigen and Peter Schuster suggests the following sort of scenario: Imagine four independent, RNA-based replicators labeled A, B, C, and D (Figure 11.15). Suppose that these replicators are all found in the same environment and in close proximity to one another, and that they interact in a cycle—that is, a closed loop of the form A → B → C → D → A, where an arrow from A to B indicates that A facilitates the replication of B, and so on (Eigen and Schuster 1977).

In a hypercycle, the rate of replication of any one replicator is a function of the concentration of the replicator that preceded it in the cycle—the more A present, the greater the replication rate of B; the more B present, the greater the replication rate of C; the more C present, the greater the replication rate of D; and, finally, to close the loop, the more D present, the greater the replication rate of A (Box 11.2). This hypercycle is a type of molecular mutualism, in that replicators affect each other's reproduction in a positive manner. Moreover, this type of mutualism may have been important in the early evolution of life, especially with respect to the evolution of the cell. And this might become an even more prominent factor in the

FIGURE 11.15 Hypercycles.
(A) In a hypercycle, each of four RNA species—indicated by the colored circles—self-replicates, and each also facilitates another's replication. These four RNA species are shown in a free solution here. As indicated by the arrows, B replicates faster when more A is present, C replicates faster when more B is present, and so on. This chain of facilitation forms a closed loop: A → B → C → D → A. Natural selection favors variants that increase their own rate of replication, but not those that increase the rates of other species' replication. **(B)** If a hypercycle is enclosed in a membrane and the replication rate of the membrane-bound ensemble depends on the replication rates of all of the RNA species within, natural selection now favors any reaction in which one RNA species increases the rate at which another species replicates.

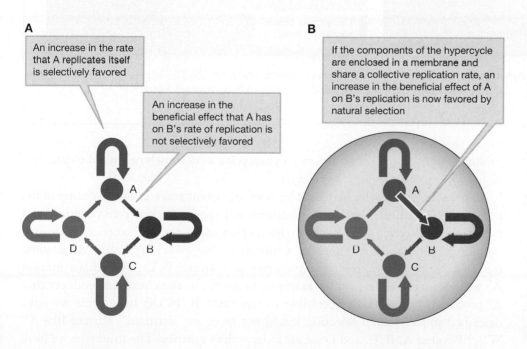

A

An increase in the rate that A replicates itself is selectively favored

An increase in the beneficial effect that A has on B's rate of replication is not selectively favored

B

If the components of the hypercycle are enclosed in a membrane and share a collective replication rate, an increase in the beneficial effect of A on B's replication is now favored by natural selection

BOX 11.2 Hypercycles in Modern Organisms

Although the ancient molecular hypercycles we discussed may seem abstract, it is important to recognize that hypercycles abound in the modern world. To see this, consider the relationship between earthworms and oak trees (Maynard Smith and Szathmary 1997). Earthworms benefit oak trees in multiple ways: They aerate the soil, and the by-products of their digestive processes provide nutrients for the oak trees, and so earthworms accelerate the rate at which oak trees can grow and reproduce. Oak trees benefit earthworms as well. The leaves of oak trees provide food for the earthworm—the more leaves,

the more earthworms, and, again, the more earthworms, the more trees, and so on. Another example of a hypercycle involves fish, water fleas, and algae. Water fleas feed on algae, so the more algae, the more water fleas. Fish feed on water fleas, and so greater numbers of water fleas mean more fish. Fish produce nitrate as a waste product, and algae use nitrate as a food source; the more nitrate present, the faster algae can grow. This leads to more water fleas and a repetition of the hypercycle loop (Figure 11.16).

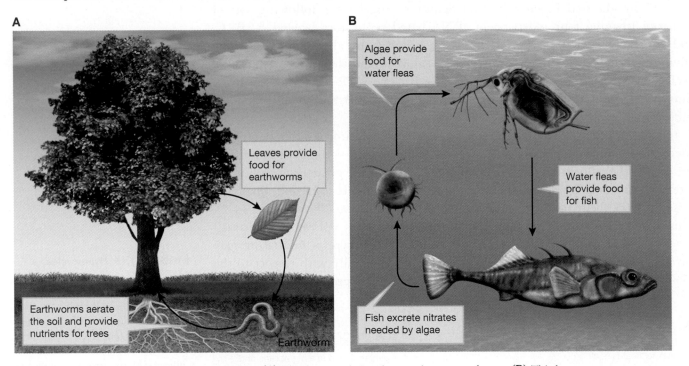

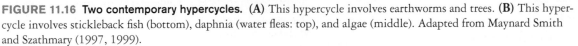

FIGURE 11.16 Two contemporary hypercycles. (**A**) This hypercycle involves earthworms and trees. (**B**) This hypercycle involves stickleback fish (bottom), daphnia (water fleas: top), and algae (middle). Adapted from Maynard Smith and Szathmary (1997, 1999).

evolution of early life if two separate hypercycles were somehow linked to produce an even more complex new hypercycle.

To see why hypercycles may have been an important stage in the evolution of the cell, we need to think about how mutations in hypercycles generate variation, and how natural selection may favor variations that lead to more complex life-forms, including cell-based life. Consider a mutation that causes replicator A, at some cost to itself, to increase the replication rate of replicator B. Let's call this mutation A′. Such a mutation might, for example, make A′, or the chemical products that A′ produces, more readily accessible to replicator B. In the hypercycle we have described so far, natural selection would not favor an "altruistic" mutant like A′. Why? Because A, B, C, and D are all independent entities. The mutation we have posited, however, would be costly only to replicator A′ and, at the same time, it would have a direct positive effect on replicator B.

Other conditions would allow our mutation to increase in frequency. Very early replicators were capable of synthesizing fatty acids. What would happen if the four replicators in our hypercycle were all enclosed within some sort of surrounding membrane made from the fatty acids they produced? That is, what would happen if the four replicators were encased in a primitive cell, sometimes called a **protocell**? For the moment, let us skip the question of how the membrane is formed—we will return to that soon. Here our critical point is that when replicators live together inside a membrane, we can speak of the entire membrane-enclosed ensemble of replicators as an organism, and so the four independent types of replicators in our original example have become a single organism with four constituent parts in a cellular hypercycle. Now when a mutant A′ increases the rate of B, this will feed back to help A′.

If the rate of cell reproduction is a function of the total concentration of all replicators within it, then this type of mutualistic relationship might be favored. In other words, natural selection *might* favor mutualistic replicators enclosed in cells over those not encapsulated in cells (Figure 11.15).

We say that natural selection *might* favor such cell life—whether selection *would* actually favor encapsulation of several cells by a membrane depends on all of the costs and benefits of our replicators encapsulated in a cell. The largest cost of such encapsulation to A, B, C, and D is that resources must now be brought across a cell membrane, which was not necessary before the replicators were enclosed together. In addition to the selective benefits previously described, other benefits of encapsulation to A, B, C, and D include controlling the microenvironment inside the cell, creating chemical gradients across membranes to let in certain chemicals and keep out others, using the cell membrane as a defensive mechanism against predatory replicators, and partitioning various functions to operate more efficiently than if they were not encapsulated in a cell (Zenisek et al. 2007). If the total benefits outweigh the costs, then selection will favor cellular life. We know that at some point this happened, because all living organisms today are composed of cells.

Reproduction in Early Cells

How would such early cells have reproduced? There are a number of hypotheses for how this occurred. Here we will look at a hypothesis that focuses on fatty acids (Chen et al. 2006). What if, in addition to making copies of themselves and being enclosed in a membrane, the replicators within the cell were also involved in producing the fatty acids that make up the cell membrane? Let's consider the case in which the hypercycle components A, B, C, and D each produce some fatty acids when they are free-living and not yet part of a cell. These fatty acids can then be used to construct the membrane that would encase early cells. If the mutualistic relationships within this cell led the replicators to be more productive in terms of fatty acid production—let's say twice as productive—then the surface area of the cell membrane would double, and the cell would start to become unstable and buckle. This is because doubling the surface area of a sphere more than doubles its volume, and this could cause the cell to swell and eventually split. A cell could split into two equal-sized daughter cells, or into two cells that were of different sizes, and initially we would expect significant variation in this trait. But the most energetically efficient way for a sphere to split is to break into two equal parts, and so natural selection should favor division into equal-sized daughter cells (Rashevsky 1938; Ganti 1978).

If the components of the original cell were divided between the daughter cells, this process would represent the early stage of cellular reproduction. Once cell reproduction—even an imperfect system of cell reproduction—exists, natural selection will favor any changes to the process of reproduction that lead to more rapid reproduction or to increased numbers of daughter cells surviving to reproduce in the next generation.

11.4 Horizontal Gene Transfer and the Evolution of Single-Celled Life

In Chapter 10, we looked at the role that horizontal gene transfer (HGT) has played in the evolution of contemporary prokaryotes and their genomes. Most likely, horizontal gene transfer was also very important in the early history of life.

There are large differences between the early cells we just described and the single-celled life we see today. How can we explain the long-term accumulations of amazing adaptations we see in modern single-celled organisms? By now, we hope the most basic answer to that question is obvious: Natural selection would have favored cells that were better suited to survive and reproduce in their environment—in this case, the environment that existed billions of years ago. But we can say more than that if we return to the distinction between modular and nonmodular cellular functions that we discussed in Chapter 10. Along these lines, Carl Woese hypothesized that during *early* cell evolution, horizontal gene transfer was a powerful force leading, in general, to more and more complex cellular organisms (Woese 1998a,b). This is because the metabolic processes within cells would have been far from integrated during this period of time, and many cell functions would have been modular. As a result, horizontal gene transfer may have been the primary means for propagating copies of a gene (Kandler 1994a,b; Woese 1998b). Even if the frequency of horizontal gene transfer was quite low relative to ordinary vertical transmission (in which genes are transferred from parent to offspring), it could have had a very big impact on the diversity of genes present and the structure of the early phylogenetic tree. Indeed, if Woese is correct, then the very idea of a single "primordial life-form" starts to unravel. Instead, early cell life would more closely have resembled a hodgepodge of different cell forms readily exchanging genetic information.

As cell structure and function became more complicated over time, they would also have become more integrated, less modular, and less likely to take up new genes by horizontal gene transfer—HGT would still occur, but its role in promoting adaptation would decrease.

The hypothesis that HGT was predominant in early evolution has ramifications for building the tree of life. The tree of life has been constructed based on patterns of common descent that presuppose that the primary way that genes are transmitted is vertically. During the early evolution of life, *if* HGT was the predominant mode by which genes were transferred—and this hypothesis is still being actively

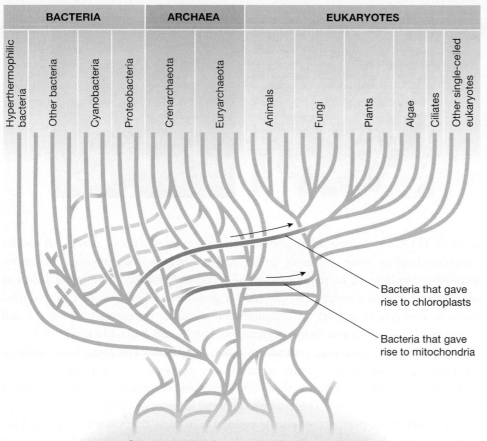

BACTERIA | ARCHAEA | EUKARYOTES

Hyperthermophilic bacteria | Other bacteria | Cyanobacteria | Proteobacteria | Crenarchaeota | Euryarchaeota | Animals | Fungi | Plants | Algae | Ciliates | Other single-celled eukaryotes

Bacteria that gave rise to chloroplasts

Bacteria that gave rise to mitochondria

Common ancestral community of primitive cells

FIGURE 11.17 Horizontal gene transfer and the early evolution of life. During the early evolution of life, *if* HGT was the predominant mode by which genes were transferred, the tree of life might have had a base that reflects a pool of early life-forms that readily swapped gene components. Adapted from Doolittle (2000).

investigated—rather than being rooted on a single universal common ancestor, the tree of life might have had a base that reflected a pool of early life-forms that readily swapped gene components (Figure 11.17). Once this period of intense HGT was complete, and vertical transmission became predominant, the three main branches of the tree of life—archaea, bacteria, and eukaryotes—would have emerged.

It is important not to misinterpret what the implications of intense HGT might have been early in the evolution of life. The hypothesis that the tree of life might have a base that reflected a pool of early life-forms that readily swapped gene components does *not* imply that all life-forms do not share a history of common descent. Instead, what we are saying is that it is not possible to delineate what ancestral species were present during the early evolution of single-celled creatures because HGT blurs the concept of a species.

11.5 Metabolic Networks, Minimal Gene Sets, and Cell Evolution

Genomic analysis provides us with another tool in our efforts to understand the early events in the evolution of life. Although we cannot make direct phylogenetic inferences about what organisms were like prior to the last universal common

FIGURE 11.18 Microbes with very small genomes may shed light on early life. (A) *Mycoplasma genitalium.* (B) *Chlamydia trachomatis.*

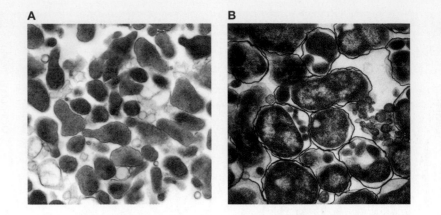

ancestor (LUCA), comparative genomic data are nonetheless useful. A better understanding of what kinds of genes are present in the genomes of extant organisms can help us to make guesses *on functional grounds* about what kinds of genes may have been present in pre-LUCA genomes. With this sort of genomic analysis, researchers can try to calculate the minimal characteristics that a cell would need to operate as a living organism (Mavelli and Ruiz-Mirazo 2007).

How might we go about estimating what constitutes the most basic cellular functions, and how many genes are necessary to code for such functions? As we noted in Chapter 10, more than 1000 prokaryotic genomes have been fully sequenced as of mid-2010. These whole-genome sequences collectively provide us with a comparative perspective on the cellular functions necessary to support life.

To pinpoint the basic and essential cellular functions, researchers have focused on a number of bacterial species with unusually small genomes (Figure 11.18). For example, the bacterium *Mycoplasma genitalium* has one of the smallest genomes of any organism that can be grown in the laboratory. This microbe is a parasite of the human urogenital system, and phylogenetic analysis suggests that a large decrease in genome size in the genus *Mycoplasma* has occurred. Indeed, *M. genitalium* has only 482 protein-coding genes and 43 RNA-coding genes. Most of the sequenced bacterial species with the smallest genomes are, like *M. genitalium*, parasitic or symbiotic in lifestyle. For example, *Wigglesworthia glossinidia* and *Buchnera aphidicola* are both *endosymbiotic* species that live in specialized organs within insect hosts; *Rickettsia prowazekii* and *Chlamydia trachomatis* are obligate intracellular parasites (that is, they live only within eukaryotic cells).

Because of their associations with eukaryotic hosts, species like *W. glossinidia, B. aphidicola, R. prowazekii,* and *C. trachomatis* face relatively stable environmental conditions, and they may have reduced metabolic requirements and hence reduced genome sizes. This makes them easy to analyze. Nonetheless, a note of caution is required as well. All of these species have small genomes as a consequence of reductive processes—that is, because of the loss of genes that were no longer needed over evolutionary time. While they may share many features in common with simple early organisms, there may also be important differences between cells that have small genomes because of loss from a more complex state and those early life-forms that had small genomes formed from the ground up by adding genes. With this caveat in mind, we turn to experimental work on the minimal set of genes needed for cellular life.

Using what is called transposon mutagenesis—a technique that allows researchers to systematically disrupt gene function—evolutionary biologists have found that all 43 RNA-coding genes and at least 380 of the 482 protein-coding genes are essential in *M. genitalium*. If any of these genes are disrupted, *M. genitalium* is not capable of growth. These essential genes are involved in energy metabolism, regulatory functions, fatty acid/lipid metabolism, synthesis of nucleotides, transcription, DNA metabolism, and protein binding.

Comparative analyses can tell us yet more about what sorts of genes appear to be essential for basic cellular life. When researchers compared the genome of *M. genitalium* with other microorganisms having small genomes—for example, *B. aphidicola*, *R. prowazekii*, *C. trachomatis*, and other microbes such as *E. coli* and *Staphylococcus aureus*—a number of interesting patterns emerged (Gil et al. 2004b). A comparison of these genomes found that certain functions were found in *all* of these organisms, and that these functions were associated with about 206 different genes—referred to as the **minimal gene set**. This gene set included 16 genes associated with DNA metabolism, 106 genes linked to RNA metabolism, 15 genes associated with the processing and folding of proteins, 56 genes linked to energetic and intermediate metabolism, and 13 genes associated with other cellular processes.

What are we to make of this minimal set of genes and the functions associated with it? To begin with, there is nothing fundamental about the actual number of genes listed. It is based on data from a small subset of species; no doubt these numbers will change when data from additional genomes are added. Rather than the absolute number of genes, it is the basic functions that are most critical. From this list, we might hypothesize that DNA and RNA metabolism, the processing and folding of proteins, and energetic and intermediate metabolism, are the central building blocks that natural selection favored during early cell evolution.

In the long run, we would really like to have comparative information about the genomes of many different creatures with small genomes so that we could search for patterns and make more general predictions. That is, if we could uncover functions that we see in all these microorganisms, then these might constitute the minimal set of functions necessary for cellular life (Koonin 2000; Gil et al. 2004b).

Let's consider another set of experiments on minimal gene sets by Csaba Pal and his colleagues (Pal et al. 2006). These researchers chose *E. coli* as their test species because a tremendous amount is known about *E. coli* cells—for example, when Pal and his colleagues began their work, it was already known that cellular metabolism in the K12 strain of *E. coli* involved approximately 904 genes and 931 unique biochemical reactions (Reed et al. 2003). The researchers ran the following experiment: They randomly selected one of the 904 genes, and they deleted it from the genome of *E. coli*. They then assayed whether this deletion decreased the fitness of the cell by measuring its rate of biomass production compared to the biomass production of an *E. coli* cell with no deletions. If the deletion did not affect fitness in any measurable way, it was permanently removed from the genome of that cell. If the gene deletion did decrease fitness, it was restored to the cell. This procedure was repeated as many times as necessary to reach the state in which deleting any of the remaining genes would decrease fitness. At the end of this deletion process, the remaining metabolic pathways were documented. This experiment was repeated 500 times to mimic 500 independent evolutionary scenarios.

Pal and his colleagues found that metabolic networks that remained in these 500 replicates were quite similar—about 77% of the metabolic pathways remaining were shared across all 500 experimental replicates. That is, of all the metabolic pathways that we know of in *E. coli*, these 77% seem to be essential to the organism—we might call this the "minimal set of metabolic pathways" in *E. coli* (K12). If an *E. coli* strain were to lose these metabolic pathways, natural selection would act against such a strain, and it would eventually be lost from the population. Pal and his colleagues used this information to make predictions about the sort of characteristics that cells might have possessed during the early stages of cell evolution. They hypothesized that, if they were to examine the cellular metabolism of species that were closely related to *E. coli* but had genomes that were much smaller than *E. coli*, they would find that metabolic networks in these species would resemble the minimal set of metabolic pathways that they had found in their *E. coli* experiment. In other words, they attempted to test ideas about very simple cell life by working backward from a modern organism that has a very long evolutionary history.

The metabolic pathway from the *E. coli* deletion experiment was compared to the metabolic pathways of two closely related species: *B. aphidicola* and *W. glossinidia.* Both *B. aphidicola* and *W. glossinidia* were chosen not only because they are evolutionarily close to *E. coli*, but also because they are endosymbiotic organisms, and so they often obtain all their resources from their hosts. What this means is that some of the genes, and the metabolic pathways typically associated with resource acquisition and processing, are not necessary in endosymbiotic organisms. Moreover, as we noted earlier, because endosymbionts cannot live outside their host, they typically experience only a narrow and controlled range of environmental conditions, further reducing the necessary set of genes for these organisms. As such, we predict that natural selection should have favored a reduction in genome size and the number of metabolic pathways in these sorts of species—and, indeed, the entire genome of *W. glossinidia* is 75% smaller than that found in *E. coli.*

Pal and his colleagues used published information on the genomes and metabolic networks in these two endosymbiotic species to examine whether their metabolic pathways were similar to the minimal metabolic pathway in *E. coli* (Gil et al. 2004a,b). They found support for their hypothesis, as those networks that were most commonly found at the end of their *E. coli* deletion experiments coincided with the metabolic networks that are present in *B. aphidicola* and *W. glossinidia.* Thus, by studying a reductive process of genome evolution, they were able to make predictions about which metabolic pathways would be present in *B. aphidicola* and *W. glossinidia*—two modern organisms that are presumably derived from reductive processes of genome evolution. This analysis also sheds light on what sorts of genes may be essential for basic cellular life, and hence on early evolution on Earth.

The approach adopted by the work we have described in this section is a powerful one, in that it allows researchers to integrate genomics, hypothesis testing, and experimental manipulations to address general questions about the evolution of early life.

In this chapter, we have outlined conceptual, theoretical, and empirical work on the origin and evolution of early life on Earth. Of course, much of what interests biologists has happened since LUCA and the early evolution of cellular life. Indeed, many major evolutionary transitions have taken place since the evolution of the prokaryotic cell. We will explore these major transitions in the next chapter.

SUMMARY

1. Understanding the origin of life requires interdisciplinary collaboration among biologists, chemists, geologists, and atmospheric scientists.

2. At the base of the tree of life is the last universal common ancestor (LUCA). LUCA was not a single organism, but a population of organisms. LUCA was not the first life-form, or the only life-form present at the base of the tree of life. But, by definition, it is the only one that left any descendant lineages that remain to this day.

3. When we use phylogenetic analysis, we cannot see back beyond LUCA, for LUCA is a common ancestor to any group of living species that we might choose to analyze. LUCA represents a phylogenetic event horizon: a point in the history of life beyond which phylogenetic analysis cannot possibly see.

4. Properties of living organisms include homeostasis, structural organization, metabolism, growth and reproduction, and the ability to respond to environmental conditions or stimuli. In addition, all life is subject to the process of evolution by natural selection.

5. The origin of life was more than just the origin of self-replicating entities; heritable variation for natural selection to operate on was also necessary.

6. In the 1920s, Oparin and Haldane proposed the prebiotic soup theory for the origin of life. Miller and Urey tried to simulate the conditions outlined by Oparin and Haldane. From their simple experimental protocol, Miller and Urey produced some of the building blocks of life: amino acids.

7. From about 4 billion to 3.5 billion years ago, life may have been based on RNA rather than on DNA. Ideas on the RNA world have been experimentally tested. Work in this area made a huge leap forward with the discovery of the first RNA enzymes: ribozymes.

8. Evolutionary biologists have built mathematical models and conducted experiments to simulate the conditions of the RNA world, in part to find a bridge from the RNA world to a world in which life is dominated by DNA and protein.

9. Molecular mutualisms may have been important among replicators in the RNA world and may have been critical in the evolution of early cells. The hypercycle model was constructed to address this possibility.

10. The origin and early evolution of bacteria were accelerated by what is known as horizontal gene transfer (HGT). HGT of genes or gene clusters may be especially important with respect to modular cell functions—those not extensively integrated with other functions in a cell. Depending on the extent of HGT, early cell life might resemble a hodgepodge of different cell forms readily exchanging genetic information.

11. Using genome analysis and experimental manipulations, scientists are attempting to understand early cellular evolution by calculating the minimal characteristics that a cell would need to operate as a living organism.

KEY TERMS

hypercycle model (p. 393)

last universal common ancestor (LUCA) (p. 381)

minimal gene set (p. 399)

molecular mutualism (p. 393)

mutualism (p. 393)

phylogenetic event horizon (p. 382)

prebiotic soup hypothesis (p. 384)

protocell (p. 395)

ribozymes (p. 388)

RNA world (p. 384)

REVIEW QUESTIONS

1. When we discussed the role of horizontal gene transfer in the evolution of early life, we primarily focused on how HGT could lead to *increased* cellular complexity. But now consider the earliest cellular parasites. How might horizontal gene transfer have led to the evolution of less complex cellular parasites?

2. Can you think of two other examples of things besides fire that appear to be alive, but aren't? For each case, what characteristics make these things appear to be alive, and what experiments could you design to test these hypotheses?

3. If extraterrestrial objects such as carbon-rich meteorites and comets deposited purines and pyrimidines on Earth during the earliest stages of the evolution of life, what implications would that have on the search for life on other planets? That is, how could our understanding of the evolution of early life on Earth guide our search for life elsewhere in the universe?

4. In 2010, Craig Venter and his colleagues published a paper in *Science* entitled "Creation of a Bacterial Cell Controlled by a Chemically Synthesized Genome," in which they describe work that comes very close to synthesizing the first man-made living cells (Gibson et al. 2010). Will such work on laboratory-created microorganisms help us understand the evolution of early life? If so, in what sense? If not, why not?

5. If life originated once on Earth, we might expect that it should be able to originate again. But we don't observe this happening. Why do you think we do not see life originating again and again from scratch?

6. All known life (a) uses nucleic acid sequences to encode inherited genetic information, and (b) translates this genetic code into proteins in order to make enzymes and other cellular components. Do you think these are necessary properties of any living things, or happenstance of the way that life evolved on Earth?

7. In what ways do you think the first life to arise on Earth was similar to LUCA? In what ways was it likely to have been different?

8. In the Pal et al. (2006) experiment we discussed in the chapter, why was it important for the researchers to randomly select the order in which they deleted genes in their 500 trials? Suppose they hadn't randomly selected the order, and they had instead used a predetermined order for deletions. What sort of problems might they have faced in terms of understanding minimal gene sets?

9. What additional experiments might follow the "one gene at a time" gene deletion experiments that are now being used to study minimal gene sets? Think about the epistatic interactions we discussed in Chapter 9. How might these sorts of interactions inform the next generation of minimal gene set experiments?

10. What sort of errors are you likely to run into when you misinterpret what evolutionary biologists mean by LUCA? Consider the following possible errors and discuss how each could lead to problems in terms of understanding the evolution of early life: (a) LUCA was an individual organism, and (b) LUCA was the first life-form on Earth.

SUGGESTED READINGS

Joyce, G. F. 2002. The antiquity of RNA-based evolution. *Nature* 418: 214–221. A review of the RNA world hypothesis.

Lazcano, A. 2006. The origins of life. *Natural History* 115: 36–43. An article on the origin of early life on Earth written for the general science reader.

Maynard Smith, J., and E. Szathmary. 1999. *The Origins of Life*. Oxford University Press, New York. A wonderful book on the origin and early evolution of life by two pioneers in the field.

Paul, N., and G. F. Joyce. 2004. Minimal self-replicating systems. *Current Opinion in Chemical Biology* 8: 634–639. A technical piece on how evolutionary biologists (and others) have attempted to mimic the processes of replication in the earliest life-forms.

Spiegelman, S. 1970. Extracellular evolution of replicating molecules. In F. Schmitt, ed., *The Neurosciences: A Second Study Program,* pp. 927–945. Rockefeller University Press, New York. A chapter about Spiegelman's work on how natural selection may have operated on the earliest life-forms.

Ⓢ **Visit StudySpace at wwnorton.com/studyspace.**

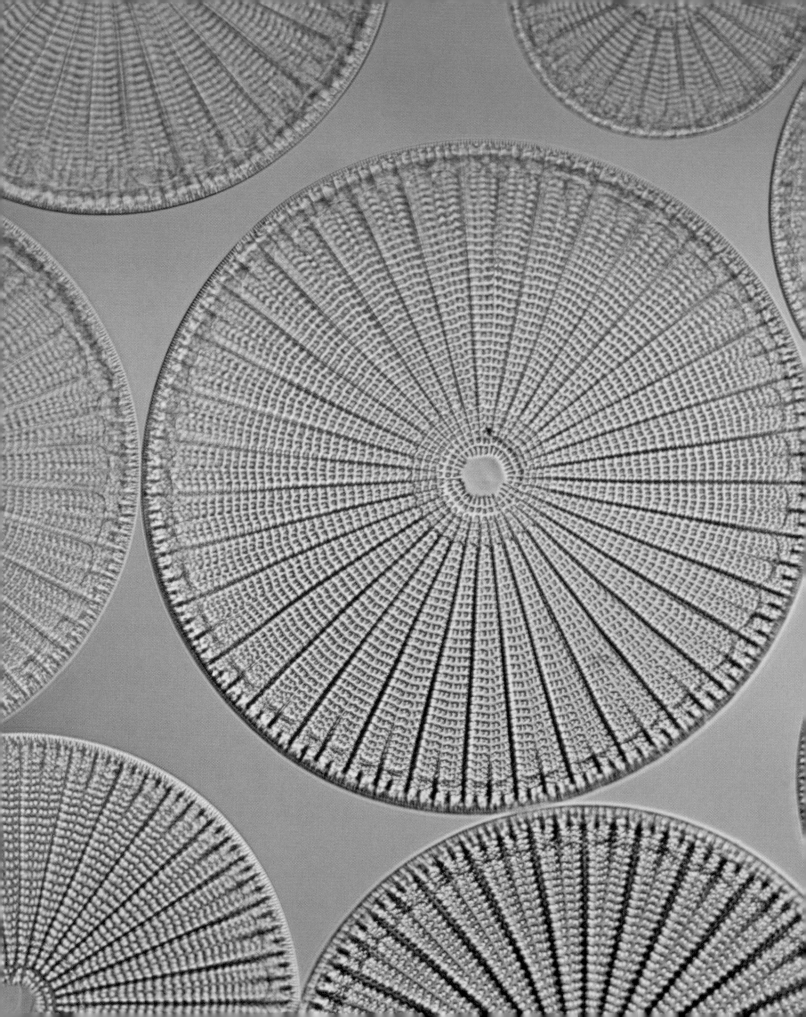

12

Major Transitions

◄ Unicellular diatom algae such as these have evolved elaborate cell walls made of silica.

Cellular slime molds—also known as social amoebas—spend much of their lives as single-celled creatures. There is nothing unusual about that. But then they undergo a radical developmental shift, in which thousands of these free-living cells come together to form a multicellular group called a "slug" (Bonner 1957, 2000, 2003; Buss 1999; Kessin 2001). This is unusual. Rarely do we see free-living, single-celled organisms relinquish their autonomy like this and become one of many cells in what amounts to a sort of primitive multicellular creature. Because of this feature of their development, slime molds, which first appeared about 1 billion years ago, are a model system for looking at what are called "major transitions" in evolution—fundamental organizational changes in the history of life. Slime molds provide some hints about one of these major transitions: from single-celled organisms to multicellular organisms.

The best studied of the slime molds is *Dictyostelium discoideum* (Raper 1935; Kessin 2001). In this species, the earliest developmental stage is a *single-celled* individual—often referred to as an amoeba—that feeds on bacteria living in the soil. A single small patch of soil may contain millions

A

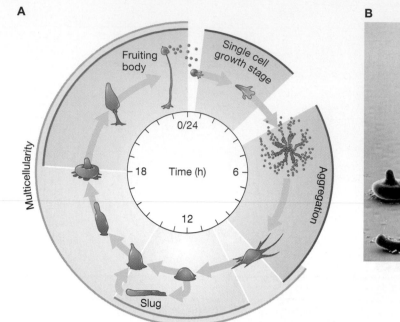

B

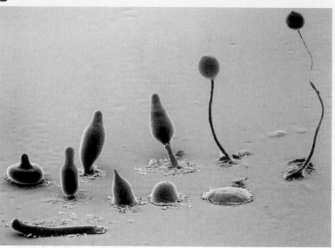

FIGURE 12.1 **Stages of development in slime mold.** (**A**) Developmental stages in *Dictyostelium discoideum*. Multicellular stages are in the green arc. Adapted from Fey et al. (2007). (**B**) Electron micrograph of the different developmental stages in *D. discoideum*.

of *D. discoideum* feeding independently of one another. Although individual cells can reproduce asexually, they rarely do so. Instead, once an area of soil is depleted of available food supplies, between 8000 and 500,000 single-celled *D. discoideum* in that area come together and form a multicellular slug. The newly formed slug then migrates to an area closer to the surface of the soil, where reproduction occurs.

After the slug form of *D. discoideum* has moved to the surface of the soil, it breaks up into a collection of what are called fruiting bodies. These fruiting bodies are also multicellular, although they contain fewer cells than do the slugs (since some of the cells in the slug die in the process of forming the fruiting body). Fruiting bodies are composed of cells that form a stalk section—these cells anchor and secure the fruiting body in place, but they sacrifice the ability to reproduce—and other cells that produce the reproductive spores of the fruiting body. Spores then detach from the stalk and are dispersed when an invertebrate predator, such as a tiny roundworm, disturbs the fruiting body, or when the fruiting body sticks to the invertebrate, or when the soil is flooded. When spores mature they form individual amoebas, and the cycle begins again in a new generation (Figure 12.1). During the fruiting body stage, the once solitary amoebas are part of a multicellular creature. The cells in its stalk behave as **somatic cells** that are responsible for growth and maintenance, while the spore cells act as **germ cells** that are specialized for reproduction. The transformation of the single-celled amoebas into multicellular slugs and fruiting bodies provides hints as to how one of the major transitions in evolutionary history may have occurred. By combining forces, individual cells receive benefits—the ability to move quickly toward light and nutrients and to be protected against predators. Some even sacrifice themselves so that other cells can reproduce. We will explore this transition in greater depth later in this chapter.

In the previous chapter, we considered the origin and early evolution of life, and we described some of the very simple organisms that may have represented the first steps in the evolution of life on Earth. In this chapter, we will examine some of the major transitions that have occurred since those first steps.

To conceptualize the idea of major transitions, compare your own physiology to the earliest organisms: Compare your body to ensembles of autocatalytic molecules, to protocells, to primitive prokaryotes, to a slime mold. Your body is composed of approximately 10^{13} cells, organized into extensive and elaborate organs and tissues. Each cell contains within it the detailed intracellular organization that we observe in eukaryotes. Within the cell nucleus, we find more than just a random collection of genes: We find a highly structured genome, arrayed along 23 pairs of homologous chromosomes. In short, the structure of our bodies is vastly more *complex* than any early life-form. The same is true along any number of branches of the tree of life; it is an astonishingly long way from autocatalytic cycles and protocells to plants and animals, forests, coral reefs, and dolphin pods (Figure 12.2).

Yet there is nothing in the process of evolution by natural selection that should *necessarily* entail a buildup over time of complexity—that is, trends such as the increase in body size, number of cell types, number of protein-coding genes, number of regulatory elements, or total genome size. Indeed, along some branches of the tree of life, we have seen very little increase in complexity for billions of years. Modern bacteria and archaea may be scarcely more complex than their ancestors that lived before the origin of multicellular life. Sometimes we even see complexity evolve, only to be lost again later.

In this chapter, we will explore the following:

- What are the major transitions in evolution?

- What are explanations for some of the major transitions?

- Why was the evolution of the eukaryotic cell a major transition?

- How can the evolution of multicellularity be understood as a second example of a major transition?

- Why did the evolution of individuality constitute a major transition?

- How can the shift from solitary to group living be seen as a major transition?

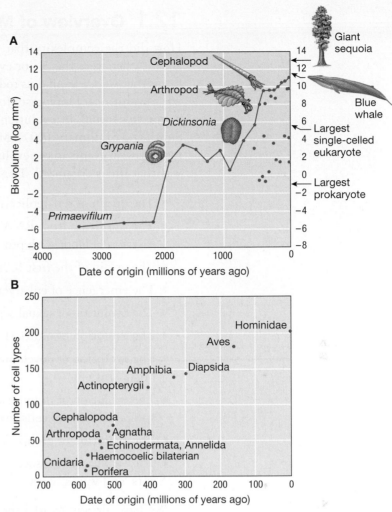

FIGURE 12.2 Organisms have become larger and more complex over evolutionary time. (A) The body size, measured as total volume, of the largest living organisms has increased over evolutionary time. The largest living things today are 10^{18}—that is, a million trillion—times larger than the earliest life-forms. Adapted from Payne et al. (2009). **(B)** The complexity of multicellular organisms—here as measured by the number of cell types—has also increased. Adapted from Valentine et al. (1994).

12.1 Overview of Major Transitions

How did the complexity that we observe along some branches of the tree of life arise? What were the major events in the history of life that led to the elaborate forms that we see around us today? To answer such questions, we will focus here on what evolutionary biologists John Maynard Smith and Eors Szathmary have called major transitions in evolution (Szathmary and Maynard Smith 1995; Maynard Smith and Szathmary 1997). Maynard Smith and Szathmary looked at some of the most critical events in the evolution of life on Earth—events that have changed the way that life is organized. These include:

- The origin of self-replicating molecules capable of heredity.
- The transition from RNA as both catalyst and genetic material to a division of labor with protein as catalyst and DNA as genetic material.
- The origin of the first cells.
- The emergence of eukaryotic cells.
- The evolution of sexual reproduction.
- The evolution of multicellular organisms from single-celled ancestors.
- The evolution of developmental complexity within multicellular organisms.
- The evolution of individuality, including the evolution of germ cells, a specialized line of cells that became gametes.
- The evolution of groups, including complex societies.
- The evolution of eusocial societies, like those seen in some species of bees, ants, and wasps, with a division of labor and sterile workers.

We have already treated the first three items in the previous chapter. We will treat the evolution of developmental complexity in Chapter 13. We will treat the evolution of sex in Chapter 16. Because the evolution of eusociality requires a background in relatedness and kin selection theory, we will postpone that topic until Chapter 18. In this chapter, we will look at the remaining evolutionary transitions.

At first glance, each of the transitions listed above appears to be a unique and perhaps highly contingent event in the history of life, but Maynard Smith and Szathmary note that many of the major transitions in evolution share a common structure and lead to common consequences. Each major transition encompasses some of the following processes, and most feature all of them.

1. *Individuals give up the ability to reproduce independently, and they join together to form a larger grouping that shares reproduction.* For example, early in the history of life, independently replicating molecules joined together within a lipid membrane to form protocells. Later, independently and along numerous branches on the tree of life, unicellular organisms joined together to form multicellular creatures. Similarly, repeatedly and along numerous branches of the tree of life, solitary individuals started living together in colonial groups, sometimes even giving up the possibility of independent replication, as we see in many species of social insects.

2. *Once individuals aggregate into higher-level groupings, they can take advantage of economies of scale and efficiencies of specialization.* Economies of scale arise when groups can perform tasks more efficiently than single individuals, or when groups can do things that lone individuals cannot do at all. For example, groups of social insects such as ants and bees can acquire food in ways that individuals working alone cannot. Collectively, ants can capture prey that are far larger than any single individual by working together (Figure 12.3). They can even engage in a sort of agriculture, as we see with leaf-cutter ants and their fungal gardens.

Once groups are collectively engaged in a task, they can benefit not only from larger numbers, but also from a division of labor, allowing different individuals to specialize in different tasks. We see this sort of task specialization in social insects, but it also occurs in organisms such as the slime mold we discussed at the start of this chapter. Within a single multicellular body, different cells may specialize in generating movement, digesting food, processing information, or other tasks. Perhaps most critically, we see a division of labor between reproductive functions and growth/maintenance functions—the *germ–soma* distinction.

3. *Aggregation and specialization facilitate changes in information technologies. Organisms develop new and increasingly efficient ways to acquire, process, transmit, and store information.* For example, once simple cells form and protein replaces RNA as a catalytic molecule, the fundamental method of storing biological information and passing it across generations can change. Single-stranded RNA with low replication fidelity is replaced as an informational molecule by double-stranded DNA with high replication fidelity. As another example, to facilitate cell differentiation in multicellular organisms, a second layer of information is added atop the DNA by what is called epigenetic tagging. As a third example, multicellular organisms have evolved a suite of positional cues used to regulate the development and differentiation of their component cells. As a final example, once group living becomes commonplace, organisms can learn about their environments through social learning rather than merely by trial-and-error learning.

FIGURE 12.3 Economies of scale in ants. *Formica hemorrhoidalis* ant workers attacking a caterpillar. This is a benefit of economies of scale, as a single worker could not capture such a prey item by itself.

Explaining Major Transitions

Given the huge advantages that come with economies of scale, division of labor, and advances in handling information, we might think it obvious that natural selection would favor these major transitions. So why do they pose a puzzle to evolutionary biologists? As we learned in Chapter 3, if we want to invoke natural selection as an explanation, we need to explain each change by the "immediate selective advantage to individual replicators," rather than by turning to group-level benefits (Maynard Smith and Szathmary 1997, p. 8). This is tricky in the case of major transitions, because it means that we have to identify advantages at the individual level, not only once the transition is complete, but also during the transition so that it can proceed.

Once the transition takes place, and individual units group together to form a higher-level individual, we also have to be able to explain why this higher-level individual continues to exist and doesn't break down (Figure 12.4). Why don't

FIGURE 12.4 Steps in a major transition. Individual "replicators" (left) band together to form a new, high-level individual (center) which can then replicate more effectively (right). To provide an adequate evolutionary explanation for this transition, we need to be able to explain both how the process of banding together is beneficial to the individual replicators, and why there is not an incentive for the components of the higher-level individual in the middle panel to cheat and revert to independent replication.

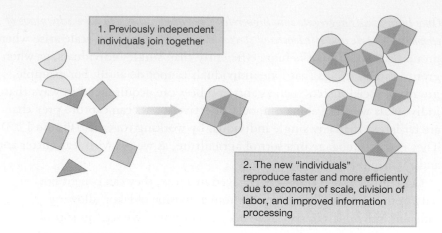

1. Previously independent individuals join together

2. The new "individuals" reproduce faster and more efficiently due to economy of scale, division of labor, and improved information processing

cooperation and coordination collapse in the face of individual incentives for selfish replication? In other words, when cells first band together to form multicellular organisms, why don't individual cells cheat, exploiting the other cells of the organism, or alternatively why do they not revert to reproducing themselves alone to avoid being exploited? When sociality arises in social insects, why don't individual workers cheat and try to produce their own offspring instead of caring for the offspring of another?

Part of the explanation is that, over evolutionary time, the higher-level individuals get "locked in" by some detail of their biology and cannot easily revert to their previous states. Sexual reproduction is a classic example. Why are there no parthenogenic mammals—that is, why are there no asexually reproducing mammals? Why doesn't sexual reproduction break down due to "cheaters" who reproduce parthenogenically, and hence pass down their entire genome intact (Chapter 16)? The process of **genetic imprinting**—in which alleles are differentially expressed according to whether they are inherited from the mother or from the father—seems to be one contributing factor. Once genetic imprinting evolved in mammals, potentially parthenogenic females faced a new and major barrier: Any parthenogenically produced offspring would have a mother but not a father, and they would thus fail to express a number of important genes that are expressed only from the paternally derived copy (Szathmary and Maynard Smith 1995; Maynard Smith and Szathmary 1997).

While parthenogenesis is common among plants, there are no parthenogenic conifers. This is due not to imprinting, but instead to the way that organelles are inherited. In conifers, unlike most other plants, the chloroplasts are transmitted through the pollen rather than through the seed. As a result, a parthenogenically produced conifer would lack chloroplasts. Of course, neither imprinting nor pollen-derived transmission of chloroplasts evolved as safeguards to prevent sexual females from reverting to parthenogenesis, but once present, these traits serve this purpose. And as a result, we see parthenogenic animals and plants of many sorts—but no parthenogenic mammals or conifers.

The point is that while there may be later developments that inhibit reversion to the pretransition state, these are not adequate explanations for the *initial occurrence and stability of the transition itself*. Here we have to look at factors that would have been present at the time of the transition. We will do this in a number of the examples in this chapter.

12.2 Major Transition: The Evolution of the Eukaryotic Cell

In the previous chapter, we discussed the evolution of prokaryotic cells—cells that lack complex membrane-bound organelles and whose DNA is not enclosed within a nucleus. Prokaryotic cells are ancient, having originated on the order of 3 billion years ago (Schidlowski 2001). The second basic cell type is the eukaryotic cell, which has membrane-bound organelles—for example, chloroplasts or mitochondria—and a distinct nucleus containing the genomic DNA. Eukaryotic cells evolved between 1 and 2 billion years after prokaryotes, with six major groups of eukaryotes now recognized by evolutionary biologists. In many ways, eukaryotic cells are more complex than prokaryotic cells, since they have very complicated within-cell communication networks that coordinate interactions among organelles, cytoplasmic elements, and the nucleus, and that target the appropriate proteins and other resources to the appropriate substructures within the cell (Knoll 2006) (Figure 12.5). How did the major transition from prokaryotes to eukaryotes

A

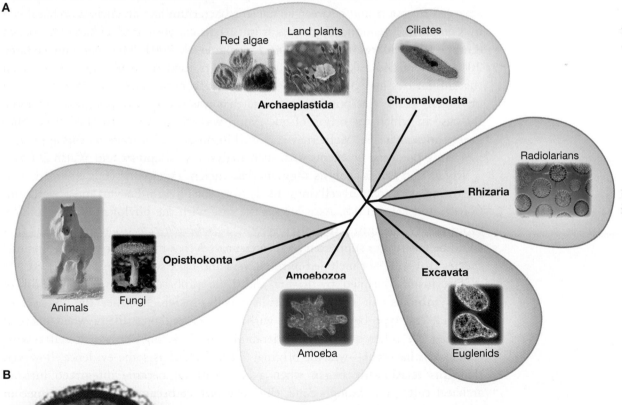

B

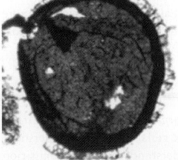

FIGURE 12.5 Evolution of eukaryotes. (A) The six hypothesized major groups of eukaryotes. Adapted from Lane and Archibald (2008). **(B)** A light micrograph of a fossil of *Shuiyousphaeridium macroreticulatum*, one of the oldest known eukaryotes. This species, uncovered in China, may date back 1.8 billion years (diameter ≅ 300 μm) (Knoll 2006).

unfold? The answer is, of course, complex, and we are still in the process of building a complete picture of this major transition. Here we highlight some of the leading theories proposed to explain such a transition.

Several early evolutionary studies on RNA, enzymes, and ribosomes show a strong phylogenetic link between prokaryotic and eukaryotic cells. Some of this early work suggested that eukaryotes shared a common ancestor with species in the prokaryotic domain Archaea (Woese et al. 1990; Gribaldo and Brochier-Armanet 2006), while other work suggested that eukaryotes traced their evolutionary roots to the other prokaryotic domain, Bacteria (Eubacteria) (Martin et al. 1996; Brown and Doolittle 1997; Feng et al. 1997; Gupta 1998). Subsequent work has shown that the situation is more complicated. Phylogenetic analyses indicate that eukaryotic "informational" genes—genes associated with transcription and translation—are most closely related to archaeal genes, whereas "operational" genes associated with metabolic processes, cell membrane formation, and amino acid production are most closely related to bacterial genes (Rivera et al. 1998). How could this be?

Maria Rivera and James Lake tested the hypothesis that eukaryotes may have emerged from a fusion between an ancient bacterium and an ancient archaeal cell by comparing genomic sequences from 10 prokaryotic and eukaryotic species (Gupta 1998; Margulis et al. 2000; Horiike et al. 2001; Hartman and Fedorev 2002; Rivera and Lake 2004). The researchers used molecular genetic data on the similarities and differences between these genetic sequences to construct a phylogeny using tree-building software that was specifically designed to handle the case in which the origin of one group was the result of the fusion of other groups in such a tree (McInerney and Wilkinson 2005; for more on this approach, including potential problems with such analyses, see Bapteste and Walsh 2005).

Rivera and Lake's analysis suggests that ancient eukaryotic cells emerged from the fusion of an archaeal cell (most likely from the phylum Eocyta) and a bacterium (Rivera and Lake 2004). As is always the case with the phylogenies produced by evolutionary biologists, this phylogeny is a *working hypothesis*—a hypothesis that could be falsified or supported by future analyses that might, for example, include species that were not included in the Rivera and Lake study (Cox et al. 2008).

The fusion outlined by Rivera and Lake probably involved some sort of **endosymbiosis**—or symbiosis within a cell—in which either the archaeal or bacterial cell type began residing within the other, most likely when one cell engulfed the other but did not metabolize it. It is unclear which cell type—archaeal or bacterial—was the original "host." There is some evidence, however, that this relationship began when a bacterial cell became integrated into an archaeal cell, and, through time, this relationship became a mutualistic one in which each provided benefits to the other (Timmis et al. 2004). Nonetheless, more work remains to be done in this area before a better resolution to the "original host question" can be obtained (Esser and Martin 2007; Pisani et al. 2007).

Endosymbiosis and the Evolution of Eukaryotic Organelles

Because the presence of complex membrane-bound organelles is one of the critical traits that separates eukaryotic cells from prokaryotic cells, we need to understand where these structures came from if we are to understand the major transition associated with the evolution of the eukaryotes. In 1970, Lynn Margulis proposed the endosymbiotic theory to explain the origin and evolution of two eukaryotic organelles: the mitochondria and chloroplasts (Margulis 1970).

Margulis argued that mitochondria and chloroplasts did not evolve de novo as internal components of a eukaryotic ancestor, but rather through a long-term symbiotic relationship. She proposed that an independent bacterial species capable of energy production and photosynthesis began to reside within early eukaryotic cells. These so-called "endosymbionts" provided their hosts with critical resources such as energy and food and, in return, they were protected from various dangers in the environment by residing inside another organism. Over time, this facultative symbiotic relationship became so strong that it developed into an obligate relationship: The endosymbionts, which evolved into organelles of the host cells, were no longer able to live on their own (Figure 12.6).

Margulis' endosymbiont hypothesis was bolstered by the fact that both mitochondria and chloroplasts each have their own genomes distinct from that found in the cell nucleus. These organelles have circular chromosomes that resemble those found in bacteria. Furthermore, phylogenetic analyses based on molecular genetic data have shown that the chloroplast RNA is more closely related to that

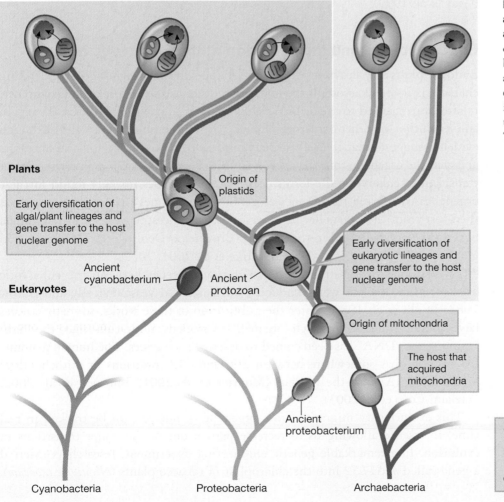

FIGURE 12.6 Endosymbiosis and the evolution of mitochondria and plastids. This phylogenetic diagram illustrates the endosymbiosis hypothesis for mitochondria and plastids such as chloroplasts in eukaryotes. Arrows within the cells indicate gene transfer from organelle to nuclear genome. Adapted from Timmis et al. (2004).

FIGURE 12.7 Chloroplast ribosomal RNA is closely related to cyanobacteria. A phylogenetic analysis based on 16S ribosomal RNA shows that the chloroplasts of plants (as well as a related algal plastid called the cyanelle) are closely related to cyanobacteria. The liverwort chloroplast is shown in red, and the cyanelle of the algae *Cyanophora paradoxa* is shown in orange. All other phyla are cyanobacteria. Adapted from Giovannoni et al. (1988).

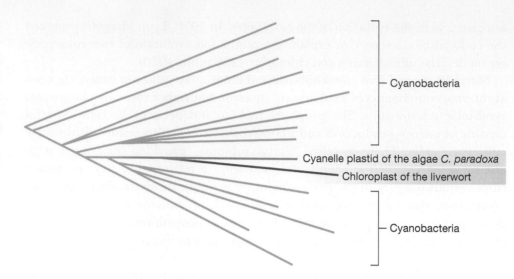

Cyanobacteria

Cyanelle plastid of the algae *C. paradoxa*
Chloroplast of the liverwort

Cyanobacteria

of the cyanobacteria than to that of other eukaryotes (Figure 12.7). This suggests that chloroplasts were once free-living photosynthetic cyanobacteria before they formed a symbiotic relationship with an ancestral eukaryotic species (Giovannoni et al. 1988). In a similar vein, mitochondrial genes in eukaryotes more closely resemble the genes in proteobacteria than other genes in their eukaryotic hosts (Gray et al. 1999).

Endosymbiosis and the Evolution of the Eukaryotic Nucleus

Endosymbiosis may also have played a role in the evolution of another of the defining characteristic of eukaryotes—the nucleus. Indeed, the same sort of endosymbiotic relationship that led to the origin of eukaryotes also may shed light on the origin and evolution of other structures found within the eukaryotic cell. Although evidence suggests that the cell nucleus may have evolved from archaeal ancestors and that the organelles may have evolved from bacterial ancestors, after the major transition to eukaryotic life-forms occurred, many genes initially found in these organelles transferred to the nuclear genome. Sometimes the ancestral gene was then lost from the organelle; in other cases, the ancestral gene was maintained both in the organelle and in the nucleus (Brown and Doolittle 1997; Ribeiro and Golding 1998; Rivera et al. 1998; Horiike et al. 2001, 2002).

Early studies demonstrating the migration of genes between the eukaryotic organelles and nucleus were conducted on maize and yeast (Farrelly and Butow 1983; Jacobs et al. 1983). After the publication of these works, so many studies have found evidence for such organelle-to-nucleus migration that the term "promiscuous DNA" has been coined to describe such genes. The human genome, for example, has somewhere between 296 and 612 insertions of mitochondrial DNA (mtDNA) into the nucleus (Mourier et al. 2001; Tourmen et al. 2002; Hazkani-Covo et al. 2003).

This evolutionary migration from organelle to nucleus can be tracked in real time, in essence allowing us to recreate part of one of the major transitions in evolution. In a remarkable genetic engineering experiment, researchers inserted a gene called *neoSTLS2* into the chloroplast of tobacco plants (*Nicotiana tabacum*).

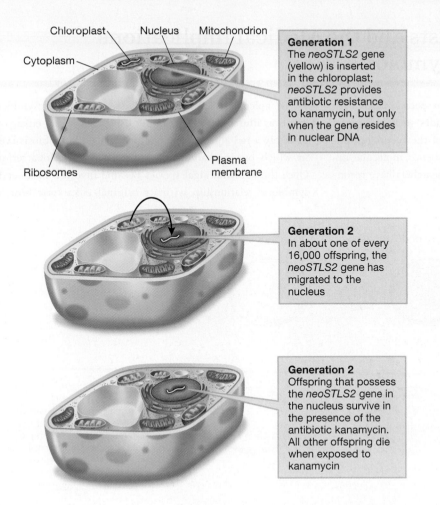

Chloroplast Nucleus Mitochondrion

Cytoplasm

Generation 1
The *neoSTLS2* gene (yellow) is inserted in the chloroplast; *neoSTLS2* provides antibiotic resistance to kanamycin, but only when the gene resides in nuclear DNA

Plasma membrane

Ribosomes

Generation 2
In about one of every 16,000 offspring, the *neoSTLS2* gene has migrated to the nucleus

Generation 2
Offspring that possess the *neoSTLS2* gene in the nucleus survive in the presence of the antibiotic kanamycin. All other offspring die when exposed to kanamycin

FIGURE 12.8 Gene migration from chloroplast to nucleus.
The *neoSTLS2* gene is shown in yellow. This gene was initially inserted in the chloroplast of tobacco cells. The *neoSTLS2* gene confers resistance to the antibiotic kanamycin, but only when it resides in nuclear DNA. Offspring that possess the *neoSTLS2* gene in the nucleus survived in the presence of the antibiotic kanamycin. All other offspring died when exposed to kanamycin. In about 1 out of every 16,000 offspring, the *neoSTLS2* gene had migrated to the nucleus.

The *neoSTLS2* gene confers resistance to kanamycin, an antibiotic that also inhibits seedling growth, but only when it is found as a *nuclear gene*. That is, the only way that the *neoSTLS2* gene in this experiment could protect against kanamycin is if it were transferred from the chloroplast to the nucleus of the tobacco plant. When offspring from plants that had the *neoSTLS2* gene inserted into their chloroplasts were tested in the presence of kanamycin, Chun Huang and his colleagues found that 16 of the 250,000 offspring they examined survived in the presence of kanamycin—in about 1 out of every 16,000 offspring produced by the tobacco plant, there was evidence that a gene initially found only in the chloroplast had migrated to the nucleus *in a single generation* (Huang et al. 2003, 2004) (Figure 12.8).

The evolution of these sorts of endosymbiotic relationships can have important medical implications (Roos et al. 2002; Huang 2004; Ralph et al. 2004). We discuss this in more detail in Box 12.1.

The major transition leading to eukaryotic cells and their complicated within-cell communication networks, then, centered on a series of endosymbiotic mergers, the first of which had either an archaeal or bacterial cell type residing within the other, and was followed by endosymbiosis associated with the nucleus and organelles within eukaryotes.

BOX 12.1 Apicoplasts and the Medical Implications of Endosymbiosis

The apicoplast is an organelle found only in species in a phylum called Apicoplasta—a phylum that includes such eukaryotic pathogens as *Plasmodium falciparum*, one of the agents responsible for malaria. Using morphological evidence, molecular genetic tools, and phylogenetic analysis, researchers have reconstructed the history of the apicoplast.

The apicoplast arose through a secondary endosymbiosis event, as illustrated in Figure 12.9. First an initial eukaryote, probably a red algae, arose by a primary endosymbiosis event in which one prokaryotic host engulfed a cyanobacterium. Once the ancestral algal species became involved in an endosymbiotic relationship with its original eukaryotic host, its

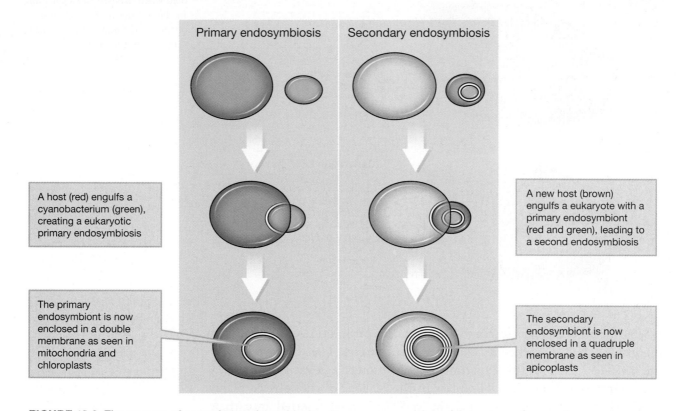

Primary endosymbiosis

Secondary endosymbiosis

A host (red) engulfs a cyanobacterium (green), creating a eukaryotic primary endosymbiosis

A new host (brown) engulfs a eukaryote with a primary endosymbiont (red and green), leading to a second endosymbiosis

The primary endosymbiont is now enclosed in a double membrane as seen in mitochondria and chloroplasts

The secondary endosymbiont is now enclosed in a quadruple membrane as seen in apicoplasts

FIGURE 12.9 The process of secondary endosymbiosis. First, a primary endosymbiosis arises; subsequently, a secondary endosymbiosis occurs when the primary endosymbiont is itself engulfed by a new host. Adapted from Gschloessl et al. (2008).

12.3 Major Transition: The Evolution of Multicellularity

Our focus in this chapter thus far has been on the evolution of single-celled organisms because these made up the earliest communities found on Earth. Now we will turn to the evolutionary transition from single-celled to multicelled organisms. Such a transition has occurred independently many times, in many taxa, over evolutionary history (Figure 12.11) (Michod 1997, 2007; Bonner 2000; Grosberg and Strathmann 2007; Herron and Michod 2008). We most often think of **multicellularity** as an obligate condition—that is, something that cannot be turned on or off. Worms, for example, don't break apart into single-celled creatures

photosynthetic properties appear to have been lost (Funes et al. 2002, 2004; Waller et al. 2003). Subsequently, that eukaryote was itself engulfed in a secondary endosymbiosis event. The original cyanobacterium, now surrounded by four membranes as illustrated in Figure 12.9, became the apicoplast (Lim and McFadden 2010). The apicoplast plays a very important role in the cells of such organisms as *Plasmodium falciparum*, where it is involved in the production of at least 500 different gene products (Figure 12.10).

How can we use this knowledge of the endosymbiotic history of the apicoplast, together with information on its modern function, to improve the medical treatment of malaria? The answer to that question revolves around what metabolic pathways in malaria should be targeted by antimalarial drugs. Think about it like this: Most metabolic pathways in *Plasmo-*

dium falciparum are similar to pathways found in other eukaryotes, because *P. falciparum* is a eukaryote. When we target these pathways with our antimalarial drugs, we risk disrupting similar pathways in eukaryotic hosts of malaria—in particular, in humans. Sometimes such risks must be taken to combat deadly diseases. But because humans lack the apicoplast organelle and because we know the evolutionary history of the apicoplast, we have a safer route we can take for targeting metabolic pathways in *Plasmodium falciparum*: We can target the pathways associated with protein production by the apicoplast. Targeting these pathways, because they have prokaryotic evolutionary roots, reduces the chance of disrupting similar pathways in human hosts (Ralph et al. 2004). Ongoing work suggests that this may be a productive line of research in developing antimalarial drugs (Dahl and Rosenthal 2008).

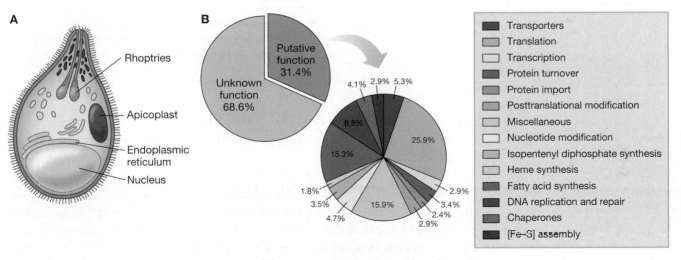

FIGURE 12.10 Apicoplasts and their functions. (**A**) The apicoplast found in a *Plasmodium* cell. (**B**) Functions of apicoplast genes in *Plasmodium*. Adapted from Ralph et al. (2004).

for a period of time, and then form back into worms. But in the early evolution of multicellularity, cells may very well have often joined together and then disbanded, forcing us to view early multicellularity as a temporary, rather than a fixed, condition. Work on the slime molds we discussed at the start of the chapter illustrates this point.

Slime Molds and Multicellularity

In a moment we will examine the selective advantages of multicellularity in the slime mold *Dictyostelium discoideum*, but first we need to establish how, during the migratory slug stage of development, slugs respond to environmental cues

FIGURE 12.11 The phylogenetic distribution of multicellularity. The distribution of multicellularity across the eukaryotes. Adapted from Grosberg and Strathmann (2007).

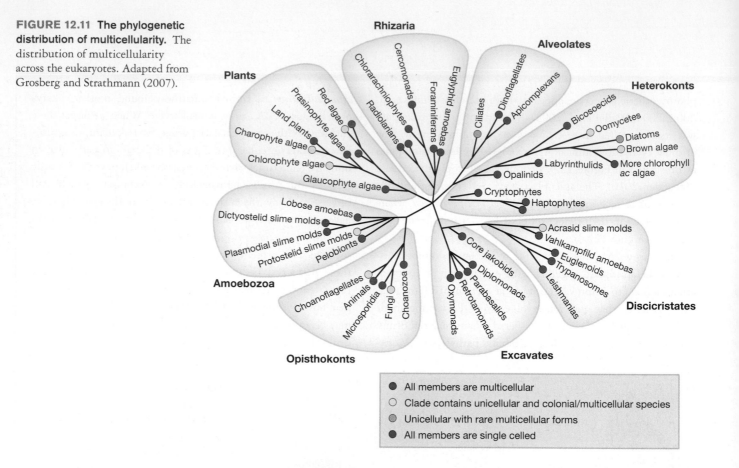

when moving about (Bonner 2000, 2003). How does the slug—composed of cells that were completely independent before its formation—orient itself in its environment?

In *D. discoideum*, the key to slug responses to the environment is communication via a chemical called cAMP (cyclic adenosine monophosphate). In the densest section of cells in the soil, cAMP is released, and this signals cells "downstream" of this point to orient in the direction from which the cAMP was emitted (Figure 12.12). Although it is unclear why some cells produce more cAMP than other cells, slime molds are extremely sensitive to differences in cAMP concentrations. Once the individual cells arrive where cAMP is being emitted, they adhere to each other, surrounding the cells that have emitted the cAMP and forming the multicellular slug. The cells produce proteins that enable them to stick to each other to form the slug. The slug is then able to orient itself and to move toward stimuli such as light and bacterial nutrients very quickly and efficiently.

The cAMP-signaling system does more than inform us about *how* a multicellular slug moves; it also helps us to understand the *benefits* of multicellularity in the life of a slime mold, and hence it provides insight into the evolution of the early stages of multicellularity. The signaling system allows the slug to orient to ambient environmental cues such as light, temperature, and ammonia and oxygen gradients that it uses to move up toward the surface of the soil, where reproduction will occur (Yamamoto 1977; Sternfeld and David 1981; Fisher 1997; Bonner et al. 1998; Kessin 2001). The slug can sense and respond to information in the form of environmental cues in a way that individual cells cannot.

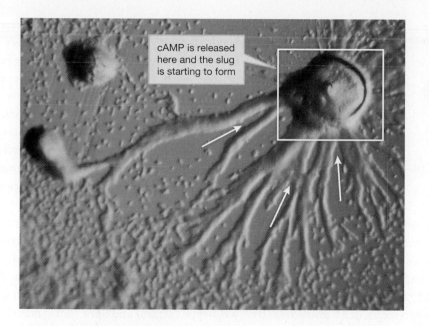

cAMP is released here and the slug is starting to form

FIGURE 12.12 cAMP and slug formation. Single-celled *D. discoideum* move (as indicated by the white arrows) toward an area where cAMP is being released (white square).

The benefits of multicellularity in slugs are not limited to orienting to stimuli and migrating to the surface for reproduction. The slug is also able to form a slime "sheath" around itself that helps protect it from nematode predators (Wang et al. 2001). This sheath is made up of cellulose and protein-rich substances, and it coats only the *surface* of the slugs; it is completely absent in the amoeba stage of slime molds. This antipredator benefit is above and beyond what slime mold cells would get by just moving together in groups (Figure 12.13). That is, many organisms travel together in groups, and such grouping per se often provides protection against predators. But slime mold cells do more than just move together in groups; they produce a protective layer around the group, and all cells that make up a slug share the same fate at this stage since they are all encased in the sheath. Producing a protective layer around the slug is also an example of an economy of scale, as the surface-to-area ratio makes such a slug sheath much less expensive to produce than many individual sheaths would be. This slime sheath thus provides additional benefits to a multicellular developmental stage in slime molds.

A

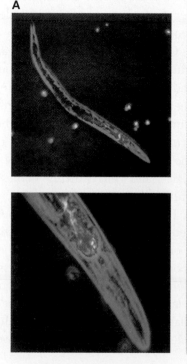

B

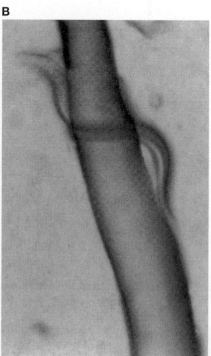

FIGURE 12.13 Slugs form a slime sheath that provides protection. (A) Micrographs of a nematode (*Caenorhabditis elegans*), shown in green, feeding on a *Dictyostelium* amoeba, shown in reddish-orange. **(B)** A nematode (top center) wraps itself around a *Dictyostelium* slug but cannot ingest or harm the much larger slug (the slug is shown running from top to bottom of image) (Kessin et al. 1996). This is an economy of scale; the group can make itself impervious to nematode predators in a way that a single individual cannot.

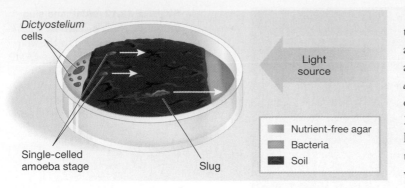

Dictyostelium cells

Single-celled amoeba stage

Slug

Light source

Nutrient-free agar
Bacteria
Soil

FIGURE 12.14 One benefit of slug formation. The experimental plate used to test whether slugs travel through their environment more efficiently than the single-celled amoeba. They do. Adapted from Kuzdzal-Fick et al. (2007).

After observing slime molds in her lab navigate through the soil, Joan Strassmann and her team at Rice University predicted that another benefit associated with this multicellular stage in *D. discoideum* was that slugs were able to move more quickly than were single cells (Kuzdzal-Fick et al. 2007). To test this hypothesis, Strassmann's team had to construct an experiment that allowed them to separate the effects of (1) single-cell movement versus slug movement and (2) developmental stage per se. To understand why both of these were necessary, remember that in normal *D. discoideum*, the amoeba stage precedes the slug stage. So, if slugs were able to navigate faster than amoebas, it could be because they were multicellular (rather than single-celled amoebas) *or* it might be that in later developmental stages—whatever those stages might be—slime molds could move more quickly. To test between these alternatives, Strassmann used a standard strain of *D. discoideum* (the "wild-type" strain) and a mutant strain (labeled CAP2 mutants) that was similar to the wild strain but did *not* form slugs in later stages of development—CAP2 mutants remained single-celled amoebas.

Wild-type cells or CAP2 mutant cells were placed on the left side of a plate as shown in Figure 12.14. On the right side of the plate was a light and a bacterial food source. *Dictyostelium discoideum* typically migrate toward such resources. To get to the food and the light, however, the slime molds had to cross a soil barrier. When wild-type *D. discoideum* were placed on the plate, in 10 of 10 trials they formed a slug and successfully migrated across the soil barrier and toward the resources. But CAP2 mutants never formed slugs, and they were only able to migrate across the soil barrier in 2 of 10 trials, suggesting a selective advantage in terms of migration for the multicellular slug stage in slime molds. By aggregating to form a slug and coordinating their behaviors, the slime mold cells were able to benefit from an economy of scale, the ability to move more efficiently in a large, coordinated group.

Once a *D. discoideum* slug reaches the soil surface, the slug breaks apart into fruiting bodies, each of which consists of a stalk, made up of nonreproductive cells, and spores (reproductive cells). In other words, the cells of the slug reassort to form yet a new set of multicellular structures (Figure 12.15). The spores at the tip of the fruiting bodies are raised from the soil surface on the stalks, and they are dispersed primarily by invertebrates that touch them as they pass by. Being elevated from the soil increases the chances of dispersal by invertebrates—another economy of scale—and thus fruiting bodies provide another set of selective advantages to multicellularity in slime molds.

Fruiting bodies also pose something of a mystery in their division of labor. What determines which slug cells become part of the stalk and which slug cells become spores? What we know is that cells that are rich in resources generally become spore cells, and those that are less well nourished become stalk cells (Kessin 2001; Bonner 2003). Because slime mold cells in a fruiting body are highly genetically related to one another, it may be in the interest of such related cells to have those that are best fed serve as reproductive spores. In Chapter 18, we will discuss the role of genetic relatedness in promoting such altruism.

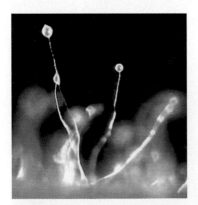

FIGURE 12.15 Fruiting bodies. A close-up photo of fruiting bodies. The stems support a "spore head" that contains the spores.

12.4 Major Transition: The Evolution of Individuality

When we look at multicellular creatures in the world around us, we tend to think of them as "individuals." Slime molds force us to think again. When does a group of cells become an individual? It is true that slime mold cells are part of a multicellular slug at some stages in their life cycle, but they are also independent, free-living creatures, capable of but rarely undergoing (asexual) reproduction at an earlier stage of development. So, is a multicellular slug an individual, or is it a group of individuals temporarily acting in concert? To answer this question, we need to have an evolutionary definition of "individual." Here we adopt a definition suggested by Rick Michod, one of the leading evolutionary biologists studying multicellularity and individuality. Michod argues that "individuals" are "integrated and indivisible wholes" that can reproduce and pass on to their offspring heritable variations (Michod 2007).

If we adopt this definition of evolutionary individuality, groups of slime mold cells are not individuals, as those very cells are free-living, independent cells capable of independent reproduction at a different life stage. Michod immediately follows up on this definition with a point that we have emphasized a number of times now: Natural selection can facilitate transitions from one level of individuality to another, by the same sort of gradual process with incremental improvements that Darwin proposed for the evolution of other complex traits.

How then did the evolutionary transition to a new level—multicellular individuals—occur? The answer entails an understanding of how fitness is transferred from one level of organization—the individual cell, for example—to a higher level of organization—the multicellular organism. In the case of the evolution of multicellular individuals, a critical component of the transfer of fitness from lower to higher levels of organization involves the differentiation of cell lines into those specialized in reproduction (germ cells) and those specialized in maintenance and growth of the organism (somatic cells, or soma). This is a differentiation that is, by definition, impossible in single-celled organisms.

Volvocine Algae and the Evolution of Individuality

To better understand how individuality, with germ and soma lines, has evolved, we will focus on volvocine algae. This group of green algae diverged from a unicellular ancestor about 230 million years ago (Herron et al. 2009), making the transition to multicellularity far more recently in evolutionary history than did most other multicellular lineages. Volvocine algae are ideal for studying the evolution of individuality, not only because of their relatively recent transition to multicellularity, but also because of the exceptional variation found within this group. Some volvocine species are unicellular; some species are made of cells that live in groups but do not have specialized germ and soma lines; and some species, such as *Volvox aureus*, show well-differentiated germ and somatic cell lines (Figure 12.16). Indeed, the division of labor between germ and soma lines has evolved on at least three separate occasions in this group. But how? How has the transition to individuals with germ and somatic cell lines taken place?

To address these questions, Michod and his colleagues focused on one species of volvocine algae, *Volvox carteri*, a species in which there are both germ and somatic

FIGURE 12.16 Cell number and germ cell specialization in volvocine algae. Six species of volvocine algae that differ in cell number and germ cell specialization (Michod 2007). **(A)** Unicellular *Chlamydomonas reinhardtii*. **(B)** *Gonium pectorale*, a sheet of 8–32 undifferentiated cells. **(C)** *Eudorina elegans*, a colony of 16–64 undifferentiated cells. **(D)** *Pleodorina californica*, a colony with between 30 and 50% somatic cells. **(E)** *Volvox carteri*, with thousands of flagella-bearing somatic cells and only a handful of germ cells. **(F)** *Volvox aureus*. In D–F, when two cell types are present, the somatic cells are smaller and the reproductive cells are larger.

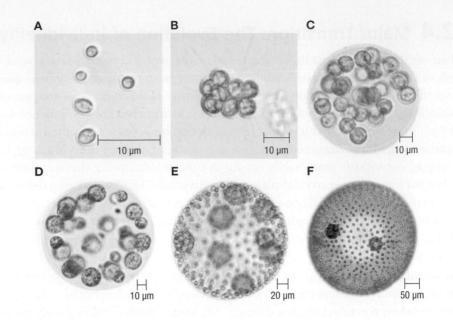

cells. Individual *V. carteri* are typically made up of about 2000 small somatic cells and as many as 16 large reproductive cells (Kirk et al. 1999). Each somatic cell has two flagella, which are long, hair-like projections from a cell that produce motion. In this species, moving by flagellar motion is critical to survival because most of the nutrients, such as phosphorus, as well as the sunlight for photosynthesis, are found close to the water surface, and *V. carteri* uses flagellar motion to avoid sinking in the water. Flagellar motion also mixes the water around individuals, and it helps them to take up nutrients and to release waste. Experimental work with mutant strains of *V. carteri*, in which somatic cells do not produce flagella, for example, shows that such mutants fare very poorly in terms of competition and reproduction (Solari et al. 2006a). Thus, the small somatic *V. carteri* cells specialize in survival and growth functions; they never divide and reproduce to form new *Volvox*, but they are critical for the survival of a colony of *V. carteri*.

The larger germ cells of *V. carteri* lack flagella and specialize in reproduction. Such large cells are necessary for reproduction because of the unusual nature of cell divisions during reproduction. Rather than doubling in size and then dividing, germ cells in *V. carteri* undergo up to 13 rounds of cell division, with almost no cell growth during these divisions. As such, a reproductive cell has to be very large from the start.

How is the fate of a cell—large germ or small soma cell—determined? The answer to this question centers on the expression of a gene known as *regA* (Meissner et al. 1999; Short et al. 2006; Solari et al. 2006a,b). When this gene is expressed, it suppresses a number of nuclear genes that code for chloroplast proteins. Since cell growth is dependent on these chloroplast proteins, and cell division depends on cells reaching a critical size, cells in which *regA* is expressed remain small and produce flagella, becoming the soma cells. If cells are above a critical size, *regA* is not expressed, and these cells photosynthesize, grow larger, and lose the ability to produce flagella. These larger cells go on to form the germ line.

We can do more than link *regA* with the evolution of individuality. Evolutionary biologists have been able to trace the evolutionary history of *regA* itself back to a unicellular volvocine ancestor of modern-day volvocine unicellular species, such as *Chlamydomonas reinhardtii*. In this species a flagellated cell first grows in size and then absorbs its flagellum and produces daughter cells. A gene that is very similar to *regA* has been found in *C. reinhardtii*: This *regA*-like gene is expressed as a function of environmental cues, and its expression determines when a cell absorbs its flagellum and begins reproduction. In essence, this single cell moves from somalike activity to germ cell–like activity. It appears that over evolutionary time, this *regA*-like gene has been co-opted from a gene that regulates the timing of cell division in unicellular organisms such as the unicellular ancestor of *C. reinhardtii* to a gene that regulates the differentiation into germ and soma of cells within a multicellular organism such as *V. carteri* (Nedelcu and Michod 2006).

We now we turn to the major transition from solitary individuals to group-living individuals.

12.5 Major Transition: Solitary to Group Living

Group living provides a suite of benefits, including benefits associated with foraging and safety from predators. But living in groups requires a degree of sociality that is not required for solitary living, and this also often entails new levels of coordination and communication between individuals to obtain such benefits. We will examine each of these components of group life in more detail in a moment, but we begin with a definition.

We can define a group as a set of conspecific individuals who affect each other's fitness (Wilson 1980). Yet there is tremendous variation in the extent to which individuals living in groups can be found throughout nature. In some ungulate species, such as the Japanese serow (*Capricornis crispus*), we see individuals spending the majority of their lives, aside from times of mating, living solitarily. In other species, such as the honeybee, individuals spend virtually all of their time in some sort of group, and they go to extreme lengths to defend others in their hive (Seeley 1985; Kishimoto and Kawamichi 1996). We need not compare such dramatically different creatures to see such variation in group living. Within spiders, for example, most species are solitary, but group living has evolved multiple times in this clade (Figure 12.17) (Agnarsson et al. 2006).

The Benefits of the Transition to Group Living

A plot of the "time budget" of almost any animal would show that organisms spend most of their time either searching for food or engaging in some sort of antipredator behavior. As such, we will focus here on the foraging and antipredator-related benefits to group living.

Foraging in Groups

Economies of scale are common in the search for food, and living in groups provides individuals with numerous foraging-related benefits. For example, consider the foraging behavior of the bluegill sunfish (*Lepomis macrochirus*). Bluegills feed

A

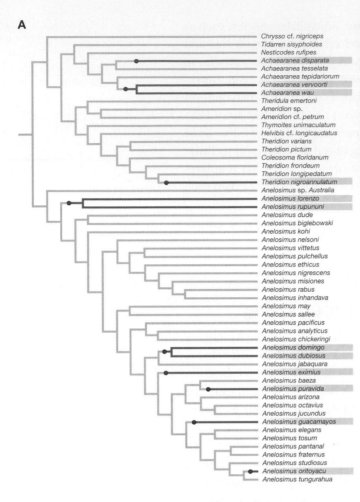

B

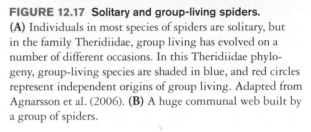

FIGURE 12.17 Solitary and group-living spiders.
(A) Individuals in most species of spiders are solitary, but in the family Theridiidae, group living has evolved on a number of different occasions. In this Theridiidae phylogeny, group-living species are shaded in blue, and red circles represent independent origins of group living. Adapted from Agnarsson et al. (2006). **(B)** A huge communal web built by a group of spiders.

primarily on small, aquatic insects that live in underwater vegetation (Figure 12.18). Aquatic insect prey are quite difficult to catch in such vegetation, but when bluegills forage together in a small area, they are able to flush out many more prey than do solitary individuals, and so foraging success *per fish* often increases as a function of group size (Morse 1970; Bertram 1978; Mock 1980).

Gary Mittlebach examined this benefit of group foraging by experimentally manipulating the group size of bluegills in a controlled laboratory setting (Mittlebach 1984). Mittlebach placed 300 aquatic prey in a large aquarium containing juvenile bluegill sunfish, and he recorded the feeding rates of bluegills that were foraging alone, in pairs, and in groups of three to six bluegills. He uncovered a positive relationship between foraging group size and individual foraging success. That is, the average amount of food that a fish received increased as its group size increased up to a certain number. This sort of relationship between group size and foraging success has been found in many different species (Creel 2001).

The bluegill example illustrates what we might call a "passive" benefit of group foraging. By passive, we mean that each bluegill in a group is foraging just as it would forage if it were alone. The fish are not behaving differently in groups; rather, the aggregate impact of their actions creates a flushing effect from which each animal benefits. Nonetheless, in other species, the benefits of foraging in groups may go beyond passive benefits and involve much more coordination and

Prey in open water are more easily captured

Prey are more difficult to capture in vegetation, but can be flushed out into the open

FIGURE 12.18 Foraging benefits of group living. Bluegills foraging in a lake feed on insects that are flushed from the vegetation. Flushing is more common in groups and leads to greater foraging success for members of a group. Adapted from Dugatkin (2009).

communication—important processes involved in major transitions. For example, Christophe and Hedwige Boesch have found that groups of chimpanzees in the Tai Forest hunt for prey in a coordinated fashion (Boesch and Boesch 1989; Boesch 2002, 2005), and that four different hunting roles are involved in the capture of a single prey (Figure 12.19). Having observed thousands of group hunts in the Tai Forest, Boesch describes the process:

> The *driver* initiates the hunt by slowly pushing the arboreal prey in a constant direction, *blockers* climb trees to prevent the prey from dispersing in different directions, the *chaser* may climb under the prey and by rapidly running after them try a capture, and the *ambusher* may silently climb in front of the escape movement of the prey to block their flight and close a trap around the prey. (Boesch 2005, p. 692)

As chimpanzee hunting groups increase in size, group members increase their per capita food intake. In addition to these group-size effects, the Boesches have found clear evidence of cooperation in Tai chimp hunting behavior (Boesch 1994). Complex but subtle social rules regulate access to fresh kills, and they provide those that are involved in a hunt greater access to prey than those who failed to join a hunt.

We can also see the benefits associated with complex, coordinated group foraging in other species. Indeed, one of the most remarkable cases of such coordinated group-level foraging is found in the communication of information provided by the waggle dance of the honeybee.

On returning to the nest with food, a worker bee that has discovered a new source of food begins the waggle dance in which it "dances" up and down a vertical honeycomb within the hive, while other foragers in the hive make physical contact with the dancer as she moves. While waggling its body vigorously, the dancer is conveying important information about the food she has found. Her dance provides directional information for finding the food source from which she has just returned—the angle at which

FIGURE 12.19 Group foraging in chimps. In the Tai Forest (Ivory Coast), chimps cooperate in both capturing and consuming prey. Once a prey is caught, subtle rules for food distribution are invoked.

A

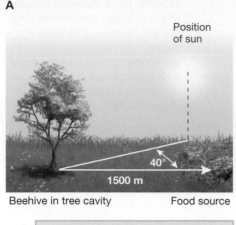

Position of sun

40°

1500 m

Beehive in tree cavity   Food source

For a resource 1500 m from the hive, the waggle dance would last approximately 1 second. The waggle dance may be repeated dozens of times in a row

B

40°

C

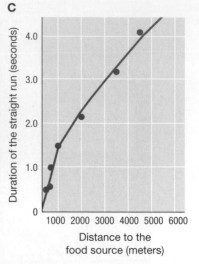

Duration of the straight run (seconds)

4.0

3.0

2.0

1.0

0

1000 2000 3000 4000 5000 6000

Distance to the food source (meters)

FIGURE 12.20 Honeybee waggle dances. **(A)** A patch of flowers that is 1500 meters from a hive, at an angle 40° to the right of the sun. **(B)** When a forager returns, the bee dances in a figure-eight pattern. In this case, the angle between a bee's "straight run" (up and down a comb in the hive) and a vertical line is 40°. **(C)** The duration of the straight run portion of the dance translates into the distance from the hive to the food source. Adapted from Dugatkin (2009) and Seeley (1985).

the forager dances shows the position of the food source of interest in relation to the hive and to the sun. In addition, the longer the waggle dance lasts, the farther away is the bounty. Every extra 75 milliseconds of dancing translates into the resource being approximately an additional 100 meters from the hive (Figure 12.20). The waggle dance thus provides bees living in groups with information about foraging sites that would not be available if they lived solitarily.

Increased Protection from Predators

Living in groups also provides benefits with respect to detecting and avoiding predators. In species in which individuals scan their environments for predators, the more "eyes" in a group searching for predators, the less likely it is that a predator will be able to capture any member of the group. Consider a single bird that lifts its head and stops feeding every five seconds to scan for a predator. Now, imagine 10 such birds that are doing the same thing. Even if the scanning behavior of each bird is completely independent of the scanning behavior of the others in its group, the probability that a predator will successfully approach and capture any of the 10 birds is dramatically lower than the probability of capturing a solitary bird, because the odds are very high that one of the 10 birds will spot the predator, and respond—perhaps by flying away—in a manner that will cause all the birds to head for safety (Pulliam 1973). The bird in our group of 10 that has detected the predator is not responding any differently than it would if it were foraging alone, but its response produces a benefit for all group members. This idea has been dubbed the **many eyes hypothesis**, but of course, it is not restricted to the case in which predators are detected visually: The same principle applies if predators are detected by sound, scent, or other sensory modalities.

As described above, the many eyes model assumes that a predator takes just a single prey item, no matter how large the group of prey. It also assumes that a predator is just as likely to spot and attempt to attack a solitary prey individual and a prey individual in a group, no matter how large the group. But what if that isn't

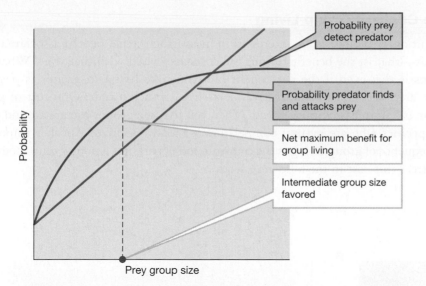
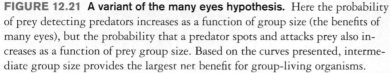

FIGURE 12.21 **A variant of the many eyes hypothesis.** Here the probability of prey detecting predators increases as a function of group size (the benefits of many eyes), but the probability that a predator spots and attacks prey also increases as a function of prey group size. Based on the curves presented, intermediate group size provides the largest net benefit for group-living organisms.

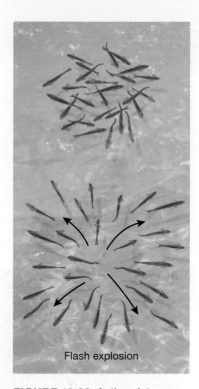

Flash explosion

FIGURE 12.22 **Antipredator benefits of group living.** During a flash explosion, fish in a school move and confuse predators by swimming off in many different directions. Adapted from Dugatkin (2004) and Pitcher and Wyche (1983).

the case? What if, for example, larger groups are more likely to draw the attention of predators? Does group living still produce a net advantage? In Figure 12.21 we show that the net benefit of group living is often maximized at intermediate group sizes under such conditions.

Above and beyond the effects of a group having "many eyes," a transition to group living can be facilitated by other benefits to group members (Hamilton 1971). For example, consider work on antipredator behaviors in schooling species of fish (Pitcher 1986). Swimming in a school produces a hydrodynamic effect that allows for faster movement than when swimming alone. This hydrodynamic effect alone can increase the chances of escaping from a predator. Fish in schools also use a number of antipredator tactics that are simply not possible for solitary individuals, including "flash explosions," in which individuals in a school swim off in all directions, which has the effect of confusing predators and facilitating escape (Figure 12.22).

Even in the absence of flash explosions, the very presence of a school of prey can "confuse" a predator by overloading the amount of information it must process and making it difficult for the predator to home in on a single target and follow it (Milinski 1979). Three-spined stickleback (*Gasterosteus aculeatus*) predators, for example, showed reduced foraging success as the group size of one of their prey—water fleas (*Daphnia*)—increased (Ioannou et al. 2008). When a model of the neural system of the stickleback was simulated using a computer model, results indicated that an increase in *Daphnia* group size caused a decrease in the ability of the stickleback to target any one specific prey item in its field of sight—that is, increased prey group size "confused" the predator, and increased the survival rates of the group-living prey (Figure 12.23).

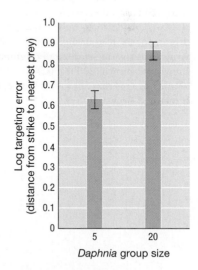

FIGURE 12.23 **Group size and the confusion effect.** Targeting errors of a predator (three-spined stickleback) increase as group size in prey (*Daphnia*) increases. Adapted from Ioannou et al. (2008).

The Costs of Group Living

Evolutionary biologists are interested in measuring the net benefit associated with a trait—that is, the benefits minus the costs associated with that trait. What this means is that even if the net benefit of a trait like living in groups is positive, there are likely to be some costs associated with such a trait. One cost of group living is a simple proximity effect: When you live in a group, you are around other conspecifics who are natural competitors for food resources. Another important consequence of group living is the transmission of parasites among group members. We treat this cost in the following section.

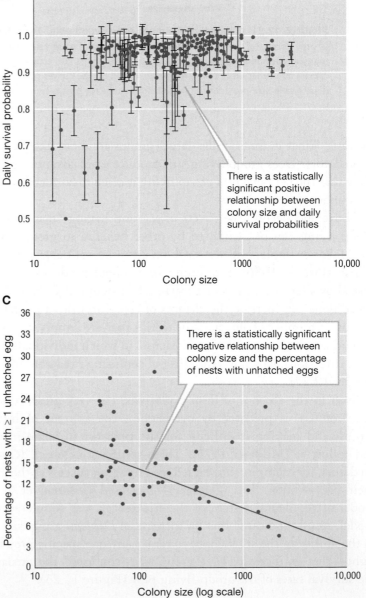

FIGURE 12.24 Group living in cliff swallows. (A) Cliff swallow nests are often clustered together on the side of cliffs. **(B)** Survival probability as a function of group size in cliff swallows. Although there is a great deal of scatter on this graph, the relationship between colony size and survival is positive and statistically significant. Adapted from Brown and Brown (2004b). **(C)** As colony size increases, the number of eggs that fail to hatch decreases. Adapted from Brown and Brown (2001).

Parasite Transmission as a Cost of Increased Group Size

As we have seen, individuals in groups transmit information about foraging, predators, and so on. But members of a group also transmit something else to one another—pathogens and parasites. Because members of a group live in close proximity to one another, parasites that infect individuals who live in groups can move from one group member to another much more easily than they can move between solitary-living hosts. This is true for many different types of parasites, including what are known as *ectoparasites*, which cling to the outside of a host and hence easily move from one host to another.

The cost of parasite transmission is nicely illustrated in Charles and Mary Brown's long-term field study of cliff swallow birds (*Petrochelidon pyrrhonota*). First, let's go into a bit of background. Cliff swallows build their nests in colonies that vary widely in size, and behavioral genetic work has found that preference for small or large groups is a heritable trait (Brown and Brown 2000). Over the last two decades, the Browns have individually marked (tagging their legs with identification numbers) over 160,000 cliff swallows in 239 different colonies, and they have recorded data on such critical evolutionary variables as the probability that eggs will hatch and the survival probabilities of swallows of all ages. Overall, these data show a clear net positive effect of living in groups. As group size increases, the probability that eggs will hatch increases, as does the survival probabilities for birds of all ages (Figure 12.24).

Yet living in groups comes with a price for cliff swallows. Swallows are parasitized by a blood-sucking insect known as the swallow bug (*Oeciacus vicarius*). This ectoparasite, which often clings to the feet of birds, can move from swallow to swallow within colonies, and it is responsible for most of the nest failures and juvenile mortality in these birds (Brown and Brown 1996). The effects of swallow bugs can be experimentally measured by fumigating some swallow nests and leaving other nests untreated by pesticides. When the Browns did this, mortality was much higher in the unfumigated nests, providing strong experimental evidence for the costs of parasitism.

But it is not just the fact that parasites have negative fitness consequences that matters for our discussion of the evolutionary transition to group living. If we are interested in the costs of group living, we need to see evidence that, as group size increases, the cost of parasitism increases. And, indeed, it does—as colony size increases, the number of swallow bugs *per nest* also increases (Figure 12.25). So, while the overall fitness effect of living in groups is positive for swallows, group living does not come cost free, and such costs are important to understand when analyzing the major transition from solitary to social living.

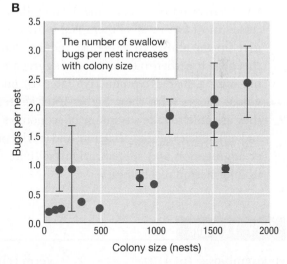

FIGURE 12.25 Cost of group living. (A) A swallow bug. **(B)** The number of bugs per nest per week increases with colony size. Part B adapted from Brown and Brown (2004a).

Following the pathbreaking work of John Maynard Smith and Eors Szathmary, we have outlined the framework biologists use to understand major evolutionary transitions—transitions such as the evolution of multicellular organisms from single-celled ancestors; the evolution of individuality, including the evolution of a specialized line of cells that become gametes; and the evolution of groups, including complex societies. We have already dealt with other transitions (the origin of self-replicating molecules, the origin of the first cells) in earlier chapters, and we will return to additional examples of major transitions throughout the remainder of the book.

SUMMARY

1. Major transitions in evolution include: (a) the origin of self-replicating molecules capable of heredity, (b) the transition from RNA as the catalyst and genetic material, to protein as the catalyst and DNA as genetic material, (c) the origin of the first cells, (d) the emergence of eukaryotic cells, (e) the evolution of sexual reproduction, (f) the evolution of multicellular organisms, (g) the evolution of developmental complexity within multicellular organisms, (h) the evolution of individuality, (i) the evolution of groups, including complex societies, and (j) the evolution of eusocial societies, with a division of labor and sterile workers.

2. Many of the major transitions in evolution share a common structure and lead to common consequences. Each transition possesses some of the following processes (most feature all of them): (a) individual agents give up the ability to reproduce independently, and they join together to form a larger aggregate ensemble with a shared reproductive fate; (b) once individual agents form these higher-level aggregations, they are able to take advantage of economies of scale and efficiencies of specialization; and (c) the processes of aggregation and specialization facilitate changes in information technologies.

3. Eukaryotes may have emerged from a fusion between an ancient bacterium and an ancient archaeal cell. This fusion likely involved some sort of endosymbiosis.

4. Endosymbiosis may also have been involved in the evolution of organelles, including mitochondria and chloroplasts, as well as some components of the cell nucleus.

5. Early on, during the evolution of multicellularity, cells may very well have joined together and disbanded often, forcing us to view early multicellularity as a temporary, rather than a fixed, condition.

6. The evolution of individuality involved the transfer of fitness from the individual cell to the multicellular organism. This transfer of fitness involved the differentiation of cell lines into those specialized in reproduction (germ cells) and those specialized in maintenance and growth of the organism (somatic cells).

7. Living in groups requires a degree of sociality that is not required for solitary living, and this also often entails new levels of coordination and communication between individuals to obtain such benefits. Group living typically imposes costs in addition to providing benefits.

KEY TERMS

endosymbiosis (p. 412)
genetic imprinting (p. 410)
germ cells (p. 406)
many eyes hypothesis (p. 426)
multicellularity (p. 416)
somatic cells (p. 406)

REVIEW QUESTIONS

1. We dated specific major transitions in specific taxa, but we did not try to create a general timeline for major transitions. Why would we not expect to be able to generate a general timeline for all major transitions?

2. Based on the common themes that underlie most major transitions, why would we *not* include the following important evolutionary changes as major transitions: (a) the shift from aquatic to terrestrial life, or (b) the evolution of flight?

3. Can you think of two more examples of economies of scale that may have played an important role in the evolutionary process?

4. How have both phylogenetic and adaptationist (cost/benefit) approaches proved important in the study of major transitions?

5. The genetic engineering experiment we discussed in the text, in which researchers inserted a gene called *neoSTLS2* into the chloroplast of tobacco plants (*Nicotiana tabacum*), illustrates the power of molecular genetics to help us understand major transitions. What sort of other molecular genetic manipulations can you imagine in the future that may shed light on major transitions in evolution? What are some of the constraints on inferring past major transition events from the results of such genetic manipulations?

6. Figure 12.2B, with its linear hierarchy of complexity placing humans at the top, bears a disquietingly strong similarity to Aristotle's *scala naturae*, or great chain of being. One possibility is that humans really do have an exceptionally high number of cell types. Can you think of any alternative reasons why there might be more known cell types in humans than in the other species shown in the figure?

7. Economist Paul Romer has argued that most important innovations for stimulating economic growth are not technological innovations such as steam power or the internal combustion engine, but rather innovations in the rules by which economic activities are conducted, such as systems of intellectual property rights or contract law. Can you make an analogous argument for the growth of biological complexity? Why might the major transitions covered in this chapter and elsewhere in the book be more important than "technological innovations" such as the evolution of wings or of the placenta?

8. In Section 11.3, we discussed one possible evolutionary explanation for the origin of protocells. Explain why this would be considered a major transition in evolution.

9. In this chapter, we considered the antipredator and foraging benefits of group living. What other benefits of group living can you think of?

10. Think about an ant colony, and list the nested levels at which an ant colony is made up of previously independent replicating individuals—that is, the major transitions through which its lineage has passed.

SUGGESTED READINGS

Kuzdzal-Fick, J. J., K. R. Foster, D. C. Queller, and J. E. Strassmann. 2007. Exploiting new terrain: an advantage to sociality in the slime mold *Dictyostelium discoideum*. *Behavioral Ecology* 18: 433–437. Experimental work on the benefits of the major transition to sociality.

Margulis, L. 1970. *Origin of Eukaryotic Cells*. Yale University Press, New Haven, CT. An early presentation of the theory of endosymbiosis.

Maynard Smith, J., and E. Szathmary. 1997. *The Major Transitions in Evolution*. Oxford University Press, New York. The book in which Maynard Smith and Szathmary presented their view of the major transitions in evolution.

Michod, R. E. 2007. Evolution of individuality during the transition from unicellular to multicellular life. *Proceedings of the National Academy of Sciences of the United States of America* 104: 8613–8618. A review of the evolution of individuality in multicellular algae.

Pulliam, R., and T. Caraco. 1984. Living in groups: is there an optimal group size? In J. Krebs and N. Davies, eds., *Behavioral Ecology*, pp. 122–148. Sinauer Associates, Sunderland, MA. A review of the costs and benefits of group living.

⑤ Visit StudySpace at wwnorton.com/studyspace.

13

Evolution and Development

◀ Developmental processes are responsible for the beautiful colors and patterns of feathers, such as these from the wing of a snow goose, *Chen caerulescens*.

W illiam Bateson (1861–1926) was an entomologist, evolutionary biologist, developmental biologist, and the man who both translated Mendel's works into English for the first time and named the science of "genetics." In the latter part of the nineteenth century, Bateson observed a number of bizarre abnormalities in insects and vertebrates. In his studies on the developmental biology of insects, Bateson documented cases in which one body part had replaced another—for example, one insect specimen had legs that developed where the antenna normally would be. Bateson found similar phenomena in vertebrates, where one vertebra had replaced another, or where there were duplicate sets of ribs in the same individual. He called these sorts of developmental changes homeotic transformations (Figure 13.1) (Bateson 1894).

For Bateson, it was not the extraordinary appearance of the homeotic transformations that made them fascinating. Bateson was interested in what homeotic transformations could tell us about evolutionary change. Today we would call his approach **evo–devo** (short for **evolutionary developmental biology**), which incorporates developmental biology into evolutionary

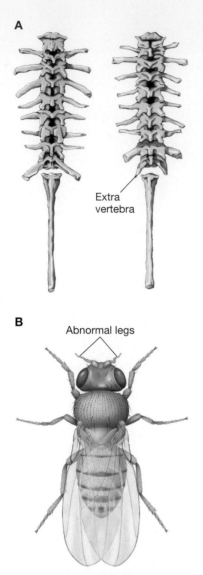

A

Extra
vertebra

B

Abnormal legs

FIGURE 13.1 Homeotic transformations. (**A**) In *Materials for the Study of Variation*, Bateson showed the rib cage of a normal *Rana temporaria* frog (left) and that of an individual with an extra vertebra (right). Adapted from Bateson (1894). (**B**) A homeotic transformation in fruit flies. In fruit flies with the *Antennapedia* mutation, legs develop in place of antenna on the fly's head. Adapted from Exploratorium (2010).

biology and is the fusion of the two disciplines (Raff and Kaufman 1983; Carroll et al. 2005; Carroll 2008). Bateson noted that the homeotic transformations he observed seemed to be most common in parts of the body that were either repeated (appendages, ribs, and so on), segmented, or both. These parts are important for an understanding of the variation we see in animal body form. In his book, *Materials for the Study of Variation*, Bateson hypothesized that homeotic transformations would eventually allow scientists to decipher the evolution of animal body plans (Bateson 1894). We will see that indeed they have.

Here we will examine evolutionary developmental biology. In an animal, this means studying the development of the animal from an egg into multiple cells and then into an embryo with various incipient organs and tissues and finally into the adult form with full-grown internal and external structures such as legs and arms in primates or wings and antennae in insects. In a flowering plant, this means studying the development of the plant from a seed into multiple cells and ultimately into roots, stems, leaves, and other structures in the developing organism. Variations in structures are based in part on when and where genes are expressed, affecting the structures that are produced and their placement in the organism.

In this chapter, we will delve more deeply into evolutionary developmental biology by addressing questions such as:

- How did the field of evo–devo emerge?
- How do homeotic genes map out a body plan, and how do changes in these genes lead to the evolution of new forms?
- How can an understanding of molecular genetics help us gain insight into the evolution of development and the variation of body shapes and forms that we see in nature?
- What is the role of gene duplication in the evolution of development?
- How does evo–devo help us to explain the evolution of novel, complex traits?

13.1 Evo–Devo: A Brief History

Although the sciences of evolutionary biology and developmental biology would not formally come into existence for millennia, the seeds of evo–devo can be found in the work of the ancient Greek philosophers and their concept of what one day would be called the *scala naturae*, or the "great chain of being" (Bonnet 1769) (Chapter 2).

In the *scala naturae*, species can be classified from "lowest" to "highest," with humans at the summit. The ancient Greeks noted a parallel between this scale—which involves the relationships between species—and the developmental stages of organisms. They argued that the development of an individual over its lifetime—its **ontogeny**—stepped through "simple" traits early in development to more complex traits later in the developmental process. In both cases—the *scala naturae* and individual ontogeny—the Greek philosophers noted that the process moved from what seemed "simple" to "complex." All life, at all scales, it seemed to the ancient Greeks, moved from simple to complex.

This idea—that developmental stages mirror the *scala naturae* in moving from the simple to the complex—became known as parallelism, and its first major spokesman after the Greek philosophers was the German anatomist J. F. Meckel (1781–1833). Meckel added a critical evolutionary slant to the concept first suggested by the ancient Greeks, as he hypothesized that the developmental stages of the individual paralleled the evolutionary history of the species being studied (Meckel 1821). In particular, Meckel argued that the developmental stages of an organism step through all the animal species that came before it on the *scala naturae* (note that Meckel's ideas were pre-Darwinian, and so he didn't use the language of evolutionary biology in his writings).

Similar ideas were put forth by the French physician and embryologist Etienne Serres (1786–1868). The "Meckel–Serres" law, as it came to be known, was quickly modified in a subtle, but important manner. While Meckel and Serres argued that embryos display characteristics of embryos from species that preceded them on the *scala naturae*, many people began to claim that the embryos of organisms step through the *adult stages* of species that preceded them. Again, although these ideas were pre-Darwinian, they clearly attempted to tie together developmental biology and what today we would call evolutionary history.

Karl Ernst von Baer (1792–1876), a German naturalist, biologist, and embryologist, rejected both the *scala naturae* and the Meckel–Serres law (von Baer 1828). Von Baer proposed that "the embryos of the Vertebrata pass, in the course of their development, through the permanent forms of no known animals whatsoever," and that the same held true for groups besides the vertebrates. Instead, what is known as "von Baer's law" states that the general characteristics of embryos in closely related species develop before specific characteristics, and embryos of higher taxa do not resemble the adult form of ancestral lower-taxa species.

For von Baer, the most general characteristics that unite embryos from closely related species appear early on in embryonic development, while specialized traits—those that start to distinguish embryos of different species from one another—appear later in development (Figure 13.2). This is a radically different approach to tying

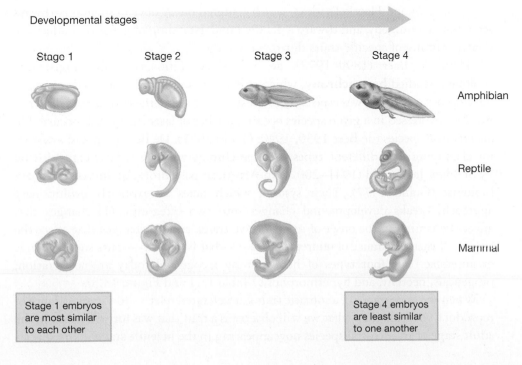

Developmental stages

Stage 1   Stage 2   Stage 3   Stage 4

Amphibian

Reptile

Mammal

Stage 1 embryos are most similar to each other

Stage 4 embryos are least similar to one another

FIGURE 13.2 Von Baer's law. Karl von Baer argued that embryos in closely related species resemble each other, and not the adult form of some ancestral species. He posited that the general characteristics that unite embryos from related species appear early in embryonic development, while specialized traits—those that start to distinguish embryos of different species from one another—appear later in development. Adapted from Horder (2006).

together developmental processes and evolutionary history than that proposed by Meckel and Serres. What von Baer was saying is that those traits that appear early in development are extremely resistant to evolutionary change, and hence they are very similar across many taxa. Presumably this is because changes at early stages in development have consequences that are enormous in magnitude and often fatal. It is only in the later stages of development, von Baer argued, that specific traits emerge that distinguish between closely related groups of organisms. Although von Baer himself argued against Darwin's ideas on evolution, in modern terminology we would say that his ideas suggest that evolutionarily novel traits tend to appear late in development, and are good diagnostics for separating closely related species.

Ernst Haeckel (1834–1919), a German biologist and naturalist, disagreed with von Baer and further expanded on the Meckel–Serres law with his *biogenetic law* (also known as Haeckel's theory of recapitulation). The biogenetic law proposes that "ontogeny is a precise and compressed recapitulation of phylogeny." Haeckel was arguing that the developmental progress of an organism (its ontogeny) replays (recapitulates) its evolutionary history (its phylogeny). This is the first theory that formally tied development to evolutionary theory by explicitly mentioning phylogeny.

For Haeckel, the evolutionary process produced a new species by tacking on something new and novel to the terminal part of the development of the ancestral species, which was an idea first proposed by Fritz Müller (1821–1897). Although this is certainly recognized today as one way that evolution and development can be linked, it is by no means the only way, as Haeckel believed.

Timing of Development

Experimental work in the 1930s and 1940s demonstrated that genes not only code for physical traits, but also control the *rate* of development, and thus the timing at which developmental stages occur (Morgan 1934; Goldschmidt 1938, 1940; Golubovsky and Gall 2003). When this was recognized, the focus moved away from von Baer and Haeckel's ideas, which centered on *all* developmental pathways seen during ontogeny, and toward a detailed understanding of genes that affect the relative timing of specific traits during ontogeny.

Gavin de Beer (1899–1972), an English evolutionary embryologist and zoologist, studied **heterochrony**, which refers to changes in the rate and timing of development. De Beer was especially interested in whether the time at which a trait was first expressed in a given species occurred earlier or later than it had occurred in an ancestral species (de Beer 1930, 1940; Gould 1977). De Beer proposed a system for classifying four different types of heterochrony, which was later expanded on by Stephen Jay Gould (1941–2002), an American paleontologist and evolutionary biologist (Gould 1977). Their system, which takes an explicitly evolutionary approach, breaks developmental changes into two categories: (1) changes that affect the timing of the onset of reproductive traits, and (2) changes that affect the timing of the appearance of nonreproductive—that is, *somatic*—traits such as wings or antennae. The four types of heterochrony recognized today are acceleration, progenesis, neoteny, and hypermorphosis (Table 13.1 and Figure 13.3).

When the appearance of a somatic trait is accelerated relative to the appearance of reproductive traits, then what we will observe is a trait that was formerly seen in the adult stage of an ancestral species now appearing in the juvenile stage in the species

TABLE 13.1		
Four Types of Heterochrony		
Appearance of Somatic Traits	Appearance of Reproductive Traits	Type of Heterochrony
Accelerated	Unchanged	Recapitulation via acceleration
Unchanged	Accelerated	Paedomorphosis via progenesis
Retarded	Unchanged	Paedomorphosis via neoteny
Unchanged	Retarded	Recapitulation via hypermorphosis

Adapted from Gould (1977) and Raff and Kaufman (1983).

derived from that ancestral species. This is referred to as **recapitulation**. Because recapitulation deals with *relative* timing, it can occur in two very different ways: (1) the somatic trait can appear earlier in development (this is called *acceleration*), or (2) the reproductive trait can appear later (be retarded) in development (this is referred to as *hypermorphosis*).

Paedomorphosis refers to the appearance of traits formerly seen in the juvenile stage of a species during the adult stage in a descendant species. Paedomorphosis, too, can occur in two very different ways: (1) reproductive traits appear earlier (*progenesis*), or (2) the onset of somatic traits is retarded (*neoteny*).

The best-studied case of heterochrony is the neoteny seen in a suite of traits in the Mexican axolotl salamander, *Ambystoma mexicanum*. Most species of salamanders live in water during the juvenile stage and live on land as adults, but the axolotl remains in the water for its entire life. Developmentally, the axolotl matures into a

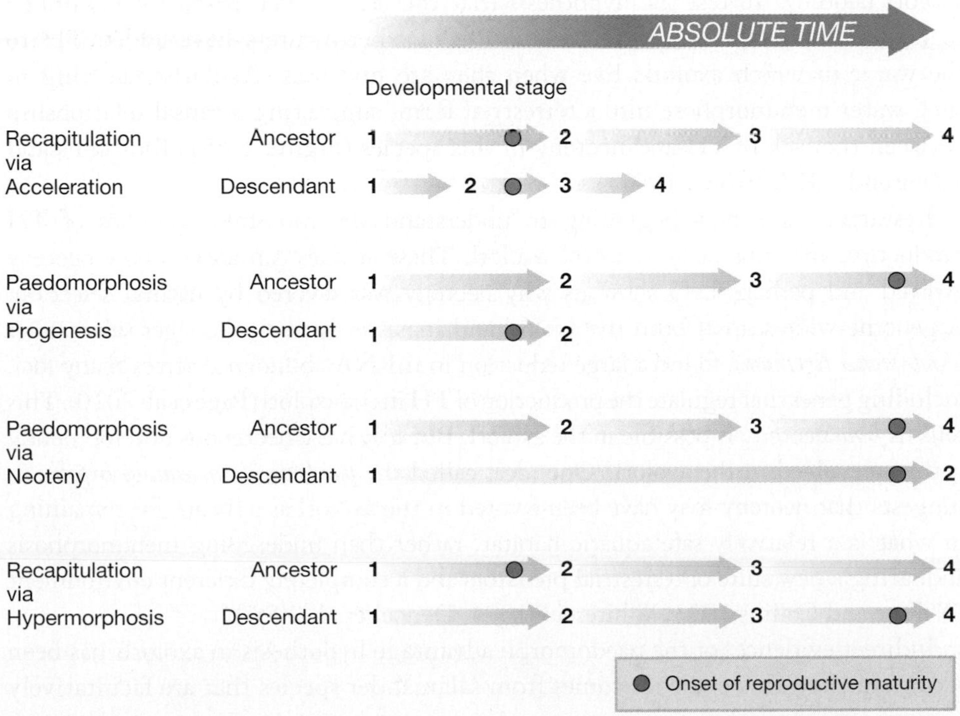

FIGURE 13.3 **Heterochrony.** Four types of heterochrony as a function of developmental stage and onset of reproductive maturity. Adapted from Ridley (2004)

A

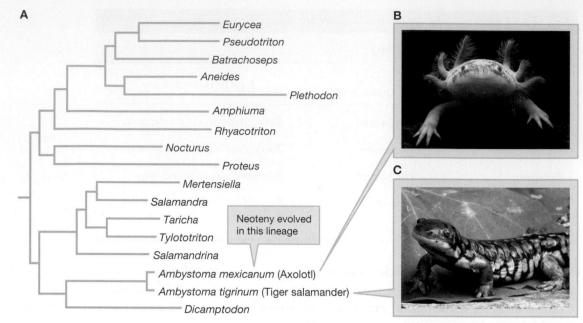

FIGURE 13.4 Neoteny in the axolotl. (A) A phylogeny of salamander families based on complete mitochondrial genomes. Adapted from Zhang and Wake (2009). (B) An adult Mexican axolotl salamander, *Ambystoma mexicanum.* Note the gills seen in this adult. (C) The tiger salamander, *Ambystoma tigrinum*, which is the sister species to the axolotl. The larval stage of the tiger salamander is aquatic, but the adult stage (seen here) is terrestrial.

normal, reproductively active adult, except that it never loses the traits associated with its aquatic existence, such as gills and a flattened tail. This represents an extreme form of neoteny in that *reproductive* traits appear at the same time in the axolotl as in most salamanders that metamorphose into land forms, but adult somatic traits (the loss of gills and the less flattened tail) are so retarded in the axolotl that they never appear at all (Figure 13.4).

Thyroid hormone (TH)—more specifically, the lack of TH—plays a role in neoteny in the axolotl. Most salamander species produce a burst of TH when they move from the water to the land. Axolotls never show this spike in TH. Researchers have set up experiments to examine whether or not TH is linked to axolotl neoteny. To test the hypothesis that the lack of TH production is linked to neoteny and to examine cause and effect, experimenters have added TH to the water in which axolotls live when they are juveniles. Axolotls maturing in such water metamorphose into a terrestrial form, suggesting a causal relationship between the lack of TH and neoteny in this species (Figure 13.5) (Tompkins and Townsend 1977; Brown 1997).

Researchers are now beginning to understand the molecular genetics of TH production and thus neoteny in the axolotl. These studies can answer how neoteny evolved and perhaps help show us why neoteny was favored by natural selection. A genome-wide scan of both the axolotl and its sister species, the tiger salamander (*Ambystoma tigrinum*), found a large reduction in mRNA abundance across many loci, including genes that regulate the production of TH in the axolotl (Page et al. 2010). This tells us *how* neoteny is possible in the axolotl. But *why* has a neotenous developmental pathway evolved in the axolotl? One idea, called the *paedomorph advantage hypothesis*, suggests that neoteny may have been favored in the axolotl as a means for remaining in what is a relatively safe aquatic habitat, rather than undergoing metamorphosis and facing a new suite of terrestrial predators and a completely different environment (Wilbur and Collins 1973; Whiteman 1994; Denoel et al. 2005).

Indirect evidence for the paedomorph advantage hypothesis in axolotls has been accumulating. This evidence comes from salamander species that are facultatively neotenous—that is, species in which some individuals, in some environments, exhibit neotenous development and remain in the water all their lives, while other individuals mature into terrestrial adult morphs. Experimental work has found that the proportion of neotenous individuals increased in facultatively neotenous salamanders when (1) pond levels were constant (as opposed to variable, with some ponds drying quickly), (2) there was a low density of conspecific competitors in ponds, and (3) predation rates were relatively low in aquatic environments relative to terrestrial ones (Harris 1987; Semlitsch 1987; Jackson and Semlitsch 1993). Such ecological factors may also have favored obligate neoteny in the axolotl lineage.

To complete our brief historical overview of evo–devo, we note that after work on heterochrony, the next historical watershed in evolutionary developmental biology was the discovery of the genes responsible for the homeotic transformations that we mentioned earlier in the chapter and which we will discuss in more detail below.

13.2 Regulation, Expression, and Switches

Consider two amazing facts about multicellular creatures: (1) every multicellular creature *develops* from a single cell and (2) except for sperm and eggs, every cell in the body of a multicellular creature contains the same set of genes. Yet skin cells look, feel, and function very differently than do the cells in muscles, cells in the liver, and so on. How cells function depends on the developmental pathways along which they progress. As we will see, this has important ramifications for the evolutionary process. Very early on in the developmental process, each cell in an embryo is totipotent—that is, it could differentiate into any of the cell types that make up the adult organism. It could potentially function as a skin cell, a muscle cell, a liver cell, and so forth. Which of these it becomes depends on the complex and fascinating ways that genes are regulated and expressed within the environment of a cell.

To understand the evolution of development, we need to recognize that the development of an organism is a dynamic process. During the developmental process when a single cell develops into a multicellular organism and then into an adult, cells receive information from local cues—that is, cues from the nearby cellular environment—and this information guides their development. What determines whether a cell will function as a liver cell or a skin cell or as any other specific type of cell depends on what is happening in the environment around that cell. In the next two sections, we will examine **homeotic genes**, master-switch genes that encode proteins that activate or repress gene expression, and **regulatory enhancers**, DNA that turns on and off the expression of particular genes. Doing so will help us to understand how development of forms and structures in plants and animals unfolds from an evolutionary perspective. We will see that development is guided by the turning on and off of genetic switches in a cascade that affects the production

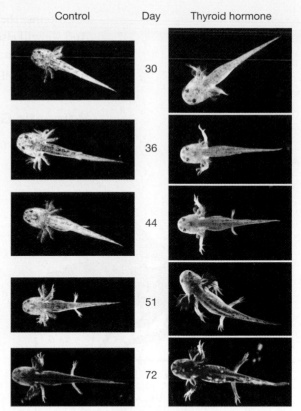

Control   Day   Thyroid hormone

30

36

44

51

72

FIGURE 13.5 Thyroid hormone causes maturation into an adult form in the normally neotenous axolotl. Adding thyroid hormone (T4) to the water in which axolotl individuals were reared causes them to mature earlier and to develop into more "adult" forms than control individuals, as shown on these photos at the specified days after fertilization occurred (Brown 1997).

of proteins, the growth of cells, and the overall body plan of plants and animals. And we will also see that evolutionary changes in developmental pathways between species, genera, and so on, are largely a function of where and when these genetic switches are flipped on or off (Carroll 2005).

Homeotic Genes, Development, and Evolution

Specific homeotic genes affect specific regions in the developing organism by delineating where morphological structures will grow within an embryo, as well as playing a key role in the development of these structures. Homeotic genes encode proteins that control the switching on and off of a cascade of other genes in a set sequence and thereby affect cell size, shape, and division, and the positioning of the cells within the organism's body plan. Homeotic genes were first recognized by Edward Lewis, Christiane Nüsslein-Volhard, and Eric Wieschaus, who shared a 1995 Nobel prize for their work (Lewis 1978; Nüsslein-Volhard and Wieschaus 1980). These researchers studied mutant fruit flies, and they hypothesized that mutations to specific genes affected the body plan of the fruit fly. They found that mutations of genes along the anterior-to-posterior (front to rear) body segments of the insect were responsible for the unusual phenotypes that could be observed in the mutant fruit flies.

Researchers have found that gene products produced from combinations of homeotic genes act as signals that create a sort of instructional map for where structures should develop. The signals occur locally and indirectly specify what structures other genes should form in those particular local regions. As such, homeotic genes play a critical role in the construction of an organism's phenotype, and it is phenotype on which natural selection acts.

Homeotic genes, and their effects on the developmental process, have been studied extensively in fruit flies, where they regulate the overall development of the insect's body regions, as well as segments within its body regions. Between 8 and 13 specific types of homeotic genes, called the ***Hox*** **genes**, affect the anterior-to-posterior positioning of structures on the embryo's body by encoding transcription factors, which are proteins that bind to DNA and that thereby influence gene expression.

Hox genes determine the ultimate fate of various cells in the head, thorax, and abdomen regions in a developing fruit fly and in other organisms (Figure 13.6). For example, the *Hox* gene called *labial* (*lab*) is expressed in cells that develop into mouth parts, while the gene *Abdominal B* (*Abd-B*) is expressed in abdominal body parts near the rear end of the fruit fly. A mutation in a *Hox* gene can lead to the type of abnormality that Bateson found, in which antennae are replaced by legs. This is known as the *Antennapedia* (*Antp*) mutant, which is affected by the *Antennapedia* (*Antp*) gene, which controls leg formation. Nevertheless, changes in *Hox* genes need not lead to such bizarre outcomes. As we will see, small changes in homeotic genes like *Hox* genes can produce large amounts of the phenotypic variance that fuels natural selection.

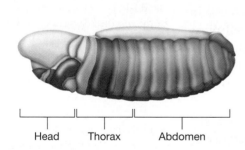

A

B

| *lab* | *pb* | *Dfd* | *Scr* | *Antp* | *Ubx* | *Abd-A* | *Abd-B* |

Drosophila Hox genes

C

Head   Thorax   Abdomen

FIGURE 13.6 *Hox* **genes and co-linearity.** At least eight different *Hox* genes are critical for the development of different body segments of fruit flies. In this diagram, *Hox* genes are color coded, revealing **(A)** the body segments in adults that are affected by each *Hox* gene, **(B)** the *Hox* genes as arrayed on a chromosome, and **(C)** the body segments in larvae that are affected by each *Hox* gene. Note the colinearity mentioned in the text. Adapted from Carroll et al. (2005, p. 24).

A remarkable feature of the *Hox* gene complex is that the position on the chromosome of each *Hox* gene corresponds to the relative position on the animal body part that the *Hox* gene regulates. This phenomenon is known as **colinearity**. For example, the genes associated with the development of mouth parts and eyes are found on the near end of the chromosome, genes associated with the thorax are found in the middle section of a chromosome, and genes associated with development of abdominal sections are found on the far end of the chromosome. How colinearity per se evolved remains a fascinating mystery, but the importance of this colinearity will become more evident in a moment, when we compare homeotic genes across very different organisms.

In the fruit fly example, we examined how homeotic gene expression affects spatial positioning during development, asking: Why do cells with the same genetic content diverge into legs, antennae, and so on? We can also examine how spatial development and homeotic gene expression are related in flowering plants, asking: Why do some cells become parts of the stamen, some parts of the carpel, and so on? In fact, researchers have found that homeotic genes also play an important role in the developmental processes in plants, particularly with respect to the structures of flowering plants (Ng and Yanofsky 2001; Krizek and Fletcher 2005). Work on *MADS-box* genes—homeotic genes that affect plant development—has shed important light on these sorts of questions in plants. For example, expression of genes from the *MADS-box AG* group is involved in the positioning of different types of plant cells (Figure 13.7). Because petals, carpels, and stamens play an important role in plant reproduction, a small change to the *MADS-box* genes underlying the development of these structures can have a large impact on the phenotype and reproductive success of an individual. Sufficiently large changes could even drive speciation.

As new molecular genetic tools became available in the 1990s, evolutionary geneticists and developmental biologists began to search for homeotic genes across a wide spectrum of plants and animals and to hypothesize how they affected their development. They discovered that the system of building organisms dynamically with homeotic genes as position-setters is extremely powerful, not only for concisely and robustly specifying how to build an organism, but also for creating a vast diversity of body forms. The rich diversity of life that we have discussed throughout the book— the diversity that Darwin tried to explain in *On the Origin of Species*—is largely a result of random mutations, subsequently acted on by natural selection, in these dynamic programs for assembling organisms. This is a remarkable statement, so let us examine the issue in a bit more detail.

The same 180-base-pair sequence, called the **homeobox**, is found in homeotic genes in several widely differing species. With this information in hand, molecular geneticists were able to search for and identify additional homeotic genes, including *Hox* genes, in species from frogs and mice to humans. As in fruit flies, the expression of *Hox* genes is often associated with delineating which cells become which body segments. And, again, as in the case of fruit flies, we see colinearity when we study *Hox* genes in vertebrates. Perhaps most remarkably, the ordering of *Hox* genes on vertebrate chromosomes parallels the ordering of *Hox* genes on fruit fly chromosomes. This means that homologous *Hox* genes in invertebrates and vertebrates not only

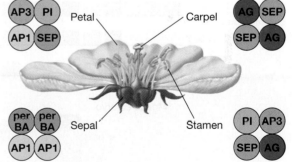

FIGURE 13.7 *MADS-box* **genes and flowering plants.** Expression of homeotic *MADS-box* genes helps explain the developmental pathways of different sections of flowering plants. The *MADS-box* transcription factor proteins (colored circles) are hypothesized to form complexes as shown, which jointly determine which structure—sepal, petal, carpel, or stamen—is formed in which location. Other *MADS-box* genes may also be involved in the development of these sections of flowering plants. Adapted from Thieben and Saedler (2001).

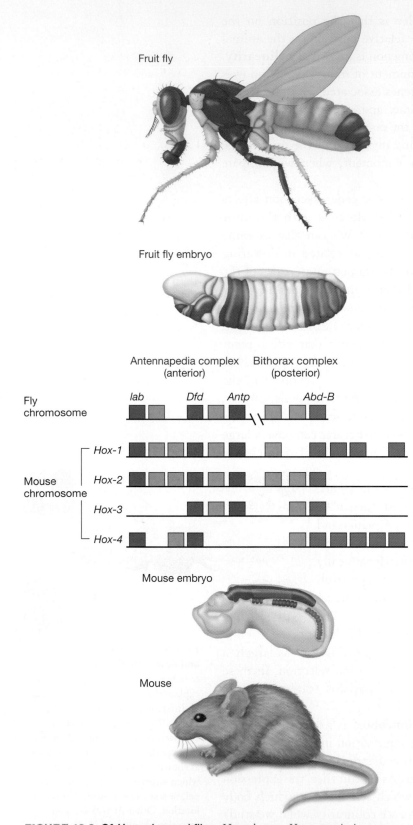

Fruit fly

Fruit fly embryo

Antennapedia complex (anterior)   Bithorax complex (posterior)

Fly chromosome — *lab*   *Dfd*   *Antp*   *Abd-B*

Mouse chromosome — *Hox-1*

Hox-2

Hox-3

Hox-4

Mouse embryo

Mouse

FIGURE 13.8 Of *Hox*, mice, and flies. Homologous *Hox* genes in invertebrates and vertebrates are not only similar in DNA sequence, but they are ordered on chromosomes in a similar way across vertebrates and invertebrates. Here we see the colinear arrangement of *Hox* genes in both fruit flies and mice. Adapted from Taubes (2010).

have similar DNA sequences, but they are also ordered on chromosomes in a similar way in both vertebrates and invertebrates (Figure 13.8). Yet, such animals are built in very different ways. This suggests that differences in the expression patterns of regulatory genes—such as *Hox* genes—in their *local cellular environment* are, at least partially, responsible for the very different sorts of body plans we see in vertebrates and invertebrates.

Researchers have developed a suite of molecular genetic techniques that allow evolutionary biologists to test hypotheses about homeotic genes and developmental pathways. For example, a specific gene can be deactivated (knocked out), allowing researchers to test hypotheses about what sorts of changes in development should then occur. A second technique is to experimentally transfer homologous *Hox* and *Hox*-like genes from one species to another. In a remarkable experiment in 1990, Bill McGinnis and his colleagues tested their hypothesis that the mouse *Hox-2.2* gene, which is structurally very similar to the fruit fly *Antennapedia* (*Antp*) gene—both genes are coded in the homeobox domain—would have the same developmental effects as *Antp* on *Drosophila*. Recall that mutations in *Antp* cause adult fruit flies to develop legs in place of antennae (see Figure 13.1B). When *Hox-2.2* from mice was experimentally inserted into the fruit fly genome and expressed in the head area of developing flies, adults produced legs in place of antennae (Figure 13.9), just as they do when *Antp* is expressed in the head area (Malicki et al. 1990; McGinnis et al. 1990; Akam 1991).

Because some *Hox* genes are so highly conserved evolutionarily, a *Hox* gene from one species can sometimes substitute for that from another, despite vast phylogenetic distances between them. In a 1997 study, Lutz and his colleagues found that by inserting a *Hox* gene from chickens into a variety of fruit fly that had a defective labial *Hox* gene, they could enable the normal phenotype of the fly to develop (Lutz et al. 1997). In other words, the appropriate *Hox* gene from a chicken worked perfectly well in regulating development in a fruit fly

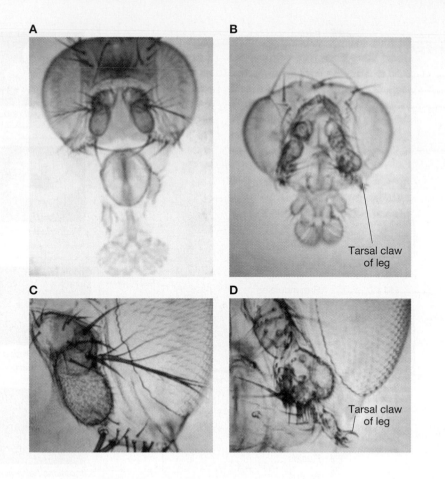

A

B

Tarsal claw of leg

C

D

Tarsal claw of leg

FIGURE 13.9 *Hox* gene transplants. When the mouse *Hox-2.2* gene is experimentally inserted into the fruit fly genome and expressed in the head of developing fruit flies, adults produce legs in place of antennae, just as they do when *Antp* is expressed in the head area (Malicki et al. 1990). **(A)** Wild-type head phenotype of an adult fruit fly. **(B)** Head of a fruit fly with the transplanted mouse *Hox-2.2* gene. **(C)** Close-up of the wild-type antenna. **(D)** Close-up of the thoracic leg parts that develop in place of the antennae in fruit flies with the transplanted mouse *Hox-2.2* gene.

that originally lacked the appropriate *Hox* gene. This is all the more surprising in that the researchers found that the fruit fly and chick proteins expressed in this experiment differed widely *except* in their homeobox regions. The key to understanding how this could be is to recognize that the homeobox regions of these *Hox* genes encode the DNA binding regions of the transcription factor proteins. Thus, the strong conservation of the homeobox regions is sufficient to allow one species' transcription factor to function in the other species, as it switches on the expression of genes in that species.

Both the comparison between mouse and fruit fly *Hox* genes and the ability to transfer *Hox* genes between species have important evolutionary implications. These results suggest that homeotic genes display deep (ancient) homologies. Homologous *Hox* genes have been uncovered in groups as diverse as polyps, mollusks, earthworms, and octopuses and, in each case, these genes are involved in constructing the anterior, central, and posterior body parts of these creatures.

Deep homology of homeotic genes is also seen in plant *MADS-box* genes. As we have seen, *MADS-box* genes play a role in flower development, but they also are instrumental in nonflowering plant species, where they are involved in developmental pathways in leaf and root systems (Kim et al. 2004; Frohlich and Chase 2007) (Figure 13.10).

Why should we see such deep homology in homeotic genes? Why do we see *Hox* genes affecting segmentation patterns early in development across the whole animal kingdom? Why are similar *MADS-box* genes important in the early developmental pathways of both flowering and nonflowering plants? One hypothesis is that while

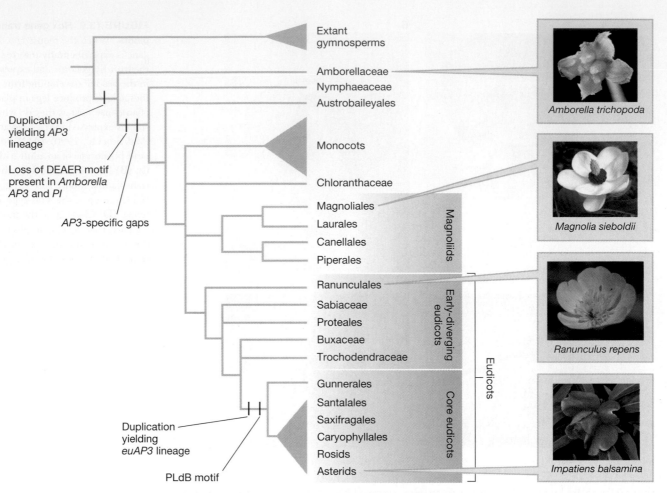

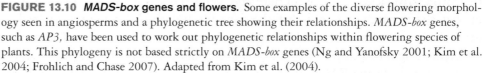

FIGURE 13.10 *MADS-box* **genes and flowers.** Some examples of the diverse flowering morphology seen in angiosperms and a phylogenetic tree showing their relationships. *MADS-box* genes, such as *AP3,* have been used to work out phylogenetic relationships within flowering species of plants. This phylogeny is not based strictly on *MADS-box* genes (Ng and Yanofsky 2001; Kim et al. 2004; Frohlich and Chase 2007). Adapted from Kim et al. (2004).

developmental changes can and do lead to radical new body plans, the dynamic programs that underlie the *early stages of development* are extraordinarily resistant to change. Mutations that change the structure of genes that affect early development are very likely to be lethal, and homeotic genes appear to be fundamental in establishing body plans early in development (as in the anterior-to-posterior axis, as directed by the *Hox* genes that we have described). As a result, we expect homeotic genes to be highly conserved over evolutionary time, and indeed they are. Still, much work remains to be done to understand deep homology in *Hox* genes.

Regulatory Enhancers as Switches

We have seen that homeotic genes encode transcription factor proteins that guide ontogeny. But transcription factors do not act in isolation; rather, they operate by binding to stretches of DNA known as regulatory enhancers (Figure 13.11). A regulatory enhancer of a gene is a section of DNA that lies outside of that gene but is involved in regulating the timing and level of that gene's expression. In

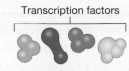

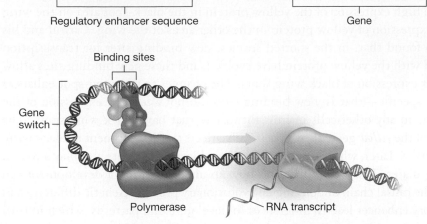

Transcription factors

DNA

Regulatory enhancer sequence

Gene

Binding sites

Gene switch

Polymerase

RNA transcript

FIGURE 13.11 Gene switches.
Gene expression is controlled in part by regulatory enhancer sequences in DNA. Transcription factor proteins bind to the regulatory enhancer, and the result is like a switch being turned on—the switch triggers RNA polymerase to start transcribing an RNA copy of the gene. Adapted from Carroll et al. (2008).

a sense, regulatory enhancers act as switches that turn genes on and off, and they affect the amount of product (primarily proteins) produced by a gene. A single gene can have numerous regulatory enhancers associated with it, and these regulators can operate independently of one another on that gene. Indeed, a gene affected by multiple regulatory enhancers can be expressed differently in different parts of the body and at different points in time. Variation in the expression of regulatory enhancers, in other words, can increase morphological variation and hence the amount of variation that natural selection has to act on.

As with homeotic genes, regulatory enhancers have been a major focus for researchers interested in understanding the construction and patterning of animal bodies. Much of this work has concentrated on traits such as size, shape, and color, as these traits are so fundamental both to the process of development and to the phenotypic variation we see around us in nature.

Extensive work has been done on the regulatory enhancers that affect development of pigmentation pattern in insects. In some species of fruit flies, males have black spots on the edge of their wings, and they use these spots for visual displays during courtship dances with females. In other fruit fly species, the black wing spots are completely absent. Why is there a difference between species? At one level of analysis, the difference between spotted and nonspotted species can be attributed to a gene called *yellow*, and the protein it codes for, which is referred to as yellow protein. In species that have black spots on the edge of their wings, the yellow protein is produced at high levels, but only in the wing cells that produce black spots. In species of fruit flies that lack black spots, the yellow protein is produced in all wing cells, but at levels much lower than those found in the black-spot cells of spotted fruit flies.

But that answer only gets us so far. We want to know why the *yellow* gene is expressed differently across different fruit fly species. The key to the differences in the amount and spatial distribution of yellow protein and wing spots across different species of fruit flies lies in the effects of regulatory enhancers on the *yellow* gene. Sean Carroll and his colleagues uncovered the role of a regulatory enhancer in this system by examining the genetic sequence around the area of the *yellow* gene. They found that in fruit fly species without spots there is a regulatory enhancer that causes the *yellow* gene to express the yellow protein at low levels all over the wing. This same enhancer in spotted species of fruit flies was associated with both high expression of the yellow protein in the black-spot area on the wing and low expression of yellow protein in the other areas of the wing. Carroll and his team also found that, in the spotted species, new binding sites for transcription associated with the yellow protein have evolved, and these new binding sites allow for greater expression of black wing spots. The changes to the wing-spot enhancer were very specific—that is, new binding sites did not affect the expression of the *yellow* gene in any other cells in fruit fly species that have black wing spots. The enhancer of the *yellow* gene, then, specifically affects the development of black wing spots in males. Later, when males mature, these black wing spots also play a role in obtaining mates. This work illustrates how an understanding of development can uncover the causal chain underlying an evolutionary process: Genetic differences in a regulatory enhancer lead to differences in developmental patterns, which in turn lead to differences in traits associated with mating (Figure 13.12).

In addition to differences in the presence of black wing spots, fruit fly species also differ in whether males have dark black coloration on their abdomen. And just as with black wing spots, coloration of the abdomen is affected by a regulatory enhancer during development and plays a role in mate choice during courtship

FIGURE 13.12 Gains, losses, and multiple enhancers. Multiple enhancers control color expression in different parts of a fruit fly's body. **(A)** A hypothetical case of a DNA sequence leading to light wings and a dark abdomen. **(B)** A new binding site produces dark wings. **(C)** A lost binding site leads to the loss of black coloration on the abdomen. Adapted from Carroll et al. (2008).

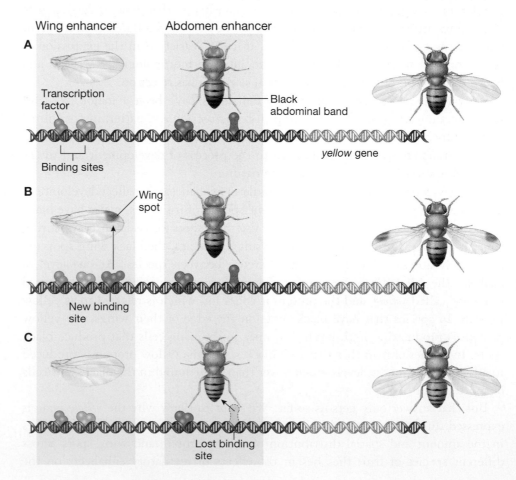

in fruit flies. But it is a second, and separate, regulatory enhancer that affects the expression of the *yellow* gene and that appears to be associated with the presence or absence of black abdomens in fruit flies. The evolutionary history of this enhancer differs from that of the wing-spot enhancer, where new binding sites facilitated dark wing spots. In the case of black abdomens, the key developmental change associated with the yellow protein is that binding sites *were lost* (as opposed to gained) in fruit fly species that lack abdomen coloration.

13.3 Evo–Devo and Gene Duplication

In Chapter 10, we briefly described some of the processes responsible for gene duplication—the establishment of multiple copies of one or more genes within the genome (Ohno 1970; Zhang 2003; Taylor and Raes 2004). Here we will see how duplicate genes play an important role in the evolution of developmental pathways.

Once a gene duplication event occurs, a number of different fates can befall the duplicate copy of a gene. It may be lost by the process of natural selection if the duplication comes at a cost, or it may evolve into a functionless copy known as a pseudogene, as described in Chapter 8. Yet, what makes gene families so important for work in evo–devo is that not all duplicate genes are lost or converted into pseudogenes. Such genes are known as **paralogs**. Paralogs of homeotic genes open the door for new developmental pathways to emerge, as two paralogs may evolve differently if they undergo different mutations or if one undergoes mutation and the other does not. These new pathways may help explain the diversity of form that we see in nature (Figure 13.13).

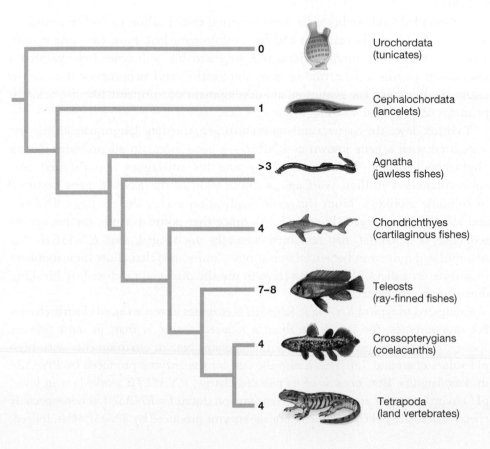

0 — Urochordata (tunicates)

1 — Cephalochordata (lancelets)

>3 — Agnatha (jawless fishes)

4 — Chondrichthyes (cartilaginous fishes)

7–8 — Teleosts (ray-finned fishes)

4 — Crossopterygians (coelacanths)

4 — Tetrapoda (land vertebrates)

FIGURE 13.13 *Hox* **gene clusters and chordate phylogeny.** The number of *Hox* gene clusters mapped onto a chordate phylogeny. The increase in some clades is due, in part, to *Hox* gene duplications. Adapted from Wagner et al. (2003).

FIGURE 13.14 Douc langur.
Douc langurs feed on leaves rather than insects and fruits.

Duplicate genes—both the original gene and the duplicate copy—may be maintained in a population for at least three different reasons:

1. Duplicate genes may influence gene expression levels, increasing production of some critical substances such as ribosomal RNA and histones.

2. After duplication, paralogs may diverge by dividing the work initially undertaken by the gene before duplication. This is referred to as **subfunctionalization**.

3. Duplicated genes may diverge, thus allowing for some new, but related, function to evolve. This process is called **neofunctionalization**.

Gene Duplication of *RNASE* in Colobine Monkeys

Let's examine evolution, development, and neofunctionalization in a bit more detail. Ideally, we would like to work in a system in which we know when gene duplication has occurred, how duplicate genes diverged, and what new function(s) arose as a result of the duplication event. Work on duplication of the gene *RNASE1* in colobine monkeys provides evolutionary biologists with just this sort of information (Figure 13.14).

The colobine monkeys are unique among primates in having a diet of leaves instead of insects and fruit. To derive nutrition from leaves, colobine monkeys, like ruminants, use bacteria in their gut to break down the leaves, and then they digest some portion of the symbiotic bacterial community. The RNA from these bacteria provide the monkeys with a critical supply of nitrogen (Hughes 2002; Zhang et al. 2002).

In a detailed study of both the developmental tract leading to this new mode of digestion (of leaves) in primates and the evolutionary history of the douc langur, a species of colobine monkey, Jianzhi Zhang and his colleagues have painted a fascinating picture of the role of gene duplication and neofunctionalization in understanding both the evolution and development of ruminant-like digestion in primates (Zhang et al. 2002).

To break down the bacteria and obtain nitrogen, the douc langur uses an enzyme associated with a gene known as *RNASE*—a gene found in all primates. Using phylogenetic dating techniques, Zhang and his colleagues hypothesized that approximately 4 million years ago, a duplication of the *RNASE* gene occurred in colobine monkeys. From this gene duplication event, the paralogs *RNASE1* and *RNASE1B* emerged (Figure 13.15). Since that point in time, the nucleotide sequence of *RNASE1* has remained virtually unchanged, but *RNASE1B* has accumulated numerous beneficial mutations—mutations that allow these monkeys to survive on a diet of leaves, as well as to use the nitrogen produced by breaking down symbiotic bacteria.

Compared to typical *RNASE1*, *RNASE1B* enzymes have a reduced electric charge. For our purposes, the key point about a reduced charge is that, in most primate species, the enzyme produced by *RNASE* operates best in environments with high pH values of around 7.4. This is also the case for the enzyme produced by *RNASE1* in douc langurs. But, because of its reduced charge, *RNASE1B* works best in lower pH environments of around 6.3. The enzyme produced by *RNASE1* is not especially effective in such a pH environment, but the enzyme produced by *RNASE1B* is. Indeed, at a pH of 6.3, the *RNASE1B* enzyme is 10 times more efficient than the *RNASE1* enzyme. This is important because the foregut of douc langurs—the place where leaves are broken down—has a pH value that is between 6 and 7. The gene duplication of *RNASE*, and the subsequent divergence of *RNASE1B* from *RNASE1*, set the stage for one paralog to specialize in digesting the nitrogen-rich symbiotic bacteria associated with life as a leaf-eater. This neofunctionalization of a duplicated gene opened up a completely novel route of foraging in colobine monkeys.

Gene Duplication of *OEP16* in Land Plants

The effects of gene duplication on the developmental process have also been implicated in the evolutionary diversification of plants. For example, the gene *OEP16* has been identified in all major lineages of land plants. OEP16 proteins are involved in activating enzyme reactions in the presence of light. Phylogenetic analysis suggests that a duplication event involving *OEP16* took place in the ancestral lineage leading to land plants. This duplication produced two genes, labeled *OEP16L* and *OEP16S*. In flowering plants (angiosperms), these paralogs have diverged from one another in a process of neofunctionalization. *OEP16L* is expressed primarily in leaves, and its expression is very sensitive to temperature change. *OEP16S*, which in flowering plants appears to have gained between 20 and 27 amino acids after the gene duplication event, operates in a very different manner than does *OEP16L*. *OEP16S* is expressed during the maturation period of seeds and pollen grain, and its primary function appears to be associated with tolerating desiccation (Drea et al. 2006). By providing additional protection to developing seeds, the neofunctionalization of the *OEP* gene may have been partly responsible for the explosion of plant diversity associated with the evolution of flowering land plants.

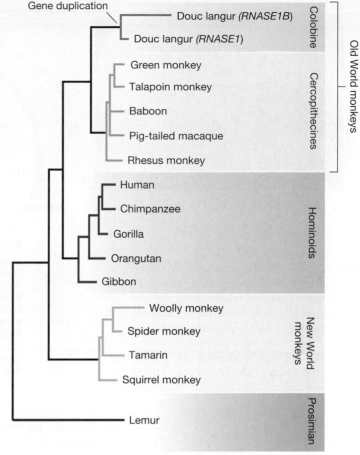

FIGURE 13.15 Gene duplication, *RNASE1*, and *RNASE1B*. A gene tree of *RNASE1* and *RNASE1B* in primates. The gene duplication that led to *RNASE1B* in the douc langur is indicated near the top of the phylogeny. *RNASE1* and *RNASE1B* are paralogs. Adapted from Zhang et al. (2002).

13.4 Evo–Devo and the Evolution of Complex Traits

In other chapters, we have discussed the evolution of complex traits. Here we will use an evo–devo approach to address the origins of complexity by examining neural crest cell development and its effects on the morphology of bird beaks.

Neural crest cells, first described by Swiss anatomist and embryologist Wilhelm His (1831–1904) in 1868, are cells that are initially positioned near the neural tube during early ontogeny and then migrate to new locations during subsequent embryological stages. Neural crest cell development is controlled by a set of homeotic genes (for example, *Hox, Snail, Dlx*), and after these cells migrate during ontogeny, they become critical for the development of sensory neurons, adipose

A

B

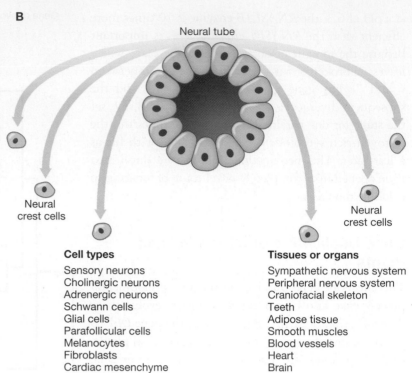

Neural tube

Neural
crest cells

Neural
crest cells

Cell types
Sensory neurons
Cholinergic neurons
Adrenergic neurons
Schwann cells
Glial cells
Parafollicular cells
Melanocytes
Fibroblasts
Cardiac mesenchyme
Adipocytes

Tissues or organs
Sympathetic nervous system
Peripheral nervous system
Craniofacial skeleton
Teeth
Adipose tissue
Smooth muscles
Blood vessels
Heart
Brain

FIGURE 13.16 Neural crest cells. (A) Neural crest cells first appear at the neural plate (indicated by the white outline) and then migrate along the vertebrate central nervous system during early development. **(B)** After neural crest cells migrate and gene expression occurs, they can affect development in a wide array of cell, tissue, and organ types, some of which are listed here. Part B adapted from Trainor et al. (2003).

(fat) cells, craniofacial development, and a suite of other characteristics (Figure 13.16) (Trainor et al. 2003).

To look at the dramatic effects that neural crest cells have on vertebrate craniofacial development, we begin with an observation made by Darwin in *On the Origin of Species*. Darwin noted that the beak proportions of birds are often constant throughout life, and that these proportions "appeared at an extremely early period [during development]…from causes of which we are wholly ignorant." Richard Schneider and Jill Helms hypothesized that beak proportions were determined early in ontogeny by the expression of neural crest cells and that this expression differed between species with different beak proportions. Schneider and Helms ran an elegant transplant experiment involving ducks and quail to test their hypothesis (Schneider and Helms 2003). In quail, neural crest cells are typically responsible for the development of narrow, short beaks, whereas in ducks, these cells are involved in the production of long, flat beaks (Figure 13.17). In both cases, the morphology of the beak is a key phenotypic component of life, having important effects on foraging, aggression, mate choice, and other aspects of life that are relevant to fitness. When embryonic neural crest cells from a duck were transplanted into a quail embryo, the quail developed a ducklike beak. The reciprocal transplant resulted in the development of a duck with a quail-like beak.

Neural crest cells were thought to have evolved during the early stages of vertebrate evolution (Santagati and Rijli 2003; Trainor et al. 2003). Indeed, it was long thought that since early vertebrate evolution coincided with the emergence of the neural crest, neural crest cells represented a fundamentally new vertebrate cell type. Recent work on *Amphioxus* (the closest living relative to vertebrates), however, indicates that this group, too, possesses cell types

A Duck

B Quail

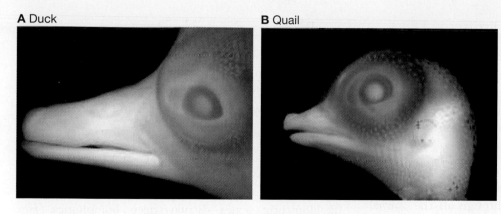

FIGURE 13.17 **Neural crest cells and beak morphology.** (**A**) In ducks, neural crest cells are involved in the production of long, flat beaks. (**B**) In quail, neural crest cells lead to the development of narrow, short beaks. These differences are seen very early in development.

that are reminiscent of neural crest cells in that they migrate over the neural tube during development and, like neural crest cells, are under the control of a series of homeotic genes (Holland et al. 1996). In addition, cells similar to these have been found in the ascidians, a close outgroup to the vertebrates (Mackie 1995; Powell et al. 1996; Shimeld and Holland 2000; Holland and Holland 2001; Wada and Satoh 2001). Based on this and other evidence, evolutionary biologists have hypothesized that neural crest cells evolved from ancestral cells similar to those found in ascidians and *Amphioxus* (Wada and Satoh 2001) (Figure 13.18).

As a second example of both the importance and complexity of neural crest cell development, let's briefly consider the role these cells play in marsupials. Marsupials are born after a relatively short gestation period, and hence they must possess the ability to suck their mother's milk at a much younger age than eutherian mammals. The jaw structure in mammals is primarily under the control of neural crest cells. Analysis of marsupial embryos found that neural crest cells begin their migration much earlier in marsupials than in other mammal groups (Vaglia and Smith 2003), allowing marsupials to begin using their mother's milk at an earlier age than do other mammals. As with our other examples, here again we see how a deeper understanding of changes in developmental processes helps us explain the myriad forms of phenotypic diversity we see in the world around us.

Evo–devo is a broad field that covers many aspects of evolution and organismal development. In this chapter, we have seen that homeotic genes and regulatory enhancers control much of the development process. We have also learned that small changes in timing or spatial positioning during ontogeny can lead to large-scale phenotypic effects, and so may be under strong selection. Some of these changes may be involved in the formation of new species, a subject to which we turn in the next chapter.

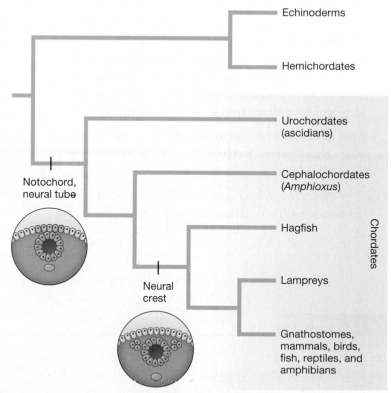

FIGURE 13.18 **Phylogeny of neural crest cells.** A phylogenetic tree mapping the emergence of neural crest cells and their possible progenitors. Adapted from Trainor et al. (2003).

SUMMARY

1. Early ideas on what today is called evo–devo can be found in the work of the ancient Greek philosophers and the concept of the "great chain of being." They noted a parallel between what would one day be called the *scala naturae*, which involves the relationships between species, and the *developmental stages* of organisms.

2. J. F. Meckel hypothesized that the developmental stages of an organism step through all the animal species that came before it on the *scala naturae*. Ernst Haeckel expanded on Meckel's ideas with his *biogenetic law*, which states "ontogeny is a precise and compressed recapitulation of phylogeny."

3. von Baer's law states that embryos in closely related species resemble each other, and not the adult form of some ancestral species. The most general characteristics that are shared among embryos from closely related species appear early in embryonic development, while specialized traits appear later in development.

4. Gavin de Beer coined the term *heterochrony* to describe changes in the rate of development, and he focused on whether the time at which a trait was first expressed in a given species was accelerated or decelerated relative to that of an ancestral species.

5. One of the key players associated with the dynamic rules that govern development are the *homeotic genes* that specify position within an embryo.

6. The position of *Hox* genes on a chromosome corresponds to the position on the anterior–posterior axis of the body part that the *Hox* gene regulates. This phenomenon is known as colinearity. Homologous *Hox* genes are ordered on chromosomes in a similar way across vertebrates and invertebrates.

7. The expression patterns of regulatory genes (such as *Hox* genes) in their local cellular environment are responsible for the very different sorts of body plans that we see in vertebrates and invertebrates.

8. Homeotic genes display deep homologies. *Hox* genes have been uncovered in polyps, mollusks, earthworms, and octopuses. In each case, the genes are involved in constructing the anterior, central, and posterior body parts of these creatures.

9. Ontogeny is also guided by regulatory enhancers. A regulatory enhancer of a gene is a section of DNA that lies outside of that gene but is involved in *regulating* the timing and level of that gene's expression.

10. A gene can have numerous regulatory enhancers associated with it. A gene with multiple regulatory enhancers can be expressed differently in different parts of the body and at different points in time. Regulatory enhancers increase morphological variation, and hence the amount of variation that natural selection has to act on.

11. Duplicated genes can evolve into *paralogs*. Paralogs of homeotic genes allow new developmental pathways to emerge, and these new pathways may help explain the diversity of forms that we see in nature.

KEY TERMS

colinearity (p. 441)

evo–devo (evolutionary developmental biology) (p. 433)

heterochrony (p. 436)

homeobox (p. 441)

homeotic genes (p. 439)

Hox genes (p. 440)

neofunctionalization (p. 448)

ontogeny (p. 434)

paedomorphosis (p. 437)

paralogs (p. 447)

recapitulation (p. 437)

regulatory enhancers (p. 439)

subfunctionalization (p. 448)

REVIEW QUESTIONS

1. Early mammalian development occurs in the female uterus. A fetus developing in the uterus encounters a uterine environment that has abiotic components (temperature, acidity, and so on) and biotic components (competition from siblings in a clutch, parasites found in the uterus, food sources from the mother). How might selection in utero make it difficult to study ontogeny as a recapitulation of phylogeny as initially suggested by Haeckel?

2. Briefly contrast the Meckel–Serres law and von Baer's law. Which of these better corresponds to a branching, phylogenetic view of species?

3. Haeckel's biogenetic law, that ontogeny recapitulates phylogeny, is generally not supported by the evidence. However, in some cases we may see a developmental sequence that appears to conform to the law. Explain how heterochrony could yield an organism that during development appears to go through the adult stage of a closely related organism.

4. Researchers have hypothesized that winglessness in some insect species represents a case of paedomorphosis via progenesis. Why might that be? Also, in general, why might natural selection favor progenesis?

5. We have focused on multicellular organisms in our discussion of evo–devo. How might evo–devo shed light on evolution in single-celled organisms? How might we use knowledge from developmental and evolutionary studies in microbes to improve our understanding of medical problems such as cancer?

6. How does an evo–devo approach to gene duplication support and strengthen the argument that evolution is like a tinkerer, building new gadgets from whatever is available at the time?

7. Stephen Jay Gould once wrote a playful essay on how Walt Disney kept making Mickey Mouse's features more and more paedomorphic, and how this seemed to increase the character's popularity (Gould 1979). Based on what you know about paedomorphosis, why do you suppose that strategy worked?

8. How does the conserved nature of homeotic genes allow for the sorts of cross-species, or even cross-genus, experiments we discussed in the *Drosophila* and mouse study involving the *Antennapedia* (*Antp*) gene?

9. A single gene can have numerous regulatory enhancers associated with it, and a gene with multiple regulatory enhancers can be expressed differently in different parts of the body and at different points in time. How does this generate potentially huge amounts of variation for natural selection to act upon?

10. Which process is more likely to facilitate evolution of novel traits or structures: subfunctionalization or neofunctionalization? Explain.

SUGGESTED READINGS

Carroll, S., J. Grenier, and S. D. Weatherbee. 2005. *From DNA to Diversity: Molecular Genetics and the Evolution of Animal Design*. Blackwell, Malden, MA. An accessible, book-length treatment of basic evo–devo concepts.

Carroll, S. B., B. Prud'homme, and N. Gompel. 2008. Regulating evolution. *Scientific American* 298: 60–67. A popular science article on regulatory enhancers and the critical role they play in the evolutionary process.

Hughes, A. L. 2002. Adaptive evolution after gene duplication. *Trends in Genetics* 18: 433–434. A review paper on how gene duplication leads to new developmental pathways and may promote adaptive change.

Pearson, J. C., D. Lemons, and W. McGinnis. 2005. Modulating Hox gene functions during animal body patterning. *Nature Reviews Genetics* 6: 893–904. A technical paper on the importance of *Hox* genes in the study of evo–devo.

Raff, R., and T. C. Kaufman. 1983. *Embryos, Genes and Evolution*. MacMillan, New York. Although somewhat dated now, this is arguably the first modern book on the subject of evo–devo and is full of interesting material.

Ⓢ **Visit StudySpace at wwnorton.com/studyspace.**

14

Species and Speciation

D uring World War II, the citizens of London were often forced to take refuge in the Underground subway tunnels during Nazi bombing raids. In addition to all of the other discomforts associated with spending long periods of time in an underground labyrinth, Londoners complained of the mosquitoes, *Culex pipiens*, that continuously harassed them (Shute 1951). This was an especially irksome problem, because above ground in England, mosquitoes preferred to bite birds rather than humans, while in the Underground subway tunnels, they showed a strong inclination to bite mammals, including humans. This underground population has the rather ominous scientific name of *Culex pipiens molestus*.

When biologists began examining the aboveground and underground forms of *Culex pipiens*, they found that mosquitoes from these populations looked remarkably similar. But, in many ways, their life histories were dramatically different (Figure 14.1 and Table 14.1). It wasn't just that mosquitoes in these populations preferred to bite different sorts of animals. There was a strong seasonal component to breeding in the aboveground populations. But in the moister, warmer underground setting, *Culex pipiens molestus* populations bred all year round.

◀ A black-browed albatross (*Thalassarche melanophrys*) soars over the South Atlantic.

455

FIGURE 14.1 Speciation in mosquitoes in the London Underground. A large, deep shelter built alongside London's Underground subway system. This is an ideal breeding ground for the mosquito *Culex pipiens molestus.*

The difference between these populations spurred evolutionary biologists to ask whether they might in fact be different species. But how could they test this? As we will see throughout the course of this chapter, especially when we discuss what is called the biological species concept, one way that evolutionary biologists diagnose whether two populations are different species is based on gene flow: the movement of genes between populations. When gene flow is absent, two populations are often diagnosed as being members of different species—or on their way to becoming such. Given the dramatic life history differences between above- and belowground London mosquito populations, Katharine Byrne and Richard Nichols had reason to suspect these were different species, or at least that they were on the evolutionary path toward becoming separate species. Byrne and Nichols hypothesized that when they examined gene flow between populations, they would find little if any gene flow (Byrne and Nichols 1999).

The researchers examined 20 populations, and they found *no* gene flow between the above- and belowground populations. This held true when comparing an aboveground group to a distant underground population (many kilometers away), but there was also no gene flow even when the comparison was between aboveground and underground populations that were very close to one another (on the scale of just 100 meters apart). And, when Byrne and Nichols undertook laboratory breeding experiments, they found that *all* mating crosses between mosquitoes from underground populations produced viable and fertile offspring, while crosses between aboveground and underground individuals produced *no* offspring at all.

The aboveground and belowground forms are genetically distinct, and they fail to produce viable offspring when crossed. Aboveground populations prefer to bite birds, and they don't breed in the winter, whereas underground populations prefer to bite mammals and breed all year long. Using gene flow as a diagnostic, the above- and belowground populations are different species, or at the very least on the path to becoming such. This work on *C. pipiens* also demonstrates nicely that

TABLE 14.1		
Differences in the Biology of *Culex pipiens* and *Culex pipiens molestus*		
Trait	*C. pipiens molestus*	*C. pipiens*
Breeding site	Underground	Above ground
Mating	In confined spaces	Not in confined spaces
Host preference	Bites mammals	Bites birds
Egg production	No blood meal needed to lay eggs	Blood meal needed to lay eggs
Life cycle	Active all year	Dormant in winter

Adapted from Byrne and Nichols (1999).

when evolutionary biologists observe the speciation process in real time, they can develop and test hypotheses about speciation using tools that are readily available. As we will see, other methods have been developed to test hypotheses about speciation events in the more distant past.

In this chapter, we will examine:

- What is a species?

- How does speciation occur?

- What creates reproductive isolation among populations?

- What do evolutionary biologists know about the genetics of speciation?

- How can we trace the evolutionary history of humans, and what has been the history of speciation in the hominid lineage?

14.1 The Species Problem

Charles Darwin chose the title of his classic book, *On the Origin of Species*, with some care. It is sometimes easy to forget that Darwin developed his theory of natural selection largely in an effort to understand what is often referred to as the "species problem"—namely, how can we account for the vast array of different life-forms that have inhabited Earth for the last 4 billion years? This requires us to answer two separate questions. The first is the question of what a species is. The second is the question of how we identify species and delineate species boundaries in nature (de Queiroz 2007).

What Is a Species?

When evolutionary biologists refer to a group of organisms as a species, the fundamental underlying notion is that this group forms a lineage that has a distinct evolutionary fate from other lineages.

This is the **evolutionary species concept**, first proposed by George Gaylord Simpson and then modified by E. O. Wiley (Simpson 1961; Wiley 1978): "A species is a lineage of . . . populations which maintains its identity from other such lineages and which has its own evolutionary tendencies and historical fate" (Wiley 1978, p. 17). This definition puts evolution front and center. The key attributes that make a group of populations into a species are their shared past evolutionary history and their common future evolutionary fate—at least until this species itself bifurcates to form new descendant species. Notice that this definition is inherently phylogenetic: A species is a group of populations that have a shared past and will have a shared future on a phylogenetic tree.

The evolutionary species concept defines what a species is, and what role species play in evolutionary history, but it does not offer particularly useful practical advice on how we should go about identifying species and drawing species boundaries in the study of natural populations. To that end, evolutionary biologists have developed a number of diagnostic approaches to decide whether populations are or are not members of the same species. These approaches include the phenetic species concept, the biological species concept, and the phylogenetic species concept.

FIGURE 14.2 The phenetic species concept. The phenetic species concept exploits the clustering of individuals or populations in phenotype space to draw species boundaries between clusters. The vertical and horizontal axes may each represent a single phenotypic trait or multiple phenotypic traits.

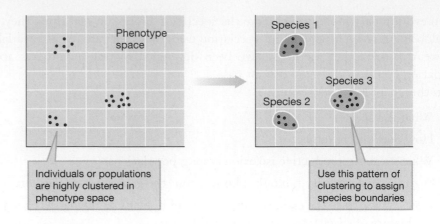

Identifying Species

As a general empirical observation, organisms are clustered together in phenotype space. If you sampled a large number of big cats, you would find many individuals that look like what we call lions, and many that look like what we call tigers, but few, if any, that look like something midway between the two. The **phenetic species concept** takes advantage of this fact, drawing species boundaries around clusters of phenotypically similar individuals or populations (Figure 14.2) (Michener and Sokol 1957; Cain and Harrison 1960; Sokol and Sneath 1963). A similar process can be applied at higher levels of taxonomic organization to delineate genera, families, orders, and other taxonomic levels.

Historically, the phenetic species concept was used by *numerical taxonomists*— scientists who use statistical analyses of multiple traits to classify organisms (Gilmour 1937; Sturtevant 1939; Rogers and Tanimoto 1960; Sokol 1985). At the most basic level, numerical taxonomists examined large data sets composed of measurements of many traits in many individuals, over many populations, and searched for patterns in these data. In particular, they used sophisticated computational algorithms to search for statistically meaningful groupings or clusters, and then they used such clusters to delineate species boundaries. Figure 14.3 shows an example in which this approach was used to classify shrubs from nine populations into three species.

The phenetic species concept remains in common use today, especially in the classification of plants and microorganisms (Sneath 1995). Paleontologists, who primarily work with fossil remains, also use this method when analyzing their data. Although fossil remains are often very fragile, it is possible to take measurements on various traits of fossilized remains (for example, tooth height, depth of ridges on teeth, and so on).

One of the challenges associated with the phenetic species concept is how to weigh the relative importance of the characters or traits used to delineate species boundaries. Should all traits be viewed as equally important in classifying organisms, or should some traits be weighed more heavily because they are particularly important? Early numerical taxonomists tended to assign equal weights to all characters they measured, but this approach was quickly abandoned by some, in favor of weighing certain characters more heavily than others (Cain and Harrison 1960).

The **biological species concept**, first introduced by Ernst Mayr, takes a very different approach to identifying species. Under the biological species concept, a species is composed of "groups of actually or potentially interbreeding populations

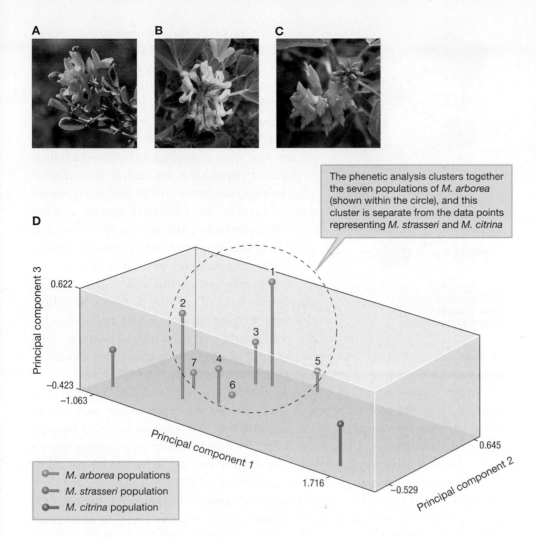

A **B** **C**

D

The phenetic analysis clusters together the seven populations of *M. arborea* (shown within the circle), and this cluster is separate from the data points representing *M. strasseri* and *M. citrina*

Principal component 3

0.622

−0.423
−1.063

1
2
3
7 4
6
5

Principal component 1

1.716 −0.529

0.645

Principal component 2

M. arborea populations
M. strasseri population
M. citrina population

FIGURE 14.3 Applying the phenetic species concept. Researchers measured twelve traits on individual plants from nine populations of shrubs from the genus *Medicago:* seven populations of (**A**) *M. arborea* (a widespread species), and a single population from each of the less widely distributed *M. citrina* (**B**) and *M. strasseri* (**C**). (**D**) They then distilled the twelve traits into what are called principal components—a statistical measure that groups a large number of different traits into a small number of variables. The *x*-, *y*-, and *z*-axes are the principal components. Part D adapted from González-Andrés et al. (1999).

which are reproductively isolated from other such groups" (Mayr 1942, 1982, 2002; Beurton 2002). Thus, under the biological species approach, it is the pattern of gene flow, rather than the pattern of phenetic similarity, that determines species boundaries (Figure 14.4). In diagnosing what constitutes a species, the biological species approach looks directly to the evolutionary mechanism—gene flow—responsible for the "shared evolutionary fate" that is fundamental to the concept of species. As a result, the biological species concept is not based on attributes of the individuals, but rather it delineates species by properties possessed by *populations*.

If individuals in one population are capable of mating with individuals in another population, then individuals in both populations are part of the same species, and they are said to share the same gene pool. If populations are reproductively isolated from one another—recall our discussion of the mosquito populations in London—then the individuals in such populations are not considered to be part of the same species.

A major practical difficulty with the biological species concept is that it is very hard to apply this concept to extinct species that are known only from paleontological evidence. Although reproductive isolation can sometimes be inferred from the distribution and form of fossils, this is not often the case.

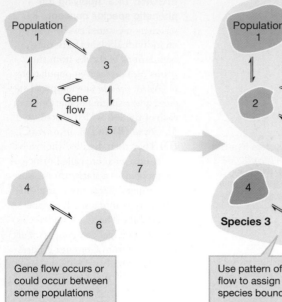

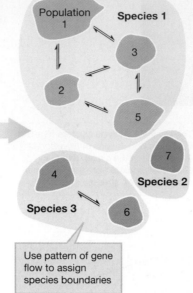

FIGURE 14.4 The biological species concept. The biological species concept uses the presence or absence of gene flow among populations to delineate species boundaries.

Gene flow occurs or could occur between some populations

Use pattern of gene flow to assign species boundaries

Another problem for the biological species concept is the occasional hybridization events between individuals in populations that are, for all practical purposes, reproductively isolated. If individuals in population 1 consistently mate with those in population 2, individuals in these populations are classified as part of the same species. But what if matings between individuals in different populations are rare or nonexistent? The two populations can still be part of a common species, because the biological species concept allows the populations to be *potentially interbreeding*. What if the offspring produced by cross-population matings are nonviable or infertile? In this case, we clearly have two species. But what if the offspring merely have reduced viability or reduced fertility? How rare do cross-population matings have to be, and how poorly must the hybrid offspring fare, before we can say that the two populations are two separate species? The answers to questions such as these are murky, and no clear consensus exists on this issue.

Another major limitation of the biological species definition is that it is restricted to sexual species. With its emphasis on the reproductive isolation of populations, the biological species concept makes little sense as a species concept for asexual organisms. As Ernst Mayr notes, "[i]n an asexually reproducing species every individual and every clone is reproductively isolated. It would be absurd to call each of them a separate species" (Mayr 1982, p. 283).

A third species concept is the **phylogenetic species concept**. Like the phenetic species concept, this approach looks to character differences in order to distinguish among species, but it does so in a different way. The basic problem in distinguishing species remains the same: How do we determine whether two groups are behaving as evolutionary species that are able to maintain distinct identities so that they have their own evolutionary histories? If two groups have been separated long enough to have diverged and produced distinguishing characters, they must have been reproductively isolated from one another and, as evidenced by these distinguishing characters, they must have already experienced unique evolutionary histories.

But what characters are the right characters to use in making such distinctions? The phylogenetic species concept proposes that we look to phylogeny to answer this question. According to this approach, we draw species boundaries using shared derived characters that are unique to one monophyletic group and absent from all other populations in the phylogeny. These characters can then be used to distinguish among species. In particular, we define a phylogenetic species as *the smallest monophyletic group distinguished by a shared derived character*. Figure 14.5 illustrates the basic way in which shared derived characters can be used to distinguish among species.

By looking at shared derived characters that distinguish monophyletic groups, the phylogenetic species concept selects appropriate characters for classifying species. Characters

FIGURE 14.5 The phylogenetic species concept. The phylogenetic species concept uses shared derived traits to draw species boundaries between monophyletic groups. Each tick mark represents a trait that is a shared derived character—this allows us to diagnose five different species here.

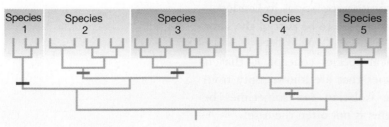

that are polymorphic within a population will not form monophyletic clades, and therefore they should not be used to define species boundaries under the phylogenetic species concept (Figure 14.6). In contrast, characters that are unique to a population or set of populations and that are also ubiquitous within those populations are ideal for drawing species boundaries; these characters will define monophyletic groups, and thus they can be employed by the phylogenetic species concept in assigning species boundaries.

By requiring that a species be the *smallest* distinguishable monophyletic clade, the phylogenetic species concept also determines an appropriate taxonomic level at which to draw species boundaries. The trait "has fur and mammary glands" is a shared derived trait of the monophyletic clade of mammals, but we certainly would not want to say that all mammals are members of the same species. Instead, we can look for shared derived traits that distinguish smaller monophyletic groups—for example, spoken language and a dearth of body hair distinguish the monophyletic clade of humans from other primates.

Whereas the biological species concept requires information about gene flow to diagnose species, the phylogenetic species concept has no such requirement. In most cases, we would expect that a breakdown of gene flow would have occurred in populations that diverged enough that we can identify shared derived characters, but in some instances this may not yet have occurred (Cracraft 1989).

One of the major critiques of the phylogenetic species concept is that the traits it uses to distinguish among species do not have to be ecologically or physiologically significant. Thus, distinguishing traits can be minor characters of minimal significance. As a result, the phylogenetic species concept often divides up organisms into more fine-grained species categories than may seem appropriate, resulting in a far greater number of species than would be delineated by other species concepts. Moreover, the phylogenetic species concept does little to ensure that species considered separate at present will have separate evolutionary fates in the future. Because there is no requirement of restricted gene flow, members of two distinct phylogenetic species may be able to interbreed readily, which would enable the two species to fuse back into one species at some point in the future. Such events run strongly counter to our intuitions about what a species is under the evolutionary species concept.

Clearly, no one species concept will work for all organisms. But it is important to recognize that once we adopt an evolutionary species concept to define what a species fundamentally is, we can then use the phenetic, biological, and phylogenetic species concepts to delineate species in nature. Each takes a somewhat different diagnostic approach: The phenetic species concept looks for clusters of phenotypic characters; the biological species concept looks at the presence or absence of gene flow; the phylogenetic species concept relies on shared derived traits of monophyletic groups. But most of the time, all three species concepts will readily agree on species boundaries. Populations that belong to different species typically show large phenotypic differences, absence of gene flow, *and* shared derived traits. These species concepts will give different answers only in relatively special

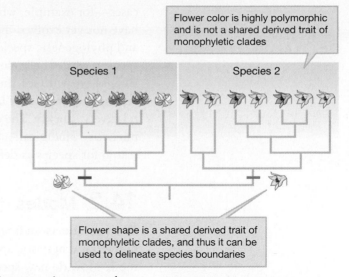

FIGURE 14.6 Polymorphic characters are not used by the phylogenetic species concept. Characters that are polymorphic within populations are not used by the phylogenetic species concept because they are not shared derived characters of monophyletic clades.

cases—for example, when populations have had time to diverge in characters but have not yet evolved mechanisms that prevent gene flow. In these cases, phenetic and phylogenetic species concepts will tend to classify the populations as separate species, while the biological species concept will classify them as a single species. But, as we will see shortly, such cases should be transient; there are numerous reasons to expect that barriers to gene flow between such populations will evolve relatively quickly. The overarching point here is that the tools associated with the phenetic, biological, and phylogenetic concepts provide us with great power to search for species as defined by the evolutionary species concept.

14.2 Modes of Speciation

With an understanding of how evolutionary biologists define a species, and of the numerous diagnostic approaches used to identify species, we now move to a related topic: How do new species originate? In other words, how can we understand the origin of species—the question that occupied, indeed tormented, Darwin for so many years? All around us we see an astonishing array of different life-forms. How could such a diversity of different species come to be? What models of speciation have evolutionary biologists developed? What predictions do these models make, and how have the models been tested (Otte and Endler 1989; Coyne and Orr 2004)?

We answer these questions by examining three models of speciation—allopatric, parapatric, and sympatric speciation. These three models are distinguished from one another by the relative geographic positions of populations undergoing speciation. In **allopatric speciation**, the process of speciation takes place in populations that are geographically isolated from one another. In **parapatric speciation**, incipient species—diverging populations on the path to speciation—have distributions that abut one another. In **sympatric speciation**, populations diverge into new species while in the same location (Figure 14.7).

FIGURE 14.7 Different types of speciation. Two forms of allopatric speciation are shown in the first and second column, parapatric speciation is shown in the third column, and sympatric speciation is represented in the last column.

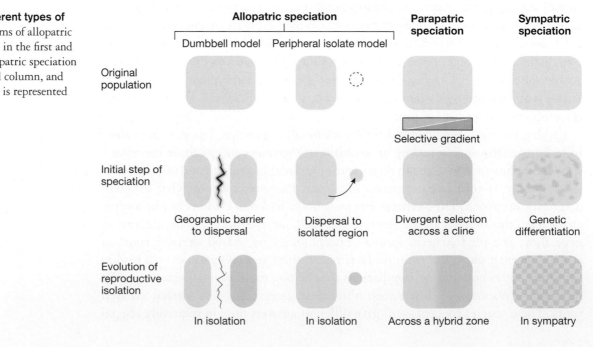

Allopatric Speciation

The central premise underlying allopatric speciation is that reproductive isolating mechanisms evolve in populations while they are geographically isolated. This geographic isolation may be a consequence of a physical barrier such as a mountain range, a river, an ocean, a desert, or some other barrier. In a moment, we will subdivide allopatric speciation into two related models, but for now let us examine one feature that *all* allopatric models of speciation have in common.

In all allopatric speciation models, the processes of genetic drift, mutation, and natural selection cause populations to diverge from one another. In Chapter 8, we saw how genetic drift leads to divergence between two populations as different alleles are fixed by chance in the two populations. Different mutations are likely to arise by chance in two allopatric populations, further augmenting their differences. Finally, no two geographically isolated populations will experience exactly the same selective conditions, and any differences in selective conditions can cause the populations to diverge by natural selection. In the long run, these processes may lead to multiple forms of reproductive isolation between these populations. This is because gene flow between geographically isolated populations may be permanently eradicated when the members of one population lose the ability to breed successfully with members of the other population due to differences in geographic range, genetics, behavior, or reproductive physiology. In Section 14.3, we will look at some of the mechanisms by which reproductive isolation occurs. For now, the key point is that once gene flow becomes impossible, the populations no longer share a common evolutionary fate, and thus this process can result in the formation of new species.

Allopatric speciation is often subdivided into a **dumbbell model** and a **peripheral isolate model**. In the dumbbell model of allopatric speciation, an initially large population is subdivided into new populations that are themselves still relatively large. In the peripheral isolate model, the populations that are geographically isolated from one another differ in size, with one large population and one or several smaller populations. A classic example of this form of allopatry would be a mainland and surrounding islands, when islands are populated by individuals who have dispersed from the mainland across some barrier like a body of water. One of the most important differences between dumbbell and peripheral isolate models pertains to the role of genetic drift in driving divergence between the populations. In the dumbbell model, the descendant populations are each relatively large in size, making it unlikely that drift dramatically affects divergence. By contrast, in the peripheral isolate model, a peripheral population may be founded by a relatively small number of individuals, resulting in strong founder effects. Moreover, the net population size in the peripheral population may be much smaller than that of the progenitor population, resulting in accelerated genetic drift.

Allopatry via the Isthmus of Panama

Allopatric speciation has been studied in the shrimp genus *Alpheus*. Approximately 3 million years ago, the Isthmus of Panama isolated populations of aquatic organisms in the Caribbean Sea from those in the Eastern Pacific. In a series of studies, Nancy Knowlton and her colleagues studied pairs of sister species of *Alpheus* snapping shrimp (recall from Chapter 4 that sister species share an immediate common ancestor on a phylogenetic tree). In each of these sister species pairs, members

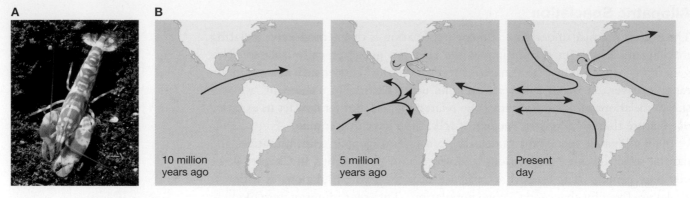

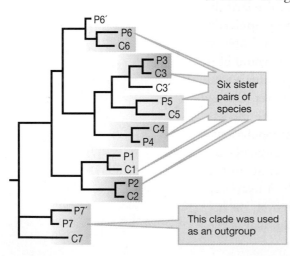

FIGURE 14.8 Allopatric speciation in shrimp. (A) A shrimp from the genus *Alpheus*. **(B)** The Isthmus of Panama: 10 million years ago, 5 million years ago, and in the present. Arrows indicate ocean currents. The Isthmus of Panama separated sister species of shrimp on the Caribbean and Pacific sides of the Isthmus of Panama, leading to allopatric speciation. Part B adapted from Haug et al. (2004).

of one of the pair lived on the Caribbean side of the isthmus, while members of the other pair lived on the Pacific side, and so we refer to these as *trans-isthmus sister pairs* (Knowlton 1993; Knowlton et al. 1993; Knowlton and Weigt 1998) (Figure 14.8).

Knowlton's team used two different molecular genetic estimates of the divergence time for a sister pair of species, and they found that sister species varied widely in their divergence times—ranging from 18 million years ago (before the Isthmus of Panama began forming) through 9 million years ago (when terrestrial mammals first began crossing from North America to South America) to 3 million years ago (when the Isthmus of Panama was complete) (Figure 14.9).

If reproductive isolation is linked to how long sister pairs have been geographically isolated, then we would expect to see a greater degree of reproductive isolation in sister pairs that had been separated 18 million years ago than those that had been separated 3 million years ago. To test this, Knowlton's team used a series of aggressive behaviors as an indicator of reproductive isolation: the more aggression displayed, the more reproductive isolation was assumed. Similarly, the more "tolerance" shown in the presence of a shrimp from a sister species, the less reproductive isolation between populations was assumed. They found that tolerance decreased and aggression increased in trans-isthmus sister pairs as a function of how long they had been geographically isolated from one another by the Isthmus of Panama. Equally important, they found that, although sister species were phylogenetically closely related, only 1% of matings between trans-isthmus pairs produced viable clutches of offspring compared to 60% of matings between different species on the same side of the isthmus (which served as control pairs).

These findings strongly suggest that speciation in snapping shrimp has largely occurred among populations in allopatry—the longer that sister species had been geographically isolated from one another, the greater the extent of behavioral and genetic divergence.

FIGURE 14.9 Sister species of shrimp. A phylogenetic tree of the genus *Alpheus* based on mitochondrial DNA (mtDNA) data. P = species from the Pacific side of the isthmus; C = species from the Caribbean side of the isthmus. Adapted from Knowlton et al. (1993).

The Peripheral Isolate Model in Black Spruce and Red Spruce Trees

In Chapter 8, we examined Isabelle Gamache's work on genetic variation and founder effects *within* populations of black spruce (Gamache et al. 2003). In the original study we discussed, Gamache and her colleagues examined mitochondrial

FIGURE 14.10 Progenitor–derivative species. The closely related red spruce (**A**) and black spruce (**B**) are thought to be a progenitor–derivative pair, with black spruce the progenitor species and red spruce the derivative species.

DNA (mtDNA) migration via wind-dispersed seeds, and they found that mtDNA distribution was more restricted and localized than nuclear DNA migration via pollen (and seed) dispersion. Here we will look at the same research group's follow-up study, which examined genetic variation in both black spruce (*Picea mariana*) and red spruce (*Picea rubens*) (Jaramillo-Correa and Bousquet 2003) in the context of allopatric speciation.

In many plant species, allopatric speciation leads to so-called progenitor–derivative species pairs. The derivative species forms when a small subgroup of the progenitor species becomes geographically isolated and begins to diverge from the original population through the process of peripheral isolate speciation (Gottlieb 1973; Gottlieb et al. 1985; Witter 1990). In such pairs, the progenitor species typically does not change very much through time, but the derivative species does change significantly.

A number of lines of evidence led evolutionary biologists to hypothesize that black spruce and red spruce form a progenitor–derivative pair (Figure 14.10). The derivative species, red spruce, seems to have arisen from a southern population of black spruce, the progenitor species, which became geographically isolated from other black spruce populations at some point during the Pleistocene glaciations. There are a number of lines of evidence for this. First, the progenitor species, black spruce, has a much broader geographic distribution than the derivative species, red spruce. Second, both nuclear and mitochondrial DNA work show that red spruce has low genetic diversity when compared to black spruce (Hawley and Dehayes 1994; Jaramillo-Correa and Bousquet 2003). Third and most critically, researchers found no unique mitochondrial haplotypes in red spruce—all mitochondrial genetic variation in red spruce is a subset of that found in black spruce. This is what we would expect if red spruce evolved from a geographically isolated population of black spruce (Perron et al. 1995, 2000; Jaramillo-Correa et al. 2003) (Figure 14.11).

Parapatric Speciation

Parapatric speciation occurs when two adjacent populations diverge into separate species without a geographic barrier to dispersal (Mayr 1970). The core of the parapatric speciation idea is that some sort of **cline**—a spatial gradient in the frequency of phenotypes or genotypes—exists in nature because adjacent populations experience somewhat different selective conditions. A **hybrid zone**—

FIGURE 14.11 Progenitor–derivative species in spruce.
Black spruce populations (1–3, with black rim on circles) show much more genetic variation than red spruce populations (4–8, with red rim on circles). All mtDNA variation in red spruce is a subset of the mtDNA variation in black spruce. Adapted from Jaramillo-Correa and Bousquet (2003).

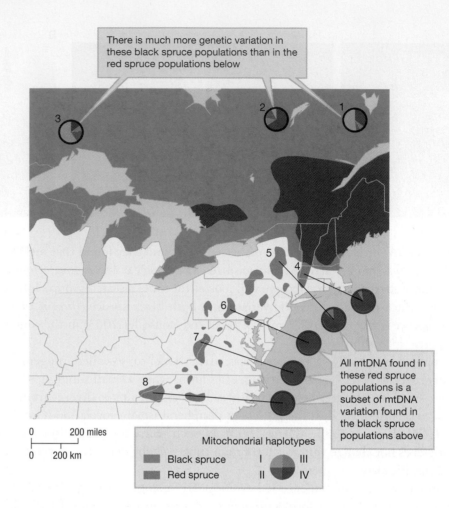

There is much more genetic variation in these black spruce populations than in the red spruce populations below

All mtDNA found in these red spruce populations is a subset of mtDNA variation found in the black spruce populations above

Mitochondrial haplotypes
Black spruce   I   III
Red spruce   II   IV

0   200 miles
0   200 km

an area in which diverging populations encounter each other, mate, and produce hybrid offspring—is essential in the parapatric speciation process (Harrison and Rand 1989; Hewitt 1989). By contrast, the allopatric and sympatric models do not include a hybrid zone during the speciation process.

The Hybrid Zone in Parapatric Speciation

In most parapatric speciation models, it is assumed that the hybrid zone between populations will eventually disappear, completing the speciation process (Bayzkin 1969; Moore 1977; Barton and Hewitt 1985). This can occur for many different reasons, the most common of which is that hybrid offspring may be at a selective disadvantage compared to offspring that come from within-population matings. This is because hybrid offspring possess a suite of traits that are not particularly well suited to life in any section of the cline, while offspring from within-population matings are well adapted to their respective environments. This generates selection for genetic, physiological, or behavioral **reproductive isolating mechanisms** that deter hybridization between the two populations, and which once in place may lead to the completion of the speciation process.

Yet, not all parapatric models assume that hybrid individuals are at a disadvantage. To see this, let's examine the work of Han Wang and his team on hybrid zones and parapatric speciation in big sagebrush (*Artemisia tridentata*) (McArthur et al. 1998; Wang et al. 1998, 1999; Byrd et al. 1999). Wang and his colleagues attempted to distinguish between two theories associated with hybrid zones. The *ecologically neutral dynamic equilibrium model* suggests that hybridization produces hybrids that are always inferior to nonhybrids. In contrast, the *ecologically dependent bounded hybrid superiority model* assumes a genotype-by-environment interaction, such that *in hybrid zones*, hybrids may have superior fitness to nonhybrids.

Wang and his team studied two parapatric subspecies of big sagebrush (Wang et al. 1997). In the mountains of Utah, basin big sagebrush (*Artemisia tridentata tridentata*) grows up to elevations of about 1800 meters, while mountain big sagebrush (*Artemisia tridentata vaseyana*) can be found at elevations above 1900 meters. Between 1800 and 1900 meters, the two subspecies form a narrow hybrid zone. To distinguish between the dynamic equilibrium and the bounded hybrid superiority models, Wang and his colleagues ran a series of reciprocal transplant experiments. In these experiments, mountain big sagebrush, basin big sagebrush, and hybrid sagebrush (taken from the hybrid zone rather than created as first-generation hybrids) were each raised in three different environments— below 1800 meters, above 1900 meters, and in the hybrid zone between 1800 and 1900 meters.

Wang and his team found strong support for the bounded hybrid superiority model. In experiments on seed survivorship, size, and flower number, they found a fascinating genotype-by-environment interaction. While hybrid individuals generally fared poorly in environments below 1800 m and above 1900 m, they had a higher fitness than either subspecies when all types were raised in the *hybrid zone* (Figure 14.12).

Although it is difficult to pinpoint why hybrids have higher fitness in the hybrid zone, it may in part be related to the fact that soil in the hybrid zone is not just a simple blend of soils from the mountain and basin areas; rather, this soil has its own unique, novel characteristics, suggesting that selection may have favored hybrids that have been produced in such soil for many, many generations (Wang et al. 1998). The parapatric speciation process under way in the big sagebrush may, over evolutionary time, result in three species rather than two—a basin species, a mountain species, and an intermediate-elevation species, where we currently see a hybrid zone.

Ring Species

It is sometimes difficult to make a clear distinction between allopatric and parapatric speciation. Consider the case in which individuals live in a series of populations that are connected to one another in a ringlike fashion, forming what is known as a ring species (Stebbins 1949; Dobzhansky 1958; Irwin et al. 2001) (Figure 14.13).

Theodosius Dobzhansky and Robert Stebbins described a beautiful example of a ring species

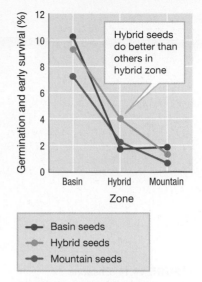

FIGURE 14.12 Bounded hybrid superiority. Mountain big sagebrush, basin big sagebrush, and hybrid sagebrush were each raised in three different environments: below 1800 meters, above 1900 meters, and in the hybrid zone between 1800 and 1900 meters. Germination and early survival rates in the seeds are shown here. Adapted from Wang et al. (1997).

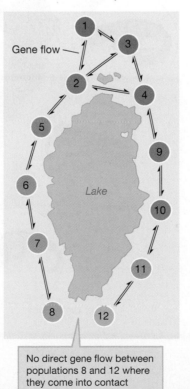

FIGURE 14.13 Ring species concept. An ancestral population (1) spreads down along both the east and west shores of a vast lake, resulting in a series of descendant populations (2–12). As a result of both selection and drift, populations along the west shore diverge from those along the east shore. Gene flow occurs between adjacent populations on each shore, but at the southern edge of the lake where west-shore population 8 comes into contact with east-shore population 12, these two populations have diverged so much that no direct gene flow occurs between them.

of *Ensatina eschscholtzii* salamanders. This species of lungless salamanders occurs in a series of populations that range from British Columbia in Canada to Baja California in Mexico (Stebbins 1949). These populations have subsequently been studied in great depth by David Wake and his colleagues (Wake 1997; Kuchta et al. 2009).

Ensatina eschscholtzii originated in northern California and southern Oregon and then, approximately 21.5 million years ago, they began to spread south along two separate but parallel fronts. One group of populations moved south along the coastal mountain range; further inland, a second group expanded south along the Sierra Nevada mountain range. These two groups were separated from one another by the hot, dry Sacramento and San Joaquin valleys, resulting in the ring distribution that we see today (Figure 14.14). Along these two ranges, salamanders show an impressive degree of phenotypic variability in skin

FIGURE 14.14 *Ensatina eschscholtzii* salamanders have been studied as a ring species. (**A**) *Ensatina eschscholtzii*. (**B**) Along the two ridges they inhabit in California, these salamanders show a tremendous amount of phenotypic variability in skin color and blotch pattern. Part B adapted from Thelander (1994).

A

B

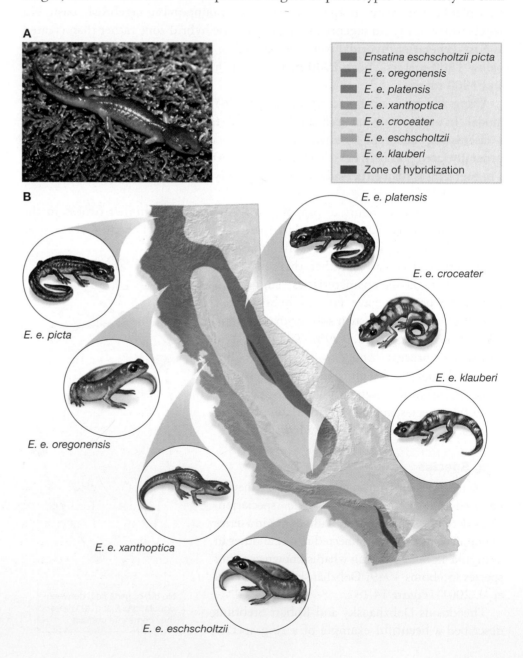

Legend:
- *Ensatina eschscholtzii picta*
- *E. e. oregonensis*
- *E. e. platensis*
- *E. e. xanthoptica*
- *E. e. croceater*
- *E. e. eschscholtzii*
- *E. e. klauberi*
- Zone of hybridization

E. e. picta
E. e. oregonensis
E. e. xanthoptica
E. e. platensis
E. e. croceater
E. e. klauberi
E. e. eschscholtzii

coloration—differences in hue, blotchiness, the number of colored stripes, and other characters.

To measure gene flow across salamander populations, David Wake and his colleagues collected skin samples from salamanders and extracted DNA from the samples so that they could compare mitochondrial DNA sequence data. They compared DNA in 385 individuals from 224 different populations along both ranges that the salamanders inhabit (Kuchta et al. 2009). Their results suggest that, while there is some gene flow between populations near one another along either ridge, gene flow is not continuous along the ring. Indeed, as one would expect if the ring originated at the northern tip, the amount of gene flow decreases along both ranges as one moves south, leading to southern populations at the end of each ridge being more genetically distinct from one another than from other populations in the ring (Wake and Yanev 1986; Wake et al. 1986; Kuchta et al. 2009). When the DNA results are mapped onto a phylogeny, separate coastal and inland clades emerge.

Sympatric Speciation

Sympatric speciation occurs when no geographic boundary exists between diverging populations. For evolutionary biologists, sympatric speciation is the most difficult of the three forms of speciation to understand. The difficulty stems from the fact that, without some sort of geographic barrier or some sort of gradient in selective conditions, some other mechanism must drive a single species to split into two species. One possibility is that speciation may be driven by resource competition; we explore this mechanism in Box 14.1. Other alternatives involve some form of reproductive isolation that arises without geographic separation. We consider some of these possibilities below.

Sympatric Speciation in Cichlids

Here we will examine two species of cichlid fish found in Nicaragua, at a site called Lake Apoyo (Figure 14.16). Lake Apoyo is a small lake, with a diameter of about 5 km, that is fairly shallow and quite homogeneous in appearance throughout. Geological data suggest that this lake is also young—it originated about 23,000 years ago.

Lake Apoyo contains two species of cichlids—the Midas cichlid (*Amphilophus citrinellus*) and the Arrow cichlid (*Amphilophus zaliosus*). Whereas the Midas cichlid is found in many Nicaraguan lakes, the Arrow cichlid is only found in Lake Apoyo.

Marta Barluenga and her colleagues hypothesized that the Arrow cichlid arose sympatrically from an ancestral population of the Midas cichlid at Lake Apoyo (Barluenga et al. 2006). To test this hypothesis, they used an array of phylogeographic, population-genetic, ecological, and morphological tools to examine whether the two species diverged sympatrically. They began by comparing 840 base pairs of mitochondrial DNA in hundreds of Arrow and Midas cichlid fish. This comparison revealed two remarkable pieces of information: (1) The Midas and Arrow cichlids form a monophyletic clade, suggesting that the Arrow cichlid arose sympatrically in Lake Apoyo, and (2) not even one

BOX 14.1 Sympatric Speciation: A Resource Competition Model

Building on earlier mathematical models, Jon Seger hypothesized that resource competition may facilitate sympatric speciation (Levene 1953; Maynard Smith 1966; Rosenzweig 1978; Bengtsson 1979; Gibbons 1979; Seger 1985). Consider a phenotypic trait that is important in resource competition—following Seger's lead, let's make this trait beak size in birds,

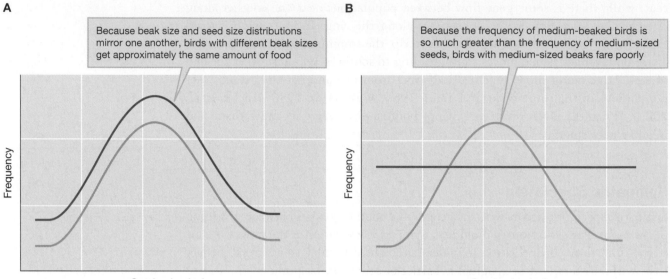

A Because beak size and seed size distributions mirror one another, birds with different beak sizes get approximately the same amount of food

B Because the frequency of medium-beaked birds is so much greater than the frequency of medium-sized seeds, birds with medium-sized beaks fare poorly

FIGURE 14.15 Two distributions of seed size and beak size. **(A)** Here both seed size (red) and beak size (blue) are normally distributed, and all birds receive approximately the same amount of food. **(B)** Here beak size (blue) is normally distributed, but seed size (red) has a flat distribution. Birds with large and small beaks get more food per bird than birds with average-sized beaks.

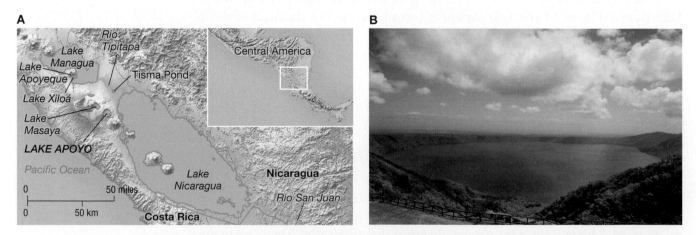

FIGURE 14.16 Lake Apoyo. **(A)** Map of the Nicaraguan lakes (including Lake Apoyo, Lake Managua, and Lake Nicaragua). Adapted from Barluenga et al. (2006). **(B)** A photograph of Lake Apoyo showing its steep wall.

and let's assume that the size of seeds consumed by birds is a function of beak size. If beak size is heritable and controlled by many genes, we can expect a normal distribution of beak size in our population of birds. If we assume that the distribution of seed size follows a matching normal distribution, individuals get about the same amount of food regardless of what beak size they have (Figure 14.15A).

But what if our *seed distribution* is flat rather than bell-shaped (normally distributed)? If beak size is still normally distributed, birds with either very large or very small beaks will get more food, and so both very large and very small beaks will be favored by natural selection. This is because, if our seed distribution is flat, there are approximately the same number of small, medium, and large seeds. But because beak size is normally distributed, there are fewer large and small-beaked birds than birds with average-sized beaks.

Large- and small-beaked birds then get more food per bird than birds with average-sized beaks, and disruptive selection will favor the extreme beak phenotypes (Figure 14.15B). Our population will then start to diverge into large- and small-beaked individuals. This is the first step toward sympatric speciation in the Seger model. The second step involves the emergence of reproductive isolation between very large- and very small-beaked birds. For speciation to occur, large-beaked birds need to mate with large-beaked birds and small-beaked birds need to mate with small-beaked birds—that is, there must be some sort of *positive assortative mating*, where like mate with like. This assortative mating could complete the process of disruptive selection and result in sympatric speciation. The result would then be a large-beaked and a small-beaked species of birds.

The problem with the above scenario is that it requires assortative mating to emerge at just the right time—that is, after disruptive selection has begun to pull apart our two types of birds (small- and large-beaked birds). The odds of assortative mating coming about at just this time are quite small. There is, however, a much simpler process that can allow sympatric speciation.

Suppose that large and small seeds are spatially segregated. For the sake of argument, imagine that plants producing large seeds prefer shady habitats, whereas plants that produce small seeds prefer more sun. Our birds would then distribute themselves according to where the food is, and we would end up with large-beaked birds spending their time in shady habitats, and small-beaked birds preferring sunny habitats. Now, even if birds don't have a preference to mate with others with similar genotypes to their own, and instead they simply choose mates randomly from those individuals around them, assortative mating occurs. This allows the sympatric speciation process to proceed.

mitochondrial haplotype was found in any other Nicaraguan lake that was the same as those found in the two Lake Apoyo species. This suggests that there was most likely a single colonization of Lake Apoyo and no further contact between fish in this lake and other lakes.

A number of lines of evidence converge to suggest that the speciation event responsible for forming these two species occurred in sympatry rather than allopatry. First, as mentioned earlier, Lake Apoyo is small, shallow, and homogeneous. Taken together, these geological and geographic characteristics of the lake make it unlikely that there is a physical boundary that Barluenga and her colleagues did not observe. Because Lake Apoyo is young, it is unlikely that such a boundary once existed, but has since disappeared. In addition, population-genetic data provide evidence that these two species were somehow dividing up Lake Apoyo in ways that were not obvious, even after Barluenga and her team observed these two species across the entire lake. But how? If there is no physical boundary to prevent gene flow between these two species, what does prevent gene flow?

FIGURE 14.17 Sympatric specia-tion, habitat specialization, and ecological specialization in two species of Lake Apoyo fish. (**A**) The Midas cichlid (*Amphilophus citrinellus*) and the Arrow cichlid (*A. zaliosus*) are morphologically quite different from one another. (**B**) The Midas and Arrow cichlids have different diets as a function of their different habitat preferences. Adapted from Barluenga et al. (2006).

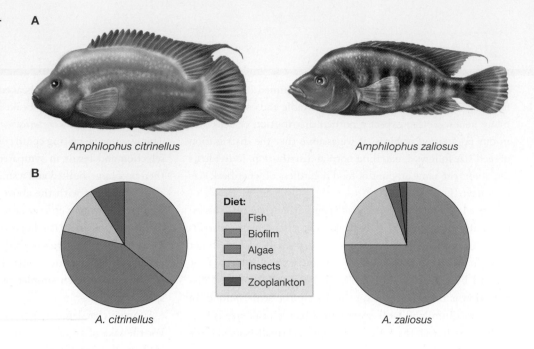

The answer appears to center on habitat and ecological specialization (Figure 14.17). The Midas cichlid is "high-bodied" and relatively short in length. This species shows the morphology classically associated with living at the bottom of lakes (known as a *benthic* body form). Diet analysis of the Midas cichlid supports the idea that this species spends its time primarily near the bottom of the lake. The Arrow cichlid is low-bodied and longer, and has a morphology associated with foraging in open water (known as a *limnetic* body form). Again, diet analyses support this contention. Moreover, Barluenga and her colleagues used behavioral experiments to demonstrate that both Arrow cichlids and Midas cichlids prefer mates from their own species, suggesting reproductive isolation between the Arrow and Midas cichlids. This partitioning of the lake into bottom-of-the-lake and open-water areas appears to be the mechanism by which sympatric speciation occurred in Lake Apoyo.

Sympatric Speciation in the Apple Maggot Fly

Perhaps the best-studied case of sympatric speciation is that of the apple maggot fly, *Rhagoletis pomonella*. Just 5 years after Darwin published *On the Origin of Species*, Benjamin Walsh suggested that sympatric speciation was common in insects such as *R. pomonella* (Walsh 1864, 1867). Walsh noted a dramatic shift in the host species of *R. pomonella* from hawthorn shrubs/trees (*Crataegus*) to both hawthorn and domesticated apple trees (*Malus pumila*) (Figure 14.18). These different forms of *R. pomonella*—known as the hawthorn "race" and the apple "race"—are arguably different species, or at the very least diverging and on the path to becoming different species.

Speciation in the apple maggot fly appears to have occurred sympatrically, based on which trees were hosts to the apple maggot fly's eggs. The domestic apple tree was introduced into North America about 400 years ago, and it has

FIGURE 14.18 Divergence in apple maggot flies. (A) The apple maggot fly, (B) a hawthorn tree, and (C) an apple tree.

always occurred sympatrically with the hawthorn tree; both trees have been hosts to apple maggot fly larvae. Because of their economic importance, apple trees have been closely monitored, and so we know that *R. pomonella* only began using the apple tree as a host about 140 years ago—before that period it was only found on hawthorn trees (Bush 1969, 1975; Berlocher and Feder 2002). Evolutionary biologists wanted to test whether differences between the races of *R. pomonella*—in particular, differences in the breeding seasons of the different races of flies—resulted from different selective conditions imposed by their host trees. Over the years, a series of experiments by numerous research groups have addressed this question. The key to understanding both the differences between races of these flies and how these differences are tied to sympatric speciation in *R. pomonella* is the different fruiting times of their host—apple trees produce fruit 3 to 4 weeks earlier than hawthorn trees.

Researchers hypothesized that the difference in the host trees' fruiting times causes the maggot flies in apples and hawthorn fruit to emerge at different times, which reduces the gene flow between the two populations. This in turn produces significant genetic differences between the hawthorn and apple races of the apple maggot (Boller and Prokopy 1976; Feder et al. 1997; Feder and Filchak 1999; Filchak et al. 2000). Indeed, host specificity reduces gene flow between the apple and hawthorn races to 4–6% each generation (Feder et al. 1994), suggesting that sympatric races of *R. pomonella* are indeed diverging, and potentially on the path to becoming separate species.

14.3 Reproductive Isolating Mechanisms and the Genetics of Speciation

Evolutionary biologists and ecologists have devoted considerable effort to understanding how reproductive isolation can arise and lead to the origin of new species (Du Rietz 1930; Mayr 1942; Dobzhansky 1970). What drives reproductive isolation between populations? Why don't two separate species merge back together if they overlap in range (Box 14.2)? In his classic book, *Genetics of the Evolutionary Process*, Theodosius Dobzhansky divided reproductive

BOX 14.2 Secondary Contact

What happens when two populations have been diverging from one another in allopatry, but are then reunited before reproductive isolation is complete? For example, imagine that population 1 and population 2 have been geographically isolated for many generations, during which time natural selection and genetic drift have caused significant, but not complete, reproductive isolation between the individuals in these populations. When the geographic isolation comes to an end, will our two populations complete the divergence process or remain part of a single species?

The answer depends on the extent of the reproductive isolating mechanisms that have evolved during allopatry. If the reproductive isolating mechanisms are sufficiently weak as to allow free interbreeding across our reunited populations, and such mating produces offspring that are not at some fitness disadvantage when compared to offspring derived from matings between individuals from within either population, then the speciation process halts and a single species remains. But if the reproductive isolating mechanisms that developed during allopatry are such that offspring from matings between population 1 and population 2 individuals are at a selective disadvantage, then the speciation process may continue, and over time we may end up with two different species. This process is referred to as secondary reinforcement (Figure 14.19).

One clue that evolutionary biologists can use to infer that secondary reinforcement has occurred is reproductive character displacement (RCD). RCD is defined as the case in which a reproductive trait, such as genital length, is less similar when two incipient species overlap (in areas of sympatry) than when these two species do not overlap (areas of allopatry). The basic premise underlying RCD is that, if hybrids are at a disadvantage, natural selection should act more intensely on the ability to mate with conspecifics in areas of sympatry than in areas of allopatry (Brown and Wilson 1956).

A fascinating case of RCD has been documented in two Japanese species of land snails, *Satsuma eucosmia* and *Satsuma largillierti*. In some areas of Japan, these two species live sympatrically; in other areas of Japan, populations of the species live in allopatry. Importantly, prior work has suggested that

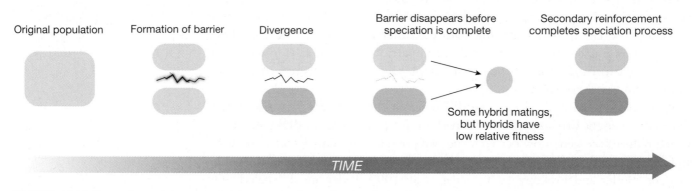

FIGURE 14.19 Secondary reinforcement. The process of secondary reinforcement in populations that diverge in allopatry may complete the speciation process.

isolating mechanisms into two categories: **prezygotic isolating mechanisms** and **postzygotic isolating mechanisms** (Table 14.2) (Dobzhansky 1970).

Prezygotic isolating mechanisms prevent or deter individuals from different populations from mating with one another, or prevent fertilization from occurring if such a mating does take place. Postzygotic isolating mechanisms operate after fertilization and conception. With postzygotic mechanisms in place, a mating between individuals from different populations may lead to successful fertilization, but the embryo may not survive. If it does, it may be either sterile or have dramatically reduced fitness. From an evolutionary perspective, even areas of modern sympatry in these species appear to represent secondary contact between *S. eucosmia* and *S. largillierti* (Kameda et al. 2007). Because of its obvious implications for mating, Yuichi Kameda used penis length as the reproductive trait of interest, and he measured this character in individual snails who lived in either sympatric or allopatric *Satsuma* populations. Average penis length showed greater differences between these species when they lived in areas of sympatry, suggesting RCD in this trait, and indicating that penis length may have been a key trait in the process of secondary reinforcement in *S. eucosmia* and *S. largillierti* (Kameda et al. 2009) (Figure 14.20).

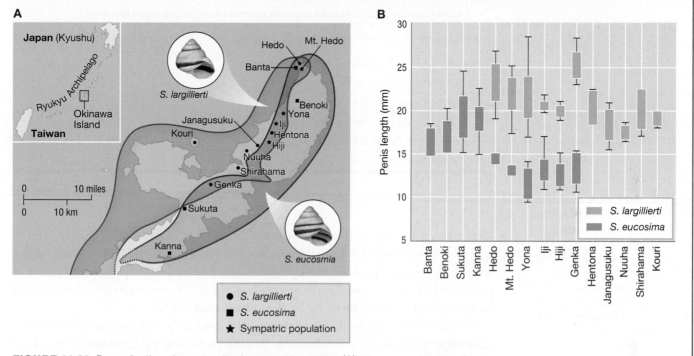

FIGURE 14.20 Reproductive character displacement in snails. **(A)** The geographic distributions of *Satsuma* snails in Japan. The blue line indicates the overall area occupied by *Satsuma largillierti*, and the red line indicates the overall area occupied by *Satsuma eucosmia*. Particular localities inhabited by *S. eucosmia* (squares) and *S. largillierti* (circles) are also shown on the map. Sympatric populations of the two species are depicted by stars. **(B)** Penis length of *S. eucosmia* (blue) and *S. largillierti* (gold) across study sites in Japan. Differences in penis length are greater in areas of sympatry than allopatry, suggesting RCD. Adapted from Kameda et al. (2009).

though matings can occur across these populations, the populations in question are functionally reproductively isolated from one another (Sobel et al. 2010).

To better understand reproductive isolating mechanisms, let us look at two examples—one involving delivery isolation in plants and one a study of reproductive isolation and shell coiling patterns in snails.

Reproductive Isolating Mechanisms: Delivery Isolation in Plants

The effects of pollinators on reproductive isolation in plants have been documented for many different plant species (Grant 1994; Hodges et al. 2004; Whittall and Hodges 2007; Widmer et al. 2009). In many plant taxa, such as the genus *Aquilegia*

TABLE 14.2

Dobzhansky's Reproductive Isolating Mechanisms

Prezygotic Isolating Mechanisms

Potential mates live in the same place, but do not encounter one another.

Habitat isolation

Temporal isolation (by time of day or time of year)

Potential mates interact, but do not mate (behavioral isolation).

Individuals copulate, but male gametes (sperm or pollen) are not transferred.

Male gametes are transferred, but the egg is not fertilized (gametic incompatibility).

Postzygotic Isolating Mechanisms

Zygote dies early in embryogenesis.

F_1 hybrids are inviable.

F_1 hybrids survive, but are sterile.

Backcross or F_2 hybrids are inviable or sterile.

Adapted from Barton et al. (2007).

FIGURE 14.21 Flowers in the genus *Aquilegia*. Floral morphology coevolves with the morphology of mouth parts in pollinators. Pictured here are **(A)** an *Aquilegia* species pollinated primarily by bumblebees, **(B)** a species pollinated primarily by hummingbirds, and **(C)** a species pollinated primarily by hawkmoths.

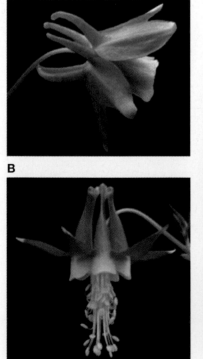

(columbine) and the genus *Ipomopsis* (skyrocket and its relatives), closely related species are pollinated by very different organisms (Figure 14.21). In *Aquilegia*, six species are primarily pollinated by hummingbirds, and four species are primarily pollinated by eastern hawkmoths; in the *Ipomopsis* group, seven species are pollinated by hummingbirds and seven by hawkmoths (Grant 1992).

Species pollinated by hummingbirds have markedly different floral structures than those pollinated by hawkmoths. In taxa pollinated by hummingbirds, floral tubes (spurs) are trumpet shaped and fairly long (16–24 mm); the hummingbirds' mouth parts used during pollination average about 23 mm in length. In contrast, the floral tube of species pollinated by hawkmoths is longer and more slender, ranging in length from 30 to 70 mm, which corresponds to the length of the proboscis of the species of hawkmoths used during pollination (Grant 1992; Whittall and Hodges 2007) (Figure 14.22). These differences in floral structures between hawkmoth and hummingbird pollinated species minimize gene flow across plant species that rely on different pollinator groups.

The difference between pollinators need not be as dramatic as the difference between insects and birds. In many closely related species of orchids, for example, reproductive isolation occurs because, although each is pollinated by insects, different species of orchids are associated with different species of bee pollinators (Van der Pijl and Dodson 1966; Nilsson et al. 1987; Armbruster et al. 1992).

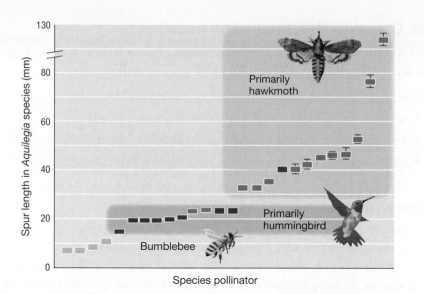

FIGURE 14.22 Differences in floral structures. The distribution of spur lengths (floral tube lengths) among *Aquilegia* ranked by size and color coded by pollinator. Adapted from Whittall and Hodges (2007).

Reproductive Isolation and Shell Coiling Patterns in Snails

Although evolutionary biologists do not assume that reproductive isolation will typically be attributed to a single point mutation, there is evidence that reproductive isolation is sometimes linked to evolutionary change in a single gene. Recall the case of land snails that have shells that either coil to the right or to the left, which we first discussed in Chapter 7. In some snail species, because of physical constraints, mating can only take place between individuals whose shells coil in the same direction: Individuals whose shells coil to the right cannot mate with individuals whose shells coil to the left.

The directionality of the coil (chirality) in snails is controlled by a single gene. Homozygous dominant and heterozygous individuals exhibit right-hand (dextral) coiling, while homozygous recessive individuals exhibit left-hand (sinistral) coiling. Because a single gene controls coil direction, and coil direction determines whether individuals are physically capable of mating with one another, evolutionary biologists have hypothesized that changes in the frequency of the dominant allele may be linked to reproductive isolation. And, in fact, phylogenetic analysis on *Euhadra* snail species suggests numerous instances in which speciation can be tied to a reversal of coil directionality (especially from left coiling to right coiling) (Figure 14.23) (Ueshima and Asami 2003).

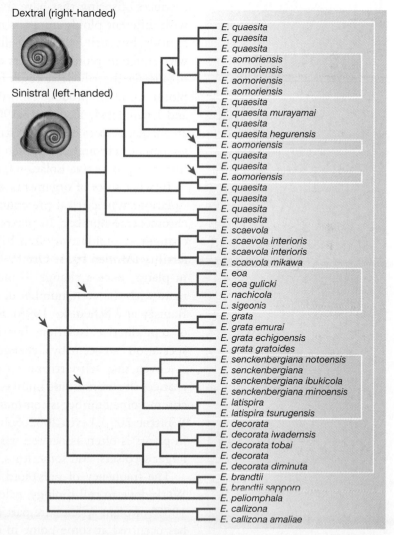

Dextral (right-handed)

Sinistral (left-handed)

FIGURE 14.23 Partial phylogeny of *Euhadra* snails and the evolution of coil direction. Blue and red indicate right-hand coiling and left-hand coiling, respectively. Arrows indicate possible points at which coil direction was reversed. This tree was constructed using maximum likelihood, based on mtDNA data. Adapted from Ueshima and Asami (2003).

The Genetics of Reproductive Isolation

In this section, we will expand our discussion of genetic/genomic mechanisms that can cause reproductive isolation. In particular, we will explore

- reproductive isolation via changes in chromosome number,
- reproductive isolation via chromosomal rearrangement,
- reproductive isolation via Dobzhansky–Muller incompatibility,
- Haldane's rule, sex chromosomes, and reproductive isolation.

Reproductive Isolation via Changes in Ploidy

Change in the number of complete sets of chromosomes that an organism possesses—also known as a change in ploidy—can lead to reproductive isolation. For example, imagine a diploid organism with three pairs of chromosomes. Suppose that, during gamete production, a breakdown in the normal process of meiosis produces offspring that have six pairs of chromosomes. In plants, two individuals with different ploidy numbers may successfully fertilize one another and produce hybrids, but such hybrids are often sterile. In animals, mating between individuals with different ploidy numbers almost always produces infertile offspring. This is because fertilizations that result from gametes produced by individuals with different ploidy numbers produce embryos that cannot properly undergo meiosis (Fowler and Levin 1984; Rodriguez 1996). If individuals with six pairs of chromosomes are viable, but can only self-fertilize or mate with another individual that possesses six pairs of chromosomes, then the change in chromosome number will result in instant reproductive isolation for the individuals with six pairs of chromosomes.

In what sorts of organisms should we expect to see this form of reproductive isolation? One critical prerequisite is the ability to survive a dramatic change in chromosome number. In plants, for reasons that are not completely understood, changes in ploidy can often be tolerated, and have minor effects on survival or fertility (Muller 1925; Orr 1990). In addition, since self-fertilization is common in plants, once a change in ploidy has occurred, an individual need not find a mate with the same number of chromosomes as itself (Stebbins 1938; Bell 1982; Ramsey and Schemske 1998). Self-fertilization is not a prerequisite for this type of reproductive isolation, but it should increase the likelihood that sympatric speciation can occur by a change in ploidy.

Given that self-fertilization is common in plants, as is the ability to survive changes in chromosome number, it is not surprising that speciation via changes in chromosome number is common in plants (Ramsey and Schemske 1998; Otto and Whitton 2000; Levin 2002; Soltis et al. 2004). At the phenotypic level, polyploidy in plants is often associated with increased cell volume, larger pollen grains, and larger seed sets, and sometimes, but not always, larger plant size.

The frequency of polyploid speciation in many plant lineages is quite high. Work done in cell biology, paleontology, and genomics suggests that from 47 to 100% of plant species are part of a lineage in which a polyploid speciation event has occurred at some point in evolutionary history. In particular, the same data show that approximately 31% of all fern species have originated as a direct result of polyploid speciation (Wood et al. 2009) (Figure 14.24).

Before leaving the subject of reproductive isolation via changes in ploidy, it is important to mention that while this phenomenon is most often seen in plants, it does occur in animals, albeit at a much reduced rate. In some cases, and for reasons that we do not yet understand, changes in chromosome number in some species of animals do not cause death or sterility—for example, speciation by changes in chromosome number may have occurred in shrimps, frogs, insects, and fish, as well as bivalves, crustaceans, and coral (Otto and Whitton 2000).

Reproductive Isolation via Chromosomal Rearrangement

Reproductive isolation may be initiated when genes or clusters of genes become rearranged on a chromosome. Such rearrangements include chromosomal fusion (the joining together of chromosomes or parts of chromosomes), chromosomal fission (the splitting of chromosomes), and chromosomal inversions and translocations.

When some individuals in a population possess the original chromosome arrangement, and others have the rearranged chromosome set, reproductive isolation between such *individuals* can occur. More importantly, when individuals in one *population* have the original chromosome arrangement and individuals in another population have the rearranged version, reproductive isolation between these *populations* may ensue.

The reproductive isolation that results from chromosome rearrangement is thought to emerge for at least two reasons (Rieseberg 2001; Ayala and Coluzzi 2005). First, hybrids formed by individuals with different chromosome arrangements will often produce dysfunctional gametes and, as such, these hybrids will have reduced rates of genetic recombination and fewer or no offspring. And so selection will lead to the production of fewer hybrids. The theoretical problem with this argument involves how the chromosomal rearrangement becomes common in the population in the first place. A chromosomal rearrangement will first appear as a mutation in a single individual, who will have to mate with another population member who has the original chromosomal arrangement. Yet, since selection acts against the resulting hybrid offspring, the mutation should quickly disappear.

A second way that chromosomal rearrangement may lead to reproductive isolation is that the reduced rates of genetic recombination found in hybrids will lead to an increase in linkage disequilibrium in their descendants. Researchers hypothesize that, if linkage disequilibrium associates traits involved with mating behaviors in hybrids and their descendants, differences between populations with respect to such mating behaviors may increase. This, in turn, can result in reproductive isolation over evolutionary time (Rieseberg 2001; Ayala and Coluzzi 2005).

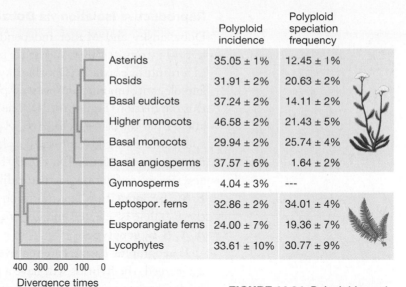

	Polyploid incidence	Polyploid speciation frequency
Asterids	35.05 ± 1%	12.45 ± 1%
Rosids	31.91 ± 2%	20.63 ± 2%
Basal eudicots	37.24 ± 2%	14.11 ± 2%
Higher monocots	46.58 ± 2%	21.43 ± 5%
Basal monocots	29.94 ± 2%	25.74 ± 4%
Basal angiosperms	37.57 ± 6%	1.64 ± 2%
Gymnosperms	4.04 ± 3%	---
Leptospor. ferns	32.86 ± 2%	34.01 ± 4%
Eusporangiate ferns	24.00 ± 7%	19.36 ± 7%
Lycophytes	33.61 ± 10%	30.77 ± 9%

400 300 200 100 0
Divergence times
(millions of years)

FIGURE 14.24 Polyploid speciation in plant lineages. Polyploid incidence and speciation frequencies across major groups of vascular plants. Less is known about polyploid speciation in gymnosperms than in other taxa. Adapted from Wood et al. (2009).

Reproductive Isolation via Dobzhansky–Muller Incompatibility

Dobzhansky and Muller independently developed a conceptual model for how hybrid incompatibility, and hence postmating reproductive isolation, might evolve as a result of epistasis (Dobzhansky 1937; Muller 1942). The basic idea is simple and elegant. Imagine an ancestral population, and consider two loci in members of this population (Figure 14.25). The population is fixed for allele A_1 at one locus, and allele B_1 at the other locus (A_1B_1/A_1B_1). Suppose now that the ancestral population splits into two geographically isolated populations. In population 1, a mutation from A_1 to A_2 occurs. On the B_1 background, the A_2 allele is selectively favored and sweeps to fixation, so that individuals in this population have genotypes A_2A_2 B_1B_1. In population 2, a mutation from B_1 to B_2 occurs. On the A_1 background, the B_2 allele is selected and goes to fixation. Thus, the populations are made up of $A_1A_1 B_2B_2$ genotypes.

If geographic barriers are removed and population 1 and population 2 can again interbreed, the hybrids will have a previously untested gene combination: A_2 with B_2. If there are epistatic interactions (Chapter 9) between the A and B loci, there may be fitness costs associated with this combination, despite the fitness advantages of having A_2 on B_1 background or B_2 on an A_1 background. The hybrids will then be selected against. This, in turn, can select for mechanisms of reproductive isolation between the two populations. Furthermore, if additional substitution differences occur at other loci with similar patterns of epistasis—for example, an allele C_2 could replace C_1 in population 1, while D_2 could replace D_1 in population 2—the fitness costs of hybridization can be compounded.

The Dobzhansky–Muller model helps explain hybrid infertility in crosses between two species of fruit flies, *Drosophila simulans* and *Drosophila melanogaster* (Brideau et al. 2006). The ancestor to both species is thought to have possessed two genes known as *Lhr* and *Hmr*. In *D. simulans*, the *Lhr* gene has diverged from the ancestral form, while in *D. melanogaster*, the *Hmr* gene has diverged from the ancestral form (Figure 14.26). When matings occur between female *D. melanogaster* fruit flies and male *D. simulans*

FIGURE 14.25 A schematic of the Dobzhansky–Muller model. Allele A_2 emerges and sweeps to fixation in population 1, while allele B_2 emerges and sweeps to fixation in population 2. Epistatic interactions between the A and B loci result in a fitness cost to individuals with both the A_2 and B_2 alleles. As a result, hybrids are selected against, driving reproductive isolation between individuals in population 1 and population 2. Adapted from Wu and Ting (2004).

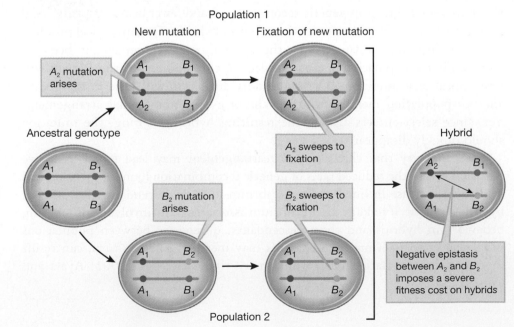

fruit flies in the laboratory, the number of offspring, compared to the number of offspring produced from within-species matings, is quite low—that is, hybridizing individuals suffer a huge fitness loss.

Follow-up genetic crosses strongly suggested that epistatic interactions between *Lhr* and *Hmr* are responsible for the fitness costs suffered by hybrid individuals.

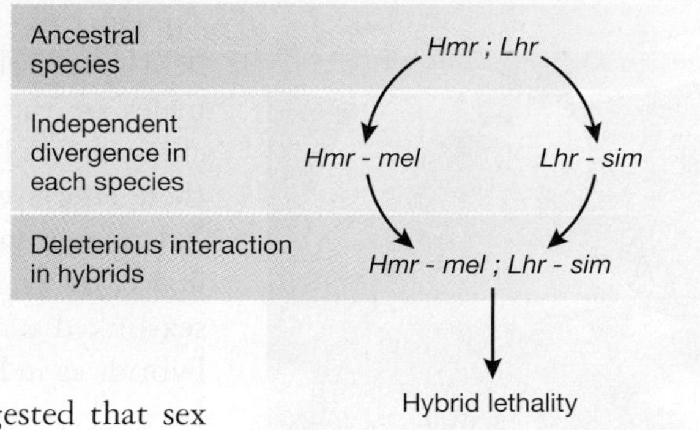

Haldane's Rule, Sex Chromosomes, and Reproductive Isolation

In the early part of the twentieth century, J. B. S. Haldane suggested that sex chromosomes may play a special role in the genetics of reproductive isolation (Haldane 1922). In discussing hybrids that are formed between incipient species, Haldane made an observation about hybrid viability and fertility that is now known as Haldane's rule. He noted that, if among hybrid offspring "one sex is absent, rare, or sterile, that sex is the heterozygous one." By heterozygous—or what today we would call *heterogametic*—Haldane meant the sex that has two different sex chromosomes as, for example, human males do with their XY sex chromosomes.

In mammals and fruit flies, males (XY) are heterogametic and females are homogametic, and so Haldane's rule predicts that, if one sex formed from the hybridization of incipient species is at a fitness disadvantage, it will be males. In birds and butterflies, on the other hand, females are heterogametic and males are homogametic, and so female hybrids should be at a disadvantage. In terms of the genetic underpinnings of reproductive isolation, Haldane's rule suggests that the decreased fitness of the heterogametic hybrids is what creates the basis for reproductive isolation.

The evidence that has amassed over the last 75 plus years overwhelmingly supports Haldane's rule. Across birds, butterflies, mammals, and fruit flies, if one sex is absent, rare, or sterile, that sex is the heterogametic sex an impressive 97.6% of the time (Table 14.3). But why is it the heterogametic hybrid sex that is at a disadvantage? While a number of theories have been put forth to

FIGURE 14.26 The Dobzhansky–Muller model and fruit flies. The Dobzhansky–Muller model has been applied to the study of hybrid infertility in crosses between two species of fruit flies, *Drosophila simulans* and *Drosophila melanogaster*. From Brideau et al. (2006).

TABLE 14.3

Evidence for Haldane's Rule

		Cases in Which One Hybrid Sex Is at a Fitness Disadvantage	Cases in Which Haldane's Rule Is Observed
Heterogametic Males			
Drosophila	Sterility	114	112
	Inviability	17	13
Mammals	Sterility	25	25
	Inviability	1	1
Heterogametic Females			
Lepidoptera	Sterility	11	11
	Inviability	34	29
Birds	Sterility	23	21
	Inviability	30	30

Adapted from Coyne and Orr (2004).

FIGURE 14.27 Raven and the First Men. The Haida people of coastal British Columbia told of humanity's origins in a clamshell that was discovered and opened by Raven.

explain this, the "dominance" theory is the most widely accepted. The idea here is straightforward. Consider the case of mammals and fruit flies, where males are the heterogametic (XY) sex. Suppose that a recessive gene on the X chromosome has negative fitness effects on hybrids. Because the gene is recessive these effects will be absent in XX females. But since males only have a single X chromosome, the effects of the recessive X-linked gene are always expressed: hence, the negative effect on male hybrids. A similar argument can be made for sex-linked genes and their effects on fitness when females are the heterogametic hybrids as in birds and butterflies.

14.4 The Evolutionary History of Humans

In this section, we turn our attention from the genetics of speciation to the story of species and speciation in our own evolutionary lineage. Throughout human history, people have struggled with the questions "who are we, and where have we come from?" The Haida people of the American Pacific Northwest, for example, tell a story of how the first humans washed up in a clamshell on the beach and were discovered by the mythological trickster, Raven (Figure 14.27). Evolutionary biology provides science's answer to the question of human origins. Genetic and genomic evidence reveal the relation between humans and other primates. Fossil

FIGURE 14.28 A phylogeny of the primates. Humans are part of the superfamily Hominoidea and are most closely related to the chimpanzee and bonobo. Also shown are the family Cercopithecidae (Old World monkeys) and the superfamily Ceboidea (New World monkeys). This phylogeny is presented as a chronogram, with the branch lengths indicating the divergence times in millions of years (mya) before the present. Adapted from Enard and Pääbo (2004).

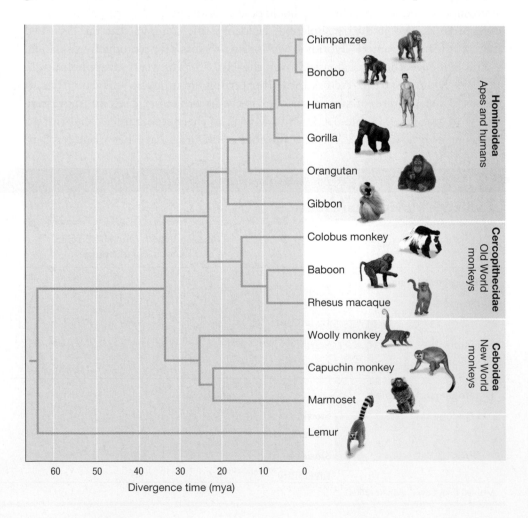

evidence helps us reconstruct the evolutionary events that occurred after the *Homo* lineage diverged from the lineage leading to chimpanzees about 5.5 million years ago. It then enables us to reconstruct how our species, *Homo sapiens*, emerged. Advances in genetic sequencing and phylogenetic inference have enormously refined our ability to reconstruct the history of how current human populations came to span the globe. In this section, we will trace out each of these stories.

Relationships among Humans and Great Apes

Genomic-scale sequence data leave little question as to the relation between humans and other living primate species. Humans are part of a superfamily known as the Hominoidea (Figure 14.28). This clade consists of eight living genera: orangutans (*Pongo*), gorillas (*Gorilla*), chimpanzees (*Pan*), humans (*Homo*), and four gibbon genera. Within the Hominoid superfamily, humans are most closely related to the two species of chimpanzee: the common chimpanzee (*Pan troglodytes*) and the bonobo (*Pan paniscus*). Gorillas are more distantly related, orangutans yet more so, and gibbons are the most distant. Figure 14.29 provides further detail on taxonomic nomenclature for humans and our near relatives.

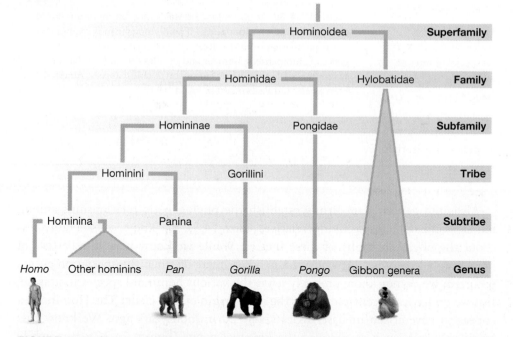

FIGURE 14.29 Hominoid nomenclature. A phylogeny of the primates, with labels for the branch tips and a few select clades. Each clade along this branch carries a name based on the Latin root *Homo* or *homin*, meaning "man." Because our understanding of hominoid phylogeny has been in flux until recently, the nomenclature for these clades has been forced to change a number of times; moreover, clade names are still used in conflicting ways in the current literature. The superfamily Hominoidea consists of humans and apes (chimpanzees, gorillas, orangutans, and gibbons), which are collectively known as hominoids. The family Hominidae comprises humans and great apes (chimpanzees, gorillas, and orangutans); the subfamily Homininae includes humans, chimpanzees, and gorillas; the tribe Hominini consists of humans and chimpanzees; the subtribe Hominina consists of all Hominini more closely related to modern humans than to chimpanzees, including both the genus *Homo* (humans) and other archaic hominin genera such as *Australopithecus* and *Praeanthropus*. In the current literature, members of the subtribe Hominina are typically referred to as hominins; in older literature, the term hominid is often used instead. Adapted from Mann and Weiss (1996).

BOX 14.3 Species Trees and Gene Trees

At the majority of loci, humans are more closely related to chimpanzees than to any other hominid. At approximately 25% of loci, however, humans appear to be more closely related to gorillas than to chimpanzees (Chen and Li 2001). How could this be, given that humans and chimpanzees share a more recent common ancestor? In Chapter 4, we learned of one possible reason that two distinct branches might share common characters: The similarities in gorillas and humans could be homoplasies or symplesiomorphies. But there is another process that could also generate this pattern.

Figure 14.30 shows this process. The locus shown is polymorphic for A_1 and A_2 at the time of the split between the human–chimp and the gorilla lineages, and this polymorphism is maintained until the split between the human and chimp lineages. By chance the same allele—A_1—is fixed in the lineages leading to humans and in gorillas, while the other allele—A_2—is fixed in the lineage leading to chimpanzees. The result is that, at this particular locus, humans and gorillas will share a common allele that is different from that which we observe in chimpanzees. This phenomenon is sometimes called *deep coalescence*, because the coalescent event between the alleles at this locus predates—that is, it is deeper than—the speciation event separating the species of interest.

This example highlights the distinction that we drew in Chapter 8 between *species trees* and *gene trees*. Often gene trees will closely reflect species trees. But along branches of the species tree that are short—that is, those that represent a relatively small number of generations—and wide—that is, those

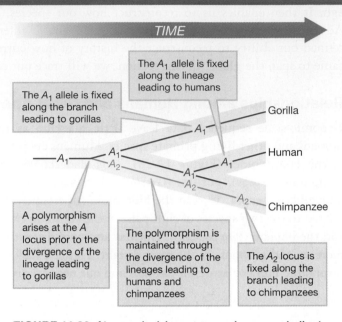

FIGURE 14.30 At some loci, humans may be more similar to gorillas than to chimpanzees. If a polymorphism is maintained in the population from the divergence of gorillas to the divergence of chimpanzees, humans and gorillas can end up sharing a common allele that differs from that in chimpanzees. Adapted from Enard and Pääbo (2004).

Although genetic data firmly establish the phylogenetic relationships among humans and our closest living relatives (Box 14.3), we know considerably less about the phylogeography of these species. While we know what the pattern of branching ancestry is, and we have a good estimate of when the speciation events occurred, we do not know precisely *where* the various hominoid species first arose. But we do have a decent picture of the *early* origin of this clade: The Hominoidea appear to have arisen in East Africa about 20 million years ago. We know this based on fossils such as those of the *Proconsul* genus (Figure 14.32). But, while *Proconsul* was likely closely related to the common ancestor of all living apes, this genus of tailless, arboreal primates is anatomically primitive compared to extant ape species. Most notably, *Proconsul* lacked the limb mobility that characterizes modern apes and allows them to swing by their arms from tree limbs—and it probably lacked a tail as well (Larsen 2008).

Details of the origin of the great apes and humans, the Hominidae, are less clear. Fossils dating to around the time of divergence of this clade—14 million years ago—have been found in Africa, Asia, and Europe alike. Where did these species arise? A sparse fossil record from this period until about 5 million years ago has hampered our progress in uncovering the answers. Many investigators suspect an

that reflect a large population size—we are particularly likely to observe deep coalescence, as in case B of Figure 14.31. Because the brown shaded branch in case A is long and narrow, coalescence along this branch is likely and deep coalescence beyond this branch is unlikely. By contrast, in case B, the brown shaded branch is short and wide, making coalescence along the branch less likely and deep coalescence beyond this branch more likely.

Deep coalescence events are particularly common in the human–chimpanzee–gorilla clade. This is because of a short, wide branch much as in case B in Figure 14.31. As illustrated in the chronogram in Figure 14.28, the split between humans and chimpanzees occurred quite soon after the split between gorillas and humans–chimps. As a result, a large number of loci remained polymorphic from the divergence between humans–chimps and gorillas, right through to the divergence between humans and chimps. As a result, 25% of loci have gene trees that do not reflect the species tree for these three species.

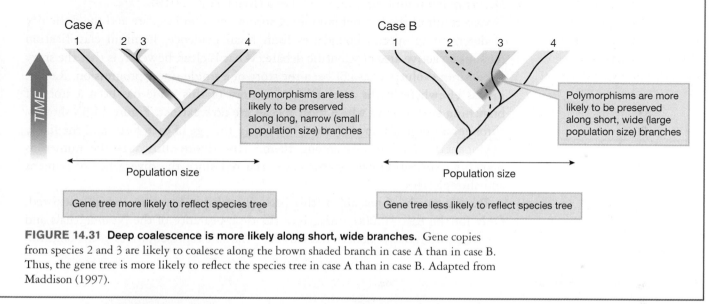

FIGURE 14.31 Deep coalescence is more likely along short, wide branches. Gene copies from species 2 and 3 are likely to coalesce along the brown shaded branch in case A than in case B. Thus, the gene tree is more likely to reflect the species tree in case A than in case B. Adapted from Maddison (1997).

East African origin for this clade as well, followed by subsequent migration into Europe and Asia (Andrews 1992). But other recent evidence is consistent with a Eurasian origin, so the topic remains an active area of investigation (Moya-Sola et al. 2009).

A

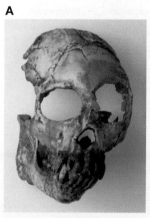

B

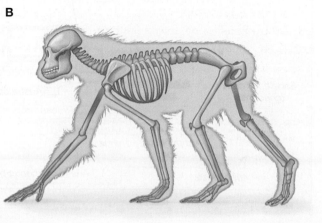

FIGURE 14.32 *Proconsul africanus.* ** *Proconsul* is considered by most researchers to be an early genus within the hominoid clade. **(A) *Proconsul* skull, and **(B)** *Proconsul* skeleton.

The Hominin Clade

Abundant genetic evidence from humans, gorillas, orangutans, and chimpanzees has allowed us to resolve the phylogenetic relationships of these species. It is much harder to resolve the phylogenetic history of the hominin clade—namely, humans and the extinct species more closely related to humans than to chimpanzees. The main reason is that all species in this clade are extinct except for our own, and we have very limited information about the others. Given the limited availability of DNA evidence, evolutionary biologists, paleontologists, and anthropologists must rely largely on fossil evidence to reconstruct the evolutionary history of the hominins. However, as of 2010, whole genome sequence data have been reported for humans, Neanderthals (*Homo neanderthalis*), and the recently discovered *Denisovan* hominin from southern Siberia (Reich et al. 2010).

As a result of the limited number of specimens found to date and the difficulty of determining species boundaries from fossil evidence, hominin classification remains an active area of scientific debate. What is clear, however, is that the story of hominin evolution has not been the story of a single linear progression. Rather, just as elsewhere in the tree of life, the story of this clade has been a story of branching evolution in which most species are now extinct. Figure 14.33 shows a chronogram of hominin evolution, indicating the age of each fossil and the likely phylogenetic relationships among them. This figure illustrates the numerous branching speciation events that have occurred since the divergence of humans and chimpanzees.

During much, if not all, of this period, multiple hominin species coexisted. Only for the past 30,000 years, since the disappearance of the Neanderthals and the Denisovans, have modern humans been the sole representatives of the hominin

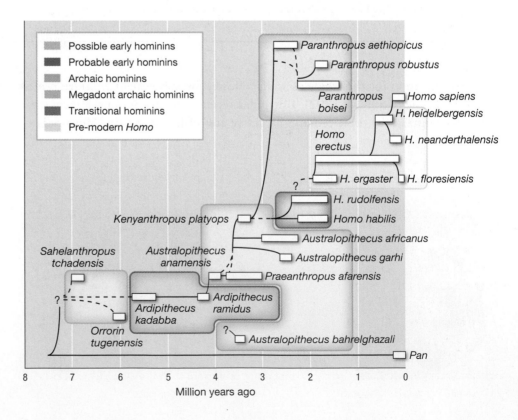

FIGURE 14.33 A chronogram of hominin evolution. Major *grades*—that is, groups of morphologically similar species that do not qualify as monophyletic clades—are indicated. Adapted from Strait et al. (2007) and Wood and Lonergan (2008).

A **B**

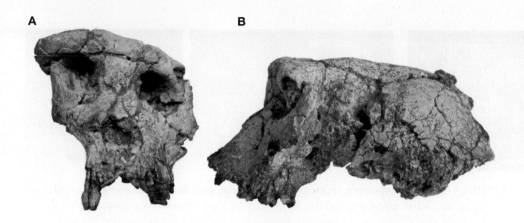

FIGURE 14.34 A fossilized skull of *Sahelanthropus tchadensis.* From the front view (**A**) the distortion of the skull is apparent. (**B**) Side view of the skull.

lineage. Perhaps this period is even briefer if skeletal remains recently discovered on the Indonesian island of Flores turn out to derive from a unique *Homo* species. This population, provisionally dubbed *Homo floresiensis*, appears to have been present as recently as 13,000 years ago.

The distinguishing features of the hominin lineage are *not* the large brains, language, and tool use that we associate with premodern and modern humans. These features evolved long after the divergence from chimpanzees. Rather, the principal common features of this lineage are changes in skeletal structure that conferred the ability for bipedal (upright) locomotion, and certain changes in dental anatomy, most notably the loss of the large shredding canines that are present in nonhuman apes (Larsen 2008).

The earliest hominin fossils, those dating to soon after the divergence of humans and chimpanzees, are the most difficult to place definitively within the hominin lineage. For one thing, these fossils tend to be in poorer condition than other hominin fossils because of their relatively great age. For example, the fossilized skull of *Sahelanthropus tchadensis*, discovered in Chad in 2001 by Brunet and colleagues (Brunet et al. 2002), has been distorted by geological forces to the point that it is challenging to reconstruct the original skull geometry (Figure 14.34).

But there is an even bigger conceptual problem: Close to the human–chimpanzee divergence, it is very difficult to tell, based on morphological characters, whether a given fossil belongs to the hominin lineage or to the panin (chimpanzee) lineage. Based on the orientation of the skull and the dental anatomy, however, it is most likely that *Sahelanthropus tchadensis* lies along the hominin lineage. Another possible early hominin in this category is *Orrorin tugenensis*, for which a complete skull has yet to be discovered.

Next to appear in the fossil record, after *S. tchadensis* and *O. tugenensis*, are a set of probable early hominins and so-called archaic hominins. Researchers have faced similar problems with the condition of these fossil remains, but advances in scanning and computer imaging have enabled sophisticated reconstructions of cranial shape based on skull fragments. Figure 14.35 shows a set of skull fragments of the early hominin *Ardipithecus ramidus* after initial physical assembly, and then as a computer-imaging reconstruction. Like *S. tchadensis* and *O. tugenensis*, *A. ramidus* had small cranial capacity, but the foot morphology of *A. ramidus* suggests that it was at least partially arboreal.

A

B

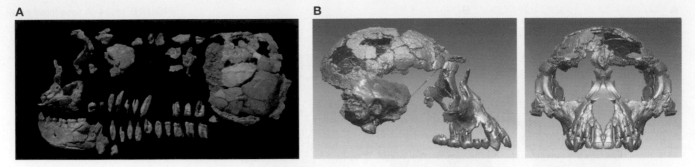

FIGURE 14.35 Reconstructing an *Ardipithecus ramidus* skull by computer imaging. (**A**) Skull fragments after initial physical reassembly. (**B**) Computer reconstructions of cranial morphology derived from these fragments. The face of *A. ramidus* is considerably flatter than that of a chimpanzee (Suwa et al. 2009).

Archaic hominins appear next in the hominin lineage (Figure 14.36) and shared an increasing number of physiological features with the genus *Homo*. They appear to have been capable of bipedal locomotion. One of these species, *Praeanthropus afarensis* (also called *Australopithecus afarensis*), is well known in the popular consciousness as the 3.2-million-year-old fossil "Lucy." Recent fossil evidence suggests that this species even fashioned stone tools and used them when scavenging for meat (McPherron et al. 2010). Later archaic hominins such as *Australopithecus garhi* certainly fashioned chipped stone tools; a large number of such tools have been dated to 2.5 to 2 million years ago (Larsen 2008).

FIGURE 14.36 Three archaic hominins. These archaic hominin species had small brain sizes compared to humans, but they appear to have walked upright at least some of the time, and they may also have fashioned stone tools. (**A**) *Kenyanthropus platyops*, (**B**) *Praeanthropus afarensis*, and (**C**) *Australopithecus africanus*.

A *Kenyanthropus platyops*   **B** *Praeanthropus afarensis*   **C** *Australopithecus africanus*

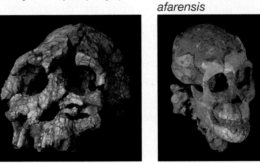
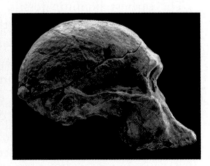

The so-called megadont archaic hominins, a group sometimes known as the *robust Australopithecines*, appear to form a proper clade: the genus *Paranthropus*. This morphologically unusual clade of hominins is characterized by huge attachment regions on the skull that confer the distinctive side and top ridges along the skulls in Figure 14.37. These attachments anchored the massive chewing muscles that *Paranthropus*

FIGURE 14.37 *Paranthropus* species. This clade was a side branch off of the hominin lineage leading to modern humans. Note the massive ridges at the top and sides of the skull; these provided attachment points for huge chewing muscles. (**A**) *Paranthropus boisei*, (**B**) *Paranthropus aethiopicus*, and (**C**) *Paranthropus robustus*.

A *Paranthropus boisei*   **B** *Paranthropus aethiopicus*   **C** *Paranthropus robustus*

possessed. The teeth of these species were correspondingly large; hence, the name *megadont*, which means "big teeth." As a result, members of this clade were well suited for grinding low-quality food sources. While an evolutionary dead-end, *Paranthropus* was contemporaneous with the later archaic hominins discussed above, and even with some early members of the genus *Homo*.

The Genus *Homo*

Approximately 2.3 million years ago, the first members of the genus *Homo* appeared. These *transitional hominins*—*Homo habilis*, and the similar, if somewhat larger, *Homo rudolfensis*—bridge the gap between some of the archaic hominins we considered in the previous subsection and the so-called premodern hominins that were the immediate ancestors to *Homo sapiens*. While both had brain sizes smaller than modern or premodern humans, *Homo habilis* and *Homo rudolfensis* had cranial capacities that were greater than the archaic hominins before them, as well as other morphological features that more closely resemble those of modern humans (Figure 14.38). They also appear to have made abundant use of tools.

Homo ergaster appeared in Africa and *Homo erectus* appeared in Africa and Asia around 1.9 million years ago. Whether these represent two separate species or regional forms of a single species, as some researchers believe, their body plans closely resembled modern humans and thus we consider them to be *premodern hominins* (Figure 14.39). Relative to archaic and transitional hominids, *Homo erectus* was taller—5 to 6 feet in height—with longer legs and a considerably larger brain. In addition to fashioning and using stone tools such as hand axes, by 600,000–400,000 years ago, *Homo erectus* was also making use of fire to cook food and to provide warmth (Larsen 2008).

Homo erectus was so successful, ecologically, that it was able to undertake a remarkable expansion, spreading from its origin in Africa throughout Europe and Asia (Figure 14.39D). Morphological differences across these regions indicate considerable regional differentiation.

The Emergence of Anatomically Modern Humans

When, where, and how did the transition from premodern hominins to modern humans take place? The **multiregional hypothesis** suggests that hominins left Africa and colonized the rest

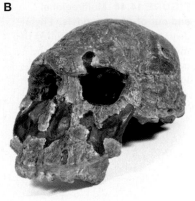

FIGURE 14.38 Transitional hominins. Skulls of the transitional hominins, **(A)** *Homo habilis*, and **(B)** *Homo rudolfensis*.

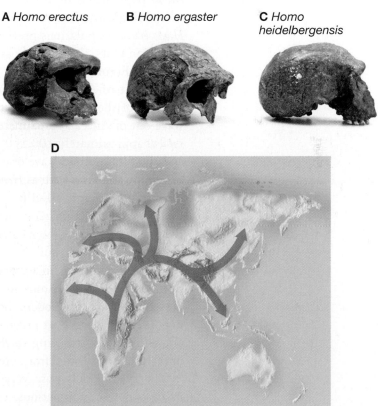

A *Homo erectus*   **B** *Homo ergaster*   **C** *Homo heidelbergensis*

D

FIGURE 14.39 Premodern hominins and the expansion of *Homo erectus*. (A) *Homo erectus*, **(B)** *Homo ergaster*, and **(C)** *Homo heidelbergensis*. *Homo heidelbergensis* appears much later in the fossil record and is a possible ancestor to both modern humans and Neanderthals. **(D)** The expansion of *Homo erectus* across modern Europe, Africa, Asia, and the South Pacific. Part D adapted from Larsen (2008).

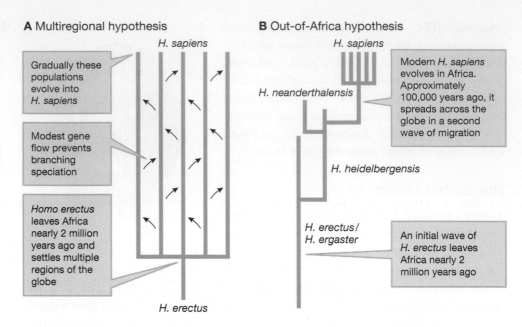

of the Old World a single time, nearly 2 million years ago, as *Homo erectus*. *Homo erectus* populations in different parts of the world then diverged from one another morphologically, but modest gene flow among these geographically separated *Homo erectus* populations prevented branching speciation. Gradually over the past 2 million years, these loosely associated populations together evolved into modern humans (Figure 14.40A).

The **out-of-Africa hypothesis** suggests that hominins left Africa and colonized the rest of the Old World in two major waves, first as *Homo erectus* in an initial wave out of Africa approximately 2 million years ago, and a second time as *Homo sapiens* approximately 100,000 years ago. Modern *Homo sapiens* emerged in Africa, and then in a second wave of colonization out of Africa, the premodern hominins of Europe and Asia such as *Homo erectus* and *Homo neanderthalensis* were replaced by *Homo sapiens* (Figure 14.40B).

The key distinction between the multiregional and out-of-Africa models involves the fate of premodern hominin populations in Europe and Asia. The multiregional hypothesis predicts humans of European ancestral origin are descended from premodern hominins in Europe, humans of Asian ancestral origin are descended from premodern hominins in Asia, and humans of African ancestral origin are descended from premodern hominins in Africa. By contrast, the out-of-Africa hypothesis predicts that premodern hominin populations in Europe and Asia died out without contributing to the modern human gene pool, and that all modern humans are descended from premodern hominins in Africa alone.

The two models pose very different explanations for the differences among modern human populations. Under the multiregional model, differences among modern populations have their origins in geographic separations that have been maintained for 2 million years. Under the out-of-Africa model, differences among modern human populations cannot predate the recent migration out of Africa around 100,000 years ago.

A number of lines of evidence support the out-of-Africa model—albeit with a surprising twist that we will discuss shortly—over the multiregional model.

First of all, fossil evidence suggests gradual divergence of the premodern *Homo* species in their various locations in Africa, Europe, and Asia, with *Homo sapiens* arising first in Africa around 130,000 years ago. Following this event, we see rapid replacement of the other *Homo* forms by the *Homo sapiens* from around the rest of the globe, sometime after 100,000 years ago. This suggests migration and replacement, as in the out-of-Africa model (Strait et al. 2007).

Genetic data provide some of the strongest evidence for the out-of-Africa hypothesis. Max Ingman and his colleagues looked at the full mitochondrial DNA sequence of 53 humans from around the world (Ingman et al. 2000). Using the molecular clock approach we discussed in Chapter 8, they estimated from these data that the coalescence time for human mitochondrial DNA is between 120,000 and 230,000 years ago, and the coalescence time for non-African mtDNA is around 38,500 years ago.

These observations tell us two important things: (1) the recent common ancestor of all human mtDNA dates to around the time of the emergence of *Homo sapiens* in Africa, not to around the time of the emergence of *Homo erectus*, and (2) non-African *Homo sapiens* share an even more recent common ancestor, dating to around the time that *Homo sapiens* replaced other *Homo* species around the globe.

Moreover, the phylogeny of modern human populations that Ingman and his colleagues created provides the crucial observation that non-Africans are a subclade of the larger phylogenetic tree for modern humans (Figure 14.41). This means that there is vastly more mitochondrial diversity within Africa than across the rest of the globe combined. Other analyses have revealed similar patterns at nuclear loci. That is again consistent with an origin of modern humans in Africa 200,000–130,000 years ago, followed by a more recent migration of *Homo sapiens* out of Africa less than 100,000 years ago.

Thus, the preponderance of the evidence points to a recent African origin followed by migration throughout the globe—that is, the out-of-Africa model. With this evidence in place, we can sketch the remainder of the story of the evolution of modern humans.

Homo erectus was contemporaneous with *Homo heidelbergensis*, a species found throughout Europe. *Homo neanderthalensis* and possible immediate predecessors were present in Europe and central Asia from around 300,000 years ago until

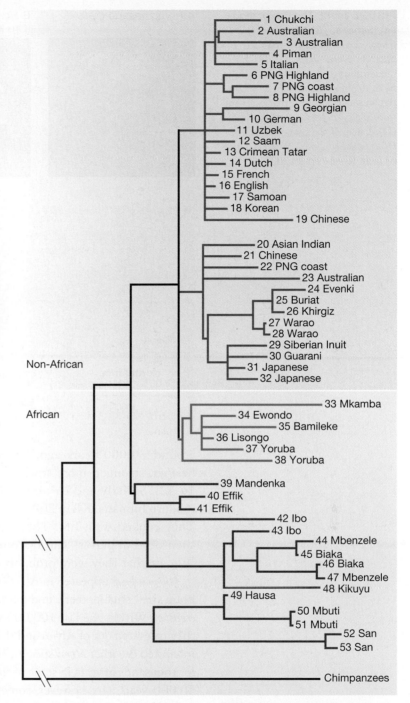

FIGURE 14.41 Modern human populations. Phylogenetic relationships among 53 modern humans based on mtDNA, with chimpanzees as an outgroup. A caveat. Because mtDNA is effectively a single locus, the tree shown here is effectively a single *gene tree* rather than an accurate representation of the relatedness of human groups. Adapted from Ingman et al. (2000).

FIGURE 14.42 The rise of modern humans. (A) Neanderthal (*Homo neanderthalensis*), (B) Cro-Magnon (*Homo sapiens*), and (C) modern human (*Homo sapiens*) skulls. Note the prominent brow ridge on the Neanderthal skull that is absent from the two *Homo sapiens* skulls. (D) A map of the proposed range of *Homo neanderthalensis*. Part D adapted from Krause et al. (2007).

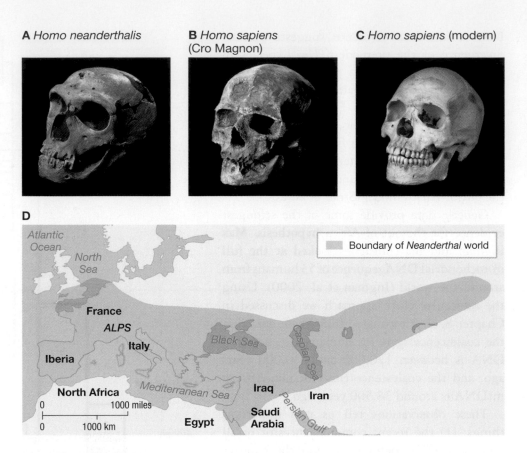

A *Homo neanderthalis*

B *Homo sapiens* (Cro Magnon)

C *Homo sapiens* (modern)

D

around 30,000 years ago. Compared to modern humans, Neanderthals were heavier, stronger, more stocky, and had a more pronounced brow ridge (Figure 14.42). Socially and culturally, however, they shared many characteristics with modern humans (Klein 2003). They made elaborate tools and hunted large game. They cooked with fire. They cared for their sick and elderly, and they buried their dead in graves. The anatomy of their vocal tract and some genetic evidence suggest that they were probably capable of some degree of speech.

Homo sapiens appeared in Africa roughly 150,000 years ago. *Homo sapiens* had larger brain sizes, smaller teeth, and reduced brow ridges compared to *Homo erectus* and *Homo ergaster*. Within the last 100,000 years and perhaps as recently as 35,000 years ago, they migrated out of Africa and throughout Europe and Asia, which at the time were inhabited by other *Homo* species. *Homo sapiens* and these other *Homo* species coexisted for thousands of years in some of these areas, as established by fossil evidence. Around 50,000 years ago, however, something changed. *Homo sapiens* rapidly developed a suite of new behavioral and cultural traits, known as the Upper Paleolithic lifestyle. This included substantially more advanced stone tools than had been seen previously, the use of bone and ivory for tools, the creation of elaborate structures for shelter, the production of art and musical instruments, the frequent hunting of larger game, ceremonial burial, and long-distance trade (Johanson 2001). Although it is more difficult to ascertain, most researchers believe that language was also fully developed by this stage. At this point, *Homo sapiens* had become, physiologically and culturally, fully modern. With this new suite of traits, *Homo sapiens* rapidly replaced the other *Homo* species around the globe. With the possible exception of *Homo floresiensis*, these other *Homo* species were gone by about 30,000 years ago.

Did Humans and Neanderthals Interbreed?

All of the evidence that we have discussed so far supports the out-of-Africa model, in which previous *Homo* species were entirely replaced by *Homo sapiens* following the migration out of Africa roughly 100,000 years ago. But if *Homo sapiens* and other *Homo* species interbred during the period in which they overlapped, some genetic variation from the original radiation of *Homo erectus* 2 million years ago could have made its way into the modern human gene pool. To determine whether this is likely, scientists have focused an extensive effort on determining whether there is any evidence of interbreeding between humans and Neanderthals in Europe. Fossil evidence suggests that interbreeding might have taken place; some fossils appear to have morphological characters intermediate between *Homo sapiens* and *Homo neanderthalensis*, for example. But it is very difficult to tell for certain whether this is indicative of interbreeding, or due to other unrelated causes. And it is yet more difficult from fossil evidence alone to ascertain whether these interbreeding events, if they did occur, have contributed to genetic variation in modern humans.

A breakthrough came with the development of techniques to extract and sequence the DNA from ancient bone material (Krings et al. 1997). In one notable study, researchers extracted mitochondrial DNA (mtDNA) from a number of Neanderthal bone fragments, and also from bones of approximately the same age from ancient *Homo sapiens* (Serre et al. 2004). They found that the Neanderthals had a unique set of mtDNA genotypes that were not represented among ancient or modern humans. From this evidence alone, the authors could rule out the possibility of *extensive* interbreeding between humans and Neanderthals. But the mtDNA is effectively only a single locus among millions of loci and, because mitochondria are maternally inherited, it represents only the history of descent through the female lineage. As a result, the researchers had no evidence of interbreeding—but they were unable to rule out the possibility that limited interbreeding had taken place.

To determine more conclusively whether any inbreeding had taken place, researchers needed a way to look at nuclear loci from Neanderthals. By sequencing the entire Neanderthal genome from ancient bone fragments, a team lead by Svante Pääbo made this possible on a large scale (Green et al. 2010).

But even with genomic information about Neanderthals in hand, it remains tricky to infer interbreeding based on nuclear DNA evidence. The problem involves the distinction between gene trees and species trees that we treated for humans, chimpanzees, and gorillas in Box 14.3. As illustrated in Figure 14.43, genetic variation in living modern humans reflects polymorphisms that date back over 500,000 years—long before the divergence of Neanderthals and *Homo sapiens*. As a result, even in the absence of interbreeding, we would expect to find that, at genetic loci with these old polymorphisms, certain individual humans may be more closely related to individual Neanderthals than they are to other individual humans (Pääbo 1999).

Given this complication, how could researchers distinguish between the results of interbreeding and the results of ancient genetic polymorphisms dating back prior to the human–Neanderthal divergence? Pääbo and colleagues reasoned that, if Neanderthals shared more alleles in common with humans that were descended from populations that overlapped with Neanderthals than with humans descended from populations that had never overlapped with Neanderthals, this would support the hypothesis that humans and Neanderthals

FIGURE 14.43 Gene trees and species trees for humans and Neanderthals. (A) The mitochondrial genome is effectively a single locus. At this locus, the gene tree reflects the species tree: As far as we know, all Neanderthal mtDNAs form one clade, and all modern human mtDNAs form another separate clade. (B) Some loci in the nuclear genome have been polymorphic in humans since long before the divergence of Neanderthals and modern humans. As a result, some human alleles at these loci may be more closely related to Neanderthal alleles than they are to other human alleles. From Pääbo (1999).

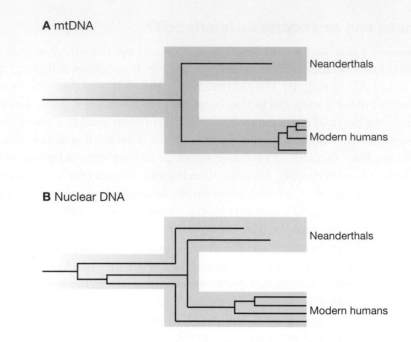

A mtDNA

Neanderthals

Modern humans

B Nuclear DNA

Neanderthals

Modern humans

had interbred. Given the European range of Neanderthals and the pattern of modern human migration out of Africa, we would expect humans descended from Europeans and Asians to share more alleles with Neanderthals than do humans descended from African populations.

This is precisely what Pääbo and colleagues found. Using a variety of statistical techniques, they estimated that 1–4% of the genome of non-African humans is derived from Neanderthals. Based on this analysis, Figure 14.44 illustrates a revised model of human evolution. It is essentially an out-of-Africa model. As in the original out-of-Africa model, modern human variation is largely derived from what was present in the population of *Homo sapiens* that first emerged in Africa less than 200,000 years ago. Moreover, all modern humans share a common ancestor that is yet more recent. This surprising twist on the out-of-Africa model is that not all genetic variation from *Homo* species prior to *Homo sapiens* has been lost. Rather, a limited amount of that variation has been passed from *Homo neanderthalensis* to Eurasian populations of *Homo sapiens* as a result of interbreeding that took place in the Near East some time within the past 100,000 years.

Evolutionary biologists today are just as interested in "the species problem" and speciation as Darwin was when he published *On the Origin of Species.* While Darwin laid out many of the questions to be addressed and provided some of the answers to these questions, much progress has been made over the last 150 years. In this chapter, we have touched on some of the major advances that have been made in the study of speciation, including speciation events that occurred within our own primate clade. With a clearer understanding of how evolutionary biologists conceptualize a species, as well as the models that have been developed to understand the process of speciation, we now move on to the topic of extinction.

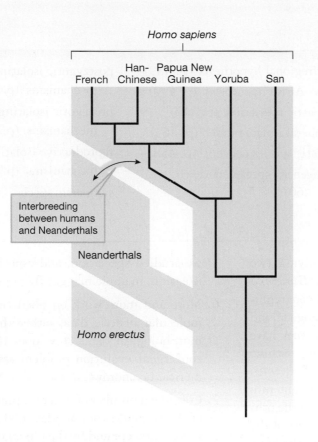

Homo sapiens

French | Han-Chinese | Papua New Guinea | Yoruba | San

Interbreeding between humans and Neanderthals

Neanderthals

Homo erectus

FIGURE 14.44 A model of gene flow between Neanderthals and modern humans. Alleles that are common to Neanderthals and modern humans have reached modern populations by two different paths: (1) Interbreeding between Eurasian *Homo sapiens* and Neanderthals, and (2) maintenance of ancient genetic polymorphisms in humans. We expect the former class of alleles to be found only in non-African populations, whereas the latter can potentially occur in any and all human populations. Adapted from Green et al. (2010).

SUMMARY

1. Evolutionary biologists have long struggled with how to classify organisms. What makes for a species? How different do two groups have to be before they are considered different species?

2. The evolutionary species concept provides an answer to what a species fundamentally *is*: a set of populations with their own distinct evolutionary history and a shared future evolutionary fate.

3. The phenetic species concept, the biological species concept, and the phylogenetic species concept each provide different diagnostic criteria for how species boundaries can be drawn in practice.

4. Major models of speciation include allopatric, parapatric, and sympatric speciation. Allopatric speciation includes a dumbbell model of speciation and a peripheral isolate model.

5. Ring species live in a series of populations that are connected to one another in a ringlike fashion. In ring species, we expect gene flow across adjacent populations, but gene flow between populations that are not adjacent, should be minimal and decrease as a function of distance.

6. Evolutionary biologists have identified and studied many types of reproductive isolation—that is, mechanisms that restrict or prevent gene flow.

7. Work on the genetics of speciation includes studies of reproductive isolation via changes in chromosome number, reproductive isolation via chromosomal rearrangement, reproductive isolation via Dobzhansky–Muller incompatibility, and reproductive isolation via Haldane's rule.

8. Evolutionary biology provides an answer to the question of human origins. Genetic and genomic evidence reveals the relation between humans and other primates.

9. Fossil evidence allows the chance to reconstruct the evolutionary events that occurred from the divergence of the *Homo* lineage from chimpanzees about 5.5 million years ago to the emergence of our species *Homo sapiens*.

10. Advances in genetic sequencing and phylogenetic inference have enormously refined our ability to reconstruct the history of how current human populations came to span the globe.

KEY TERMS

allopatric speciation (p. 462)

biological species concept (p. 458)

cline (p. 465)

dumbbell model (p. 463)

evolutionary species
   concept (p. 457)

hybrid zone (p. 465)

multiregional hypothesis (p. 489)

out-of-Africa hypothesis (p. 490)

parapatric speciation (p. 462)

peripheral isolate model (p. 463)

phenetic species concept (p. 458)

phylogenetic species concept
   (p. 460)

postzygotic isolating
   mechanisms (p. 474)

prezygotic isolating
   mechanisms (p. 474)

reproductive isolating
   mechanisms (p. 466)

sympatric speciation (p. 462)

REVIEW QUESTIONS

1. Imagine that researchers have been studying two populations of a hypothetical creature, *Darwinius huxlianus*. These two populations are geographically separated by 3000 miles, and geological and molecular genetic evidence suggests that they have been separated for at least 20 million years, with no exchange of gene flow during that time. Yet, when individuals from each population are brought into the lab, they readily mate with members of the other populations. Are the two populations part of the same species? Make the case that they are the same species, and then make the case that they are not.

2. Some evolutionary biologists have suggested that the peripheral isolate model should lead to speciation at a faster rate than the dumbbell model. Why should this be the case? Try to cast your answer in terms of population size and environmental heterogeneity on islands versus mainlands.

3. What sort of follow-up experiments would you run to further examine whether the above- and below-ground *Culex pipiens* populations we discussed at the start of the chapter are different species?

4. Why is it important for evolutionary biologists to separate the question "What is a species?" from the question "How can we distinguish among species in nature?"

5. Evolutionary biologists suggest that speciation in apple maggot flies is a case of sympatric speciation because the two host species—apple trees and hawthorn shrubs/trees—occur in the same location. But suppose that from a "fly's eye perspective," there *were* barriers that separated apple trees and hawthorn shrubs/trees—barriers that were small enough and different enough, that we, as humans, would not notice them. Would this make a difference in the way we studied speciation, and especially reproductive isolation, in apple maggot fly populations? Why?

6. More and more work on phenetic species is using molecular genetic data, rather than anatomical or morphological data. How does this minimize the convergent evolution problem associated with numerical taxonomy?

7. Consider a number of different purebred dog breeds. (a) How would they be classified by the biological species concept and by the phylogenetic species concept? (b) Which does a better job of accounting for the future evolutionary fates of these breeds? (Hint: Your answer may depend on the assumptions you make about the degree of control humans will have on these breeds in the future.)

8. Only once extensive genetic data became available were scientists able to definitively resolve the relationship between humans, chimpanzees, and gorillas. Explain why this posed a particularly difficult challenge.

9. Svante Pääbo and colleagues found evidence that a nontrivial fraction of the genome of non-African humans is derived from Neanderthals. Does this mean that *Homo sapiens* and *Homo neanderthalensis* should not be classified as two separate species? Explain your answer using one or more of the species concepts we have discussed in this chapter.

10. The highly polymorphic HLA alleles in humans reflect polymorphisms that have been conserved since before the divergence of humans and chimpanzees approximately 5.5 million years ago. Explain how this can be consistent with the observation that the most recent common ancestor of all human mitochondrial genomes—sometimes referred to as the *Mitochondrial Eve*—dates to around 200,000 years ago.

SUGGESTED READINGS

Coyne, J. A., and H. A. Orr. 2004. *Speciation*. Sinauer Associates, Sunderland, MA. An interesting book on the process of speciation.

Johanson, D. 2001. Origins of modern humans: multiregional or out of Africa? American Institute of Biological Science. www.actionbioscience.org/evolution/johanson.html. An online article on human evolutionary history by one of the world's leading experts.

Sobel, J. M., G. F. Chen, L. R. Watt, and D. W. Schemske. 2010. The biology of speciation. *Evolution* 64: 295–315. A current overview of models of speciation, with an emphasis on the ecology of speciation.

Wang, H., E. McArthur, S. Sanderson, J. Graham, and D. Freeman. 1997. Narrow hybrid zone between two subspecies of big sagebrush. *Evolution* 51: 95–102. An overview of the sagebrush peripheral isolate model we discussed in the chapter.

Wood, T. E., N. Takebayashi, M. S. Barker, I. Mayrose, P. B. Greenspoon, and L. H. Rieseberg. 2009. The frequency of polyploid speciation in vascular plants. *Proceedings of the National Academy of Sciences of the United States of America* 106: 13875–13879. An overview of sympatric speciation by polyploidy in plants.

 Visit StudySpace at wwnorton.com/studyspace.

15

Extinction and Evolutionary Trends

15.1 The Concept of Extinction

15.2 Background Extinction

15.3 Mass Extinction

15.4 Factors Correlated with Extinction

15.5 Rates of Evolutionary Change and Evolutionary Trends

◀ A fossilized *Tribrachidium heraldicum* with an unusual disk-shape that has threefold rotational symmetry. The relationship of *Tribrachidium* to other groups is still uncertain, although some researchers have suggested that it may have been distantly related to either corals and anemones or perhaps urchins and seastars. This fossil is from the Ediacaran period (approximately 630–542 million years ago).

F ifty thousand years ago, individuals from more than 150 genera of megafauna (large animals) roamed the Earth across five continents giant ground sloths, mammoths, mastodons, short-faced bears weighing 2500 pounds, saber-toothed cats, and 500-pound kangaroos, to name just a few examples (Figure 15.1). But, by 10,000 years ago, two-thirds of the 150 genera that were present 40,000 years earlier had gone extinct in what is known as the Pleistocene megafauna extinction (Barnosky et al. 2004). They were gone forever, leaving only fossil remains—and on rare occasion, a well-preserved frozen carcass—to alert us to the fact that they had ever existed.

What caused these extinctions? Evolutionary biologists continue to pursue this question, putting forward a number of possible explanations for the megafauna extinction. The evidence on islands strongly suggests that intense hunting by humans as well as less direct human impacts, such as fire, habitat fragmentation, and the introduction of exotic species, played a large role in the Pleistocene megafauna extinction (Martin and Kelin 1984; MacPhee 1999).

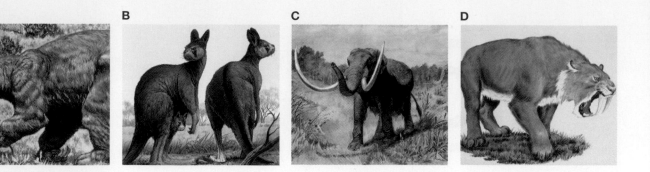

FIGURE 15.1 Pleistocene megafauna. A few examples of the megafauna that existed 50,000 years ago. **(A)** Harlan's ground sloth (*Glossotherium harlani*). **(B)** Giant short-faced kangaroo (*Procoptodon goliah*). **(C)** American mastodon (*Mammut americanum*). **(D)** Saber-toothed cat (*Smilodon gracilis*).

The disappearance of megafauna on continents is less well understood. Dramatic changes in average temperature in certain regions are correlated with many of these extinctions. For example, during the ice age about 18,000 years ago, temperatures were 2°C–5°C colder than modern temperatures at low altitudes, and 10°C–20°C colder at higher latitudes and altitudes (Kutzbach et al. 1998). These temperature changes would have had direct and indirect effects on survival by changing the food chain, and hence the diet of megafauna. As on islands, human hunting again seems to have played a role in some extinctions. Archeological remains suggest that human hunters in this period had superior weapons compared to their predecessors (Bar-Yosef 2002). Human hunting and climate change are not mutually exclusive explanations. Indeed, it may be that intense human hunting precipitated extinction in Pleistocene megafauna that were already on the decline as a result of environmental change (Figure 15.2) (Barnosky et al. 2004).

More work is clearly needed to better understand the precise causes of the Pleistocene megafauna extinction. But one thing is certain: These species are gone forever. All of the exquisite adaptations that natural selection produced in those creatures over millions of years, all of the unique genetic variation that they possessed, all of this is lost.

As we will see, many biotic and abiotic factors have been responsible for extinction over the last 600 million years. In this chapter, we will examine how evolutionary biologists tackle the conceptual, theoretical, and empirical issues associated with extinction. We will address the following questions:

- How does background extinction differ from mass extinction?

- How do processes such as competition, predation, host–parasite interaction, and even the impact of asteroids sometimes lead to the extinction of species and contribute to shaping the diversity of life?

- How does a better understanding of extinction inform us about large-scale changes in the history of life—massive pruning of the tree of life and the subsequent growth of newer, often quite different, branches?

- Are there certain attributes of a species or taxa that make them more or less prone to going extinct, either in periods of mass extinction or during background extinction?

- Is the rate of evolutionary change steady and gradual, or punctuated by periods of rapid change?

- Are there evolutionary "trends" both at the microevolutionary and the macroevolutionary scale? If so, what explains these trends?

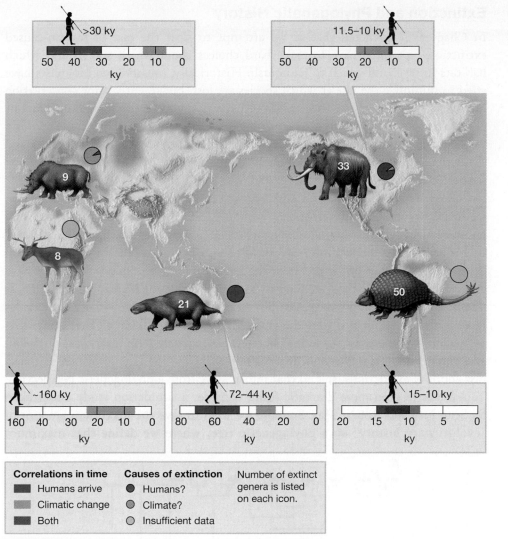

15.1 The Concept of Extinction

Although it may seem obvious, even trivial, we need to begin with a definition of extinction. When we say that a species has gone extinct, we mean that all individuals in that species have died out and left no living descendants. If all species in a genus are extinct, then that genus is extinct, and similarly for all genera in a family, and so on. Today, we know that most species that have ever lived have gone extinct. But it is worth noting that the very notion that a species could go extinct was a matter of much debate until the turn of the nineteenth century. Many philosophers, religious leaders, and even scientists ruled out the possibility of extinction, as this would suggest a less-than-perfect world, and hence an imperfect supernatural creator of that world.

When we study extinction as a process that is occurring right now in contemporary times, we can literally search for the last living representatives of a species. When we look at extinction in evolutionary time, we most often must use evidence from the fossil record to determine if a species has gone extinct.

Extinction and Phylogenetic History

In Chapter 1, we noted that as we attempt to slow the rate of human-caused extinctions, we often have to make hard choices about which species and which habitats to save, and which to relinquish. Historically, conservation biologists have tried to minimize the rate at which species are lost, but recently some conservation biologists have suggested that we should try to conserve the maximum amount of what is sometimes referred to as phylogenetic diversity (Faith 1992). There is a very general point here, above and beyond decisions relating to human activity and extinction. And that point is that not all extinctions are equal with respect to how they affect phylogenetic history (Nee and May 1997; Erwin 2008; Purvis 2008).

To see why, look at Figure 15.3, which depicts three different hypothetical extinction scenarios (Erwin 2008). In scenario A, we lose 7 species of the 21 terminal species on our tree. Although this is an extinction rate of 33%, no entire clade has gone extinct—no major branches are lost from the tree. Scenario B illustrates a single hypothetical extinction event, but had it occurred, the *largest* clade on the tree would never have evolved. In scenario C, there are three hypothetical extinction events, but this time, as a result of extinction, the three most divergent clades on the tree—that is, the clades most distant evolutionarily from the others on the tree—would never have evolved.

Which of these three scenarios is preferable is partly a philosophical question, and partly a practical question. But for our purposes, the key point here is that a raw count of extinctions isn't all that informative from a phylogenetic perspective. To see this in even more dramatic form, consider a simulation study undertaken by Sean Nee and Robert May. Suppose we wish to save the maximum amount of "evolutionary history" on a phylogenetic tree, where we define that maximum

FIGURE 15.3 The consequences of extinctions depend on the phylogenetic distribution of those extinctions. Black horizontal lines represent extinction events. Clades in red would never have evolved given prior extinctions. Three different hypothetical scenarios are depicted here. **(A)** Seven of twenty-one taxa are lost (33% extinction), but the general structure of the phylogeny is preserved. **(B)** A clade of seven taxa would never have evolved as the result of a single extinction, but the rest of the tree would have been preserved. **(C)** Three extinctions lead to seven taxa never having evolved. Here we lose the clades most divergent from the rest of the tree. Adapted from Erwin (2008).

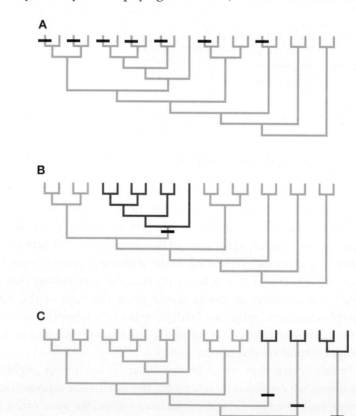

amount in terms of the number of clades that have at least one remaining species after some large-scale extinction event. Nee and May demonstrate that sometimes the loss of many species translates into the loss of much phylogenetic history, but that, under certain reasonable conditions, even if only 5% of the species survive a large-scale extinction, more than 80% of the evolutionary history of the tree can be preserved (Nee and May 1997).

Extinctions and the Fossil Record

Fossil evidence is the key to understanding the history of extinction on Earth. Paleontologists typically define a **fossil** as the remains or traces of a past-living organism, and the term is usually reserved for remains or traces that are greater than 10,000 years old (Prothero 2003; Larsen 2008). Fossil remains, once part of a living organism, are slowly transformed over time into rock. Minerals such as calcium and phosphorus, which were once part of a living organism, are slowly replaced by such minerals as iron and silica through chemical processes (on rare occasions, organisms are fossilized within a gooey tree resin called amber) (Figure 15.4). Fossils vary in the extent to which this replacement process has occurred: The longer the fossilization process has been going on, the more rocklike the fossil. In fossils formed fairly recently, bones, skin, and even remains in the organism's digestive tract are sometimes uncovered. In some instances, the biochemical substances in these once-living organisms can be extracted and analyzed.

Organic remains that have been fossilized into rock are not the only way to tap into what we call the **fossil record**, by which we mean the history of life on Earth as recorded by fossil evidence of one type or another. Sometimes water seeps into fossils and breaks down the fossil that has formed (this is referred to as dissolution), but the *shape* of the fossil is preserved in the sediment around it, providing a rough outline of the organism. In other cases, organisms fossilize as layers of thin carbon spread on sandstone and shale (this is referred to as carbonization), a process that is particularly common in plants (Figure 15.5). We can also use the fossil record to tell us about the environment that organisms lived in—and hence the selective conditions they faced—by employing geological analyses of oxygen content, acidity, and other properties of fossils and their surrounding substrates.

The process of fossilization requires just the right conditions to occur (Figure 15.6). A dead organism, or at least parts of it, must be buried—often by soil deposited by flowing water, or by volcanic ash—and the remains must stay in an anoxic (oxygen-free) environment. Although fossilization can occur in many types of rock, it most often occurs in sedimentary rock, such as chalk, limestone, sandstone, and shale, which makes up only about 5% of all rocks and occurs mostly as a thin layer on the surface.

Even when geological and abiotic conditions are right for fossilization to occur, many factors can disrupt the process. Predators and scavengers often leave little behind of dead organisms, even those that die in areas conducive to fossilization.

A

B

FIGURE 15.4 Fossils. (A) The beautiful fossilized remains of *Darwinius masillae*, a new primate genus and species found recently in Messel, Germany. This fossil is approximately 47 million years old. **(B)** A fossil of a hymenopteran insect in amber found in Ethiopia. This fossil has been dated at approximately 95 million years old.

FIGURE 15.5 Carbon layer plant fossil. A flower of *Porana oeningensis* fossilized as a thin carbon layer on rock.

FIGURE 15.6 The process of fossilization. Fossilization can occur in many ways. Here a dead gazelle lies on the shore. Soft tissues quickly decay, and only skeletal remains are left. After the water level rises, sediments settle on the remains of the gazelle, producing anoxic conditions needed for fossilization. Adapted from Larsen (2008).

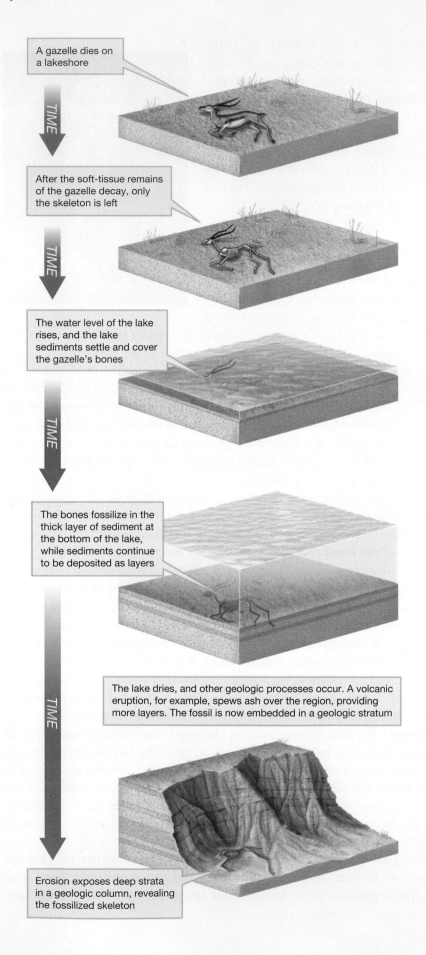

A gazelle dies on a lakeshore

TIME

After the soft-tissue remains of the gazelle decay, only the skeleton is left

TIME

The water level of the lake rises, and the lake sediments settle and cover the gazelle's bones

TIME

The bones fossilize in the thick layer of sediment at the bottom of the lake, while sediments continue to be deposited as layers

The lake dries, and other geologic processes occur. A volcanic eruption, for example, spews ash over the region, providing more layers. The fossil is now embedded in a geologic stratum

TIME

Erosion exposes deep strata in a geologic column, revealing the fossilized skeleton

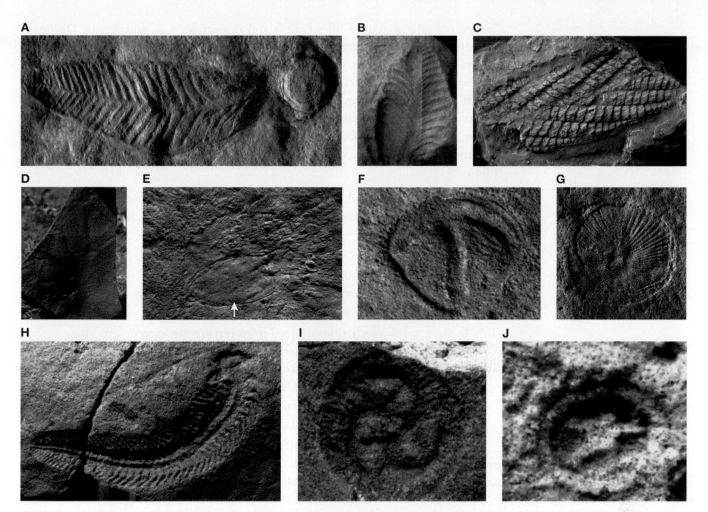

FIGURE 15.7 Ediacaran fossils. Some of the body plans represented in the Ediacara biota. **(A)** *Charniodiscus* frond. **(B)** *Rangea.* **(C)** *Charnia* frond. **(D)** *Swartpuntia* frond. **(E)** *Kimberella* (white arrow). **(F)** *Parvancorina.* **(G)** *Dickinsonia.* **(H)** *Spriggina.* **(I)** *Tribrachidium.* **(J)** *Arkarua.*

Soft tissues rarely remain long enough to fossilize, which is why so much of what we see in the animal fossil record consists of hard substances that were once teeth, bones, shells, exoskeletons, and so on. Wind, water currents, and other abiotic processes break down even those parts of an organism that could fossilize. It is not surprising, then, that paleontologists typically find evidence of only a single bone, tooth, shell, or exoskeleton of an organism, and they rarely find anything as complete as we what we see in Figure 15.4. On occasion, however, much more complete fossils—multiple bones, even full skeletons—are uncovered. On even rarer occasions, geological and biotic conditions in the past have, by chance, been such that huge numbers of fossils are found together—these are referred to as *Lagerstatten* (a German word meaning something like "resting place"). Examples of these include the exquisite Ediacara fossils from 635 to 541 million years ago— fossils that tell of a huge burst of new multicellular organisms—and the Burgess Shale fossils dating from about 520 million years ago (Xiao and Laflamme 2009) (Figure 15.7).

Paleontologists use many factors when deciding where to search for fossils of the organisms they study. Take the case of *Tiktaalik roseae*, which we discussed in Chapter 5 (Daeschler et al. 2006; Shubin et al. 2006). Researchers chose the

Canadian Arctic region of Ellesmere Island in part because, during the Devonian period 375 million years ago, the area was subtropical and replete with the shallow stream systems like those that evolutionary biologists hypothesized would be associated with the transition from water- to land-based life. Serendipity also played a role here. Not long after the researchers began work on Ellesmere Island, they found a *Tiktaalik* fossil skull literally jutting out of a stone on an icy bluff. Of course, they then intensified their search as a result.

Some of the factors paleontologists use when choosing sites include the following:

1. At a very broad level, paleontologists will focus on the sites that best match the geological and abiotic conditions in which fossilization may have occurred.

2. In most instances, a paleontologist is unlikely to be the first researcher to be searching for fossils from their organism of interest. In those instances, researchers often begin at or near sites where others have already uncovered related fossils.

3. In some cases, predictions derived from phylogenetic reconstruction, biogeography, and/or molecular genetics might guide paleontologists to a particular area. Recall the two hypotheses we discussed for patterns of speciation in our own genus, *Homo*. Both the multiregional hypothesis and the out-of-Africa hypothesis would suggest that the search for earlier *Homo* fossils be centered at sites in Africa.

These three factors are not mutually exclusive, and they often work in concert. In many cases, it is a combination of all three of these factors that lead paleontologists to choose their sites for excavating the fossil record.

Paleontologists have many techniques for determining the age of a fossil. Some of these techniques provide a measure of relative time and others provide a measure of absolute time. An example of the way relative time is gauged is the **law of superposition**, which states that fossils found lower down in the sediment at a particular locality are older than those found closer to the surface. Certain types of chemical dating also provide information on the relative age of a fossil. For example, fluorine is found in some types of soil, and it builds up in bone remains as they fossilize. At a given site, the older a bone is, the more fluorine it will have in it.

To estimate the absolute age of a fossil, paleontologists use such techniques as **radiocarbon dating** and radiopotassium dating. In 1949, Willard Libby found that one form of carbon—what is known as isotope carbon-14 (^{14}C)—decays into a second form of carbon—isotope ^{12}C—at a constant rate. Every 5730 years, half of the ^{14}C in a substance will decay into ^{12}C. This rate of decay is known as the half-life of an element, and so for ^{14}C the half-life is 5730 years. All living plants and animals absorb small amounts of ^{14}C that is in Earth's atmosphere, but once an organism dies, the intake of ^{14}C ceases. ^{14}C then begins to decay, and so we can measure the age of a fossil by looking at the $^{14}C/^{12}C$ ratio in its remains. Because ^{14}C has a short half-life, radiocarbon dating is a useful tool for measuring absolute time for about 50,000–75,000 years, after which point there is usually not enough ^{14}C remaining to use the technique.

The half-life of other elements can also be used to date fossils. Potassium-40 (^{40}K) has a very long half-life of approximately 1.3 billion years. After 1.3 billion years, half of the ^{40}K in a substance will have decayed into a gas form of argon-40 (^{40}Ar). When a volcano erupts, the heat is so intense that all the ^{40}Ar gas present is driven off, but the nongaseous ^{40}K remains. Because of this, scientists can use the ratio of gaseous ^{40}Ar (that is, ^{40}Ar formed by the decay of ^{40}K) to nongaseous ^{40}K to date the igneous rocks that form when the molten lava created by volcanic eruptions cools. Although most fossils are found in sedimentary rock, the age of a fossil can be estimated by dating igneous rock layers above and below it. Paleontologists also use the half-life of elements such as uranium-235 (which decays to lead-207 with a half-life of 700 million years) to date fossil beds and the fossils within them.

Paleomagnetic dating can also measure the age of a fossil by using changes in Earth's magnetic field (the position of true north). Over the last 6 million years, the polarity of the Earth has changed back and forth on numerous occasions. These shifts cause the metal grains in the crust to realign. By measuring the alignment of metal particles in the substrate in which their fossil was found, paleontologists can estimate a relative date for that fossil.

Finding high-quality fossilized remains is grueling and painstaking work. Despite the fact that evolutionary biologists and paleontologists have developed hypotheses for where it's best to search for fossils, and despite the fact that we now have exquisitely good techniques for dating strata (layers of rock) in the Earth's surface, the last *fossilized* remains of an organism that we find are rarely if ever from the last actual survivor of that species. And the further back in time we go, the fewer the fossils that can be recovered. Thus, the last fossilized sample of a species that can be recovered is rarely, if ever, the last true survivor of a species that has gone extinct. This time lag between the last known fossil and actual extinction is called the **Signor–Lipps effect**, named after Jere Signor and Philip Lipps, who came up with this idea (Signor and Lipps 1982). The Signor–Lipps effect is a form of "backward smearing": Its effect is to make us date an extinction earlier than it actually occurred.

A second problem associated with dating extinction from the fossil record is called "forward smearing." A common cause of forward smearing is the fact that burrowing animals move fossilized remains up through layers of earth and distort the fossil record. Worms and shrimp, for example, stir up sand and sediment and, in so doing, they push fossilized remains into a strata that is more recent than the one in which the now-extinct species perished. This then makes it appear that a species that is actually extinct still shows fossilized remains well after its extinction. One way that paleontologists minimize forward smearing effects is to search for evidence of extensive burrowing in the strata from which they obtain their fossils (Jin et al. 2000).

Magnitude of Extinction: Background Extinction versus Mass Extinction

The fossil record shows that rates of extinction vary over time, and that extinction rates sometimes spike in what are referred to as **mass extinctions**. When extinction occurs outside a period of mass extinction, it is referred to as part of **background extinction**. The distinction between background and mass extinction can, in principle, be fuzzy, but as we will see in this chapter, in practice these are usually easy to distinguish.

Although there is no hard-and-fast definition adopted by all evolutionary biologists, a mass extinction usually refers to a series of events that cause large-scale loss—on the order of 40 to 50% of all species in many major taxa—over a broad geographic range (Benton 2003a). Evolutionary biologists have evidence for between five and eight mass extinctions over the course of the last 600 million years, with some of these extinctions wiping out 90% of all species that existed before that point. But mass extinctions are few and far between, and of all the extinctions that have ever occurred, about 95% have *not* been associated with a mass extinction; rather, they represent background extinction (Raup 1986, 1992; Pimm et al. 1995). And so it is with background extinction that we begin to delve more closely into the phenomenon of the extinction of species.

15.2 Background Extinction

Our discussion of background extinction will focus on testing hypotheses about extinctions caused by predation, competition, disease, and climate change. Many of our examples involve species that are **endemic**—that is, native to only one area (Figure 15.8). The reason for focusing on endemic species is twofold: (1) extinctions are common in such species (IUCN 2001; Jansson 2003), and (2) it is much easier to study extinction in endemic species because, in these cases, local extinction becomes synonymous with global extinction.

Extinction and Predation

Many organisms manifest antipredator adaptations, both morphological and behavioral. For example, groups of squirrels will often mob a snake predator, biting and harassing it until the snake is forced to leave the area (Owings and Coss 1977; Coss and Owings

FIGURE 15.8 Endemic hot spots. Endemic hot spots for vertebrates around the world. The number of endemic mammal, reptile, and amphibian species in each location is indicated. Adapted from MHHE (2010) and based on data from Myers (1988) and World Conservation Monitoring Center (1992).

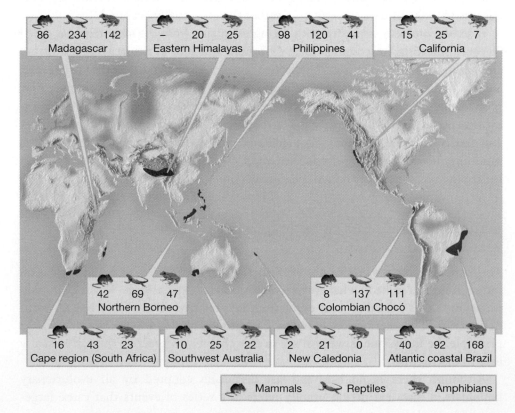

1985; Coss 1991). Their antipredator behavior also includes kicking dirt and rocks at predators, as well as emitting alarm calls that specifically signal that snakes, as opposed to other predators, are present (Owings and Leger 1980) (Figure 15.9).

Not only does natural selection operate on prey to favor behaviors that help them avoid predators, it also operates on their predators, improving their ability to capture their prey. We will examine this sort of *coevolutionary arms race* further in Chapter 19. For now, our point is that natural selection can favor traits in predators that make them very efficient at capturing their prey. Such efficient foraging behavior may result in the extinction of the prey species. In addition, when new predators enter an area—either through migration or some form of human introduction—they may cause extinction of the prey species that they feed on. Here we will look at two examples in which human-introduced predators have led directly or indirectly to the extinction of native fauna. Of course, many background extinction events are not tied to human introduction of new predators to an ecosystem, but these human introductions have been especially well studied, which is why we focus on them here.

Cats and Seabird Extinctions

The California Channel and Northern Baja Islands are home to many seabird species. These islands are also home to an array of mammals that have been introduced by humans since the late 1800s (Anthony 1925). Human-introduced mammals include such potential predators as cats, dogs, and black rats.

Black rats, which occur on at least 7 Channel Islands, are responsible for a sharp decline in small, hole-nesting birds, although no extinctions are as yet attributed to rat predation. Feral cats, which now live on 10 of the Channel Islands, have had a devastating impact on a number of local seabirds, including dramatically lowering population size in black-vented shearwaters (*Puffinus opisthomelas*), Cassin's auklets (*Ptychoramphus aleuticus*), and Xantus's murrelets (*Synthliboramphus hypoleucus*). Predation by cats has also been responsible for the extinction of the endemic Guadalupe storm-petrel (*Oceanodroma macrodactyla*) (Jehl and Everett 1985; McChesney and Tershy 1998) (Figure 15.10).

Human introductions of cats and rats have caused the most dramatic declines in seabird populations in areas of the Channel Islands that have no native mammals. In those areas, there has been no selection on birds to maintain or develop mammalian antipredator behaviors, and these species were consequently hardest hit when predatory mammals were introduced.

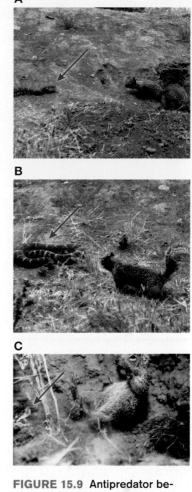

FIGURE 15.9 Antipredator behavior in squirrels. (A) Ground squirrels emerging from their burrow recognize snakes as predators. (B) Confrontations with rattlesnakes (red arrow) are common, and (C) they sometimes lead a squirrel to kick dirt and rocks at the snake (note the snake's head at the red arrow) to defend itself.

FIGURE 15.10 Extinction on the Channel Islands. (A) The Channel Islands. (B) Three species that have had their populations almost brought to extinction by cats: black-vented shearwaters (*Puffinus opisthomelas*), Cassin's auklets (*Ptychoramphus aleuticus*), and Xantus's murrelets (*Synthliboramphus hypoleucus*). (C) The endemic Guadalupe storm-petrel (*Oceanodroma macrodactyla*) was driven to extinction by cats.

Predation, Extinction, and Indirect Effects

Continuing our focus on the California Channel and Northern Baja Islands, we will examine the near-extinction of a carnivorous mammalian predator there (Roemer et al. 2002). First, let's consider some demographic background to set the stage for what has happened: Six of the Channel Islands are inhabited by an endemic fox species, the island fox (*Urocyon littoralis*)—a predator that feeds primarily on mice, insects, and fruit (Roemer et al. 2001b). Two of these six islands are also inhabited by the endemic Western spotted skunk (*Spilogale gracilis*) (Crooks 1994). In addition, one island—Santa Cruz—is home to a human-introduced population of feral pigs (*Sus scrofa*). Until the early 1990s, when foxes and skunks co-occurred on some of the islands, foxes outcompeted skunks for prey, and the foxes were found in high numbers. But then, in about 1992, a series of indirect events led to the near-extinction of foxes by a new predator—the golden eagle (*Aquila chrysaetos*).

What happened is illustrated in Figure 15.11. First, the feral pig population on Santa Cruz grew large enough in the early 1990s to attract a colony of golden eagle

FIGURE 15.11 Interspecific interactions on Santa Cruz Island. Increased pig number led to more eagles, and then to almost no foxes. At the same time, the decrease in fox number led to an increase in the number of skunks. **(A)** Pictured are the feral pig, the golden eagle, the island fox, and the Western spotted skunk. **(B)** A schematic of indirect effects, predation, and extinction on Santa Cruz Island.

A

B

Humans bring to the island pigs...

that escape and form wild populations,

and attract golden eagles,

which also eat the foxes

that were previously keeping skunks in check,

and the skunk population increases dramatically

predators. Before this, golden eagles had been seen in the Channel Islands, but no established colonies of golden eagles had existed in these islands since the 1950s (Roemer et al. 2001a). The eagles then began feeding on pigs, which became a staple of their diet. At the same time, once they became established, the golden eagles also began attacking foxes, with eagles driving fox populations to near-extinction. That is, initial colonization of the eagles occurred as a result of an increased population size of pigs, but once the colonization occurred, it also caused the near-extinction of the island fox.

Not only did the increase in the number of pigs lead to more eagles and then a drastic decline in the fox population, but at the same time, the decrease in the number of foxes led to an increase in the number of skunks (Roemer et al. 2002; Coonan et al. 2005; Knowlton et al. 2007). In this instance, human introduction (of pigs) led to the establishment of a new predator population (of eagles), a near-extinction (of foxes), and an increase in the number of another species (skunks). Extinction often occurs as a result of such tangled ecological and evolutionary interactions.

Indeed, predation may be particularly likely to lead to extinction in cases like this. When a predator relies heavily on a single prey species, a decline in the prey population can restrict the predator's food source enough to reduce or eliminate the predator population before extinction of the prey species can occur. But when a predator relies on one food source (in this case, pigs) and incidentally catches another food source (foxes), the decline in the number of foxes will have relatively little impact on predator numbers, since the predator can always forage on their primary food source. It is also worth noting that in cases in which the predator population is very dense and in which their survival in a location depends on their ability to forage for both the original and the incidental populations of prey, foraging may drive both the primary and the incidental prey to extinction.

Extinction and Disease

Over the course of the last 30 to 40 years, there has been a major decline in amphibian populations worldwide, including the extinction of many amphibian species (Houlahan et al. 2000; Collins and Storfer 2003; Storfer 2003; Wake and Vredenburg 2008) (Figure 15.12). Recent analyses suggest that the current rate of extinction in amphibians is much higher than typical background extinction rates in this taxon, and some have argued that these rates approach those seen in mass extinctions (McCallum 2007). Although there are many factors that contribute to this steep increase

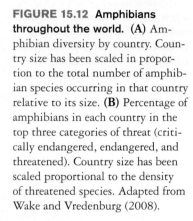

FIGURE 15.12 Amphibians throughout the world. (A) Amphibian diversity by country. Country size has been scaled in proportion to the total number of amphibian species occurring in that country relative to its size. **(B)** Percentage of amphibians in each country in the top three categories of threat (critically endangered, endangered, and threatened). Country size has been scaled proportional to the density of threatened species. Adapted from Wake and Vredenburg (2008).

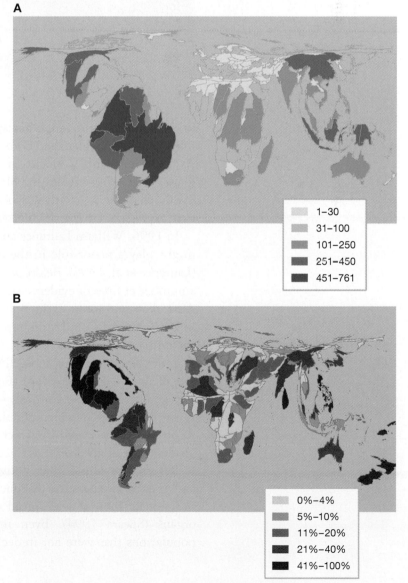

FIGURE 15.13 Disease and amphibian decline. (A) Tropical rain forests in eastern Australia. (B) Spread of amphibian population declines in eastern Australia measured from the distance of the outbreak of disease. Circles indicate where population size declined dramatically in less than 12 months. Declines spread across eastern Australia at a rate of roughly 100 km/year. Adapted from Laurance et al. (1996).

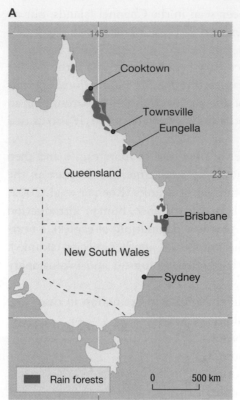

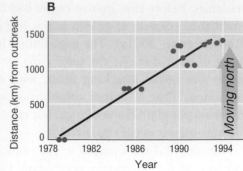

in extinction rates in amphibians, here we will focus on one of the major culprits, infectious disease (Hero and Gillespie 1997; Berger et al. 1998; Daszak et al. 1999). As a case in point, we focus on 14 species of frogs native to the Australian rain forest (Tyler 1991; Ingram and McDonald 1993; Laurance et al. 1996). Since the late 1970s, as many as 7 of these species may have gone extinct in the wild, and others have experienced a dramatic decrease in their population size.

In 1996, William Laurance and his colleagues suggested that infectious disease might play a major role in the decline of the Australian rain forest frog species (Laurance et al. 1996). Before we identify the specific infectious agent, let's look at a number of lines of evidence suggesting infectious disease as a major cause of the Australian rain forest extinctions. To begin with, the dramatic declines occurred in a specific order, and quite rapidly. At first, populations in the southern part of the rain forest began to decline. Next, populations farther north were affected—a map of the populations in decline shows a wavelike pattern moving north, at a rate of about 100 km per year (Figure 15.13). Once a population began to decline in size, it dropped precipitously—often by 80% or more—in a matter of months. Both the manner in which populations declined in a wavelike pattern from south to north and the speed by which populations declined are classic signatures of a virulent infectious disease as the cause.

Researchers found that individuals in affected populations were lethargic, and they showed motor dysfunction and anemia. Histological analysis of dead individuals showed widespread damage to the kidneys, liver, skin, and other organs (Speare 1994). Even more striking, when healthy individuals from populations that were not in decline were introduced into populations that were

A   **B**

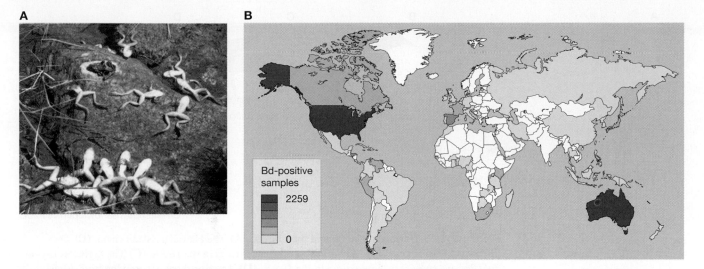

FIGURE 15.14 **Chytridiomycosis and amphibian decline.** **(A)** Consequences of a chytridiomyco-sis outbreak in the Sixty Lake Basin. **(B)** Global data on the prevalence of *Batrachochytrium dendroba-tidis* (Bd). Part B from Bd-Maps (2010).

in decline, the healthy individuals often soon began to display the symptoms and pathologies described above (Laurance et al. 1996). While Laurance and his colleagues suggested a number of possible diseases as the culprit behind the population decline and extinction of many of the Australian rain forest frog species, a definitive diagnosis of which specific disease was responsible was not made for another 2 years, when the disease was identified as chytridiomycosis (caused by a chytridiomycete fungus, *Batrachochytrium dendrobatidis*) (Laurance et al. 1996; Berger et al. 1998). *B. dendrobatidis*, which interferes with the ability of amphibians to transport chemicals across the epidermis, has been found not only in frogs from the Australian rain forest, but also in sick individuals in declining populations of Panamanian frogs. Moreover, the distribution of the disease itself is broad (Voyles et al. 2009) (Figure 15.14).

Multiple Causes of Background Extinction

Our examples above are not meant to suggest that predation, competition, and disease are mutually exclusive explanations for background extinction. Indeed, in many cases, two or all three of these causes may be connected to background extinction. To see this, let's consider the bird extinctions on the Hawaiian Islands.

The Hawaiian Islands are home to a diverse array of plants and animals, many of which are native to the islands. But a majority of the bird species that existed on the Hawaiian Islands just a few thousand years ago have gone extinct, and most of the species we see today are the result of human introduction (Figure 15.15).

The Pacific Islands, including the Hawaiian Islands, have gone through at least two waves of human colonization (Diamond 1984a,b; Milberg and Tyrberg 1993; Smith et al. 1993; Pimm et al. 1995). From about 4000 to 1500 years ago, people in the first wave of human colonization emigrated from the East Indies and settled on the Hawaiian Islands. The second wave, which was primarily led by European explorers, began with Magellan and essentially came to an end when Captain Cook

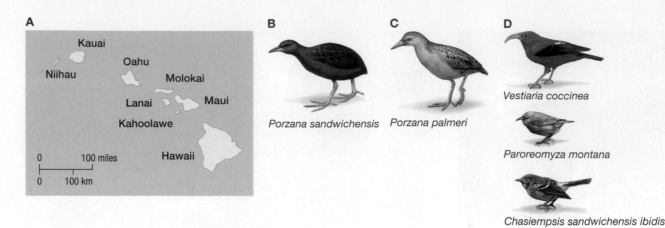

FIGURE 15.15 Hawaiian Islands and their birds. (**A**) The Hawaiian Island chain. (**B**) The flightless Hawaiian rail (*Porzana sandwichensis*) went extinct in the 1890s. (**C**) The flightless Laysan rail (*Porzana palmeri*) became extinct in the 1940s. (**D**) An assortment of surviving bird species endemic to the Islands.

died on the Hawaiian Islands in 1779 (Pimm et al. 1995). Stuart Pimm and his colleagues found evidence that 30 species of land-dwelling birds went extinct during the first wave of human colonization (Pimm et al. 1994, 1995; Boyer 2008). And just since 1800, at least 19—and probably many more—Hawaiian bird species have gone extinct.

Of the approximately 125 to 145 bird species that once inhabited the Hawaiian Islands before human colonization, 90 to 110 are now extinct, many as the result of direct and indirect interactions with humans. Disease introduced by humans, predation by humans who hunted birds for food as well as for decorative feathers, predation by human-introduced species, competition with human-introduced species, and destruction of the native habitat by humans have acted together to lead to the background extinction that has occurred on these islands (Pimm et al. 1995, 2006).

When Alison Boyer analyzed the data on the Hawaiian Islands extinctions, she uncovered some fascinating patterns with respect to which species were most likely to survive human colonization (Boyer 2008). During the first round of colonization—the prehistoric colonization—bird species that were large, flightless, and nested on the ground suffered much higher rates of extinction than other bird species. Evidence suggests that this was largely a result of humans who were hunting the less mobile targets. In the second human colonization wave, many large species of birds were already extinct (Figure 15.16). At that time, bird species that fed on insects and nectar were especially susceptible to extinction. Why? The evidence suggests that habitat destruction by humans and human-introduced predators devastated the lowland forests of the Hawaiian Islands during the second wave of colonization, and those forests were home to many birds that fed on insects and nectar.

FIGURE 15.16 Hawaiian bird extinctions. Extinction of Hawaiian bird species during the first (prehistoric) wave of human migration and the second (historic) wave. Data shown in relation to body mass. Adapted from Boyer (2008).

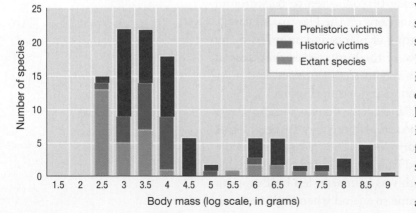

15.3 Mass Extinction

Think about this for a moment: If there had not been a mass extinction about 65 million years ago, reptilian dominance of the land would likely have continued for some indefinite amount of time, which in turn would likely have kept mammals as they were—small, mouse-sized, nocturnal creatures. There would have been no mammalian expansion, no primates, no humans. As paleontologist David Jablonski has written:

> To the conservation biologist, there is little positive to be said about extinction. From an evolutionary perspective, however, extinction is a double-edged sword. By definition, extinction terminates lineages and thus removes unique genetic variation and adaptation. But over geological time scales, it can reshape the evolutionary landscape in more creative ways, via the differential survivorship of lineages and the evolutionary opportunities afforded by the demise of dominant groups and the postextinction sorting of survivors. (Jablonski 2001, p. 5393)

Although no precise technical definition exists, a mass extinction typically refers to the wholesale loss of many, many groups of organisms over a broad geographic range. At least five, and perhaps as many as eight, such mass extinctions have occurred over the last 600 million years—at the end of the Ordovician, in the Late Devonian, the Late Permian, at the end of the Triassic, and at the Cretaceous–Tertiary boundary (Figure 15.17). The numbers associated with mass extinctions can be staggering—estimates for the late Permian extinction have somewhere from 80 to 96% of all marine species going extinct (Raup and Sepkoski 1979; Stanley and Yang 1994).

And it isn't just the sheer numbers of species that are lost that make the effects of mass extinction so dramatic. For example, Douglas Erwin found that mass extinction is not only associated with decreasing the total number of surviving species and genera (loss of taxonomic diversity), but it also decreases diversity with respect to morphology (form and structure of organisms), behavior (measured in the fossil record by "traces" left by burrowing organisms, herds of animals, and so on, indicating movement of organisms), the number of different types of niches inhabited by organisms, and developmental patterns (for example, number of body parts) (Erwin 2008).

The effects of mass extinction can be far-reaching in time. Jablonski has coined the phrase "dead clade walking" to describe clades that survived a period of mass extinction, only to go extinct some time in the following geological time period (Jablonski 2002). These groups were found in substantial numbers in four of the five mass extinctions recorded (the one exception being the mass extinction occurring at the end of the Triassic period). Jablonski found that the geological time periods immediately following mass extinctions were often marked by the subsequent loss of 10–20% of the orders of marine invertebrates that had made it through

FIGURE 15.17 Extinction rates of marine families over time. Spikes represent mass extinctions.

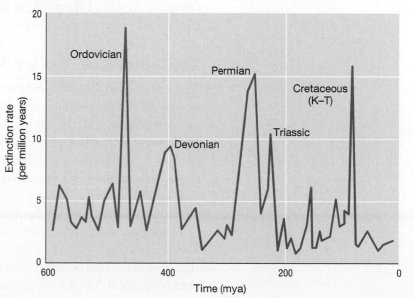

FIGURE 15.18 Dead clades walking. Many clades survive periods of mass extinction, only to go extinct some time in the following geological time period. In four **(A, B, C, E)** of the extinctions shown here, the survival rate for genera during the period before the mass extinction (yellow dot) was higher than the survival rate during the period that followed the mass extinction (blue dot). Only **D** shows higher survival after the mass extinction. The abbreviations on the *x*-axis indicate geological time periods going forward in time from left to right. Adapted from Jablonski (2002).

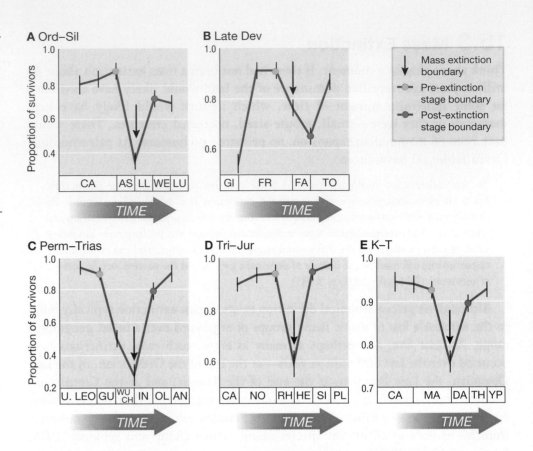

the mass extinction (Jablonski 2002). This rate of extinction was significantly greater than that seen in the geological time period immediately preceding the mass extinction, providing evidence that the post–mass extinction deaths were higher than normal background extinction in the marine invertebrate orders in question (Figure 15.18).

As we have seen, evolutionary biologists have gathered data of various sorts on a number of the mass extinctions that have occurred over the last 600 million years. But for a whole suite of reasons, including access to fossil beds and clues to causation, much of the work on mass extinctions has focused on the Permian and Cretaceous–Tertiary extinctions.

The Cretaceous–Tertiary (K–T) Mass Extinction

The most well-known, and well-studied, of the mass extinctions occurred approximately 65 million years ago, at the boundary of the Cretaceous and Tertiary periods (Figure 15.19). This most recent of mass extinctions—often called the **K–T mass extinction** after the German words for Cretaceous and Tertiary—had profound effects on many different taxa, both in the water and on the land, flora and fauna, invertebrate and vertebrate. The most famous victims were the dinosaurs. Conservative estimates report that half of all the genera alive before the end of the Cretaceous period died off during this mass extinction.

Geologists and evolutionary biologists first worked under the assumption that, although the K–T extinction was a mass extinction in terms of its effect on diversity, this mass extinction occurred gradually over the course of millions of years, likely as a result of gradual changes in temperature, humidity, sea level, and other environmental properties. This initial assumption of slow, gradual change makes good sense. Recall our discussion of Lyell's theory of uniformitarianism in Chapter 2. Geologists and evolutionary biologists since Lyell's day have gathered enormous amounts of data suggesting that change—both geological and biological—is often slow and gradual. But in the case of the K–T extinction, the more data that were collected, the less it looked like the mass extinction was the result of slow gradual processes. The alternative—some sort of catastrophic event—had to be reconsidered.

The first serious attempt to do so was made by Dale Russell and Walter Tucker, who suggested that if a giant supernova had exploded near Earth, the radiation produced and the climate changes triggered by such an explosion would be massive enough to cause extinction on the scale seen in the K–T mass extinction (Russell and Tucker 1971). This extraterrestrial hypothesis for the K–T mass extinction has not stood the test of time, but another closely related idea—that the K–T extinction was the result of a large asteroid colliding with Earth—has.

Walter Alvarez, a paleontologist, began discussing the idea of an extraterrestrial cause for the K–T mass extinction with his father, physicist Luis Alvarez, in 1976. If this mass extinction was the result of some large extraterrestrial event, the evidence should indicate that the mass extinction happened rapidly—on the order of years or decades—rather than over millions of years as in the gradualist theories. They started examining the K–T boundary—the rock strata that denote the end of the Cretaceous period and the start of the Tertiary—to see if there was evidence that events surrounding the K–T mass extinction had occurred on the timescale of years or decades. What they found stunned them.

At sites in Italy and Denmark, Alvarez and his colleagues examined a 1-cm-thick layer of clay that demarcates the boundary between the Cretaceous and Tertiary periods. They measured the concentration of 28 elements in the clay sections right above and below the K–T boundary (Alvarez et al. 1980). At their Italian site, 27 of the 28 elements examined were found at similar concentrations above and below the layer of clay demarcating the Cretaceous–Tertiary boundary. The one exception was iridium, which showed a dramatic 30-fold increase. Similar results were found at a Danish test site, with iridium that was just above the K–T boundary measuring 160 times greater than baseline levels. Subsequent work has overwhelmingly confirmed the findings from the sites in Italy and Denmark: This work has found such increased levels of iridium at the K–T boundary at 50 other sites around the world (Figure 15.20) (Ganapathy 1980; Kyte et al. 1980; Smit and Hertogen 1980; Orth et al. 1981; Alvarez et al. 1990).

But so what? What does it matter if the amount of one element in the crust increased, even if the change was dramatic? The answer to that question lies in the

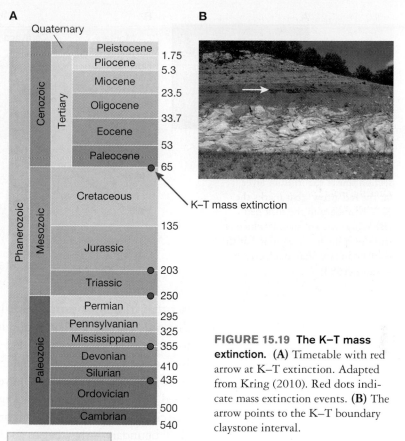

FIGURE 15.19 The K–T mass extinction. (A) Timetable with red arrow at K–T extinction. Adapted from Kring (2010). Red dots indicate mass extinction events. **(B)** The arrow points to the K–T boundary claystone interval.

A

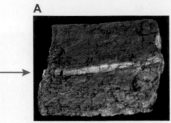

FIGURE 15.20 Iridium at the K–T boundary. **(A)** A stratigraphic section shows the iridium layer (at the red arrow) that marks the K–T boundary in the rock bed. **(B)** World map of sites at which an iridium spike has been found at the K–T boundary. Part B adapted from Alvarez (1983).

B

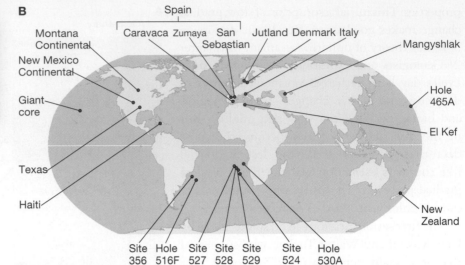

fact that iridium is extraordinarily rare in Earth's crust. Geological work has shown that much of the iridium found in Earth's crust is extraterrestrial in origin, brought to Earth on comets, meteors, and asteroids. Perhaps the spike in iridium found in the K–T clay was evidence that the mass extinction was the result of the collision of such an object with Earth 65 million years ago. As Alvarez and his colleagues methodically ruled out alternative explanations for a spike in iridium—for example, under very rare circumstances, iridium might have built up in seawater and been deposited at the K–T boundary—they began to think that the iridium indeed was extraterrestrial in origin (Alvarez 1983; Alvarez et al. 1980, 1984a,b, 1990; Kastner et al. 1984).

Alvarez and his team hypothesized that an asteroid, approximately 10 km in diameter, struck Earth 65 million years ago. Models (that were initially built to examine the impact of the use of nuclear weapons) indicate that if a 10-km asteroid struck the surface of Earth, it would form a huge crater (Figure 15.21). Giant tsunamis would strike, wiping out everything in their paths. Huge amounts of particulate matter from the crater would shoot into the atmosphere, spread around the globe, and block out sunlight. Photosynthesis would plummet, causing a collapse of the food chain and mass extinction.

Are there asteroids out in space large enough to cause such an impact? Astronomers have calculated that at any given time, there are about seven asteroids of 10 km in diameter or larger that are orbiting Earth (Chapman et al. 1978; Wetherill 1979), and that, on average, one of these asteroids collides with Earth approximately every

FIGURE 15.21 Asteroid impact craters. **(A)** The 1.13-km-diameter Pretoria Saltpan impact crater. **(B)** Arizona's Barringer Meteor crater. The meteor that produced this crater was only 150 feet in diameter, *much* smaller than the asteroid that likely was the major cause of the K–T extinction.

A

B

30 million years. The shutdown of photosynthesis that Alvarez and his colleagues predicted would follow the impact of such an asteroid is consistent with what we know from the huge eruption of the Krakatoa volcano in 1883 (Symons 1888). That eruption shot 18 km^3 of dust into the stratosphere, causing a significant decrease in the average temperature of Earth for over 2 years and a subsequent drop in photosynthesis. Simulations indicate that the impact of a 10-km asteroid would have dwarfed the effects of the Krakatoa eruption. As such, the decreased primary productivity seen in at least some taxa, such as plankton, at the K–T boundary was likely a consequence of the particulate matter that was shot into the atmosphere when an asteroid hit Earth about 65 million years ago (Stuben et al. 2002).

Iridium in the clay at the K–T boundary is only one of the many sources of evidence that now suggest that an asteroid collision was the primary cause of the K–T mass extinction. Subsequent work shows the following:

- There are amino acids of extraterrestrial origin in the clay at the K–T boundary (Bada et al. 1986; Zhao and Bada 1989). Researchers have uncovered both alpha-aminoisobutyric acid and racemic isovaline—two amino acids that are very rare on Earth but that are major amino acids found on comets, meteors, and asteroids—at the K–T boundary.

- At the K–T boundary there are hard glassy minerals, called spinels. These appear to have crystallized from the vapor associated with the impact of a large asteroid (Smit and Kyte 1984).

- Tiny impact diamonds were formed as a result of a large collision approximately 65 million years ago (Carlisle and Braman 1991; Carlisle 1992).

- Evidence from over 3000 fossils, from 340 genera of bivalves (clams, oysters, scallops, and others) indicates that extinction occurred on a global level, as predicted from an impact of a huge asteroid (Raup and Jablonski 1993).

- Recent work from astrophysics indicates that the breakup of a huge asteroid in the solar system's inner asteroid belt would have produced large fragments, a shower of which would likely have struck Earth about 65 million years ago (Bottke et al. 2007). Nonetheless, there remains an active debate as to whether this event was associated with the K–T mass extinction (Reddy et al. 2009).

The asteroid hypothesis for the K–T mass extinction suggests that this extinction happened over a very short geological time period—on the order of dozens to thousands of years. And, indeed, there is evidence that the decline in biodiversity was not gradual, at least not for pollen-producing plants, which showed a dramatic decrease that coincides almost exactly with the iridium spike. But is this pattern of *rapid decline* that is seen in pollen-producing plants at the K–T boundary observed in other taxa?

To examine this question, Alvarez and his colleagues examined the fossil record data on four groups that were common during the Paleozoic era—ammonites (extinct organisms whose closest living relatives are cephalopods like octopuses, squids, and cuttlefish), bryozoans (small aquatic invertebrates), branchiopods (marine invertebrates that have hard shells on their upper and lower surfaces and use a fleshy, stalklike structure to burrow), and bivalves. They found clear evidence for the rapid disappearance of these four groups, almost exactly at the K–T boundary, as expected from the asteroid hypothesis. That said, Alvarez and his colleagues found that a gradual decline was in fact under way in *some* of these groups *before* the K–T boundary. Ammonites, for example, often went through cycles of abundance and decline, and they were likely in decline 65 million years ago. Nonetheless, ammonites had also gone through a dramatic decline in diversity earlier in their evolutionary history, and they had rebounded at that time. The impact of an asteroid, then, may have precipitated a mass extinction of ammonites that likely would not have occurred otherwise. Indeed, Alvarez argues that the groups that were in a period of decline, perhaps in response to normal environmental fluctuations, were generally those that were most susceptible to extinction when an asteroid hit (Alvarez et al. 1984a,b).

If there was a dramatic mass extinction as the result of an asteroid colliding with Earth 65 million years ago, where is the huge crater that would have resulted from the impact? Alvarez and his team recognized the importance of finding this crater, but they were not particularly optimistic about discovering it when they first hypothesized the asteroid theory in the early 1980s. After all, a substantial portion of pre-Tertiary ocean had long ago disappeared under the Earth's surface, which could have made it impossible to find the crater (Alvarez et al. 1980, 1990).

In 1980, when the asteroid hypothesis first came to light, there were about 100 large craters identified around the world. Alvarez and his team immediately encountered a problem—they estimated that the K–T crater should be about 150 to 200 km in diameter, and yet the available data did not suggest any crater that matched that description. A few of the 100 craters were large enough, but they were the wrong age.

An important discovery by Jody Bourgeois put the Alvarez team on the path that would eventually lead to the discovery that some would come to call the "crater of doom." The Brasos River, which empties into the Gulf of Mexico, was one of the many sites showing high levels of iridium at the K–T boundary. At that site, Bourgeois found sedimentary evidence that a giant tsunami had struck approximately 65 million years ago (Bourgeois et al. 1988). Recall that the asteroid hypothesis predicts that just such a tsunami would be produced if the impact occurred in water.

Working with data from geology, oceanography, and many other disciplines, in 1991 Alan Hildebrand and his group published the first account of the location of the crater associated with the K–T mass extinction (Hildebrand et al. 1991; Sharpton et al. 1992; Urrutia-Fucugauchi et al. 1996). The so-called Chicxulub crater was discovered in the Yucatán peninsula of Mexico (Figure 15.22). In fact, scientists at PEMEX—the state oil company of Mexico—had known of the existence of this crater for many years, but it took Hildebrand and his group to recognize that this might be the crater that Alvarez and his team were looking for. The Chicxulub crater had very high levels of iridium, it was the right size and the right age, it was located near the site of a giant tsunami that would have been generated by an asteroid impact, and it was composed of rock with mineralogical characteristics that showed evidence of a massive collision (Izett 1991; Sigurdsson et al. 1991; Smit et al. 1992; Swisher et al. 1992).

The Chicxulub crater provides us with evidence that an asteroid collision occurred at the K–T boundary that had profound effects on the worldwide

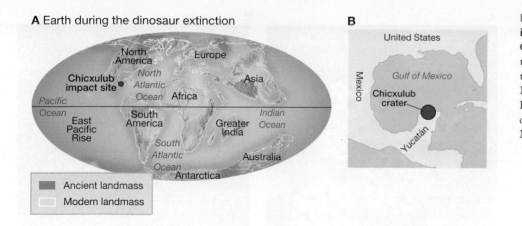

A Earth during the dinosaur extinction

North America
Europe
North Atlantic Ocean
Asia
Chicxulub impact site
Africa
Pacific Ocean
East Pacific Rise
South America
Greater India
Indian Ocean
South Atlantic Ocean
Australia
Antarctica

■ Ancient landmass
□ Modern landmass

B

United States
Mexico
Gulf of Mexico
Chicxulub crater
Yucatán

FIGURE 15.22 The Chicxulub impact site. (A) Location of the Chicxulub crater on a map of what the continents looked like at the time of the K–T extinction. From National Geographic (2010). (B) Location of the Chicxulub impact crater in the Yucatán Peninsula, Mexico.

environment and on a large number of life-forms that existed at the time of its impact. Nonetheless, there remains an active debate as to whether the asteroid that produced the Chicxulub crater was *the* asteroid that set the K–T extinction into motion, or whether it was only one of a series of such asteroids (Keller et al. 2004a,b).

The Late Permian Mass Extinction

Dramatic as the K–T mass extinction was, the greatest mass extinction on record occurred much earlier, at the end of the Permian period, approximately 250 million years ago. In the late Permian mass extinction, an astonishing 90% of all species went extinct (Figure 15.23). Many major groups of plants, animals, fungi, and so on, went completely extinct—in essence, the slate of life was *almost* wiped clean. In his book *When Life Nearly Died: The Greatest Mass Extinction of All Time*, Michael Benton suggests that if we think of the diversity of life as a tree, then during the late Permian extinction, "vast swathes of the tree are cut short, as if attacked by crazed, axe-wielding madmen. . . . After such a severe attack, the great tree of life, with over 3000 million years of history behind it at the time, might have withered away and died completely" (Benton 2003a, p. 10).

Damage in the Sea

Rather than try to categorize the approximately 90% of aquatic and 70% of terrestrial species that went extinct during the late Permian mass extinction, we will discuss some of the taxa that were most devastated. Then we will examine which, if any, characteristics were shared by the species that survived. This will then lead us into an examination of the possible causes of the late Permian mass extinction.

Plankton—small, often microscopic creatures that drift in water currents—were hit hard during the late Permian mass extinction. For example, one group of protozoan plankton called the radiolarians were nearly annihilated, despite the fact that their extinction rate prior to the end of the Permian period was relatively low (Rampino and Adler 1998; Rampino 1999).

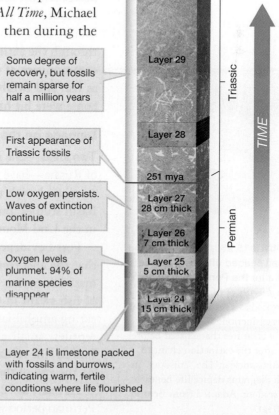

FIGURE 15.23 The late Permian mass extinction. A schematic of the Permian extinction as seen in a rock bed from China. Adapted from Benton (2003b).

Some degree of recovery, but fossils remain sparse for half a milliion years

Layer 29

First appearance of Triassic fossils

Layer 28

251 mya

Low oxygen persists. Waves of extinction continue

Layer 27 28 cm thick

Layer 26 7 cm thick

Oxygen levels plummet. 94% of marine species disappear

Layer 25 5 cm thick

Layer 24 15 cm thick

Layer 24 is limestone packed with fossils and burrows, indicating warm, fertile conditions where life flourished

Triassic

Permian

TIME

FIGURE 15.24 Two groups affected by the late Permian extinction. (A) Radiolara (Phaeodaria). **(B)** Fossilized fusulinids cover the surface of this rock slab collected in Kansas.

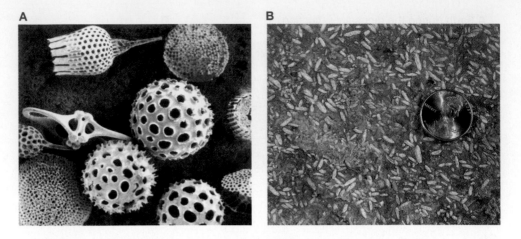

Another plankton group called the fusulinids, which were first thought to have gradually declined before the end of the Permian period now seem to have died off en masse about 251 million years ago—5000 species of plankton existed prior to the late Permian mass extinction, and scarcely any of them remained after it (Kozur 1998; Bragin 2000; Isozaki et al. 2004). Since plankton are at the base of the aquatic food chain, their destruction during the late Permian mass extinction had effects on the survival of many organisms higher in the food chain (Figure 15.24).

Fossil evidence suggests that hundreds of species inhabited the many coral reefs that thrived during the Permian period—mollusks, starfish, shrimp, fish, and so on, abounded near these often massive structures. And then, 251 million years ago, the fossil record shows a "reef gap"—that is, for the 7 to 8 million years that followed the late Permian mass extinction, reef life was remarkably depleted (Figure 15.25). Moreover, the reefs that evolved subsequent to the Permian extinction looked very different from the vibrant communities that had existed before. These reefs appear to have been strung together by the few surviving species that managed to make it through the extinction, and they took millions of years more to develop the diverse communities of organisms associated with them. Other aquatic creatures—such as fish and bivalve mollusks—were also hit by the late Permian mass extinction. Some were hit harder than others, but all suffered significant losses.

Damage on the Land

Two areas—one in the Karoo Basin of South Africa and one in the Ural Mountains of Russia—have been the primary repositories for information on land extinctions associated with the end of the Permian period. Here we will focus on the Karoo Basin studies, but the patterns of change found in South Africa were also mirrored in the Ural Mountains (Benton et al. 2000).

Fossil evidence from the Karoo Basin before the late Permian extinction shows that the area teemed with centipedes, spiders, cockroaches, and beetles before the mass extinction (Benton 2003a). There is also fossil evidence that many species of fish swam in the lakes and rivers, as well as fossil remains of 2 species of fish-eating amphibians and 72 species of reptiles. Herbivorous reptiles and carnivorous reptiles, small and large, were plentiful before the end of the Permian period. Indeed, the remains from the Karoo Basin show evidence for a complex ecosystem. At the end of the Permian period, this Karoo ecosystem disappeared.

As many as 72 of the 74 sampled species of vertebrates present before the end of the Permian period went extinct. At the global level, 36 of the 48 families of amphibians

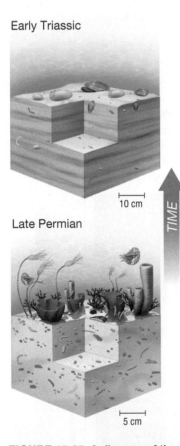

Early Triassic

10 cm

Late Permian

5 cm

FIGURE 15.25 A diagram of the ocean seabed off China before and after the Permian extinction. The *diversity* of life dramatically decreased as a result of the Permian mass extinction. Whether the *density* of creatures that existed before and after the extinction changed remains unclear. The "cut outs" in both diagrams depict life beneath the seafloor. Adapted from Benton (2003a).

and reptiles vanished (Maxwell 1992; Benton 1993, 1997). There was also a massive dieoff of plant species, with huge amounts of excess sedimentation. Many species of woody trees and bushes went extinct, leaving mostly low-lying mosses and lycopsids (small vascular plants) on an otherwise fairly barren landscape (Eshet et al. 1995; Retallack et al. 1996; Retallack 1999; Looy et al. 2001; Smith and Ward 2001; Twitchet et al. 2001).

FIGURE 15.26 The Siberian Traps. A map of the world during the Permian period shows the area of the massive volcanoes in the Siberian Traps. Adapted from Benton (2003b).

What Caused the Greatest Mass Extinction of All Time?

Having laid out the consequences of the late Permian extinction, we now turn to its cause. Paul Wignall has put together a multilayered hypothesis to explain what happened. Wignall begins with a well-established geological fact—251 million years ago, there was a series of huge volcanic eruptions in Siberia in an area known as the Siberian Traps (Campbell et al. 1992) (Figure 15.26). It is estimated that these eruptions spewed between 2 and 3 million km^3 of lava into the air, and that this lava covered almost 4 million km^2 of Siberia to a depth of somewhere between 400 to 3000 m. The resulting cycle of extreme cooling and extreme heating of the planet that these eruptions produced may have brought about the greatest mass extinction ever recorded.

This cycle began with a brief period of global cooling followed by global heating of the planet. The huge amounts of carbon dioxide released from the volcanic eruptions produced a massive "greenhouse effect," raising the temperature around the planet by as much as 6°C, as well as likely causing an increase in ocean acidity. Vast quantities of sulfur dioxide and chlorine were also spewed into the atmosphere. These gases—carbon dioxide, sulfur dioxide, and chlorine—created an atmosphere that was very low in free oxygen. This has led Anthony Hallam and Paul Wignall to hypothesize that creatures well adapted to low-oxygen environments may have been predisposed to survive the Permian mass extinction. Indeed, a detailed analysis of the survivors and victims of the mass extinction at the end of the Permian period by Hallam and Wignall found that, the better that individuals in a species were adapted to low oxygen (hypoxia), the more likely that species was to survive the Permian mass extinction (Hallam and Wignall 1997). The sulfur dioxide and chlorine also created a worldwide acid rain problem that devastated plant life. In addition, one consequence of the greenhouse effect was the melting of polar ice caps and the release of large quantities of methane gas buried around these ice caps.

This deadly mix of increased temperature, global hypoxia, massive amounts of acid rain, and the release of methane gas may have combined in just the right manner to produce the late Permian mass extinction (Figure 15.27).

15.4 Factors Correlated with Extinction

Evolutionary biologists have long been interested in whether certain attributes of a species or taxa make them more or less prone to going extinct, both in periods of mass extinction and during background extinction. In general, we hypothesize that any factor that allows a species to better endure environmental

FIGURE 15.27 Direct and indirect effects of the Siberian Traps eruptions on the Permian extinction. A complex combination of direct and indirect effects resulted from the Siberian Traps eruptions. Adapted from Benton (2003a).

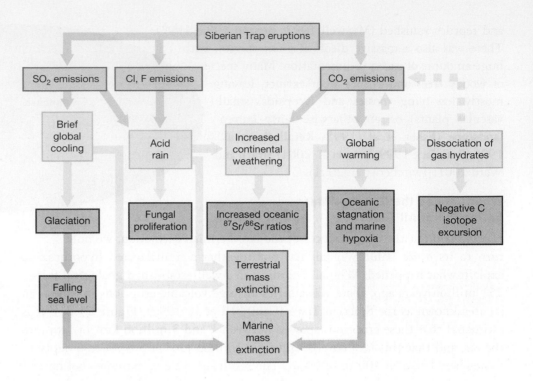

perturbations—changes in temperature, oxygen availability, and so on—might reduce the probability of extinction. We have seen one example of this during the late Permian mass extinction and the differential survival of species that fared well in low oxygen environments. A second example comes from a 2009 study of 4536 species of modern mammals, which found that species that went into hibernation or that used burrows or hiding places of some sort—all means of escaping environmental perturbations—were at lower risk of extinction. They were less likely to be classified as "endangered" on the International Union for Conservation of Nature Red List than were species that did not hibernate or use burrows or hiding places (Liow et al. 2009).

In this section, we will look at two other factors that have been examined as possible correlates with the probability of extinction: (1) species' longevity, and (2) species' geographic range.

Species' Longevity and Extinction Probability

What do we expect when we compare the length of time a taxon has existed and its probability of going extinct in a subsequent time interval? One hypothesis is that the longer a taxon has existed, the less likely it is to go extinct at any given point in time, because long-lived taxa have demonstrated the ability to adapt to their environment. An alternative hypothesis is that species might have some constraints on their life span; the longer they lived, the more we might expect local conditions, both biotic and abiotic, to have changed, making it less likely that the species will survive much longer. Or perhaps the age of a species is irrelevant to its chances of extinction; maybe extinction has nothing to do with how long a species has existed.

To distinguish among these possibilities, Leigh van Valen plotted the probability of extinction as a function of species' longevity in a wide array of different taxa. He found that species' longevity had no effect on the probability of extinction in virtually any of the taxa he examined (Figure 15.28). Why? Van Valen suggested that this was because how well species in a taxon had adapted in the past was irrelevant to the probability of extinction in the future (Van Valen 1973). That is, the biotic and abiotic environments are always changing, and extinction is a function of how well individuals in a species adapt to the *current* environment, not how well they adapted to *past* environmental conditions.

Species' Geographic Range and Extinction Probability

Whatever the cause or causes of extinction may be, evolutionary biologists hypothesize that the broader the geographic range of a species, the less likely that species will go extinct (Manne et al. 1999; Foote 2003; Jones et al. 2003). The logic here is straightforward. The broader a species' geographic range, the less likely that each and every population of which it is composed will be extirpated.

Jonathan Payne and Seth Finnegan analyzed fossil data from 12,300 marine invertebrate genera spanning from the Middle Cambrian (about 500 million years ago) through the Middle Miocene (about 14 million years ago), and they subdivided their samples in 10-million-year periods of time (Payne and Finnegan 2007). They found very strong support that broad geographic ranges reduced rates of extinction of species within genera (Figure 15.29). Such an effect was statistically significant in 44 of 47 10-million-year time frames analyzed by Payne and Finnegan.

The exceptions to the rule that wide geographic range is positively correlated with the probability of survival were clustered around times of mass extinctions, at which time the correlation was weaker. This makes some sense, since mass extinctions themselves are very broad geographically, which would dampen the generally positive effect that a species' geographic range would normally have on survival. But for some taxa, geographic range is correlated with species' survival even around periods of mass extinction. David Jablonski and his colleagues' work on geography and extinction in gastropods (slugs and snails) provides a good example of this.

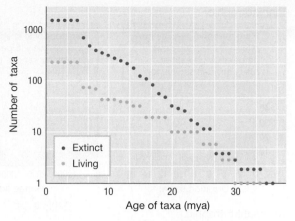

FIGURE 15.28 The probability of extinction is not a function of how long a taxon has already existed. Here we see data on mammalian genera. An extinct genus is one in which all species are extinct, while a living genus has species that are not extinct, as well as extinct species in it. On the vertical axis is the number of surviving taxa, and on the horizontal axis is the age of a taxon. The vertical axis is measured on a logarithmic scale. Because of this scaling, when the probability of extinction is unrelated to the age of taxa, we expect a linear relationship, similar to the one seen here. Adapted from Van Valen (1973).

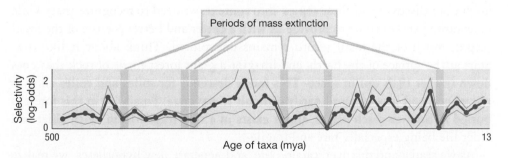
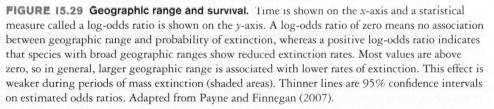

FIGURE 15.29 Geographic range and survival. Time is shown on the *x*-axis and a statistical measure called a log-odds ratio is shown on the *y*-axis. A log-odds ratio of zero means no association between geographic range and probability of extinction, whereas a positive log-odds ratio indicates that species with broad geographic ranges show reduced extinction rates. Most values are above zero, so in general, larger geographic range is associated with lower rates of extinction. This effect is weaker during periods of mass extinction (shaded areas). Thinner lines are 95% confidence intervals on estimated odds ratios. Adapted from Payne and Finnegan (2007).

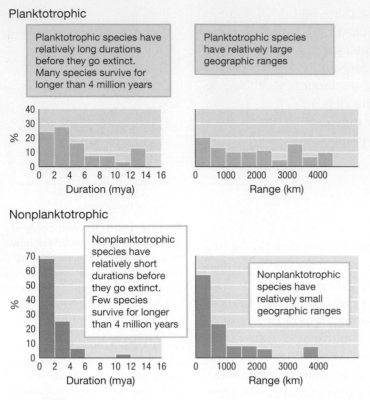

Planktotrophic

Planktotrophic species have relatively long durations before they go extinct. Many species survive for longer than 4 million years

Planktotrophic species have relatively large geographic ranges

Nonplanktotrophic

Nonplanktotrophic species have relatively short durations before they go extinct. Few species survive for longer than 4 million years

Nonplanktotrophic species have relatively small geographic ranges

FIGURE 15.30 Larval development and survival rates. The relationship between larval development mode (planktotrophic versus nonplanktotrophic), species' geographic range, and extinction in gastropods of the late Cretaceous era. Adapted from Jablonski and Lutz (1983).

Jablonski and his co-workers found that, for gastropods of the Late Cretaceous period, the key to a broad geographic range at the species level—and hence to increasing the chances of surviving the mass extinction at the K–T boundary—is a specific form of larval development. Planktotrophic larvae feed in the open water on very small prey (zooplankton and phytoplankton), and they develop into adults at a relatively slow pace. Because they are small sized for a long period of time and live in the open water, planktotrophic larvae are often dispersed long distances, leading to a broad geographic range for species with such larvae compared to species with nonplanktotrophic larvae. When Jablonski examined the larval development patterns, extinction rates were half as high in planktotrophic species (Figure 15.30) (Jablonski and Lutz 1983).

15.5 Rates of Evolutionary Change and Evolutionary Trends

In the last section of this chapter, we will discuss both rates of evolutionary change and evolutionary trends. These topics will allow us to connect our discussion of speciation in the previous chapter with our discussion of extinction in the current chapter and will help us to bridge the gap between microevolutionary and macroevolutionary approaches to the study of evolution.

Rates and Patterns of Evolutionary Change

In this section, we address questions about *rates* and *patterns* of evolutionary change. To put our discussion of these topics into context, we need to recognize that, while paleontologists continue to provide us with a better and better picture of the fossil record, much of the fossil record remains unexplored. Think about it like this: Start at the surface of the Earth, and imagine a giant forest made of rock that goes down (rather than up) in space. This forest descends for miles and miles all over the planet, and it is extraordinarily difficult to navigate. Fossil life in this forest is buried in hard-to-reach places, and it exists in a very fragile form, with older and older life being especially rare and fragile.

As we search and discover, catalog life, and generate new hypotheses, we realize that our inverted forest, spanning the Earth's circumference and going miles and miles down, is a vast area, and much of it remains dark and hidden. This poses some difficult challenges, especially as we aim to infer the rates and patterns of change that have occurred over evolutionary time. In earlier chapters, we have seen cases where paleontologists can trace a lineage back through evolutionary time and obtain a fairly detailed record of the changes that have occurred in that lineage, including the birth and death of new species. But in most instances, we get a more fragmentary view of evolutionary change. We see snapshots at various points in time, with little information about what has occurred between these snapshots.

Fossil snapshots pose an interesting problem. Suppose that we find evidence for one particular species in the fossil record from 5 million years ago. Then the fossil record from 4 million years ago shows two new species, both of which appear to be descended from that prior species. We may not have much in the way of fossil evidence from the interval spanning 5 to 4 million years ago, either because we have not had the opportunity to investigate this stratum or we have done some investigation but found few fossils. In either case, we might say that there is a gap in the fossil record.

With respect to rate and pattern of change, at least two interpretations of this sequence are possible. On the one hand, the fossil record may be hiding a series of small-scale changes that lead from the original species to its two descendant species. On the other hand, very little change may have occurred during most of the period 5 to 4 million years ago, and the changes that led to the diversification of species we uncovered may have happened quite rapidly, perhaps so rapidly that we would not expect them to be captured in the fossil record.

Presumably, if we had extensive fossil information from the period 5 to 4 million years ago in our hypothetical case, we could distinguish between these two possibilities. The problem for evolutionary biologists is that these gaps in the fossil record are common, and they will likely remain so for the foreseeable future (recall our forest analogy). As a result, in the early 1970s, two conceptual schools developed around how to interpret the sorts of data on rates and patterns of evolutionary change when such gaps exist. One of these schools of thought is called phyletic gradualism and the other is known as punctuated equilibrium (Eldredge and Gould 1972; Gould 1985, 2002).

The **phyletic gradualism model** of evolutionary change can be traced to Darwin and his argument that adaptations that arise within a species are the result of a slow, gradual process, where any variant that provides the slightest net benefit slowly increases in frequency. Perhaps more critically, on many occasions, Darwin argued that this very same slow, gradual process eventually led to the formation of new species.

The theory of phyletic gradualism hypothesizes that new species arise from a gradual transformation of an ancestral species through slow, constant change. A classic example of phyletic gradualism is the case of equine evolution that we discussed in Chapter 5, in which slow, gradual changes in the skull and limb morphology led to new equine species over evolutionary time (Figure 15.31). According to this view, the new forms that appear in the fossil record may arise either through branching speciation events—a process known as **cladogenesis**—or through gradual modification of form over evolutionary time without branching speciation, a process known as **anagenesis**. Paleontologists sample from the fossil record. When the lineage being studied has changed enough—through the slow, gradual accumulation of adaptive and nonadaptive changes—it is considered a new species. The earlier forms no longer occur in the fossil record, and it appears as if this earlier species has gone extinct. We call this phenomenon

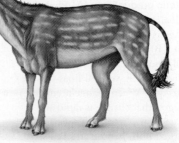

FIGURE 15.31 Phyletic gradualism and equine evolution. The fossil record of equine evolution reveals a gradualist pattern, with slow and continuous physiological change in equine species over time. Shown here is an early equid of the genus *Eohippus*.

FIGURE 15.32 Cladogenesis, anagenesis, and pseudoextinction. This diagram illustrates the evolutionary fate of a lineage over geological time. Labels A–F indicate species as classified by paleontologists from the fossil record. In this diagram, species A undergoes cladogenesis, giving rise to species B and C by a process of branching speciation. The lineage that includes species B goes truly extinct, whereas the lineage that includes species C changes gradually and by the process of anagenesis is eventually classified as species D. We say that species C has undergone pseudoextinction. Species D then undergoes another round of branching speciation, resulting in species E and F.

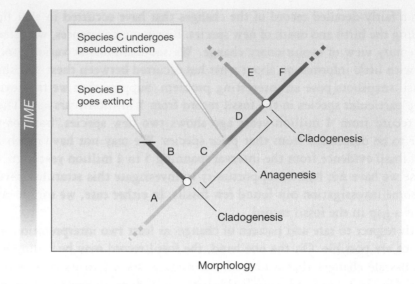

Morphology

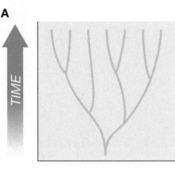

Morphology

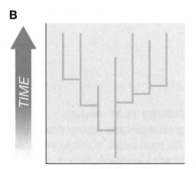

Morphology

FIGURE 15.33 A schematic of evolutionary trees predicted by punctuated equilibrium and phyletic gradualism. (A) Phyletic gradualist models hypothesize that change is slow, gradual, and constant. **(B)** According to punctuated equilibrium theory, stasis—represented by long, vertical lines—is the status quo. When change occurs in lineages, that change is rapid and associated with branching speciation. Adapted from Benton and Pearson (2001).

pseudoextinction (Smith et al. 2001), because the lineage has not actually died out; rather, its members have changed so much that they are now reclassified as a new species (Figure 15.32).

Eldredge and Gould's **punctuated equilibrium** theory provides an alternative to phyletic gradualism. This view posits that major evolutionary changes do not occur through a slow, gradual process. Instead, while *some* minor degree of change is always occurring within lineages, stasis—the *absence of change*—is the rule during the vast majority of a lineage's history. When evolutionary change does occur in lineages, it is not only rapid, but it typically leads to branching speciation—that is, cladogenesis. Thus, periods of rapid morphological change coincide with bursts of rapid branching speciation (Figure 15.33).

As an extreme example of a period of rapid evolutionary change, consider what is known as the **Cambrian explosion**. Fossil evidence from the Cambrian period, from approximately 543 to 490 million years ago, shows a huge spike, not just in the number of marine species, but in the number of genera, families, and other taxonomic units, as well as in an exquisite array of new multicellular creatures with new body forms and shapes. Indeed, most of the animal groups that have ever lived appeared in the fossil record for the first time during a 5–25-million-year interval in the early Cambrian period (Conway Morris 1998).

Much of the evidence for the Cambrian explosion comes from an extraordinary bed of fossils known as the **Burgess Shale** (in British Columbia, Canada), which, for a complicated set of geological reasons, contains samples from soft-bodied species that elsewhere tend to fossilize poorly if at all. Active debate continues about the causes of the Cambrian explosion. Our point here is not that the Cambrian explosion is best explained by punctuated equilibrium models—there is scant evidence one way or the other on this—but just that the fossil record does show periods in which evolutionary change seems to be rapid, such as during the early Cambrian period, and other times when this is not the case.

The theoretical underpinnings of punctuated equilibrium are tied to Mayr's peripheral isolate model of allopatric speciation (Chapter 14) (Eldredge and Gould 1972; Gould and Eldredge 1993). If speciation most often occurs in small peripheral populations, we are likely to see large-scale evolutionary changes arising out of rapid punctuated bursts of change. But why? To answer that question, imagine a large mainland population of animals surrounded by an archipelago of small islands. Imagine that these islands are colonized by individuals from the mainland. (The argument applies equally well to other forms of geographic isolation.) As we have learned, when small genetically isolated subpopulations are adjusting to local conditions, natural selection combined with genetic drift may lead to rapid evolutionary change. If such change is occurring on many small islands, all populated from some larger mainland, the rate of branching speciation may be rapid.

But there is more to it than that. If populations on islands diverge rapidly, then sometimes one of the new species on an island will migrate back to the mainland and coexist with its ancestral species. If we sample the fossil record, we will see the following: Initially we will see our ancestral mainland population. If change occurs rapidly on the surrounding islands, we are unlikely to catch that speciation in the fossil record. When we next look back at the mainland, we may find our ancestral species and one of its descendant species that seems to have appeared from nowhere. But, of course, it didn't appear from "nowhere." Rather, the descendant species migrated from small, isolated populations that are unlikely to be sampled from the fossil record.

An example of punctuated equilibrium in the fossil record was documented by Alan Cheetham in his work on speciation and evolutionary change in aquatic invertebrates called bryozoans (Cheetham 1986). Using the fossil record from the last 20 million years, Cheetham tracked speciation patterns in one genus of bryozoans (*Metrarabdotos*) by measuring change in 46 morphological characters—including the size and shape of various cells in a colony—in fossil bryozoans from the entire geographic range of genus *Metrarabdotos* (Figure 15.34)

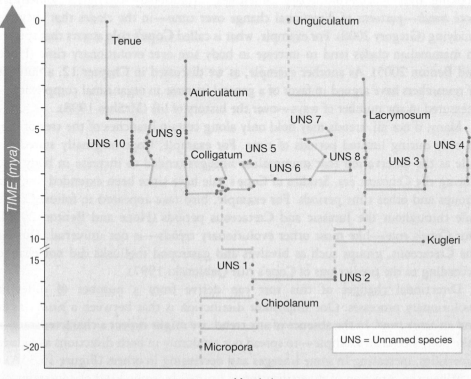

FIGURE 15.34 Punctuated equilibrium in bryozoans. A phylogeny of colonial bryozoan species from the genus *Metrarabdotos*. Time runs along the *y*-axis, and morphological difference is depicted on the *x*-axis. Stasis is shown in long vertical lines, while punctuated speciation is depicted in horizontal lines. UNS = unnamed species. Adapted from Cheetham (1986).

The pattern of speciation uncovered by Cheetham resembles the pattern hypothesized by punctuated equilibrium theory. Most bryozoan species showed little change for long stretches of time (represented by the long vertical lines). Then in a punctuated burst of change, speciation occurred (represented by the horizontal lines). The question of what, if anything, caused rapid speciation to occur in many different lineages of bryozoans at about the same time—6 to 8 million years ago—remains unresolved, although there is some suggestion that "pulses" of speciation in this group might be tied to the rise and fall of oceanic boundaries over time (Jackson and Cheetham 1999).

What are we to make of this work on punctuated equilibrium, phyletic gradualism, and rates of evolutionary change? One review of nearly 60 studies found evidence for *patterns* of punctuated equilibrium in some lineages and phyletic gradualism in others. This review also found numerous studies in which phyletic gradualism best explained change for some period of evolutionary time in a given lineage, while change in other periods of time, for the same lineage, were better described by punctuated equilibrium (Erwin and Anstey 1995).

Evolutionary biologists want more than number estimates of the frequency of punctuated equilibrium, of phyletic gradualism, or of some model that combines the other models. What we seek is to understand whether the predictions from one of these theories are better supported by the data. Is punctuated change, for example, found more often in island archipelagos or in environments that change often and dramatically? As of today, we can't answer these questions, but research on this topic continues.

Evolutionary Trends

Looking over macroevolutionary timescales, evolutionary biologists sometimes note *trends*—patterns of directional change over time—in the clades that they are studying (Gregory 2008). For example, what is called **Cope's rule** asserts that species in mammalian clades tend to increase in body size over evolutionary time (Hone and Benton 2005). As another example, as we discussed in Chapter 12, a number of researchers have argued in favor of a general increase in organismal complexity—measured in any number of ways—over the history of life (McShea 1998).

Many, if not all, trends may hold only along certain branches of the tree of life, or only during limited periods of time. For example, Cope originally stated his rule as the observation that mammalian lineages tended to increase in body size during the Cenozoic era. Studies of Cope's rule have since been extended to other groups and other time periods. For example, bird taxa appeared to follow Cope's rule throughout the Jurassic and Cretaceous periods (Hone and Benton 2005). But Cope's rule—like most other evolutionary trends—is not universal. During the Cretaceous, groups such as bivalves and gastropod mollusks did not change according to the predictions of Cope's rule (Jablonski 1997).

Directional changes of this sort can derive from a number of different evolutionary processes. One important distinction is that between a *passive trend* and an *active trend*. In the absence of any trend, we might expect a character state—take body size as an example—to spread out randomly in both directions as a clade diversifies, increasing in some lineages and decreasing in others (Figure 15.35A).

When the direction of diversification is limited by some kind of constraint on evolution—for example, when the ancestor was already the minimal viable size—

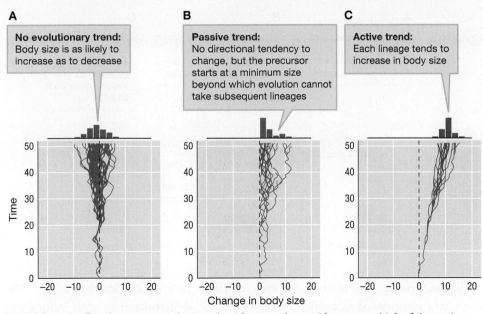

A No evolutionary trend: Body size is as likely to increase as to decrease

B Passive trend: No directional tendency to change, but the precursor starts at a minimum size beyond which evolution cannot take subsequent lineages

C Active trend: Each lineage tends to increase in body size

FIGURE 15.35 Passive versus active trends. These graphs provide a way to think of change in a clade over time. The *y*-axis represents time, while the *x*-axis represents a character state that we are interested in—body size, in this example. In each, a species starts at some point in character space, and over time, evolutionary changes and speciation events lead to increasing diversity in the character states represented within the clade. (**A**) Here we have no evolutionary trend. Different lineages diffuse out in both directions from the original ancestor. (**B**) Here again there is no directional tendency to the evolution of any individual lineage, but our precursor species starts at a boundary—a minimum value—beyond which a lineage cannot go. As a result, all derived lineages have body sizes at least as large, and often larger, than that of the ancestor. This generates a "passive trend" in which the mean size within the clade increases over time, even though at any point away from the boundary, the evolutionary process is equally likely to lead to an increase or a decrease in size. (**C**) Here we see an active trend, in which each lineage tends to increase in the character state. Adapted from McShea (1994, 1998).

variations in body size again will spread out throughout the clade, but only in the unconstrained direction (Figure 15.35B). The result is a trend in the sense that we see an increase in the mean body size within the clade, but we call this a *passive trend* because, away from the boundary, evolution is as likely to lead to a decrease in body size as to an increase in body size (Stanley 1973). In other cases, we may see a tendency for the entire distribution of body sizes within the clade to increase. Such an *active trend* results in an increase in mean body size even without relying on a boundary, as Figure 15.35C illustrates.

Ideally, evolutionary biologists would be able to link an active trend of this type to the underlying selective conditions that generate such a trend. Kingsolver and Pfennig did precisely this in a large-scale analysis of body size evolution in plants, invertebrates, and vertebrates (Kingsolver and Pfennig 2004). They aggregated results from 42 studies of selection on 854 traits in 39 species, and they found that, on average, selection favored traits associated with increased body size. The direction of selection was evenly distributed between increasing or decreasing the magnitude of other traits (Figure 15.36). This kind of work fashions a bridge between microevolutionary studies of local selective conditions and macroevolutionary studies of long-term trends across taxa.

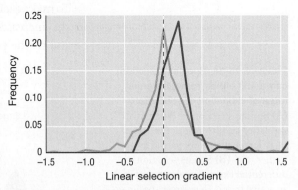

FIGURE 15.36 Selection for increased size. Kingsolver and Pfennig examined selection gradients—a measure of the strength and direction of selection—for 854 traits from 39 species. They found that selection on increased body size tended to be positive (red line), whereas selection on other traits averaged zero (blue line). These results suggest a selective mechanism behind Cope's rule as an active trend. Adapted from Hone and Benton (2005).

FIGURE 15.37 An active trend can arise from either of two evolutionary processes. (A) An active trend—for example, an increase in body size—results in an across-the-board change in the value of the trait of interest; here the entire distribution shifts to the right. **(B)** Species selection occurs when the trait value influences extinction or speciation probabilities. Here smaller subclades go extinct and new subclades tend to be larger in body size. **(C)** Alternatively, an active trend can occur when each subclade undergoes a parallel evolutionary change. Here that change is toward larger body size. Adapted from Adamowicz et al. (2008).

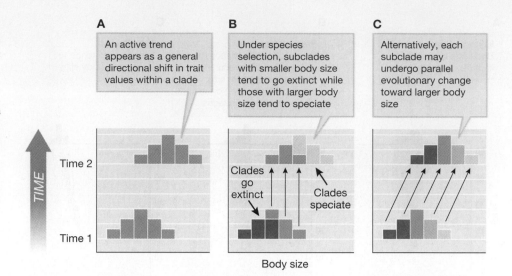

Active trends can arise from two distinct processes (Figure 15.37). The most straightforward is a process by which the distribution of trait values (for example, size) in a clade shifts because the trait values within each subclade shift in parallel. Alternatively, the average trait value may increase due to **species selection**: speciation and extinction rates that vary according to the value of the trait in question. For example, if species with larger body size are more likely to speciate and/or less likely to go extinct, species selection can result in a shift of the distribution of trait values across species (Figure 15.38).

The distinction between parallel evolution within subclades and species selection recalls the distinction we made in Chapter 2 between transformational and variational processes of evolution. But here we are applying these concepts at the level of species. If each subclade goes through a parallel process of evolutionary change, we have a transformational process at the species level. If instead species vary in some trait, and species with certain trait values are more likely to speciate (reproduce) or go extinct (die), we have a variational process at the level of species, in which species are *sorted* according to their trait values.

To make each of these concepts concrete, consider the evolutionary trend manifested in the increase in morphological complexity of the Crustacea from the Cambrian period to the present. Focusing on a trait that has been well preserved in the fossil

FIGURE 15.38 Species selection can result either from differential rates of extinction or speciation. (A) Species selection by differential speciation results in an active trend. Here larger species speciate at a higher rate. **(B)** Species selection by differential extinction also results in an active trend. Here larger species go extinct at a lower rate; we can infer this from the fact that branches of the phylogenetic tree are shorter in smaller species. Although the distinction between differential speciation and differential extinction is conceptually useful, both processes can occur simultaneously. Adapted from Gregory (2008).

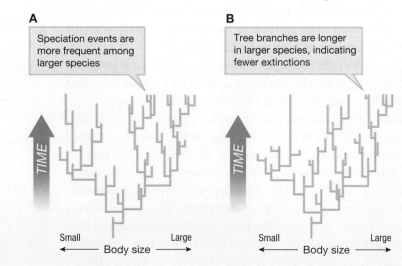

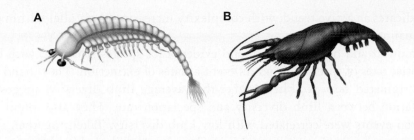

record—the morphological structure of the limb—researchers have noticed that, over the past 500 million years, there has been a trend toward increasing differentiation of limbs, with different morphology on different body segments (Figure 15.39). But why? Is this an example of a passive trend, leading some but not all taxa away from an absorbing barrier (namely, having only one limb type)? Is it an active trend due to parallel evolution toward increasing limb complexity in each of many taxa? Or is it an active trend driven by some kind of species selection?

To resolve this question, Sarah Adamowicz and her colleagues compiled data on 66 crustacean orders from extant organisms and from the fossil record (Adamowicz et al. 2008). For each order, they computed an overall indicator of limb morphological complexity known as the Brillouin index, and they established the time at which the order was present in the fossil record. Plotting the Brillouin index as a function of time (Figure 15.40) and testing for statistical significance, they found that complexity has increased significantly from 500 million years ago to the present. The dearth of minimally complex orders at the present time suggests that this pattern is not simply the result of a passive trend.

But is this increase in complexity due to the differential success of more complex orders, or is it due to a trend toward the development of increasing complexity within individual orders, or to both? To check for parallel evolutionary change within multiple subclades, Adamowicz and her colleagues selected 12 paired comparisons between fossil orders and their closest relatives, chosen so as to make each comparison phylogenetically independent from the others. They found that in 10 to 11 of these cases, the present-day orders exhibited greater complexity than did the fossil orders to which they were related (Figure 15.41).

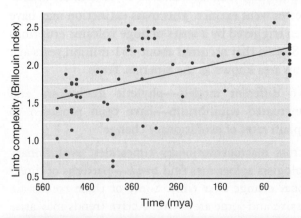

FIGURE 15.40 A trend toward increasing limb complexity in the crustaceans. The complexity of limb morphology in observed fossils increases as we move from 500 million years ago to the present. Note the dearth of low values for present-day and recent fossil species. Adapted from Adamowicz et al. (2008).

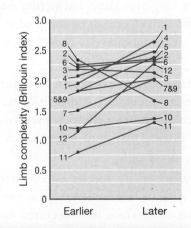

FIGURE 15.41 Increasing limb complexity using phylogenetically independent data points. Paired comparisons (chosen for phylogenetic independence) of fossil crustacean taxa with closest current relatives shows increasing complexity in 10–11 of 12 comparisons. This reveals that there has been parallel evolutionary change in the various subclades. Adapted from Adamowicz et al. (2008).

This indicates an active trend, with complexity increasing in parallel within each individual order.

Adamowicz and her team also found evidence for species selection, with both differential rates of speciation and differential rates of extinction. They found that newly originated taxa exhibited higher-than-average limb diversity, suggesting a correlation between limb diversity and speciation rate. They also found that extinction events were correlated with low limb diversity. Taken together, these observations suggest that species selection has also contributed to the pattern of increasing limb diversity in the Crustacea.

Both because of the effect of extinction on the tree of life, and the alarming rate at which species are going extinct at the hands of modern humans, it is critical that we understand the dynamics of extinction. In this chapter, we have examined background extinction, mass extinction, and evolutionary trends, and how evolutionary biologists study both background extinction and mass extinction for clues as to causes and for information on which to build a conceptual framework for understanding extinction.

SUMMARY

1. A species is said to be extinct when all members of that species have died out and left no living representatives. If all species in a genus are extinct, then that genus is extinct. Most species that have ever lived have gone extinct.

2. When estimating extinction dates from the fossil record, evolutionary biologists must be aware of both backward smearing and forward smearing effects.

3. Rates of extinction vary over time. Extinction rates sometimes spike far above normal, or above what is sometimes called background levels. These spikes in extinction rate are called mass extinctions.

4. Many causes for background extinction have been studied, among them, predation, competition, and disease. Both direct and indirect effects of predation, competition, and disease may lead to background extinction.

5. Mass extinctions affect many species over a broad geographic range. At least five (and perhaps as many as eight) such mass extinctions have occurred over 600 million years—at the end of the Ordovician, in the Late Devonian, at the end of the Permian, at the end of the Triassic, and at the Cretaceous–Tertiary (K–T) boundary.

6. Mass extinction not only leads to fewer species and genera, but it also decreases diversity with respect to morphology, behavior, the number of different types of niches inhabited by organisms, and developmental patterns.

7. The best studied and most famous of the mass extinctions is the K–T mass extinction. A large asteroid that collided with Earth approximately 65 million years ago initiated this mass extinction.

8. The Permian extinction occurred approximately 250 million years ago, and 80 to 96% of all marine species went extinct. This mass extinction may have been triggered by a series of huge volcanic eruptions in Siberia that occurred about 251 million years ago in an area known as the Siberian Traps.

9. Two different models—phyletic gradualism and punctuated equilibrium—have been proposed to explain rates of evolutionary change.

10. Across macroevolutionary timescales, evolutionary biologists sometimes find *trends*—patterns of directional change over time. Some of these trends are passive and some are active. Active trends may arise when the distribution of trait values in a clade shifts because the trait values within each subclade shift in parallel. Alternatively, the average trait value may increase because of species selection.

KEY TERMS

anagenesis (p. 527)

background extinction (p. 507)

Burgess Shale (p. 528)

Cambrian explosion (p. 528)

cladogenesis (p. 527)

Cope's rule (p. 530)

endemic (p. 508)

fossil (p. 503)

fossil record (p. 503)

K–T mass extinction (p. 516)

law of superposition (p. 506)

mass extinction (p. 507)

paleomagnetic dating (p. 507)

phyletic gradualism model (p. 527)

pseudoextinction (p. 528)

punctuated equilibrium (p. 528)

radiocarbon dating (p. 506)

Signor–Lipps effect (p. 507)

species selection (p. 532)

REVIEW QUESTIONS

1. What are the pros and cons of the following argument: The tree of life is simply a summary of speciation and extinction events that have occurred over the last several billion years.

2. Let's return to the following quote from David Jablonski cited in this chapter: "To the conservation biologist, there is little positive to be said about extinction. From an evolutionary perspective, however, extinction is a double-edged sword. By definition, extinction terminates lineages and thus removes unique genetic variation and adaptation. But over geological time scales, it can reshape the evolutionary landscape in more creative ways, via the differential survivorship of lineages and the evolutionary opportunities afforded by the demise of dominant groups and the postextinction sorting of survivors." What do you suppose Jablonski meant by "reshaping the evolutionary landscape in creative ways?"

3. Why do you think that the study of the Pleistocene megafauna extinction was just the sort of work that would bring about collaborations between evolutionary biologists and researchers in archeology, anthropology, and even sociology?

4. Why would you expect extinction rates to be much higher in species that are endemic to islands than to those found more broadly distributed?

5. Has your introduction to the evolutionary study of extinction changed the way you think about issues relating to conservation programs? How so?

6. Think about our discussion of evolutionary trends. How is a scientific understanding of such trends critical in helping to dispel the common misperception that the evolutionary process has "purposes" and "goals"?

7. If Charles Lyell (Chapter 2) had lived to see the debate between the proponents of phyletic gradualism and the proponents of punctuated equilibrium, which side do you think he would have taken? Why?

8. Make the following argument: Because the distinction between what constitutes mass extinction versus background extinction is always going to be subjective, this distinction should be abandoned. Now make the counterargument.

9. Consider the following hypothetical scenario: You are studying Genus G at two time periods: 310 to 300 million years ago and 100 to 90 million years ago. Between 310 and 300 million years ago, you have data on 10 species in G, of which 5 went extinct during this 10 million year period. Between 100 and 90 million years ago, you have data on 100 species, of which 30 went extinct. Which was more severe: the extinctions between 310 and 300 million years ago, where 50% of the species you were studying went extinct but this amounted to a loss of only 5 species, or the extinctions between 100 and 90 million years ago, where 30% of the species in G went extinct but this amounted to 30 different species going extinct?

10. How might the Signor–Lipps effect make what was a mass extinction appear to be the gradual accumulation of smaller extinction events spread over time?

SUGGESTED READINGS

Alvarez, W. 1998. *T. Rex and the Crater of Doom*. Vintage, New York. A popular science book on the K–T extinction.

Benton, M. 2003. *When Life Nearly Died: The Greatest Mass Extinction of All Time*. Thames & Hudson, London. A very nice, but long, general discussion of the Permian extinction.

Boyer, A. G. 2008. Extinction patterns in the avifauna of the Hawaiian islands. *Diversity and Distributions* 14: 509–517. An overview of the two rounds of extinction that have hit Hawaiian bird species.

Gregory, T. R. 2008. Evolutionary trends. *Evolution: Education and Outreach* 1: 259–273. An article that summarizes work on evolutionary trends.

McCallum, M. L. 2007. Amphibian decline or extinction? Current declines dwarf background extinction rate. *Journal of Herpetology* 41: 483–491. A review of recent spikes in amphibian extinction rates.

 Visit StudySpace at wwnorton.com/studyspace.

PART IV

Evolutionary Interactions

Two polar bears (*Ursus maritimus*) spar on the ice of Canada's Hudson Bay.

16

The Evolution of Sex

◀ A single coho salmon female (*Oncorhynchus kisutch*) produces thousands of bright orange eggs.

The New Zealand mud snail, *Potamopyrgus antipodarum*, looks fairly ordinary. It is small—about 4 to 7 mm—very common, and it serves as host to dozens of different parasites that lay eggs between the snail's body and its shell (Figure 16.1). But for the evolutionary biologist, there is something quite extraordinary about these little snails, and it is that different snails in the same lake reproduce in dramatically different ways.

In Lake Alexandria, New Zealand, populations of *P. antipodarum* are composed of two very different types of females. Some females reproduce sexually, mating with and being fertilized by males in the population. Other females reproduce asexually; these *P. antipodarum* females produce unfertilized eggs that mature into the next generation of females. In these asexual lineages, each offspring is a clone of its parent.

Among multicellular eukaryotic organisms, the vast majority of species reproduce only sexually. All species of birds and mammals, for example, reproduce sexually. Other species, such as aphids reproduce sexually some of the time and asexually other times. Some other species reproduce only

FIGURE 16.1 Snails and sex.
(A) The New Zealand mud snail, *Potamopyrgus antipodarum*. **(B)** Lake Alexandria, South Island, New Zealand, one of the primary study sites for work on *P. antipodarum*.

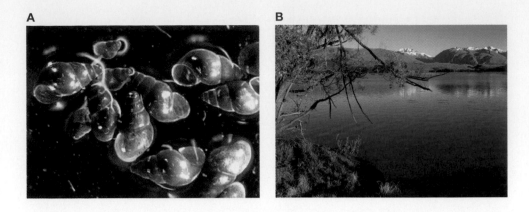

asexually. But it is quite rare to discover a species of animal such as *P. antipodarum*, in which some females only reproduce sexually (obligate sexual reproduction), while others only reproduce asexually (obligate asexual reproduction).

How can both asexual and sexually reproducing females coexist in the same population? After all, asexual females produce genetic clones of themselves, while sexually reproducing females produce offspring that contain genomes that are a mixture of their own and their mate's genes. And as we will see in subsequent sections, all else being equal, asexual lineages should multiply at twice the rate of sexual lineages. So why haven't the asexually reproducing lineages of *P. antipodarum* replaced the sexual lineages? Why do we still see sexual reproduction at all in the New Zealand populations of *P. antipodarum*? Beginning in the late 1980s, Curt Lively and his colleagues set out to answer this question (Lively 1987).

Lively tested three models that had been developed to better understand the conditions that favor sexual reproduction (Figure 16.2). He tested the *environmental unpredictability hypothesis* by examining whether sexual reproduction was seen at higher frequencies in rapidly changing and unpredictable environments (Lively 1987). The basic idea is that, because of recombination, sexual reproduction generates a great deal of genetic variation, and this variation may allow sexual lineages to adapt to unpredictable environments faster than do asexual lineages. Prior work suggested that lakes in New Zealand had changed less over time than had streams. Thus, Lively could compare the frequency of sexual reproduction in lakes versus sexual reproduction in streams and test the prediction that sexual reproduction occurred more frequently in streams, the more unpredictable of the two environments.

The lakes and streams that Lively studied also differed in another fundamental way: Lakes had a greater number of distinct ecological niches—habitats with particular, well-defined resources, competitors, predators, parasites, and pathogens—that could be occupied by snails. Evolutionary theory predicts that sexual reproduction should be favored over asexual reproduction in environments with more niches, because the huge number of genotypes generated by sexual reproduction may include genotypes

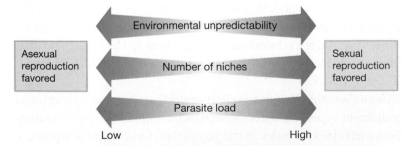
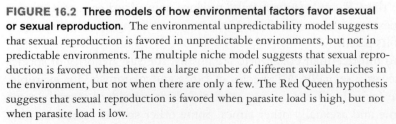

FIGURE 16.2 Three models of how environmental factors favor asexual or sexual reproduction. The environmental unpredictability model suggests that sexual reproduction is favored in unpredictable environments, but not in predictable environments. The multiple niche model suggests that sexual reproduction is favored when there are a large number of different available niches in the environment, but not when there are only a few. The Red Queen hypothesis suggests that sexual reproduction is favored when parasite load is high, but not when parasite load is low.

that are able to colonize niches unavailable to individuals from asexual lineages. This idea is referred to as the multiple niche hypothesis. If this model applies to the New Zealand populations of *P. antipodarum*, sexual reproduction should be more common in lakes than in streams. Note that this is the exact opposite prediction from that of the environmental unpredictability model.

Lively also tested a third model, the **Red Queen hypothesis**, named after the Red Queen in Lewis Carroll's *Through the Looking-Glass*. We will discuss this hypothesis in more depth later in the chapter, but for now, note that this hypothesis predicts that the frequency of sexual reproduction will be related to the level of parasitic infection. In particular, the Red Queen hypothesis predicts that *if* parasites infect an asexual lineage, the parasites are likely to be very successful, since in each generation, their host's genome remains largely unchanged as a result of asexual reproduction. The situation with sexual lineages is quite different. Offspring still resemble their parents, but because of recombination they are not genetic clones. Over generations, the genetic variation produced by sexual reproduction creates a moving target for parasites. Even when parasites can successfully infect sexual lineages, the genetic variation produced by sexual reproduction—including the occasional production of new parasite-resistant genotypes—may favor sexual reproduction when parasites are abundant. Sexual reproduction may not enable the organisms to totally outrun the parasites, but it will at least enable the hosts to keep pace with the parasites rather than being totally overwhelmed by the parasites, as are organisms in asexual lineages.

Lively collected samples of 40–100 snails from each of a number of different lakes and streams across New Zealand, assayed parasitic infections in these snails, and used the frequency of males in each sample to measure the prevalence of sexual reproduction. He found that sexual reproduction was more common (1) in lakes than in streams, as predicted by the multiple niche hypothesis, and (2) in populations that had high parasite loads, as predicted by the Red Queen hypothesis (Figure 16.3). To distinguish between the multiple niche and the Red Queen hypotheses, Lively analyzed the data using statistical tools that allowed him to ask the following question: If we control for differences in the frequency of parasites, do lakes and streams still differ in the frequency of sexually reproducing snails? The answer was "no." Next, Lively asked whether there was still a positive correlation between the frequency of parasites and the frequency of sexual reproduction when he controlled for whether the data came from streams or lakes. This time, the answer was "yes." The Red Queen model for the evolution of sex best explained the frequency of sexual reproduction in populations of *P. antipodarum*.

The literature on the evolution of sexual and asexual reproduction includes other such elegant tests of models. In this chapter, we will examine both asexual and sexual reproduction and talk about the costs and benefits of each type of reproduction. We will also discuss the role of environmental unpredictability and variation and how they can affect the evolution of asexual versus sexual reproduction. Then we will briefly examine a hypothesis for the *origin* of sexual reproduction. We will seek to answer the following questions within this chapter:

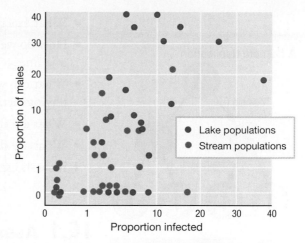

FIGURE 16.3 Parasites and sexual reproduction. Snails are infected with many different parasites. On the horizontal axis is the percentage of individuals infected with *Microphallus* or *Stegodexamene* parasites. The proportion of males in a population—a measure of the prevalence of sexual reproduction in that population—is shown on the vertical axis. Both axes are displayed on a logarithmic scale. Red circles represent data from lakes, and blue circles are data from streams. Adapted from Lively (1987).

A Gamete duplication

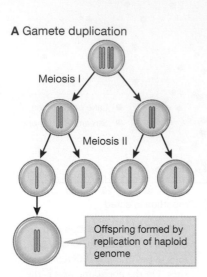

Offspring formed by replication of haploid genome

B Terminal fusion

Offspring formed when meiotic products containing sister chromatids fuse

C Central fusion

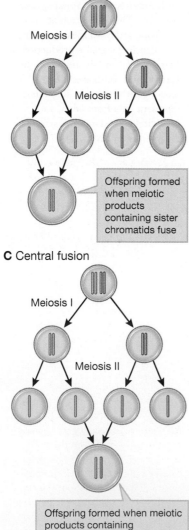

Offspring formed when meiotic products containing homologous chromatids fuse

- What exactly are asexual and sexual reproduction?
- How do we know that sexual reproduction is occurring, or has occurred, in a lineage?
- What can we infer about sexual and asexual reproduction from their phylogenetic distribution?
- What are the costs of sexual reproduction?
- What are the benefits of sexual reproduction?
- How did sexual reproduction originate?

16.1 Asexual and Sexual Reproduction

In order to understand the evolution of sexual and asexual reproduction, we will begin by clearly defining what we mean by each. We will then briefly look at the modes of asexual and sexual reproduction.

Asexual Reproduction

In multicellular eukaryotes, **asexual reproduction** is typically defined as the production of offspring from unfertilized gametes (Schurko et al. 2009). There are two basic forms of asexual reproduction: apomixis and automixis. In both cases, reproduction involves a single, female parent (we do not include selfing, in which hermaphrodites self-fertilize, as asexual reproduction). In **apomixis**, an unfertilized gamete undergoes a single mitosis-like cell division, producing two "daughter cells." Each daughter cell has an unreduced number of chromosomes and is genetically identical to its mother. In plants, apomixis is sometimes referred to as apogamy. **Automixis** involves the production of haploid gametes via meiosis, but diploidy is usually restored by the fusion of haploid nuclei from the same meiosis (some biologists consider this a form of sexual reproduction). Figure 16.4 illustrates several modes of automixis, and their differing consequences for the kinds of offspring that are produced. Offspring from automictic asexual reproduction are genetically different from their parent and their siblings, but much less genetic variation is generated here than in sexual reproduction.

Sexual Reproduction

In the broadest sense, **sexual reproduction** involves the joining together of genetic material from two parents to produce an offspring that has genes from each parent (Barton and Charlesworth 1998). More specifically, the process of sexual reproduction is characterized by amphimixis, which involves alternating phases of

FIGURE 16.4 Three modes of automixis. (A) In gamete duplication, a full round of meiosis is followed by replication of the haploid genome. Even in the presence of recombination, the resulting offspring will be entirely homozygous. **(B)** In terminal fusion, meiotic products containing two sister chromatids fuse to form an individual that is homozygous except for any recombination events that occur during meiosis. **(C)** In central fusion, meiotic products containing homologous chromatids fuse, leading to a heterozygous offspring that would be identical to the parent were it not for recombination events that may have occurred during meiosis. Adapted from Engelstädter (2008).

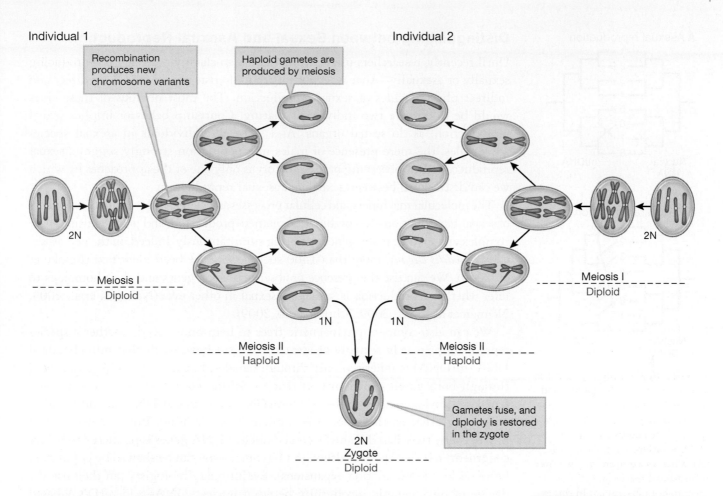

FIGURE 16.5 **Amphimixis.** In sexual reproduction, diploid individuals produce haploid gametes via meioses. Gametes fuse (syngamy), producing a diploid offspring. Adapted from Schurko et al. (2009).

meiosis and gamete fusion (syngamy) (Kondrashov 1993). There are three steps in the process of amphimixis (Figure 16.5):

1. Recombination: the crossover between homologous chromosomes, which produces new chromosomal variants.

2. Gamete production: the production of haploid gametes by diploid individuals via reductive meiotic division.

3. Gamete fusion: the gametic exchange between (usually) unrelated individuals, in which haploid gametes fuse to produce a diploid offspring.

As we will see, the vast majority of multicellular eukaryotes reproduce only sexually, and virtually all eukaryotes reproduce sexually either at some point in their life cycle or periodically across generations. Mechanisms of genetic exchange among bacteria, such as transduction, conjugation, and transformation, are sometimes also referred to as "sex" (Figure 16.6) (Redfield 2001; Franklin 2007; Michod et al. 2008; Vos 2009). But these mechanisms do not satisfy the definition we have presented above (Barton and Charlesworth 1998; Otto and Lenormand 2002).

FIGURE 16.6 **Genetic exchange in bacteria.** Three types of genetic exchange occur in bacteria: transduction, conjugation, and transformation. In all three, a DNA fragment from a chromosome of a donor cell is transferred to a recipient cell, and then it is integrated into the DNA of the recipient cell. Adapted from Redfield (2001).

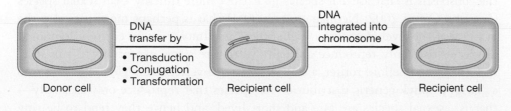

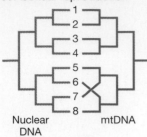

A Asexual reproduction

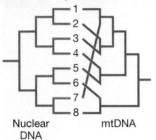

B Sexual reproduction

Nuclear   mtDNA
DNA

FIGURE 16.7 Using phylogenetic incongruence to infer sexual and asexual reproduction. Four hypothetical phylogenetic trees of eight species, based on either nuclear DNA or mtDNA, are shown here. **(A)** When we compare the two trees in an asexual species, we expect them to be quite similar. In this example, only the resolution of the 5,6 and 7,8 subclades has changed. **(B)** When we compare them in sexual species, we expect a greater degree of incongruence between nuclear and mtDNA trees. In this case, the structures of the two trees differ in numerous ways.

Distinguishing between Sexual and Asexual Reproduction

Until recently, researchers used natural history to classify a species as reproducing sexually or asexually—that is, evolutionary biologists would search for direct and indirect physical clues of sexual reproduction. The most obvious of these clues would be to observe two individuals mating. Courtship behavior implies sexual reproduction, as do sexual organs. And since all individuals in asexual species are females, the mere presence of males in a population strongly suggests sexual reproduction. But observing natural history is only one of the approaches by which we can distinguish between sexual and asexual reproduction.

The molecular machinery and cellular processes associated with various components of sexual reproduction—recombination, gamete production, and gamete fusion—are complex and involve many genes operating simultaneously. Indeed, in the nematode, *Caenorhabditis elegans*, more than 1400 such genes have been identified (Reinke et al. 2000). We can use the presence or absence of such genes and their homologs to infer whether reproduction is sexual or asexual in other taxa (Normark et al. 2003; Nciman et al. 2005, 2009; Schurko et al. 2009).

We can also compare phylogenetic trees to help understand whether a species reproduces primarily sexually or asexually. To see how, recall that mitochondrial DNA (mtDNA) is inherited only through females, but nuclear DNA is inherited through both parents. This means that in asexual species—which contain only females—phylogenetic trees based on mtDNA and nuclear DNA should be fairly congruent—that is, they should be similar to one another. But a comparison of phylogenetic trees based on mtDNA and nuclear DNA genes is predicted to be less congruent in sexual species (although this comparison can be skewed by population bottlenecks and population expansions). Evolutionary biologists can then use the degree of phylogenetic incongruity between nuclear DNA- and mtDNA-based trees to infer mode of reproduction (Figure 16.7).

A Phylogenetic Overview of Sexual and Asexual Reproduction

In a moment, we will examine the relative costs and benefits of sexual versus asexual reproduction, but before we do so, let's survey the rather striking phylogenetic distribution of these forms of reproduction in eukaryotes. In eukaryotes, very few species reproduce only asexually. Among vertebrates, for example, of the more than 42,000 species recognized, only 22 species of fish, 23 species of amphibians, and 29 species of reptiles reproduce exclusively by asexual reproduction (Vrijenhoek et al. 1989) (Figure 16.8).

But simply tallying the numbers of asexual versus sexual species will only get us so far. Evidence also suggests that asexual taxa are short-lived compared to sexual taxa. Although there is much debate as to how to calculate "short-lived" in absolute time—that is, how many thousand years is considered short?—in general, the consensus is that asexual species go extinct more quickly than sexual species (Law and Crespi 2002; Neiman et al. 2009). What is perhaps more critical is the fact that at the level of genus or higher there are almost no taxa entirely composed of species that only reproduce asexually (there are a few exceptions to this rule, including the bdelloid rotifer, a tiny freshwater invertebrate). This translates into a "twiggy" phylogenetic distribution for species that reproduce only asexually—that is, asexual species are rare and short-lived, and hence they tend to be tiny

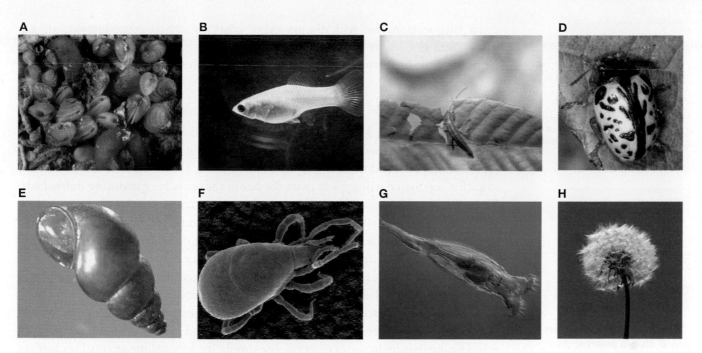

FIGURE 16.8 Examples of asexual species from a variety of taxa. Although asexual species are rare among eukaryotes, they can be found in a number of taxa as shown here. (**A**) *Lasaea australis*, marine clam; (**B**) *Poecilia formosa*, Amazon molly; (**C**) *Timema douglasi*, stick insect; (**D**) *Calligrapha suturella*, leaf beetle; (**E**) *Potamopyrgus antipodarum*, snail; (**F**) *Archegozetes longisetosus*, oribatid mite; (**G**) *Philodina roseola*, bdelloid rotifer; and (**H**) *Taraxacum officinale*, dandelion.

twigs on phylogenetic trees (Figure 16.9). Indeed, to date, evidence suggests that all species of eukaryotes that reproduce only asexually are derived from an ancestral sexual species, strongly suggesting that sexual reproduction is the ancestral state in eukaryotes (Malik et al. 2008).

16.2 The Costs of Sexual Reproduction

Our phylogenetic survey reveals that sexual reproduction is the norm among eukaryotes. This poses something of a challenge to evolutionary biologists, because sexual reproduction has a number of significant costs associated with it. Perhaps the most obvious of these costs is that diploid sexual females produce haploid gametes containing only one of the two sets of chromosomes that they possess. As such, any haploid gamete, should it successfully fuse with another haploid gamete, will produce a diploid offspring that contains only one set of its mother's chromosomes (and one set of its father's chromosomes). A diploid asexual female produces offspring that possess two sets of chromosomes from the mother. Another way to say this is that, assuming no inbreeding, asexual females are twice as genetically related to their offspring as are sexual females (Figure 16.10).

FIGURE 16.9 A hypothetical phylogenetic distribution of asexual species. This figure represents a typical animal phylogeny. In animals, asexual species (blue) compared to sexual species (gold) are rare, making up less than 0.1% of all animal species. Asexual lineages are also relatively short-lived on an evolutionary timescale. Adapted from Rice (2002).

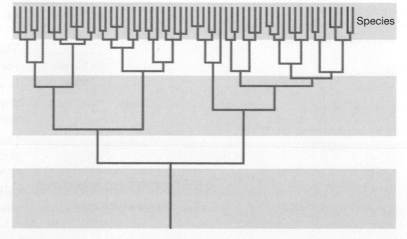

Species

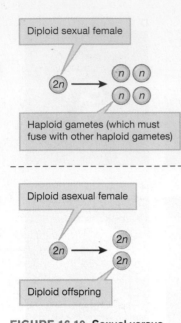

FIGURE 16.10 Sexual versus asexual diploid parents. As a result of meiosis, each gamete produced by a sexual female has only one set of the mother cell's chromosomes. An asexual female passes two sets of chromosomes to each offspring cell. As such, diploid asexual females are twice as related to each of their offspring as are sexual females.

In this section, we will discuss other costs to sexual reproduction; in the subsequent section, we will describe some potential compensating benefits that may be responsible for natural selection favoring the evolution of sexual reproduction.

The Twofold Cost of Sex

In the 1970s, John Maynard Smith made the following argument: Consider a population that is made up of asexually reproducing females, as well as sexually reproducing males and females (Maynard Smith 1971, 1978). The number of asexuals in such a population will grow at twice the rate of the sexually reproducing individuals. The reason for this is that, from a demographic perspective, a male's only function is to inseminate females; males never produce offspring directly. Asexual females avoid the "cost of producing males" by producing offspring that do not need to be inseminated in order to reproduce themselves—that is, by producing only females.

To see how the Maynard Smith model works, consider a sexually reproducing population, into which a small number of asexually reproducing females are introduced by mutation or migration. Let n = the number of asexually reproducing females in our population, and let N_m = the number of males and N_f = the number of sexually reproducing females in our population. For simplicity, let's assume an equal sex ratio in the sexually reproducing population, so that $N_m = N_f = N$. In generation t, let k be the number of offspring produced by a female, and assume that this number is not affected by whether the female reproduces sexually or asexually. Finally, let s = the probability that an offspring will survive and eventually breed. Maynard Smith calculated the number of adults in generation $t + 1$ (Maynard Smith 1971, 1978) as follows:

	Adults in t Generation	Offspring	Adults in $t + 1$ Generation
Asexually Reproducing Females	n	kn	skn
Sexually Reproducing Males	N	$kN/2$	$skN/2$
Sexually Reproducing Females	N	$kN/2$	$skN/2$

Now let's see what has happened to the proportion of asexual females in our population. At generation t, the proportion of asexual females—that is, the number of asexual females divided by the total number of individuals—was $n/(2N + n)$. In the next generation, this proportion is now:

$$\frac{skn}{skN + skn} = \frac{sk(n)}{sk(N + n)} = \frac{n}{N + n}$$

The proportion of asexually reproducing females has gone from $n/(2N + n)$ to $n/(N + n)$. When n is small compared to N—that is, when we have a population composed mostly of sexually reproducing individuals and only a few asexual females—the proportion of asexual females approximately doubles each generation, from approximately $n/(2N)$ to approximately $n/(N + n)$. Maynard Smith called this the **twofold cost of sex**. As n gets larger, the proportion of asexual females still increases each

FIGURE 16.11 The twofold cost of sex. If sexual and asexual females each produce four offspring (four females in an asexual population, two males and two females in a sexual population), the population size increases twice as fast in asexual versus sexual populations.

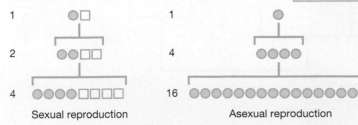

generation, but not at so fast a rate. In Figure 16.11, we show how to conceptualize the twofold cost of sex when we begin with one asexual population and one sexual population, rather than one population with both asexual and sexual individuals.

The twofold cost of sex is a consequence, not of sex itself, but rather of **anisogamy** (Box 16.1), which is the production of two different kinds of gametes—generally sperm and eggs (Figure 16.13A) (Bell 1982). Imagine that the growth of a lineage is constrained by the amount of resources that parents can invest in the biomass of their gametes, and thus in the biomass of their offspring. In the case of sex with anisogamy, a female produces large gametes, each with sufficient biomass to develop into an adult. For example, in Figure 16.13A, a female can produce two large gametes. In anisogamous sexual reproduction, males do not invest resources

BOX 16.1 The Evolution of Different-Sized Gametes: Anisogamy

Here we examine when natural selection should favor anisogamy, the production of different-sized gametes—namely, small sperm and large eggs. Our model is based on work by Geoff Parker and his colleagues (Parker et al. 1972; Bulmer and Parker 2002) and on discussions of Parker's model by Maynard Smith (1978) and Randerson and Hurst (2001).

We begin by imagining an ancestral marine organism that sheds its gametes into the water, where these gametes then fuse with gametes from other parents to produce offspring. Imagine a population of individuals in which a wide range of gamete sizes are produced, and suppose that (1) there is a trade-off between the size of gametes and the number of gametes, so that the larger the gamete, the smaller the number of gametes an individual can produce, (2) the larger the size of a gamete, the less mobile it is, and (3) the probability that a zygote survives increases with its size, where the size of the zygote is a function of the sizes of the fusing gametes (Parker et al. 1972; Bulmer and Parker 2002).

Because individuals that produce very small proto-sperm can produce many such gametes, most zygotes come from the fusion of two very small proto-sperm. But these zygotes have low survival rates compared to zygotes that were formed by the fusion of large gametes (proto-eggs). Clearly, selection favors proto-sperm that fuse with proto-eggs. All else being equal, proto-eggs should be favored to fuse with other proto-eggs, but all else is not equal. Proto-eggs, because of their large size, are relatively rare. As such, selection may favor proto-eggs that differentially fuse with proto-sperm, producing disruptive selection for proto-eggs and proto-sperm. Intermediate-sized gametes begin to decrease in frequency. And researchers have predicted that intermediate-sized gametes will decrease to a frequency of zero—leaving just proto-eggs and proto-sperm—when the relationship between zygote size and zygote fitness is as shown in Figure 16.12 (Parker et al. 1972; Bulmer and Parker 2002).

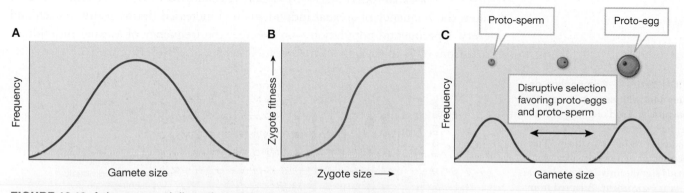

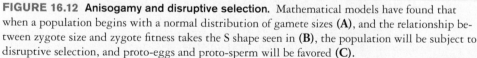

FIGURE 16.12 Anisogamy and disruptive selection. Mathematical models have found that when a population begins with a normal distribution of gamete sizes (**A**), and the relationship between zygote size and zygote fitness takes the S shape seen in (**B**), the population will be subject to disruptive selection, and proto-eggs and proto-sperm will be favored (**C**).

FIGURE 16.13 The twofold cost of sex arises only in anisogamy. (A) With anisogamous sexual parents, only the female invests in offspring biomass; the male invests in sperm that mostly go to waste so far as the growth rate of the lineage is concerned. (B) In asexual reproduction, all parents are female and invest entirely in offspring production. (C) With isogamous sexual parents, investment again goes to biomass rather than "wasted" sperm, and the lineage is able to grow at the same rate as an asexual lineage. Males are represented by squares; females by circles. Each offspring in (C) has white and green color to indicate investment from both parents. (In isogamous mating systems, biologists often refer to "mating types" rather than males and females.)

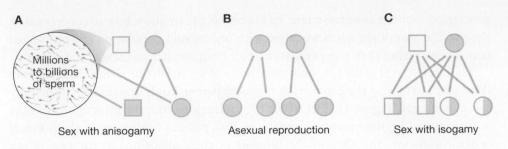

A   B   C

Sex with anisogamy   Asexual reproduction   Sex with isogamy

in offspring biomass. Rather, they produce millions to billions of tiny sperm, of which only a few will pass genes—and essentially no other biomass—to the next generation. In Figure 16.13A, two parents produce two zygotes. So far as the growth rate of the *lineage* is concerned, the male reproductive effort is wasted on sperm, most of which do not ever fertilize eggs.

In the case of asexual reproduction, all individuals in the population are female, and thus all reproductive effort is invested in large gametes and thus in offspring biomass. In Figure 16.13B, each of two females produces two offspring. Two parents produce four zygotes, and thus this asexual lineage grows at twice the rate of the sexual lineage in Figure 16.13A. This is the twofold cost of sex that we have described above.

In the case of sex with **isogamy**—when individuals produce one type of gamete— each parent produces mid-sized gametes that, when they fuse, are together the size of the large gametes produced by anisogamous females. In Figure 16.13C, each parent can produce four mid-sized gametes. These eight gametes fuse to produce four zygotes. These two sexual isogamous parents produce as many offspring as two asexual parents—they do not pay the twofold cost of sex. Thus, we see that the twofold cost of sex arises because under anisogamy males invest in sperm—most of which are wasted—rather than in biomass that goes to the offspring. For this reason, the twofold cost of sex is sometimes called the "cost of males."

The twofold cost of sex is supported empirically as well as theoretically. For example, Curt Lively and his colleagues tested the prediction that the proportion of asexual females should increase in mixed populations of asexual and sexual individuals of *Potamopyrgus antipodarum*, the snail species that we discussed at the opening of this chapter. They created 14 replicate experimental populations, each composed of 120 sexual snails and 65 asexual snails—that is, each tank was made up of 35% asexual individuals—from Lake Alexandrina in New Zealand. One year later, the frequency of asexual individuals had increased dramatically in each and every experimental population—on average, the frequency of asexual individuals rose from 35% to 62% in a single year (Jokela et al. 1997) (Figure 16.14).

FIGURE 16.14 Competition experiment with asexual versus sexual snails. Each of 14 experimental replicates began with 35% asexual and 65% sexual individuals. The percentage of asexually reproducing snails significantly increased in this 1-year experiment. Adapted from Jokela et al. (1997).

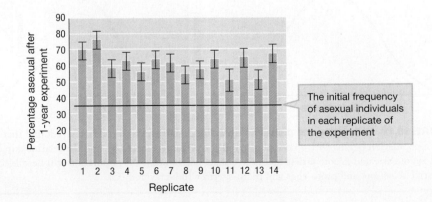

The initial frequency of asexual individuals in each replicate of the experiment

Sex Can Break Up Favorable Gene Combinations

As we will discuss in more detail soon, there are many benefits to the genetic variation that is created as a result of the outcrossing and recombination that take place during sexual reproduction (Felsenstein 1974, 1988; Felsenstein and Yokoyama 1976). But recombination has its costs, in that it can potentially break up associations between gene combinations that have been favored by natural selection. When we speak of a favorable gene combination, we mean that an allele at one locus is favored when it occurs in the presence of a specific allele at another locus, but not otherwise.

To see how sexual reproduction breaks up a favorable gene combination, let's look at a hypothetical example of a species in which individuals use their claws to fight for resources. Imagine two loci, each with two alleles, in our hypothetical population. At locus 1, we have either allele A_1, which codes for large claws (useful for fighting), or allele A_2, which codes for small claws; and we assume A_1 is dominant. At locus 2, our two alleles, B_1 and B_2, code for aggressiveness and meekness, respectively, and we assume B_1 is dominant. Genotypes with large claws and aggressive tendencies are favored by natural selection. Natural selection may also favor genotypes with small claws and meek tendencies, because although these genotypes don't win many fights for resources (as a consequence of smaller claws), they also do not pay the energetic cost of maintaining large claws, and their meekness keeps them out of aggressive interactions they would likely lose anyway. All other genotypes—large claw/meek and small claw/aggressive—should fare poorly in our population.

Sexual reproduction can break up our favorable gene combinations (large claw + aggressive, small claw + meek), because when combinations of our favored genotypes mate with one another, they will produce some proportion of offspring with genotypes that are not favored by selection (Figure 16.15). This is not a

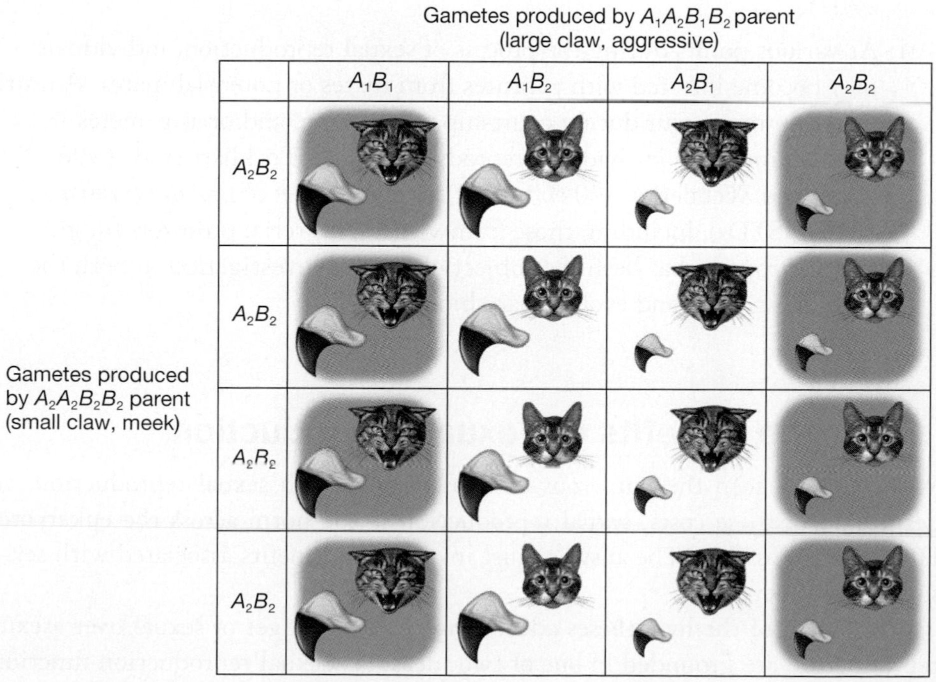

Gametes produced by $A_1A_2B_1B_2$ parent
(large claw, aggressive)

Gametes produced by $A_2A_2B_2B_2$ parent
(small claw, meek)

	A_1B_1	A_1B_2	A_2B_1	A_2B_2
A_2B_2				
A_2B_2				
A_2B_2				
A_2B_2				

FIGURE 16.15 Sexual reproduction can break up favorable gene combinations. A Punnett square showing how sexual reproduction can break up favorable gene combinations (such as large claw + aggressive and small claw + meek). Allele A_1 codes for large claws, allele A_2 codes for small claws (A_1 is dominant). B_1 and B_2 code for aggressiveness and meekness (B_1 is dominant). Here we see the offspring produced by a mating between an $A_1A_2B_1B_2$ and an $A_2A_2B_2B_2$ individual. Offspring in the far left and far right columns (orange) have a favorable gene combination; others do not.

cost paid in asexual populations, because asexual individuals produce offspring with their own genotypes: Individuals with large claws and aggressive behavior produce offspring with large claws and aggressive behavior, and individuals with small claws and meek behavior produce offspring with small claws and meek behavior.

Other Costs of Sex

Compared to asexual reproduction, sexual reproduction has costs above and beyond those that we just detailed. These costs include, but are not limited to, the following:

1. The search for potential mates requires time and energy. For example, John Byers and his team tracked individually marked pronghorns (*Antilocapra americana*). These researchers then compared the energy used by pronghorns actively searching for mates, and those not actively searching for mates. Over a two-week period, the difference in energy expended between these groups equaled approximately the energy used by an average pronghorn in half a day (Byers et al. 2005).

2. It takes time and energy to court potential mates. These costs have been examined in detail in many frog species in which males form choruses to court females and sing for hours each evening—sometimes for weeks at a stretch—to attract females (Wells and Schwartz 2007).

3. When individuals are searching for and courting potential mates, they are less vigilant for predators in the environment. For example, experimental manipulations have found that individual *Littorina plena* snails that are part of "mating pairs" are more prone to be attacked and captured by predators than are other individual snails, and that snails respond to predation threat by decreasing mating in the presence of putative predators (Koch et al. 2007).

4. At various points during the process of sexual reproduction, individuals may become infected with parasites from mates or potential mates. Parasitic infection may occur during courtship, copulation, and/or as gametes travel through an individual's reproductive tract (Lockhart et al. 1996; Knell and Webberley 2004). Indeed, an entire class of *sexually transmitted diseases* (STDs), including those from viruses, bacteria, protozoa, fungi, and arthropods, has been the subject of intense investigation in both the medical sciences and evolutionary biology.

16.3 The Benefits of Sexual Reproduction

We have now seen the numerous costs associated with sexual reproduction, yet despite all of these costs, sexual reproduction is the norm across the eukaryotes. How is this possible? The answer must involve the benefits associated with sex—so what are they?

Almost all of the hypotheses addressing the advantages of sexual over asexual reproduction are grounded in one of two ideas: (1) sexual reproduction functions to purge deleterious mutations, and (2) sexual reproduction generates genetic variation that is favored by natural selection. We will look at these hypotheses in turn now, but note first that they are not mutually exclusive, and that a "pluralist" hypothesis for the evolution of sex that combines some of the advantages of each is developing within evolutionary biology (West et al. 1999; Cooper et al. 2005; Meirmans and Neiman 2006). It may be that no single one of the benefits that we describe is by itself able to overcome the twofold cost of sex to the individual; hence, there is the appeal of explanations that rely on these benefits acting in concert.

Before we begin to explore the advantages of sex, it is useful to be explicit about what it takes to explain how sex is maintained over evolutionary time.

Sex Purges Deleterious Mutations

Sexually produced offspring are genetically different from their parents. One consequence is that when a deleterious mutation arises in a sexual population, individuals with this mutation can produce offspring without it. This is not the case with asexual reproduction, in which whole genomes are passed on from parent to offspring. As a result, deleterious mutations can't be purged as readily in asexual species as in sexual species.

The irreversible buildup of deleterious mutations in asexual populations was first discussed by population geneticist Herman Muller and has come to be known as **Muller's ratchet** (Muller 1932, 1964). Muller's basic idea amounts to this: Imagine a population of asexual organisms, in which deleterious mutations occur, but back mutations—mutations from deleterious to wild type—do not. Let the smallest number of deleterious mutations present in any individual's genome be some number j. For the sake of illustration, let's say that j is 1. Eventually, a new deleterious mutation or set of mutations will emerge in each and every genome in which $j = 1$. When that happens, j increases to 2. The "ratchet" has clicked one turn, so that the minimal number of deleterious mutations in our asexual population is now 2. Or it may be that any initial individuals with $j = 1$ die or fail to reproduce, leaving no descendants. Again, j will now be 2 (or more). Without the recombination that occurs in sexual reproduction, j can only increase; it can never decrease (Figure 16.16).

Note that looking at Muller's ratchet is not the same as looking at the fixation of deleterious alleles. The ratchet can turn without any particular deleterious mutation becoming fixed (Figure 16.17A). Similarly, a deleterious mutation can be fixed without causing the ratchet to turn (Figure 16.17B).

Muller noticed that one function of recombination is that it can reverse the ratchet effect we have been discussing. By recombining different segments of the chromosome, a region with few mutations from one parent can be combined with a region with few mutations from another parent to generate a new genome

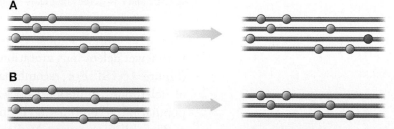

FIGURE 16.16 Muller's ratchet. (A) In the population on the left, most of the genomes (lines) have two deleterious mutations (red circles), but the highlighted genome (red line) has only a single deleterious mutation. When a new deleterious mutation (shown in dark red) arises in that genome, all genomes in the population have at least two deleterious mutations, and Muller's ratchet has turned. **(B)** Muller's ratchet can also turn if the genome or genomes with the fewest deleterious mutations fail to leave any descendants in the next generation. Here, the ratchet turns because the genome indicated by the red line fails to produce offspring, and thus all genomes in the subsequent generation have two deleterious mutations.

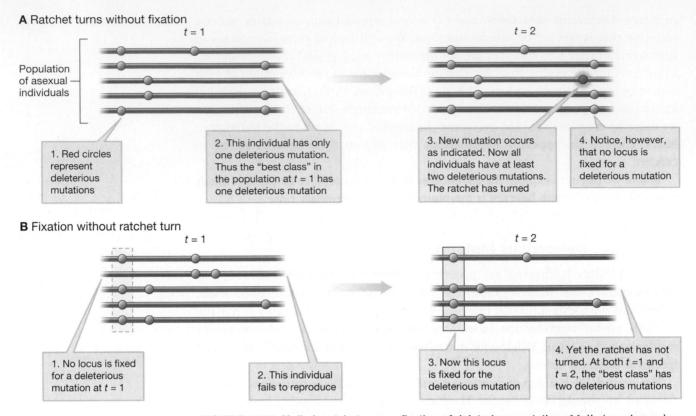

FIGURE 16.17 Muller's ratchet versus fixation of deleterious mutation. Muller's ratchet and the fixation of deleterious mutations are two processes by which deleterious mutations accumulate in an asexual population. In each image, the lines indicate chromosomal segments and the red circles represent the positions of deleterious mutations. **(A)** Muller's ratchet turns without fixation of any particular deleterious mutation. At time $t = 1$, the "best class" in the population features one deleterious mutation only. All other individuals have two. At $t = 2$, this individual's offspring picks up a novel deleterious mutation. Now all individuals in the population have at least two deleterious mutations. Muller's ratchet has turned. Notice that fixation has not occurred; no locus (at $t = 1$ or at $t = 2$) is fixed for the deleterious mutation. **(B)** A deleterious mutation is fixed without a turn of Muller's ratchet. At $t = 1$, no locus is fixed for a deleterious mutation. But the sole individual without the deleterious mutation at the indicated locus fails to reproduce and leaves no offspring at $t = 2$. Now the indicated mutation has become fixed. Notice that the ratchet has not turned; both at $t = 1$ and $t = 2$, the "best class" in the population has two deleterious mutations.

with fewer deleterious mutations than were present in any single individual parent (Figure 16.18). But recombination cannot in and of itself reverse the fixation of deleterious mutations; once a particular deleterious mutation is fixed in the population, recombination cannot undo this fact.

The ability to reverse Muller's ratchet and thereby purge deleterious mutations is a beneficial consequence of sexual reproduction. But is there empirical evidence for this theoretical result? One way to answer this question is to directly compare sexual populations with asexual populations. When Maurine Neiman and her colleagues asked this question by comparing sexual and asexual populations of the snail *P. antipodarum* (discussed at the beginning of the chapter), they found increased mutation accumulation in recently derived asexual populations of this much-studied species (Neiman et al. 2010). Similar work using sexual and asexual lineages of the water flea, *Daphnia pulex*, reveals that deleterious mutations accumulate at four times the rate in asexual versus sexual lineages (Paland and Lynch 2006).

Comparable sexual and asexual populations can be hard to come by, but there is another way to test the theory on species that are entirely sexual. The key to this approach is that the Y chromosome has no homolog, it does not undergo recombination, and thus it functions much as if it were in an asexually reproducing species. By comparing the Y chromosome to the other chromosomes in a sexual species, we can get another view on how recombination affects the accumulation of deleterious mutations.

Comparisons of the Y chromosome to other chromosomes show that the Y chromosome is not only significantly smaller, with few functional genes, but also that this chromosome has "degenerated" and accumulated nonfunctional, likely deleterious genes at a faster rate than other chromosomes (Rice 1994, 2002; Lahn et al. 2001; Tilford et al. 2001; Wilson and Makova 2009). These results are consistent with mutation accumulation models of sex, although other processes may also contribute to the degeneration of the Y chromosome.

Alexey Kondrashov has expanded Muller's ratchet model to consider how *epistasis*—interactions between the effects of alleles at different loci—influences the accumulation or purging of deleterious mutations. He considered the case in which the effects of mutations are synergistic, in the sense that two mutations that occur together have a stronger detrimental effect than the summed effect of each mutation alone (Kondrashov 1982, 1988, 2001). Kondrashov's model found that such synergistic epistasis strongly favors recombination, and hence sexual reproduction over asexual reproduction. Indeed, Kondrashov found that under synergistic epistasis, when the deleterious mutation rate per diploid genome per generation is greater than 1, sexual reproduction is favored over asexual reproduction.

The evidence available to date—primarily obtained from work on *Escherichia coli* and *Saccharomyces cerevisiae*—suggests that such synergistic epistasis among mutations occurs, but whether it is prevalent enough to explain the maintenance of sex remains unclear (Elena and Lenski 1997; He et al. 2010). As evolutionary geneticists and molecular biologists gather more data on mutation rates and on the extent of synergistic epistasis among mutations, we will be better able to test this idea (Kondrashov 1988; Kondrashov and Kondrashov 2010).

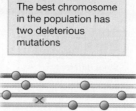

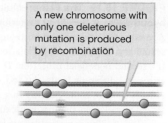

FIGURE 16.18 Recombination reverses Muller's ratchet. Prior to recombination, every chromosome in the population has at least two deleterious mutations, indicated by the red circles. Recombination then occurs as indicated by the orange crossover point. This creates two new chromosomal variants, one with only a single deleterious mutation, and one with three deleterious mutations. Recombination has driven the ratchet backward.

Sex Accelerates Adaptive Evolution: The Fisher–Muller Hypothesis

One advantage of sexual reproduction is that recombination allows natural selection to operate at a quicker rate than is possible in asexual species. This idea was first proposed by R. A. Fisher (1930) and later discussed by Herman Muller (1932), both of whom made the following argument: Compare two large populations, one sexual population and one asexual population. Imagine that a beneficial mutation, *A*, arises, and increases in both of our populations. Now suppose that a second beneficial mutation, *B*, arises. In an asexual population, *AB* individuals can only come about if the *B* mutation occurs in an individual with *A*. But, in sexual

FIGURE 16.19 The Fisher–Muller hypothesis. A beneficial mutation, *A*, arises, and increases in both large sexual and large asexual populations. If a second beneficial mutation, *B*, arises at a different locus, *AB* individuals emerge more quickly in a sexual population than in an asexual population. The same holds true for a third beneficial mutation *C*. This occurs because, in an asexual population, *AB* and *ABC* individuals can only arise if the *B* mutation occurs in an individual with *A*, and the *C* mutation then occurs in *AB* individuals (or if the *C* mutation occurs in an individual with *A*, and the *B* mutation then occurs in *AC* individuals). But, in sexual populations, recombination brings together beneficial mutations from separate lineages. Adapted from Crow and Kimura (1965) and Maynard Smith (1988), after Muller (1932).

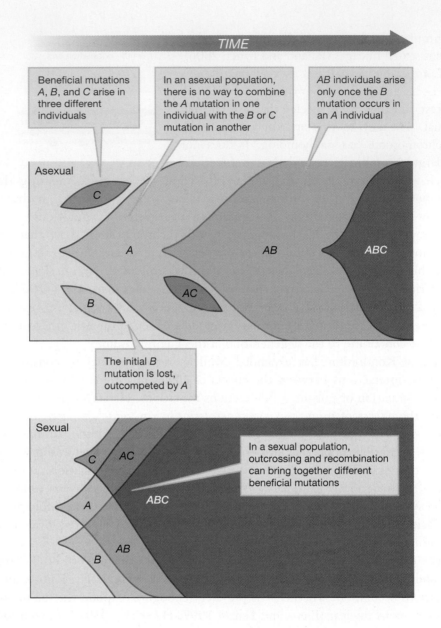

populations, recombination can bring the two beneficial mutations together even when the *B* mutation does not arise in an individual that already has *A*: an *AB* individual can be the product of a mating between one individual with *A* and one with *B*. If a third beneficial mutation, *C*, now arises, we can use the same argument for how the frequency of *ABC* individuals can increase more quickly in sexual populations (Figure 16.19). The Fisher–Muller hypothesis, then, predicts that sexual reproduction will accelerate the speed at which evolution operates. In our example, we get high frequencies of *ABC* individuals more quickly in large sexual populations than in asexual populations. Notice that the key thing that sex is doing in this model is breaking down linkage disequilibrium (Felsenstein 1988). When beneficial mutations *A*, *B*, and *C* arise in different individuals, there is initially linkage disequilibrium among these loci—that is, the presence of *A* guarantees the absence of *B* and *C*, and so on. Sex makes it possible to have *A* and *B* or *A* and *C*, and so on, in the same individual, breaking down this nonrandom association.

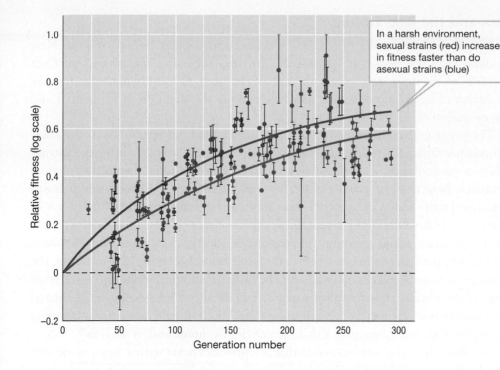

In a harsh environment, sexual strains (red) increase in fitness faster than do asexual strains (blue)

FIGURE 16.20 Relative fitness in asexual versus sexual populations. The relative fitness of asexual and sexual lineages of yeast was measured in a benign environment and a harsh environment by measuring population growth rates. In the benign environment (not shown), there was no difference in the fitness of asexual and sexual populations, but in the harsh environment, a difference was observed. Adapted from Goddard et al. (2005).

Experimental evidence comparing fitness in asexual versus sexual populations of yeast and green algae (*Chlamydomonas reinhardtii*) supports the prediction of the Fisher–Muller hypothesis (Zeyl and Bell 1997; Greig et al. 1998; Colegrave 2002; Goddard et al. 2005). In one of these experiments, Austin Burt and his colleagues experimentally created asexual lines of yeast by deleting two genes (*SPO11* and *SPO13*) associated with meiosis and recombination in yeast cells from a line of yeast that has a typical sexual phase during its reproductive cycle (Goddard et al. 2005). They replaced one of these genes with a neutral marker—a marker gene that did not otherwise affect function in the yeast—allowing them to distinguish asexual lineages. These manipulations enabled them to compare populations of yeast that differed only with respect to whether reproduction was sexual or asexual.

Burt and his colleagues next compared the growth rate of the asexual and sexual lineages of yeast. Lineages were raised in one of two types of environments: (1) a "benign" environment, in which there were relatively high levels of glucose sugar that the yeast could use as a resource and a temperature that facilitated yeast growth, or (2) a "harsh" environment, in which glucose was more limited and the temperature was above optimal for yeast growth. In the benign environment, in which selection for favorable mutations was presumably weak, there was no difference in the fitness of asexual and sexual populations when fitness was measured by growth rate. But, in the harsh environment, in which selection was more intense, sexual lineages had significantly higher growth rates than asexual lineages (Figure 16.20). Although the precise mutations involved, and the order in which they occurred, could not be measured in this experiment, these findings are consistent with the Muller–Fisher hypothesis for the accelerated rates of natural selection in sexual populations.

Sex and the Red Queen

Let's again return to our discussion of sexual and asexual lineages of the New Zealand snail, *Potamopyrgus antipodarum*. In that discussion, we noted that Curt Lively's work supported what we called the Red Queen hypothesis for the evolution of sexual reproduction. Here we explore the Red Queen hypothesis and the evolution of sex in more detail.

The key to understanding how the Red Queen hypothesis operates is to recognize a fundamental aspect of the relationship between parasites and their hosts. Natural selection is constantly favoring hosts that can better defend themselves against parasites. Selection, of course, also favors parasites that can overcome a host's defenses. In this sense, the parasite–host relationship is an arms race (Chapter 19). But parasites have a built-in advantage in this arms race because, in almost all cases, their generation times are orders of magnitude shorter than those of their hosts. As a result, pathogens can often evolve far faster than their hosts. For example, the generation time of many bacteria is on the order of an hour, while the generation time of their human host is on the order of two decades. One generation of natural selection in humans corresponds to more than 100,000 generations in a bacterial pathogen. Can hosts overcome this inherent disadvantage and, if so, how?

The Red Queen hypothesis posits that sex provides hosts with a way around this disadvantage, and this hypothesis predicts that host lineages that rely on sexual reproduction will outcompete host lineages that rely on asexual reproduction. When a new asexual lineage of host emerges, it may initially be resistant to parasites in its environment, and it may quickly grow to much higher frequencies in the overall population. But this increase in frequency is a double-edged sword for asexual hosts because, as they become prevalent, natural selection quickly favors adaptations in parasites to overcome the defenses of their now-prevalent asexual hosts. Since asexual individuals in a lineage are genetically identical, when effective adaptations to circumvent a host's defense system emerge in parasites, they will be particularly effective against asexual hosts.

If hosts reproduce sexually, the genetic variability generated by recombination makes it much more difficult for a parasite to home in on vulnerabilities and breach the host's defenses. As in the Fisher–Muller hypothesis, sex breaks down linkage disequilibrium; here the advantage to the host is that, with reduced linkage disequilibrium, the pathogen has a harder time tracking the host genotype. This argument has been dubbed the Red Queen hypothesis for sexual reproduction because the continual generation of new genotypes by sexually reproducing species makes them akin to the Red Queen in Lewis Carroll's stories: Carroll's Red Queen has to keep moving just to stay in place (Bell 1982). Similarly, sexual lineages must keep producing new genotypes, and keep breaking down linkage disequilibrium, to "keep up" with adaptations in parasites. The Red Queen hypothesis predicts that we will observe:

1. Oscillations in the relative frequency of asexual lineages when parasites are present: An asexual clone may be resistant at first, and it may increase in frequency in the population relative to sexual lineages. But, for the reasons we have discussed, this relative advantage will dissipate as an asexual lineage becomes more common.

2. Time lags: Suppose an asexual host evolves an effective defense against parasites. Initially this lineage will increase in frequency. But, as it

becomes common in the host population, natural selection will act strongly on the parasite population to favor parasite variants that can evade the defenses of this now-common host strain. Because of these dynamics, we expect a short time lag between the emergence of an effective host defense and the evolution of pathogen traits that can counter that defense (Figure 16.21).

3. Correlation between parasite load and sexual reproduction: Sexual reproduction will increase in frequency relative to asexual reproduction when the level of parasitism in an environment is high (Jaenike 1978; Hamilton 1980; Hamilton et al. 1990; Salathe et al. 2008).

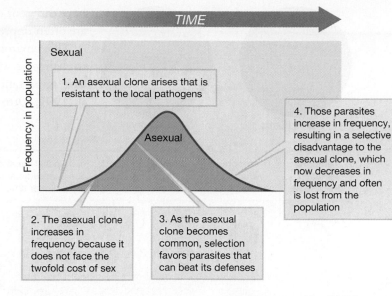

In the opening section of this chapter, we presented evidence for the third of these predictions in the snail *P. antipodarum*, but there is also evidence in this system that supports the first and second predictions. When Lively and his colleagues collected snail samples from lakes in New Zealand for four consecutive years, they found that the frequency of different asexual clones varied across years. As a clonal lineage reached higher frequencies, the proportion of individuals of that clone that were infected by parasites increased (prediction 1). Moreover, a time lag was present, such that the proportion of a particular clonal lineage infected by parasites tended to increase the year after the frequency of that clonal lineage had increased (prediction 2). This fieldwork was supplemented by laboratory experiments in which Lively and his colleague Mark Dybdahl raised uninfected snails from many different asexual clonal lineages and then experimentally exposed them to parasites from the wild. Lively and Dybdahl then recorded the proportion of individuals in each clonal lineage that became infected. As expected under the Red Queen hypothesis, the clonal lineages that were most common in nature had the highest proportion of individuals infected, while those clonal lineages that were rare in nature had a lower proportion of individuals that were infected (Figure 16.22) (Dybdahl and Lively 1998).

FIGURE 16.21 The Red Queen hypothesis. When a single parasite-resistant asexual clone arises within a sexual population, it increases in frequency quickly and reduces the relative proportion of the sexual genotypes in the population. As this clone increases in frequency, natural selection favors parasites that can infect this clone, which prevents its further increase. Adapted from Jokela et al. (2009).

FIGURE 16.22 Parasite infection and frequency of asexual clones in nature. The frequency of two common asexual clones—clones 22 and 19—were plotted (gold bars) against the proportion of those clones infected by *Microphallus* parasites (blue), and by all parasite species present (red). Adapted from Dybdahl and Lively (1998).

Sex, Environmental Unpredictability, and Variation among Offspring

Environments vary in both time and space. Organisms in the same location may experience very different conditions—temperature, humidity, predators, parasites, and so forth—across time. What's more, *during* a specific time interval, different patches in the same area may

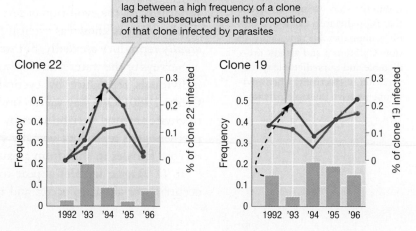

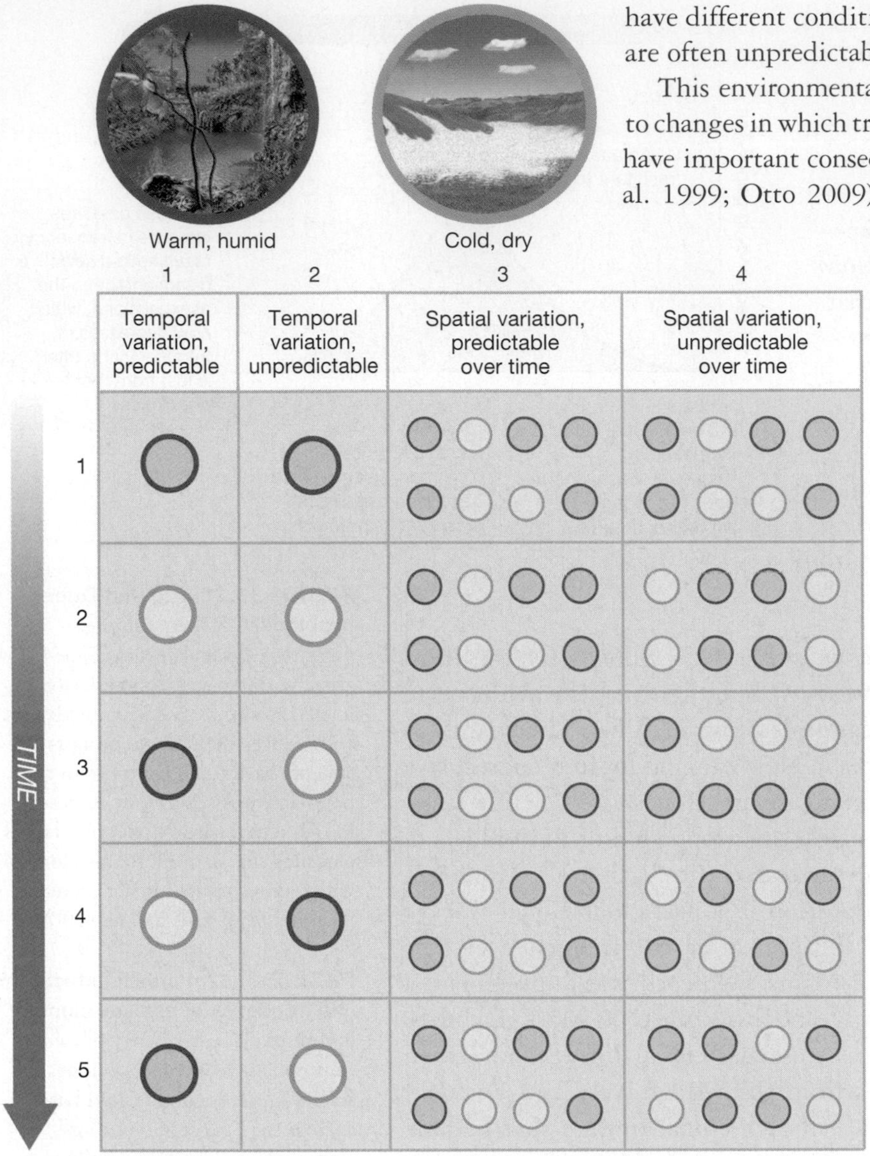

	Warm, humid	Cold, dry	
1	2	3	4
Temporal variation, predictable	Temporal variation, unpredictable	Spatial variation, predictable over time	Spatial variation, unpredictable over time

FIGURE 16.23 Temporal and spatial variability. Environments may show temporal variability, spatial variability, or both. Variability may be predictable or unpredictable. Column 1 shows temporal variation that is predictable. Column 2 displays unpredictable temporal variation. Columns 3 and 4 show predictable and unpredictable spatial variation, respectively, when environments have multiple patches.

have different conditions. Both spatial and temporal variability are often unpredictable (Figure 16.23).

This environmental variability and unpredictability can lead to changes in which traits are favored by natural selection and may have important consequences for the evolution of sex (Robson et al. 1999; Otto 2009). If environments are unpredictable, then natural selection may favor individuals that reproduce sexually. This is because they produce a diverse set of genotypes among their offspring as a result of both crossing over and the fact that offspring are made up of parts of the genomes of two different parents. This is akin to "bet hedging," in which individuals attempt to maximize the chance that some of their lottery tickets (offspring) are winners (survive to reproduce). Reproducing sexually in such environments provides two different kinds of immediate benefits to individuals (Kondrashov 1993):

1. It increases the chance that at least one of an individual's offspring will be a good match to the environment in which the offspring find themselves (Williams and Mitton 1973; Ghiselin 1974; Williams 1975).

2. Because individuals with similar genotypes are likely to be competitors for the same resources, the genetic variability that exists among offspring produced sexually creates the opportunity for these offspring to specialize in different niches in an environment: Sexual reproduction reduces competition between siblings (Maynard Smith 1978; Bell 1982; Price and Waser 1982).

One way that evolutionary biologists have examined the role of environmental variability on the evolution of sexual reproduction is by examining organisms that display what is known as cyclical parthenogenesis, in which cyclical parthenogens *usually* reproduce asexually, but *occasionally* reproduce sexually.

Surveys of the water flea, *Daphnia magna*, have found some populations in which individuals reproduce by cyclical parthenogenesis (Hebert 1974; Hebert and Crease 1980). One way we can understand how sexual reproduction may be linked to environmental variability is by examining the conditions under which cyclical parthenogens reproduce sexually versus asexually. In particular, what factors cause diploid females to shift from asexual reproduction to sexual reproduction?

In *D. magna*, females appear to respond to cues that environmental change is occurring or about to occur, and these cues trigger a shift from asexual to sexual reproduction. Introduction of a new predator—which represents a substantial environmental change—triggers sexual reproduction in some *Daphnia* populations (Pijanowska and Stolpe 1996; Slusarczyk 1999). A decrease in the quality of food may also trigger sexual reproduction, as do cues that a temporary pond may be drying up (sexual eggs are very resistant to drying up) (Carvalho and Hughes 1983; Koch et al. 2009). All of these findings suggest that sexual reproduction in cyclically parthenogenic *Daphnia* is favored as an adaptation to environmental variability and unpredictability (Figure 16.24).

FIGURE 16.24 Environmental change and reproductive mode. Cues of environmental change lead to a shift from an asexual to a sexual cycle.

16.4 The Origin of Sexual Reproduction

We conclude this chapter by examining one hypothesis for the possible origin of sexual reproduction. Notice that the question of *origin* is distinct from the question of *maintenance*, in that we need to understand not only why the fully formed system of sexual reproduction is maintained by natural selection operating on individuals, but also how the process got started in the first place. The idea is to be able to explain how the components of sexual reproduction—recombination, gamete production, and gametic fusion—could have arisen in a series of individually beneficial steps. A number of hypotheses for the origin of sexual reproduction have been proposed—including one in which sex originated as a means for transposable genetic elements to spread from cell to cell (Hickey 1993)—but we will focus on the hypothesis that DNA repair was integrally involved in the origin of sex.

One hypothesis for the origin of sexual reproduction is that it served as a mechanism for DNA repair (Dougherty 1955; Bernstein et al. 1981, 1985a,b; Bernstein and Bernstein 1991). To understand the DNA repair hypothesis for the evolution of sex, we begin by considering DNA damage in a haploid asexual organism. If such damage occurs on a single strand of DNA, then many repair mechanisms exist to fix such damage. These mechanisms often use the intact, nondamaged strand of DNA as a repair template. But what if there is damage to both strands of DNA—often referred to as double-stranded damage? In this case, no template is available to use for repair in haploid asexual organisms.

The DNA repair hypothesis postulates that, in such scenarios, diploid asexual reproduction would be favored by natural selection over haploid asexual reproduction, since double-stranded damage on one chromosome could potentially be repaired by using the homologous chromosome as a repair template. So far, so good, and we note that what we have here is a theory for the evolution of diploidy per se. But in asexual diploids, a basic problem remains: Deleterious alleles are fully exposed to selection, just as they are in haploids, because the two homologous chromosomes in an asexual diploid are highly similar, if not identical.

The DNA repair hypothesis postulates that diploid sexual reproduction is now favored because it introduces variation: By pairing homologous chromosomes from two different parents, deleterious recessive alleles from one parent can be masked by functioning copies of those alleles from the other parent. In other words, the gametes produced by the process of meiosis and the fusion of such gametes derived from different parents will restore heterozygosity in offspring.

In this chapter, we have examined the evolution of sexual reproduction. Sexual reproduction is the predominant reproductive mode of eukaryotes, while asexual reproduction in eukaryotes is relatively rare and has a very twiggy phylogenetic distribution. Yet, sexual reproduction has many costs associated with it, not the least of which is that parents and offspring are no longer genetically identical (as they are in asexual reproduction). The compensating benefits associated with sexual reproduction have helped evolutionary biologists understand the origins and maintenance of this fascinating form of reproduction.

SUMMARY

1. Asexual reproduction involves the production of offspring from unfertilized gametes. Sexual reproduction involves the joining together of genetic material from two parents to produce a progeny that has genes from each parent.

2. Mode of reproduction—sexual versus asexual—can be inferred by observation, but also by molecular genetics and phylogenetic comparisons.

3. In eukaryotes, species that reproduce only asexually are rare and short-lived on an evolutionary timescale. Their phylogenetic distribution has a "twiggy" appearance.

4. Despite its ubiquity, sexual reproduction has many costs associated with it, including the "twofold cost of sex" and the breaking up of favorable gene combinations.

5. The advantages of sexual reproduction can be divided into two general categories: (a) sexual reproduction is more efficient at purging deleterious mutations from a genome than asexual reproduction, and (b) natural selection favors the production of more variable offspring through the processes of recombination and gametic fusion.

6. The repair of double-stranded DNA may have played a role in the origin of sexual reproduction.

KEY TERMS

anisogamy (p. 547)
apomixis (p. 542)
asexual reproduction (p. 542)

automixis (p. 542)
isogamy (p. 548)
Muller's ratchet (p. 551)

Red Queen hypothesis (p. 541)
sexual reproduction (p. 542)
twofold cost of sex (p. 546)

REVIEW QUESTIONS

1. In 2010, NASA researchers found a new, shrimplike creature, living 600 feet below the ice sheets of the Antarctic. We know almost nothing about this new species—including how it reproduces (sexually or asexually). Why might deep under the ice sheets of the Antarctic be the sort of environment especially likely to house asexual creatures?

2. Assuming that it is not feasible to observe the reproductive behavior of the shrimp species described in question 1, how else could researchers determine its mode of reproduction?

3. Following our discussion of the evolution of anisogamy in Box 16.1, why do you suppose we almost never find species with three sexes, four sexes, and so on?

4. Recent work (Neiman et al. 2009) suggests that some asexual lineages may be much more ancient than originally thought. What sort of variables or properties would you first look at in these lineages to better understand their longevity? Defend your choice.

5. Add a fourth beneficial mutation (call it D, since our first three were A, B, and C) to the example we showed in Figure 16.19. What does the figure look like now?

6. What sort of practical applications can you envision based on Muller's ratchet? Could we somehow, at least in principle, use this against microbes that are dangerous to humans and that reproduce asexually?

7. We mentioned evidence that every eukaryote species that reproduces strictly asexually is derived from an ancestral sexual species, suggesting that sexual reproduction is the ancestral state in eukaryotes. Does this surprise you? Why or why not?

8. If you had unlimited time and money, what sort of experiments on the evolution of sexual and asexual reproduction would you like to undertake in the *Potamopyrgus antipodarum* snail system we discussed many times in this chapter?

9. One of the costs of sexual reproduction is that it breaks up favorable gene combinations. How might assortative mating minimize this cost in sexually reproducing species?

10. In addition to those discussed in this chapter, can you think of any costs of sexual reproduction associated with courting potential mates? List such costs and briefly discuss how they might apply to a species in which courtship has been studied.

SUGGESTED READINGS

de Visser, J., and S. F. Elena. 2007. The evolution of sex: empirical insights into the roles of epistasis and drift. *Nature Reviews Genetics* 8: 139–149. A review of recent empirical work testing population genetic models for the evolution of sex.

Kondrashov, A. S. 1993. Classification of hypotheses on the advantage of amphimixis. *Journal of Heredity* 84: 372–387. A conceptual approach to classifying models for the evolution of sex.

Michod, R. E., and B. Levin, eds. 1988. *The Evolution of Sex*. Sinauer Associates, Sunderland, MA. An edited volume on many aspects of the evolution of sexual and asexual reproduction across taxa.

Otto, S. P. 2009. The evolutionary enigma of sex. *American Naturalist* 174: S1–S14. A nice overview of the evolution of sex and problems that remain to be solved in this area.

Schurko, A. M., M. Neiman, and J. M. Logsdon. 2009. Signs of sex: what we know and how we know it. *Trends in Ecology & Evolution* 24: 208–217. A review of the advantages and disadvantages of approaches used by evolutionary biologists to distinguish between asexual and sexual lineages.

Ⓢ **Visit StudySpace at wwnorton.com/studyspace.**

17

Sexual Selection

17.1 Overview of Sexual Selection

17.2 Intersexual Selection

17.3 Intrasexual Selection

17.4 Postcopulatory Sexual Selection

17.5 Conflicts of Interest between Males and Females

◀ Male king bird of paradise tail feathers, *Cicinnurus regius*, Papua New Guinea.

espite the considerable interest that natural historians took in animal behavior, for decades after Darwin published *On the Origin of Species* they paid very little attention to *evolutionary questions* regarding behavior. Such disinterest is puzzling, as Darwin himself was quite clear that natural selection should operate on behavior in the same manner as it acts on anatomical traits, physiological traits, and so on. Indeed, Darwin focused specifically on the effects of selection on sexual behavior in his book *The Descent of Man and Selection in Relation to Sex* (Darwin 1871). Nonetheless, the birth of ethology—the scientific study of animal behavior—would not come about until the 1930s and 1940s, when Konrad Lorenz (1903–1989), Niko Tinbergen (1907–1988), and Karl von Frisch (1886–1982) began their pathbreaking field and laboratory studies on aggression, mate choice, territoriality, and other aspects of behavior. This work turned out to be so fundamental to all of biology that these researchers were all eventually awarded a Nobel Prize (Hinde 1973; Marler and Griffin 1973).

In this chapter, we will focus on the evolution of traits and behaviors related to selecting mates, and we begin with one of the most dramatic examples: the peacock.

During the breeding season, male peafowl—referred to as peacocks and characterized by their dramatic and elaborate tails (often called trains)—set up and defend small arenas called leks that contain no apparent resources such as food or shelter. Females then come to these leks and select mates from among the males present.

Females often visit many leks, and they prefer leks that contain the most males (Alatalo et al. 1992; Hoglund and Alatalo 1995; Kokko et al. 1998). On these leks, females choose among males that have elaborate, colorful trains that often contain beautiful "eyespots." Early work had shown that females preferred males with longer, more elaborate trains. Indeed, experimental reduction of the number of eyespots on a train reduced a male's attractiveness to females (Petrie et al. 1991). But why do females prefer males with ornate trains?

Marion Petrie and her colleagues hypothesized that because elaborate trains are costly to produce and maintain, train length and elaboration are signals of male genetic quality. To test this hypothesis, Petrie and her team ran a series of controlled mating experiments in natural enclosures that they built and in a field in Whipsnade Park, England (Petrie 1994). They placed eight males in experimental pens (one male per pen). They randomly selected eight sets of females—with four females per set—and placed one set of females in the pen of each of the eight males. After mating, a total of 349 offspring hatched, and the researchers took measurements on the chicks at numerous points during their development. When Petrie's team used the weight of the chicks as an indication of chick health, they found a positive correlation between the weight of the chick (a sign of health) and both the train length of the chick's father and the number of eyespots on the train of its father (a measure of the degree of elaborateness of a train), suggesting that females who chose males with longer, more elaborate trains were indeed choosing males with good genes. The researchers next released the chicks that had been born during the experiment into the field at Whipsnade Park, and they checked survival rates of these individuals over the course of 2 years. They found a positive correlation between survival of the released birds and the size of the eyespots on that individual's father (Figure 17.1), in part because the chicks from fathers with large eyespots were particularly good at fending off infectious diseases (Hale et al. 2009). This provides even stronger evidence that females were selecting males with high quality as their mates, and that the females were using length and elaborateness of the peacock's train as indicators of male quality when making such choices.

FIGURE 17.1 Good genes, offspring health, and survival. (A) Male and female peafowl. Males with more elaborate trains had **(B)** healthier chicks, as measured by weight at day 84 after birth, and **(C)** offspring with higher survival rates, as found from tracking them for 2 years in the field. Parts B and C adapted from Petrie (1994).

A

B

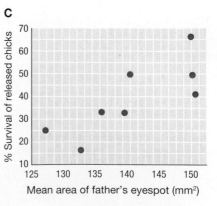

In this chapter, we will pursue this theme of how selection operates on traits related to mating. We will examine the following questions:

- What are the components of the process that Darwin dubbed sexual selection?
- What are the evolutionary models of female mate choice?
- How does male–male competition affect sexual selection?
- How does postcopulatory sexual selection occur?
- What are conflicts of interest between the sexes and how do they affect sexual selection?

17.1 Overview of Sexual Selection

Darwin was initially puzzled by extravagant traits such as the beautiful plumage of many male birds, the melodic songs of many species across the animal kingdom, and the giant horns found on males in many mammals and insects. How, he wondered, could such traits ever be favored by natural selection? In *The Descent of Man and Selection in Relation to Sex*, Darwin proposed that such traits evolved via **sexual selection**—a process that "depends on the advantage which certain individuals have over other individuals of the same sex and species *in exclusive relation to reproduction*" (Darwin 1871) (Box 17.1).

BOX 17.1 A Primer on Mating Systems

Here we provide a quick survey of the different types of mating systems seen in nature (Figure 17.2).

A **monogamous mating system** is usually defined as a mating system in which a male and female mate with each other, and only each other, *during a given breeding season*. As such, we can have animal societies in which we see lifetime monogamy, or a system in which pairs mate only with one another during a single season but in subsequent years find new mates (serial monogamy). Indeed, the latter sort of mating system is very common in territorial animals.

A **polygamous mating system** is defined as one in which either males or females have more than one mate during a given breeding season or cycle. Polygamy includes **polygyny**, in which males mate with more than one female per breeding season, and **polyandry**, in which females mate with more than one male per breeding season. Polygamy can be simultaneous or sequential. Simultaneous polygamy refers to the case in which an individual maintains numerous mating partners in the same general time frame, whereas sequential polygamy involves individuals forming many short-term pair bonds in sequence during a given breeding season.

When polyandry and polygyny are occurring in the same population of animals, the breeding system is said to be promiscuous. There are two very different kinds of promiscuity, which vary dramatically as a function of the presence or absence of pair bonds between mating individuals. In one form of promiscuity, both males and females mate with many partners, and no pair bonds are formed. In the second type of promiscuous breeding system, **polygynandry**, several males form pair bonds with several females simultaneously.

A Monogamy

B Polygyny

C Polyandry

D Polygynandry

FIGURE 17.2 Four basic mating systems. These are **(A)** monogamy (1 male, 1 female), **(B)** polygyny (1 male, more than 1 female), **(C)** polyandry (1 female, more than 1 male), and **(D)** polygynandry (more than 1 male, more than 1 female). Each class of mating system can be further subdivided in a number of ways. From Dugatkin (2009).

Following Darwin's lead, evolutionary biologists often divide sexual selection into (1) **intersexual selection**, in which individuals of one sex select among individuals of the other sex as mates, and (2) **intrasexual selection**, in which members of one sex, most often males, compete with each other for mating access to the other sex.

Different Selective Processes Operate on Males and Females

Competition between members of one sex for mating access to the other sex—intrasexual selection—is often much stronger among males than among females. This is, in part, due to a fundamental difference between the sexes. By definition, females produce fewer, but larger, gametes than males. Compared to sperm, each egg is extremely valuable, because of both its size and its relative scarcity. Each sperm, on the other hand, requires much less energy to produce, and sperm are usually produced in prolific quantities.

This means that male reproductive success is limited by the comparatively few eggs that are available to fertilize, causing severe competition for this scarce resource (Trivers 1985). By this logic, evolutionary biologists have hypothesized the following:

1. Because eggs are the limiting resource, males should compete for access to mating opportunities with females.

2. Because eggs are expensive and rare, females should be the choosier sex.

The huge number of sperm produced by males means that males chosen by multiple females may have extraordinarily high reproductive success, while males not chosen by any females or by only a few females may have very low reproductive success. The situation is different for females. Because of the relatively high costs related to egg production, and the relative scarcity of eggs, variation in female reproductive success should be fairly low. This is especially the case in species in which females have internal gestation and devote resources to a developing embryo. In such instances, females cannot become pregnant again until after they give birth, further reducing the variance between females in reproductive success. Indeed, evolutionary biologists observe much greater variation in reproductive success of males than of females (Bateman 1948) (Figure 17.3).

From the time of Darwin until about the 1970s, the majority of research on evolution and mate selection focused on male–male competition and intrasexual competition, rather than on mate choice and intersexual selection (Andersson 1994; Andersson and Simmons

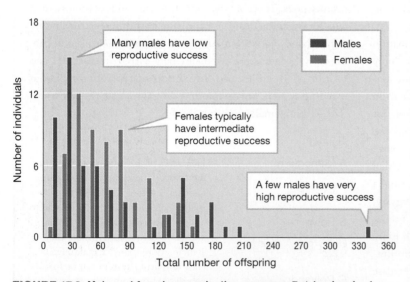

FIGURE 17.3 Male and female reproductive success. Reichard and colleagues measured lifetime reproductive success for male and female bitterling fish (*Rhodeus amarus*). Reproductive success in the fish follows the typical pattern: Males have a greater variance in reproductive success, with many males leaving no offspring and a few having a large number of offspring. Females have a smaller variance in reproductive success. Because each offspring has one male and one female parent, the mean reproductive success of each sex is equal. Adapted from Reichard et al. (2009).

2006; Clutton-Brock and McAuliffe 2009). This bias may have resulted from the fact that male–male competition is very easy to observe in nature, as well as from the fact that some early, prominent evolutionary biologists had dismissed mate choice as unimportant, thus directing research toward male–male competition (Huxley 1938).

In principle, intersexual selection includes both female choice of a male mate, as well as male choice of a female mate—both occur in nature. In practice, many mating systems involve *mutual mate choice,* in that both sexes are selective in their choice of partners (Bergstrom and Real 2000). Nonetheless, theory predicts that females will tend to be more discriminating about who has access to their gametes because females stand to lose more than males by making bad choices of mates. There are at least two reasons for this: (1) eggs are larger and more energetically expensive than are sperm and so they have what we might call a higher replacement value, and (2) in species with internal gestation, females are usually the only sex to devote energy to offspring before they are born, and so females are under strong selection to choose good mates.

17.2 Intersexual Selection

Male–male competition, to which we will return later in this chapter, held center stage in the sexual selection theater until approximately 30 years ago. But the research focus has shifted, and now the majority of studies done on sexual selection involve intersexual selection whereby females choose among males. Thus, we will begin with a discussion of female mate choice.

We will examine four evolutionary models of female mate choice: the "direct benefits," "good genes," "runaway selection," and "sensory bias" models. We will begin by outlining the logic of the models, and then we will look at case studies. Our focus is primarily on case studies in which the evolution of a sexually selected trait is best explained by just one of our four models. Nonetheless, in many species, the evolution of sexually selected traits might best be explained by a combination of two or more models.

Direct Benefits and Mate Choice

The direct benefits model of sexual selection is fairly straightforward. Selection favors females who have a genetic predisposition to choose mates that provide them with some resources—above and beyond sperm—which increase their fecundity and/or survival (Kirkpatrick and Ryan 1991; Price et al. 1993; Andersson 1994; Møller and Jennions 2001). For example, Randy Thornhill has found that female hanging flies, *Hylobittacus apicalis*, prefer males that bring them "nuptial gifts" of large prey items during courtship because such gifts increase the amount of resources available to the female to expend on growth and future reproductive effort (Figure 17.4) (Thornhill 1976). In the next section, we will examine the direct benefits model when the benefit provided by males is protection from predators.

FIGURE 17.4 Direct benefits to hanging fly females. Male hanging flies present females with food items that the females eat during courtship or mating. Females prefer males that provide larger prey items. The red arrow points to a prey captured by a male hanging fly.

Direct Benefits and Safety from Predators

In numerous species of amphibians, insects, and crustaceans, males and females perform premating rituals that include what is called *amplexus*, in which males and females are physically joined together, often with one individual on the back of the other. This form of mate guarding involves males defending the female from other males, and positioning themselves so that they are present when females become receptive to copulation.

Such premating behavior is seen in crustaceans from the genus *Hyalella* (Cothran 2008). In *Hyalella* amphipods, males carry females using large, clawlike appendages called gnathopods. Such appendages are found in both sexes, but they are larger and more muscular in males. Females mate more often with larger males who have larger gnathopods, and behavioral work has shown that this is not a function of male–male competition (Strong 1973). Moreover, females are not forced to mate by larger males. Rather, the distribution of matings is the result of female preference for mates with larger gnathopods.

Why should females prefer mates with larger gnathopods? What benefits, if any, do females obtain as the result of such a preference? Rickey Cothran hypothesized that females may be safer from predators while in amplexus with larger males, and he set out to address this question using *Hyalella* populations around the University of Oklahoma. Cothran collected a large sample of *Hyalella* and brought them into the laboratory. He then randomly paired one male and one female and placed the pair in a large jar filled with lake water. Once a male and female were in amplexus, Cothran added a predator—a larval dragonfly—and noted whether the predator attacked the *Hyalella* pair, whether such attacks were successful, and, if they were, which individual was taken by the predator (Cothran 2008).

When Cothran analyzed the data on attack rate and predator success, he found that the size of a *female* in amplexus did not affect the probability that a predator attacked. But a female's probability of survival increased dramatically as a function of her partner's size. Females that mated with larger males were much less likely to be eaten by predators (Figure 17.5). In the *Hyalella* system, then, one reason that selection favors females who mate with larger males is that

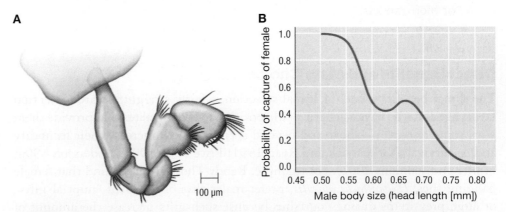

FIGURE 17.5 *Hyalella* **females receive a direct benefit in the form of safety. (A)** The gnathopod of *Hyalella azteca*. Adapted from Gonzalez and Watling (2002). **(B)** The probability that a female is captured by a predator during mating decreases with the size of the male she chooses as a mate. The curve shown here has been estimated using a statistical procedure known as cubic spline estimation. Adapted from Cothran (2008).

such females receive direct benefits—in the form of safety from predators—as a result of their choices.

Good Genes and Mate Choice

What if males do not provide females with direct resources such as food and shelter? How does female mate choice evolve in species in which the female receives nothing but sperm from her mate? In such systems, selection may favor females who choose mates that possess "good genes"—that is, genes that code for some suite of favorable traits (Fisher 1915; Kodric-Brown and Brown 1984; Andersson 1994; Kokko et al. 2003; Mays and Hill 2004). This sounds straightforward enough, but how can females assess whether males have good genes? Wouldn't selection favor males who have traits that indicate that they possess appropriately good genes, even if they don't? The answer is yes. This, in turn, selects for females who can avoid being fooled. As a result, evolutionary biologists suggest that only traits that are accurate and honest indicators of male genetic quality should be used by females when choosing mates. One theory is that honest indicator traits should be generally costly to produce or, more specifically, costly to fake. The more costly a trait is to fake, the more likely it is that the trait is a true indicator of good genes. We will return to honest indicators in a discussion of signaling behavior in Chapter 18.

Good Genes, Parasites, and the Hamilton–Zuk Hypothesis

The idea that females use honest signals to choose males with good genes has been studied in a number of traits associated with mating behavior. Perhaps the most well-known set of studies centers on bright color as an honest indicator of a male's ability to fend off parasite infection.

The argument about color as an indicator of male resistance goes like this. There are two types of parasites: ectoparasites that live on the surface of an individual (skin, fur, feathers, and so on), and endoparasites that reside inside an organism (in the gut, blood, and so on). Females may be able to visually gauge a male's resistance to ectoparasites, and as such there is little selection for males to signal their ability to resist ectoparasites, and there is little selection for females to look for such male signals. But when resistance to parasites cannot be visually gauged, as in the case of endoparasites, which live inside the organism, how can females figure out which males have good resistance to the parasites?

The answer to this problem must in some way lie in the female's ability to identify some *other* male trait that correlates with the ability of a male to avoid being parasitized. If possessing a given trait also means that males are good at fighting endoparasites, females can use that trait as a proxy for judging what they really need to know. One hypothesis is that body coloration is a good proxy. This is labeled the Hamilton–Zuk hypothesis after the researchers who first came up with the idea (Hamilton and Zuk 1982). Healthy males tend to be very colorful, while diseased males have much drabber colors, in part because it is difficult for males to shift resources to the production of bright colors when they are fighting endoparasite infection. Thus, coloration is costly, and hence it is a good candidate for an honest signal (Milinski and Bakker 1990). Although this hypothesis remains the subject of some heated debate, many studies on mate choice and coloration find

FIGURE 17.6 Color, good genes, and mate choice in sticklebacks. (A) Males and females around a stickleback nest. Note the variation in male color (red). Redder-colored male sticklebacks were in better health **(B)** and were preferred by females **(C)**. Milinski and Bakker were able to show that it was color that females preferred, not other traits that bright-colored males might possess. **(D)** When females were given a choice between colorful and duller males, but red color was filtered out (using a filter sheet), females did not significantly prefer the brighter of the two males. One reason stickleback females may prefer redder-colored males is that this color indicates resistance to parasites. Part A adapted from Dugatkin (2009); parts B–D adapted from Milinski and Bakker (1990).

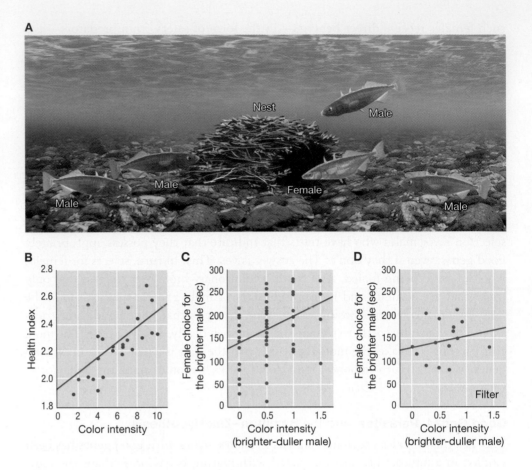

that females often choose the most colorful, least parasitized males (Figure 17.6). If bright color in males is a signal that a male is especially resistant to parasites, females who choose such males as mates may receive indirect benefits by producing offspring with good genes.

Runaway Sexual Selection

The **runaway sexual selection model** was first proposed by Sir Ronald Fisher (Fisher 1915, 1958). In this model, a gene that codes for a particular trait in males over time becomes associated with a gene that codes for a particular behavioral mating preference in females—that is, linkage disequilibrium arises between alleles for male phenotype and female preferences. In order to follow how this occurs, let's consider a population in which some fraction of the females have a heritable preference for brightly colored males, while the remainder of the females choose males randomly with respect to color. Suppose that, in this population, the degree of male coloration is also a heritable trait—some males are more colorful than others. So, we have a group of females, some of whom prefer brightly colored males and some of whom have no preference for any particular male coloration, as well as a group of males, some of whom are more colorful than others. Further suppose that the genes coding for male color and female preference are present in both males and females, but each is expressed only in the appropriate sex—preference genes in females; color genes in males.

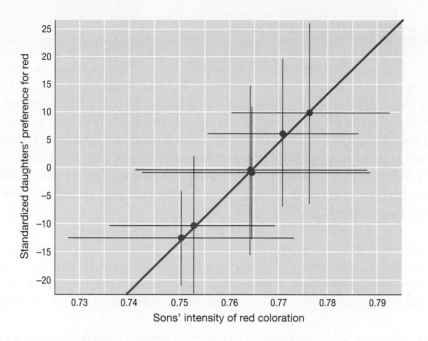

FIGURE 17.7 Genetic correlation in male trait and female preference. A number of studies have found a genetic correlation between female preference and male trait. In sticklebacks, for example, Theo Bakker examined the sons and daughters from six males and found a genetic correlation between red color in males and preference for red color in females. The horizontal lines show the standard deviation in red color among sons from a given father, and the vertical lines show the standard deviation in preference for red color among daughters from a given father. From Bakker (1993).

If the genetic correlation between trait and preference is extremely high, this system can "run away" in a positive feedback loop, like a snowball rolling down a snowy mountain and accelerating as it becomes larger and larger. Across generations, selection may produce increasingly exaggerated male traits—for example, exaggerated male color pattern. As the male trait increases in frequency, the female preference also increases because of the genetic association between these traits. This leads to stronger and stronger female preferences for such exaggerated traits, even if the female preference changes to a point where it has a detrimental impact on female fitness. Similarly, a selective advantage due to female preference can drive an increase in the frequency of an associated male trait.

The Fisher runaway selection model has been notoriously difficult to test. The reason for this is that one of its key predictions—linkage disequilibrium between preference and male traits—is also predicted by the good genes model (Lande 1980; O'Donald 1980; Pomiankowski 1988; Houde 1997). This means that evidence of a genetic correlation between female preference and male trait—a correlation that has been demonstrated in guppies, sticklebacks (Figure 17.7), stalk-eyed flies, and field crickets—is not sufficient to distinguish between good genes and runaway models (Houde and Endler 1990; Bakker 1993; Wilkinson and Reillo 1994; Gray and Cade 1999).

One way to distinguish between the runaway sexual selection model and the good genes model is with information on whether the female preference has evolved to a point that it is detrimental to female fitness. This can happen under some conditions in the runaway sexual selection model, but it is never predicted to occur under the good genes model. Ruling out all possible benefits that females may receive by choosing mates is extremely difficult. Indeed, in all systems in which a genetic correlation between female preference and male trait has been demonstrated, there is some evidence that females benefit either directly or indirectly as a consequence of their mate preferences.

The Sensory Bias Hypothesis

The last of the evolutionary models of female mate choice we consider is known as the **sensory bias model** (also known as the sensory exploitation model or preexisting bias model) (West-Eberhard 1979, 1981; Endler and McLellan 1988; Ryan 1990). This model hypothesizes that females initially prefer a certain male trait—let's call it M1—but not because of any mating benefit that is associated with that male trait. Instead, the sensory bias hypothesis speculates that female nervous systems respond to M1 either because it is associated with some benefit outside of mate choice, or simply as an artifact of how they are "wired." Males are simply tapping into a preexisting sensory bias for trait M1, such that males with M1 are now preferred as mates.

For example, suppose that red berries are the most nutritious food source available to a fruit-eating species of birds. Selection will favor individuals who are best able to search out and consume red berries—that is, natural selection will fine-tune the neurobiology of the birds so that they are acutely aware of the color red and will hone in on red things in their environment (Kirkpatrick and Ryan 1991).

Suppose that, after selection has favored a nervous system that is especially adept at picking up red-colored objects in the environment, red feathers randomly arise in some individuals in our population. Red-feathered males may be chosen as mates because the female's nervous system is already designed to preferentially respond to red objects. Males with red feathers, then, are exploiting the preexisting neurobiologically based preferences of females—preferences that evolved as a result of selection for other functions.

The sensory bias model is unique among models of mate choice in that it leads to a very clear, very simple prediction regarding the phylogenetic history of male traits and female preferences. When we look at a phylogeny that includes information on both female preference and the male trait preferred by females across closely related species, the female preference trait should predate the appearance of the male trait. In our case of red berries, the preference for red should be in place *before* red feathers are present.

Sensory Bias and Female Mate Choice in Frogs

One of the earliest studies of sensory bias involved two closely related species of frogs, *Physalaemus pustulosus* and *Physalaemus coloradorum* (Ryan et al. 1990). Males of both species employ advertisement calls to attract females. Males in both *pustulosus* and *coloradorum* species begin this advertisement with what is referred to as a high-frequency "whine." But *pustulosus* males add a low-frequency "chuck" sound to the end of their call (Figure 17.8). When *pustulosus* females choose between *pustulosus* males who produce a chuck and those who do not produce a chuck, they consistently prefer to mate with the former. Ryan and his colleagues hypothesized that the female preference for chucks was the result of a sensory bias in favor of such low-frequency sounds, which are detected by the amphibian papilla section of the inner ear.

Phylogenetic and behavioral evidence support the contention that the female preference for chucks in *pustulosus* is due to sensory bias. Recall that the chuck part of the call is absent in the call of *coloradorum* males. Indeed, when Michael

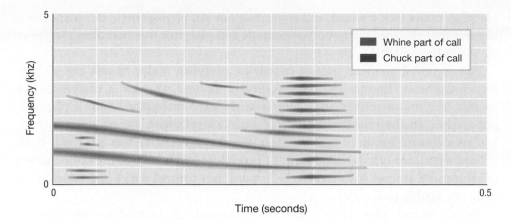

FIGURE 17.8 A whine and chuck call from the frog *Physalaemus pustulosis.* Other *Physalaemus* species produce the whine without the chuck. Adapted from the Ryan Lab (2011).

Ryan and his colleagues used molecular and morphological data to reconstruct the evolutionary history of the genus *Physalaemus*, they inferred that the common ancestor of *coloradorum* and *pustulosus* did not use a chuck call (Figure 17.9). Yet, when computer audio technology is used to add a chuck call to the end of pre-recorded *coloradorum* male calls, *coloradorum* females show a preference for calls that include a chuck—as soon as chucks appear in a *coloradorum* population, females prefer males who produce such calls. These studies suggest the auditory circuitry in *Physalaemus* frogs is built in such a way as to produce a preference for a certain class of low-frequency calls like chucks. The *coloradorum* studies provide evidence that the preference for chucks has predated the actual production of chucks in this species, in accordance with the sensory bias hypothesis.

Sensory Bias and Female Mate Choice in Swordtails

As a second example of the sensory bias hypothesis, let's examine Alexandra Basolo's work on two closely related species of tropical fish: the green swordtail (*Xiphophorus helleri*) and the platyfish (*Xiphophorus maculatus*). Female swordtails prefer males with long "swords"—an elongation of the tail fin—as mates. But

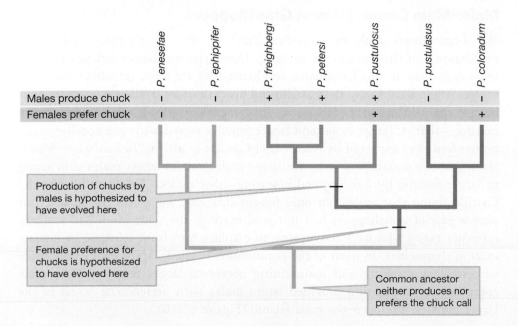

FIGURE 17.9 Preference for chuck calls arose before production of the calls themselves. The chuck call (or a similar "squawk") is thought to have evolved in the common ancestor of *P. pustulosus*, *P. petersi*, and *P. freighbergi*. Males of other species do not produce the chuck. Female preference for the chuck call has been observed in *P. pustulosus* and *P. coloradum*, but is not present in *P. enesefae*. Ryan and colleagues therefore hypothesize that the preference predated the chuck call itself. Adapted from Ryan and Rand (2003).

platyfish males lack a sword, and some, but not all, phylogenetic reconstructions suggest that the ancestor to both swordtails and platyfish was swordless (Meyer et al. 1994; Basolo 1995).

Basolo sewed artificial plastic swords on platyfish males to test the sensory bias hypothesis (Basolo 1990)—specifically, to test whether a preference for swords was found immediately after swords were introduced into a platyfish population. In the treatment group of males, a yellowish sword was attached, and in a control group of males a clear, see-through sword was attached. Both groups went through the same surgical procedure and both had the same burden of swimming around with an attached sword, but control males had swords that were not visible to females.

Basolo found that female platyfish showed an immediate and strong attraction to treatment males with visible swords over the control males. Even though there had been no evolutionary history of platyfish females choosing males with swords, females viewed this elaborate male trait as attractive as soon as it appeared in the population, suggesting a sensory bias for such elaborate traits.

17.3 Intrasexual Selection

Male–male competition can take many forms. Males may fight among themselves: for example, male stag beetles (*Lucanidae cervus*) use their "horns" to engage in combat, and male red deer (*Cervus elaphus*) battle each other with their antlers. The winners of such contests mate more often than do the losers.

In this section, we will explore the various ways—sometimes obvious; other times more subtle—in which males compete with each other for access to females, and the evolutionary consequences of such competition.

Male–Male Competition in Grasshoppers

We begin with male–male competition in the grasshopper, *Sphenarium purpurascens*. In this species, as in many insect species, larger females produce more eggs than do smaller females, and hence they are more valuable to males as mates. When Raul Cueva del Castillo and his colleagues studied the grasshopper both in the wild and in the laboratory, they uncovered strong, positive assortative mating—that is, larger males and larger females mated with one another much more often than expected by chance (del Castillo et al. 1999). But why? Was it the result of a female preference for larger males, or did larger males gain access to larger females by fighting and defeating other males? The team lead by del Castillo found that males not only fought intensely for access to females, but once a pair of grasshoppers had initiated mating, the male often had to fight off other males who were attempting to displace him from a female to initiate mating themselves. In most of the populations studied, larger males were more successful at securing and maintaining access to larger females. Male–male competition via fighting provided larger males with preferential access to the larger females, who were the most fecund (Figure 17.10).

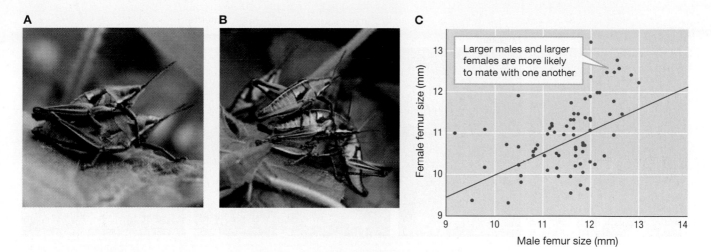

A

B

C

Larger males and larger females are more likely to mate with one another

FIGURE 17.10 Male–male competition and assortative mating in grasshoppers. (**A**) A male (on top) mating with a female (below) and guarding her from other males. (**B**) Two males try to displace a male mating with a female. (**C**) Positive assortative mating between larger males and larger females. Part C adapted from del Castillo et al. (1999).

Male–Male Competition by Cuckoldry

Male–male competition can be more subtle than direct fights between individuals. In many species, a single population may contain numerous different male reproductive morphs that are distinct in physiological, endocrinological, and behavioral traits. These morphs compete with each other for access to mating opportunities in indirect and often complicated ways (Gross and Charnov 1980; Gross 1985). For example, in bluegill sunfish (*Lepomis macrochirus*), three male morphs—known as parental, sneaker, and satellite morphs—coexist within many populations (Gross 1982; Neff et al. 2003).

In bluegills, parental males are light-bodied in color, but they have dark yellow-orange breasts. They build nests, and they are highly territorial, chasing off any other males that come near their territory. Parental males also invest substantial amounts of energy in caring for their offspring—fanning the eggs to oxygenate them and defending the nest against predators during nesting (Coleman et al. 1985).

Sneaker bluegill males are smaller and less aggressive than parental males, and they do not hold territories. They camouflage themselves in hiding places near a parental male's territory. When they see a parental male and female spawning, they quickly swim toward the pair, shed their own sperm, and swim away, all within about 10 seconds (Gross 1982). Using molecular paternity analysis, researchers have found that, depending on their relative numbers in a population, sneaker males fertilized up to 58.7% of all bluegill eggs laid in Lake Opinicon, Canada (Philipp and Gross 1994).

A third male reproductive morph, called a satellite male, is also found in some bluegill populations. Satellite males look like females, and they often swim between a spawning pair that contains a parental male and a female. If the parental male attempts to spawn with both the female and the satellite male that is posing as a female, the satellite male will release his own sperm (Figure 17.11).

FIGURE 17.11 Parental, sneaker, and satellite males. Bluegill morphs, from left to right: a bluegill parental male preparing a nest, sneaker males hiding behind plants awaiting a chance to quickly swccp into a parental nest, a satellite male (outlined in white oval) about to swim over a nest containing a parental male and a female, and a satellite male swimming between a parental male and a female. Adapted from Gross (1982).

The continued coexistence of parental, sneaker, and satellite males nicely demonstrates how males compete for mating opportunities in complex ways. We will return to this case again in just a moment.

17.4 Postcopulatory Sexual Selection

Sexual selection acts not only on behavior and external morphological traits, but also on traits that affect a sperm's ability to reach and ultimately fertilize an egg. In such cases, competition occurs *after* a female has mated with numerous males. If sperm from numerous males are present, sperm may compete with one another over access to fertilizable eggs. When such sperm competition exists, selection can operate directly on various attributes of sperm, such as sperm size and shape. Sperm competition is one form of what is known as **postcopulatory sexual selection** (Eberhard 2009).

Sperm Competition in Bluegill Reproductive Morphs

Let's now return to the case of the bluegill male reproductive morphs. Because the three different morphs exhibit very different reproductive behaviors, Bryan Neff and his colleagues reasoned that there might also be differences in sperm production and sperm quality across morphs (Neff et al. 2003). In particular, they hypothesized that, because of their "hit-and-run" mating strategy, sneaker males might invest most heavily in sperm *production*. The results of their investigations are consistent with this prediction. Although parental males are larger than sneakers and have testes that are larger, when Neff and his colleagues examined the ratio of testes size to body size—that is, the relative investment in testes—sneaker males had the highest ratio, followed by satellites and then parentals.

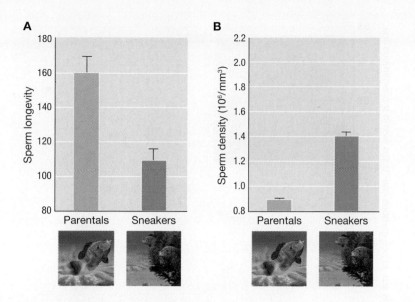

FIGURE 17.12 Sperm production in bluegill morphs (parental and sneaker). Sperm longevity **(A)** and ejaculate sperm density **(B)** in parental and sneaker morphs. Adapted from Neff et al. (2003).

The relative investment in testes size is an *indirect* measure of sperm production. A more direct measure would be the number of sperm produced per ejaculate. Given that the sperm produced by sneakers are always competing with parental sperm, but parental sperm are not always competing directly with sneaker sperm, a high density of sperm per ejaculate should be more strongly favored in sneaker males. When Neff and his team looked directly at the density of sperm per ejaculate, they found that sneakers indeed produced more sperm per ejaculate.

Nonetheless, sneakers do pay costs for investing so heavily in sperm production. First, their sperm survive shorter periods of time than do parental sperm. Second, when Neff and his colleagues removed sperm from both sneakers and parentals, and then released the same number of parental and sneaker sperm over eggs, parental sperm were more likely to fertilize eggs than were sneaker sperm. Sneakers invest in producing many short-lived, lower-quality sperm, while parentals invest in producing fewer, but higher-quality sperm (Figure 17.12).

Sperm Competition and Testis Size in Beetles

In the dung beetle, *Onthophagus taurus*, sperm competition is intense, as females are often inseminated by numerous males. This has led to the evolution of large testis size and greater sperm production in *O. taurus* relative to other species of dung beetles. Indeed, in general, comparative analyses of the genus *Onthophagus* show a positive relationship between size of testes and the frequency of sperm competition (Simmons et al. 2007).

If large testes in *O. taurus* are an adaptation to sperm competition, then experimentally enforcing monogamy—one male mating with one female per breeding cycle—should decrease sperm competition. If there is genetic variation in testis size and sperm production, reduced sperm competition could, over multiple generations, select for a decrease in testis size. Leigh Simmons and Francisco Garcia-González tested this prediction by experimentally enforcing monogamy for 21 generations in three replicate lines of *O. taurus* (Simmons and Garcia-González 2008). In these lines, the researchers placed a female and a randomly chosen male together and allowed them to breed. They also created three control lines in which

FIGURE 17.13 Sperm competition and mating systems in dung beetles. Testis size of the experimentally enforced monogamous line and the polygamous line of dung beetles diverged through the course of a 21-generation experiment. Adapted from Simmons and Garcia-González (2008).

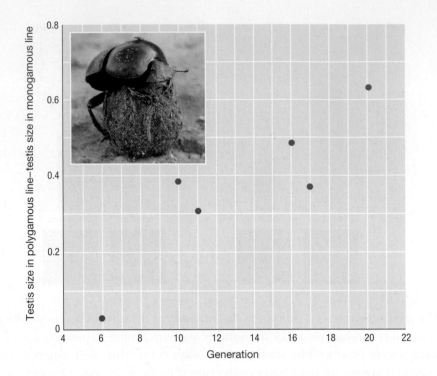

male and female beetles were allowed to mate polygamously as they do in the wild. They then compared testis size between males from such experimental lines with those of males in the control lines.

Figure 17.13 shows the results of this experiment. Compared to the control lines, over the course of 21 generations, the testis size of male beetles in lines with enforced monogamy decreased significantly. To get a more direct measure of the effect of enforced monogamy on sperm competition, at generations 11 and 16 of their experiment, Simmons and Garcia-González inseminated females with sperm from males from both the control and the experimental monogamy lines. They found that sperm from males in the control line outcompeted sperm from males in the enforced monogamy line in the sense that they fertilized a greater proportion of a female's eggs. Testis size and investment in sperm production do indeed appear to be an adaptation to sperm competition in *O. taurus*.

17.5 Conflicts of Interest between Males and Females

As we have seen throughout this chapter, selection operates differently on males and females with respect to mating behavior. When these differences are sufficiently strong, **sexual conflict** may result: traits that evolve in one sex may be detrimental to individuals of the other sex. As an extreme example of this type of sexually antagonistic coevolution, consider the case of the yellow dung fly, *Scathophaga stercoraria*. Male dung flies fight for access to females, and these fights are so intense at times that females are drowned in the melee (Parker 1979).

At a more general level, sexual conflict between males and females will emerge over the type of mating system in place. For example, in the dunnock bird (*Prunella modularis*), some males and females are monogamous, while others mate

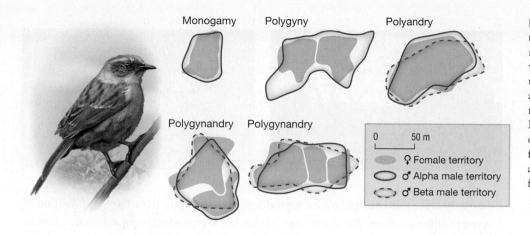

with multiple opposite-sex partners. Underlying much of the variance in mating systems, including that of the dunnock, is the fact that the fitness of males and females is affected in different ways by the mating system.

In general, a conflict of interest between the sexes exists with respect to what constitutes the optimal breeding system (Arnqvist and Rowe 2005; Rowe and Day 2006; Fricke et al. 2009). For a male, *potential* reproductive success will often be lowest when he has shared access to a single female (polyandry), and then male reproductive success will increase in the following order: sole access to a single female (monogamy), joint access to two females (polygynandry), and access to numerous females (polygyny) (see Box 17.1 for more).

Potential reproductive success in females increases in precisely the opposite direction to male reproductive success, with polyandrous and polygynandrous females having the highest reproductive success. In dunnocks, females appear to be winning this battle of the sexes over breeding system (at least for now), as over the course of 10 years, 75% of females and 68% of males were involved in either polyandrous or polygynandrous mating groups (Davies 1992) (Figure 17.14). It is difficult to say precisely why females currently are winning the sexual conflict, but it may be because, early in the breeding season, females compete with each other to establish territories, and such female territories are chosen independently of the position of males. Males may then attempt to delineate their own territories so that they can overlay as many female territories as possible. This pattern of dispersal and territoriality may allow females a degree of control over the mating system because female territories are already established when males try to establish their territories.

Conflicts of Interest among Red Deer

Evolutionary biologists have hypothesized that conflicts of interest between males and females should be most pronounced when selection favors different traits in each sex. In red deer (*Cervus elaphus*), males are much larger than females, and they have weaponry (their antlers) that they use during fights during the reproductive season (Figure 17.15). Females lack such weaponry. Males and females also differ in that females invest heavily in parental care, whereas males do not. In addition, most females in a population produce offspring. In contrast, because of intense male–male competition, some males may sire many offspring, but many males may never sire any offspring.

FIGURE 17.15 Red deer during mating season. (A) During the reproductive season, male deer battle with their antlers for access to females. **(B)** Males are much larger than females. Here a male is seen with multiple females.

Clearly, the traits that make for a reproductively successful male (size and weaponry) are very different from the traits that make for a reproductively successful female (maternal investment in offspring) (Foerster et al. 2007). What sort of impact does this sexual conflict have on the evolutionary dynamics of the red deer population? Tim Clutton-Brock and his team, who have been studying the life history of red deer on the Island of Rum, Scotland, for more than two decades, wanted to know.

Clutton-Brock's team had records on the lifetime reproductive success of hundreds of red deer on Rum. They examined correlations between an individual's lifetime reproductive success, and the lifetime reproductive success of its male and female offspring. Of the four possible correlations—male fitness/son fitness, male fitness/daughter fitness, female fitness/son fitness, female fitness/daughter fitness—only one was significant. Males whose lifetime reproductive success was relatively high tended to produce daughters whose lifetime reproductive success was relatively low. In the case of red deer, sexual conflict produces an antagonistic relationship between traits that confer high male reproductive success and high female reproductive success.

Drosophila Seminal Fluid and Sexual Conflict

In the case of red deer, different suites of traits were favored in males and females, and sexual conflict led to an antagonistic relationship between male reproductive success and female reproductive success. But that is not the only way that conflicts of interest can manifest themselves in sexual selection. It is also possible that a single trait, expressed in only one sex, can have positive effects on individuals of that sex but negative effects on members of the opposite sex. An excellent example of this sort of antagonistic relationship is seen in the evolution of proteins in the seminal fluid of *Drosophila* (fruit fly) males (Chapman et al. 1995, 2003; Chapman 2001).

The seminal fluid of *Drosophila* males contains at least 80 different proteins produced primarily in what are called accessory glands. The genes that code for proteins produced in the accessory gland proteins are all found on autosomes but they are only expressed in males, and they evolve at relatively quick rates compared to other autosomal genes (Swanson et al. 2001). The accessory gland proteins have a remarkable number of different functions. They facilitate egg production and laying in females, decrease a female's receptivity to other males, and form part of what is called a mating plug—a gelatinous mass that temporarily blocks a female's reproductive tract so that she cannot successfully mate with other males (Chapman 2001).

The benefits to males of such accessory gland proteins are obvious, but what are the benefits to females? Do they receive any benefits associated with proteins transferred in male seminal fluid? It might seem that the increased egg production and laying

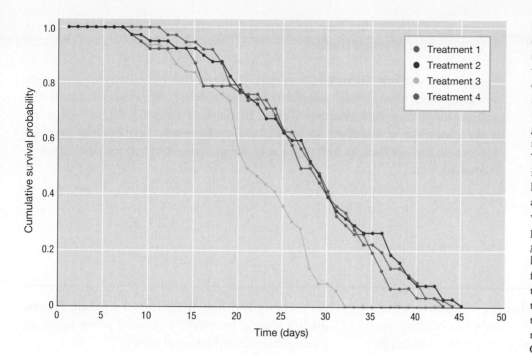

FIGURE 17.16 Cost to females of accessory gland proteins in male seminal fluid. Females survived shorter periods of time when exposed to accessory gland proteins in male seminal fluids. Treatment 1: Males with no sperm or accessory gland proteins; these males behaved normally and mated with females. Treatment 2: Males similar to males in Treatment 1, except that they had their genitalia experimentally altered so that they could not mate with females. Treatment 3: Males produced seminal fluid and accessory gland proteins, but not sperm; males behaved normally and mated with females. Treatment 4: Males similar to males in Treatment 3, except that they had their genitalia experimentally altered so that they could not mate with females. Adapted from Chapman et al. (1995).

associated with the accessory gland proteins transferred in seminal fluid would be beneficial to females as well. This would be the case if eggs were cost-free to produce. But, of course, they aren't. And when a female increases her egg production as a result of contact with the accessory gland proteins transferred to her by an initial male, this may affect her chances of surviving and producing offspring with other males in the future. Tracey Chapman and her colleagues hypothesized that this might impose a mating cost on female fruit flies, and they designed an ingenious experiment to test the idea (Chapman et al. 1995). They used four experimental, transgenic lines of fruit flies. Males in these lines differed from one another as follows:

- Treatment 1: Males produced no sperm or accessory gland proteins. These males behaved normally and mated with females.

- Treatment 2: Males produced no sperm or accessory gland proteins. These males behaved normally, except that they had their genitalia experimentally altered so that they could not mate with females.

- Treatment 3: Males produced seminal fluid and accessory gland proteins, but not sperm. These males behaved normally and mated with females.

- Treatment 4: Males produced seminal fluid and accessory gland proteins, but not sperm. These males behaved normally, except that they had their genitalia experimentally altered so that they could not mate with females.

This combination of treatments allowed Chapman's team to isolate the possible negative effects of male accessory gland proteins on females, while holding constant other attributes of sperm, as well as male behavior outside of courtship behavior. *If there was a cost to females associated with accessory gland proteins transferred by males during mating, then that cost should show up in treatment 3 but not in the other treatments.* When Chapman and her team looked at survival probabilities of females across the four treatments, they found that treatment 3 females died earlier than females from other treatments, suggesting a significant cost to females from exposure to male accessory gland proteins (Figure 17.16). Thus, accessory gland proteins benefit males, but they are detrimental to females. Together these effects are the signature of sexual conflict.

Ever since Darwin first discussed the subject, few topics have fascinated evolutionary biologists more than sexual selection. In this chapter, we have examined the underpinnings of sexual selection theory. In the next chapter, we will move on to the evolution of sociality, including the evolutionary relationship between cooperation and conflict, as well as the role of communication in promoting social interaction.

SUMMARY

1. In *The Descent of Man and Selection in Relation to Sex*, Darwin hypothesized that some traits evolved via sexual selection, which "depends on the advantage which certain individuals have over other individuals of the same sex and species in exclusive relation to reproduction."

2. Sexual selection can occur via (1) intersexual selection, wherein individuals of one sex choose individuals of the other sex as mates, and/or (2) intrasexual selection, in which members of one sex compete for mating access to the other sex.

3. The four main evolutionary models of female mate choice are the "direct benefits," "good genes," "runaway selection," and "sensory bias" models.

4. In the direct benefits model, selection favors females who have a genetic predisposition to choose mates that provide them with some tangible resource—for example, protection from predators.

5. In systems in which females do not receive direct benefits, selection may favor females that choose mates that possess so-called "good genes." Evolutionary biologists have hypothesized that traits that are accurate and honest indicators of male genetic quality should be used by females when selecting among males based on good genes.

6. In the runaway sexual selection model, there is a genetic correlation between a male trait and a female preference for that trait. If the frequency of one increases—for example, by selection—the other increases as well due to this correlation. This process can "snowball," resulting in dramatic male traits and strong female preferences.

7. In the sensory bias model of mate choice, females initially prefer a trait in males because the female nervous system already responds to that trait outside the context of mate choice.

8. Males compete with each other for access to females in direct ways, such as one-on-one fights, but also in more indirect, complex ways, such as via "sneaker" strategies and sperm competition.

9. When selection operates differently on males and females, sexual conflict may result, where the traits that evolve in one sex are detrimental to individuals of the other sex.

KEY TERMS

intersexual selection (p. 566)
intrasexual selection (p. 566)
monogamous mating system (p. 565)
polyandry (p. 565)

polygamous mating system (p. 565)
polygynandry (p. 565)
polygyny (p. 565)
postcopulatory sexual selection (p. 576)

runaway sexual selection model (p. 570)
sensory bias model (p. 572)
sexual conflict (p. 578)
sexual selection (p. 565)

REVIEW QUESTIONS

1. Imagine a group of males that are engaged in a series of fights, and suppose male 1 wins most of those fights. Further suppose that a female watches the fights and prefers male 1 as a mate because he wins them. How does this example blur the distinction between intrasexual selection and intersexual selection?

2. Sensory bias models of sexual selection examine the origins of female preference, but not their subsequent evolution. Outline a system in which sensory bias establishes a preference but direct benefits or good genes lead to the evolution of the female preference.

3. Why would you predict that females in species with internal fertilization and long periods of gestation would be choosier in selecting their mates than females in other species?

4. Besides food and protection from predators, what might be a few other examples of direct benefits that males could provide females during courtship and mating?

5. In many species, there is much greater variance in male reproductive success than in female reproductive success. This could result from females independently choosing a select few mates as partners. But what if females copied the mate choice of others?

How might that also lead to higher variance in male than female reproductive success?

6. Do you think that sexual selection occurs in plants? If so, how? If not, why not?

7. In many (although not all) species of parrots, the male is more brightly colored than the female. Propose two hypotheses for this phenomenon, one of which involves sexual selection and one of which does not.

8. Figure 17.3 shows the distribution of reproductive success for male and female bitterling fish. As noted in the figure caption, the mean reproductive success is the same for males and females because each offspring has one male and one female parent. However, the median reproductive successes need not be the same. Based on the figure, which sex has the higher median reproductive success? Do you think this pattern is typical for other vertebrate species?

9. Why do you think endoparasites might be more important in the "good genes" model of sexual selection, but ectoparasites might play a more prominent role in relation to direct benefits models?

10. One could argue that postcopulatory sexual selection is the last line of defense: a female's last chance to express choice in terms of which male fathers her offspring. What implications follow from this perspective on postcopulatory sexual selection?

SUGGESTED READINGS

Clutton-Brock, T., and K. McAuliffe. 2009. Female mate choice in mammals. *The Quarterly Review of Biology* 84: 3–27. A review of intrasexual selection.

Davies, N. B. 1992. *Dunnock Behavior and Social Evolution*. Oxford University Press, Oxford. A lovely, short book with information on multiple mating strategies in birds.

Eberhard, W. G. 2009. Postcopulatory sexual selection: Darwin's omission and its consequences. *Proceedings of the National Academy of Sciences of the United States of America* 106: 10025–10032. A review of work on sexual selection and postcopulatory competition.

Kokko, H., M. D. Jennions, and R. Brooks. 2006. Unifying and testing models of sexual selection. *Annual Review of Ecology, Evolution, and Systematics* 37: 43–66. A somewhat technical article that tries to bring together theoretical and empirical work in the field of sexual selection.

Ryan, M. J., J. H. Fox, W. Wilczynski, and A. S. Rand. 1990. Sexual selection for sensory exploitation in the frog *Physalaemus pustulosus*. *Nature* 343: 66–67. An early study of sensory bias as one driver of frog mating call evolution.

18

The Evolution of Sociality

18.1 Cooperation

18.2 Conflict

18.3 Information and Communication

◄ Lichen on rock on Steeple Jason Island, Falkland Islands.

I f you were to undertake a search for unselfish behavior, you would probably not begin by looking at slime molds. But slime molds might be just the place to start. In Chapter 12, we described how slime mold cells join together to form mobile slugs. The slugs are ensembles of individual cells that unite to form a sort of pseudo-multicellular creature. Slime mold slugs produce fruiting bodies as their primary reproductive structures. Within each fruiting body, some of the cells are part of a stalk that holds up a capsule full of reproductive spores, while others are the reproductive spores (Figure 18.1). We noted in Chapter 12 that we would return to the possible benefits that a stalk cell might receive (Brannstrom and Dieckmann 2005; Santorelli et al. 2008). We do so here.

Evolutionary biologists have found that slime mold cells that are large and well fed generally become spore cells, and those that are less well nourished become stalk cells (Kessin 2001; Bonner 2003). But this does not tell us how selection could ever favor a once free-living cell that gives up its opportunity for reproduction and instead becomes part of the nonreproducing stalk. To resolve this puzzle, we need to understand that

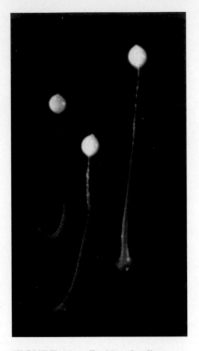

FIGURE 18.1 Fruiting bodies of the slime mold, *Dictyostelium discoideum*. The fruiting body is composed of a stalk and a capsule on the top of the stalk which holds reproductive spores.

slime mold slugs are composed of cells that are close genetic relatives. By acting to increase the reproductive success of their close genetic relatives in the capsule, cells in the stalk may increase the number of copies of their own genes that make it into the next generation, albeit indirectly through their genetic kin.

In this chapter, we will examine the important impact that genetic relatedness can have on social behaviors such as cooperation in organisms ranging from microbes to vertebrates. But our goals are broader than this, as we wish to provide something of an overview of important topics in the area of social behavior. By social behavior, we mean the interactions that organisms have with others—most often, their conspecifics. When we look at behavior in a social context, the actions taken by one individual affect not only its own fitness, but also the fitness of those around it.

In particular, we will address the following questions in this chapter:

- What are the evolutionary processes leading to cooperation?
- How do evolutionary processes lead to conflict?
- How has signaling behavior—involved in both cooperation and conflict—evolved?

18.1 Cooperation

We define cooperation as follows: **cooperation** occurs when two or more individuals each receive a net benefit from their joint actions. Each cooperator may pay an immediate cost for its action, but the overall effect on fitness is positive. Even when everyone benefits from cooperating, however, it is not obvious that natural selection will favor cooperative behavior. The reason is that *free riding*—receiving benefits but not generating them for others—may be possible, and may be even more beneficial than cooperation to an individual free rider. To understand when natural selection favors cooperation, we need to understand how natural selection has solved the following two related problems:

1. The **altruism** problem. Why would natural selection favor an individual who performs an action that has the immediate consequence of reducing its own fitness while increasing the fitness of another?

2. The free-rider problem. In many cases, groups of individuals cooperate, each investing time, energy, and other resources in activities that benefit the entire group. Why are individuals selected to do so, when they could instead *free ride* on the efforts of others, receiving the public benefits while shirking their own duties?

To answer these questions, we explore three evolutionary "paths" to cooperation: (1) kinship, (2) reciprocity, and (3) group selection.

Path 1: Kinship and Cooperation

Most of us feel a special loyalty to our familial kin. "Blood is thicker than water, is it not? If cousins are not friends, who can be?" asks Anthony Trollope in *The Belton Estate* (1866). While among humans this sentiment may largely be a learned cultural convention, there are strong evolutionary reasons why we might expect to see cooperation and altruism among close relatives of any species. The basic reason is that genetic relatives are likely to share common genes that they have inherited from common ancestors—parents, grandparents, and so on. Here we will explore how this observation relates to cooperation and altruism among kin.

Common Ancestry and Shared Alleles

Alleles that are shared because of common ancestry are referred to as identical by descent. For example, you and your sibling share some of the same copies of alleles that you both inherited from common ancestors—in this case, your mother and father. In a similar vein, you and your cousins are genetic kin because you share genes in common; in this case, your most recent common ancestors are your grandparents. In general, a *most recent common ancestor* is the most recent individual through which two (or more) organisms can trace gene copies that they share by descent. Full siblings share the same mother and father, cousins share some subset of the same grandparents, and so on.

Inclusive Fitness and Genetic Relatedness

Notice that there are two ways for an individual to increase the probability that copies of her alleles will reach the next generation. The straightforward way is to produce surviving offspring of her own. The less direct way is to act in a manner that increases the number of offspring produced by her genetic relatives, as these relatives share her genes with some probability, and they may pass them on to *their* offspring. Thus, an individual can get copies of her genes into the next generation either by producing more offspring herself or by helping her kin in their reproductive endeavors. This observation forms the basis for the notion of **inclusive fitness**.

To formalize this intuition, British evolutionary biologist W. D. Hamilton (1936–2000) proposed that we broaden our definition of fitness (Hamilton 1963, 1964). He proposed that an individual's total fitness can be viewed as the sum of (1) its **direct fitness**, which is the number of viable offspring that it produces, and (2) its **indirect fitness**, which is the incremental effect that the individual's behavior has on the (direct) fitness of its genetic relatives. The latter quantity reflects the fact that when an individual increases the number of its genetic kin that survive and reproduce, it is indirectly getting copies of some of its own genes into the next generation. Hamilton termed the sum of the two components the inclusive fitness of an individual.

Since genetic relatedness is one path to the evolution of cooperation, we would like to have a way of quantifying the relatedness between two individuals. Fortunately, there is a straightforward procedure for doing so. To calculate the **coefficient of relatedness** (often denoted r) between two individuals "A" and "B," we follow these steps:

1. We locate the most recent common ancestor or ancestors of A and B. This may be a single individual as in Figure 18.2A, or it may be a mated pair, as illustrated in Figure 18.2B.

2. For each most recent common ancestor, we calculate the probability that a given allele copy in that ancestor has been passed on to *both* A and B. This computation is straightforward. In sexual diploid organisms, the process of

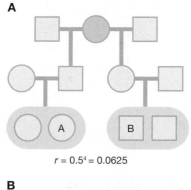

$r = 0.5^4 = 0.0625$

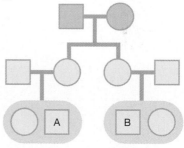

$r = 0.5^4 + 0.5^4 = 0.125$

FIGURE 18.2 Pedigrees for calculating relatedness. Squares indicate males; circles indicate females. Individuals A and B may have one or two most recent common ancestors (dark shading). **(A)** A and B have the same grandmother but different grandfathers. Thus, their grandmother is their sole most recent common ancestor. **(B)** A and B have the same maternal grandmother and the same maternal grandfather. Thus, both maternal grandparents are the most recent common ancestors.

meiotic segregation occurs once per generation. Thus, for any given allele copy in the parent, there is a 50% chance that this allele will be passed on to each offspring, and a 50% chance that the homologous allele will be passed on instead. To compute the coefficient of relatedness (r), we simply tally the number of meiotic divisions that occur along the paths from the common ancestor to A and from the common ancestor to B. Going through multiple generations is straightforward; each generation reduces the probability of obtaining a particular allele that is identical by descent by one-half. According to the rules of probability, if there are two meiotic divisions, the probability is then $0.5 \times 0.5 = 0.25$. If there are four meiotic divisions, the probability is $0.5 \times 0.5 \times 0.5 \times 0.5 = 0.0625$. In general, if there are k meiotic divisions separating A and B, the probability that they share an allele through a single most recent common ancestor is 0.5^k. In Figure 18.2A, individuals A and B share a single ancestor—a grandmother—and are separated by four meiotic divisions (A's grandmother to A's father, A's father to A, B's grandmother to B's mother, B's mother to B). Thus, the coefficient of relatedness (r) between A and B is $0.5^4 = 0.0625$.

3. If A and B have only one most recent common ancestor, we are done. The probability that we have computed is the coefficient of relatedness (r) between them. If A and B have two most recent common ancestors, they could share a given allele through either of those ancestors (but not both). Therefore, again following the rules of probability, we *add* the probability that A and B share an allele through one of the most recent common ancestors to the probability that A and B share an allele through the other most recent common ancestor. In Figure 18.2B, individuals A and B have two most recent common ancestors—their maternal grandparents. The chance they share an allele through one specific grandparent is $0.5^4 = 0.0625$, so the total chance they share an allele through either grandparent is $r = 0.5^4 + 0.5^4 = 0.125$.

In Box 18.1, we show how to calculate the coefficient of relatedness (r) for other sets of genetic relatives.

There is a straightforward way to use genetic relatedness to predict whether an allele for helping one's relatives is favored by natural selection. As a rule of thumb, Hamilton showed that an allele X for helping a relative increases in frequency whenever

$$rb - c > 0$$

where b is the benefit that the genetic relative receives from traits associated with allele X, c is the cost accrued to the individual expressing the trait, and r is the coefficient of relatedness (Lush 1948; Grafen 1984). The equation is often called Hamilton's rule after W. D. Hamilton, the founder of inclusive fitness theory.

Hamilton's rule shows that the extent to which natural selection favors assisting family members depends on how related individuals are, and how high or low the associated costs and benefits turn out to be. When relatedness r is high, benefit b to the recipient is high, and cost c to the actor is low, then natural selection should strongly favor individuals who help their kin.

Some actions, such as issuing an alarm call in response to a predator, may help several relatives at the same time. Hamilton's rule is easily extended to those cases

BOX 18.1 Calculating Genetic Relatedness

Let us work through a few more examples of calculating genetic relatedness. In Figure 18.3A, individuals A and B are half siblings, with the same mother but different fathers. To compute the coefficient of relatedness (r) between A and B, we first must find the most recent common ancestor or ancestors. In this case, there is one: their mother. Second, we compute the probability that a given allele copy in the mother is passed to both offspring. The probability is 0.5 that the allele will be passed to A, and the probability is 0.5 that it will be passed to B, so the probability that it will be passed to *both* is $0.5 \times 0.5 = 0.25$. Because the mother is the sole most recent common ancestor, this is the total coefficient of relatedness (r).

In Figure 18.3B, individuals A and B are full siblings, with the same mother and the same father. Thus, both parents are the most recent common ancestors. For each, we compute the probability that a given allele copy will be passed to both offspring. The calculation is as above. With probability $0.5 \times 0.5 = 0.25$, a given allele in the mother will be passed to both offspring, and by similar logic, with probability 0.25, a given allele in the father will be passed to both offspring. The total coefficient of relatedness will be the sum of these two paths: $0.25 + 0.25 = 0.5$.

In Figure 18.3C, A and B have a single most recent common ancestor who is A's maternal grandmother and B's mother. The chance that a given allele copy in this ancestor reaches A is 0.25, because there is a 0.5 chance that it will reach A's mother, and if it does, there is an additional 0.5 chance that it will go on to reach A, for a net chance of 0.25. The chance that a given allele will reach B is 0.5, since only a single meiosis separates

B from the common ancestor. Thus, the chance that the given allele copy will reach *both* A and B is $0.25 \times 0.5 = 0.125$. The coefficient of relatedness between A and B is therefore 0.125. (If B had been a full sibling to A's mother, the coefficient of relatedness between A and B would have instead been 0.25). Similar calculations allow us to compute the genetic relatedness between any pair of individuals with a known pedigree.

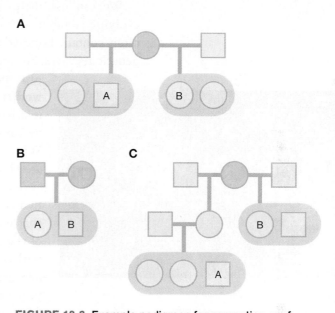

FIGURE 18.3 Example pedigrees for computing coefficients of relatedness. (A) A and B are half siblings. **(B)** A and B are full siblings. **(C)** A more complicated scenario, with A and B coming from different generations.

as well. If relatives 1, 2, . . . , n are related by $r_1, r_2, \ldots, r_n$, and receive benefits $b_1, b_2, \ldots, b_n$, an action with cost c will be favored by natural selection if

$$\sum_1^n r_i b_i - c > 0$$

Inclusive Fitness and Eusociality

Inclusive fitness theory has played a prominent role in understanding the transition from solitary to group living (Chapter 12). In particular, inclusive fitness theory has been employed to examine the evolution of eusocial behavior and group living (Batra 1966). Although debate continues over how best to define **eusociality**, it is most often defined as a social system with the following properties (Alexander et al. 1991):

1. Reproductive division of labor. Only a fraction of the population is actively breeding at a given time; others are infertile as a result of some form of reproductive suppression.

2. Cooperative rearing. Multiple individuals, beyond the immediate parents, work together to feed and care for the young.

3. Overlapping generations. Not only do the generations of a eusocial species overlap (unlike annual plants or many annual insect species), but the members of different generations also live together and work together in a single group.

Eusociality has evolved in termites, beetles, aphids, thrips, shrimps, and mammals such as the naked mole rat. But this extreme form of sociality is most often associated with ants, bees, and wasps. This group is part of the insect order Hymenoptera. Eusociality has evolved independently on at least nine separate occasions in hymenopterans (Hughes et al. 2008) (Figure 18.4).

We can use inclusive fitness theory to understand why we see eusociality evolving so often in ants, wasps, and bees. First, bee, ant, and wasp nests, which often contain hundreds or thousands of individuals, are composed primarily of genetic relatives. So, the altruistic acts associated with eusociality may benefit not just one but many, many genetic relatives—for example, when a worker bee defends the hive, she may save hundreds of genetic relatives by her act. But this logic applies to any colonial species in which relatives live near one another, not just the Hymenoptera. While not all hymenopteran species are eusocial, many are. Why are Hymenoptera particularly prone to evolve eusociality?

The answer may lie in the unusual genetic architecture of the hymenopterans. Ants, bees, and wasps are haplodiploid species—that is, all males are haploid and all females are diploid. Because of the genetics of haplodiploidy, when a queen in a colony mates with a single male, sister workers are related to one another on average by a coefficient of relatedness (r) of 0.75—that is, the probability that a given parental allele ends up in both sisters is 0.75. Here's why. The probability that the sisters share a given allele copy through their mother is 0.25 (as in the case of diploid species), but because all males are haploid, the probability that sisters share an allele copy through their father is 0.5. Adding these probabilities gives us our genetic relatedness value of 0.75.

A genetic relatedness of 0.75 between sisters has the remarkable effect of making females more related to their sisters than to their own offspring! Think about it like this: The queen of the hive produces both females (workers) and males (drones). If the female workers produce their own offspring, they have a genetic relatedness of 0.5 to such offspring, but if they help their mother produce more workers—more sisters—they have a genetic relatedness of 0.75 to such new sisters. Females are not as closely related to their male sibs. A female worker and a male sibling share an r of 0.25, whereas r between siblings in a diploid species is 0.5. Because of the asymmetries in genetic relatedness, we predict that eusocial behaviors should be displayed by female workers but not by male drones. And, indeed, that is what we see: Females tend to the brood at the hive, they defend the nest (even at the cost of their lives), and they do the foraging for the nest (Figure 18.5).

A

B

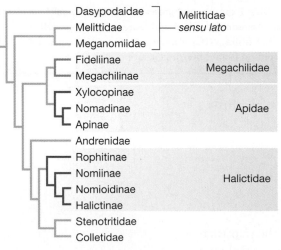

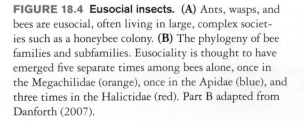

FIGURE 18.4 Eusocial insects. (A) Ants, wasps, and bees are eusocial, often living in large, complex societies such as a honeybee colony. **(B)** The phylogeny of bee families and subfamilies. Eusociality is thought to have emerged five separate times among bees alone, once in the Megachilidae (orange), once in the Apidae (blue), and three times in the Halictidae (red). Part B adapted from Danforth (2007).

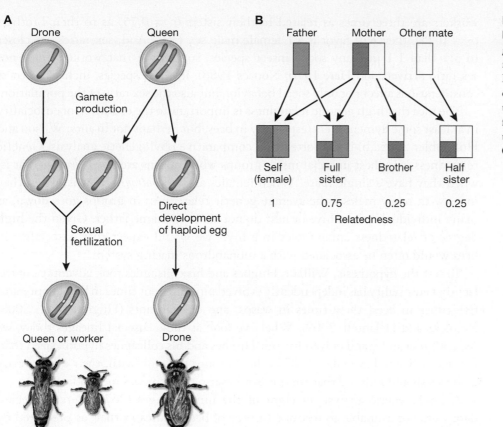

A

Drone | Queen

Gamete production

Sexual fertilization

Direct development of haploid egg

Queen or worker | Drone

B

Father | Mother | Other mate

Self (female) | Full sister | Brother | Half sister

1 | 0.75 | 0.25 | 0.25

Relatedness

FIGURE 18.5 Relatedness in haplodiploids. (**A**) Haplodiploid genetics in bees. (**B**) Coefficients of relatedness among haplodiploids. Blue represents the paternally derived alleles; red represents maternally derived alleles in the focal female labeled "Self." Relatedness to self is always 1. Adapted from Queller (2003).

We are not suggesting that eusociality in hymenopterans is *completely* explained by the high genetic relatedness between workers that comes about via their haplodiploid genetics. After all, *all* hymenopteran species are haplodiploid, but only some hymenopteran species are eusocial. What's more, there are also examples of eusociality in diploid species such as naked mole rats and termites. So, haplodiploidy *alone* is neither necessary nor sufficient for the evolution of eusociality, but it does help partly to explain why eusociality is overrepresented in hymenopterans.

The unusual genetic architecture of hymenopterans also has a dramatic effect on sex ratios in these species. To see why, we need to return to Fisher's original argument for sex ratio evolution, which we explored in Chapter 1. Recall Fisher's conclusions: For most systems of genetic inheritance, natural selection will favor a 1:1 sex ratio, assuming that the cost of producing a male is the same as the cost of producing a female. If the costs differ, an analogous argument reveals that selection will favor parents who invest an equal *amount of resources* in offspring of each sex. As a result, parents will produce more of whichever sex is less expensive to produce.

This result can be extended yet further, to treat the curious genetics of haplodiploid species. An extension of Fisher's argument reveals that, when relatedness varies by sex, as in haplodiploid species, natural selection will favor individuals who invest in kin of a given sex proportional to their relatedness to kin of that sex.

As we have seen, haplodiploid mothers are equally related to their sons ($r = 0.5$) and their daughters ($r = 0.5$). Assuming an equal cost of producing male and female offspring, queens are expected to favor a 1:1 sex ratio. But because female workers are three times as related to their sisters ($r = 0.75$) as to their brothers ($r = 0.25$), they will favor a 3:1 female:male sex ratio. And, sex ratios are closer to 3:1 than 1:1 in many social insect species, suggesting that workers influence sex ratio (Trivers and Hare 1976; Nonacs 1986). In those species, inclusive fitness considerations affect not only social behavior, but also the sex ratio of the population.

The idea that high genetic relatedness is important to the evolution of eusociality in at least some hymenoptera (especially in bees, but perhaps not in ants; Wilson and Holldobler 2005a,b) is supported by a comparative phylogenetic analysis. Genetic relatedness is highest in social insect groups when queens are *monandrous*—that is, when they have a single mate. When females are *polyandrous*—that is, when they mate with many males—the average genetic relatedness in groups goes down, as many individuals in the hive or nest do not share the same father. Given the high degree of relatedness among bees in a hive, we would expect that eusociality in bees would often be associated with a monandrous mating system.

To test the hypothesis, William Hughes and his colleagues took advantage of the fact that eusociality has independently evolved nine different times in hymenopterans: five times in bees, three times in wasps, and once in ants (Hughes et al. 2008; Ratnieks and Helantera 2009). When we look at these eusocial lineages today, we see both monandry and polyandry. But Hughes and his colleagues hypothesized that, for eusociality to have taken hold in these groups to begin with, their evolutionary histories should indicate that the ancestral mating system was monandrous.

A phylogenetic analysis of eight of the nine lineages (267 different species; data were not available to test one lineage of bees) indicates that, as predicted by inclusive fitness theory, monandry was the ancestral state in *all* eusocial lineages examined (Figure 18.6). This suggests that eusocial species that are not currently monandrous (about one-third of all eusocial species) evolved from monandrous ancestors after eusociality was already in place. Why the evolution to polyandry occurred in some hymenopteran species has not been fully explained.

Path 2: Reciprocity

In 1971, Robert Trivers hypothesized that if individuals benefited from *exchanging* acts of altruism, then this sort of reciprocal exchange system—which Trivers called **reciprocal altruism**—might be favored by natural selection (Trivers 1971). If individual A pays some cost to help individual B, but the cost is recovered at some point in the future (when B helps A), then natural selection might favor behaviors that lead to this type of reciprocity. Reciprocal altruism might be especially likely to occur among individuals that live in stable groups because they are likely to have ongoing interactions with the same set of partners.

The Prisoner's Dilemma

Trivers addressed the question of the evolution of reciprocity using a theoretical framework known as *game theory*. Game theory allows us to analyze decision making in a social context. It is useful when dealing with strategic situations, in which the results of one participant's actions depend on the behaviors that other participants adopt. In particular, Trivers, with some help from W. D. Hamilton, suggested that the evolution of cooperation could best be understood by employing a mathematical game called the prisoner's dilemma.

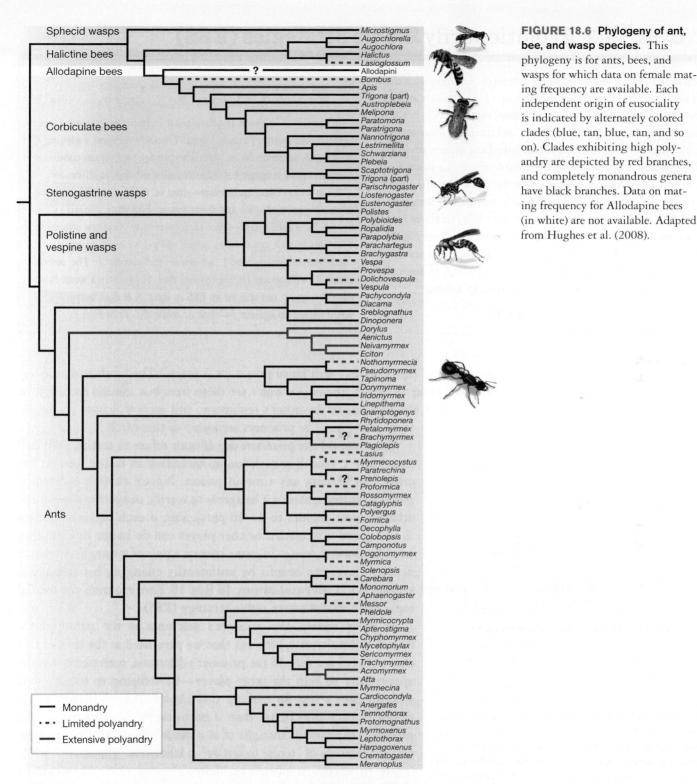

FIGURE 18.6 Phylogeny of ant, bee, and wasp species. This phylogeny is for ants, bees, and wasps for which data on female mating frequency are available. Each independent origin of eusociality is indicated by alternately colored clades (blue, tan, blue, tan, and so on). Clades exhibiting high polyandry are depicted by red branches, and completely monandrous genera have black branches. Data on mating frequency for Allodapine bees (in white) are not available. Adapted from Hughes et al. (2008).

The prisoner's dilemma game is based on a scenario in which two criminal suspects are caught by the police. They are taken to two different rooms and interrogated separately. The police have enough circumstantial evidence to put each suspect in prison for 1 year, even without a confession from either. In an effort to get the two suspects to testify against one another, the police offer each suspect the following deal: "If you testify against the other guy, you'll walk away a

BOX 18.2 Evolutionarily Stable Strategies (ESS)

An evolutionarily stable strategy is defined as "a strategy such that, if all the members of a population adopt it, no mutant strategy can invade" (Maynard Smith 1982). Here "mutant" refers to a new strategy introduced into a population, and successful invasions center around the relative fitness of established and mutant strategies. If the established strategy is evolutionarily stable, the payoff from the established strategy is greater than the payoff from the mutant strategy. To see this more formally, let's consider two strategies, I and J (for example, I might be to cooperate, while J might be not to cooperate). We will denote the expected payoff of strategy I against strategy J as $E(I, J)$, the payoff of J against I as $E(J, I)$, the payoff of I against I as $E(I, I)$, and the payoff of J against J as $E(J, J)$. Strategy I is an ESS if for every possible alternative strategy J, either

$$E(I, I) > E(J, I) \qquad (18.1)$$

or

$$E(I, I) = E(J, I), \quad \text{but} \quad E(I, J) > E(J, J) \qquad (18.2)$$

If the first condition (Equation 18.1) holds true, then I does better against other I's than J does. Thus, if everyone is playing I, no one can do better by unilaterally shifting to J. Thus, condition 1 ensures that strategy I is what is called a Nash equilibrium.

If I is *a strict Nash equilibrium*—that is, if $E(I, I) > E(I, J)$ for all other strategies, J—we are done. I is an ESS. But if strategy I is a *weak Nash equilibrium*—that is, if there is at least one strategy J that does as well against I as I itself does—we need an additional condition to ensure that I can resist invasion by J. The second condition (Equation 18.2) provides this. If I is only a weak Nash equilibrium, it can still be an ESS so long as it does better than J when paired up against J—that is, when $E(I, J) > E(J, J)$.

free man and the other guy will go to prison for 5 years." The catch is that if *both* prisoners agree to testify, the police won't set them free, but instead each will be convicted on the grounds of the other's testimony, and each will have to serve 3 years in prison. The prisoners are aware of this catch.

What should the prisoners do? If both refuse to testify, each will serve only 1 year. But each has an incentive to testify against the other: not serving any time in prison. Notice that an individual serves a shorter sentence if he agrees to testify, *irrespective* of what the other suspect decides to do. In particular, if each player's strategy in this game is to testify, neither player can do better by changing what he alone is doing. In game theory, a pair of strategies in which neither player can benefit by unilaterally changing his strategy is known as a Nash equilibrium. In Box 18.2 we examine the related topic of an evolutionary stable strategy (ESS).

In game theory, the prisoner's dilemma is the paradigmatic model of the altruism problem that we presented at the start of this chapter. Let's see why. In the prisoner's dilemma, each player has the opportunity to help the other player—by refusing to testify—but a player pays a cost for making that choice. If both players refuse to testify, each does better than if both testify. For this reason, the prisoner's dilemma is thought of as a model of cooperation, and the players' strategy of "refuse to testify" is labeled as "cooperate," while "agree to testify" is labeled as "defect" (notice that, when defined this way, to cooperate means to cooperate with one's codefendant, not with the authorities, and to defect is to no longer cooperate with one's codefendant).

Figure 18.7 depicts the payoffs—measured as years in prison—to each suspect as a function of what he decides to do, and what the other suspect decides to do. If both suspects cooperate, they both

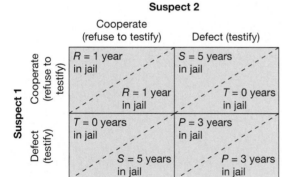

FIGURE 18.7 The prisoner's dilemma game. In this game, each player can either cooperate or defect. To cooperate is to refuse to testify; to defect is to testify. For the matrix to qualify as a prisoner's dilemma game, it must be true that $T > R > P > S$, where T is "temptation to defect" payoff, R is "reward for mutual cooperation" payoff, P is "punishment for mutual defection" payoff, and S is "sucker's" payoff. Each cell shows the payoff to suspect 1 (above the dashed diagonal line) and the payoff to suspect 2 (below the dashed diagonal line). For example, in the lower left cell, when suspect 1 defects and suspect 2 cooperates, the former gets no time in jail, while the latter gets 5 years in jail. Technically, in order for the game to be a prisoner's dilemma, it must also be true that the payoff for mutual cooperation ($2R$) is greater than the sum of the payoffs received by two players in a cooperator–defector interaction—that is, $2R > T + S$. Adapted from Dugatkin (2009).

receive a payoff of R (the reward for mutual cooperation; 1 year in jail), but if they both defect, each receives P (the punishment for mutual defection; 3 years in jail). If suspect 1 defects, but suspect 2 cooperates, the former receives a payoff of T (the temptation to defect; no time in jail), and the latter receives S (the sucker's payoff; 5 years in jail). If we order the payoffs in this matrix from high to low, we see that $T > R > P > S$. It is this series of inequalities that defines our game as a prisoner's dilemma—that is, for a game to be a prisoner's dilemma, the payoff structure of the matrix must be $T > R > P > S$.

With the game laid out in this way, we can explore the strategic problem facing our two suspects: Suspect 1 will receive a higher payoff individually (serving fewer years in prison) if he defects, without considering what suspect 2 does. As such, suspect 1 should always defect, assuming that he prefers to minimize the length of his prison sentence. The same holds true individually for suspect 2, and he should also always defect. So, if both subjects want to minimize the lengths of their prison sentences, each should defect and agree to testify against his codefendant. The *dilemma* in the prisoner's dilemma is that, while each suspect receives P (3 years in prison) when they testify against one another, both suspects would have received better payoffs (R, which is only 1 year in prison) if they had both refused to testify—that is, if they had cooperated with each other. The seemingly intractable problem that the prisoners face is that, once taken to their separate interrogation rooms, each has no way to ensure that the other will cooperate if he does so himself—and in fact by the logic above, each has every reason to suspect the other will defect instead.

So, why would we ever see cooperative behavior in games that take the form of the prisoner's dilemma? In *one-shot games*—that is, in circumstances in which the game is played only once—the answer is that we should not expect to see cooperation. Defection is a Nash equilibrium (neither player can benefit by changing his strategy and cooperating), and it is in fact the only Nash equilibrium.

But what if the game is played repeatedly? Then perhaps the logic of reciprocal altruism may lead to cooperative behavior. Indeed, it can, but only under certain conditions. The key insight is that each player can "demand" cooperation from the other, using the promise of future cooperation and the threat of future defection as carrot and stick. One notable strategy of this sort is known as tit for tat (TFT). In the tit-for-tat strategy, an individual cooperates on the initial encounter with a partner and subsequently copies its partner's previous move. This means that, after the first move, TFT operates under an if–then rule: if the partner cooperated in the previous round, then cooperate now; if the partner defected in the previous round, then defect now. That is, TFT reciprocates both acts of cooperation and acts of defection. Playing repeatedly against a TFT player, one can defect now, but at the cost of being defected against in the next round. So, can this make cooperation advantageous?

To see, let's first suppose that players 1 and 2 know that they are going to play the prisoner's dilemma game with one another 10 times in a row. We might imagine that each would cooperate in the early rounds, so that the other would continue to cooperate throughout the series of games. But does this really work? Think about what each player should do on the tenth and final round of the game. In this final round, as in any single round of the prisoner's dilemma game, he will do strictly better by defecting. Moreover, there are no further rounds to worry about, so each player may as well defect on the final round. Now step back to the second-to-last round. Knowing that the other player is likely to defect on the final round, by the logic above there is no harm in defecting on the second-to-last round, because there is no cooperation to preserve. So, each should defect on the second-to-last round as well. By the same logic, each should defect on the third-to-last round, the fourth-to-last round, and so on, all the way back to the first round of the game. Thus, in a repeated prisoner's dilemma where the players know there will be some *fixed number of interactions*, the only Nash equilibrium that remains is to defect throughout. Simply playing repeatedly does not necessarily solve the altruism problem.

But the altruism problem can be solved with a bit of uncertainty about how many times the game will be played. If neither player knows when the game will end, neither can apply the logic described above. There is no definitive "last round" in which defection is the obvious choice. Instead, at any present time, each player must cooperate now so as to ensure cooperation by the other player in the future.

Robert Axelrod and W. D. Hamilton used both analytical techniques and computer simulations to examine what sorts of behavioral strategies fared well in an iterated (repeated) prisoner's dilemma game (Axelrod and Hamilton 1981; Axelrod 1984). They found that, while the strategy "always defect" is the only Nash equilibrium in the single-shot prisoner's dilemma, the tit-for-tat strategy was one Nash equilibrium in the iterated prisoner's dilemma that has an uncertain end point. This work established the basic theoretical foundation for reciprocal altruism.

Numerous studies have examined reciprocity in animals (Dugatkin 1997). Here we examine one such study that addresses reciprocity in the context of predator mobbing by birds.

FIGURE 18.8 Crows mobbing an owl. Evidence from some species suggests that this sort of antipredator behavior may involve reciprocity among the mobbers.

Reciprocity and Mobbing in Birds

Along with the altruism problem, we described a closely related problem known as the free-rider problem. The gist of the free-rider dilemma is that it may be hard to establish costly cooperation in groups because each individual has an incentive to "free ride" on the efforts of the others. The behavior of mobbing a predator provides a good example.

Mobbing behavior is an antipredatory tactic, in which one or more individuals approach, chase, and sometimes even attack a potential predator that may be much larger than individuals of the mobbing species. This sort of behavior is common among birds, where mobbing behavior often causes a potential predator to leave an area as a result of continual harassment (Sordahl 1990) (Figure 18.8).

Mobbing behavior can be costly, both in terms of the time and energy invested, and because mobbing individuals are occasionally caught by the predator they are trying to mob (Sordahl 1990; Krama and Krams 2005). But once a predator is driven away, *all* of the prey individuals in that area benefit, not just those that were involved in mobbing. So, why do individual birds join a mobbing group? Why don't they simply let others take on the cost and risk? Indrikis Krams and his colleagues designed an experiment to examine whether reciprocity played a role in the mobbing behavior of the pied flycatcher (*Ficedula hypoleuca*) (Krams et al. 2008). The researchers examined whether flycatchers were more willing to risk the danger associated with mobbing when they had partners who had helped them in the past. To test this, they set up three nestboxes that each housed a pair of flycatchers. They placed the nestboxes about 50 m from one another, and birds in each nest could see all the other nestboxes.

To begin the experiment, a stuffed "model" predator was placed near nestbox 1 (Figure 18.9). Birds from nestbox 1 mobbed the predator, and they were joined by birds from nestbox 3. But the experimenters had placed the birds in nestbox 2 in a cage, so that they could not join the mobbing event at nestbox 1. As a consequence, the birds in nestbox 1 had the experience of being aided by those at nestbox 3, but not by those at nestbox 2. In two follow-up experiments in which a stuffed predator was placed at nestboxes 2 and 3, birds from nestbox 1 joined birds at nestbox 3 in mobbing a predator—they reciprocated the aid they had received—but they did not join a mob when a predator was placed near nestbox 2. Together, these experiments suggest that pied flycatchers may use a reciprocal altruistic strategy when mobbing dangerous predators in their environment.

A Phase 1

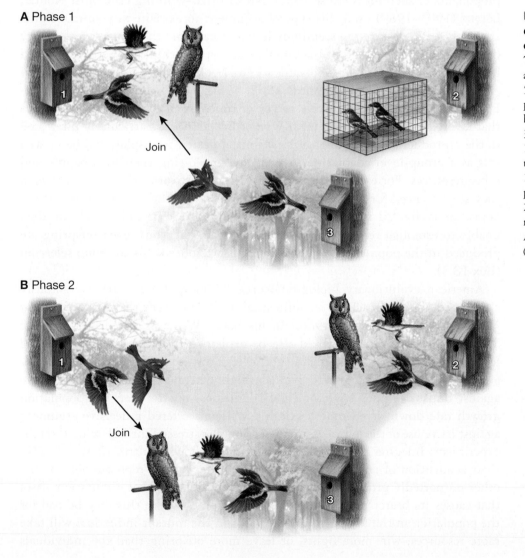

Join

B Phase 2

Join

FIGURE 18.9 The experimental design for examining reciprocal mobbing in pied flycatchers. Three nestboxes were placed on a triangular grid spaced roughly 50 m apart. **(A)** Phase one: A stuffed predator (owl) was placed near nestbox 1. Birds from nestboxes 1 and 3 mobbed the predator at nestbox 1, but birds in pair 2 could not join this mob. **(B)** Phase two (conducted 1 hour after phase one): A stuffed predator was placed at nestboxes 2 and 3. Pair 1 joined the mob at nestbox 3, but not at nestbox 2. Adapted from Wheatcroft and Price (2008).

Path 3: Group Selection

A third evolutionary path to cooperation is via group selection. Ideas about group selection have a long history (Wilson 1980; Sober and Wilson 1998; Wilson and Wilson 2007). Although still quite controversial (Lehmann et al. 2007; Reeve and Holldobler 2007), *modern* group selection models—sometimes called trait-group selection models—of cooperation are conceptually straightforward. Before treating these, however, we will briefly review the history of group selection thinking and the critiques that brought earlier group selection approaches into disfavor.

Group Selection and "Good of the Species" Logic

From the time that Darwin and Wallace laid out their theory of evolution by natural selection and through the 1960s, evolutionary biologists would sometimes attempt to explain certain aspects of animal behavior or physiology as adaptations that had arisen "for the good of the species" or "for the good of the population"—that is, adaptations that would minimize the chances that the species or population as a whole would go extinct. Wallace himself and, to a lesser extent, Darwin were proponents of such ideas (Ruse 1980). Nobel prize–winning ethologist Konrad Lorenz (1903–1989) used this type of argument to explain why animal fights are rarely fatal, despite the seemingly lethal armaments that many species carry (Lorenz 1966). The "good of the population" type of thinking perhaps reached its pinnacle in V. C. Wynne-Edwards's 1962 book *Animal Dispersion in Relation to Social Behavior* (Wynne-Edwards 1962, 1986, 1993). In his book, Wynne-Edwards presented an exhaustive survey of traits that he felt to be adaptations that favored the survival of groups. Wynne-Edwards was particularly interested in the reproductive restraint that organisms appeared to display, and he viewed this as a group-level adaptation to avoid overexploiting their food supply and other resources. For example, individuals defend territories that are larger than they seem to need for survival and reproduction, with the consequence that the landscape is divided into fewer breeding territories. Some individuals are then unable to establish territories on which to breed, and thus fewer offspring are produced in the population. Wynne-Edwards attributed this to group selection (Box 18.3).

American evolutionary biologist George Williams (1926–2010) vigorously challenged this approach in an influential 1966 book entitled *Adaptation and Natural Selection* (Williams 1966). In his book, Williams noted that most of Wynne-Edwards' examples could also be explained by natural selection at the level of the individual, rather than at the level of the group (Bergstrom 2002). For example, he hypothesized that individuals might defend large territories as a hedge against unusually poor environmental conditions, not to keep the population growth rate down. Even more critically, Williams offered a decisive argument against naive use of the logic of group selection. He stressed the following thought experiment: Imagine a population of individuals showing altruistic restraint (in their acquisition of resources, severity of fighting, rate of reproduction, or any other purportedly group-level adaptation). Now imagine that a mutation arises that causes its bearer not to exercise such restraint. While this may be bad for the population in the long run, in the short run the mutant individual will take more resources, win more fights, or leave more offspring than the individuals

BOX 18.3 The Tragedy of the Commons

In a famous 1968 essay, Garrett Hardin presented a metaphor for the overexploitation of natural resources, which he called "The Tragedy of the Commons" (Hardin 1968). Hardin describes the following pastoral fable:

> Picture a pasture open to all. It is to be expected that each herdsman will try to keep as many cattle as possible on the commons. Such an arrangement may work reasonably satisfactorily for centuries because tribal wars, poaching, and disease keep the numbers of both man and beast well below the carrying capacity of the land. Finally, however, comes the day of reckoning, that is, the day when the long-desired goal of social stability becomes a reality. At this point, the inherent logic of the commons remorselessly generates tragedy. (Hardin 1968, p. 1244)

Hardin proceeds to explain why this leads to tragedy:

> As a rational being, each herdsman seeks to maximize his gain. Explicitly or implicitly, more or less consciously, he asks, "What is the utility to me of adding one more animal to my herd?" (Hardin 1968, p. 1244)

Hardin points out that adding one additional goat to his personal herd—grazed on communal land—offers both benefits and costs to the individual herdsman. The benefit accrues to the individual herdsman alone; he now has one more animal that he can use or sell. This brings him a net benefit of one goat.

The cost of adding one more goat to his personal herd comes in the form of the further overgrazing to the commons that is caused by the added goat. This cost is shared among all of the people who graze goats on the commons, and thus even if that cost is quite large, the part that the individual herdsman must pay is only a small fraction of one goat. Based on this logic, Hardin explains,

> The rational herdsman concludes that the only sensible course for him to pursue is to add another animal to his herd. And another; and another. . . . But this is the conclusion reached by each and every rational herdsman sharing a commons. Therein is the tragedy. Each man is locked into a system that compels him to increase his herd without limit—in a world that is limited. Ruin is the destination toward which all men rush, each pursuing his own best interest in a society that believes in the freedom of the commons. Freedom in a commons brings ruin to all. (Hardin 1968, p. 1244)

Hardin's tragedy of the commons is yet another form of the altruism problem or free-rider problem. In this case, the cooperative or altruistic thing to do would be to show restraint and limit one's own herd, but this creates group benefits at an individual cost. In the context of natural selection, we can then ask why natural selection would favor such moderation; indeed, this is precisely the question that led Wynne-Edwards to advocate the form of group selection thinking that he did.

---

who exhibited restraint. As a result, the frequency of the mutation will increase over time within the population, as its bearers outcompete the more restrained wild type, and natural selection will eliminate restraint. Williams' point is that natural selection typically acts more strongly on individual-level traits than on group-level traits. For this reason, he argued that appeals to group-level selection should be an absolute last resort for evolutionary biologists.

Modern Approaches to Group Selection: Trait-Group Selection Models

Although Williams' arguments against group selection are sound, they do not entirely rule out the possibility of selection acting at the level of a group. Modern *trait-group models* address this, and they specify the precise circumstances in which selection can favor group-beneficial traits even when such traits impose individual-level costs.

A *trait group* is defined as a group in which all individuals affect one another's fitness. Many such trait groups make up a population. The essence of trait-group selection models is that natural selection operates at two levels: within-group selection and between-group selection. In the context of cooperation, within-group selection acts *against* cooperators who pay some cost that others do not. Selfish free riders—those who do not cooperate—are always favored by within-group selection because they receive any benefits that accrue through the actions of cooperators, but they pay none of the costs.

As opposed to within-group selection, between-group selection favors cooperation if groups with more cooperators outproduce other groups—for example, by producing more total offspring or being able to colonize new areas faster. Consider alarm calls. Alarm callers pay a cost within groups, as they will be the most obvious target of a predator alerted by such a call. But their sacrifice may benefit the group overall, as other individuals—including other alarm callers, as well as those that don't call—are able to evade predators because of the alarm call. Thus, groups with many alarm callers may outproduce groups with fewer alarm callers. For such group-level benefits to be manifest, groups must differ in the frequency of cooperators within them, and groups must be able to "export" the productivity associated with cooperation (for example, by having more total offspring, by moving more quickly to colonize newer areas, and so on).

Many evolutionary biologists argue that group selection models (including trait-group selection models) can be translated mathematically into "classic" models of natural selection—that is, they claim that group selection models simply partition the effect of a trait into within- and between-group components, but that if you sum up the effects over all groups making up a population, you get the same solution as a classic model would produce by tracking gene frequency in an entire population (Queller 1992; Lehmann et al. 2007; Reeve and Holldobler 2007). This is absolutely correct. We can always take a group selection model and translate the mathematics into a model of alternative alleles in which natural selection favors one allele over another in a given population. Such mathematical equivalence, however, does not mean that group selection models do not shed new light on behavior, as trait-group selection models focus attention on what is happening within and between groups, and this is not necessarily the case for more classic models (Dugatkin and Reeve 1994; Kerr and Godfrey-Smith 2002). Thus, under certain conditions, trait-group selection models may spur investigators to conduct experiments or pursue lines of research that would not have been obvious had they been using classic models.

Within- and Between-Group Selection in Ants

Cooperative colony foundation occurs in a number of species of ants where cooperating cofoundresses are *not* closely related (Holldobler and Wilson 1990; Bernasconi and Strassmann 1999). This type of cooperative foundation has been especially well studied in the desert seed harvester ant *Messor pergandei*, in which nests are often initiated by two unrelated queens (cofounders). Cofounding queens in a nest assist in excavating their living quarters, and each produces approximately the same number of offspring.

Steve Rissing and his colleagues have found a positive correlation between the number of cooperating foundresses in a nest and the number of initial workers produced by that colony (Rissing and Pollock 1986, 1991) (Table 18.1). The number of workers produced by a nest is important for nest survival, because *brood raiding* is common in this species. Brood-raiding ants attack nearby colonies and capture their larvae and pupae. The stolen brood are brought to the nests of the victorious ants. Colonies that lose their brood in such interactions die; such a fate often befalls colonies that are just starting up. This competition between nests favors cooperation at the level of the group (Wheeler and Rissing 1975; Ryti and Case 1984). Nests with more cooperating foundresses—and thus with more workers—are more likely to win brood raids (Rissing and Pollock 1987, but see Pfennig [1995] for a critique of this work).

TABLE 18.1

Cooperating Cofoundresses in Unrelated Queens of *Messor pergandei*

Nest Number	Number of Eggs	PERCENT LAID BY[a]			
		W	Y	B	O
1	22	—	32	41	27
2	24	38	25	38	—
4	32	41	28	25	6
5	11	45	—	55	—
8	29	28	34	38	—
9	44	34	16	27	23
10	36	31	19	50	—
11	21	43	—	24	33
15	29	38	—	21	41

[a]W, queen marked with white paint; Y, queen marked with yellow paint; B, queen marked with blue paint; O, queen marked with orange paint. The reproductive output of queens within a nest tends to be approximately equal. From Rissing and Pollock (1986).

Until workers emerge, queens within a nest do not fight, and no dominance hierarchy exists (Figure 18.10). After workers emerge and the between-group benefits of having multiple foundresses are already set in place with the presence of brood raiders, all that remains is within-group selection, which always favors noncooperative behavior. It is at this juncture that queens within a nest often fight to the death.

One of the strongest cases for group selection comes from Rissing's work on another ant, *Acromyrmex versicolor* (Rissing et al. 1989). In this species, nests are often founded by multiple queens, there is no dominance hierarchy among queens,

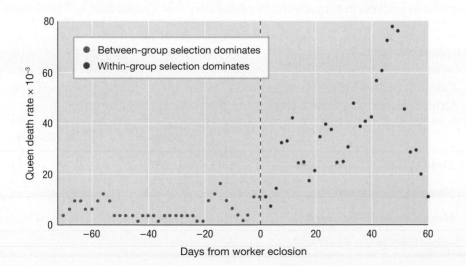

FIGURE 18.10 From cooperation to aggression. Cofounding *Messor pergandei* queens are cooperative during worker production, with very little queen–queen aggression during this phase of colony development. But once workers are produced—known as "worker eclosion," starting at day 0—aggression between queens escalates, as does the queen death rate. Adapted from Rissing and Pollock (1987).

FIGURE 18.11 Cooperation by foraging queens. In the ant *Acromyrmex versicolor*, a single queen (shown in the blowup circle) is the forager for a nest. Such foraging is very dangerous, but all food collected is shared equally among (unrelated) queens. Adapted from Dugatkin (2009).

and all *A. versicolor* queens produce workers. As was the case for *M. pergandei*, brood raiding among starting nests is common, and the probability that a nest survives the brood-raiding period is a function of the number of workers it has produced.

In *A. versicolor*, a single queen in the nest takes on the role of forager for that entire nest (Figure 18.11). Foraging entails bringing vegetation back to the nest, where this resource is added to a "fungus garden" from which the ants feed. As a result of increased predation pressure outside of the nest, foraging is a dangerous activity for a queen. Yet, once a queen takes on the role of forager, she remains in that role. The queen that is the sole forager for her nest shares all the food she brings into her nest with her cofoundresses. This means that the forager assumes both the risks and the benefits of foraging, while the other queens in her nest reap the benefits without paying the costs (Table 18.2). Once again, however, cooperation within nests—in this case, on the part of the forager—appears to lead to more workers. The increase in workers in turn affects the probability that a given nest will be the one to

---

TABLE 18.2

Cooperation among Forager and Nonforager *Acromyrmex versicolor* Queens Leads to Equal Reproduction by All the Queens

	Forager	Nonforager
Mean Number of Primary Eggs	8.6	8.5
Mean Primary Egg Length	0.52	0.54
Mean Number of Total Eggs	20.37	18.94

Adapted from Rissing et al. (1989).

survive the period of brood raiding, thus providing the between-group component necessary for cooperation to evolve (Rissing et al. 1989; Seger 1989).

Rather than separating these two ant examples into within- and between-group selection, we could have analyzed the cooperation described in both examples in terms of the relative success of alternative alleles, and we would have come to the same conclusions we arrived at from the trait-group perspective. For example, in the case of *A. versicolor*, we could say that foraging by the foraging specialist is favored over not foraging because decreased survival rates associated with foraging are, on average, made up for by the increased expected survival of her reproductive brood owing to enhancement of the worker defense force. Here we have averaged survival rates over all groups, rather than separating our example into what happens within and between groups. Both explanations are correct in that they are mathematically equivalent, but in both *M. pergandei* and *A. versicolor* we see systems in which population biology and demographics match those postulated in trait-group models. The multiple nests, intense competition between nests, and multiple foundresses in these species make them ideal for an analysis at the within-group and between-group levels.

Within- and Between-Group Selection in Microbes

Work on testing group selection models of cooperation has not been limited to insects. Indeed, microbial systems have become a model for testing group selection models of cooperation. Microbial systems afford the ability to work with both large populations and rapid generation times so as to provide a system with relative ease of manipulation.

Most microbes live in structured colonies, and hence they seem like good candidates for analysis at the level of within- and between-group selection. For example, in *E. coli*, most cells produce β-lactamase, a substance that breaks down certain antibiotics, but some cells do not. In addition, some of the cells that produce β-lactamase secrete (or can be experimentally induced to secrete) a portion of it into their environment and hence protect other cells. From the perspective of trait-group selection models, we could say that, within groups, cells that do not produce β-lactamase have an advantage over cells that produce and secrete β-lactamase. They receive the protection afforded by β-lactamase when it is produced by other cells in their colony, but they do not pay the costs of producing β-lactamase. These costs have been measured in the absence of any antibiotics and the growth rate of cells that secrete β-lactamase is lower than the growth rate of cells that do not secrete β-lactamase. On the other hand, at the between-group level, groups with "secretors" should outcompete groups without secretors. The relative strength of within- and between-group selection will determine the frequency of secretors (Dugatkin et al. 2005).

Just as with the foraging queen example above, we could cast our explanation so that it refers to within- and between-group selection. Or we could note that, averaged over all groups, the frequency of secretors is determined by the cost of secreting versus the benefits of being in the area with other secretors who provided added protection against antibiotics (Dugatkin et al. 2003). Regardless of which perspective we adopt, the empirical data suggest that secretors and cells that do not produce β-lactamase can coexist at equilibrium in species such as *E. coli* (Dugatkin et al. 2004).

18.2 Conflict

Thus far, our discussion of the evolution of sociality has centered on cooperation. But prosocial behavior such as cooperation is only one type of social behavior that evolutionary biologists study. Indeed, much of the work on the evolution of behavior focuses not on prosocial behavior but instead on the behaviors associated with conflict. In one sense, conflict behavior is less interesting theoretically, as its evolution is easier to understand. When resources are limited, sometimes they are worth contesting.

Conflict manifests itself in many ways in nature. The most obvious form is aggressive behavior, such as when two rams butt horns or when two male elephant seals fight for access to mating opportunities. But conflict can also occur in unexpected places, such as between genetic relatives, where we would generally expect cooperation. Finally, conflict is not limited to conflict between individuals. Conflict, in the broadest sense, can occur at many levels, including among genes in the same genome. In this section, we will work through examples of each—conflict among nonkin, conflict within families, and conflict within genomes.

Conflict among Nonkin

In the previous chapter, we considered various types of sexual selection. We noted that intrasexual selection involves direct competition among members of the same sex—typically, although not always, males—for territory or access to members of the opposite sex. This sort of male–male competition is a major source of conflict in nature (Figure 18.12). Many of the conflicts are resolved by direct fights, which in turn have a strategic dimension. For example, when should an individual risk a fight, and when should it flee? As mentioned earlier in this chapter, evolutionary biologists can use game theory models as a tool for thinking about social interactions and their fitness consequences. For example, if an individual is willing to fight for a contested resource, the outcome will depend on whether its opponent opts to fight or simply to flee, and so we can model fighting behavior using game theory.

The hawk–dove game is a classic model of the evolution of aggression, and it was among the first applications of game theory in evolutionary biology (Maynard Smith and Price 1973; Maynard Smith 1982). John Maynard Smith and his colleagues wanted to understand why in contests among organisms with lethal armaments—sharp teeth, claws, horns, and so forth—one individual often backed down, thereby avoiding a fight that might lead to lethal injury.

FIGURE 18.12 Conflict can lead to fights. Here two oryx lock horns in a struggle over access to mates.

The modern form of the hawk–dove game posits two individuals contesting a single resource with a value *v*. They face off over the resource, and they can adopt one of two behavioral strategies when contesting the resource: Each can play the aggressive "hawk" strategy, or the cautious "dove" strategy. If both select hawk, they end up in a damaging fight incurring total cost *c*. After fighting, each gets half of the resource (or, alternatively, we can think of the probability that a given individual gets the resource as 0.5). If one individual selects hawk and the other selects dove, the hawk gets the resource, while the dove retreats and gets nothing. If both select dove, they share the resource. We can write down the payoffs for this game as in Figure 18.13. We assume here that the cost of a fight *c* is greater than the benefit of the resource *v*.

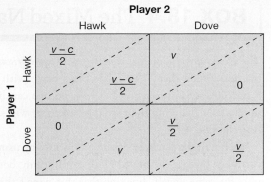

FIGURE 18.13 Payoffs for the hawk–dove game. The row indicates the strategy of player 1, the column indicates the strategy of player 2. Each cell indicates the payoffs to player 1 (above the diagonal line) and player 2 (below the diagonal line). For example, we look in the upper right box to see what happens if player 1 plays hawk and player 2 plays dove. There we see that player 1 gets a payoff of *v* and player 2 gets a payoff of 0.

Just as when we analyzed the prisoner's dilemma game, we are interested in finding the Nash equilibrium (or Nash equilibria, if there are more than one) for the hawk–dove game. Thus, we want to find a pair of strategies for player 1 and player 2 such that neither player can benefit from unilaterally changing his strategy. In the hawk–dove game, there are two such strategy pairs: If player 1 always plays hawk and player 2 always plays dove, neither player can benefit by switching his strategy alone. If player 1 switched to dove, he would have to share the resource with player 2 instead of getting it all for himself. If player 2 switched to hawk, he would end up in a costly fight against player 1, who was also playing hawk. We see this kind of Nash equilibrium in some territorial interactions in nature: Often a territory holder will be willing to fight to keep the territory (thereby playing a hawklike strategy), and an invader will flee immediately when challenged by the territory holder (thereby playing a dovelike strategy). The second Nash equilibrium, which is equivalent to the first, occurs when player 1 plays dove and player 2 plays hawk.

But what if the two individuals don't *know* who is player 1, and who is player 2—that is, what if there is not any salient cue, such as the status of territory owner or invader, that distinguishes the roles of the two players? Then it is impossible to play either of the Nash equilibria described above because players cannot condition their strategy on whether they are player 1 or player 2. In this case, no strategy by itself is a Nash equilibrium in the hawk–dove game. But there is a Nash equilibrium in which each player plays hawk some fraction of the time, with probability *p*, and plays dove the rest of the time, with probability $1 - p$. Because the players mix up their actions, sometimes playing hawk and sometimes playing dove, this type of equilibrium is called a *mixed Nash equilibrium*. Box 18.4 shows how we calculate the mixed Nash equilibrium for the hawk–dove game.

Conflict over Parental Investment

As we discussed earlier in this chapter, genetic relatedness plays a pivotal role in understanding the evolution of cooperation. Inclusive fitness theory can also be used to understand conflict within families.

A major source of familial conflict is *parental investment*: the resources—food, shelter, defense—that parents provide to their offspring. At first glance, it might seem that parental investment should be a straightforward matter: Natural selection favors the parents who leave the most surviving offspring, so where is the potential for conflict? But as we look more closely within an inclusive fitness framework, areas of potential

BOX 18.4 The Mixed Nash Equilibria for the Hawk–Dove Game

The hawk–dove game has a *mixed Nash equilibrium*: There is a fraction p such that, if everyone plays hawk with probability p and dove with probability $1 - p$, no one can benefit from unilaterally changing their strategy. Here we will show how to find the value of p.

We can find the mixed Nash equilibrium by using a trick. It turns out that, at the mixed Nash equilibrium, both strategies give the same payoff. Imagine that this wasn't the case. Then one strategy would provide a higher payoff than the other, and a player could shift to playing only the higher-paying strategy and unilaterally increase his payoff. But, by definition, at any Nash equilibrium, players *cannot* unilaterally increase their own payoff. So, we know that at a mixed equilibrium, the two strategies cannot give different payoffs.

To find a mixed Nash equilibrium, then, we look for a point where both strategies give the same payoff. Suppose that everyone else in the population is playing hawk with probability p and dove with probability $1 - p$. Then we can work out the

payoff if an individual plays hawk: With probability p our individual plays against another hawk and gets payoff $(v - c)/2$, and with probability $1 - p$ he plays against a dove and gets payoff v. This gives an expected payoff of $p(v - c)/2 + (1 - p)v$. We can also calculate the payoff if an individual plays dove. In that case, he plays against a hawk with probability p and gets the payoff of 0, and he plays against a dove with probability $1 - p$ and gets the payoff $v/2$. This gives an expected payoff of $(1 - p)v/2$. At a mixed Nash equilibrium, the payoff from playing hawk must equal the payoff from playing dove. So, at the mixed Nash equilibrium, the following equation must hold:

$$p(v - c)/2 + (1 - p)v = (1 - p)v/2$$

Solving this equation for p, we get

$$p = v/c$$

This is the mixed Nash equilibrium frequency of playing hawk. The frequency of playing dove is then $1 - p = 1 - v/c$. Notice that the lower the cost of fighting c, and the higher the value of the contested resource v, the more often individuals will play hawk.

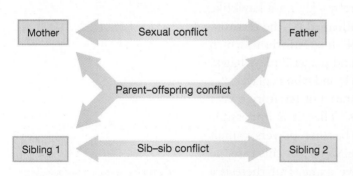

FIGURE 18.14 Familial conflicts. Familial conflicts include sexual conflict between parents, parent–offspring conflict, and sib–sib conflict. All can influence parental investment in offspring. Adapted from Parker et al. (2002).

conflict rise to the surface. Even within a family, individual interests vary, creating the potential for multiple conflicts (Figure 18.14). Parents face sexual conflict over issues such as who should provide how much parental care. Each parent is selected to hand off as much of the parental care as possible to the other. Siblings face sib–sib conflict over which sibling receives the most resources from the parents. Each is selected to try to obtain more than an even share of the total (Mock and Parker 1997). Parents and offspring face **parent–offspring conflict** over how parents allocate resources to their offspring. All else being equal, parents are selected to invest equally in all of their offspring. But individual offspring seek more for themselves, even at the expense of their siblings. In the following section, we will examine parent–offspring conflict.

Parent–Offspring Conflict

Because in diploid species parents and their offspring have a coefficient of relatedness (r) of 0.5, inclusive fitness theory predicts that parents should go to great lengths to help their offspring. And, indeed, they generally do just that. Hundreds of studies have shown that parents—mothers in particular—provide aid in many forms to their offspring.

Natural selection favors individuals who produce the most surviving offspring, and thus selection often favors parents who provide food, shelter, and other sorts of aid—collectively called *parental care*—to their offspring. Yet, there are limits to how much aid parents are selected to provide. These limits were first conceptualized by Robert Trivers in his parent–offspring conflict model (Trivers 1974). From the perspective of the parent, these decisions are affected by how much energy the parent has available to help current offspring, and by how many offspring the parent is likely to have in the future.

In principle, a parent could use every bit of energy it has to provide one particular offspring with all the benefits at its disposal. But if such an effort kills the parent or severely hampers the parent from producing other offspring in the future, then natural selection may not favor such behavior, as it might not maximize the *total number of offspring* that the parent is able to produce over the course of his or her lifetime. So, there are limits on parental investment with respect to any given child.

Now, let's look at parental investment from an offspring's perspective. The offspring will receive some inclusive fitness benefits when its parent provides aid to both current and future siblings; if they are full siblings, these individuals are related to the offspring in question by $r = 0.5$. Yet, the individual offspring is more related to itself ($r = 1$) than to any of its siblings. As such, in terms of inclusive fitness, the offspring values the resources it receives from its parent more than the resources that its parent provides to its current or future siblings. The conflict between parent and offspring arises because, although each offspring will value the resources it receives more than those dispensed to its siblings, all offspring are equally valuable to a parent. These different valuations set up a zone of conflict between how much an offspring would optimally receive from a parent, and how much a parent would optimally provide to an offspring (the former always being greater than the latter). This zone is where parent–offspring conflict takes place (Figure 18.15).

Parent–Offspring Conflict and Mating Systems in Primates

The degree of parent–offspring conflict predicted in any population is in part a function of the mating system that exists in that population (Long 2005; Hain and Neff 2006). To see why, recall that in any parent–offspring conflict situation,

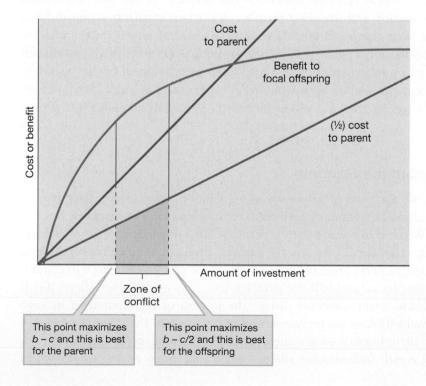

Cost or benefit

Cost to parent

Benefit to focal offspring

(½) cost to parent

Amount of investment

Zone of conflict

This point maximizes $b - c$ and this is best for the parent

This point maximizes $b - c/2$ and this is best for the offspring

FIGURE 18.15 Parent–offspring conflict. A parent can either allocate resources to a "focal" offspring, or redirect those resources to other current or future offspring. The *x*-axis represents the amount of resources that the parent invests in the focal offspring, and the *y*-axis represents fitness costs (*c*) to the parent or benefits (*b*) to the offspring. Benefits here refer to increases in the fitness of the focal offspring, whereas costs are quantified in terms of decreases in fitness of other offspring. The more resources that a parent invests in the focal offspring, the greater the benefits to that offspring—albeit with decreasing returns—but the greater the costs as well. The parent is equally related to all of its offspring, and so it is selected to maximize fitness—that is, to maximize the difference between benefit and cost. But the offspring is only half as related to its full siblings as it is to itself, and thus by the logic of inclusive fitness, it is selected to maximize the difference between benefit, and cost divided by 2. As a result, parent and offspring prefer different amounts of resource allocation. This zone of conflict is shaded in the figure. To the left of the zone, parents and offspring alike benefit from increasing allocation to the offspring. To the right of this zone, parents and offspring alike benefit from decreasing allocation to the offspring.

natural selection favors offspring that balance (1) the inclusive fitness benefits associated with receiving continued parental assistance versus (2) the inclusive fitness benefits of curtailing the degree of parental assistance received and thereby leaving a parent with more resources to produce future offspring.

The degree of relatedness between current offspring and future offspring is not fixed, but rather it is a function of the mating system. In a strictly monogamous species, current offspring and future offspring will have an average genetic relatedness of $r = 0.5$, because they are likely to be full siblings (they have the same mother and the same father). But suppose the mating system is polyandrous, with a female mating with many males. Then the genetic relatedness between current and future offspring will be somewhere between 0.5 (for full siblings) and 0.25 (for half siblings). Assume that the mother provides the majority of the parental care. Then compared to the case of monogamous mating systems, in polyandrous mating systems, natural selection will favor an offspring who attempts to extract more in the way of parental assistance—the siblings from which it is effectively taking resources are not as closely related as they would be in a monogamous system. Thus, parent–offspring conflict should be more intense in polyandrous than in monogamous mating systems (Trivers 1974; Mock and Parker 1997).

Tristan Long tested the hypothesis that offspring will attempt to extract more resources from parents in polyandrous systems than in monogamous systems. He did this by asking whether there was evidence that fetuses grow faster in utero—taking more maternal resources—in polyandrous primate species.

Long used the method of independent contrasts (Chapter 5) to examine whether strong parent–offspring conflict was more likely in polyandrous or monogamous primate species. He began by employing a phylogenetic tree for primates. From the tree, he was able to find 16 pairs of primates to use in his independent contrast analysis. Each pair was made up of species that had diverged from a recent common ancestor—one member of the pair was from a monogamous species, and the other member of the pair was from a polyandrous species. Long then compared already published data on fetal growth rates for each of the species in his pairwise comparison (Long 2005). He predicted that in polyandrous mating systems, a fetus would attempt to sequester more resources during development, and hence it would show faster rates of growth than would a fetus from a species that was monogamous. Long's analysis found just such a relationship.

Conflict within the Genome

In Chapter 6, we reviewed Mendel's law of segregation, which states that the two alleles at each locus segregrate at meiosis so that each gamete receives one but not both alleles. We tend to think of this process as "fair," in the sense that each allele is equally likely to make it into a viable gamete. Thus, we tend to expect that, on average, half the gametes produced by a heterozygote at a given locus will contain one allele at that locus, and half the gametes will contain the other allele. But if a particular allele could somehow distort the process of segregation in its own favor—if it could increase its representation to being more than half the gametes produced by an individual—that allele would be favored by natural selection, all else being equal. Indeed some alleles can do that. Such alleles are known as

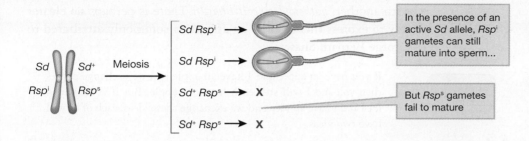

In the presence of an active *Sd* allele, *Rsp*ⁱ gametes can still mature into sperm...

But *Rsp*ˢ gametes fail to mature

FIGURE 18.16 Meiotic drive in *Drosophila*. The segregation distorter system in *Drosophila* involves two linked loci, *Sd* (*Sd* or *Sd*⁺) and *Rsp* (*Rsp*ⁱ or *Rsp*ˢ). *Sd* and *Rsp*ⁱ are often found together, as are *Sd*⁺ and *Rsp*ˢ. In the presence of an *Sd* allele on either of the homologous chromosomes, sperm that have the *Rsp*ˢ allele break down: In a double heterozygote, 99% of the sperm are *Rsp*ⁱ and, because of linkage, 99% are *Sd*. Adapted from Hurst and Werren (2001).

segregation distorters (or **meiotic drive alleles**). When these alleles are in place, we can speak of a genetic conflict of interest within individuals.

Segregation distortion has best been studied in fruit flies and mice, but it has also been found in many other species (Hartl 1972; Lyttle 1991; Hurst and Werren 2001). Moreover, its evolution has been modeled mathematically by evolutionary biologists (Dunn et al. 1958; Haig 2010). In *Drosophila melanogaster,* one of the best-studied cases of meiotic drive involves two linked loci (Hartl et al. 1967). A segregation distorter locus houses either the active allele *Sd* or the inactive allele *Sd*⁺, while a different responder locus houses what is known as the responder gene (*Rsp*), which is either *Rsp*ⁱ (response insensitive) or *Rsp*ˢ (response sensitive). The loci are in linkage disequilibrium, in that *Sd* and *Rsp*ⁱ are typically found together, as are *Sd*⁺ and *Rsp*ˢ. In the presence of the active *Sd* allele, sperm that have the *Rsp*ˢ (response sensitive) allele break down: 99% of surviving sperm in such individuals are *Rsp*ⁱ and, because of genetic linkage, 99% of these sperm are *Sd* (Merrill et al. 1999) (Figure 18.16). In this extreme example, rather than observing an allele in half the gametes produced by heterozygotes, we see it in virtually all of the gametes.

This raises a question: If segregation distortion is so strongly favored by selection, why do we see segregation distorters such as *Sd* at intermediate frequencies in populations? The answer is that many segregation distorters probably go to fixation very quickly, and we do not see them because the disadvantaged allele is quickly lost. The ones that we do see are special cases in which the segregation advantage to a segregation distorter is balanced by a severe fitness cost paid by the distorter when it is found in homozygotes—that is, in individuals with two copies of the "driving" allele (Hartl 1972). For example, in the *t*-allele meiotic drive system in mice, individuals that are homozygous for the driving *t*⁺ allele have greatly reduced survival and fertility. This can lead to an evolutionary equilibrium in which the driving *t*⁺ allele remains in the population but is unable to reach high frequency (Dunn and Bennett 1967).

18.3 Information and Communication

Regardless of whether social behavior involves cooperation, conflict, or the mating decisions we discussed in Chapter 17, information is likely being transferred. Signals are involved in virtually all social interactions. And so we need to understand the evolutionary pressures associated with signaling.

As resources go, *information* is remarkably well suited for sharing. Compared to a nest or a heavy carcass, information is easy to transport from place to place. More importantly, unlike food or shelter or mates, one individual can share information

Private information

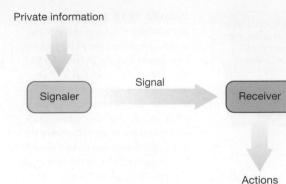

FIGURE 18.17 **Signaling.** A *signaler* with private information sends a signal to a *receiver*, informing the receiver about the state of the world. The receiver can then act on the information.

FIGURE 18.18 **The greater honeyguide** *Indicator indicator.* The honeyguide forages on beehives disturbed by honey badgers (*Mellivora capensis*), and it also leads human foragers to hives.

with another, *without losing it himself.* There is perhaps no clearer way to express this than by an aphorism commonly attributed to George Bernard Shaw:

> If you have an apple and I have an apple and we exchange apples then you and I will still each have one apple. But if you have an idea and I have an idea and we exchange these ideas, each of us will have two ideas.

Of course, not all information can be shared without cost. If I tell you where an indivisible food resource is located, you may collect it at my expense. If I show you a safe hiding place, you can take it before I do. But when I give you information, I do not give up the information itself (Figure 18.17).

Because of this unique property, information sharing is ubiquitous in nature. In many cases, it is relatively straightforward to understand how information sharing might evolve. If two individuals have entirely coincident interests, it is straightforward to see why both would benefit from communication. One striking example occurs between humans and a bird known as a honeyguide, *Indicator indicator* (Figure 18.18). This African species has been documented to lead human hunters to bees' nests, where the hunters can use smoke and other techniques to extract the honey that would otherwise be inaccessible to the birds. In the process, the birds obtain some of the honey. Here both sides benefit from honest communication. The humans are led to a food source; the birds gain access to resources they could not otherwise have exploited (Isack and Reyer 1989). In this case, there is no incentive for birds to mislead humans about where the honey is located. Mathematical models reveal that signaling systems can readily evolve under such circumstances (Skyrms 2010).

Honest Signaling

Matters get substantially more complicated when, despite some commonality of interests, signalers have incentives to deceive. The key problem can be summarized very simply: Two individuals have access to different information. They could both gain if they could honestly share this information. But their interests do not coincide entirely, and so each has an incentive to deceive the other. How can honest communication be ensured? Evolutionary biologists have proposed a number of solutions to this puzzle. We will treat each of them in turn.

Mind Reading versus Manipulation

One possibility is that honest communication is *not* ensured. Rather, the signals and responses that we observe may result from an ongoing antagonistic coevolutionary process. In the *mind reading versus manipulation* view of communication proposed by Richard Dawkins and John Krebs, signaling arises when receivers attempt to gain an edge by closely observing the cues—not necessarily meant as signals—sent by another individual (Dawkins and Krebs 1978; Krebs and Dawkins 1984).

Krebs and Dawkins illustrate this idea with the example of a dog baring its teeth. If a dog is to bite a rival without severing its own lip in the process, it must pull its lip back prior to striking. This motion, however small, can tip off the rival that there will be an impending attack. By watching for such a cue, a rival can *mind read*, pushing an antagonistic interaction up to the point that an attack is imminent, and then fleeing before actual harm is done. Where mind reading aids in avoiding injury, it will be favored by natural selection.

But once a rival attends to the cues that the angry dog is sending, the dog has a "handle" by which to *manipulate* its rival's behavior. It can now influence its rival's behavior by altering the type or timing of the cues that it sends. For example, the angry dog can cause a rival to flee simply by baring its teeth, even when it does not actually intend to bite. Such a behavior will be selected when it confers an advantage in antagonistic interactions.

According to this view, signals emerge not as cooperative solutions to exchanging information, but rather through a process of antagonistic coevolution. Receivers attempt to obtain an edge by mind reading; signalers respond by sending cues to manipulate receiver behavior; receivers counter by adjusting their responses, and so forth.

Costly Signaling Theory

The mind reading versus manipulation view presents signals and responses as tactics in a coevolutionary arms race. If this view is correct, we would not expect the *same* signals to be maintained over long stretches of evolutionary time. To explain cases in which the same signals are maintained over evolutionary time, we would need some other explanation of how signaling evolves and is maintained. One such explanation arises from costly signaling theory. The basic structure of costly signaling arguments is as follows: Suppose that signals are costly, and that for one reason or another, dishonest signals cost more than honest signals. If telling the truth is cheap enough and telling a lie is costly enough, it may be worthwhile to communicate honestly and not to lie.

Without further exposition, it is not easy to see exactly how or why this might work. To explain, we will look at a pair of examples: the long tails of widowbirds, and the begging behavior of baby birds. First, though, a bit of background. In the early 1970s, natural historian Amotz Zahavi struggled to understand a problem that had puzzled researchers since Darwin: Why do animals often produce costly and extravagant displays or physical ornaments? Why do peacocks have such spectacular plumage? Why do baby birds beg so loudly? Why do gazelles jump up and down when they see a lion?

Zahavi proposed that physical extravagances are signals to other individuals (Zahavi 1977). For example, a peacock's tail may be a signal used by prospective mates in order to estimate the individual's overall condition and/or genetic quality. Zahavi named his hypothesis the **handicap principle**, and he suggested that there is something about costly behaviors or physical features that makes for inherently reliable signals.

SEXUAL SELECTION SIGNALS The long-tailed widowbird of Africa exhibits extraordinary sexual dimorphism (Figure 18.19). Not only do the male and female differ strikingly in coloration, but the males also have very long tails, whereas female tails are short and compact.

In a classic study, Malte Andersson wanted to see whether the long tails of males are a product of sexual selection (Andersson 1982). If so, he reasoned that females should display a preference for males with longer tails. Andersson artificially

FIGURE 18.19 Long-tailed widowbird (*Euplectes progne*). (A) Male widowbirds have long flamboyant tails that are roughly 50 cm in length, whereas (B) female widowbirds have short tails of roughly 7 cm in length.

A

B

FIGURE 18.20 Female widowbirds prefer males with artificially lengthened tails. As a proxy for reproductive success, Andersson recorded the number of new nests that females built on the territories occupied by males in four groups: (1) males with artificially shortened tails, (2) males with tails cut and reglued into place, (3) males with tails left unaltered, and (4) males with tails that had been artificially lengthened by gluing on additional plumage. Compared to males in the three other groups, males with artificially lengthened tails had significantly higher reproductive success. Adapted from Andersson (1982).

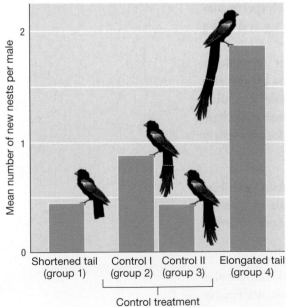

shortened or lengthened males' tails, and he recorded their subsequent success at obtaining mates. His results, illustrated in Figure 18.20, strongly supported the hypothesis that females prefer males with longer tails.

But *why* do female widowbirds prefer males with longer tails? In Chapter 17, we considered a number of possible explanations: direct benefits, good genes, runaway sexual selection, and sensory bias. It seems unlikely that tail length provides any direct benefits to females and other studies have implicated the good genes model as the most likely explanation (Pryke et al. 2001).

Zahavi's handicap principle explains why tail length can be a reliable signal of good genes. His argument goes something like this: Female widowbirds choose mates from a pool of suitors. Because the females cannot judge a male's genetic quality directly, they instead attend to signals that the male provides: males advertise their quality with a long flamboyant tail. This advertisement is a handicap in the sense that it is energetically costly to produce and maintain, and it may reduce the male's ability to maneuver as well. The cost of producing long tails varies among males. A weak and sickly male cannot afford to divert energetic resources from basic metabolism to the production of ornaments. Moreover, an unhealthy bird would have a hard time flying if he were also hindered by a long tail. By contrast, a strong and healthy male can afford the additional costs of producing a long tail, and he may be able to fly reasonably well even when hampered by a lengthy tail.

Because only high-quality males can afford long tails, females prefer mates with those characteristics. High-quality males, for their part, produce the extravagant plumes to ensure that they are chosen as mates. Low-quality males cannot afford to do so, and so they will produce shorter tails.

Thus, among widowbirds, the bright colors and long tails may be honest signals of male quality that are used by females to choose their mates. This is the basic idea behind the use of the costly signals in a sexual selection context. Of course, the costly signal need not involve extended tail feathers; bright colors, a large rack of antlers, an elaborate song, a captured prey item offered as a gift, or any number of other expensive ornaments or displays could serve equally well. Nor, for that matter, must the male sex be the signaling sex. In some cases, females may use costly signals to advertise their own qualities to male suitors.

A number of authors, most notably Alan Grafen, have used mathematical models to demonstrate that the costly signaling mechanism can indeed allow the evolution and maintenance of honest communication (Grafen 1990). Although the mathematics involved get rather complicated, we can capture a good fraction of the intuition behind the models with a simple graphical illustration.

Figure 18.21 shows the cost of producing a tail of a given length for widowbird males of three different underlying genetic qualities: low, medium, and high. Producing a longer tail always costs more than producing a shorter tail, but producing a longer tail is comparably less expensive for high-quality individuals. On the same graph, the fitness benefits that result from improved mating success are indicated as the black curve. The longer the tail, the greater the mating success. In this graph, a male's fitness is maximized when his tail length maximizes the difference between the fitness benefits and the fitness costs. Individual males do not consciously choose their tail length; rather natural selection will favor an appropriate norm of reaction for tail length. On the graph, low-quality, medium-quality, and high-quality males, respectively, produce tails of short, medium, or long length. Thus, each type of male maximizes its fitness with a different tail length: long tails for high-quality males, short tails for low-quality males, and intermediate-length tails for medium-quality males. Zahavi's predictions are met. Tail length as a signal will be (1) *honest*, in that the higher the male's quality, the longer the tail, and (2) *costly*, in that all males produce tails that impose significant fitness costs.

In this model, signaling honesty comes down to a sort of cost–benefit analysis. Signalers send the signals that they do because the cost of doing otherwise exceeds the benefits of doing so. Here, the costs come from the act of producing the signal, and the benefits come from the response of the signal receiver.

SIGNALS OF NEED If you have ever located a bird's nest by listening to the begging nestlings within, you've recognized another type of costly signal. The loud begging calls that nestlings make are thought to be costly signals of hunger or need. Consider the strategic problem that the mother bird faces when she returns with a morsel of food. Arriving at the nest, she finds herself faced with an array of gaping beaks. Which of them should she feed? Natural selection favors efficient allocation of the food among her offspring. Therefore, a mother bird would benefit from knowing precisely how much food each nestling needs.

But will the nestlings be willing to signal their true hunger levels? Here we have another example of parent–offspring conflict. The parent would prefer to feed the hungriest chick, but each offspring would like to receive the food itself. As a result, nestlings may exaggerate the signals they emit regarding their levels of hunger, unless some mechanism prevents deception.

Fortunately for the mother, costly signals can provide a way out of this dilemma. Suppose that nestlings must signal their hunger by squawking loudly—the louder a chick squawks, the hungrier the mother infers it to be. And suppose

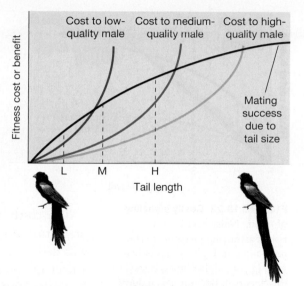

FIGURE 18.21 Costly signaling of male quality. Longer tails cost more to make and maintain, but they cost relatively less for higher-quality males. Females prefer longer tails, and thus mating success increases as a function of tail length. Low-, medium-, and high-quality males maximize fitness (as indicated by the black curve) by producing tails of the lengths L, M, and H, respectively. These optimal signals are both honest—higher-quality males produce longer tails—and costly, in that all males expend a substantial fitness cost on tail production.

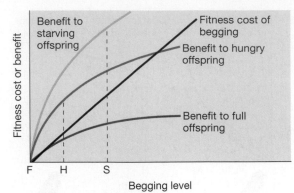

FIGURE 18.22 Costly signaling of need. Noisy begging carries a risk of attracting predators to the nest, and thus higher levels of begging impose higher fitness costs (black curve). Begging also induces the parents to feed the offspring. This creates substantial benefits for starving offspring, intermediate benefits for hungry offspring, and minimal benefits for satiated or "full" offspring. Full, hungry, and starving offspring therefore optimize fitness by begging at levels F, H, and S, respectively.

FIGURE 18.23 Parents respond to begging calls. Both male and female parents brought more food to treatment nests where begging calls were played back from hidden speakers than to control nests with no playback. Adapted from Price (1998).

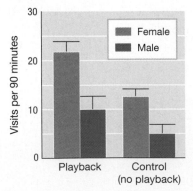

that squawking in this way is not without its risks. Among other things, the begging calls may attract predators to the nest.

Under these conditions, the nestlings may end up honestly revealing their hunger levels. If a nestling's hunger is satiated, the risk of predation will outweigh any potential gain from begging. By contrast, if a nestling is starving, then the predation risk is overshadowed by the need for food. As a result, the hungry chicks will beg, the satiated ones will stay silent, and the mother will receive honest information about each offspring's condition. Because the begging signal is costly in terms of predation, it ends up being honest as well (Figure 18.22).

The costly signaling explanation of begging makes at least three empirical predictions. If begging calls are costly signals, we would expect that (1) parents will deliver more food in response to stronger begging, (2) begging intensity will reflect the hunger level of nestlings, and (3) begging will be costly (Searcy and Nowicki 2005). Each of these predictions has been tested extensively. Here we briefly consider a few studies that test each of the predictions above.

To determine whether parents heed begging calls and deliver more food in response to more intense begging, Katie Price recorded the begging calls of yellow-headed blackbird (*Xanthocephalus xanthocephalus*) nestlings (Price 1998). She then divided a set of blackbird nests into two groups. For each of the nests in the treatment group, she played back the begging calls from concealed speakers near the nest. For each nest in the control group, a concealed speaker was placed nearby and turned on, but no begging calls were broadcast. Price then compared the rate at which parents brought food to the treatment nests to the rate at which parents brought food to the control nests. She found that the rate at which parents brought food to the treatment group was significantly greater than the rate at which they brought food to the control group. Both male and female parents approximately doubled their rate of provisioning in response to calls played back from the hidden speakers (Figure 18.23). The added provisioning translated into weight gain for the nestings. Price also found that nestlings in the treatment nests gained significantly more weight than those in the control nests.

To establish that begging accurately reflects hunger levels, Rebecca Kilner and her colleagues fed a group of reed warbler nestlings (*Acrocephalus scirpaceus*) until they were satiated; then they measured the begging rate as they withheld food over the subsequent 110 minutes (Kilner et al. 1999). They found that for both 3- to 4-day-old chicks and 6- to 7-day-old chicks, begging rate increased with the time since last feeding and thus presumably with hunger (Figure 18.24). From these results, Kilner and her colleagues concluded that begging intensity is an honest signal of hunger in reed warblers.

To explore whether begging calls are costly, researchers measured two different potential costs associated with begging: the metabolic cost of the begging, and the predation risk associated with the begging. Results from the metabolic cost studies suggest that begging only slightly raises metabolic rate above baseline levels. Given that nestlings are not begging continually, this minor increase during a small fraction of time confers minimal metabolic cost (Searcy and Nowicki 2005). But the predation costs associated with begging behavior appear to be more substantial. To estimate those costs, David Haskell placed a set of artificial nests, each baited with a quail egg, in a New York state park (Haskell 1994). Half of the nests were placed on the ground, and half were placed in trees. Each nest contained a two-way radio, which could be used to broadcast prerecorded begging calls of the western bluebird (*Sialia mexicana*). Haskell then divided the nests into two groups: (1) a treatment group from which begging calls were broadcast, and (2) a control group from which no sounds were broadcast. He checked twice daily for 5 days to see whether each nest had been raided by predators. Haskell found that begging calls increased the rate of predation significantly for nests on the ground but not for nests in the trees (Figure 18.25). In a second study, Haskell also found that increasing begging led to increased predation. Follow-up work of similar design by Susan Leech and Marty Leonard found that predation rates also increased for tree-nesting birds (Leech and Leonard 1997). Overall, there is clear evidence that begging is costly in terms of predation.

From this set of studies, researchers have amassed considerable evidence consistent with the hypothesis that begging behavior is costly. In addition to the begging example considered here, costly signaling theory has been applied in many other domains as well—from threat displays to antipredator signals. While not all of these cases have been tested as rigorously as has the begging case, costly signaling is an important explanation for how honest signaling can evolve. The brilliance of Zahavi's solution was that he took two major puzzles in evolutionary biology—"Why are signals honest despite conflicting interests?" and "Why are signals extravagant despite selection for efficiency?"—and recognized that these puzzles, when coupled, resolve one another. Signals are honest because they are extravagant (in the right way); signals are extravagant because such extravagance may be required to ensure honesty.

Conventional Signals

Although costly signaling may be important in explaining many examples of honesty, it cannot be the only mechanism that serves this purpose. The words that you are reading now do not have the sort of production costs associated with them that make widowbirds' tails or begging calls honest. Moreover, costly signaling can be an extremely wasteful way of communicating. Indeed, in some cases, costly signaling can be so costly that both signaler and signal receiver end up worse off than if they had not communicated in the first place (Bergstrom and Lachmann 1997).

To see how signals can be honest without extravagant cost, we turn to the house sparrow (*Passer domesticus*). Members of this species, like many other sparrow species, use subtle variations in plumage coloration to signal fighting ability and social dominance. House sparrows, for example, signal fighting ability by the size of their black throat patches (Figure 18.26). The larger the throat patch, the less likely a bird is to be challenged and the more likely it is to win in a fight if it is challenged.

The sparrow's throat badge is inexpensive to produce, as it entails only a small color change in a small number of feathers. Signals of this type are known as

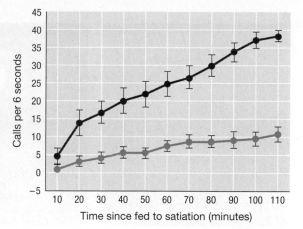

FIGURE 18.24 Begging intensity reflects hunger. Begging rate of 6- to 7-day-old nestlings (blue) and 3- to 4-day-old nestlings (green) as a function of time since feeding. Adapted from Kilner et al. (1999).

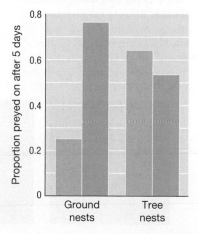

FIGURE 18.25 Begging is costly. In an experimental study using artificial nests, begging calls significantly increased the rate of predation for nests on the ground but not for nests in trees. Predation on nests from which begging calls were broadcast is shown by the blue bars, and predation on silent nests (those from which no begging calls were broadcast) is shown by the gold bars. Adapted from Haskell (1994).

A

B

FIGURE 18.26 Badges of status. House sparrows signal fighting ability by means of black throat badges. **(A)** This bird has a small badge, indicating low fighting ability, and **(B)** this bird has a large badge, indicating high fighting ability.

conventional signals—that is, their meaning is established by a convention, rather than intrinsically connected with their structure. But what keeps conventional signaling systems honest? Why, for example, don't sparrows who are poor fighters adorn themselves with deceptively large throat patches? The answer appears to be social enforcement: If they are discovered, birds that have exaggerated their condition with a large throat badge, but are poor fighters, tend to be attacked by more dominant sparrows (Rohwer 1977).

Evolutionary biologist Elizabeth Tibbetts has demonstrated that paper wasps (*Polistes dominulus*) use a similar type of conventional signal to communicate their fighting abilities, and that these signals are kept honest by social punishment (Tibbetts and Dale 2004; Tibbetts and Lindsay 2008; Tibbetts and Izzo 2010). *Polistes dominulus* wasps have variable black facial patterns. In an initial study, Tibbetts and Lindsay demonstrated that the "brokenness" (fragmentation) of the black facial patterning signals dominance (Figure 18.27). Brokenness could be assesssed by noting the number of black facial spots on the wasps: 0 spots was correlated with low fighting ability and low dominance; 2 spots were correlated with high fighting ability and high dominance. The researchers manipulated facial patterns of individual wasps, adding spots with paint, and they found that wasps preferred to contest food resources with other wasps that had fewer black facial spots (which signaled lower quality and therefore lower fighting ability).

In a follow-up study, Tibbets and Izzo explored why these signals were honest. They wanted to know why wasps of low fighting ability didn't fake dominance by producing broken facial patterns. They hypothesized that wasps could recognize when signals were not honest—that is, when the wasp's facial pattern indicated dominance and high fighting ability while the wasp actually had low fighting ability. To test this, they manipulated either the wasp's facial pattern using paint, the wasp's dominance behavior by applying an artificial hormone (which increased aggressive behavior), both, or neither. They found an increased incidence of aggressive behavior toward wasps whose facial patterns indicated dominance (and high fighting ability) but whose behavior did not (Figure 18.28). From their results, the authors argue that wasps detect dishonest signals as mismatches between markings and behavior. Wasps impose a social cost to such dishonest signals, attacking those individuals with facial markings that falsely indicate dominance and high fighting ability.

FIGURE 18.27 Conventional signaling of fighting ability by paper wasps. Moving from left to right, we see increasing "brokenness" of the black patterning on the face (from 0 black facial spots to 2 black facial spots), and thus signals of increased fighting ability.

While the logic of conventional signals—such as those displayed by sparrows and paper wasps—seems at first glance quite different from that of costly signals—such as those displayed by begging birds—we can apply the same sort of cost–benefit framework to understand why conventional signals are honest. In doing so, we learn something important about how signal cost relates to signal honesty as illustrated by Figure 18.29. In this figure, signalers pay no cost unless they overstate their quality. If they do overstate their quality, they will face social punishment, and thus they will pay substantial costs. Here each individual does best to signal its true fighting ability, so the signals will be honest (Lachmann et al. 2001).

Conventional signals are honest, but they are not costly. *Deviations* from these signals—namely, exaggerations of fighting ability—would be costly, however, and it is this cost of deviation that keeps signals honest. Thus, we see that it is not the cost of the signal per se, but rather the cost of shifting to a dishonest signal, that keeps signaling honest in each example we've treated.

One question remains: Why do some communication systems rely on costly signals, while others use conventional signals? Why do chicks produce expensive alarm calls to signal their hunger, while wasps can use inexpensive conventional signals to indicate their fighting ability? The difference between the two cases is that in the begging chick case, the signal receiver—the mother—cannot readily assess the honesty of the message. Was the chick that was begging the loudest actually the one that needed food the most? The answer to that question is difficult for the parent bird to ascertain. In the case of the wasps, the signal receiver can directly probe the accuracy of the signal by instigating a fight. Thus, we might expect that conventional signals can be used when communicating about verifiable traits, whereas costly signals will be required otherwise.

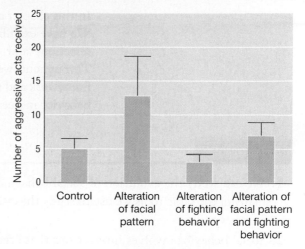

FIGURE 18.28 Aggression toward wasps with exaggerated facial patterns. Wasps with facial patterns that falsely indicated high fighting ability (dishonest signal) suffered from higher levels of aggression from other wasps than did wasps in any of the other experimental treatments. Adapted from Tibbetts and Izzo (2010).

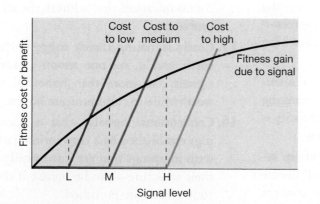
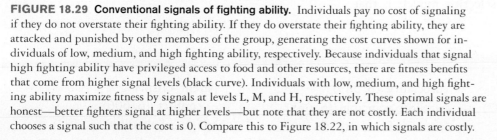

FIGURE 18.29 Conventional signals of fighting ability. Individuals pay no cost of signaling if they do not overstate their fighting ability. If they do overstate their fighting ability, they are attacked and punished by other members of the group, generating the cost curves shown for individuals of low, medium, and high fighting ability, respectively. Because individuals that signal high fighting ability have privileged access to food and other resources, there are fitness benefits that come from higher signal levels (black curve). Individuals with low, medium, and high fighting ability maximize fitness by signals at levels L, M, and H, respectively. These optimal signals are honest—better fighters signal at higher levels—but note that they are not costly. Each individual chooses a signal such that the cost is 0. Compare this to Figure 18.22, in which signals are costly.

In this chapter and the preceding one, we have focused on the evolution of behavior. We have examined sexual selection, including intrasexual selection and intersexual selection, and the evolution of cooperation, conflict, and signaling behavior. Throughout, we cast our evolutionary questions within a conceptual and theoretical framework and then examined empirical studies on both the costs and benefits of the behavior in question and the phylogenetic history of the subject matter.

SUMMARY

1. We can study the evolution of social behavior using many of the same tools we use to study the evolution of other traits.

2. Social behavior involves interactions that organisms have with others—most often, their conspecifics. In these interactions, the actions taken by one individual affect not only its own fitness, but also the fitnesses of those around it.

3. Cooperation occurs when two or more individuals each receive a net benefit from their joint actions, even though individuals may pay a cost for interacting cooperatively.

4. At least three different paths can lead to the evolution of cooperation: (1) kinship, (2) reciprocity, and (3) group selection. All three paths are susceptible to cheaters—those who receive the benefits of cooperation, but do not pay the costs.

5. Evolutionary theory predicts that cooperation and altruism should be common among close relatives, because relatives are likely to share common genes that they have inherited from common ancestors—parents, grandparents, and so on. This idea has been formalized in inclusive fitness theory.

6. Another path to cooperation is via reciprocal altruism in which individuals benefit from exchanging acts of altruism. One formal model for reciprocity is called the repeated prisoner's dilemma game.

7. A third path to cooperation may be via group selection, although this is a matter of heated debate among evolutionary biologists. The core concept underlying modern group selection models is that natural selection operates at two levels: within-group selection and between-group selection.

8. Conflict can occur between unrelated individuals and, under certain conditions, between related individuals. Evolutionary biologists have developed and tested models predicting when and where such conflict should occur.

9. Segregation distorters have been examined to study evolutionary conflict within genomes.

10. Signals of one sort or another are involved in virtually all social interactions, whether they revolve around cooperation or conflict.

11. In some cases, it is straightforward to understand how information sharing might evolve. If two individuals have entirely coincident interests, it is easy to see why both would benefit from communication.

12. Despite some commonality of interests, signalers often have incentives to deceive. Evolutionary biologists have developed and tested many models of communication that address the incentive-to-cheat problem.

13. Costly signaling theory suggests that if signals are costly and if, for one reason or another, dishonest signals cost more than honest signals, it may be worthwhile to communicate honestly and not to lie.

14. Conventional signals—that is, signals with meanings established by a convention, rather than signals with meanings that are intrinsically connected with their structure—can be honest if those who violate conventions are punished.

KEY TERMS

altruism (p. 586)
coefficient of relatedness (p. 587)
conventional signals (p. 616)
cooperation (p. 586)
direct fitness (p. 587)

eusociality (p. 589)
handicap principle (p. 611)
inclusive fitness (p. 587)
indirect fitness (p. 587)
meiotic drive alleles (p. 609)

parent–offspring
    conflict (p. 606)
reciprocal altruism (p. 592)
segregation distorters (p. 609)

REVIEW QUESTIONS

1. When it comes to inclusive fitness theory, why is it important to distinguish between genetic kinship and kinship in the everyday sense of "family"?

2. Why would you expect that individuals in many species are able to gauge their genetic relatedness to those around them? When would such behavior be favored by natural selection?

3. Why might you expect the "zone of conflict" between parents and offspring to decrease as a parent's age increases?

4. Microbes don't have neurons, let alone brains. Given this, how could they possibly be involved in cooperative, altruistic, or competitive interactions? When answering this question, explain why evolutionary definitions of behavior work best when they are cast in terms of costs and benefits.

5. In the iterated prisoner's dilemma model that we considered, each player was able to perfectly ascertain what the other player did on the previous round. In this case, the tit-for-tit strategy proved very effective. Now imagine that players were occasionally mistaken about what their opponent had done on the previous round. How would tit-for-tat fare?

6. Can you think of three examples of costly signals that humans use? Why do you think these signals are not readily replaced with cost-free conventional signals?

7. Pelicans have clutches of two, with each nest having two eggs and two babies. Suppose among pelicans a new allele arises that causes a nestling to share its food with its nestmate if it is not particularly hungry. This gene imposes a fitness cost of 0.2 on those who carry it, while conferring a 0.5 benefit on the sibling who receives the additional food. Will this gene increase in frequency if nestmates are always full siblings, sharing the same mother and father? What if nestlings are always half-siblings, sharing the same mother but different fathers?

8. In a 1974 review paper on social evolution, Richard Alexander minimized the importance of parent–offspring conflict by making the following argument: Imagine a "rotten kid" allele that drives selfish behavior on the part of an offspring toward its parents. This rotten kid allele may be beneficial to the offspring while it is young, but any benefits that an individual receives from being selfish as a juvenile will be countered by the increased risk of having selfish offspring of its own. Critique Alexander's argument.

9. In Box 18.3, we considered Garrett Hardin's tragedy of the commons. How does Hardin's logic apply to the problem of air pollution? How does it apply to antibiotic resistance?

10. Many species of bacteria, including *Streptococcus pneumoniae*, *Clostridium difficile*, and *Salmonella typhimurium*, produce toxins that deter competing species or cause damage to a eukaryotic host and thereby make additional resources available to the bacterial colony. These toxins are released only when a bacterial cell lyses (bursts). The bursting cell dies while the other bacterial cells in the colony benefit; thus, toxin production and release is a form of altruism. Could this trait be explained by reciprocal altruism? Under what conditions might natural selection favor such lysing behavior?

SUGGESTED READINGS

Axelrod, R., and W. D. Hamilton. 1981. The evolution of cooperation. *Science* 211: 1390–1396. A classic paper on the use of evolutionary game theory to model cooperation.

Mesterton-Gibbons, M., and L. A. Dugatkin. 1992. Cooperation among unrelated individuals—evolutionary factors. *The Quarterly Review of Biology* 67: 267–281. A review of models for the evolution of cooperation when individuals are not related.

Reeve, H. K., and B. Holldobler. 2007. The emergence of a superorganism through intergroup competition. *Proceedings of the National Academy of Sciences of the United States of America* 104: 9736–9740. An attempt to bridge the gap between classic natural selection models and group selection models of social behavior.

Robinson, G. E., C. M. Grozinger, and C. W. Whitfield. 2005. Sociogenomics: social life in molecular terms. *Nature Reviews Genetics* 6: 257–271. A concise overview of how genomics can inform our understanding of social behavior.

Velicer, G. J. 2003. Social strife in the microbial world. *Trends in Microbiology* 11: 330–337. A review of the evolution of cooperation and conflict in microbes.

Ⓢ Visit StudySpace at wwnorton.com/studyspace.

19

Coevolution

◀ The Iiwi, or scarlet Hawaiian honeycreeper, *Vestiaria coccinea*, feeding on flowers of the 'ohi'a lehua plant (*Metrosideros polymorpha*) in Hawaii. The curved beak of this bird is the result of coevolution with lobelioid plants (not shown here).

All over the planet, lichens grow on rocks and trees. Everything about these lichens—the way they look, the way they reproduce, the way they respond to environmental change—would make the casual observer think that lichens are well-integrated multicellular organisms. And they are, but not in the usual sense: Every lichen is made up of two different species (Brodo et al. 2001).

There are thousands of different kinds of lichens, each of which is composed of one fungal species and one species of either photosynthetic algae or cyanobacteria. In the case of fungal–algal lichens, fungal cells typically surround the algal cells to form the body (the thallus) of a lichen. Each species derives benefits from the other. The fungi use sugars produced by photosynthesis in the algae. The algae benefit from the fungi's ability to retain water, and they also use some of the resources that fungal cells extract from soil. The algae and fungi in a lichen live in a mutualistic relationship—each benefits the other. The codependency between algae and fungi is so complete that, for most lichens, neither the fungal nor the algal species can survive in the absence of its partner. As a result, the fungi and algae have evolved to disperse together. One form of reproduction in lichens is via the spread of *diaspores*, which contain both algal and fungal cells.

When species interact, as fungi and algae do in lichen, the action of natural selection on one species may cause selection to operate in new ways on the other. Evolutionary biologists say that coevolution occurs when changes to heritable traits in species 1 drive changes to heritable traits in species 2, which in turn feed back to affect heritable traits in species 1, and so on, back and forth. When the interaction of the two species increases the fitness of both species, this is called a mutualism.

If one species is a fungus and a second species is an alga, for example, we can examine how the two species can coevolve in lichen. Above and beyond their remarkable natural history, lichens are an excellent model system for formulating and testing hypotheses about coevolution, particularly molecular genetic and phylogenetic questions regarding coevolution. This is because: (1) biologists have the tools to make molecular genetic comparisons among many species involved in lichen formation, and (2) many species of fungi that are part of a lichen have sister species that are not in a lichen association, which allows us to use the comparative method to address coevolutionary questions. For example, evolutionary biologists have hypothesized that the transition to, and the maintenance of, a mutualistic relationship like that seen in algae and fungi in lichens must be complex and require many changes to the genomes of both species. We can then ask: Is there evidence for such changes to the genomes of algae and fungi that associate to form lichens?

To answer that question, François Lutzoni and Marc Pagel compared the rate of nucleotide substitution in free-living versus mutualistic fungi (Lutzoni and Pagel 1997). They compared 1550 nucleotide sites in 16 species of mutualistic fungi (primarily in lichens, but some in liverworts) and 13 species of free-living fungi that are closely related to the mutualistic species found in lichens. They found that there was a faster rate of molecular evolution in the mutualistic fungi. Specifically, the rates of nucleotide substitution were much higher in fungal species involved in mutualistic relationships with algae and liverworts than were the rates in the closely related, free-living fungal species. Moreover, the researchers found evidence consistent with the hypothesis that the transition to mutualism was responsible for accelerating the rate of molecular evolution. They found that the increased rate of nucleotide substitution occurred only during and after the transition to the mutualistic relationship, not before. Finally, they also discovered that the increased rate of nucleotide substitution in mutualistic species was not constrained to one specific area of the genome, but rather it was widespread across many sections of the genome (Figure 19.1). Not only are the fungal and algal species that are in a lichen association coevolving, but the process of coevolution has also quickened the pace of evolutionary change throughout the genome of at least one of the partners in this mutualistic relationship.

At a very general level, if we consider the long-term evolutionary dynamics of coevolution, two basic scenarios emerge: (1) mutualistic interactions, such that evolutionary changes in each species benefit the other species (Boucher 1985; Bronstein 1994; Connor 1995; Thompson 2005), and (2) **antagonistic coevolution**, in which evolutionary changes in each species decrease the fitness of the other species. A classic example of antagonistic coevolution is the relationship between predators and their prey, where selection for antipredator traits in prey (faster escape time, camouflage ability, and so on) favors traits in predators that produce better success

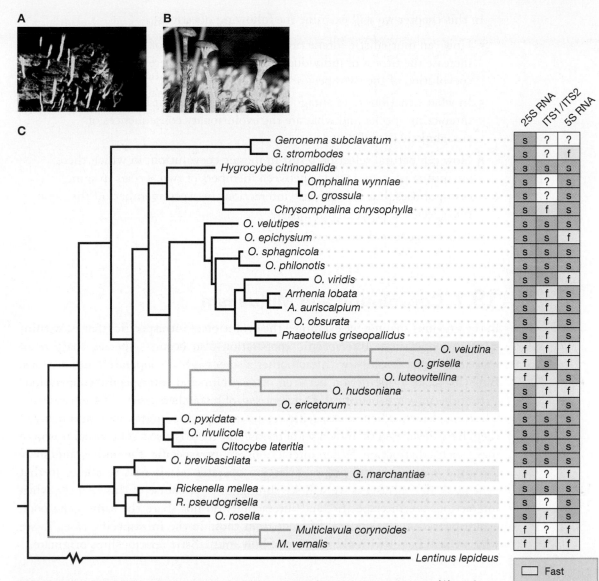

FIGURE 19.1 Nucleotide changes in mutualistic and free-living species of fungi. (A) *Multiclavula mucida*, which forms a lichen with the green alga *Coccomyxa*. **(B)** *Omphalina velutipes*, a free-living species of fungi. **(C)** A phylogenetic tree of the fungal genus *Omphalina* and related species reveals a more rapid pace of molecular evolution in the species involved in mutualistic associations. On the phylogenetic tree, clades associated with lichen-forming green algae are indicated in green; those involved in mutualistic associations with liverworts are shown in blue. In the columns on the right, the rate of molecular evolution for each species is shown using three different molecular genetic measures: 25S RNA, ITS1/ITS2, and 5S RNA. The columns on the right indicate slow (in pink) and fast (in yellow) rates of nucleotide substitution in the fungi. Where the rate of molecular evolution is unknown, there is a question mark. Fast evolutionary change tends to be associated with a mutualistic lifestyle. Part C adapted from Lutzoni and Pagel (1997).

at catching these prey, which selects for new antipredator behavior in the prey, and so on. Antagonistic coevolution produces an evolutionary "arms race" between predator and prey which may go on indefinitely, producing a wide array of both mechanisms by which prey can protect themselves against predators, and systems by which predators can find and capture prey. We will discuss such evolutionary arms races in depth later in this chapter.

In this chapter we will examine the following questions:

- How can mutualistic interactions between members of different species increase the fitness of individuals in each species and result in the coevolution of the two species?

- In what situations does antagonistic coevolution occur between interacting species and what are the evolutionary consequences of antagonistic coevolution?

- How can natural selection result in mosaic coevolution, in which there are mutualistic interactions between members of two species in some communities, but antagonistic interactions between members of the same two species in other communities?

- What is gene–culture coevolution?

19.1 Coevolution and Mutualism

In the previous chapter, we examined the evolution of intraspecific (that is, within-species) cooperation. Interspecific cooperation also occurs—species that are in mutualistic relationships with each other also engage in cooperative interactions. But why make a distinction between intraspecific and interspecific cooperation? Part of the answer is historical. Different sets of researchers, with different research questions, have studied intraspecific versus interspecific interactions, and many of them have developed their own set of terms. But there is also a conceptual reason to make such a distinction. When interactions are intraspecific, the interactants share the same gene pool, and natural selection operates on alternative alleles in that gene pool. In contrast, as we will see throughout the course of this chapter, when interactions are interspecific, and interactants do not share the same gene pool, evolutionary interactions are different from those in the intraspecific case. To see why, we will begin with a discussion of how mutualistic relationships originate.

The Origin of Mutualisms

When we study a specific case of mutualism, we are looking at a snapshot of one point in evolutionary time. But we can also ask how such a mutualism might have evolved. The answer is that there is no one set path by which a mutualism originates and evolves. In some cases, mutualisms may have evolved from initially neutral interactions between species, in which neither party initially affected the other's fitness. Some mutualisms may have evolved from initial interactions in which one species benefited and the other species was unaffected. Or mutualisms may have evolved from an initially parasitic relationship when the costs and benefits of that parasitic relationship changed and favored mutualism. And yet in still other instances, the relationship between species may have been mutualistic all along.

In Table 19.1, we present a few of the numerous systems in which the evolution of mutualism has been studied. While this gives us a sense of the wide array of mutualisms that exist in nature, to fully understand the exquisite adaptations that result from the evolutionary dynamics of mutualisms, and to better comprehend

TABLE 19.1			
Examples of Mutualisms			
Example	Partner 1	Partner 2	Context
Survival and Growth			
Mitochondria	Eukaryotes	Bacteria	Cellular energy
Chloroplasts	Eukaryotes	Cyanobacteria	Photosynthesis
Marine reefs	Corals	Dinoflagellates	Photosynthesis
Lichens	Fungi	Green algae/ cyanobacteria	Nutrition
Mycorrhizae	Plants	Fungi	Plant nutrition
Rhizobia	Plants	Bacteria	Nitrogen fixation in soil
Gut symbionts	Animals	Bacteria	Digestion in animals
Gut symbionts	Termites	Protozoa, bacteria	Ability to digest cellulose
Fungus gardens	Ants	Fungi	Agriculture by ants
Chemosymbiosis	Bacteria	Invertebrates	Colonization of deep sea vents
Reproduction			
Pollination	Plants	Animals	Sexual reproduction in plants
Seed dispersal	Plants	Animals	Sexual reproduction in plants

Adapted from Thompson (2010).

the complex, often indirect interactions between the parties in such mutualisms, we need to delve more deeply into some well-studied systems. We will begin with a mutualism between ants and fungus.

Ant–Fungus Mutualisms

Approximately 30 million years ago, ants began cultivating their own food by entering into a mutually beneficial relationship with certain species of fungi (Mueller and Rabeling 2008) (Figure 19.2). They continue to do so today, and they are one of the few taxa on the planet that grow their own food. The ants promote the growth of the fungi, while eating some of the vegetative mycelium—threadlike hyphae that absorb nutrients from the soil and break down plant material—produced by their fungal partners.

Cameron Currie and his colleagues have found a fascinating adaptation in ants such as the leaf-cutter ant (*Acromyrmex octospinosus*), which is involved in an ant–fungus mutualism. Remarkably, the leaf-cutter ants not only provide a safe

FIGURE 19.2 A phylogeny of attine fungus-growing ants. The phylogenetic history of the five known ant agricultural systems: lower agriculture, coral fungus agriculture, yeast agriculture, higher agriculture, and leaf-cutter agriculture. Adapted from Schultz and Brady (2008).

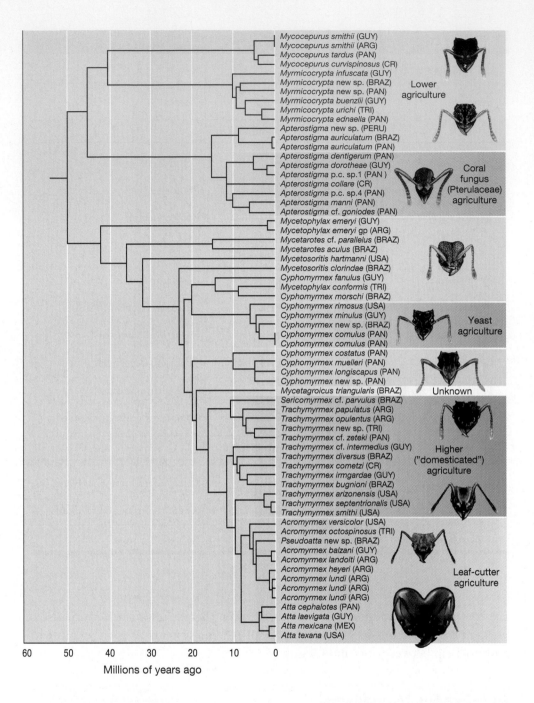

haven for fungi to grow, but they also protect the fungi from disease (Currie et al. 1999a,b; Cafaro and Currie 2005; Mangone and Currie 2007; Clardy et al. 2009; Cafaro et al. 2011). Researchers who study ants with fungal food gardens have long known of a whitish-gray crust found on and around many of the ants, but they did not know what this crusty substance—found only on ants in mutualistic relationships—actually was. Recent work has demonstrated that it is, in fact, a mass of bacteria—primarily *Pseudonocardia* and *Streptomyces* bacteria (Figure 19.3).

Currie and his colleagues hypothesized that ants use the antibiotic substances produced by the *Pseudonocardia* and *Streptomyces* bacteria to kill parasites that grow in their fungal gardens, thereby protecting their fungal food supply. A number of lines of evidence support this claim: (1) all 20 species of the fungus-growing ants that Currie and his team examined had *Streptomyces* bacteria associated with them, (2) the bacteria found on fungus-growing ants produce antibiotics that wipe out *certain* parasitic diseases, (3) ants transmit the bacteria across generations, with parents—primarily mothers—passing the bacteria on to offspring, and (4) when male and female reproductive ants are examined before their mating flights, only females are covered in *Streptomyces*; this is important, as only females start new nests that will rely on the bacteria to produce antibiotics, and only females are involved in "cultivating" fungal gardens.

One thing that makes the use of antibiotics by the ants in this mutualism so remarkable is that the antibiotics are specifically targeted toward diseases that are dangerous to the fungus growing in the ants' garden. When Currie and his colleagues tested the antibiotics produced by the bacteria that made up the white crust on the ant, they found that these antibiotics were potent only against the parasitic *Escovopsis* fungus—a serious threat to the ants' fungal garden. Other parasitic fungal species (those not a danger to fungus-growing ants) were unaffected by the antibiotics produced by *Streptomyces*, suggesting that selection has favored the use of the *Streptomyces* bacteria by the ants.

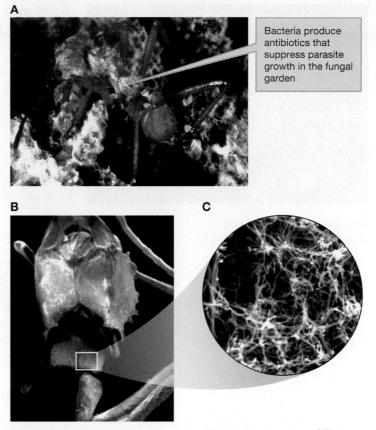

Bacteria produce antibiotics that suppress parasite growth in the fungal garden

FIGURE 19.3 Leaf-cutter ants protect their fungal garden. (A) A worker of the leaf-cutter ant (*Acromyrmex octospinosus*) tending a fungal garden. The thick whitish-gray coating on the worker are bacteria that produce the antibiotics that suppress the growth of parasites in the fungal garden. **(B)** Scanning electron microphotograph (SEM) of a worker, showing the location of the bacteria. **(C)** Detail of SEM in part B.

Other work by Currie and his team has uncovered even more subtle components to the ant–fungus mutualism. In addition to directly using the antibiotics produced by *Streptomyces* to protect their fungal gardens, the ants meticulously groom these gardens and physically remove fungus from their garden that has been infected with *Escovopsis* (Mangone and Currie 2007). Ants pick up parasitic fungal *Escovopsis* spores and hyphae and place them in areas of their body called infrabuccal pockets. Inside these pockets, the spores and hyphae are killed by the antibiotics that are also present in the infrabuccal pockets. The ants then take the dead spores and hyphae and deposit them in a separate pile away from the fungal garden (Little et al. 2003, 2006).

The relationship between ants and fungi demonstrates that not only do mutualistic relationships involve increases in the fitness of individuals of each species, but such relationships often are also characterized by complex and subtle interactions that may involve even additional species such as the bacteria producing the antibiotics in the ant–fungus example. Recent work in other bacteria–insect mutualisms is also beginning to shed light on the genomics of coevolutionary change (McCutcheon et al. 2009; McCutcheon and Moran 2010).

FIGURE 19.4 Butterfly–ant mutualism. Butterflies and ants in a mutualistic relationship. In the mutualism between the butterfly *Jalmenus evagoras* and the ant *Iridomyrmex anceps*, butterfly larvae cannot survive in the absence of ants, and ants receive some of their food from the nectar produced by the butterfly larvae.

Ants and Butterflies: Mutualism with Communication

Communication involves the transfer of information from a signaler to a receiver (Chapter 18), and natural selection will favor communication between individuals from different species that are involved in a mutualistic relationship if such communication increases the fitness of the individuals in each species. To examine the role of communication in mutualistic relationships, we will focus on the work of Naomi Pierce and her colleagues, who have been studying a mutualistic relationship between the imperial blue butterfly (*Jalmenus evagoras*) and the ant *Iridomyrmex anceps* (Pierce et al. 2002) (Figure 19.4). The benefits to both parties in this mutualism are enormous: The butterfly larvae and pupae secrete a sugary nectar composed of sucrose and fructose that nourishes the ants, while the ants protect the larvae and pupae from predators such as wasps. Pierce and her colleagues have found that butterfly larvae have reduced survival rates when ants are experimentally removed (Figure 19.5). While ants can survive in the absence of the nectar that they consume from larvae and pupae, they nonetheless obtain a significant portion of their nutrients from their butterfly larvae partners (Pierce et al. 1987; Fiedler and Maschwitz 1988).

This ant–butterfly mutualism involves costly investment by both parties. To see this investment, consider this: Butterfly larvae raised in a predator-free laboratory environment develop into much larger pupae than butterfly larvae raised in the wild (Pierce et al., 1987). Why? This is because larvae raised in the laboratory are able to pupate later than larvae raised in the wild, since they do not experience the threat of predation. As such, they can reduce the amount of nectar they secrete for use by the ants, and the larvae can use the nutrients normally provided to ants for their own development. Since size in both male and female butterflies is related to reproductive success, pupating early in the wild leads to lower reproductive success for the butterflies, and hence it represents a significant investment in the mutualistic relationship (Elgar and Pierce 1988; Hill and Pierce 1989; Hughes et al. 2000). There is probably also a cost to ants for protecting butterfly larvae, but it has not yet been quantified by researchers. Ants involved in a mutualistic relationship with butterflies likely have an increased risk of detection by their own predators and parasitoids, as well as bearing metabolic costs that are associated with defense of the butterfly larvae (Pierce et al. 1987).

FIGURE 19.5 Butterflies benefit from their ant partners. The probability of survival of *Jalmenus evagoras* larvae and pupae when faced by predation was much higher when ants were present than when they were experimentally excluded at two Australian field sites: **(A)** Mt. Nebo site and **(B)** Canberra site. Adapted from Pierce et al. (1987).

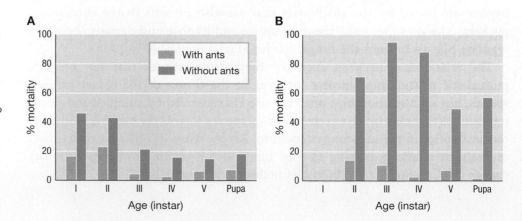

Given that ants and butterflies are tied together in a mutualism that is costly to maintain, researchers hypothesized that communication between the two species would be beneficial to both parties. They decided to test whether such communication was indeed taking place. Travasso and Pierce found that ants are almost deaf when it comes to airborne sounds, but they are quite sensitive to vibrational signals traveling through solid substrates (Travasso and Pierce 2000; Cocroft 2001; Cocroft and Rodriguez 2005). In examining the role of vibrational communication between ants and butterflies, Travasso and Pierce found that larval stridulation (vibrational signals produced when the larvae rubbed stridulatory organs together) was higher when ants were in the vicinity, suggesting that the larvae used such vibrational signals as a way to communicate with their ant guards.

In a follow-up experiment, Travasso and Pierce examined pairs of butterfly pupae, "muting" one of the pair by applying nail polish to its stridulatory organs and allowing the other member of the pair to stridulate normally. Then, using a preference testing device that included two bridges on which the ants could move about, Travasso and Pierce tested whether ants were more attracted to the muted individual in a pair or to the individual who was free to produce vibrational communication. They discovered that ants demonstrated a clear preference for associating with the pupae that could and did produce vibrations, providing evidence that vibrational communication plays a role in this ant–butterfly mutualistic relationship (Figure 19.6).

In this experiment, Travasso and Pierce did not directly measure whether the butterfly pupae stridulate more when their ant partners are present. Nonetheless, it appears that the fitness benefits accrued by both parties in the ant–butterfly mutualism are valuable enough that a form of vibrational communication has evolved between these mutualistic partners.

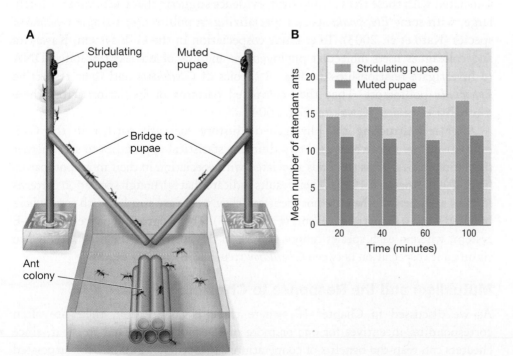

A

Stridulating pupae

Muted pupae

Bridge to pupae

Ant colony

B

Mean number of attendant ants

- Stridulating pupae
- Muted pupae

20

15

10

5

0

20   40   60   100

Time (minutes)

FIGURE 19.6 Communication between mutualists. (A) The apparatus used in preference tests. Ants from a colony could choose to move along either bridge. One of the pupae at the top was "muted." **(B)** Stridulating attracts ants. Stridulating *J. evagoras* pupae attracted more ants than *J. evagoras* pupae that had been experimentally muted. Differences between treatments were significant at all time intervals (20, 40, 60, and 100 minutes). Adapted from Travasso and Pierce (2000).

A

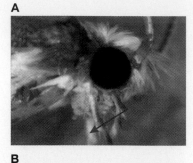

B

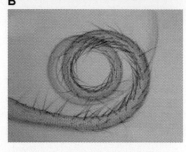

FIGURE 19.7 Pollination of *Glochidion* tree flowers by *Epicephala* moths. (A) Pollen on the proboscis of a female moth is shown at the tip of the red arrow. **(B)** One section of the proboscis of a female *Epicephala* moth. The hairlike projections (sensilla) in the females of pollinating *Epicephala* have likely evolved as a specialized trait associated with pollination.

Mutualism and Cospeciation

When the benefits of mutualism to both species are high and the mutualistic relationship has been in place over long periods of evolutionary time, the link between mutualists may result in **cospeciation**, in which speciation in one species leads to speciation in the other (Page 2003). But how might a simultaneous breakdown of gene flow *within each* of two mutualistic species occur and allow for cospeciation? Geographic separation provides one possible mechanism (Chapter 14). If some physical barrier separates communities that contain both mutualistic species, then a pair of species involved in a mutualistic relationship in community 1 could evolve independently of that pair of species in community 2, leading to cospeciation in allopatry.

As an example of cospeciation, we will examine the mutualism that exists between *Glochidion* trees and the *Epicephala* moths that pollinate them (Kawakita et al. 2004). The *Glochidion* trees and *Epicephala* moth study system (which we will label as the G–E system) is an **obligate mutualism**, in which each partner can only survive and reproduce successfully in the presence of the other (Thompson 1994, 2005). The strong reciprocal reliance of each species on the other in such an obligate mutualism suggests that this system may be one in which cospeciation is occurring.

In the G–E system, a female moth transports pollen between *Glochidion* tree flowers on a tubelike mouthpart called a *proboscis*. A comparison of the proboscis of male and female moths shows that female moths possess specialized hairlike projections called sensilla that play a role in pollination (Figure 19.7). Females lay their eggs in the flower's style, and their larvae feed on developing seeds, destroying a small portion of the seeds in the process. The larvae rely on these seeds as their food source, and the trees rely on the moths for pollination (Kawakita et al. 2004; Kawakita and Kato 2006).

The G–E system is made up of 300 species of *Glochidion* distributed across Asia, Australia, and Polynesia. Whereas the exact number of *Epicephala* moth species associated with these trees is unknown, evidence suggests that the number is likely large, with some *Epicephala* species specializing in pollinating a single *Glochidion* species (Kato et al. 2003). To examine cospeciation in the G–E system, Kawakita and colleagues used molecular phylogenetic analysis of nuclear ribosomal DNA to investigate relationships among 18 species of *Glochidion* and their respective *Epicephala* pollinators. They then compared patterns of speciation across these mutualistic species (Kawakita et al. 2004).

After reconstructing the phylogenetic history for each partner in the G–E mutualism, the researchers used two different statistical approaches to see whether speciation in *Glochidion* trees was associated with speciation in their moth pollinators (Page 1994; Ronquist 1995). Their results indicate that, although speciation patterns in trees and their moths were not identical, they were very similar, with somewhere between 6 to 10 cospeciation events (Figure 19.8). The reciprocal reliance in the G–E system, wherein each species cannot survive in the absence of the other, has led to significant cospeciation between *Glochidion* trees and *Epicephala* moths.

Mutualism and the Response to Cheaters

As we discussed in Chapter 18, where there is cooperation, there are often corresponding incentives for one or more of the parties involved to cheat, since cheaters can reap the benefits of cooperation without having to pay the associated

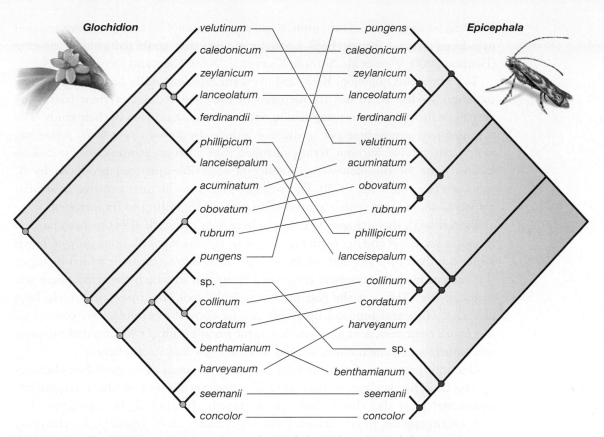

Glochidion

velutinum — pungens *Epicephala*
caledonicum — caledonicum
zeylanicum — zeylanicum
lanceolatum — lanceolatum
ferdinandii — ferdinandii
phillipicum — velutinum
lanceisepalum — acuminatum
acuminatum — obovatum
obovatum — rubrum
rubrum — phillipicum
pungens — lanceisepalum
sp. — collinum
collinum — cordatum
cordatum — harveyanum
benthamianum — sp.
harveyanum — benthamianum
seemanii — seemanii
concolor — concolor

FIGURE 19.8 Cospeciation. Phylogenetic trees for *Glochidion* (left) and *Epicephala* (right) showing tree–moth associations. Species of *Glochidion* are also designated by their species names ("sp." indicates an unnamed species). The *Epicephala* moths in this study were all undescribed species, so each is indicated here by the species name of its host tree. Lines connect moth species and tree species that are associated with them. Nodes associated with cospeciation are indicated with colored circles. Adapted from Kawakita et al. (2004).

FIGURE 19.9 Soybean–rhizobial bacterium mutualism. Soybean legumes (*Glycine max*) are involved in a mutualistic relationship with rhizobial bacteria (*Bradyrhizobium japonicum*). Pictured here is a soybean with nodules containing *Bradyrhizobium japonicum*.

costs. How do interspecific mutualists handle the "cheater problem"? Does one partner in a mutualism respond when the other cheats? To address this question, Toby Kiers and her colleagues examined the mutualism between a soybean legume, *Glycine max*, and a rhizobial bacterium, *Bradyrhizobium japonicum*, which is a soil bacterium that forms nodules on the roots of the soybean plant (Kiers et al. 2003) (Figure 19.9). In a process known as nitrogen fixation, *B. japonicum* converts inorganic N_2 in the root nodules of the plant into an organic form of nitrogen, providing a critical resource that the plant uses for growth and synthesis. On the other end of the mutualism, the soybean plant provides carbohydrates and other energetic resources to *B. japonicum*, which they use for their growth and maintenance.

What would happen if one party in this mutualism cheated? For example, nitrogen fixation is costly for *B. japonicum*, as the resources used to fix nitrogen could instead be used by the

rhizobial bacterium for its own growth and reproduction. Can soybean plants respond to reduced nitrogen fixation by *B. japonicum* in a way that would reduce such cheating (Denison 2000; West et al. 2002a,b; Kiers et al. 2006; Kiers and Denison 2008)?

To address this question, Kiers and her team experimentally forced *B. japonicum* to "cheat"—that is, to not fix nitrogen in the root nodules of their hosts—by creating a nitrogen-free atmosphere in one treatment condition of their study. The nitrogen-free atmosphere per se did not reduce the growth rate of *B. japonicum*, which can survive without fixing nitrogen, but such an atmosphere created an indirect cost for the plant—the absence of accessible nitrogen produced by *B. japonicum* (Layzell et al. 1979). Did soybean plants in the nitrogen-free treatment respond and, in some sense, "punish" *B. japonicum* for failing to fix nitrogen?

Even though the nitrogen-free atmosphere does not itself affect growth rates in *B. japonicum*, Kiers and her team found that *B. japonicum* populations grew to much larger numbers in plant nodules in an experimental treatment in which nitrogen was present in the atmosphere versus in a treatment in which the atmosphere was nitrogen-free. This was the case even in a "split root" treatment in which, by a clever experimental protocol involving precise control of atmospheric conditions in growth chambers, a single plant had some nodules subject to a normal nitrogen atmosphere and some nodules subject to a nitrogen-free atmosphere.

One interpretation of these results is that the soybean plant punished cheating by the bacteria, leading to decreased *B. japonicum* growth in the nitrogen-free treatment. How did the soybean plant punish cheating in *B. japonicum*? The mechanism appears to be curtailing the O_2 available to *B. japonicum* by changing the permeability of the nodule membrane, which in turn reduces *B. japonicum*'s growth rate. The split nodule control treatment condition also demonstrated that it isn't just that plants with *B. japonicum* that fail to fix nitrogen have lower levels of O_2 themselves, and hence have less to put into nodules. Rather, the results indicated that plants differentially allocated O_2 to nodules with nitrogen-fixing *B. japonicum* over nodules containing experimentally created *B. japonicum* cheaters.

19.2 Antagonistic Coevolution

In addition to mutualism, coevolution may also occur when each of two species has a negative effect on the other. We refer to this as antagonistic coevolution, and here we will examine the two most common forms of this type of coevolution: (1) between predator and prey, and (2) between parasite and host.

Predator–Prey Coevolution

Consider a simple predator–prey system in which a predator feeds on only one species of prey, and this species of prey is preyed on by only this one predator. Selection favors any trait in prey that increases their chances of escaping predation. When such a trait evolves in prey, this immediately intensifies selection on predators for traits that increase their probability of capturing and consuming their now better-adapted-to-escape prey. Such a trait in predators will then favor any trait in prey that allows them to escape their now better-adapted-to-kill predators, and so on. This coevolutionary dynamic is known as an evolutionary arms race.

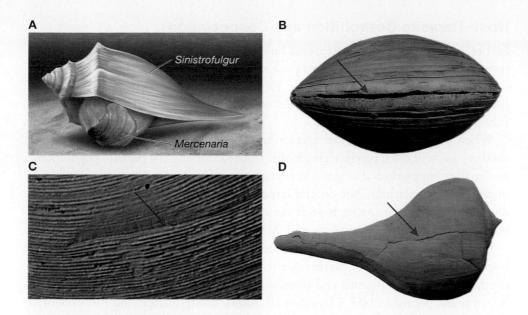

FIGURE 19.10 Predator–prey
interactions and coevolution.
(A) A predatory whelk *Sinistrofulgur* mounts its bivalve prey *Mercenaria* and chips away at its shell. Adapted from Dietl (2003a,b). **(B)** Evidence of a successful attack on *Mercenaria* (red arrow). **(C)** Evidence of an unsuccessful attack on *Mercenaria*. The *Mercenaria* shell is worn down (red arrow) but not cracked by *Sinistrofulgur* shell chipping. **(D)** A whelk occasionally breaks its own shell while trying to open its prey. Damage is indicated by the red arrow (Dietl 2003a,b).

To better understand such evolutionary arms races, let's examine predator–prey interactions between the predatory whelk *Sinistrofulgur* and its bivalve prey *Mercenaria* (Dietl 2003a,b). In this system, the fossil record is detailed enough that it is possible to record both successful and unsuccessful attempts at predation over significant periods of time. During an attack, a whelk "mounts" its prey, and it uses its shell lip to chip away at the bivalve shell. When it is successful, it kills the prey, but even when it is unsuccessful, the telltale chips and cracks from a failed predation attempt are preserved in the fossil record. The cost of predation can also be documented in the fossil record, as the whelk occasionally breaks its own shell while trying to open its prey, and such damage and subsequent repair can be seen when examining whelk shells (Figure 19.10).

Evidence for an evolutionary arms race can be seen in the fossil record of the *Sinistrofulgur–Mercenaria* system. Over evolutionary time, selection has favored an increased shell size and shell thickness in *Mercenaria* prey, which would reduce its probability of being eaten by *Sinistrofulgur*. As *Mercenaria* evolved a thicker shell, selection then favored any trait in *Sinistrofulgur* that allowed it to kill its thicker-shelled *Mercenaria* prey. The fossil record shows that, over the same time period that *Mercenaria* were evolving a thicker shell, *Sinistrofulgur* predators were also increasing in size. Larger *Sinistrofulgur* predators would have been able to penetrate the shells of their *Mercenaria* prey more easily (and would have been safer from their own predators). This would also produce selection for increased size in *Mercenaria* prey, and back and forth in an evolutionary arms race with *Sinistrofulgur* with respect to size.

Is it possible that, rather than a predator–prey arms race, natural selection acted on size, outside the context of predator–prey interaction, and independently in each species? Could this explain the increase in size in both *Sinistrofulgur* and *Mercenaria*? While this is possible, evidence suggests that an evolutionary arms race is a more likely explanation. The fossil record also shows that, over evolutionary time, *Sinistrofulgur* predators changed the typical position they assumed during an attack in such a way as to increase the probability of successfully killing their *Mercenaria* prey. The positional change recorded in the predator is consistent with the hypothesis that the adaptive change was in response to adaptations in its prey.

Host–Parasite Coevolution and Cospeciation

Earlier in this chapter, we discussed how mutualistic interactions can result in cospeciation. Cospeciation between parasites and hosts may also occur, as initially suggested by Kellogg and later by Fahrenholz, both of whom hypothesized that phylogenies of parasites and hosts often change in parallel (Kellogg 1896; Fahrenholz 1909; Klassen 1992).

When populations of a host species become geographically isolated from one another, this will often produce geographic isolation among the parasite populations. As the host populations diverge, selection acts in new ways, not only on individuals in the host populations, but on the respective parasite populations they carry. If divergence in the host species is great enough, and host speciation occurs, this could lead to speciation in the parasite as well (Moran and Baumann 1994; Wade 2007).

Dale Clayton and his colleagues examined the role of parasite–host coevolution and cospeciation in ectoparasitic feather lice (*Columbicola*) that complete their life cycle on their bird hosts—pigeons and doves—feeding on the bird's abdominal feathers (Clayton and Johnson 2003; Clayton et al. 2004). Using nuclear and mitochondrial DNA sequences, these researchers constructed phylogenies of both lice and their hosts, and then they compared these phylogenies to test whether cospeciation had occurred. Their analysis uncovered eight cospeciation events (Figure 19.11).

FIGURE 19.11 Parasite–host cospeciation. Phylogenies of pigeons and doves and their lice (genus *Columbicola*). Lines connect host–parasite associations. Cospeciation events are color coded: Matching colors on each side indicate cospeciation events. Adapted from Clayton et al. (2003).

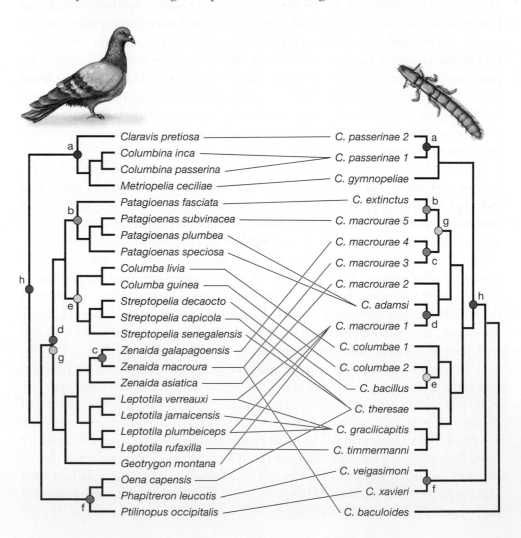

What drove cospeciation in this system? One clue came when Clayton and his team found that the lice species in which individuals were large tended to live on larger species of pigeons and doves (Figure 19.12). The researchers hypothesized that there were benefits to lice if they stayed on size-matched species (small lice with small host species, large lice with large host species). If this was correct, then when speciation occurred in birds, their lice were constrained to remain on their hosts because of the benefits of size matching, and this led to a tight linkage between host and parasite, leading to cospeciation events. But what exactly were the benefits that lice received for size matching with their hosts?

Clayton and his team experimentally examined whether body size matching allowed lice to remain attached to their hosts more efficiently. Lice were placed on feathers from either a host species or a nonhost species, and these feathers were attached to a fan to test the ability of the lice to remain on their hosts. Results indicate that size matching did not improve the ability of lice to attach to the feathers. Other work also found that body size matching did not affect the feeding ability of lice. Body size matching did, however, have a significant effect on the ability of lice to escape the defensive preening behavior of their host species. Compared to the case of lice on their natural hosts, when lice were experimentally placed on nonhost species, they were unable to evade preening (self-cleaning) acts by a host, and they were eaten by birds at high rates; as such, lice could not establish populations on hosts that differed in size from their normal host.

The process of cospeciation between birds and their parasitic lice in this example appears to unfold as follows: After a speciation event occurs in a bird group, lice are constrained to remain on their host species because they often fare poorly when switching hosts. Such switches might involve living on a new host that is a different size than their original host, which could potentially make the lice susceptible to significant predation by the new host. Constrained to remain on their original bird host, when natural selection acts on the bird host, their lice also experience selection operating in new ways. This can lead to new adaptations by the parasites as well, and cospeciation may occur.

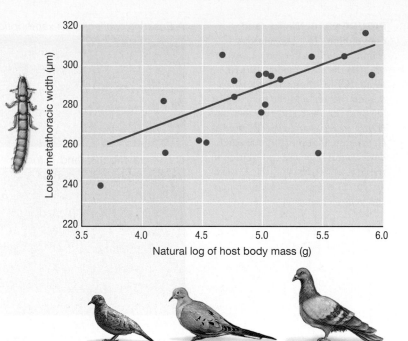

FIGURE 19.12 Body size in parasite and host. Parasite body size in relation to host body size across the associations shown in Figure 19.11. Adapted from Clayton et al. (2003).

Mimicry and Coevolution

Predation can affect the coevolutionary process both directly and indirectly. To see how, recall the *Ensatina* salamanders we discussed as a classic example of a ring species in Chapter 14. Here we return to these salamanders, but we will focus on one particular subspecies, *Ensatina eschscholtzii xanthoptica,* also known as the yellow-eyed salamander. This subspecies displays striking colors—an orange ventral region and yellow eyes, neither of which are found in other *Ensatina* salamanders. Why have these dramatic traits evolved in *Ensatina eschscholtzii xanthoptica*? One hypothesis is that they

FIGURE 19.13 Mimicry and coevolution. (A) The nontoxic mimic, *Ensatina eschscholtzii xanthoptica*, **(B)** the toxic model, *Taricha torosa*, **(C)** the mimic (*E. e. xanthoptica*, on the left) and the toxic model (*T. torosa*, on the right) together, and **(D)** *E. e. oregonensis*, a species used as an experimental control in research done on mimicry and coevolution in the *E. e. xanthoptica* and the *T. torosa* system. *E. e. oregonensis* lacks orange and yellow coloration.

A *Ensatina eschscholtzii xanthoptica*

B *Taricha torosa*

C *E. e. xanthoptica* and *T. torosa*

D *E. e. oregonensis*

are warning signals to predators that *E. e. xanthoptica* is unpalatable. Such **aposematic coloration** is quite common in salamanders and other animals—but *E. e. xanthoptica* is *not* unpalatable to predators. Instead, orange body color and yellow eyes in *E. e. xanthoptica* appear to be the result of coevolution. In the case of *E. e. xanthoptica*, the second species in this story of coevolution is the California newt, *Taricha torosa*, which lives sympatrically with *E. e. xanthoptica* populations (Figure 19.13).

The California newt, in addition to possessing orange body coloration and yellow eyes, also produces a neurotoxin called tetrodotoxin in its skin. While this toxin is potent, predators whose attacks fail but who ingest a small dose of tetrodotoxin often survive and learn to avoid this potential prey type.

More than 60 years ago, American evolutionary biologist George Ledyard Stebbins (1906–2000) hypothesized that the orange body color and yellow eyes of *E. e. xanthoptica* had been selected because they mimic the coloration of the California newt. Such mimicry would protect *E. e. xanthoptica* from predators who might confuse it with the toxic California newt (Stebbins 1949). This sort of mimicry, in which one species is palatable and the other is not, is called **Batesian mimicry** (Bates 1862), and it is different from **Müllerian mimicry** (Müller 1879), in which multiple unpalatable species evolve similar phenotypes to reinforce warning signals that predators can pick up.

But how can evolutionary biologists test whether *E. e. xanthoptica*'s brilliant colors are a result of coevolution via Batesian mimicry? Shawn Kuchta first attempted to answer this question by setting up an experiment in which he placed in the field clay salamander models that either looked like *E. e. xanthoptica* or lacked the orange body color and yellow eyes of *E. e. xanthoptica*. His results showed that predators attacked the *E. e. xanthoptica* models significantly less often than models without the *E. e. xanthoptica* coloration (Kuchta 2005).

In a follow-up experiment, Kuchta and his colleagues examined rates of predation on *E. e. xanthoptica* in a controlled laboratory setting in which a predator was provided with the opportunity to feed on live salamanders (Kuchta et al. 2008). Western Scrub Jays taken from the field were used as predators because they

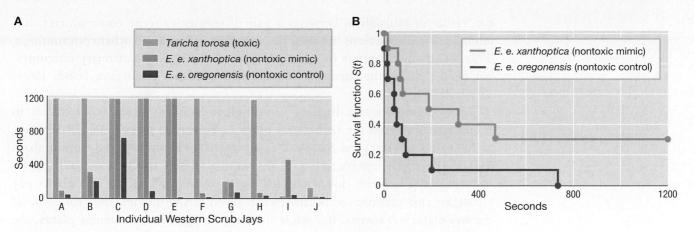

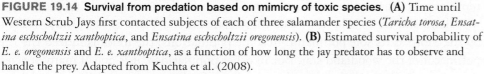

FIGURE 19.14 Survival from predation based on mimicry of toxic species. (**A**) Time until Western Scrub Jays first contacted subjects of each of three salamander species (*Taricha torosa, Ensatina eschscholtzii xanthoptica*, and *Ensatina eschscholtzii oregonensis*). (**B**) Estimated survival probability of *E. e. oregonensis* and *E. e. xanthoptica*, as a function of how long the jay predator has to observe and handle the prey. Adapted from Kuchta et al. (2008).

had experience with toxic California newts. But the Western Scrub Jays had no experience with *E. e. xanthoptica*, which are not usually found in the Western Scrub Jay habitat. The researchers first presented a scrub jay with a California newt, and then they presented the jay with either an *Ensatina eschscholtzii xanthoptica* salamander, or an individual from the closely related subspecies *Ensatina eschscholtzii oregonensis*, which is morphologically similar to *E. e. xanthoptica*, except that it lacks orange and yellow coloration. Kuchta and his team found that, after experience with the California newts—some of which the jays attacked—the jays took more time to approach *E. e. xanthoptica* than *E. e. oregonensis* individuals.

From their encounters with the California newts, the scrub jays had learned to avoid creatures that had orange body color and yellow eyes and a newtlike body. As a result, jays were very hesitant to approach *E. e. xanthoptica*, even though that species does not possess the neurotoxin found in the newts (Figure 19.14). In an encounter in the wild, such extra time could make the difference between survival or being eaten by a jay.

In this case, we have seen how traits in one species (the toxic California newt) influence the operation of selection on another (the nontoxic *Ensatina eschscholtzii* salamander), but, unlike many of the other examples, we do not know yet whether the *E. eschscholtzii* has a reciprocal influence on the species that is being mimicked. There is, however, a testable prediction here: Because predators will occasionally eat *E. e. xanthoptica*, they also will occasionally eat a *Taricha torosa* newt, and in so doing create new selection pressures on newts to signal their toxicity in a slightly different way. Future research could look for hints of such a change.

19.3 Mosaic Coevolution

Up to this point we have examined the dynamics of coevolution leading to mutualism (the G–E example) or antagonistic coevolution (whelks and bivalves) between a pair of species. But it is important to understand that, depending on the ecology and behavioral interactions between two species, natural selection can result in mutualism between a pair of species in some communities, but antagonistic interactions between the *same* pair of species in other communities. This idea, which centers on geographic *variation* in coevolutionary outcomes, has been dubbed the theory of **mosaic coevolution** (Thompson 1982, 1994, 1999, 2009).

John Thompson and Bradley Cunningham studied mosaic coevolution in interactions between the herbaceous plant *Lithophragma parviflorum* (also known as the woodland star) and the moth *Greya politella* (Thompson and Cunningham 2002). The moth lays its eggs into developing flowers of the woodland star by inserting its ovipositor down into the floral ovaries. Woodland star plants pay a cost for this pollination, because when moth larvae mature, they eat some of the woodland star's seeds. But while inserting its eggs into numerous plants, the moth also pollinates the woodland star. *Greya politella* is completely reliant on the woodland star as its sole host. But the woodland star plant is not always reliant on the moth as its sole pollinator. In some populations, the moth is indeed the *sole* pollinator of the woodland star. But in other populations, it is one of many species that act as pollinators for this plant, and many of these other pollinators do not produce larvae that eat the plant's seeds, and so they are less costly to the woodland star. This means that there is geographic variation in the costs and benefits of the plant–moth interactions. Thompson and Cunningham tested whether this geographic variation in costs and benefits was correlated with geographic variation in mutualism versus antagonistic coevolution. They hypothesized that the more reliant the plant was on this moth species as a pollinator, the more mutualistic the coevolutionary dynamics would be (Figure 19.15).

In four populations in which *G. politella* acted as the *sole* pollinator, and the woodland star was the *sole* host for *G. politella*, Thompson and Cunningham found a mutualistic relationship between plants and moths. In these populations, the woodland star rarely aborted flower capsules that contained moth eggs (doing so would also kill the moth larvae) compared to capsules that had no moth eggs. But, depending on the costs and benefits to the plants and moths, mutualism need not be the outcome of coevolution in this system. In four other populations, in which the

FIGURE 19.15 Mosaic coevolution in plants and moths. (A) A female *Greya politella* moth on the flower of the woodland star (*Lithophragma parviflorum*). The moth is laying eggs as she pollinates the flower. **(B)** Mosaic coevolution must be studied across many sites. In their work on the interaction between the woodland star and its moth pollinator, Thompson and Cunningham sampled sites (shown with red dots) in Oregon, Washington, and Idaho. Part B adapted from Thompson and Cunningham (2002).

woodland star had numerous pollinators besides *G. politella*, the researchers found evidence of an antagonistic relationship between plants and *G. politella*. The plants *selectively* aborted flower capsules that contained moth eggs. In these populations, where alternative pollinators were present, the costs of having the moth pollinator outweighed the benefits, and selection favored an antagonistic, rather than a mutualistic, response from the plant.

The woodland star–*Greya politella* moth system is a good example of mosaic coevolution in nature. Depending on the costs and benefits to each party, the coevolutionary process may lead to mutualism—as in the case where the moth is the sole pollinator for the plant—or antagonistic interaction. What's more, with a solid understanding of the natural history of the species involved, evolutionary biologists can derive hypotheses about which populations will head down one coevolutionary path and which will head down the other.

19.4 Gene–Culture Coevolution

In addition to studying coevolution between species, over the last 30 years evolutionary biologists and anthropologists have begun examining **gene–culture coevolution**, the coevolutionary process between genes and cultural traits, both within and between species. Throughout this book, we have examined how allele frequencies change over time. In that sense, we have already delved deeply into the "gene" part of gene–culture coevolution. Let's briefly examine how cultural evolution operates, and then we will move on to discuss gene–culture coevolution.

In Chapter 3, we noted that, for natural selection to act on a trait, a mechanism for transmitting that trait across generations is required. Once Mendel's work on genetics was rediscovered in the early 1900s, it became clear that genes are one means of transmitting traits across generations. Culture provides another means, and recently evolutionary biologists have become interested in this phenomenon as well.

Cultural transmission is often defined as the transfer of information from individual to individual through social learning. A slightly different way of saying this is that cultural transmission is a system of information transfer that affects an individual's phenotype via social learning (Bonner 1980; Cavalli-Sforza and Feldman 1981; Boyd and Richerson 1985).

Although cultural and genetic transmission each provide a means of passing traits down from one generation to another, there are a number of unique aspects of cultural transmission. When individuals learn from others—that is, when social learning occurs—information can be spread through a population very quickly. As a consequence, the behavior of a single individual can dramatically shift the behavior patterns of an entire group. Cultural transmission can change the frequency of behavioral traits, not only across generations, when younger individuals learn from older individuals, but also within a single generation, when individuals from the same cohort learn from one another (Boyd and Richerson 1985, 2004). When cultural transmission leads to changes in the frequency of traits within or between generations, we call this **cultural evolution**.

FIGURE 19.16 Scavenging and cultural transmission. (A) A scavenging rat often encounters new food items while foraging. **(B)** Smelling another rat provides olfactory cues about what it has eaten. This transfer of information from one rat to another about safe foods is a form of cultural transmission.

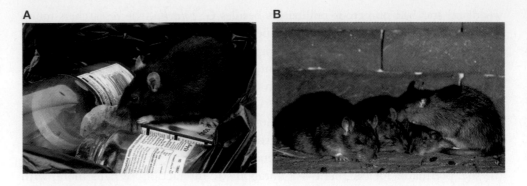

There are different types of cultural evolution. Vertical cultural transmission refers to the scenario in which information is transmitted between generations from parent(s) to offspring. Horizontal cultural transmission involves the transfer of information between individuals who are in the same age cohort. Oblique cultural transmission involves the transfer of information across generations outside the context of interactions between parents and their offspring—that is, young receive information from adults that are not their parents.

For an interesting case study illustrating the importance of cultural evolution and social learning in animals, let's examine foraging behavior in rats. As scavengers, rats sample many new foods. Yet, scavenging can present a dilemma. A new food source may be an unexpected bounty for a rat. But new foods can also be dangerous. They may contain elements (such as poisons) that are inherently bad for rats or, because a rat doesn't know how a new food should smell, it is difficult to tell if a novel food is fresh (and will serve as nourishment) or spoiled (and may make it sick).

Jeff Galef and his colleagues have studied the role of cultural transmission in the scavenging behavior of Norway rats (*Rattus norvegicus*) (Figure 19.16) (Galef and Wigmore 1983; Galef and Laland 2005; Galef and Whiskin 2006). To test whether cultural transmission via social learning plays a role in rat foraging, the researchers examined whether observers could learn about a new, distant food source simply by interacting with a demonstrator that had experienced such a new addition to its diet.

After two rats had been caged together for days, one rat was removed and taken to another experimental room, where it was given one of two new diets—either rat chow flavored with Hershey's cocoa or rat chow mixed with ground cinnamon: The rats had never experienced either of these two additions to their diets. This "demonstrator" rat was then brought back to its home cage and allowed to interact with the observer rat for 15 minutes. The demonstrator rat was removed from the cage. For the next 2 days, the observer rat was given two food bowls, one with rat chow and cocoa, the other with rat chow and cinnamon. Although the observer rat had no direct experience with either of the novel food mixes and it had not *seen* the demonstrator rat eating these new food items, it was more likely to eat the food that the demonstrator rat ate, both when the demonstrator had eaten rat chow with cocoa and when it had eaten rat chow with cinnamon, strongly suggesting that Norway rat foraging behavior was affected by cultural transmission via olfactory cues (Figure 19.17).

With the growing realization that cultural transmission affects many different types of behavior, evolutionary biologists have become interested in the ways in which genetic and cultural transmission can interact. Do genetic changes affect cultural evolution? Do changes in cultural evolution affect genetic evolution? Do they both affect each other? Researchers are actively looking into these questions, and the answers have promise for helping us to understand the coevolution of genetic and cultural transmission.

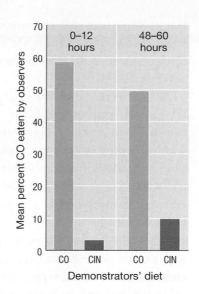

FIGURE 19.17 Cultural transmission across generations. Social learning and foraging in the Norway rat. Observer rats had a "tutor" (demonstrator) who was trained to eat rat chow containing either cocoa (CO) or cinnamon (CIN) flavoring. Rats with "cinnamon tutors" preferred cinnamon-flavored food (not shown here), and rats with "cocoa tutors" preferred cocoa-flavored food (shown here). Adapted from Galef and Wigmore (1983).

Gene–Culture Coevolution in Darwin's Finches

The medium ground finch (*Geospiza fortis*) and the cactus finch (*G. scandens*) live on the Galápagos island of Daphne Major (Figure 19.18). These species are capable of interbreeding. Hybrid offspring produced from *G. fortis* by *G. scandens* matings do not appear to suffer a decrease in fitness as compared to offspring from *G. fortis* by *G. fortis* matings or *G. scandens* by *G. scandens* matings. Nevertheless, interbreeding between *G. fortis* and *G. scandens* remains a rare

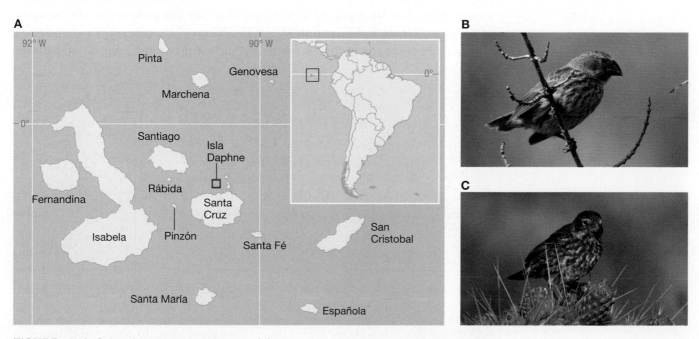

FIGURE 19.18 Cultural evolution in birdsong. (**A**) The Galápagos Islands, including Daphne Major Island (in red). (**B**) *Geospiza fortis*. (**C**) *Geospiza scandens*.

event. Why is it that the two species seldom interbreed? Here we will examine work that suggests a role for cultural transmission in inhibiting such matings (Nelson et al. 2001; Slabbekoorn and Smith 2002; Freeberg 2004; Lachlan and Servedio 2004).

When Peter and Rosemary Grant examined the songs of *G. fortis* and *G. scandens* during the birds' mating season, they found evidence of cultural transmission across generations (Grant and Grant 1996). To understand what was happening, the Grants compared the songs of sons, fathers, and grandfathers. One hypothesis for why father and son finches have very similar songs is that the song is a genetic trait passed from father to son; another hypothesis is that cultural transmission of the song from father to son takes place when the son hears and learns the song sung by his father. To test the two hypotheses, we can compare the songs of sons to those of their paternal and maternal grandfathers, and we can distinguish between these possibilities. If song types are genetically controlled, we would expect the songs of the sons to be similar to the songs of both their paternal and maternal grandfathers, since the son inherits genes from both grandfathers. But if cultural transmission from father to son is the mechanism, then the songs of the sons should resemble those of their paternal grandfather, not those of their maternal grandfather, since it is the paternal grandfather who would have transmitted the song to the father, who in turn would have transmitted the song to the son. Comparison of the son's song to those of the maternal and paternal grandfathers shows that the son's song resembles the song of the paternal grandfather, not the song of the maternal grandfather—suggesting that birdsong is culturally transmitted (Figure 19.19).

In studying the birdsong of the two finch species, the Grants found that the songs varied significantly from one another: The birdsong of *G. scandens* has shorter components that are repeated more often than the components of the birdsong of *G. fortis* (Grant and Grant 1994, 1997). These differences in their songs—a culturally transmitted trait—have a dramatic impact on gene flow between

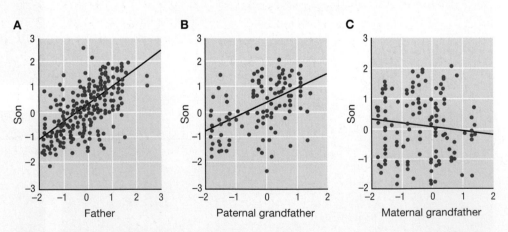

FIGURE 19.19 Finch songs across generations. Male finches' songs are positively correlated with those of their fathers (**A**) and those of their paternal grandfathers (**B**), but not with those of their maternal grandfathers (**C**). The horizontal axis and vertical axis are in units that summarize numerous components of the songs. Adapted from Grant and Grant (1996).

ground finches and cactus finches. The researchers sampled 482 females and found that over 95% of them mated only with males who sang the song typically produced by males of their own species. This suggests that cultural transmission plays a large role in why ground and cactus finches rarely mate, even though hybrid offspring suffer no fitness costs. The song sung by males—a culturally transmitted trait—provides females with a means to recognize individuals of their own species, which in turn leads to few between-species matings. Further support for this interpretation comes from the fact that the Grants uncovered 11 cases in which the male of one species sang the song of another species; most of these males mated with females from the other species. In such cases of cross-species breeding, viable hybrid offspring were produced. Remove the normal pattern of cultural transmission, and the barrier to breeding across species disappears.

In this chapter, we have seen the many ways that coevolution has shaped the diversity of life around us, and the complex interactions that often define between-species relationships, be they mutualistic, parasitic, or predator–prey interactions. In the final chapter of this book, we will move to a subject—evolutionary medicine—that ties together many of the concepts that we have covered throughout the book.

SUMMARY

1. Coevolution occurs when evolutionary changes to traits in one species cause selection to act in new ways on traits in another species, which in turn feed back to alter the nature of selection on traits in the first species, and so forth.

2. Mutualisms may evolve from initially neutral interactions between species or from interactions in which one species initially benefits and the other is initially unaffected in any way. A mutualism can even evolve from an initially parasitic relationship when the costs and benefits of that parasitic relationship change over time to favor the mutualism.

3. Under certain conditions, natural selection may favor communication between individuals of different species involved in a mutualistic relationship.

4. If the benefits of mutualism to both species are high and the mutualistic relationship has been in place over a long period of evolutionary time, the link between mutualists may lead to cospeciation, in which

speciation in one species is associated with speciation in the other.

5. The two most common forms of antagonistic coevolution are that between predator and prey, and that between parasite and host.

6. The dynamics of antagonistic coevolution can take the form of an evolutionary arms race.

7. Cospeciation may occur in parasite–host systems. Speciation in hosts can drive speciation in their parasites.

8. Natural selection can result in mutualism between a pair of species in some communities but antagonistic interactions between the same species in other communities. This leads to geographic variation in coevolutionary outcomes, or mosaic coevolution.

9. Evolutionary biologists and anthropologists have begun examining gene–culture coevolution and its consequences, both within and between species.

KEY TERMS

antagonistic coevolution (p. 622)

aposematic coloration (p. 636)

Batesian mimicry (p. 636)

cospeciation (p. 630)

cultural evolution (p. 639)

cultural transmission (p. 639)

gene–culture
  coevolution (p. 639)

mosaic coevolution (p. 638)

Müllerian mimicry (p. 636)

obligate mutualism (p. 630)

REVIEW QUESTIONS

1. Why distinguish coevolution from other evolutionary processes? Why not treat species 2 as a part of species 1's environment in exactly the same way that temperature is a part of species 1's environment?

2. Consider the following argument: In environments that are especially harsh, with little food and much competition, mutualism is unlikely to evolve, as it is not in the best interest of individuals of any species to act in a way that increases the fitness of individuals of other species. Now make the counterargument—that is, that such environments are where mutualisms are most likely to evolve.

3. Why might microbes found in the guts of a series of host species and the host species themselves be an especially likely system in which to find cospeciation?

4. Why is an understanding of the natural history of the species being studied so critical to using the mosaic theory of coevolution to make specific, testable predictions?

5. How might genetic drift, along with or in the absence of natural selection, shape the coevolutionary process?

6. Many species of ants and acacia trees have evolved a mutualistic relationship, in which the ants protect the acacia from mammalian and insect predators, and the acacia trees provide food to the ants through extrafloral nectary glands full of a carbohydrate-rich liquid that the ants eat, as well as through globules, called Beltian bodies, containing proteins that the ants consume. How might you design an experiment to test the hypothesis that if the protection that the ants provide becomes less necessary, natural selection will favor acacia trees that produce fewer resources for ants?

7. When we discussed mosaic coevolution, we framed that discussion in terms of two species. But how might yet other species in a community—their absence, presence, or behavior—affect mosaic coevolution of the primary two species a researcher may be studying?

8. How might the rapid speed at which cultural evolution operates affect the rate of genetic evolution? Is it possible, in certain situations, that the rapid speed at which cultural evolution operates could slow down the rate of genetic evolution? How so?

9. Why do you suppose that communication between mutualistic species might speed up the pace of coevolutionary change? Can you think of an example in which this has occurred?

10. Recall our definition of Batesian mimicry and Müllerian mimicry. What sort of differences would you expect in coevolutionary dynamics between these two types of mimicry?

SUGGESTED READINGS

Boyd, R., and P. J. Richerson. 1985. *Culture and the Evolutionary Process*. University of Chicago Press, Chicago. A book-length treatment of the coevolution of genes and culture.

Dawkins, R., and J. Krebs. 1979. Arms races between and within species. *Proceedings of the Royal Society B: Biological Sciences* 205: 489–511. One of the first detailed treatments of evolutionary arms races.

Pellmyr, O. 2003. Yuccas, yucca moths, and coevolution: a review. *Annals of the Missouri Botanical Garden* 90: 35–55. There are many wonderful examples of coevolution, and the yucca–yucca moth system described in this review is a classic case.

Proctor, H., and I. Owens. 2000. Mites and birds: diversity, parasitism and coevolution. *Trends in Ecology & Evolution* 15: 358–364. A summary on the coevolutionary dynamics seen between mites and their bird hosts.

Thompson, J. N. 2010. Four central points about coevolution. *Evolution: Education and Outreach* 3: 7–13. A well-crafted, general review of coevolution.

 Visit StudySpace at wwnorton.com/studyspace.

20

Evolution and Medicine

◀ A traditional mask from the Tsimshian people of coastal British Columbia.

olklore has it that fever can bring about prophetic dreams, visions, and epiphanies. One could even argue that, in a sense, the subject of this entire book—the theory of biological evolution—arose in part through a feverish epiphany. Alfred Russel Wallace, who developed the theory of evolution by natural selection in parallel with Charles Darwin, described how he came to realize the role of natural selection:

> At that time I was suffering from a rather severe attack of Intermittent [malarial] fever at Ternate in the Moluccas, and one day while lying on my bed during the cold fit, wrapped in blankets, though the thermometer was at 88°F... there suddenly flashed upon me the idea of the survival of the fittest—that the individuals removed by [disease, famine, and the like] must be on the whole inferior to those that survived. In the two hours that elapsed before my ague fit was over I had thought out almost the whole of the theory, and the same evening I sketched the draft of my paper, and in the two succeeding evenings wrote it out in full, and sent it by the next post to Mr. Darwin. (Wallace 1891, p. 20)

But whether fever brings epiphanies—or more likely bizarre dreams—fever makes one feel miserable. Fever can even be life threatening if body temperature rises too high. Fortunately, fever is usually easy to remedy. A number of common over-the-counter drugs have *antipyretic* effects: aspirin, acetaminophen, and ibuprofen all reduce or eliminate fever with a minimum of side effects in most patients. Thus, it makes perfect sense, on first consideration, that we should treat fever whenever we can. Doing so is easy, and it relieves suffering or worse.

But if we consider what fever is *for* in the first place, we might begin to wonder about the wisdom of treating it. There is overwhelming evidence that fever, like cough and diarrhea, is one of the body's evolved defenses against infection by pathogens (Nesse et al. 2006). Might there be times when it is best to let a fever run its course rather than to treat it? Or is it reasonable to block the fever response even if it has evolved as a defense against pathogens? We begin by describing one particularly striking observation, and we will return to answer these questions in more detail later in this chapter.

Fever is not restricted to *endotherms,* the so-called warm-blooded species that actively regulate their body temperatures. In a classic experiment, Linda Vaughn and her colleagues demonstrated that a cold-blooded *ectotherm,* the desert iguana (*Dipsosaurus dorsalis*), induces fever behaviorally in response to bacterial infection by moving to warmer locations within its habitat (Vaughn et al. 1974). To demonstrate this, Vaughn and her colleagues constructed environmental chambers with regions that were kept at two different temperatures: one below the thermal optimum for the lizards, and one above the thermal optimum. They found that healthy lizards thermoregulate by moving between the two areas, as shown in Figure 20.1.

Vaughn and her colleagues then stressed a group of lizards by injecting them with bacteria. To avoid causing infections that would kill the lizards, the researchers killed the bacteria with heat prior to injecting the lizards. Although the dead bacteria were not able to reproduce, they could nonetheless stimulate an immune response in the injected lizards. To make sure that any change in the behavior of the lizards was due to the bacteria rather than to the handling process and the physical injection, the researchers compared the individuals injected with bacteria to a control group that had been injected with saline solution. They found that both groups used behavioral thermoregulation, but the lizards injected with bacteria had a shift in their preferred temperature. Compared to the controls, lizards injected with bacteria stayed on the warm side of the enclosure until reaching a higher body temperature. Similarly, the lizards injected with bacteria were quicker to leave the cool side of the enclosure than were the controls.

This work provides compelling evidence that the lizards respond to bacterial infection with what is called behavioral fever. Subsequent studies have shown behavioral fever in other ectothermic vertebrates, including a wide range of reptiles, fish, and amphibians (Monagas and Gatten 1983; Reynolds et al. 1976; Covert and Reynolds 1977). A number of invertebrate species have also been shown to elevate their body temperatures in response to infection (Thomas and Blanford 2003).

These results suggest that there is something about fever that may be advantageous in dealing with infection. Both endothermic and ectothermic vertebrates induce fever, albeit by very different mechanisms, in response to

A

B

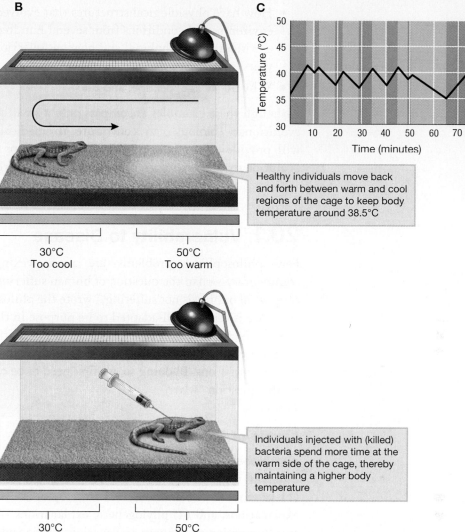

Healthy individuals move back and forth between warm and cool regions of the cage to keep body temperature around 38.5°C

30°C
Too cool

50°C
Too warm

Individuals injected with (killed) bacteria spend more time at the warm side of the cage, thereby maintaining a higher body temperature

30°C

50°C

infection. This might give us pause as we think about the purpose of fever and the medical implications of treating it. We will return to the issue of fever in Section 20.2. There we will see why, defensive role notwithstanding, treating fever may be advisable in most cases.

In this chapter, we will explore the new and rapidly growing field of evolution and medicine. Throughout, a unifying theme will be that evolutionary biology informs medical science by providing explanations for how vulnerabilities to disease have evolved—and that this in turn helps us both to understand the proximate mechanisms responsible for disease and to generate hypotheses about how disease can be treated. We will look at six different classes of explanation for disease vulnerability, and we will explore case studies for four of them. In the course of discussing these four cases, we will resolve the following questions:

- Why can we safely treat most fevers despite the fact that fever appears to be an evolved defense?

- What is the role of the immune system in host–pathogen coevolution?

FIGURE 20.1 Thermoregulatory behavior in the desert iguana *Dipsosaurus dorsalis.*

(A) *Dipsosaurus dorsalis.* **(B)** Vaughn observed how *D. dorsalis* thermoregulated by moving between the cool end and the warm end of its enclosure. **(C)** A graph showing temperature regulation behavior for an uninfected lizard. The curve goes up and down as the lizard thermoregulates, moving to the cool end of the chamber when it reaches too high a body temperature and to the warm end of the chamber when it reaches too low a temperature. Part C adapted from Vaughn et al. (1974).

- How have physiological structures that evolved in response to environmental conditions from several hundred million years ago left us vulnerable to choking, and how has selection within the past million years exacerbated this problem?
- Why do we age, decline, and die?

Although these examples encompass only a small sampling of the ways in which evolutionary biology can contribute to medical research and practice, they will provide a sense of the intimate connections between evolutionary history, evolutionary processes currently in operation, and the practice of human medicine.

20.1 Vulnerability to Disease

Few philosophical problems are more vexing—or have attracted more commentary—than the question of human suffering. "If the immediate and direct purpose of our life is not suffering," wrote the philosopher Schopenhauer, "then our existence is the most ill-adapted to its purpose in the world" (Schopenhauer 1851, cited in Nesse 2001). *Why* does the human condition involve such heavy doses of pain, misery, anxiety, and sadness? In this chapter, we will look at science's answers to these questions. In doing so, we first need to be clear as to exactly what we mean by the question "why?"

Levels of Explanation

The Nobel prize–winning biologist Niko Tinbergen distinguished among four different types of answers that can be given to a "why" question in biology: (1) proximate explanations, (2) developmental explanations, (3) evolutionary explanations, and (4) phylogenetic explanations. Although Tinbergen developed this distinction to apply to explanations for behaviors, we can apply them equally well to explanations for illness or disease. Proximate explanations tell us about the immediate mechanism that precipitated a particular pathology. Developmental explanations tell us how the pathology came about over the course of the organism's lifetime. Evolutionary explanations tell us how natural selection and other evolutionary processes interact to leave the body vulnerable to a particular pathology. Phylogenetic explanations look at a species' evolutionary history and explain where in this evolutionary history such vulnerabilities came about.

Medicine largely deals with the first two levels of explanation, for good reason: Proximate and developmental explanations associate disease with factors *that we have the power to change.* Much of clinical medicine is *reactive:* It aims to respond to a problem, and to correct that problem to whatever degree is possible. This aim places a significant premium on a proximate understanding of disease. If we are to intervene to eliminate or eradicate disease, we must understand the proximate contributors to that disease, and then alter them. Similarly, *preventative medicine* commonly considers the developmental explanations of disease in order to intervene before illness begins. How does a patient's lifestyle and life experience shape disease vulnerability? How can changes in lifestyle reduce the probability of illness later in life?

Beyond the obvious importance of proximate and developmental explanations for medicine, researchers and clinicians alike are beginning to recognize the utility of evolutionary and phylogenetic explanations as well. An understanding of these factors deepens our understanding of the defenses and weaknesses of the human body. This in turn can suggest strategies for treatment or prevention. Moreover, explanations at the evolutionary and phylogenetic levels can suggest new hypotheses about mechanisms and consequences at the proximate and developmental levels. Evolutionary thinking can guide medical research on mechanisms of disease.

We also need to consider what is the *object* of evolutionary explanation. What is it about disease that we aim to explain? Illness itself is rarely an adaptation; rather, we aim to understand the *vulnerability* to illness. How did these vulnerabilities evolve, and why have such vulnerabilities persisted despite the operation of natural selection?

Six Explanations for Vulnerability to Disease

Randy Nesse and George Williams proposed six classes of evolutionary explanation for vulnerability to disease in their 1994 book *Why We Get Sick* (Nesse and Williams 1994). Here we will consider a slightly reformulated list of six reasons, as presented in a paper that Nesse wrote a decade later (Nesse 2005). All of these reasons are based on principles we have already studied in the book. But in this chapter we will use them to understand human vulnerabilities to disease. It is important to recognize that these reasons are not mutually exclusive. As we will see explicitly in our discussion of choking, multiple explanations may contribute to a single vulnerability.

The first two explanations revolve around the fact that evolution by natural selection may not be fast enough to solve certain problems.

1. Humans are locked in a coevolutionary arms race with their pathogens, most of which evolve much more rapidly than do humans.

2. Natural selection has not had time to catch up with rapid changes in the environment.

In each of these cases, there may be heritable genetic variation for disease susceptibility in the population, and less susceptible variants may be favored. Yet, susceptibility to disease may remain in the population simply because selection has not had time to eliminate it. In the first case, vulnerabilities remain because pathogens provide a moving target, evolving rapidly to escape whatever mechanisms evolve in the host to prevent or eliminate infection. In the second case, vulnerabilities remain because of relatively recent environmental changes. This may be a particularly important mechanism for disease vulnerabilities in humans because human cultural innovations have radically changed human life and human diets over the past 30,000 years.

The next two explanations pertain to limits on what evolution can do, even given huge amounts of time.

3. The laws of physics and the nature of biology impose trade-offs on what an organism can do.

4. Natural selection lacks foresight, so that sometimes we are stuck with historically contingent relics of our past.

In the first of this pair of reasons, we note that many aspects of our physiology reflect compromises in function. For example, thicker bone structure would reduce the number of fractures that we suffer, but it would come at the expense of nimbleness and speed. Higher metabolic rates might improve numerous aspects of physiological function, but they would do so at the cost of increased nutrient demands. In the second of this pair of reasons, we note that evolution is a historically contingent process; our current anatomy evolved by a gradual process of modification that occurred without the benefit of foresight. For example, when the basic tetrapod body plan shared by all terrestrial vertebrates was evolving, evolution had no way to plan ahead and ensure that this body plan would be an appropriate foundation for some future bipedal hominid.

The last two explanations for vulnerability to disease are focused on what it is that natural selection actually favors: not health and well-being, but rather reproductive success.

5. Natural selection favors reproductive success, even at the expense of vulnerability to disease.

6. Some defenses, such as fever, nausea, and anxiety, may be unpleasant to experience, but they are beneficial adaptations rather than maladies.

In each of these explanations, reactions or symptoms that we label as disease because they are unpleasant may not be maladaptive from a fitness perspective. In the first of this pair, we need to recognize that natural selection does not maximize health at age 70 or even survival at age 16—rather, it maximizes *expected lifetime reproductive success*. Thus, phenomena such as physical decline associated with old age or risk-taking behavior associated with adolescence may be adaptive if the alleles responsible also contribute to reproductive success. In the second of this pair, we need to distinguish between symptoms that are unpleasant but beneficial as defenses—vomiting, fever, or itching—and truly maladaptive defects such as chronic pain.

20.2 Fever

Having laid out the big picture regarding why we are vulnerable to disease, we now turn to a set of case studies illustrating some of the reasons for these vulnerabilities. In this section, we return to the example with which we began this chapter: the phenomenon of fever and the issue of how we should treat it.

The proximate mechanisms responsible for triggering mammalian fever are relatively well understood. When immune cells recognize the presence of a pathogen, they release signaling chemicals known as *cytokines*. Among many other functions cytokine signals stimulate the brain region known as the hypothalamus, which is responsible for regulating many of the body's physiological systems. The signals induce a shift in the body's thermal setpoint for temperature regulation, driving an increase in body temperature and inducing fever (Figure 20.2). For example, when components of the bacterial cell wall are bound by immune cells known as macrophages, the macrophages produce the fever-inducing signals. The

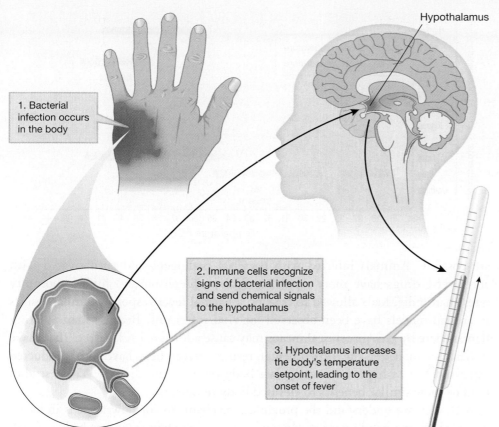

1. Bacterial infection occurs in the body

2. Immune cells recognize signs of bacterial infection and send chemical signals to the hypothalamus

3. Hypothalamus increases the body's temperature setpoint, leading to the onset of fever

Hypothalamus

FIGURE 20.2 Proximate mechanism for fever. Immune cells recognize a pathogen and send chemical signals to the hypothalamus in the brain, which increases the body's thermal setpoint, leading to fever.

presence of bacterial cell wall components is a strong indicator of bacterial infection, and thus this pathway illustrates a mechanistic coupling between indications of pathogen challenge and the fever response.

Consequences of Fever

Correlational studies indicate that, in humans, higher fevers are associated with more severe infections. From studies that control for infection severity in animal models, it appears that the presence of fever is also correlated with better disease outcomes. A number of studies of human patients have shown that patients who receive fever-reducing antipyretic drugs such as aspirin, acetaminophen, or ibuprofen recover *less* quickly from viral infection (Hasday et al. 2000). This indicates that fever may play a beneficial role in shortening the duration of infection. Studies of patients with bacterial sepsis (a severe and very dangerous full-body inflammatory response induced by bacterial infection) have often, but not always, revealed a higher survival rate in patients that exhibit fever than in patients without fever. It is important to recognize that this observed correlation does not provide direct evidence of causation. Survival rates could be lower in patients without fever because, for example, the inability to mount a fever response could be indicative of more severe illness.

As we described in the introduction, numerous nonhuman species exhibit a fever response as well (Figure 20.3). In many of the species, manipulative studies

FIGURE 20.3 Basal and fever temperatures in vertebrate species. Fever has been documented in many vertebrates, including both endotherms and ectotherms. Adapted from Hasday et al. (2000).

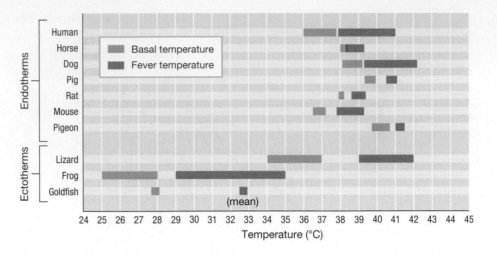

are possible. Animals infected with bacterial pathogens and then treated with antipyretic drugs have more rapid bacterial proliferation and higher mortality rates than individuals allowed to develop a normal fever response. Similar results in animal models have been observed for viral infections. Because manipulating temperature by pharmaceutical means may cause additional side effects that make it hard to interpret the experimental results, researchers have also conducted experiments in which they manipulate body temperature directly. These studies tend to show similar benefits to elevated body temperature.

Although we understand the proximate mechanisms for how fever is induced, we do not have a good understanding of precisely why fever seems to be beneficial. The temperature increase due to fever may have deleterious effects on the growth of some microorganisms, but many bacteria will grow at comparable rates whether at normal (basal) or fever temperatures. Another possibility is that temperature increases may up-regulate some immune defenses and increase the rate at which immune cells both proliferate and act against pathogens. Fever may also trigger the expression of *heat shock proteins*, which help cells deal with intracellular damage, and may thus be useful during infection (Hasday and Singh 2000).

If fever conferred only benefits, we might expect endotherms to have higher body temperatures all the time, ill or otherwise. But fever has its costs. For one thing, the metabolic costs of fever are significant. Simply running a fever of 2–3°C above normal temperature causes a 20% increase in metabolic rate, and the shivering response sometimes used to elevate temperature can increase metabolic rate up to sixfold over baseline (Kluger et al. 1996, 1998; Hasday et al. 2000). This increased metabolic load may be especially expensive in precisely the situations in which it occurs: when individuals are already stressed by infection. By up-regulating the immune system, fever may also exacerbate the tissue damage due to aspects of the immune response. Indeed, some experimental studies in mice suggest that, while increased body temperature increases the rate at which bacteria are eliminated from the bloodstream, it also increases the probability of death.

The Smoke Detector Principle

Given the costs associated with fever, why are fevers so common? And if fevers are evolved defenses, why can we treat them without severe consequences? To answer these questions, we need to think about the decision to trigger a defensive reaction.

Ideally, the body would initiate an unpleasant or costly defensive reaction only when such a reaction was absolutely necessary. For example, the body would only mount a fever response when it was challenged by a pathogen that could not be cleared without fever. But there is an information problem here. At the time a bacterial infection is detected, the body doesn't know just how severe a particular infection may turn out to be—or, in some cases, whether an infection is really present at all.

Moreover, for fever as for other defensive responses, some kinds of mistakes are far more costly than others. A *false positive*, in which a defensive response is initiated in the absence of a threat, typically imposes only a modest energetic cost, whereas a *false negative*, in which no defensive reaction is imposed even in the presence of a threat, can be extremely costly or even fatal. This is the *principle of asymmetric harm*—that is, failing to initiate a defense response when it is needed tends to be vastly more harmful than needlessly invoking a defense response when it is unnecessary. As a result, we expect evolution to tune our defensive responses so that they will be invoked too often, rather than too seldom.

Randy Nesse has compared this problem to that faced in the design of a smoke detector (Nesse 2001). Nesse points out that no one wants a smoke detector that only detects some or even the majority of fires; a smoke detector needs to raise the alarm each and every time there is a fire, even at the cost of the occasional false alarm when cooking bacon. Here again the underlying principle is that false positives are inexpensive compared to false negatives. Indeed, given that false positives are so much cheaper than false negatives, an optimally designed smoke detector may be in error (of the false positive sort) 99% of the time that it triggers an alarm, so long as it is almost never in error when it does not sound.

Applying this logic to fever, we can make a number of predictions. First, the cost of a modest fever—some energetic expense, some discomfort and associated downtime—is far less than the cost of failing to produce a fever when it is needed to clear an infection. Thus, we might expect that fevers should be relatively common, and most should be unnecessary. If so, this means that, in most cases, we should be able to safely intervene, reducing fever with antipyretic drugs. Medical technologies, most notably antibiotics and rehydration therapy, further reduce the risk of death by infection relative to what it would have been throughout much of our evolutionary history. Thus, even many of those infections that might have been lethal without a fever response prior to these medical developments can now be safely controlled without fever. This is not to say that treating fever with antipyretic drugs is always the right decision. Further research will be needed to resolve that issue. But it does provide an explanation for why we can often safely interfere with fever, despite its role as an evolved defense against pathogens.

20.3 Coevolutionary Arms Races between Pathogens and Hosts

Infectious disease medicine aims to help us deal with challenges from pathogens ranging from viruses such as influenza and HIV to bacteria such as *Staphylococcus aureus* to eukaryotic parasites such as the malaria parasite (*Plasmodium falciparum*) and parasitic helminth worms. Throughout this book, we have looked at some of the evolutionary considerations that arise in this area of medicine, most notably the evolution of antimicrobial resistance to the drugs that we use against these pathogens. In this section, we will step back and consider more generally why natural selection has not solved the problem for us already. Why are we vulnerable to pathogens in the first place? Why hasn't natural selection provided us with impenetrable immune defenses?

To answer this question, we return to a subject from the previous chapter: the phenomenon of coevolutionary arms races. Such arms races are particularly important in the evolution of hosts and pathogens. Pathogens are selected to do whatever furthers their own reproduction and transmission, and this often involves exploiting the host. Hosts are selected to minimize the harm caused by pathogens; this is often best accomplished by eradicating the pathogen entirely from the host's body.

At first glance, multicellular hosts appear to have a marked disadvantage in this coevolutionary arms race. Such hosts are typically much larger than their pathogens, with two important coevolutionary consequences: First, hosts usually have far longer generation times than do their pathogens. For example, the human generation time is on the order of 20 years, whereas many bacterial pathogens have generation times on the order of an hour or two. This is a 100,000-fold difference in generation times! As a result, natural selection can act extremely rapidly on pathogens relative to hosts. In the time it takes an individual human to go from birth to sexual maturity, bacterial pathogens can go through more generations than there have been in the entire evolutionary history of the *Homo* genus. Second, pathogens have much larger population sizes than their hosts. A single bacterial infection may consist of billions of cells. A patient infected with HIV may produce more than 100 billion HIV virions per day during the period of peak viral load. Large pathogen populations that have rapid turnover are able to generate a great deal of genetic variation by mutation, and thus they can generate ample raw material on which natural selection may sort.

With the odds stacked so badly against hosts, how can they possibly keep up? Across the tree of life, hosts have evolved immune systems that isolate pathogens, minimize the harm that they cause, and, if possible, eliminate them from the body. Perhaps the best known of these is our own, the vertebrate adaptive immune system, but there are many others. Bacteria use so-called *restriction–modification systems* to identify and eliminate viral nucleic acids. Bacterial CRISPRs (clustered regularly interspaced short palindromic repeats) may serve a similar function. RNA interference appears to have originated in early eukaryotes as a system for silencing viral gene expression. Plants have extensive systems of nonspecific or "innate" immunity, as do multicellular animals. Among social insects, colony recognition systems and the associated response to intruders can even be viewed as a colony-level immune system. In the remainder of this section, we will focus on the human immune system, but comparable analysis is possible for any of these other systems as well.

Immune Strategies

The human immune system must (1) recognize, and (2) eliminate or incapacitate microbial pathogens. Because pathogens draw on such a large pool of genetic variation and evolve so rapidly, a host population could never keep up if it had to match every adaptive substitution in the pathogen population with an adaptive substitution of its own. Instead, immune systems rely on a number of tactics that limit the ability of pathogen populations to outrun their hosts by virtue of rapid natural selection. There are at least three ways that host immune systems can do this.

Detecting Characteristic Components of Pathogens

Perhaps the most straightforward strategy for dealing with rapidly evolving pathogens is to target those components of the pathogen that cannot easily be changed. Fortunately for us, bacteria have a number of such components. Our innate immune response detects the presence of pathogens using *pattern recognition receptor* molecules that bind to common components of pathogens known as *pathogen-associated molecular patterns,* or PAMPs. These receptors recognize highly conserved components of pathogens, such as the peptidoglycan polymer that makes up bacterial cell walls (Figure 20.4), the lipopolysaccharide molecules in Gram-negative bacterial cell membranes, and the flagellin protein of bacterial flagella. These elements are structurally essential to the various bacteria that use them, and their structures appear to be so highly conserved that bacteria are unable to evolve variants that pass undetected by the pattern recognition receptors.

Finding Infected Cells

One drawback of this strategy is that it doesn't work well against viruses, for a number of reasons: First, viruses have few conserved external structures that an immune system could use to identify them. Second, many viruses are produced by budding from a host cell. Such viruses are therefore wrapped in a membrane layer that is structurally the same as the host cell membrane, save for the inclusion of a few viral transmembrane proteins. Third, viruses replicate within host cells. Therefore, it is not sufficient for the immune system to find and eliminate free viral particles or *virions*; it must also locate and deal with infected host cells that serve as sources that produce additional virions.

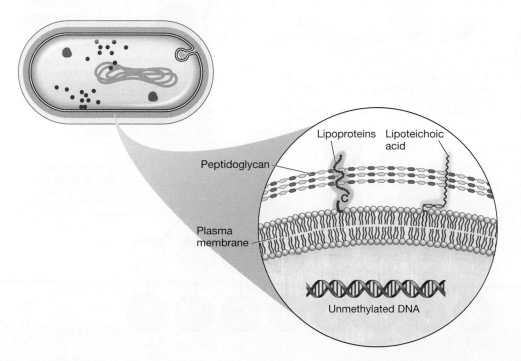

Peptidoglycan

Plasma membrane

Lipoproteins **Lipoteichoic acid**

C

Unmethylated DNA

FIGURE 20.4 Pathogen-associated molecular patterns (PAMPs). One way in which immune systems detect the presence of pathogens is with receptors that bind to common and highly conserved pathogen components known as PAMPs. Shown here are a number of PAMPs in a Gram-positive bacterium: elements of the peptidoglycan layer, membrane lipoproteins, lipoteichoic acid, and unmethylated CG sites in bacterial DNA. Adapted from Wardenburg et al. (2006).

That said, viruses are not without distinctive molecular characteristics. Most notably, properly functioning eukaryotic cells should not contain long double-stranded RNAs, whereas cells infected by viruses do. Double-stranded RNAs occur as either viral genomes or intermediates in the process of viral genome replication. As such, the innate immune system responds aggressively to such double-stranded RNAs within a cell.

To find other cues that reveal viral pathogens, host immune systems must learn, one way or another, to recognize the molecular signs associated with viral infection of a cell. Vertebrate adaptive immune systems do exactly this. They learn, during the lifetime of a single individual, to detect the cues associated with pathogens or pathogen-infected cells. The basic mechanism is known as clonal selection (Figure 20.5); we present a simplified picture of the process here. The immune system produces an enormous, highly diverse repertoire of immune cell lines. Each cell line expresses a different receptor that specifically recognizes some unique antigen, often a small section of a protein. Rather than directly encoding the million or so different receptors in the genome, organisms produce the diversity of receptors through a process known as somatic recombination. Much as a combination lock can specify

FIGURE 20.5 Clonal selection and clonal expansion. The immune repertoire features roughly a million receptor types, and the process of clonal expansion results in a 1000-fold increase in the number of pathogen-specific immune cells, but the basic process is shown here using a far smaller number of cells. Adapted from Bergstrom and Antia (2006) and Goldsby et al. (2000).

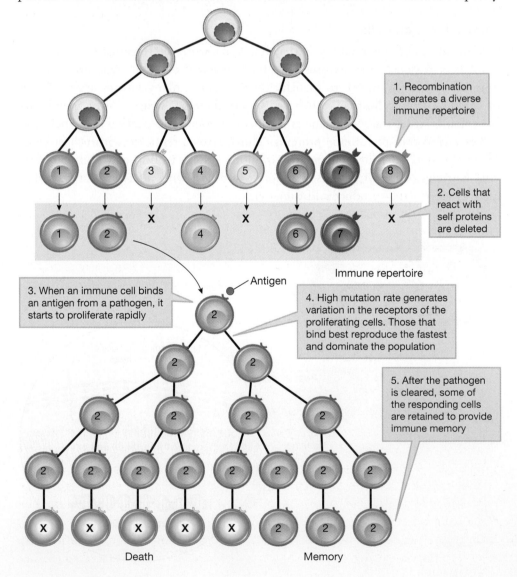

10,000 different codes using only 40 digits (four wheels, each with the numbers 0–9), the immune system can potentially create millions of different receptors by combining a relatively small number of receptor subunits in different ways.

Because these receptors have been created by a process of random recombination, many will bind to *self* proteins—that is, those formed by the host itself. In a screening process, the cell lines with receptors that bind to self proteins are deleted shortly after they are produced. The end product is a vast diversity of immune receptors, all specific for proteins that are not produced by the host itself. If one of these binds something, that something probably shouldn't be there. It is probably a component of a pathogen.

When an immune cell does bind to an antigen, it begins to proliferate rapidly in a process known as **clonal expansion**. By the process of clonal expansion, the immune system creates a large number of cells that specifically react with the antigen that has been detected; these cells can then eradicate the pathogen.

This process already has much in common with natural selection: A good fit between receptor and target is developed by randomly producing large amounts of variation that is subsequently selected on. But in the humoral branch of the immune system that is responsible for antibody production, additional rounds of selection occur as well. During clonal expansion, an improved match to the pathogen is achieved through a process known as **affinity maturation**. Affinity maturation works by the same logic we have already seen: Variation is produced and subsequently selected on. In the dividing immune cells, the coding sequence that specifies the receptor structure undergoes mutation at a much higher rate than usual, thereby generating further variation in the receptors. Those receptor variants that most effectively bind to the pathogen proliferate at a faster rate, outcompeting less effective variants of the receptor. The result is a gradual improvement in the match of the receptor to the antigen target over the course of clonal expansion.

The key point about the entire system is that vertebrates are fighting fire with fire—or rather, they are fighting selection with selection. Vertebrate *organisms* do not have short enough generation times or large enough populations to have any hope of matching pathogens in their rate of evolution. But vertebrate *cells* can reproduce rapidly and form large populations. Thus, the vertebrate adaptive immune system sets up its own internal selective process on its immune cells—that is, it stages its own selective competition. The internal selective process on individual cells operates on a comparable timescale to pathogen evolution, thereby enabling vertebrates to keep pace with their rapidly evolving antagonists.

Creating Variation through Sexual Reproduction

If there are weaknesses to a defense system such as the adaptive immune system, rapidly evolving pathogens will find them and exploit them. How do immune systems deal with this threat? According to the Red Queen hypothesis that we described in Chapter 16, sexual reproduction may be an adaptation for generating large amounts of variation in host lineages, and thus preventing pathogens from specializing on any particular host genotype. Sex may be particularly useful for dealing with pathogens that are transmitted *vertically*, from parent to offspring. If hosts reproduce asexually, their pathogens will already be well adapted to exploit their offspring, because those offspring are genetically identical to their parent. By contrast, if hosts reproduce sexually, the offspring will be genetically different from the parents, and thus less susceptible to pathogens that have been successful in the parents.

Evolution of Pathogens to Subvert Immune Systems

Pathogens for their part evolve remarkably sophisticated and effective ways to avoid being eliminated by their hosts' immune systems. Some of these mechanisms help the pathogens avoid detection by immune systems, but this is not the only way that pathogens deal with immune challenge. Another approach is to subvert the function of the immune system by sabotage or subterfuge (Bergstrom and Antia 2006).

Viruses employ a wide repertoire of subversive tactics. The human immuno-deficiency virus (HIV) not only down-regulates the expression of host MHC molecules involved in recognizing an infected cell, it also induces programmed death in uninfected immune cells (Evans and Desrosiers 2001). The poxviruses—large DNA viruses responsible for diseases such as chickenpox and smallpox—have numerous ways of tampering with the signaling molecules that immune cells use to coordinate and regulate their activity. These viruses produce enzymes that degrade the immune system's chemokine signals (chemical signals controlling the replication and migration of cells to fight infection) before they reach their destination, and they produce molecules that block the host's chemokine signal receptors. They produce false chemokine signals that stimulate some receptors, and they produce decoy receptors that lure the true signals away from their intended targets (Liston and McColl 2003). Some viruses even turn RNA-directed components of the host's immune system against the host itself, using this system to knock out certain host genes and thereby render the host more susceptible to the pathogen (Wang et al. 2004; Pfeffer et al. 2004).

Since internal pathogens have already invaded the body, they are well positioned to tamper with the immune system's communication and coordination pathways. Intracellular pathogens can go yet further; having invaded individual host cells, they can readily manipulate host gene expression. To deal with these challenges, immune systems must be able to function robustly despite targeted misinformation and other forms of "information warfare." How they do this remains an open research question, although hypotheses are starting to emerge. Multiple redundant defenses, fail-safe devices rather than feedback control, cross-validation of signals, and distributed rather than centralized decision making all appear to be mechanisms for defending against internal subversion by pathogens (Bergstrom 2009).

Effects of Immune Systems on Pathogens

Pathogens have driven the evolution of immune systems, and immune systems have driven the evolution of pathogen countermeasures. We have looked at examples of each of these. And if the Red Queen hypothesis for sex is correct, the influence of pathogens on hosts has been far greater than simply forcing hosts to have immune systems. If this theory is correct, pathogens are ultimately responsible for the large-scale patterns of host evolution: Pathogens have driven the evolution of sex, a principal mechanism by which most multicellular eukaryotes generate genetic variation on which natural selection can operate.

The influence of immune systems on the large-scale patterns of pathogen evolution is no less significant. Recent phylogenetic studies of microbial pathogens have revealed that the nature of immune selection on a pathogen population often has a strong influence on the phylogenetic structure of that pathogen.

Some viral pathogens, such as measles, generate a very strong immune response that confers lifelong immunity; these pathogens cannot readily evolve **escape variants** that can dodge the immune memory of previously infected individuals by changing a few key epitopes. Such pathogens have the sort of classic phylogenies that we might expect to see in nonpathogen species (Figure 20.6A). Other viruses, such as the human influenza A virus, can evolve escape variants capable of reinfecting previously infected individuals. These escape variants enjoy a tremendous selective advantage in a population of hosts with immunological memory against previous strains. The result is a distinctive phylogenetic pattern, known as a **cactus-shaped phylogeny**, in which most clones are lost in any given year, and the lineage is continued by one or at most a small number of escape variants (Figure 20.6B). These phylogenies typically have no deep branches, but instead each has a single trunk with only very minor twiggy lineages—the spines of the metaphorical cactus—branching off from it.

Viruses that cause long-term infection, such as HIV, provide an opportunity to compare phylogenetic structure within a single host to phylogenetic structure across a population of hosts. Within a host, strong immunological selection drives a series of escape variants to high frequency, generating a cactuslike phylogeny reminiscent of the population-level phylogeny of influenza (Figure 20.7A). Looking at the population as a whole, however, immunological selection plays very little role. Cross-immunity, in which infection by one strain prevents subsequent infection by another, is minimal, as are differences in transmission rate due to immune selection. The result is that, at the population level, HIV has a more conventional branching phylogeny, as illustrated in Figure 20.7B. The shape of the HIV phylogeny also reveals something about the recent demographic history of this pathogen. The branches are very deep compared to what we would expect in a constant-size population, reflecting the epidemic expansion of HIV over the past quarter century.

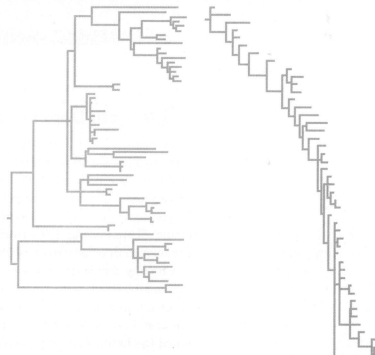

A Measles virus population phylogeny

B Human influenza A virus population phylogeny

TIME

TIME

FIGURE 20.6 Phylogenies of measles and influenza. (A) The measles virus generates lifelong immunity, so there is no immune selection for escape variants. Shown here is a gene tree for the measles nucleocapsid gene. **(B)** The influenza virus can evolve escape variants that get around immunological memory, reinfecting hosts that have previously had the disease. Such escape variants are strongly favored by immune selection; this process gives rise to the characteristic cactus-shaped phylogeny of the influenza hemagglutinin gene shown here. Adapted from Grenfell et al. (2004).

FIGURE 20.7 Within-host and population-wide phylogenies of HIV. (**A**) Within a single host, HIV has a cactuslike phylogeny. (**B**) By contrast, at the population level, the phylogeny of HIV is more conventionally shaped, though with deep branches due to its rapid epidemic expansion. Adapted from Grenfell et al. (2004).

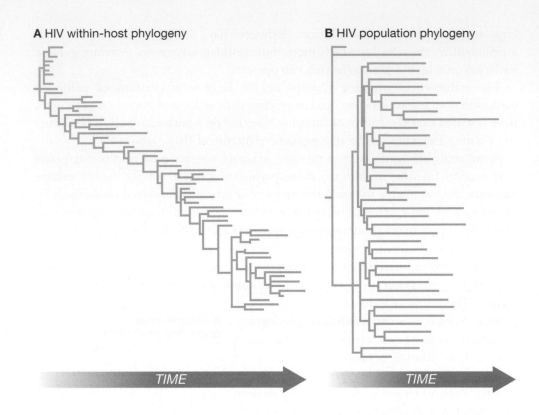

A HIV within-host phylogeny

B HIV population phylogeny

TIME

TIME

20.4 Path Dependence and Vulnerability to Choking

On January 13, 2002, the 43rd presidency of the United States of America nearly came to a premature and tragic end. The evolutionary cause was a "mistake" that evolution had made more than 300 million years previously; the proximate cause was a pretzel. President George W. Bush was watching a football game alone but for the company of his dogs Barney and Spot, when he choked on a pretzel he was eating. He passed out, fell forward, and according to his own recollection awoke on the floor to find the two dogs staring at him.

Choking accidents such as this one are surprisingly common. According to the National Safety Council, choking is the fourth leading cause of accidental death (after automobile accidents, poisoning, and falls) in the United States. Looking at our anatomy, the reason has been obvious to thinkers as far back as Aristotle, who noted the unfortunate intersection of the trachea with the esophagus as a cause of choking (Aristotle's *On the Parts of Animals* 3:3; Held 2009). Figure 20.8 illustrates that the route that air takes through the nasal cavity to the trachea and into the lungs actually intersects the route that food or water takes through the mouth to the esophagus and into the stomach.

A structure known as the *epiglottis* has evolved as a partial work-around to this problem. As shown in Figure 20.8, the epiglottis functions as a trapdoor over the larynx and trachea. When we breathe, the epiglottis is raised, allowing free passage of air into the lungs. When we swallow, the epiglottis is pushed downward over the opening and thus prevents food or water from entering the trachea and lungs.

Even with the epiglottis as a safety mechanism, this is a poor design at best. In principle, there is very little reason that the path of food and water should have to

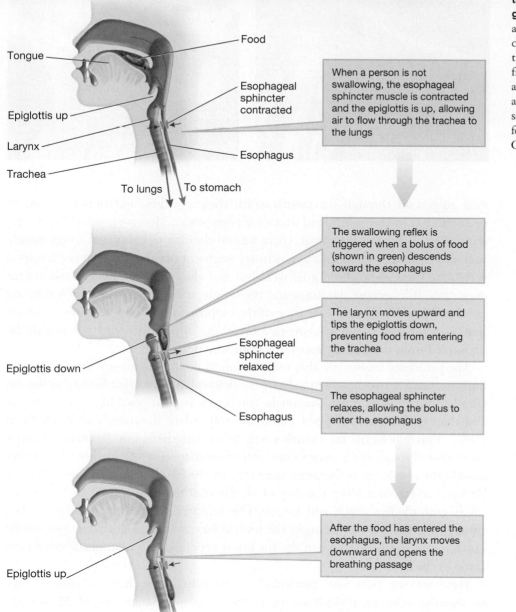

Tongue

Food

Epiglottis up

Esophageal sphincter contracted

Larynx

Trachea

Esophagus

To lungs   To stomach

When a person is not swallowing, the esophageal sphincter muscle is contracted and the epiglottis is up, allowing air to flow through the trachea to the lungs

The swallowing reflex is triggered when a bolus of food (shown in green) descends toward the esophagus

Epiglottis down

Esophageal sphincter relaxed

Esophagus

The larynx moves upward and tips the epiglottis down, preventing food from entering the trachea

The esophageal sphincter relaxes, allowing the bolus to enter the esophagus

Epiglottis up

After the food has entered the esophagus, the larynx moves downward and opens the breathing passage

FIGURE 20.8 Anatomy of the throat, trachea, and esophagus. The pathway through which air passes from the nose to the trachea and lungs crosses the pathway through which food or water passes from the mouth to the esophagus and stomach. By closing down like a trap door, the epiglottis provides a safeguard against accidently taking food into the trachea. Adapted from Othman (2010).

intersect the airway. Figure 20.9 shows a much more sensible alternative, in which the two are kept entirely separate. Admittedly, one benefit of the present structure is that the mouth provides a backup airway if the nose becomes clogged—but a broader nasal opening could readily solve the problem as well. Another, as we will see, is that the present structure facilitates complex vocalizations, including human speech.

So how did we end up in this mess? This aspect of our physiology is a legacy of our evolutionary history (Liem 1988; Held 2009). Lungs arose very early in primitive fish, as a pouch of esophagus or gut tissue that probably served to trap gas bubbles and thereby to capture additional oxygen in low-oxygen environments.

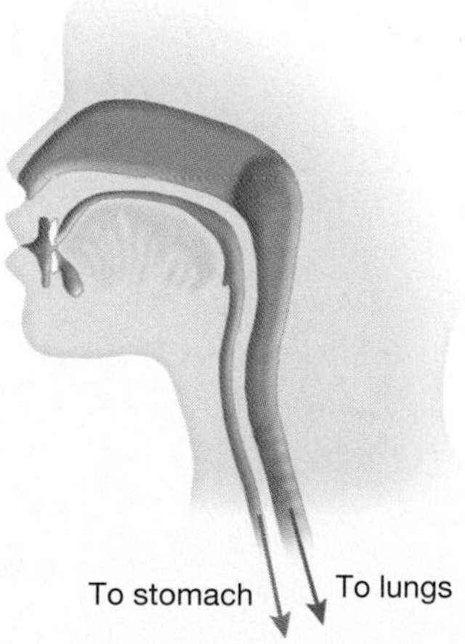

To stomach   To lungs

FIGURE 20.9 How to design a trachea and esophagus that do not intersect. Shown here is a hypothetical anatomy in which the airway and the route of food are entirely separate and choking on food is impossible. Adapted from Held (2009).

FIGURE 20.10 A descended larynx exacerbates the choking hazard in humans. (A) In nonhuman mammals, such as the dog shown here, the epiglottis reaches the soft palate, effectively sealing off the mouth cavity whenever the trachea is open. **(B)** In humans, the larynx is positioned much lower. This facilitates speech production by the passage of air through the mouth, but it comes at the expense of not blocking the flow of material from the mouth, even when the tracheal opening is exposed (Laitman and Reidenberg 1993).

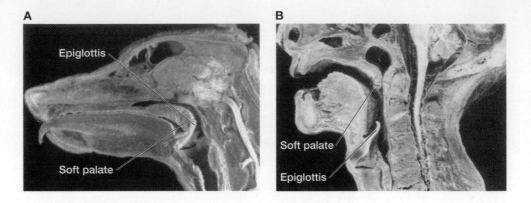

Fish gulped air through the mouth to fill these pouches, and thus they had to be connected to the mouth and digestive passageway. Moreover, when lungs first arose as an offshoot of the throat, there was no choking risk, because lungs merely provided a backup to the gills as a primary source of oxygen. Later in the tetrapod vertebrate lineage, however, gills were lost and the lungs became the sole source of oxygen. But because the lungs and the development process by which they are generated had arisen as an extension of the esophagus and its development, rather than as a separate organ system, this breathing apparatus could not readily be decoupled from the feeding apparatus from which it arose.

The problems created by this morphological configuration are exacerbated in humans, relative to other mammals, by the descended larynx that facilitates human speech. In most nonhuman mammals, the larynx is positioned high in the throat and the epiglottis when raised meets the soft palate (Laitman and Reidenberg 1993). This blocks off the mouth cavity when the trachea is exposed, allowing air to flow through the airway in one unbroken channel while food or water flows around the epiglottis at the same time (Figure 20.10). But humans are different; we vocalize by controlling the flow of air through the mouth cavity, shaping the cavity with the lips, teeth, and tongue. This requires free airflow from the trachea through the mouth. Accordingly, the human larynx and epiglottis are positioned much lower, so that even when the trachea is exposed, air has an unimpeded path through the mouth.

Here we see how one particular vulnerability—the human vulnerability to choking—can be understood by considering more than one of Nesse's six explanations. The basic vulnerability, in which the airway and path of food cross one another, arose in an early vertebrate ancestor prior to the evolution of the lungs as the primary breathing organs. This organization could not readily be reversed later in the evolutionary process, and we are stuck with this anatomical inefficiency as a relic of our evolutionary history. Humans are further vulnerable because of a trade-off: Speech requires slow controlled flow of air from the lungs through the mouth cavity, and thus humans face a trade-off between communication ability and choking risk. But perhaps we should not be too discouraged by this. For all of the problems that our feeding anatomy gives us, Lewis Held points out that we don't have it as bad as cephalopods such as octopuses and squid. In these organisms, the brain wraps around the esophagus, so that each bite of food must pass through the middle of the brain, and too large a morsel can have disastrous consequences (Held 2009).

20.5 Senescence

Senescence refers to a general decline in the physical functioning or performance of living organisms with age. Typically the process of senescence results in an increase in the mortality rate and a decrease in fecundity—that is, ability to produce offspring—with age. While the eventual consequence of senescence is death, the two are not synonymous. Rather, senescence refers to the general process of decline that ends in death. Thus, while average life span or maximum life span are correlates of the senescence rate, senescence has effects reaching beyond longevity (Ackermann and Fletcher 2008). We see these effects dramatically in the decline in human athletic performance with age (Figure 20.11).

Vulnerability to Senescence

Senescence is a very general phenomenon among multicellular organisms. Figure 20.12A shows the general increase in age-specific mortality for humans, chimpanzees, and porpoises. Figure 20.12B shows age-specific mortality for water fleas (*Daphnia*) and fruit flies (*Drosophila*). While the timescales vary over several orders of magnitude, the general pattern is the same in all cases: The rate of mortality accelerates as individuals age. Figure 20.13 shows fertility—the actual number of offspring produced—as a function of age, for humans and for *Drosophila*. Again we see a similar pattern: Fertility declines dramatically as individuals age

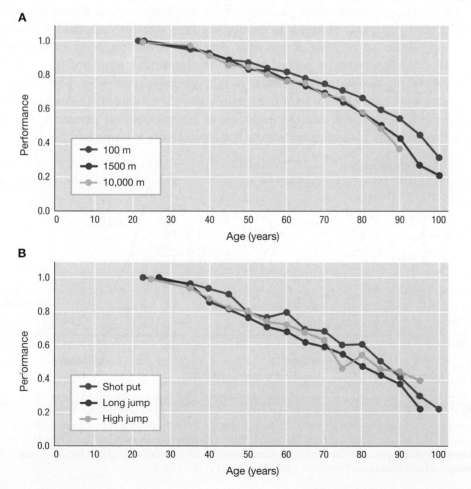

FIGURE 20.11 Decline in maximal human physical performance with age. Here maximal physical performance is as measured by world record times in **(A)** track and **(B)** field events. Performance is scaled relative to the world record time or distance for any age. In track events, performance is quantified as average speed; in field events, performance is quantified as distance or height. Note that, as world record times, these represent the limits of human performance. Thus, the falloff in performance of average individuals with age is likely to be substantially greater than that shown in the graphs. Adapted from Nesse and Williams (1994), Austad and Finch (2008), Track and Field News (2010), and World Masters Athletics (2011).

FIGURE 20.12 Age-specific mortality in five species. Age-specific mortality—that is, the per capita probability of death at a given age—shows a characteristic increase at older ages across a wide range of species. Here we show five examples. **(A)** Age-specific mortality in human females in an Ache hunter-gatherer population, chimpanzee (*Pan troglodytes*) females, and harbor porpoises (*Phocoena phocoena*) of both sexes. Adapted from Kaplan et al. (2000) and Moore and Read (2008). **(B)** Age-specific mortality in water fleas (*Daphnia*) and in fruit flies (*Drosophila melanogaster*). Adapted from Nisbet and Murdoch (1995) and Snoke and Promislow (2003).

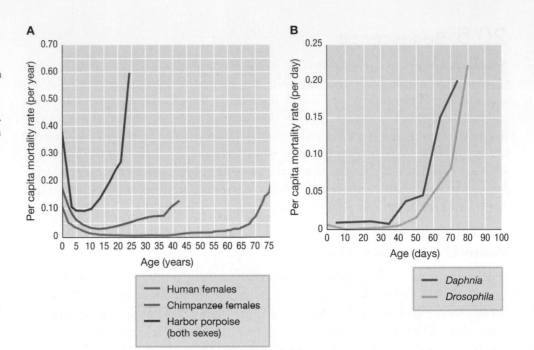

beyond reproductive maturity. (Human females go through menopause, and thus they show a particularly strong pattern in this regard; this phenomenon appears to be quite rare in other species.)

How can we explain these patterns? Why are humans, like almost all other multicellular species, vulnerable to senescence? How did this vulnerability evolve? It is important to be clear about what precisely is our target of evolutionary explanation. Senescence is not an evolved developmental program; it is not an adaptation in its own right. Rather, it is a by-product of other physiological adaptations. Thus, what we want to explain is not the evolution of senescence, but rather why we have not seen the evolution of adaptations that prevent aging. In other words, why has evolution left the body vulnerable to aging and death?

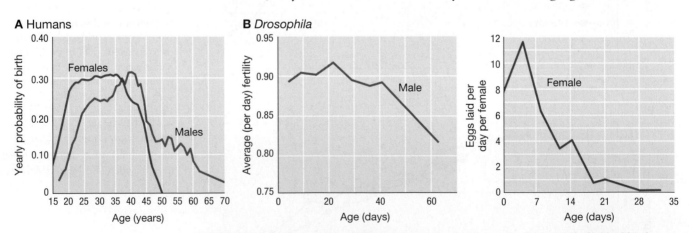

FIGURE 20.13 Age-specific fertility. Age-specific fertility—that is, the per-individual birth rate—shows a characteristic decline at older ages across a wide range of species. **(A)** In an Ache hunter-gatherer human population, female age-specific fertility drops dramatically because of menopause as individuals reach their forties. Male age-specific fertility also falls off with age, although less rapidly. Adapted from Hill and Hurtado (1996). **(B)** Age-specific male and female fertility in *Drosophila melanogaster*. Female fertility is determined by the number of eggs laid by the female. Adapted from Snoke and Promislow (2003) and Tatar et al. (1996).

Rate-of-Living Hypothesis for Senescence

One possible explanation is that senescence is simply unavoidable. Just as machines wear down over time and eventually break down completely, so the bodies of living beings do as well. Selection may result in a slowing down of senescence, but there is only so much that selection can do in this respect. This type of explanation is sometimes known as the **rate-of-living hypothesis** for senescence, because it posits that senescence is a consequence of physical wear and tear.

The rate-of-living hypothesis for senescence makes two strong testable predictions. First, if selection has already done everything possible to slow the pace of senescence, there should be little or no remaining genetic variability in the rate of senescence. But biologists have uncovered ample evidence to contradict this prediction. In such model organisms as fruit flies (*Drosophila melanogoster*), nematode worms (*Caenorhabditis elegans*), and mice (*Mus musculus*), researchers have identified scores of known *longevity mutations* that confer slower rates of senescence (Tatar et al. 2003). Even in humans, a few alleles—for example, the *APOE2* allele—are known to contribute to greater longevity (Christensen et al. 2006).

Furthermore, the heritability of life span is substantial in many species. Table 20.1 lists the observed ranges of heritabilities for humans and model organisms from a number of empirical studies. From this, we again can conclude that there is considerable genetic variation for life span in these species. Consonant with these observations, researchers have found that it is possible to increase life span through artificial selection in a number of model species (Figure 20.14).

Second, the rate-of-living hypothesis predicts a strong inverse correlation across species between metabolic rate and life span. The faster an organism's metabolism, the faster its physical structures should wear out and break down and the faster it should senesce. More specifically, intracellular damage due to oxidative stress should increase with metabolic rate. Thus, the rate-of-

TABLE 20.1

Heritability of Life Span

Species	Heritability (h^2)
Caenorhabditis elegans	0.34
Drosophila melanogaster	0.06–0.09
Mus musculus	0.29
Homo sapiens	0.17–0.35

Adapted from Austad and Finch (2008).

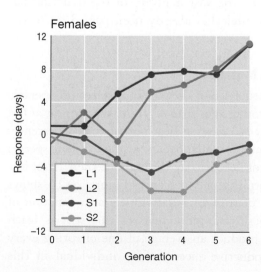
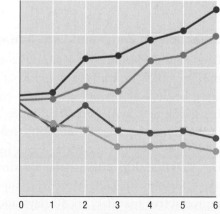

FIGURE 20.14 Artificial selection influences life span in *Drosophila*. Bas Zwaan and colleagues selected for short life span in two lines (S1 and S2) of *Drosophila melanogaster* and for long life span in two other lines (L1 and L2). The *y*-axis indicates the difference in the life span of each treatment group relative to the life span of unselected control lines. Life span changes significantly in the direction of selection in each case. This indicates the presence of substantial genetic variation associated with the rate of senescence. Adapted from Zwaan et al. (1995).

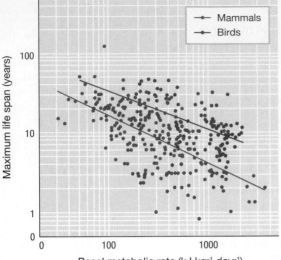

FIGURE 20.15 An inverse relationship between metabolic rate and life span. This plot of basal metabolic rate against maximum life span, on logarithmic axes, shows an inverse relationship between the two variables. But the broad scatter observed and the different curves for birds and mammals indicate that differences in metabolic rate only partially explain differences in life span. Adapted from Hulbert et al. (2007).

living hypothesis predicts that longevity and metabolic rate should be inversely correlated. Indeed, comparing across species, we do see a general trend in this direction (Figure 20.15). But a close look at the data leaves us with a number of reasons to be skeptical that the rate-of-living hypothesis provides a complete explanation for differences in senescence rates. These include the following: (1) longevity mutants—that is, mutants that live longer than do wild-type individuals of the same species—do not necessarily show reduced metabolic rate relative to wild-type individuals; (2) within a species, longevity and metabolic rate are not associated; (3) frequent exercise increases metabolic rate but does not decrease longevity; and (4) as illustrated in Figure 20.15, birds typically have much longer life spans than mammals of comparable basal metabolic rate (Hulbert et al. 2007).

Thus, the testable predictions of the rate-of-living model are not well supported by the available data. While metabolic rate may influence the rate at which oxidative stress and other forms of damage accumulate, the notion that damage is responsible for senescence provides us with only a proximate explanation. We have seen that there is substantial variation in the ability to withstand and repair such damage within and across species, and we need an evolutionary explanation for these observations.

An Evolutionary View of Senescence

For senescence, as for many other biological phenomena, J. B. S. Haldane was the first to provide a cogent evolutionary explanation (Haldane 1941). Haldane wanted to understand why Huntington's disease, a genetic disease caused by a mutation in a single gene, was so prevalent in the population—at a frequency of approximately 1 in 10,000 in the United States—despite its debilitating effects. If Huntington's disease is caused by a deleterious mutation, why isn't this mutation eliminated from the population through the action of natural selection? The answer, Haldane reasoned, has to do with the age at which Huntington's disease begins to manifest symptoms. Symptoms typically arise only once an individual is in his or her mid- to late-forties, *after* the vast majority of reproduction has already occurred, and after the disease allele has already been passed on to the next generation.

Selection on Early- and Late-Acting Mutations

In general, senescence is a simple consequence of the fact that selection operates more strongly on traits that appear at young ages than on traits that appear at old ages, because of *extrinsic mortality*—that is, causes of death other than senescence. Let's look at a hypothetical example. Suppose members of a small rodent species face a constant external mortality risk of 2% per day throughout their lives, but otherwise they do not suffer from mortality or senescence. Figure 20.16 shows the **survivorship curve**—the fraction of surviving individuals as a function of age—for this species. Moreover, suppose that individuals of this species reach reproductive maturity in 30 days and produce an average of one offspring every 2 weeks thereafter. The average reproductive success of an individual in this

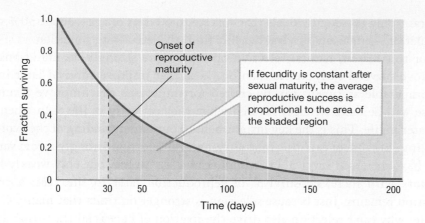

FIGURE 20.16 **A survivorship curve.** A survivorship curve indicates the fraction of individuals who survive to reach a given age. In this example, individuals reach maturity at 30 days and reproduce at a constant rate thereafter. The average reproductive success is proportional to the average reproductive life span—that is, the average number of days lived beyond 30. This quantity is indicated by the area under the shaded curve.

population is therefore proportional to the number of days beyond 30 that the individual lives. In this particular example, the average reproductive success is about 1.9 offspring produced.

Now compare two different mutations, each of which reduces mortality by 50% relative to the wild type, for a period of 30 days. An *early-acting* mutation reduces mortality by 50% from birth until reproductive maturity at 30 days, whereas a *late-acting* mutation reduces mortality by 50% from 60 days to 90 days. Which of these mutations confers a greater fitness benefit?

Because fecundity is unchanged by either mutation, we can answer the question by comparing the survivorship curves, illustrated in Figure 20.17. The average reproductive success is proportional to the shaded area in each case. As shown by the comparison figure, the shaded area—and thus the average reproductive success—is larger for individuals with the early-acting mutation than for individuals with the late-acting mutation. Individuals with the early-acting mutation have an average reproductive success of about 2.6 offspring, whereas individuals with the late-acting mutation have an average reproductive success of about 2.2 offspring.

Why do we see this difference, given that the benefit of the mutation lasts only 30 days in either case? Put simply, it is a matter of how likely an individual is to benefit from each type of mutation. The early-acting mutation takes effect immediately at birth, and thus each individual is certain to reap its benefits. By

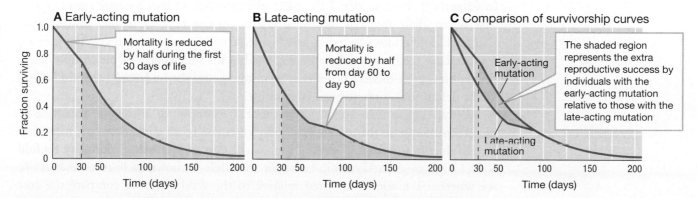

FIGURE 20.17 **Comparing the fitness benefit of an early-acting mutation with that of a late-acting mutation.** The difference in reproductive success due to (**A**) the early-acting mutation and (**B**) the late-acting mutation is given by (**C**) the difference in the shaded areas in each case.

contrast, only those individuals that survive to 60 days of age—about 30% of the population—enjoy any survival benefits from the late-acting mutation.

For this reason, natural selection operates more strongly on mutations that improve survival early in life than on those that improve survival later in life. A comparable argument can be crafted for mutations that improve fecundity: These will be more strongly favored when they act early in life than when they act later in life. This is the key insight behind our understanding of the evolution of aging. Because of extrinsic mortality, natural selection strongly favors variants that have increased survival or reproduction early in life, but only weakly favors variants with increased survival or reproduction later in life. Yet, a critical question remains: Just because selection is stronger on traits that manifest early in life, why can't selection also drive the fixation of beneficial traits that appear later in life?

The Mutation Accumulation Hypothesis

One answer is that, as we saw in Chapter 8, in a finite population natural selection is not effective at fixing or eliminating mutations that have very small fitness effects. The **mutation accumulation hypothesis** proposes that for late-life traits, selection is simply not strong enough to purge deleterious mutations (Medawar 1952). As a result, mutations that have deleterious effects later in life build up in the genome, whereas mutations with deleterious effects early in life are purged by natural selection. The consequence is that individuals who live long enough will be plagued by a suite of late-acting deleterious mutations; senescence is the consequence of these mutations' effects.

The Antagonistic Pleiotropy Hypothesis

Some mutations may have multiple effects at different points in the life cycle. Peter Medawar imagined what would happen if a single allele had beneficial effects early in life but deleterious effects later in life. He noted that such an allele could easily be favored by selection. Because natural selection acts more strongly on traits that manifest early in life, "a relatively small advantage conferred early in the life of an individual may outweigh a catastrophic disadvantage withheld until later" (Medawar 1952, p. 49). The evolutionary biologist George Williams called this explanation the antagonistic pleiotropy hypothesis (Williams 1957). In Chapter 3, we considered the basic phenomenon of antagonistic pleiotropy— that is, an allele that has beneficial effects on one trait or in one context may also have deleterious effects on another trait or in another context. Age-specific antagonistic pleiotropy might be responsible for senescence. A pleiotropic allele that confers even modest benefits at a young age might be favored despite having major deleterious consequences later in life.

We can again turn to our hypothetical example to illustrate antagonistic pleiotropy. Suppose that a new mutation gives rise to an allele that cuts mortality in half during the first 30 days of life, at the cost of increasing mortality tenfold after 120 days. A survivorship curve for the allele is shown in Figure 20.18A. To see whether the allele is favored relative to the wild type, we compare the area under the curve in Figure 20.18A to that in Figure 20.16. Figure 20.18B overlays the two curves to facilitate the comparison. The green-shaded area, where the survivorship curve for the new allele lies above the survivorship curve for the wild

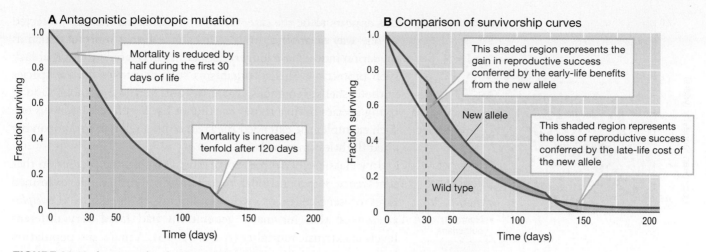

FIGURE 20.18 An example of antagonistic pleiotropy. A new allele has antagonistic pleiotropic effects, increasing survivorship early in life, but reducing survivorship later in life by a much greater fraction (**A**). Comparing the survivorship curves, we see that the early benefits of the new allele outweigh its later costs (**B**).

type, represents the increased reproductive success due to the early-life benefits of the new allele. The gold-shaded area, where the survivorship curve for the new allele lies below the survivorship curve for the wild type, represents the lost reproductive success due to late-life costs. Because the former area is much larger than the latter, the new allele has higher net reproductive success, and thus it will be favored by natural selection. In this particular example, individuals with the new allele have an average reproductive success of around 2.3 offspring, compared to 1.9 offspring for the wild type.

The antagonistic pleiotropy hypothesis states that if a number of new mutations with similar effects on fecundity or on survivorship were to be fixed in the population by natural selection, organisms would experience a large decay in function later in life because of the collective effects of these alleles. It is important to recognize, however, that the deleterious mutation accumulation hypothesis and the antagonistic pleiotropy hypothesis are not mutually exclusive. It is entirely possible, and indeed entirely likely, that populations will accumulate both late-acting deleterious alleles because of drift and antagonistic pleiotropic alleles because of selection.

Under evolutionary theories of aging, senescence should always be manifested as a generalized deterioration, rather than the result of deterioration of one single bodily system. Moreover, Williams notes that the accumulation of antagonistic pleiotropic mutations creates positive feedback in favor of further such mutations. Not only does extrinsic mortality reduce the strength of selection later in life, but also the late-life decline of survival and fecundity due to senescence further reduces any selection for late-acting benefits.

Senescence Should Be Proportional to Extrinsic Mortality

The mutation accumulation hypothesis and the antagonistic pleiotropy hypothesis agree in a number of their testable predictions. Both predict that the rate of senescence will be proportional to the rate of extrinsic mortality in a population. This is because as extrinsic mortality increases, expected life span decreases and older ages will contribute a smaller fraction of the total reproductive success. In general,

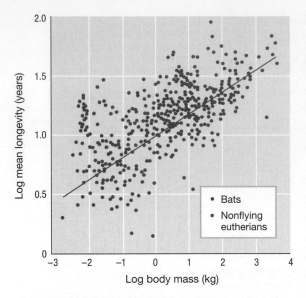

FIGURE 20.19 Bats senesce slower than do flightless mammals. Flightless mammals show a close relationship between body mass and longevity; this relation is approximately linear on a log–log plot. Bats lie well above the trendline for flightless mammals, indicating that they live much longer relative to their body size. Adapted from Austad and Fischer (1991).

this appears to be the case. Groups of organisms that are protected in one way or another from sources of external mortality such as predation indeed have longer life spans than related groups that lack such protection. Shelled organisms such as turtles outlive species without shells; venomous organisms outlive those without venom; flying species outlive those that cannot fly. Figure 20.19 shows one such example: the extended life span of bats relative to comparably sized flightless mammals.

In a classic experiment, Steve Austad tested the prediction that senescence is proportional to extrinsic mortality. He examined rates of senescence in two populations of opossums (*Didelphis virginiana*) that for many generations had faced very different levels of extrinsic mortality (Austad 1993). A mainland population had experienced high levels of predation by bird and mammal predators. By contrast, an isolated population on Sapelo Island, 5 miles off the coast of Georgia, had experienced much lower predation pressure because there were no large predators on the island. Austad reasoned that, if the evolutionary theories of aging are correct, selection should have favored slower rates of senescence in the island population than in the mainland population. By contrast, if the rate-of-living hypothesis was responsible for the phenomenon of senescence, both populations should senesce at comparable rates.

Austad attached radio collars to opossums in each population, in order to monitor their mortality rates. He also recaptured individuals intermittently to assess their fecundity and physiological condition. Austad's results provided strong evidence for the evolutionary explanation of senescence. Relative to individuals in the mainland population, island individuals had lower age-specific mortality (Figure 20.20) and lower rates of physiological decline as measured by the breaking time of the collagen fibers in the tail. Age-specific fertility also declined more quickly on the mainland than on the island, as we would expect under evolutionary explanations of senescence.

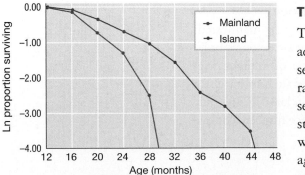

FIGURE 20.20 Senescence in island and mainland opossums. Opossums from the island population outlive those from the mainland. Mortality rate increases with age in both populations, as indicated by the increasingly steep slope of the survival curves with time. Adapted from Austad (1993).

The Disposable Soma Hypothesis

The strength of the antagonistic pleiotropy hypothesis is that, according to the hypothesis, senescence is a consequence of natural selection on correlated traits (early and late survival and fecundity) rather than simply being a side consequence of mutations on which selection is so weak that they cannot be eliminated. But with this strength comes a significant challenge. We need to be able to explain why we would expect to see antagonistic pleiotropy with respect to age, and this is not at all obvious. After all, most of the adaptations we have considered throughout this book—cryptic coloration, for example—should be beneficial irrespective of age. Williams' theory requires a large class of alleles that are beneficial early in life but harmful later in life. Why would such alleles exist?

The **disposable soma hypothesis** provides an answer to this puzzle. It suggests that these antagonistic pleiotropic relationships between beneficial effects early in life and deleterious effects later in life are the result of a fundamental trade-off. The disposable soma hypothesis was first framed narrowly, as a trade-off between growth, on one hand, and repair of the transcriptional and translational machinery within cells on the other (Orgel 1963; Kirkwood 1977). The basic idea is that, once organisms have evolved a distinction between germ-line cells and somatic cells, these two types of cells face different requirements. The transcriptional and translational machinery within the germ-line cells, which are passed from generation to generation, is selected to avoid degradation and decay. Otherwise, the genes carried in these germ-line cells would not be transmitted faithfully to future generations. But matters are different in somatic cells. At some point, energy that could be invested in ensuring transcriptional and translational fidelity might better be invested in promoting rapid growth, even at the cost of such fidelity. While the structure and function of the genetic machinery within somatic cells would then degrade over an organism's lifetime, the whole process could be reset in the next generation as new and perfectly intact somatic cells would be produced from the more carefully preserved germ-line cells.

Today the disposable soma hypothesis is typically interpreted in a more general manner. It is seen as pertaining to any trade-off between investment in *reproduction* and investment in *repair*. Any allocation of resources toward immediate reproductive benefit and away from repair and regeneration is an allocation that privileges the germ line while relinquishing the soma to the ravages of entropy. In other words, the disposable soma hypothesis focuses on the trade-offs between early fecundity and later survival. Why not preserve the soma as well as the germ line? We have already seen the answer that evolutionary theories provide: The presence of extrinsic mortality means that sooner or later any given soma's luck will run out. Given this, selection will not tend to favor the investments in repair that would confer indefinite survival of the entire soma, as there is no selective benefit to investing in repairs that the organism will not live long enough to need (Williams 1957; Kirkwood and Austad 2000).

Aging in Bacteria

In an ingenious experiment, Martin Ackermann and his colleagues showed that the trade-off between repair and reproduction occurs even in organisms that lack a germ–soma distinction (Ackermann et al. 2003). They reasoned that, if the evolutionary perspective on aging is correct, even bacteria should senesce, provided that they divide in a way that clearly differentiates a new daughter cell from an older mother cell (Partridge and Barton 1993). We usually envision bacteria as dividing symmetrically into two similar daughter cells, rather than as dividing into an older mother and a younger daughter cell. But Ackermann and his colleagues found a bacterium with an unusual life cycle: *Caulobacter crescentus* divides asymmetrically with a clear mother–daughter distinction. The bacterium begins life as a free-swimming *swarmer cell* propelled by a flagellum, but later it matures to become a *stalked cell*, anchoring itself to a surface using an attachment known as a *holdfast* (Figure 20.21). After attaching, stalked cells never return to the swarmer state; instead, they undergo repeated cell divisions, with the newly formed cell taking on the swarmer state while the original cell retains its holdfast.

FIGURE 20.21 Stalked cells of *Caulobacter crescentus* cells produce swarmer cell daughters. Here we see a swarmer cell with its flagellum about to bud off from a stalked cell, which is attached to the substrate by a holdfast.

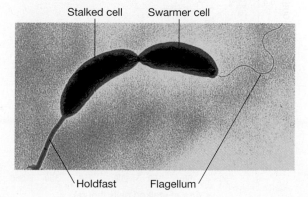

Stalked cell   Swarmer cell

Holdfast   Flagellum

FIGURE 20.22 Measuring reproductive rate in *C. crescentus*. By filming stalked cells dividing in a flow chamber, Ackermann and his colleagues could measure the rate of reproduction as a function of each cell's age—that is, time since forming a stalk.

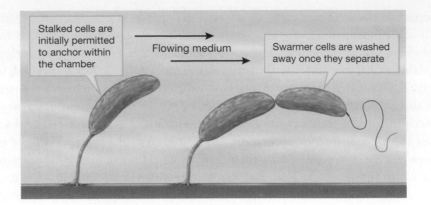

Stalked cells are initially permitted to anchor within the chamber

Flowing medium

Swarmer cells are washed away once they separate

An evolutionary perspective on aging would predict that the stalked cells should exhibit senescence. Because stalked cells face extrinsic mortality and do not revert to the swarmer form, evolutionary logic suggests that they should at some point stop investing in repair and instead invest in extra production of swarmer cell offspring. The expected consequence is senescence: Stalked cells should experience higher mortality and/or lower rates of reproduction as they age.

To test this prediction, Ackermann and his colleagues devised a way to measure the reproductive rate of stalked cells over time. They allowed cells to anchor themselves in a chamber, with liquid medium flowing past, and filmed the cells over a couple of weeks under a microscope. Because of the flowing medium, newly formed swarmer cells were washed away, and they did not clutter the chamber, allowing the researchers to note the times at which each stalked cell underwent cell division (Figure 20.22).

Ackermann filmed the bacteria over a period of more than 300 hours subsequent to anchoring, and he measured the rate at which stalked cells produced new swarmer offspring. His results provided strong evidence of senescence: Reproductive output declined substantially over the course of the experiment, and the rate of decline accelerated for older cells (Figure 20.23). To be certain that this decline was the result of senescence rather than a consequence of changing experimental conditions, he also measured the rate of reproductive output of swarmer cells produced at 250 hours into the experiment. These cells were rejuvenated: They reproduced at the rates observed at the start of the experiment, not at the reduced rates found in aged mother cells.

FIGURE 20.23 The asymmetrically dividing bacterium *Caulobacter crescentus* undergoes senescence. Age-specific reproductive output, shown here for three replicate experiments, incorporates both the probability of survival and the rate of cell division. Adapted from Ackermann et al. (2003).

With this study, Ackermann and his colleagues showed that senescence can occur in bacteria, at least when they divide asymmetrically, as does *C. crescentus*. Subsequent work indicates that this phenomenon may be much more general. Even bacteria that appear to divide symmetrically, such as *E. coli*, may actually distribute new and old components of the cytoplasm to two different cell poles during division, thereby producing an aged "parent" and a younger "daughter." Indeed, older bacteria with older cytoplasmic components appear to have decreased rates of replication (Lindner et al. 2008). Again the trade-off between repair and reproduction arises; rather than repairing old cytoplasmic components, these structures can be segregated together into a senescing parent while new components are synthesized for a rejuvenated offspring (Ackermann et al. 2007).

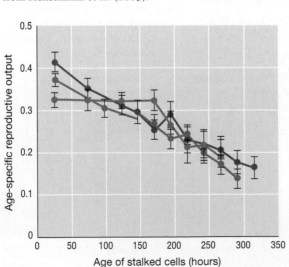

These studies have substantially advanced our understanding of what it means to grow old. The trade-off between reproduction and

repair has often been framed as a consequence of the distinction between the germ line, which is potentially eternal, and the somatic cells, which are disposed of in each generation. But this new work has demonstrated compellingly that a reproduction–repair trade-off can occur even without the germ–soma distinction, and thus even unicellular organisms such as bacteria can senesce. Ultimately, evolutionary hypotheses for senescence posit that aging occurs because—like inexpensive consumer electronics—our bodies, cells, and intracellular components are cheaper to replace than to repair.

In this chapter, we have seen a number of ways in which the principles of evolutionary biology can contribute to our understanding of medicine. By necessity, we have merely scratched the surface of this vibrant and rapidly emerging area of research. Numerous other examples and applications are being explored, and doubtless many more will be discovered as this research area continues to expand.

Having finished our discussion of evolution and medicine, we have reached the end of this volume. Thousands of observations and experiments, as well as mathematical and conceptual models, have demonstrated that the theory of evolution—descent with modification—explains the diversity of life on our planet, both present and past. No other scientific theory comes remotely close to explaining so much of what we know about the diversity of life. And so, more than 150 years after Charles Darwin wrote *On the Origin of Species,* we can think of no more appropriate way to draw to a close than in the same spirit that Darwin concluded his revolutionary book: contemplating the grandeur of a universe in which "from so simple a beginning endless forms most beautiful and most wonderful have been, and are being, evolved."

SUMMARY

1. In biology, the question "Why do we see a particular trait or phenomenon?" can be answered at multiple levels. Proximate explanations specify immediate mechanisms, developmental explanations specify changes that occur during an individual's lifetime, evolutionary explanations specify how selection and other evolutionary processes have shaped a trait, and phylogenetic mechanisms specify when and where in the history of life the trait arose.

2. An evolutionary perspective on human disease does not ask why disease is evolutionarily advantageous, but rather it asks why evolution has left the body vulnerable to disease.

3. Nesse and Williams distinguished six different evolutionary explanations for vulnerability to disease: (a) coevolutionary arms races, (b) not enough time for selection to catch up with environmental conditions, (c) trade-offs, (d) historical contingency and path dependence, (e) selection favoring reproductive success at the expense of health and well-being, (f) some symptoms possibly being defenses rather than pathologies. These explanations need not be mutually exclusive.

4. Fever appears to reduce the duration and severity of microbial infection, but usually it can be treated without major negative consequences.

5. The smoke detector principle suggests that defenses will tend to be overly sensitive because the cost of a false alarm is much less than the cost of failing to respond to a true threat.

6. Immune systems help hosts cope with pathogens that typically evolve far more rapidly, but pathogens evolve ways of subverting immune responses.

7. Selection due to immune responses can have a major impact on the phylogenetic structure of viral clades.

8. Evolution is unable to plan ahead for future contingencies; as a result, organisms may be susceptible to problems that could have been avoided by structuring the anatomy in a different way. Human susceptibility to choking provides an example.

9. Organisms senesce because natural selection is strong on traits that are manifest early in life but weak on traits that appear later in life. The mutation accumulation hypothesis suggests that drift leads to an accumulation of alleles with deleterious effects later in life. The antagonistic pleiotropy hypothesis suggests that alleles with beneficial effects early in life but deleterious effects later in life will be favored by selection and therefore accumulate in genomes. The disposable soma hypothesis focuses on a trade-off between investment in reproduction and repair.

KEY TERMS

affinity maturation (p. 659)

cactus-shaped phylogeny (p. 661)

clonal expansion (p. 659)

disposable soma hypothesis (p. 672)

escape variants (p. 661)

mutation accumulation hypothesis (p. 670)

rate-of-living hypothesis (p. 667)

senescence (p. 665)

survivorship curve (p. 668)

REVIEW QUESTIONS

1. In a 1994 paper published in the *Annals of Internal Medicine*, Philip Mackowiak attempted to explain fever (the febrile response) as follows:

> If one considers the consequence of the febrile response and its mediators only from the point of view of the host, there can be no reconciliation between its reported capacity for benefit at certain times and harm at others. However, if one views the febrile response from the perspective of the species, its salutary effects on mild to moderately severe infections and its pernicious influence on fulminating infections become less paradoxic—that is, if one accepts preservation of the species rather than survival of the individual as the essence of evolution. An evolutionary process driven by such a principle might lead to sacrifice of the individual if it poses a threat to the species. In this context, the febrile response and its mediators might have evolved both as a mechanism for accelerating the recovery of infected individuals with localized or mild to moderately severe systemic infections and for hastening the demise of hopelessly infected individuals, who pose a threat of epidemic disease to the species. (Mackowiak 1994, p. 1039)

In other words, Mackowiak proposed that fever may be harmful to individuals who manifest it, but beneficial to the species in that it prevents transmissible pathogens from spreading rapidly through the population. Based on your understanding of evolution by natural selection, critique this explanation.

2. In his classic 1957 paper on the evolution of senescence, George Williams noted that some species, such as carp, increase continually in size and also

in fecundity over the course of their lifetimes (Williams 1957). He predicted that such species would not senesce as fast as species that do not increase in fecundity beyond reproductive maturity. Explain the reasoning behind this prediction.

3. Anxiety appears to be an evolved mechanism to help us avoid or escape from dangerous situations. Apply the smoke detector principle to explain why—even if levels of anxiety have been optimized by natural selection—many actual instances of anxiety that people experience would be unnecessary. That is, explain why many or most episodes of anxiety may not be associated with real danger and can safely be treated with antianxiety medications.

4. We used the smoke detector principle to explain why many defenses may be overly sensitive in the sense that they are often triggered in the absence of threat. Use the same logic to explain why many defenses may be larger in magnitude than is usually needed. Why, for example, does the average T-cell response lead to the production of far more pathogen-specific T-cells than are needed to clear the average infection?

5. We have seen how the vulnerability of humans to choking is the consequence of path dependence in evolution. When the structures that would someday become lungs first evolved, evolution lacked the foresight to head off a future choking risk once lungs became the sole way of breathing. Can you think of another example of how path dependence and evolution's lack of foresight have left the human body vulnerable to disease, injury, or malfunction?

6. The figure to the right (adapted from Adams et al., 2006) shows a phylogeny of dengue virus serotype 3. Based simply on the structure of this phylogeny, would you guess that infection by the dengue virus confers long-term immunity? Explain.

7. Feedback control allows a response to be finely regulated by signals or cues from the environment. Explain why feedback control might be a dangerous way to regulate immune responses when pathogens are present.

8. Two mutations arise in a population of mice. The first increases fecundity by 20% during the first month after sexual maturity (but not beyond), whereas the second increases fecundity by 20% during the third month after sexual maturity (but not beyond). Which mutation will be more strongly favored by natural selection? Explain.

9. Propose (a) proximate and (b) evolutionary explanations for why we sneeze.

10. Critique the following claim: "All vertebrate organisms senesce, therefore senescence must be an adaptation for something—we just need to figure out what it is an adaptation for."

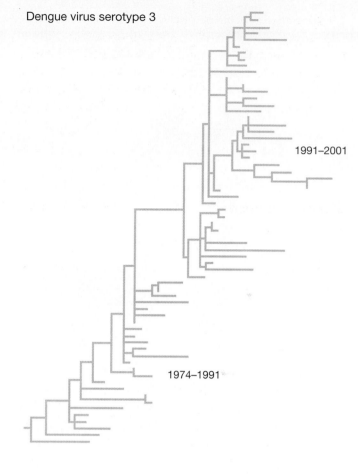

Dengue virus serotype 3

1991–2001

1974–1991

SUGGESTED READINGS

Gluckman, P., and N. Hansen. 2006. *Mismatch: Why Our World No Longer Fits Our Bodies.* Oxford University Press, Oxford. An exploration of how recent rapid changes in our environment leave our bodies mismatched with our circumstances of life, with a particular focus on early development.

Kirkwood, T. B. L., and S. N. Austad. 2000. Why do we age? *Nature* 408: 233–238. An engaging review of the evolutionary biology of aging.

Nesse, R. M., and G. C. Williams. 1994. *Why We Get Sick: The New Science of Darwinian Medicine.* Times Books, New York. The original book that spurred current interest in the relevance of evolutionary biology to medicine.

Shubin, N. 2009. *Your Inner Fish: A Journey into the 3.5-Billion-Year History of the Human Body.* Vintage, New York. In this gripping book, paleontologist Neil Shubin explains how our bodies reflect the legacy of our evolutionary history.

Sompayrac, L. 2008. *How the Immune System Works,* 3rd Ed. Wiley-Blackwell, Malden, MA. This short primer on immunology explains the fundamental logic behind the operation of the immune system without getting bogged down in the intricate details of immune function.

Visit StudySpace at wwnorton.com/studyspace.

APPENDIX

Likelihood Methods and Bayesian Methods for Phylogenetic Inference

Neither parsimony methods nor distance methods for phylogenetic inference consider an explicit probability model of how evolutionary changes occur. Maximum likelihood methods and Bayesian methods each aim to remedy this deficit. Each attempts to infer which tree or trees best explain observed character data, but likelihood and Bayesian methods differ in their interpretation of what "best explains" should mean.

A.1 Models of Sequence Change

Maximum likelihood methods and Bayesian methods both employ models of evolution change. For the purposes of statistical inference, a model of evolutionary change provides us with a way of computing the probability that a particular character changes from state 1 to state 2 along a given branch of a phylogenetic tree (or alternatively, that no change occurs along this branch). There are any number of models that we can use for this purpose; here we look briefly at a few of the models that are commonly used when building trees based on DNA sequence data.

The Jukes–Cantor Model

Here we will consider only *homogeneous* models, in which the probability of character change does not change over time. The simplest such model of DNA sequence change is known as the *Jukes–Cantor model*; in this model, any of the four bases (A, C, G, T) is equally likely to change into any of the other bases at any given time. Notice that what we are modeling here is not the occurrence of mutation alone, but rather *substitutional change*, a change in which a mutation arises and goes near or all the way to fixation in the population. Changes are assumed to occur randomly and independently, generating what is called a *Poisson distribution* of substitution events along the branches of the tree.

The Kimura Two-Parameter Model

In practice, a given base may not be equally likely to change into any other base. Rather, each of the *purines* A and G is typically more likely to change into the other purine, and each *pyrimidine* C and T is more likely to change into the other pyrimidine. These purine-to-purine and pyrimidine-to-pyrimidine changes are known as *transitions*.

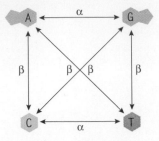

FIGURE A.1 The Kimura two-parameter model. This model of substitution events allows transitions and transversions to occur at different rates. Adapted from Elias and Lagergren (2007).

Switches from purine to pyrimidine and vice versa are known as *transversions*. The *Kimura two-parameter model*, shown in Figure A.1, captures this difference in substitution rate using two parameters: α for the transition rate, and β for the transversion rate.

Making the transition/transversion distinction can be particularly useful if some of the branches on our phylogenetic tree are so long that transitions have largely saturated—that is, some branches are long enough that there is a high probability that at least one transition has occurred at each site. Using the Kimura two-parameter model in concert with likelihood methods, we can put more weight on the rarer, and thus more informative, transversion events.

The Tamura–Nei Model

Both the Jukes–Cantor and Kimura two-parameter models assume that all four bases are equally common, occurring at a frequency of 0.25 each. This assumption may not be appropriate, because many species have base pair frequencies that differ substantially from an even distribution. For such species, we may want to elaborate on the Kimura model further. The *Tamura–Nei* model offers one way of doing this. Like the Kimura two-parameter model, the Tamura–Nei model allows transitions to occur at some rate α and transversions to occur at some rate β. But when a transversion occurs, say, from the purine A, the new base does not become either of the two pyrimidines, C or T, with equal probability. Rather, we imagine drawing the new base from a pool of pyrimidines in which the bases are represented in proportion to their frequencies in the genome. Dealing with transitions is only slightly more complicated. We again envision drawing the new base—possibly the same as the old base—from a pool in which the bases are represented in proportion to their frequencies in the genome. For example, instead of the purine A always switching to the other purine T, a transition from A goes to A (no change) or to T (change), with probabilities proportional to the frequencies of A and T in the genome, respectively. Because every new base that is added is chosen by drawing it from a pool that reflects the base composition in the germ line at large, under this model the average base frequencies do not change along the tree. The Tamura–Nei algorithm thus generates a substitution process that preserves the genomic base composition, which may be more suitable for species with significantly biased base pair compositions.

A.2 Maximum Likelihood

Once we have selected a model of character change, the next step in generating a maximum likelihood tree is to construct the so-called *likelihood function*. This function, which gives us a numerical likelihood score as a function of the character data and the tree we want to evaluate, is the quantity that we aim to maximize in maximum likelihood phylogenetic reconstruction. That is, we look for the tree with the highest likelihood score, in which the likelihood L is defined as follows:

$$L = \text{Probability}(\text{Observed Data} \mid \text{Tree})$$

Here the term "(Observed Data | Tree)" is read "observed data *given* a tree," and probabilities of this sort are referred to as a *conditional probabilities*.

To construct the likelihood function, we typically assume:

1. Each character evolves independently of each other character.

2. Evolution proceeds independently along each branch of the phylogenetic tree.

The first assumption effectively allows us to work with one character at a time. If X and Y are independent events, the probability of X and Y occurring [denoted "Prob(X and Y)"] is equal to the probability of X occurring multiplied by the probability of Y occurring [Prob(X) Prob(Y)]. So if we have characters $1, 2, 3, 4, \ldots$, all the way up to some large number of characters that we will denote as k, and the observed character data are labeled $D_1, D_2, \ldots, D_k$, the probability of observing the full suite of data that we did for all characters is:

$$L = \text{Prob(Observing } D_1 \text{ and } D_2 \text{ and so on to} \ldots D_k | \text{Tree})$$

We can go through and compute the probabilities of observing the data that we saw for each character, one by one. Next we multiply these together. This multiplication is denoted with the symbol Π. The full likelihood score is then:

$$L = \prod_i \text{Prob}(D_i | \text{Tree})$$

But we still have to figure out how to calculate the probability of observing the data that we saw for a particular character. The problem is that there are a lot of different evolutionary histories that could have led to the same combination states at the branch tips. We need to account for *all* of them. To deal with this problem, we apply another basic law of probability. If X and Y are mutually exclusive events—that is, X and Y cannot both be true—then the probability that either X or Y is occurring [Prob (X or Y)] is equal to Prob(X) + Prob(Y). This is the case when considering two mutually exclusive events (X and Y), but we can generalize to any number of events. The probability of observing one of k mutually exclusive events is equal to the sum of the probabilities of each individual event. So we compute the probability of observing a given set of character states on a given tree as the sum of the probabilities of all of the different ways that those character states could have come about. In our case, these "different ways" are the various different combinations of internal node states that led to the observed characters on the tips. Figure A.2 illustrates the different combinations of internal node states for a single DNA site on a tree with three branch tips.

So now we can write:

$$L = \prod_i \text{Prob}(D_i | \text{Tree})$$
$$= \prod_i \text{Prob} \sum_{\text{all ways } j} \text{Prob}(D_j | \text{Tree})$$

where $\sum$ is the symbol for summing up over a certain number of cases: in our examples, summing up the probabilities of all of the different ways that those character states could come about.

Finally, we need to be able to compute the probability that one particular evolutionary history along our tree is realized, namely Prob(D_j | Tree). Now we turn to our chosen model of evolution. Using that model, we can work out the probabilities that a given character change occurs along a given branch of our tree. Here we use our second independence assumption. Because we have assumed that evolution occurs

FIGURE A.2 Possible evolutionary histories of a rooted tree with three species. A rooted tree with three species has two internal nodes, and therefore, if the character states are the four DNA bases, there are $4^2 = 16$ possible histories. All 16 are shown, with the two that would typically contribute the most to the likelihood highlighted in blue. These involve fewer changes than the other trees.

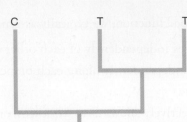

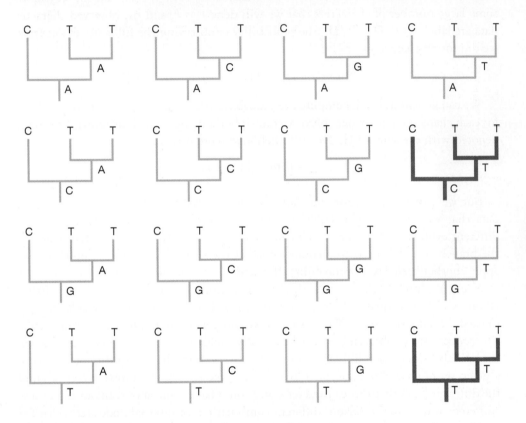

independently along each branch of the tree, the probability of all of the changes and nonchanges that we observe is equal to the product of each change or nonchange along each branch:

$$L = \prod_i \text{Prob} \sum_{\text{all ways } j} \text{Prob}(D_j | \text{Tree})$$

$$= \prod_i \text{Prob} \sum_{\text{all ways } j} \prod_{m \text{ in tree } Dj} \text{Prob}(\text{branch } m)$$

The trouble is that, even for very small trees, there are a very large number of internal nodes and thus a very large number of possible histories that could lead to the observed data. In computing the likelihood of the observed data given the tree, we have to sum over all of these possible histories as specified by the equation above. Figure A.2 shows the possible histories associated with a rooted tree of only three species for one particular DNA site. This tree has two interior nodes, and for each internal node, there are four possible bases, giving a total of 16 different histories that we have to sum across. Some of these histories may be quite unlikely; in this example, it is unlikely that both interior nodes have an A in these sites. Rather, most of the probability will be associated with a few outcomes; two of the most likely outcomes for Figure A.2 are indicated in blue. But we need to sum over *all* of the possibilities.

For larger trees, the number of possible histories will quickly become astronomical. A tree with k species will have $k - 1$ internal nodes; thus, if we are looking at DNA sequence data, for example, we are faced with the challenge of summing over $(k - 1)^4$ unique histories in order to compute the likelihood of a single tree. Fortunately, this problem is more manageable than it might seem; there are fast algorithms for computing these likelihood scores over all possible sets of internal characters.

We conclude our discussion of likelihood methods by noting that under some circumstances, there is a close relationship between maximum parsimony inference and maximum likelihood inference. As branch lengths become short and rates of character change become low, the possibility of multiple character changes along an individual branch becomes vanishingly small. In this special case, the most *likely* phylogeny will be the one that requires the fewest single character changes. That phylogeny is, of course, precisely the one selected by maximum parsimony.

A.3 Bayesian Methods

Maximum likelihood methods look for the tree from which we have the highest probability of generating the data that we actually observed; Bayesian methods look at the closely related question of which tree is *a posteriori* most probable given that we observed these data. Bayesian methods are similar in many ways to likelihood methods. Like maximum likelihood methods, they work with explicit models of character change, such as the Jukes–Cantor, Kimura, or Tamura–Nei models, in order to compute probabilities of various evolutionary histories leading up to the observed character data. The salient conceptual difference is that, in using Bayesian methods, we actually assign a true probability—rather than a likelihood score—to each possible tree that we are trying to evaluate. To do so, we need to begin with a *prior distribution:* a set of beliefs or assumptions about the probabilities of the various tree topologies that are possible. That is, we need to have an advance estimate for Prob(Tree) for each possible tree. Different Bayesian approaches have different ways of assigning these priors, but the hope is generally that, as we add more data, the influence of the prior *distribution* on the results of our analysis will be minimal. The Bayesian approach then assigns what is called a *posterior probability* to each possible tree, using Bayes' rule:

$$\text{Prob}(\text{Tree}|D) = \frac{\text{Prob}(D|\text{Tree})\,\text{Prob}(\text{Tree})}{\text{Prob}(D)}$$

The problem with Bayesian phylogenetic methods is that we are often unable to write down explicitly each of the necessary terms. Fortunately, we can sample from these distributions without being able to compute them explicitly using a computational approach known as Markov chain Monte Carlo (MCMC). This allows us to make good estimates of Prob(Tree$|D$) for any tree we wish to consider. Because these probabilities are true probabilities, they are additive. In principle, Bayesian methods allow us to do things such as sum the probabilities for all trees that include a given monophyletic clade; this summed probability then represents a probability estimate that the clade in question is indeed monophyletic. This greater flexibility to assign probability values to hypotheses is one of the major attractions of the Bayesian approach. To understand maximum likelihood and Bayesian methods better, let us work through an example.

A.4 A Conceptual Comparison of Likelihood and Bayesian Approaches

In the course of doing science, we often want to infer something about the world—say, the value of a parameter—from data that we have collected. For example, suppose an urn has been filled with 100 balls, some red and some black. We might want to infer the true number r^* of red balls in the urn, based on sampling 10 balls from that urn. Suppose that our sampling occurs *with replacement*—that is, we return each ball to the urn after it is drawn, so that successive samples do not change the frequencies of the balls in the urn. Suppose that, of our ten draws, we get six red balls and four black ones. Then what should be our "best guess" r about the true number r^* of red balls in the urn (Figure A.3)?

One good answer is that our best guess of the value of r is the *most likely* value, *given* our observation of drawing six red balls. So our guess for the total number r of red balls in the urn might be the one with the highest probability, given our observations.

In mathematical terms, we are looking for the value of r that maximizes the *conditional probability* that there are a total of r red balls in the urn, given that we observed six red balls in our ten draws. In the notation of probability, we say we are looking for the value of r that maximizes

FIGURE A.3 Sampling from an urn. The process of sampling 6 red and 4 black balls from an urn of 100 balls is shown in the diagram.

Prob(r red balls in urn | 6 of 10 draws are red)

More generally, the idea is that if we are trying to infer any parameter r, given our observations of data D, our best guess might be the value of r that maximizes Prob($r|D$). Yet, more generally, if we are trying to choose among alternative models M_0, M_1, M_2, M_3, etc., given that we observed data D, we should choose M_i that maximizes Prob($M_i|D$).

So far, this all seems straightforward: Our best guess of a parameter is the value of that parameter that is most likely, given the data we observed. But there is a problem. In our example, we don't actually have enough information to compute the value Prob($r|D$), and so it seems as though we are stuck. Maximum likelihood and Bayesian inference offer two different ways out of this dilemma. We consider these in turn.

The maximum likelihood approach is that while we can't compute the quantity we want, we can make do with a related quantity that we are able to compute. In our example, we cannot compute the value of Prob($r|D$)—the probability that there are r red balls in the urn given the data D that we observed—but we can compute the reverse, the value of Prob($D|r$)—the probability of observing the data that we did, given that there are r balls in the urn. So, the trick behind likelihood is that for our purposes, it may be good enough to maximize the term that we do know how to compute: Prob($D|r$). We call this value a *likelihood* in order to distinguish it from a probability. What is the difference? A likelihood maps from known outcomes to unknown parameters; a probability maps from known parameters to unknown outcomes. Moreover, unlike a probability density function, which must sum to one—after all, one of the possible outcomes will happen—a likelihood function need not sum to one.

We write the likelihood of a parameter value r, given data D, as $L(r|D)$, and it is defined as:

$$L(r\,|\,D) = \mathrm{Prob}(D\,|\,r)$$

In our urn case, we can work out the values of the likelihood function. Sampling n times with replacement from an urn with a fraction r red balls yields a sample with a binomial distribution of red balls. The probability of observing k red balls in this sample is:

$$\binom{n}{k} r^k (1-r)^{n-k}$$

where:

$$\binom{n}{k} = \frac{n!}{(n-k)!\,k!}$$

In our urn case, $n = 10$ and $k = 6$. So, we are interested in finding the value of r that maximizes

$$L(r\,|\,n = 10, k = 6) = \binom{10}{6} r^6 (1-r)^4$$

The *likelihood function*—that is, the value of L as a function of r—is in this case given in Figure A.4.

This likelihood function has its maximum at $r = 60$—that is, when there are 60 red balls in the original urn. As a result, we conclude that our best guess—the *maximum likelihood estimate*—is that there were 60 red balls in the urn. Maximum likelihood, then, is the inference procedure of selecting our estimate to maximize the likelihood of the parameter values, given the data that we have observed. Next, we consider an alternative approach, Bayesian inference.

In *Bayesian inference*, our problem in deciding on a best estimate r, given our data, is that we don't know how to directly compute the value of $\mathrm{Prob}(r\,|\,D)$. Likelihood deals with this problem by flipping the $\mathrm{Prob}(r\,|\,D)$ term and by working with $\mathrm{Prob}(D\,|\,r)$— which we can compute directly—instead. Bayesian inference takes a different approach, using *Bayes' rule* to rewrite $\mathrm{Prob}(r\,|\,D)$ as follows:

$$\mathrm{Prob}(r\,|\,D) = \frac{\mathrm{Prob}(D\,|\,r)\,\mathrm{Prob}(r)}{\mathrm{Prob}(D)}$$

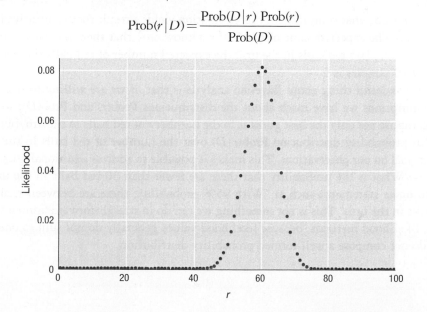

FIGURE A.4 A likelihood function. Here we illustrate the likelihood function $L(r\,|\,n = 10, k = 6)$ for our urn example.

With Bayes' rule, we can replace the problematic $\mathrm{Prob}(r|D)$ with the manageable $\mathrm{Prob}(D|r)$—but again there's a catch. The catch is that now we have to specify what the probabilities $\mathrm{Prob}(r)$ and $\mathrm{Prob}(D)$ are. The quantity $\mathrm{Prob}(r)$ is a probability distribution over possible values of our parameter r; this is called the *prior distribution* on the value of the parameter r because it reflects what we believe about the possible values of r *prior* to our observations. Similarly, the quantity $\mathrm{Prob}(D)$ is called the prior distribution in the observations D, because it reflects what we believe about the probability of observing the specific set of data that we did, prior to actually making that observation. Where do we get the values $\mathrm{Prob}(r)$ and $\mathrm{Prob}(D)$? This is the hard part of Bayesian inference; typically these values are chosen to reflect our previous knowledge of the system we are studying and/or to minimally influence the outcome of our calculations.

Once we have selected values for $\mathrm{Prob}(r)$ and $\mathrm{Prob}(D)$, however, we can go forward with probability calculations about $\mathrm{Prob}(r|D)$. For example, suppose that we choose a *flat prior* of r—that is, each value of $r = 0, 1, \ldots, 100$ is equally likely. Then $\mathrm{Prob}(r) = 1/101$, because there are 101 possible values of r from 0 to 100. The value of $\mathrm{Prob}(D)$ is slightly more difficult to calculate, but we recall that when there are r red balls in the urn, the probability of observing k red balls out of 10 draws is given by the binomial probability:

$$\binom{10}{k} r^k (1-r)^{10-k}$$

Here, r could be anything from 0 to 100, with equal probability.

To find the chance that we observe k red balls out of 10 draws, given that we don't know what the value of r is, we simply sum over all possible values of r times the probability of observing that value (which is $1/101$).

$$\mathrm{Prob}\{k \text{ red balls observed}\} = \frac{1}{101} \sum_{r=0}^{100} \binom{10}{k} r^k (1-r)^{10-k}$$

From this, we can now compute more or less whatever we want. For example, the expected number of red balls that we observe in our 10 draws is given by:

$$\sum_{k=0}^{10} k * \mathrm{Prob}\{k \text{ red balls observed}\}$$

Although somewhat complicated, this calculation yields a result that is intuitive in its symmetry: The expected value of k is 5. If we know only that there are equally likely to be $0, 1, \ldots, 100$ red balls in the urn, the expected number of red balls that we will draw in 10 samples is 5.

The wonderful thing about Bayesian analysis is that, if we are willing to live with the assumptions we have made about the distributions $\mathrm{Prob}(r)$ and $\mathrm{Prob}(D)$, we can now compute not only the best guess as to the number r of red balls in the urn, but also the full probability distribution $\mathrm{Prob}(r|D)$ over the number of red balls in the urn, conditional on our observations. This makes it possible to address additional questions such as "What is the probability that there are fewer than 30 red balls in the urn?," and to make statements such as "With 95% probability, there are between r_1 and r_2 red balls in the urn." This is not something we can do in straightforward fashion when using likelihood methods, because likelihood values generally do not sum to one and thus do not compose a well-formed probability distribution.

GLOSSARY

adaptation A trait that increases an organism's fitness and that is the result of the process of natural selection for its present function.

adaptive landscape A heuristic representation of fitness as a function of genotype or phenotype. Adaptive landscapes are commonly used by biologists to envision the course of evolutionary change. Also known as fitness landscapes.

additive genetic effects Genetic contributions to phenotype for a polygenic trait, in which the effects of each allele simply sum together to determine phenotype.

affinity maturation A selective process by which immune receptors develop an improved match to a pathogen during clonal expansion.

alleles Gene variants—that is, alternate forms of the same gene.

allopatric speciation Speciation that occurs when incipient species are geographically isolated from one another.

altruism An action that has the immediate consequence of reducing an individual's own fitness while increasing the fitness of another.

amino acids Specified by nucleotide triplets, these molecules are the building blocks of proteins.

anagenesis Gradual modification of form over evolutionary time, without branching speciation. *See also* cladogenesis.

analogous trait A trait that is similar in two different species or taxa, not because of common descent, but rather as a result of natural selection operating in similar ways along separate evolutionary lineages.

anisogamy A reproductive system in which at least two different kinds (sizes) of gametes—such as eggs and sperm—are produced. *See also* isogamy.

antagonistic coevolution An evolutionary relationship in which evolutionary changes in each species decrease the fitness of the other species.

antagonistic pleiotropy A phenomenon in which a single gene has multiple phenotypic consequences with opposing effects on fitness. *See also* pleiotropic genes.

antibiotic resistance The ability of microbes to survive and reproduce in the presence of antibiotics.

apomixis A form of asexual reproduction in which an unfertilized gamete undergoes a single mitosis-like cell division, producing two daughter cells that have an unreduced number of chromosomes and that are genetically identical to those of the mother.

aposematic coloration Warning coloration that functions to alert predators that a potential prey item is venomous or unpalatable.

artificial selection The process of human-directed selective breeding aimed at producing a desired set of traits in the selected species.

asexual reproduction The production of offspring from unfertilized gametes.

assortative mating A mating pattern in which individuals with similar phenotypes or genotypes mate with one another.

automixis A form of asexual reproduction in which haploid gametes are produced via meiosis. Subsequently, diploidy is restored when sister chromatids fuse, homologous chromatids from the same meiosis fuse, or a haploid genome is replicated.

background extinction A standard or "baseline" process of extinction occurring outside a period of mass extinction.

background selection A process by which neutral or beneficial alleles are lost because they are physically linked to nearby deleterious alleles. Background selection decreases genetic variation relative to what would be expected in a neutral model.

balanced polymorphism A stable equilibrium in which more than one allele is present at a locus.

balancing selection Natural selection that leads to an intermediate phenotype or to a stable equilibrium in which more than one allele is present. *See also* balanced polymorphism.

Batesian mimicry Mimicry in which a palatable species resembles an unpalatable species. *See also* Müllerian mimicry.

Bayesian inference A statistical approach often used to model evolutionary processes. Bayesian inference selects as "best" the tree that is most probable given both the observed data and some prior assumptions about possible trees.

biological species concept An approach to determining species boundaries in which a species is composed of actually or potentially interbreeding individuals. In the biological species concept, reproductive isolation determines species boundaries.

bootstrap resampling A statistical technique for quantifying how strongly a data set supports a given phylogeny.

breeder's equation The equation $R = h^2 S$, relating the selection response R to the selection differential S and the narrow-sense heritability h^2.

broad-sense heritability (H^2) The fraction of the phenotypic variance that can be attributed to genetic causes and thus is potentially heritable.

Burgess Shale A fossil bed in British Columbia, Canada, containing extensive fossil evidence from the Cambrian explosion.

cactus-shaped phylogeny A phylogeny with short branches off a primary "backbone." Cactus-shaped phylogenies are observed in some infectious pathogens for which most clones are lost in any given year and a lineage is continued by one or at most a small number of escape variants.

Cambrian explosion The relatively rapid evolution of extensive phenotypic diversity during the early part of the Cambrian period (543–490 million years ago).

catastrophism The theory that the geology of the modern world is the result of sudden, catastrophic, large-scale events.

centromere drive Selection at the level of the chromosome that favors mutations to centromeres that increase their chance of segregating to the oocyte instead of to the polar bodies.

characters Measurable aspects of an organism. Characters may be anatomical, physiological, morphological, behavioral, developmental, molecular genetic, and so on.

chromosomal deletion A mutation involving the loss of a section of a chromosome.

chromosomal duplication A mutation involving the duplication of a section of a chromosome.

chromosomal sex determination A sex determination system such as that of most mammals, in which the sex of an individual is determined by the combination of sex chromosomes it possesses.

chronogram A phylogenetic tree on which absolute time is denoted.

cis **regulatory elements** DNA sequences that modify the expression of other genes that are nearby on the chromosome, often by acting as binding sites for transcription factors.

clade A taxonomic group including an ancestor and all of its descendants.

cladogenesis Modification of form associated with branching speciation.

cladogram A phylogenetic tree in which cladistic (historical evolutionary) relationships are represented but in which branch lengths do not indicate the degree of evolutionary divergence. *See also* clade, phylograms.

cline A spatial gradient in the frequency of phenotypes or genotypes.

clonal expansion The process by which a specific immune cell is stimulated by an antigen and rapidly proliferates to create a large population of antigen-specific cells that can eradicate a pathogen.

clonal interference An overall reduction in the rate at which beneficial alleles are fixed in asexual populations due to competition among alternative beneficial mutations.

coalescent point The point on a gene tree that delineates the gene copy that is the most recent common ancestor of the genes being studied in a population.

coalescent theory A theory developed to study the gene–genealogical relationships in a population by tracing the ancestry of gene copies backward from the present through a finite population.

codominant Two alleles are said to be codominant if both contribute to the phenotype of the heterozygote—that is, if an A_1A_2 heterozygote manifests a phenotype intermediate between the A_1A_1 and A_2A_2 homozygotes. *See also* dominant, recessive.

codon usage bias A bias in which certain codons occur more frequently than others that specify the same amino acid.

codons A sequence of three consecutive nucleotides specifying an amino acid product.

coefficient of linkage disequilibrium (D) A measure of nonrandom association between alleles at two different loci. The coefficient of linkage disequilibrium D between two loci is defined as the difference between the actual frequency of a haplotype and the expected frequency of that haplotype if there were no association between alleles at one locus and alleles at the other locus.

coefficient of relatedness A measure of the extent to which individuals share alleles that are identical by descent.

coevolution The process in which evolutionary changes to traits in species 1 drive changes to traits in species 2, which feed back to affect traits in species 1, and so on, back and forth, over evolutionary time.

colinearity An arrangement of genes along a chromosome in which the relative position of a gene on the chromosome corresponds to the relative position of the body part that the gene regulates.

comparative anatomy The study of trait structure and function by comparing anatomical structures across species.

conservative transposon A transposable element that excises itself and moves to a new location rather than inserting a second copy into a new location while leaving the original copy in place. *See also* nonconservative transposon.

conventional signals Signals that take their meaning from an arbitrary convention rather than from the cost of producing them.

convergent evolution The process in which natural selection acts in similar ways in different taxa, driving the independent evolution of similar traits in each taxon. *See also* analogous trait.

cooperation The process by which two or more individuals each receive a net benefit from their joint actions.

Cope's rule The observation that mammalian clades tend to increase in body size over evolutionary time.

cospeciation Concurrent occurrence of speciation in both partners of an interspecific mutualism.

coupling Linkage disequilibrium in which the coefficient of linkage disequilibrium D is positive. *See also* repulsion.

crossing-over The physical exchange of segments of DNA on homologous chromosomes during meiosis.

cultural evolution The process by which culturally transmitted traits change over time.

cultural transmission The transmission of information from one individual to another by teaching or social learning.

C-value paradox The observation that differences in genome size measured in base pairs do not correlate with the number of protein-coding genes that an organism has, nor with its phenotypic complexity.

derived trait A trait that has changed form or state from the ancestral form over evolutionary time.

descent with modification The evolutionary process by which species change over time.

differential reproductive success The difference in the expected number of surviving offspring that can be attributed to having one particular genotype or phenotype instead of another. This is one component of natural selection.

direct fitness The expected number of viable offspring an individual produces. *See also* inclusive fitness, indirect fitness.

directional selection A process of selection in which selection drives phenotype in a single direction, or in which selection drives allele frequencies in a single direction toward fixation of a favored allele. *See also* balancing selection.

disassortative mating A mating pattern in which individuals with dissimilar phenotypes or genotypes mate with one another.

disposable soma hypothesis The hypothesis that senescence results from a necessary trade-off between investment in reproduction and investment in repair.

divergent evolution The process in which natural selection operates in different ways in each of two or more taxa that share a recent common ancestor, leading to different traits in these taxa.

dominance effects Interactions between two alleles at the same locus in determining phenotype.

dominant An allele A_1 is said to be dominant over another allele A_2 if its effects on phenotype cover up those of A_2—that is, if the A_1A_2 heterozygote manifests the same phenotype as the A_1A_1 homozygote.

dumbbell model A form of allopatric speciation in which a population splits into two comparably sized subpopulations separated by geographic barriers.

effective population size The size of an idealized population (no migration, mutation, assortative mating, or natural selection) that loses genetic variation due to genetic drift at the same rate as the population under study.

endemic Found in only one specific area of the world.

endosymbiosis A mutually beneficial relationship in which one organism lives within the body—often within the cells—of another.

enhancers Regulatory elements that increase the rate of transcription.

epistasis The phenomenon in which alleles at two or more loci interact in nonadditive ways to determine phenotype.

escape variants Variant forms of a pathogen that are not recognized by the immunological memory of previously infected hosts.

eusociality A form of extreme sociality involving reproductive division of labor and the cooperative rearing of offspring.

evo–devo (evolutionary developmental biology) The subdiscipline within evolutionary biology that deals with the evolution of developmental pathways and the role that developmental changes have played in the evolution of life's diversity.

evolution Broadly defined as any instance of change over time. More specifically, in a biological context, it is the process of descent with modification that is responsible for the origin, maintenance, and diversity of life.

evolutionary arms race A form of coevolution in which the species involved each evolve countermeasures to the adaptations of the others; most often associated with host–pathogen and predator–prey coevolution.

evolutionary genomics The study of how the composition and structure of genomes have evolved and are evolving.

evolutionary radiation A burst of rapid speciation in a taxon, often associated with entering a new, relatively unoccupied niche.

evolutionary species concept The basic notion in evolutionary biology of what a species is. According to this view, a species is a lineage that maintains a unique identity over evolutionary time.

evolutionary synthesis The collected efforts, primarily in the 1930s and 1940s, of evolutionary biologists, systematists, geneticists, paleontologists, population biologists, population geneticists, and naturalists in shaping modern evolutionary theory to show that a Darwinian view of small-scale and large-scale evolution alike is compatible with the mechanisms of genetic inheritance. Also known as the modern synthesis.

exaptation A trait that currently serves one function today but that evolved from a trait that served a different function in the past.

exons Stretches of DNA that code for protein products. *See also* introns.

expected heterozygosity Denoted as H_e, this is the fraction of heterozygotes expected under the Hardy–Weinberg model, given the allele frequencies in the population.

extinction The loss of all individuals in a species.

fecundity A measure of the ability to produce offspring.

fitness A measure of reproductive success relative to the average reproductive success in a population.

fitness peaks Combinations of traits associated with the greatest fitness values on an adaptive landscape.

fitness valleys Combinations of traits associated with lower fitness values on an adaptive landscape.

fixation In population genetics, an allele is said to go to fixation in a population when it replaces all alternative alleles at the same locus—that is, when its frequency reaches 1.

fossil The remains or traces of a once-living organism. This term is usually used for remains that are greater than 10,000 years old.

fossil record The history of life on Earth as recorded by fossil evidence.

founder effect A change in allele frequencies that results from sampling effects that occur when a small number of individuals derived from a large population initially colonize a new area and found a new population.

frameshift mutation A mutation in which the addition or deletion of base pairs causes a shift in reading frame. Unless an addition or deletion involves a number of base pairs that is a multiple of 3, it will cause a frameshift mutation.

frequency-dependent selection A form of selection in which the fitness associated with a trait or genotype is dependent upon the frequency of that trait or genotype in a population.

frequency-independent selection A form of selection in which the fitness associated with a trait is not directly dependent upon the frequency of that trait in a population.

gametes The sex cells of an organism. In animals, sperm and eggs are gametes.

GC content The fraction of nucleotides in a gene, chromosome, or genome that are G or C rather than A or T.

GC skew A measure of whether G nucleotides or C nucleotides are overrepresented on either the leading or lagging strand of the chromosome. GC skew is typically measured as the $(G - C)/(G + C)$ ratio of nucleotides along one strand of the chromosome.

gene A sequence of DNA that specifies a functional product.

gene–culture coevolution The interaction between genetic and cultural evolutionary change in which each drives the other.

gene duplication A new duplicate copy of a gene that is produced by mutation.

gene expression The process by which a gene produces a functional product (often a protein).

gene sharing The phenomenon in which a protein has more than one function and is expressed in more than one part of the body.

genetic distance A measure of the genetic divergence between populations.

genetic drift Random fluctuation in allele frequencies over time due to sampling effects in finite populations.

genetic hitchhiking The process by which a neutral or even disadvantageous allele is able to "ride along" with a nearby favorable allele to which it is physically linked, and thus increase in frequency.

genetic imprinting A phenomenon in which alleles are expressed differently when inherited from the mother than when inherited from the father.

genotype Either the combination of alleles that an individual has at a given locus, or the combination of alleles that an individual has at *all* loci.

genotype space A conceptual model in which similar genotypes occupy nearby points on a plane. Adaptive landscapes are often illustrated in genotype space. *See also* adaptive landscape, phenotype space.

germ cells Most multicellular organisms have germ cells, such as sperm and eggs, that are specialized for reproduction. Germ cells are the cells that pass DNA to the next generation.

G-value paradox The observation that despite seemingly large differences in organismal complexity, multicellular eukaryotes tend to have very similar numbers of protein-coding genes. *See also* C-value paradox.

handicap principle The hypothesis that the costs of producing signals ensures that they will be reliable or "honest."

haplotype A set of alleles, one at each locus.

haplotype blocks Stretches of the genome where recombination is infrequent, and linkage disequilibrium is high.

Hardy–Weinberg equilibrium Given a set of allele frequencies, the expected set of genotype frequencies that will be observed under the Hardy–Weinberg model.

Hardy–Weinberg model A null model for how genotype frequencies relate to allele frequencies in large populations and how they change over time in the absence of these evolutionary processes: natural selection, mutation, migration, assortative mating, and genetic drift.

heterochrony Changes in the rate and timing of development.

heterozygote advantage *See* overdominance.

heterozygotes Individuals with two different alleles at a given locus—for example, A_1A_2.

homeobox A conserved 180-base-pair sequence present in homeotic genes in widely differing species.

homeotic genes Genes associated with mapping out body shape and form during development.

homologous trait A trait shared by two or more species because those species have inherited the trait from a shared common ancestor.

homoplasy A trait that is similar in two species because of convergent evolution rather than common ancestry.

homozygotes Individuals with two copies of the same allele at a given locus—for example, A_1A_1 or A_2A_2.

horizontal gene transfer (HGT) The transfer of genetic material from one organism to another organism that is not its offspring. Also called lateral gene transfer.

***Hox* genes** A set of genes that direct anterior-to-posterior positioning of body structures during the developmental process in animals.

hybrid zone An area in which diverging populations encounter each other, mate, and potentially produce hybrid offspring.

hypercycle model A model for the evolution of early life involving multiple types of replicators, each of which facilitates the replication of another in cyclical fashion.

hypotheses Proposed explanations for a natural phenomenon. Scientists are interested in hypotheses that generate testable predictions.

identical by descent When two or more gene copies are identical because of shared descent through a recent common ancestor.

inbreeding Mating with genetic relatives.

inbreeding depression A decrease in fitness that results from individuals mating with genetic relatives. *See also* inbreeding.

inclusive fitness The sum of indirect and direct fitness.

independent contrasts A technique for accounting for shared common ancestry when using the comparative method to find trends and patterns in evolutionary events.

indirect fitness The incremental effect that an individual's behavior has on the fitness of its genetic relatives.

inheritance Transmission down across generations.

inheritance of acquired characteristics The hypothesis that traits acquired during the lifetime of an organism are passed on to its offspring. This idea was championed by J. B. Lamarck.

intersexual selection Processes in which individuals of one sex select among individuals of the other sex as mates.

intrasexual selection Processes in which members of one sex, most often males, compete with each other for mating access to the other sex.

introns Noncoding stretches of DNA that interrupt protein-coding regions known as exons and that are excised before translation.

inversion A mutation in which the orientation of a stretch of a chromosome is reversed.

isochores Extended stretches on chromosomes that have similar GC content.

isogamy A reproductive system in which all individuals produce gametes of the same size. *See also* anisogamy.

K–T mass extinction A mass extinction that occurred about 65 million years ago at the boundary between the Cretaceous and Tertiary periods.

last universal common ancestor (LUCA) The population of organisms at the base of the tree of life. All living things today are descended from this one lineage.

lateral gene transfer *See* horizontal gene transfer (HGT).

law of independent assortment Mendel's principle stating that which allele is passed down to the next generation at one locus is independent of which allele is passed down at other loci. This law holds only for pairs of unlinked loci, such as loci on different chromosomes.

law of segregation Mendel's principle that each individual (of a diploid species) has two gene copies at each locus and these gene copies segregate during gamete production. Thus, at each locus only one gene copy goes into each gamete, and an offspring receives one gene copy at each locus from each parent.

law of superposition The observation that fossils found lower down in the sediment at a particular locality are older than those found closer to the surface.

life history strategy The way that an organism invests time and resources into survivorship and reproduction over its lifetime.

LINEs Long INterspersed Elements. A common class of transposable elements, these autonomous transposons make up approximately 17% of the human genome.

linkage disequilibrium The presence of statistical associations between alleles at different loci.

locus The physical location of a gene on a chromosome.

long-branch attraction The tendency of some phylogenetic inference methods to incorrectly infer too close a relationship among rapidly evolving taxa.

major transitions Fundamental changes and developments in the organization of living things that have occurred over the history of life.

many eyes hypothesis The hypothesis that group living provides an advantage in the ability to detect predators.

marker gene A neutral gene with readily observable phenotypic consequences that can be used to track different experimental lines—for example, in microbial evolution experiments, such genes can be used to track different bacterial strains.

mass extinction A large-scale extinction of many taxa over a relatively short period of evolutionary time.

mating systems The mode or pattern of reproductive pairing in a population. Mating systems include monogamy, polygyny, and polyandry.

maximum likelihood A statistical approach often used to model the evolutionary process. This approach selects as "best" the phylogenetic tree that would have the highest probability of generating the observed data.

meiotic drive alleles *See* segregation distorters.

methodological naturalism An approach in which the world is explained solely in terms of natural, rather than supernatural, phenomena.

minimal gene set The hypothetical minimal number of genes thought necessary to allow for cellular-based life.

modern synthesis *See* evolutionary synthesis.

molecular clock A technique for assigning relative or absolute age based on genetic data. In their simplest form, molecular clock methods assume that substitutions at neutral loci occur in clocklike fashion, and so researchers use genetic distances between populations to estimate the time since divergence.

molecular mutualism When different molecules act such that they increase each other's rate of replication. *See also* mutualism.

monogamous mating system A mating system in which a male and female mate with each other, and only each other, during a given breeding season.

monophyletic group A group that consists of a unique common ancestor and each and every one of its descendant species, but no other species.

mosaic coevolution A situation in which the same two species interact mutualistically in some communities but antagonistically in others.

Muller's ratchet A process by which the number of deleterious mutations builds up irreversibly over time in asexual populations.

Müllerian mimicry Mimicry in which unpalatable species resemble one another. *See also* Batesian mimicry.

multicellularity The state in which a single individual is composed of multiple cells.

multiregional hypothesis The hypothesis for human origins stating that 2 million years ago hominins left Africa and colonized the Old World a single time, as *Homo erectus*. These populations in different parts of the world diverged from one another morphologically, but they did not speciate due to modest gene flow among them. Gradually over the past 2 million years, populations together evolved into modern humans.

mutation A change to the DNA sequence.

mutation accumulation hypothesis The hypothesis that senescence occurs because natural selection is not strong enough to purge deleterious mutations associated with traits that are expressed only late in life.

mutation–selection balance An equilibrium frequency of deleterious mutations in which these deleterious mutations are maintained at a positive frequency in a population due to a balance between ongoing deleterious mutation and the purging effect of natural selection.

mutualism An ecological interaction in which different individuals, often of separate species, act so as to increase each other's fitness.

narrow-sense heritability (h^2) The fraction of the total phenotypic variation that is due to additive genetic variation and thus is readily accessible to natural selection.

natural history The comprehensive study of organisms in their natural environment.

natural selection The evolutionary process by which beneficial alleles increase in frequency over time in a population due to increased survival and reproductive success of individuals carrying those alleles. Natural selection is the consequence of variation, inheritance, and differential survival.

nearly neutral theory The hypothesis that most substitutions, if not strictly neutral, are only mildly deleterious.

neofunctionalization An evolutionary process in which duplicated genes diverge, and one copy takes on a new function.

neutral mutations Mutations that do not affect fitness either because they have no effect on phenotype or because the change in phenotype they induce has no fitness consequences.

neutral theory The hypothesis that at the molecular level of DNA sequence or amino acid sequence, most of the variation present within a population and most substitutional differences between populations are selectively neutral.

niche construction The process by which an organism shapes its own environmental conditions.

node A branch point on a phylogenetic tree, representing an ancestral population or species that subsequently divided into multiple descendant populations or species.

noncoding DNA DNA that does not specify an expressed product such as a protein, tRNA, or mRNA.

nonconservative transposon A transposable element that creates a duplicate copy of itself to be inserted elsewhere in the genome rather than excising itself and moving to a new location. *See also* conservative transposon.

nonsynonymous mutation A mutation in a gene that changes the amino acid sequence of the protein that gene encodes.

norm of reaction A curve that represents the phenotype expressed by a given genotype as a function of environmental conditions.

nuclear genome The set of chromosomes contained in the eukaryotic nucleus.

obligate mutualism A mutualism in which each partner requires the other for successful survival and/or reproduction.

observed heterozygosity The fraction of individuals in the population that are heterozygous at a given locus.

odds ratio testing A statistical technique for quantifying how strongly a data set supports a particular hypothesis. Applied to phylogenetics, odds ratio testing is sometimes used to determine how strongly the data support the hypothesis that a given group represents a monophyletic clade.

ontogeny The development of an organism.

outgroup A distantly related group for which we already have information with respect to its evolutionary relationship to the taxon we are studying. Outgroups are used in rooting phylogenetic trees.

out-of-Africa hypothesis The hypothesis that hominins left Africa and colonized the Old World in two major waves, first as *Homo erectus* in an initial wave out of Africa approximately 2 million years ago, and a second time as *Homo sapiens* approximately 100,000 years ago.

overdominance A form of frequency-independent selection in which heterozygote genotypes have higher fitness than the corresponding homozygote genotypes.

paedomorphosis The appearance of traits seen in the juvenile stage of an ancestral species during the adult stage of a descendant species.

paleomagnetic dating The method of estimating fossil dates based on shifts in the Earth's magnetic field by measuring the alignment of metal particles in the substrate in which the fossil was found.

paralogs A pair of genes within a genome that share common ancestry due a gene duplication event.

parapatric speciation The process of speciation that occurs when diverging populations have distributions that abut one another.

paraphyletic group A group that includes the common ancestor of all its members but does not contain every species that descended from that ancestor.

parent–offspring conflict Conflict that arises when the genetic interests of offspring and their parents are not perfectly aligned.

parsimony An approach to selecting the best phylogenetic tree given some set of character data. Parsimony methods assume that the best tree is the one that requires the fewest character changes to explain the data.

periodic selection A process in which a series of clones carrying beneficial mutations successively go to fixation in an asexual population.

peripheral isolate model A form of allopatric speciation in which a population is split into geographically isolated populations that differ substantively in size, with one large population and one or several smaller populations. *See also* dumbbell model.

phenetic species concept An approach to determining species boundaries in which species are identified as clusters of phenotypically similar individuals or populations.

phenotype The observable physical, developmental, and behavioral characteristics of an organism.

phenotype space A conceptual model in which similar phenotypes occupy nearby points on a plane. Adaptive landscapes are often illustrated in phenotype space. *See also* adaptive landscape, genotype space.

phyletic gradualism model The hypothesis that new species arise by a gradual transformation of an ancestral species through slow, continual change. *See also* punctuated equilibrium model.

phylogenetic distance methods Methods of constructing phylogenetic trees based upon measurements of pairwise "distances" between species, where distance is a measurement of morphological or genetic differences between species.

phylogenetic diversity The amount of diversity in the evolutionary history of a taxon, sometimes measured as the sum of the branch lengths of the phylogeny of that taxon.

phylogenetic event horizon The point in the history of life beyond which phylogenetic analysis is uninformative because there are no surviving descendants from ancestors before this point. *See also* last universal common ancestor (LUCA).

phylogenetic species concept An approach to determining species boundaries in which a species is defined as the smallest monophyletic group that shares a unique derived character absent from all other groups on the phylogeny.

phylogenetic systematics An approach to classifying organisms based upon their evolutionary histories.

phylogenetic tree A visual representation, in the form of a bifurcating tree, of the evolutionary relationship between species, genera, families, and higher taxonomic units.

phylogeny The branching pattern of relatedness among populations (or occasionally, individuals) in a group or taxon.

phylogeography The use of phylogenetic and population-genetic tools to study the geographic distributions of populations or species.

phylograms A phylogenetic tree in which the length of each branch represents the amount of evolutionary change that has occurred along that branch.

physical linkage The occurrence of two or more loci on the same chromosome. Physical linkage causes alleles at linked loci to segregate together (in the absence of recombination) into the gametes.

plasmids Circular extrachromosomal genetic elements common in bacteria and some other microorganisms.

pleiotropic genes Genes that affect more than a single trait.

polarity The order in which different variants of a trait evolved over evolutionary time.

polyandry A mating system in which females mate with more than one male per breeding season.

polygamous mating system A mating system in which either males or females—or both—have more than one mate during a given breeding season.

polygenic traits Traits that are affected by many genes simultaneously.

polygynandry A mating system in which several males form pair bonds with several females simultaneously.

polygyny A mating system in which males mate with more than one female per breeding season.

polyphyletic group A group that does not contain the common ancestor of its members and/or all descendants of that common ancestor.

polytomy A node on a phylogenetic tree that has more than two branches arising from it. Polytomies are often used to represent our uncertainty about phylogenetic relationships on a phylogenetic tree.

population A group of individuals of the same species that are found within a defined area and, if they are a sexual species, interbreed with one another.

population bottleneck A brief period of small population size. Population bottlenecks reduce genetic diversity and can accelerate changes in allele frequencies due to genetic drift.

population genetics A subdiscipline in evolutionary biology that investigates how allele frequencies and genotype frequencies change over time.

positive selection Selection favoring new beneficial mutations.

postcopulatory sexual selection Sexual selection that occurs after matings have taken place. Sperm competition is one form of postcopulatory sexual selection.

postzygotic isolating mechanisms Reproductive isolating mechanisms that occur after fertilization and conception, often leading to embryos that may not develop fully to birth or to sterile offspring. *See also* prezygotic isolating mechanisms, reproductive isolating mechanisms.

prebiotic soup hypothesis The idea that the earliest life emerged in a "soup-like" liquid environment, drawing upon energy from cosmic rays, volcanic eruptions, and the Earth's own internal heat.

prezygotic isolating mechanisms Reproductive isolating mechanisms that prevent mating from occurring in the first place or that prevent fertilization from occurring if such a mating does occur. *See also* postzygotic isolating mechanisms, reproductive isolating mechanisms.

promoter A short DNA sequence before the transcribed region of a gene, to which the RNA polymerase binds to initiate transcription.

prophages Viral genomes that insert themselves into bacterial chromosomes. Prophages can subsequently be excised from the genome and initiate viral replication within the bacterial cell.

protocell A cell-like entity that predated cellular life-forms in the history of life.

pseudoextinction A phenomenon in which a population changes by anagenesis over evolutionary time, until it is so different from the ancestral population that it is reclassified as a new species.

pseudogene A nonfunctional and typically untranslated segment of DNA that arises from a previously functional gene.

punctuated equilibrium model The hypothesis that major evolutionary changes, including speciation, do not occur through a slow, gradual process. Instead, stasis—the absence of change—is the rule during the vast majority of a lineage's history. But when evolutionary change does occur in lineages, it is rapid, and typically leads to branching speciation (cladogenesis). *See also* phyletic gradualism model.

purifying selection Selection against deleterious mutations.

QTL mapping A technique for identifying the regions of the genome in which quantitative trait loci occur. *See also* quantitative trait loci (QTLs).

quantitative genetics A mathematical approach to the population genetic study of continuously varying traits.

quantitative trait loci (QTLs) Loci responsible for quantitative—that is, continuously varying—traits.

radiocarbon dating A technique for dating geological strata by using the decay rate of Carbon 14 to Carbon 12.

rate-of-living hypothesis The hypothesis that senescence is an inevitable consequence of accumulated physical wear and tear.

realized heritabilities Narrow-sense heritability values estimated by using values of the selection differential and selection response in the breeder's equation.

recapitulation The appearance of traits in the juvenile stage of a descendant species that were expressed in the adult stage of an ancestral species.

recessive An allele A_1 is said to be recessive to another allele A_2 if its effects on phenotype are covered up in the heterozygote—that is, if the A_1A_2 heterozygote manifests the same phenotype as the A_2A_2 homozygote. *See also* dominant.

reciprocal altruism The hypothesis that altruistic behavior can be maintained evolutionarily if individuals sequentially exchange acts of altruism.

recombination hotspots Small regions of the genome that are particularly prone to serving as locations of crossing over.

Red Queen hypothesis The hypothesis that sexual reproduction is an adaptation allowing hosts to generate sufficient genetic variation to keep up with their pathogens and parasites in the coevolutionary arms race. This hypothesis predicts that the level of parasitic infection will be related to the frequency of sexual versus asexual reproduction.

regulatory elements Stretches of DNA involved in controlling levels of gene expression.

regulatory enhancers A section of DNA that lies outside of a gene but is involved in up-regulating that gene's expression.

reproductive isolating mechanisms Mechanisms that prevent gene flow between populations.

repulsion Linkage disequilibrium in which the coefficient of linkage disequilibrium *D* is negative. *See also* coupling.

ribozymes RNA molecules with enzymatic function.

RNA world A hypothetical early stage in the history of life in which RNA was the fundamental unit upon which life was based, fulfilling both an informational role (much as DNA does today) and a catalytic role (much as protein-based enzymes do today).

root The basal (most ancestral) lineage on a phylogenetic tree.

rooted tree A phylogenetic tree in which the root is indicated and thus the direction of time is specified.

runaway sexual selection model A model of sexual selection in which a positive feedback loop develops between genes that code for male traits and genes that code for particular mating preferences in females, leading to exaggerated male traits and strong female preferences for them.

saltationism The hypothesis that evolutionary change occurs primarily as a result of large-scale changes.

segregation distorters Alleles that bias the process of meiotic segregation in their own favor, increasing their representation to more than half the gametes produced by an individual. Also known as meiotic drive alleles.

selection coefficient A measure of the strength of natural selection for or against a specific phenotype or genotype.

selection differential (S) In quantitative genetics, the difference between the mean trait value of the individuals who reproduce and the mean trait value of all individuals.

selection response (R) In quantitative genetics, the difference between the mean trait value of the offspring population and the mean trait value of the parental population.

selective breeding A process in which humans decide which plants or animals in a population are allowed to breed. *See also* artificial selection.

selective sweep A phenomenon in which a selected allele goes to fixation, carrying with it alleles at tightly linked loci. *See also* genetic hitchhiking.

selectively neutral Alternative alleles are selectively neutral when there is no fitness difference between them.

selfish genetic elements Stretches of DNA, such as transposons, that act primarily to ensure their own survival and replication within a genome, even at a fitness cost to the organism.

senescence General decline in the physical functioning or performance of living organisms with age.

sensory bias model Model for the evolution of elaborate traits by sexual selection, in which a preexisting bias in the perceptual system of one sex favors members of the other sex who display a particular trait.

sequence divergence A measure of the extent to which two DNA sequences differ from one another.

sex ratio The ratio of males to females in a population.

sexual conflict A phenomenon in which selection operates differently on males and females, typically with respect to mating behavior.

sexual reproduction Joining together of genetic material from two parents to produce an offspring that has genes from each parent. Typically sexual reproduction involves both recombination between homologous chromosomes and outcrossing—mating between genetically different individuals.

sexual selection A form of natural selection that refers to selection for traits and behaviors that confer mating or success (as opposed to survival).

Signor–Lipps effect The lag between the last observed fossil of an extinct species and the actual date of extinction. This effect can cause paleotologists to date an extinction earlier than it actually occurred.

silent mutation *See* synonymous mutation.

SINEs Short INterspersed Elements. A common class of transposable elements in humans, these nonautonomous transposons are incapable of independent replication but rather rely on genes encoded by autonomous transposons elsewhere in the genome.

sister taxa Two taxa that derive from the same node on a phylogenetic tree.

somatic cells Cells specialized in the maintenance and growth functions of an organism.

speciation The process by which new species arise from previously existing species. All models of speciation involve some type of breakdown of gene flow across populations.

species selection A process of differential speciation and/or extinction that may drive some of the macroevolutionary trends observed across taxa.

sperm competition Competition among sperm for access to eggs.

spontaneous generation The now-disproven hypothesis that complex life-forms can arise, de novo, from inorganic matter.

struggle for existence Darwin's idea that organisms are continually in competition for resources.

subfunctionalization A molecular evolutionary process by which gene duplication produces gene copies that diverge and divide the work initially undertaken by the gene before duplication.

substitution The process in which a new allele arises by mutation and is subsequently fixed in a population.

survivorship curve The fraction of surviving individuals as a function of age.

sympatric speciation A process of speciation in which diverging populations are not geographically separated.

symplesiomorphy A derived trait that has arisen so recently that it appears in only one of two sister taxa. Evolutionary biologists try to avoid using symplesiomorphies in phylogenetic reconstruction.

synapomorphy A derived trait that is shared in two populations because it was inherited from a recent common ancestor. Evolutionary biologists aim to use synapomorphies in phylogenetic reconstruction, as they provide useful information about the evolutionary relationships among populations.

synonymous mutation A base pair substitution that does not change the amino acid that a codon normally produces. Also known as a silent mutation.

systematics The scientific study of classifying organisms.

taxon A group of related organisms.

trade-off A situation in which constraints prevent simultaneously optimizing two different characters or two different aspects of a character.

traits Any observable characteristics of organisms, such as anatomical features, developmental or embryological processes, behavioral patterns, or genetic sequences.

***trans* regulatory elements** DNA sequences that modify the expression or activity of genes that are not nearby on the chromosome, often by coding for transcription factors.

transcription The process of copying DNA sequence into a complementary messenger RNA (mRNA).

transcription factors Proteins that bind to DNA and influence gene expression.

transduction Horizontal gene transfer that occurs when a bacteriophage packages host DNA into its capsule. If that DNA is injected into a new host, it can be incorporated into the genome.

transformational process A process of change in which the properties of a group change because every member of that group changes.

transition A mutation in which a purine (adenine or guanine) is replaced by a purine, or a pyrimidine (cytosine or thymine) is replaced by a pyrimidine. *See also* transversion.

translocation A mutation in which a section of a chromosome is moved to a nonhomologous chromosome.

transmission genetics The study of the mechanisms by which genes are passed from parents to offspring.

transposable element A self-replicating genetic unit that can move or copy itself within a genome.

transversion A mutation in which a purine is replaced by a pyrimidine or vice versa. *See also* transition.

tree of life A phylogenetic tree that depicts the evolutionary relationships among all living things.

twofold cost of sex The observation that—with all else equal—an asexual lineage introduced into a population of sexually reproducing organisms would initially double in representation in each generation.

underdominance A form of frequency-independent selection in which the heterozygote genotype has a lower fitness than either corresponding homozygote genotype.

uniformitarianism Charles Lyell's theory that the very same geological processes that we observe today have operated over vast stretches of time, and explain the geology of the past and the present.

unrooted tree A phylogenetic tree in which the root, and thus the direction of time, is unspecified.

variation In evolutionary biology, genetic variation is one of the components of the process of natural selection.

variational process A process of change in which the properties of an ensemble change, not because the individual elements change, but because of some sorting process. In evolutionary biology, the sorting process is natural selection.

vestigial traits Traits that have no known current function but that appear to have had a function in the evolutionary past.

virulence factors Specialized genes that assist bacteria in exploiting eukaryotic hosts.

Wright's *F*-statistic A statistical measure of the degree of homozygosity in a population.

Wright–Fisher model A population genetic model of evolutionary change in small populations with nonoverlapping generations.

REFERENCES

Abbo, S., D. Shtienberg, J. Lichtenzveig, S. Lev-Yadun, and A. Gopher. 2003. The chickpea, summer cropping, and a new model for pulse domestication in the ancient Near East. *The Quarterly Review of Biology* 78: 435–448.

Ackerly, D. D., and M. J. Donoghue. 1998. Leaf size, sapling allometry, and Corner's rules: phylogeny and correlated evolution in maples (*Acer*). *American Naturalist* 152: 767–791.

Ackermann, M., and L. Chao. 2006. DNA sequences shaped by selection for stability. *PLoS Genetics* 2: 224–230.

Ackermann, M., L. Chao, C. T. Bergstrom, and M. Doebeli. 2007. On the evolutionary origin of aging. *Aging Cell* 6: 235–244.

Ackermann, M., and S. D. Fletcher. 2008. Evolutionary biology as a foundation for studying aging and aging-related disease. In S. C. Stearns and J. C. Koella, eds., *Evolution in Health and Disease*, 2nd Ed., Chapter 18. Oxford University Press, Oxford.

Ackermann, M., S. C. Stearns, and U. Jenal. 2003. Senescence in a bacterium with asymmetric division. *Science* 300: 1920.

Adamowicz, S. J., A. Purvis, and M. A. Wills. 2008. Increasing morphological complexity in multiple parallel lineages of the Crustacea. *Proceedings of the National Academy of Sciences of the United States of America* 105: 4786–4791.

Adams, B., E. C. Holmes, C. Zhang, M. P. Mammen, Jr., S. Nimmannitya, S. Kalayanarooj, and M. Boots. 2006. Cross-protective immunity can account for the alternating epidemic pattern of dengue virus serotypes circulating in Bangkok. *Proceedings of the National Academy of Sciences of the United States of America* 103: 14234–14239.

Agnarsson, I., L. Aviles, J. A. Coddington, and W. P. Maddison. 2006. Sociality in Theridiid spiders: repeated origins of an evolutionary dead end. *Evolution* 60: 2342–2351.

Agris, P. F. 2008. Bringing order to translation: the contributions of transfer RNA anticodon-domain modifications. *EMBO Reports* 9: 629–635.

Ahlberg, P. E., and J. A. Clack. 2006. Palaeontology—a firm step from water to land. *Nature* 440: 747–749.

Akam, M. 1991. Wondrous transformation. *Nature* 349: 282.

Alatalo, R. V., J. Hoglund, A. Lundberg, and W. J. Sutherland. 1992. Evolution of black grouse leks: female preferences benefit males in larger leks. *Behavioral Ecology* 3: 53–59.

Alberti, S. 1997. The origin of the genetic code and protein synthesis. *Journal of Molecular Evolution* 45: 352–358.

Alexander, R. D. 1974. The evolution of social behavior. *Annual Review of Ecology and Systematics* 5: 325–383.

Alexander, R. D., K. M. Noonan, and B. J. Crespi. 1991. The evolution of eusociality. In P. Sherman, J. U. M. Jarvis, and R. D. Alexander, eds., *The Biology of the Naked Mole-Rat*, pp. 3–44. Princeton University Press, Princeton.

Alexander, R. M., A. S. Jayes, G. M. O. Maloiy, and E. M. Wathuta. 1979. Allometry of the limb bones of mammals from shrews (*Sorex*) to elephant (*Loxodonta*). *Journal of Zoology* 189: 305–314.

Alvarez, L. W. 1983. Experimental evidence that an asteroid impact led to the extinction of many species 65 million years ago. *Proceedings of the National Academy of Sciences of the United States of America. Physical Sciences* 80: 627–642.

Alvarez, L. W., W. Alvarez, F. Asaro, and H. V. Michel. 1980. Extraterrestrial cause for the Cretaceous–Tertiary extinction. *Science* 208: 1095–1108.

Alvarez, W. 1998. *T. Rex and the Crater of Doom*. Vintage, New York.

Alvarez, W., L. W. Alvarez, F. Asaro, and H. V. Michel. 1984a. The end of the Cretaceous: sharp boundary or gradual transition? *Science* 223: 1183–1186.

Alvarez, W., E. G. Kauffman, F. Surlyk, L. W. Alvarez, F. Asaro, and H. V. Michel. 1984b. Impact theory of mass extinctions and the invertebrate fossil record. *Science* 223: 1135–1141.

Alvarez, W., F. Asaro, and A. Montanari. 1990. Iridium profile for 10-million years across the Cretaceous–Tertiary boundary at Gubbio (Italy). *Science* 250: 1700–1702.

Andersson, D. I., and D. Hughes. 2010. Antibiotic resistance and its cost: is it possible to reverse resistance? *Nature Reviews Microbiology* 8: 260–271.

Andersson, M. B. 1982. Female choice selects for extreme tail length in a widowbird. *Nature* 299: 818–820.

Andersson, M. 1994. *Sexual Selection*. Princeton University Press, Princeton.

Andersson, M., and L. W. Simmons. 2006. Sexual selection and mate choice. *Trends in Ecology & Evolution* 21: 296–302.

Andolfatto, P. 2005. Adaptive evolution of non-coding DNA in *Drosophila*. *Nature* 437: 1149–1152.

Andrews, P. 1992. Evolution and environment in the Hominoidea. *Nature* 360: 641–646.

Anthony, A. 1925. Expedition to Guadalupe Island, Mexico, in 1922. The birds and mammals. *Proceedings of the California Academy of Sciences* 14: 277–320.

Aristotle. 350 B.C.E. *History of Animals*. Translated by R. Cresswell. In Schneider, J. G. 1862. *Ten Books*. H. G. Bohn, London.

Aristotle. 350 B.C.E. *On the Parts of Animals*. Translated by W. Ogle. http://classics.mit.edu/Aristotle/parts_animals.html.

Aristotle. 1986. *The Physics: Books 1–4*. Translated by P. H. Wicksteed and F. M. Cornford. Harvard University Press, Cambridge.

Arjan-G., J., M. de Visser, C. W. Zeyl, P. Gerrish, J. Blanchard, and R. Lenski. 1999. Diminishing returns from mutation supply rate in asexual populations. *Science* 283: 404–406.

Armbruster, W. S., A. L. Herzig, and T. P. Clausen. 1992. Pollination of 2 sympatric species of *Dalechampia* (*Euphorbiaceae*) in Surinam by male Euglossine bees. *American Journal of Botany* 79: 1374–1381.

Armstrong, K. 2005. *A Short History of Myth*. Canongate, New York.

Arnqvist, G., and L. Rowe. 2005. *Sexual Conflict*. Princeton University Press, Princeton.

Asami, T., R. H. Cowie, and K. Ohbayashi. 1998. Evolution of mirror images by sexually asymmetric mating behavior in hermaphroditic snails. *American Naturalist* 152: 225–236.

Austad, S. N. 1993. Retarded senescence in an insular population of Virginia opossums (*Didelphis virginiana*). *Journal of Zoology* 229: 695–708.

Austad, S. N., and C. E. Finch. 2008. The evolutionary context of human aging and degenerative disease. In S. C. Stearns and J. C. Koella, eds., *Evolution in Health and Disease*, 2nd Ed., Chapter 23. Oxford University Press, Oxford.

Austad, S. N., and K. E. Fischer. 1991. Mammalian aging, metabolism, and ecology: evidence from the bats and marsupials. *Journal of Gerontology* 46: B47–53.

Axelrod, R. 1984. *The Evolution of Cooperation.* Basic Books, New York.

Axelrod, R., and W. D. Hamilton. 1981. The evolution of cooperation. *Science* 211: 1390–1396.

Ayala, F. J., and M. Coluzzi. 2005. Chromosome speciation: humans, *Drosophila*, and mosquitoes. *Proceedings of the National Academy of Sciences of the United States of America* 102: 6535–6542.

Babbitt, G. A. 2008. How accurate is the phenotype? An analysis of developmental noise in a cotton aphid clone. *BMC Developmental Biology* 8: 19.

Bada, J. L., M. X. Zhao, and N. Lee. 1986. Did extraterrestrial impactors supply the organics necessary for the origin of terrestrial life—amino acid evidence in Cretaceous–Tertiary boundary sediments. *Origins of Life and Evolution of the Biosphere* 16: 185.

Baer, C. F., M. M. Miyamoto, and D. R. Denver. 2007. Mutation rate variation in multicellular eukaryotes: causes and consequences. *Nature Reviews Genetics* 8: 619–631.

Bakker, T. C. M. 1993. Positive genetic correlation between female preference and preferred male ornament in sticklebacks. *Nature* 363: 255–257.

Baldauf, S. L. 2003. Phylogeny for the faint of heart: a tutorial. *Trends in Genetics* 19: 345–351.

Bamshad, M., and S. Wooding. 2003. Signatures of natural selection in the human genome. *Nature Reviews Genetics* 4: 99–111.

Bapteste, E., and D. A. Walsh. 2005. Does the "Ring of Life" ring true? *Trends in Microbiology* 13: 256–261.

Barluenga, M., K. N. Stolting, W. Salzburger, M. Muschick, and A. Meyer. 2006. Sympatric speciation in Nicaraguan Crater Lake cichlid fish. *Nature* 439: 719–723.

Barnosky, A. D., P. L. Koch, R. S. Feranec, S. L. Wing, and A. B. Shabel. 2004. Assessing the causes of Late Pleistocene extinctions on the continents. *Science* 306: 70–75.

Barrier, M., R. H. Robichaux, and M. D. Purugganan. 2001. Accelerated regulatory gene evolution in an adaptive radiation. *Proceedings of the National Academy of Sciences of the United States of America* 98: 10208–10213.

Barsh, G. S. 1996. The genetics of pigmentation: from fancy genes to complex traits. *Trends in Genetics* 12: 299–305.

Barton, N., D. Briggs, J. Eisen, D. Goldstein, and N. Patel. 2007. *Evolution.* Cold Spring Harbor Press, Cold Spring Harbor, NY.

Barton, N. H., and B. Charlesworth. 1998. Why sex and recombination? *Science* 281: 1986–1990.

Barton, N. H., and G. M. Hewitt. 1985. Analysis of hybrid zones. *Annual Review of Ecology and Systematics* 16: 113–148.

Bar-Yosef, O. 2002. The Upper Paleolithic revolution. *Annual Review of Anthropology* 31: 363–393.

Basolo, A. 1990. Female preference predates the evolution of the sword in swordfish. *Science* 250: 808–811.

Basolo, A. 1995. Phylogenetic evidence for the role of a pre-existing bias in sexual selection. *Proceedings of the Royal Society of London* 259: 307–311.

Bateman, A. J. 1948. Intra-sexual selection in *Drosophila. Heredity* 2: 349–368.

Bates, H. W. 1862. Contributions to the insect fauna of the Amazon valley. *Transactions of the Linnean Society* 23: 495–566.

Bateson, W. 1894. *Materials for the Study of Variation: Treated with Special Regard to Discontinuity in the Origin of Species.* MacMillan, New York.

Batra, S. W. T. 1966. Nests and social behavior of halictine bees of India (Hymenoptera:Halictidae). *Indian Journal of Entomology* 28: 375–393.

Batzer, M. A., and P. L. Deininger. 2002. *Alu* repeats and human genomic diversity. *Nature Reviews Genetics* 3: 370–379.

Baum, D. 2008. Reading a phylogenetic tree: the meaning of monophyletic groups. *Nature Education* 1(1). Accessed November 11, 2010. www.nature.com/scitable/topicpage/reading-a-phylogenetic-tree-the-meaning-of-41956.

Bayzkin, A. D. 1969. Hypothetical mechanism of speciation. *Evolution* 23: 685–687.

Bd-Maps. 2010. Accessed December 27, 2010. www.spatialepidemiology.net/bd/.

Bean, D. C., D. M. Livermore, I. Papa, and L. M. Hall. 2005. Resistance among *Escherichia coli* to sulphonamides and other antimicrobials now little used in man. *Journal of Antimicrobial Chemotherapy* 56: 962–964.

Beer, B., E. Bailes, P. Sharp, and V. Hirsch. 1999. Diversity and evolution of primate lentiviruses. In B. Korber, C. Brander, B. F. Haynes, J. P. Moore, R. Koup, B. Walker, and D. Watkins, eds., *HIV Molecular Immunology Database 1999.* Los Alamos National Laboratory, Theoretical Biology and Biophysics (LA-UR 00-1757), Los Alamos.

Bell, G. 1982. *The Masterpiece of Nature: The Evolution and Genetics of Sexuality.* University of Berkeley Press, Berkeley.

Bengtsson, B. 1979. Theoretical models of speciation. *Zoological Scripta* 8: 303–304.

Benner, S. A., A. D. Ellington, and A. Tauer. 1989. Modern metabolism as a palimpsest of the RNA world. *Proceedings of the National Academy of Sciences of the United States of America* 86: 7054–7058.

Bennett, A. F., and R. E. Lenski. 2007. An experimental test of evolutionary trade-offs during temperature adaptation. *Proceedings of the National Academy of Sciences of the United States of America* 104: 8649–8654.

Benson, S. B. 1933. Concealing coloration among some desert rodents of the southwestern United States. *University of California Publications, Zoology* 40: 1–70.

Benton, M., ed. 1993. *The Fossil Record: 2.* Chapman & Hall, London.

Benton, M. 1997. *Vertebrate Paleontology.* Chapman & Hall, London.

Benton, M. 2003a. *When Life Nearly Died: The Greatest Mass Extinction of All Time.* Thames & Hudson, London.

Benton, M. 2003b. Wipeout. *New Scientist* 178: 38–41.

Benton M. J., and P. N. Pearson. 2001. Speciation in the fossil record. *Trends in Ecology & Evolution* 16: 405–411.

Benton, M., M. Shishkin, D. Unwin, and E. Kurochkin, eds. 2000. *The Age of Dinosaurs in Russia and Mongolia.* Cambridge University Press, Cambridge.

Berger, L., R. Speare, P. Daszak, D. E. Green, A. A. Cunningham, C. L. Goggin, R. Slocombe, M. A. Ragan, A. D. Hyatt, K. R. McDonald, H. B. Hines, K. R. Lips, G. Marantelli, and H. Parkes. 1998. Chytridiomycosis causes amphibian mortality associated with population declines in the rain forests of Australia and Central America. *Proceedings of the National Academy of Sciences of the United States of America* 95: 9031–9036.

Bergsten, J. 2005. A review of long-branch attraction. *Cladistics* 21: 163–193.

Bergstrom, C. T. 2009. Dealing with deception in biology. In B. Harrington, ed., *Deception: Methods, Motives, Contexts & Consequences.* Stanford University Press, Palo Alto, CA.

Bergstrom, C. T., and R. Antia. 2006. How do adaptive immune systems control pathogens while avoiding autoimmunity? *Trends in Ecology and Evolution* 21: 22–28.

Bergstrom, C. T., and M. Feldgarden. 2008. The ecology and evolution of antibiotic-resistant bacteria. In S. Stearns and J. Koella, eds., *Evolution in Health and Disease*, pp. 125–137. Oxford University Press, Oxford.

Bergstrom, C. T., and M. Lachmann. 1997. Signalling among relatives. 1. Is costly signalling too costly? *Philosophical Transactions of the Royal Society of London Series B—Biological Sciences* 352: 609–617.

Bergstrom, C. T., and L. A. Real. 2000. Towards a theory of mutual mate choice: lessons from two-sided matching. *Evolutionary Ecology Research* 2: 493–508.

Bergstrom, T. 2002. Evolution of social behavior: individual and group selection. *Journal of Economic Perspectives* 16: 67–88.

Berlocher, S. H., and J. L. Feder. 2002. Sympatric speciation in phytophagous insects: moving beyond controversy? *Annual Review of Entomology* 47: 773–815.

Bernal, D., K. A. Dickson, R. E. Shadwick, and J. B. Graham. 2001. Review: analysis of the evolutionary convergence for high performance swimming in lamnid sharks and tunas. *Comparative Biochemistry and Physiology* 129: 695–726.

Bernasconi, G., and J. E. Strassmann. 1999. Cooperation among unrelated individuals: the ant foundress case. *Trends in Ecology & Evolution* 14: 477–482.

Bernstein, C., and H. Bernstein. 1991. *Aging, Sex and DNA Repair*. Academic Press, New York.

Bernstein, H., H. C. Byerly, F. A. Hopf, and R. E. Michod. 1985a. Genetic damage, mutation and the evolution of sex. *Science* 229: 1277–1281.

Bernstein, H., H. C. Byerly, F. A. Hopf, and R. E. Michod. 1985b. Sex and the emergence of species. *Journal of Theoretical Biology* 117: 665–690.

Bernstein, H., G. S. Byers, and R. E. Michod. 1981. Evolution of sexual reproduction: importance of DNA repair, complementation and variation. *American Naturalist* 117: 537–549.

Berthold, P., and F. Pulido. 1994. Heritability of migration activity in a natural bird population. *Proceedings of the Royal Society B: Biological Sciences* 257: 311–315.

Bertram, B. 1978. Living in groups: predators and prey. In J. R. Krebs and N. Davies, eds., *Behavioural Ecology: An Evolutionary Approach*, pp. 64–96. Blackwell, London.

Beurton, P. J. 2002. Ernst Mayr through time on the biological species concept—a conceptual analysis. *Theory in Biosciences* 121: 81–98.

Birkhead, T., and A. Møller. 1992. *Sperm Competition in Birds*. Academic Press, London.

Birkhead, T., and A. Møller. 1998. *Sperm Competition and Sexual Selection*. Academic Press, London.

Birkhead, T. R., and T. Pizzari. 2002. Postcopulatory sexual selection. *Nature Reviews Genetics* 3: 262–273.

Bisgaard, M., K. Fenger, S. Bulow, E. Niebuhr, and J. Mohr. 1994. Familial adenomatous polyposis (FAP): frequency penetrance and mutation rate. *Human Mutation* 3: 121–125.

Bittles, A. H. 1988. *Empirical Estimates of the Global Prevalence of Consanguineous Marriage in Contemporary Societies*. Morrison Institute for Population and Resource Studies, Stanford University, Palo Alto, CA.

Bittles, A. H., H. Savithri, H. Murthy, G. Baskaran, and G. Wang. 2001. Consanguinity: a familiar story full of surprises. In H. MacBeth and P. Shetty, eds., *Health and Ethnicity*, pp. 68–78. Taylor and Francis, London.

Bjedov, I., O. Tenaillon, B. Gérard, V. Souza, E. Denamur, M. Radman, F. Taddei, and I. Matic. 2003. Stress-induced mutagenesis in bacteria. *Science* 300: 1404–1409.

Bock, W. J. 1959. Preadaptation and multiple evolutionary pathways. *Evolution* 13: 194–211.

Bock, W. J. 1969. The origin and radiation of birds. *Annals of the New York Academy of Sciences* 167: 147–155.

Bodmer, W. 1999. Familial adenomatous polyposis (FAP) and its gene, APC. *Cytogenetics and Cell Genetics* 86: 99–104.

Boesch, C. 1994. Cooperative hunting in wild chimpanzees. Animal Behaviour 48: 653–667.

Boesch, C. 2002. Cooperative hunting roles among Tai chimpanzees. *Human Nature—An Interdisciplinary Biosocial Perspective* 13: 27–46.

Boesch, C. 2005. Joint cooperative hunting among wild chimpanzees: taking natural observations seriously. *Behavioral and Brain Sciences* 28: 692–694.

Boesch, C., and H. Boesch. 1989. Hunting behavior of wild chimpanzees in the Tai National Park. *American Journal of Anthropology* 78: 547–573.

Boller, E. F., and R. J. Prokopy. 1976. Bionomics and management of rhagoletis. *Annual Review of Entomology* 21: 223–246.

Bonnell, M., and R. K. Selander. 1974. Elephant seals: genetic variation and near extinction. *Science* 184: 908–909.

Bonner, J. T. 1957. A theory of the control of differentiation in the cellular slime molds. *The Quarterly Review of Biology* 32: 232–246.

Bonner, J. T. 1980. *The Evolution of Culture in Animals*. Princeton University Press, Princeton.

Bonner, J. T. 2000. *First Signals: The Evolution of Multicellular Development*. Princeton University Press, Princeton.

Bonner, J. T. 2003. Evolution of development in the cellular slime molds. *Evolution & Development* 5: 305–313.

Bonner, J. T., L. Segel, and E. C. Cox. 1998. Oxygen and differentiation in *Dictyostelium discoideum*. *Journal of Biosciences* 23: 177–184.

Bonnet, C. 1769. *La palingenesis philosophique*. C. Philibert, Geneva.

Borges, J. L. 1952. *Other Inquisitions (1937–1952)*, Translated by R. I. Simms. University of Texas, Austin, 1964.

Borodina, I., P. Krabben, and J. Nielsen. 2005. Genome-scale analysis of *Streptomyces coelicolor* A3(2) metabolism. *Genome Research* 15: 820–829.

Bottke, W. F., D. Vokrouhlicky, and D. Nesvorny. 2007. An asteroid breakup 160 Myr ago as the probable source of the K/T impactor. *Nature* 449: 48–53.

Boucher, D., ed. 1985. *The Biology of Mutualism: Ecology and Evolution*. Oxford University Press, New York.

Bourgeois, J., T. Hansen, P. L. Wiberg, and E. G. Kauffman. 1988. A tsunami deposit at the Cretaceous–Tertiary boundary in Texas. *Science* 241: 567–570.

Boyd, R., and P. J. Richerson. 1985. *Culture and the Evolutionary Process*. University of Chicago Press, Chicago.

Boyd, R., and P. J. Richerson. 2004. *Not by Genes Alone*. University of Chicago Press, Chicago.

Boyer, A. G. 2008. Extinction patterns in the avifauna of the Hawaiian islands. *Diversity and Distributions* 14: 509–517.

Bragin, N. Y. 2000. The Permian–Triassic crisis in the biosphere as manifested in the Paleo-Pacific deep-water sequences. *Stratigraphy and Geological Correlation* 8: 232–242.

Brannstrom, A., and U. Dieckmann. 2005. Evolutionary dynamics of altruism and cheating among social amoebas. *Proceedings of the Royal Society B: Biological Sciences* 272: 1609–1616.

Brideau, N. J., H. A. Flores, J. Wang, S. Maheshwari, X. Wang, and D. A. Barbash. 2006. Two Dobzhansky–Muller genes interact to cause hybrid lethality in *Drosophila*. *Science* 314: 1292–1295.

Bridge, E. S., A. W. Jones, and A. J. Baker. 2005. A phylogenetic framework for terns (*Sternini*) inferred from mtDNA sequences: implications for taxonomy and plumage evolution. *Molecular Phylogenetics and Evolution* 35: 459–469.

Bridgham, J. T., S. M. Carroll, and J. W. Thornton. 2006. Evolution of hormone–receptor complexity by molecular exploitation. *Science* 312: 97–101.

Brodie, E. D. 1992. Correlational selection for color pattern and antipredator behavior in the garter snake, *Thamnophis ordinoides*. *Evolution* 46: 1284–1298.

Brodo, I., S. D. Sharnoff, and S. Sharnoff. 2001. *Lichens of North America*. Yale University Press, New Haven, CT.

Bronstein, J. L. 1994. Our current understanding of mutualism. *The Quarterly Review of Biology* 69: 31–51.

Brown, C. R., and M. B. Brown. 1996. *Coloniality in the Cliff Swallow: The Effect of Group Size on Social Behavior*. University of Chicago Press, Chicago.

Brown, C., and M. B. Brown. 2000. Heritable basis for choice of group size in a colonial bird. *Proceedings of the National Academy of Sciences of the United States of America* 97: 14825–14830.

Brown, C. R., and M. B. Brown. 2001. Egg hatchability increases with colony size in cliff swallows. *Journal of Field Ornithology* 72: 113–123.

Brown, C. R., and M. B. Brown. 2004a. Empirical measurement of parasite transmission between groups in a colonial bird. *Ecology* 85: 1619–1626.

Brown, C. R., and M. B. Brown. 2004b. Group size and ectoparasitism affect daily survival probability in a colonial bird. *Behavioral Ecology and Sociobiology* 56: 498–511.

Brown, D. D. 1997. The role of thyroid hormone in zebrafish and axolotl development. *Proceedings of the National Academy of Sciences of the United States of America* 94: 13011–13016.

Brown, J. H., and M. V. Lomolino. 1998. *Biogeography*, 2nd Ed. Sinauer Associates, Sunderland, MA. Accessed January 18, 2011. www.geo.arizona.edu/Antevs/ecol438/ratite.html.

Brown, J. R., and W. F. Doolittle. 1997. Archaea and the prokaryote-to-eukaryote transition. *Microbiology and Molecular Biology Reviews* 61: 456–502.

Brown, W. L., and E. O. Wilson. 1956. Character displacement. *Systematic Zoology* 5: 49–64.

Brunet, M., F. Guy, D. Pilbeam, H. T. Mackaye, A. Likius, D. Ahounta, A. Beauvilain, C. Blondel, H. Bocherens, J. R. Boisserie, L. De Bonis, Y. Coppens, J. Dejax, C. Denys, P. Duringer, V. R. Eisenmann, G. Fanone, P. Fronty, D. Geraads, T. Lehmann, F. Lihoreau, A. Louchart, A. Mahamat, G. Merceron, G. Mouchelin, O. Otero, P. P. Campomanes, M. Ponce De Leon, J.-C. Rage, M. Sapanet, M. Schuster, J. Sudre, P. Tassy, X. Valentin, P. Vignaud, L. Viriot, A. Zazzo, and C. Zollikofer. 2002. A new hominid from the Upper Miocene of Chad, central Africa. *Nature* 418: 145–151.

Bubb, K. L., D. Bovee, D. Buckley, E. Haugen, M. Kibukawa, M. Paddock, A. Palmieri, S. Subramanian, Y. Zhou, R. Kaul, P. Green, and M. V. Olson. 2006. Scan of human genome reveals no new loci under ancient balancing selection. *Genetics* 173: 2165–2177.

Buffon, G. L. 1778. *Historie Naturelle, Supplement, Epoques de lan Nature*. Imprimerie Royale, puis Plassan, Paris.

Buick, R. 2010. Early life: ancient acritarchs. *Nature* 463: 885–886.

Bulmer, M. G., and G. A. Parker. 2002. The evolution of anisogamy: a game–theoretic approach. *Proceedings of the Royal Society B: Biological Sciences* 269: 2381–2388.

Burkhardt, R. W., ed. 1984. *The Zoological Philosophy of J.-B. Lamarck*. University of Chicago Press, Chicago.

Burkhardt, R. W. 1995. *The Spirit of System: Lamarck and Evolutionary Biology*. Harvard University Press, Cambridge.

Bush, G. L. 1969. Sympatric race host formation and speciation in the frugivorous flies of the genus *Rhagoletis* (Diptera: Tephritidae). *Evolution* 23: 237–251.

Bush, G. L. 1975. Sympatric speciation in phytophagous parasitic insects. In P. W. Price, ed., *Evolutionary Strategies of Parasitic Insects and Mites*, pp. 237–251. Plenum, New York.

Buss, L. W. 1999. Slime molds, ascidians, and the utility of evolutionary theory. *Proceedings of the National Academy of Sciences of the United States of America* 96: 8801–8803.

Bustamante, C. D., A. Fledel-Alon, S. Williamson, R. Nielsen, M. T. Hubisz, S. Glanowski, D. M. Tanenbaum, T. J. White, J. J. Sninsky, R. D. Hernandez, D. Civello, M. D. Adams, M. Cargill, and A. G. Clark. 2005. Natural selection on protein-coding genes in the human genome. *Nature* 437: 1153–1157.

Byers, J. A., P. A. Wiseman, L. Jones, and T. J. Roffe. 2005. A large cost of female mate sampling in pronghorn. *American Naturalist* 166: 661–668.

Byrd, D. W., E. D. McArthur, H. Wang, J. H. Graham, and D. C. Freeman. 1999. Narrow hybrid zone between two subspecies of big sagebrush, *Artemisia tridentata* (Asteraceae). VIII. Spatial and temporal pattern of terpenes. *Biochemical Systematics and Ecology* 27: 11–25.

Byrne, K., and R. Nichols. 1999. *Culex pipiens* in London Underground tunnels: differentiation between surface and subterranean populations. *Heredity* 82: 7–15.

Cafaro, M. J., and C. R. Currie. 2005. Phylogenetic analysis of mutualistic filamentous bacteria associated with fungus-growing ants. *Canadian Journal of Microbiology* 51: 441–446.

Cafaro, M. J., M. Poulsen, A. E. Little, S. L. Price, N. M. Gerardo, B. Wong, A. E. Stuart, B. Larget, P. Abbot, and C. R. Currie. 2011. Specificity in the symbiotic association between fungus-growing ants and protective *Pseudonocardia* bacteria. *Proceedings of the Royal Society B: Biological Sciences* 278: 1814–1822.

Cain, A., and G. Harrison. 1960. Phyletic weighting. *Proceedings of the Zoological Society of London* 135: 1–31.

Callinan, P., and M. Batzer 2006. Retrotransposable elements and human disease. *Genome and Disease* 1: 104–115.

Campbell, I., G. Czamanske, V. Fedorenko, R. Hill, and V. Stepanov. 1992. Synchronism of the Siberian traps and the Permian–Triassic boundary. *Science* 258: 1760–1763.

Canchaya, C., C. Proux, G. Fournous, A. Bruttin, and H. Brussow. 2003. Prophage genomics. *Microbiology and Molecular Biology Reviews* 67: 238–276.

Cann, A. 2005. *Principles of Molecular Virology*, 4th Ed. Elsevier Academic Press, Amsterdam.

Carlisle, D. B. 1992. Diamonds at the K/T Boundary. *Nature* 357: 119–120.

Carlisle, D. B., and D. R. Braman. 1991. Nanometer size diamonds in the Cretaceous–Tertiary boundary clay of Alberta. *Nature* 352: 708–709.

Carr, D. E., and M. R. Dudash. 2003. Recent approaches into the genetic basis of inbreeding depression in plants. *Philosophical Transactions of the Royal Society of London Series B—Biological Sciences* 358: 1071–1084.

Carroll, R. 1988. *Vertebrate Paleontology and Evolution*. Freeman, New York.

Carroll, S. 2005. *Endless Forms Most Beautiful*. W.W. Norton, New York.

Carroll, S. B. 2008. Evo-devo and an expanding evolutionary synthesis: a genetic theory of morphological evolution. *Cell* 134: 25–36.

Carroll, S., J. Grenier, and S. D. Weatherbee. 2005. *From DNA to Diversity: Molecular Genetics and the Evolution of Animal Design.*, 2nd Ed. Blackwell, Malden, MA.

Carroll, S. B., B. Prud'homme, and N. Gompel. 2008. Regulating evolution. *Scientific American* 298: 60–67.

Carter, J., and V. Saunders, eds. 2007. *Virology: Principles and Applications*. Wiley, Chichester, UK.

Carvalho, G. R., and R. N. Hughes. 1983. The effect of food availability, female culture density, and photoperiod on ephippoa production in *Daphnia magna. Freshwater Biology* 13: 37–46.

Cavalier-Smith, T. 1978. Nuclear volume control by nucleoskeletal DNA: selection for cell volume and cell growth rate, and solution of DNA C-value paradox. *Journal of Cell Science* 34: 247–278.

Cavalli-Sforza, L. L., and M. W. Feldman. 1981. *Cultural Transmission and Evolution: A Quantitative Approach*. Princeton University Press, Princeton.

CDC, U.S. Department of Health and Human Services. 2007. *National Antimicrobial Systems for Enteric Bacteria (NARMS): Human Isolates Final Report, 2004*. Atlanta, GA.

Ceccatti, J. S. 2009. Natural selection in the field: insecticide resistance, economic entomology and the evolutionary synthesis, 1914–1951. *Transactions of the American Philosophical Society* 99: 199–217.

C. elegans Sequencing Consortium. 1998. Genome sequence of the nematode *C. elegans*: a platform for investigating biology. *Science* 282: 2012–2018.

Center for North American Herpetology. 2010. A modern taxonomy of chordates. Accessed November 10, 2010. www.cnah.org/taxonomy.asp.

Chambers, R. 1845. *Vestiges of the Natural History of Creation*. John Churchill, London.

Chao, L., and D. Carr. 1993. The molecular clock and the relationship between population size and generation time. *Evolution* 47: 688–690.

Chapman, C. R., J. G. Williams, and W. K. Hartmann. 1978. Asteroids. *Annual Review of Astronomy and Astrophysics* 16: 33–75.

Chapman, T. 2001. Seminal fluid-mediated fitness traits in *Drosophila. Heredity* 87: 511–521.

Chapman, T., G. Arnqvist, J. Bangham, and L. Rowe. 2003. Sexual conflict. *Trends in Ecology & Evolution* 18: 41–47.

Chapman, T., L. F. Liddle, J. M. Kalb, M. F. Wolfner, and L. Partridge. 1995. Cost of mating in *Drosophila melanogaster* females is mediated by male accessory gland products. *Nature* 373: 241–244.

Charlat, S., E. A. Hornett, J. Fullard, N. Davies, G. Roderick, N. Wedell, and G. Hurst. 2007. Extraordinary flux in sex ratio. *Science* 214: 317.

Charlesworth, B. 1996. The good fairy godmother of evolutionary genetics. *Current Biology* 6: 220–220.

Charlesworth, D., and S. I. Wright. 2001. Breeding systems and genome evolution. *Current Opinion in Genetics & Development* 11: 685–690.

Cheetham, A. H. 1986. Tempo of evolution in a Neogene bryozoan: rates of morphological change within and across species boundaries. *Paleobiology* 12: 190–202.

Chen, F. C., and W. H. Li. 2001. Genomic divergences between humans and other hominoids and the effective population size of the common ancestor of humans and chimpanzees. *American Journal of Human Genetics* 68: 444–456.

Chen, I., M. Hanczyc, P. Sazani, and J. Szostak. 2006. Protocells: genetic polymers inside membrane vesicles. In R. Gesteland, T. Cech, and J. Atkins, eds., *The RNA World*, 3rd Ed., pp. 57–88. Cold Spring Harbor Laboratory Press, Cold Spring Harbor, NY.

Christensen, K., T. E. Johnson, and J. W. Vaupel. 2006. The quest for genetic determinants of human longevity: challenges and insights. *Nature Reviews Genetics* 7: 436–448.

Christiansen, F. B. 2008. *Theories of Population Variation in Genes and Genomes*. Princeton University Press, Princeton.

Clardy, J., M. A. Fischbach, and C. R. Currie. 2009. The natural history of antibiotics. *Current Biology* 19: R437–R441.

Clark, A. B., and T. J. Ehlinger. 1987. Pattern and adaptation in individual behavioral differences. In P. P. G. Bateson and P. H. Klopfer, eds., *Perspectives in Ethology*, pp. 1–45. Plenum, New York.

Clausen, J., D. D. Keck, and W. M. Heisey. 1940. Experimental studies on the nature of species, I. The effect of varied environments on western North American plants. *Carnegie Institute of Washington Publication* 520: 1–452.

Clausen, J., D. D. Keck, and W. M. Heisey. 1948. Experimental studies on the nature of species, III. Environmental responses of climatic races of *Achillea. Carnegie Institute of Washington Publication* 581.

Clayton, D. H., S. E. Bush, B. M. Goates, and K. P. Johnson. 2003. Host defense reinforces host–parasite cospeciation. *Proceedings of the National Academy of Sciences of the United States of America* 100: 15694–15699.

Clayton, D. H., S. E. Bush, and K. P. Johnson. 2004. Ecology of congruence: past meets present. *Systematic Biology* 53: 165–173.

Clayton, D. H., and K. P. Johnson. 2003. Linking coevolutionary history to ecological process: doves and lice. *Evolution* 57: 2335–2341.

Clutton-Brock, T. H., ed. 1988. *Reproductive Success*. University of Chicago Press, Chicago.

Clutton-Brock, T., and K. McAuliffe. 2009. Female mate choice in mammals. *The Quarterly Review of Biology* 84: 3–27.

Coates, M., M. Ruta, and M. Friedman. 2008. Ever since Owen: changing perspectives on the early evolution of tetrapods. *Annual Review of Ecology and Systematics* 39: 571–592.

Cocroft, R. B. 2001. Vibrational communication and the ecology of group-living, herbivorous insects. *American Zoologist* 41: 1215–1221.

Cocroft, R. B., and R. L. Rodriguez. 2005. The behavioral ecology of insect vibrational communication. *Bioscience* 55: 323–334.

CoGePedia. 2009a. Bacteria genomic inversion E. coli K12. Accessed May 18, 2011. http://genomeevolution.org/wiki/index.php/Bacteria_Genomic_Inversion_E_.coli_K12.

CoGePedia. 2009b. SynMap syntenic dotplot between chromosome 4 of *Arabidopsis lyrata* (x-axis) and chromosome 2 of *Arabidopsis thaliana* (y-axis). Accessed March 17, 2011. http://genomevolution.org/wiki/index.php/File:SynMap-inversion.png.

Cohn, M. J., and C. Tickle. 1999. Developmental basis of limblessness and axial patterning in snakes. *Nature* 399: 474–479.

Cole, S. T., R. Brosch, J. Parkhill, T. Garnier, C. Churcher, D. Harris, S. V. Gordon, K. Eiglmeier, S. Gas, C. E. Barry, F. Tekaia, K. Badcock, D. Basham, D. Brown, T. Chillingworth, R. Conner, R. Davies, K. Devlin, T. Feltwell, S. Gentles, N. Hamlin, S. Holroyd, T. Hornsby, K. Jagels, A. Krogh, J. McLean, S. Moule, L. Murphy, K. Oliver, J. Osborne, M. A. Quail, M. A. Rajandream, J. Rogers, S. Rutter, K. Seeger, J. Skelton, R. Squares, S. Squares, J. E. Sulston, K. Taylor, S. Whitehead, and B. G. Barrell. 1998. Deciphering the biology of *Mycobacterium tuberculosis* from the complete genome sequence. *Nature* 393: 537–544.

Colegrave, N. 2002. Sex releases the speed limit on evolution. *Nature* 420: 664–666.

Coleman, R., M. Gross, and R. C. Sargent. 1985. Parental investment decision rules: a test in bluegill sunfish. *Behavioral Ecology and Sociobiology* 18: 59–66.

Collins, J. P., and A. Storfer. 2003. Global amphibian declines: sorting the hypotheses. *Diversity and Distributions* 9: 89–98.

Comas, I., J. Chakravartti, P. M. Small, J. Galagan, S. Nieman, K. Kremer, J. D. Ernst, and S. Gagneux. 2010. Human T cell epitopes of *Mycobacterium tuberculosis* are evolutionarily hyperconserved. *Nature Genetics* 42: 498–503.

Connor, R. C. 1995. The benefits of mutualism: a conceptual framework. *Biological Reviews* 70: 427–457.

Constantini, M., O. Clay, F. Auletta, and G. Bernardi. 2006. An isochore map of human chromosomes. *Genome Research* 16: 536–541.

Conway Morris, S. 1998. *The Crucible of Creation: The Burgess Shale and the Rise of Animals*. Oxford University Press, Oxford.

Coonan, T. J., C. A. Schwemm, G. W. Roemer, D. K. Garcelon, and L. Munson. 2005. Decline of an island fox subspecies to near extinction. *Southwestern Naturalist* 50: 32–41.

Cooper, G. M., and R. E. Hausman. 2009. *The Cell: A Molecular Approach*, 5th Ed. Sinauer Associates, Sunderland, MA.

Cooper, G. W., N. Kimmich, W. Belisle, J. Sarinana, K. Brabham, and L. Garrel. 2001. Carbonaceous meteorites as a source of sugar-related organic compounds for the early Earth. *Nature* 414: 879–883.

Cooper, T. F., R. E. Lenski, and S. F. Elena. 2005. Parasites and mutational load: an experimental test of a pluralistic theory for the evolution of sex. *Proceedings of the Royal Society B: Biological Sciences* 272: 311–317.

Cooper, V. S., A. F. Bennett, and R. E. Lenski. 2001. Evolution of thermal dependence of growth rate of *Escherichia coli* populations during 20,000 generations in a constant environment. *Evolution* 55: 889–896.

Cordaux, R., and M. A. Batzer. 2009. The impact of retrotransposons on human genome evolution. *Nature Reviews Genetics* 10: 691–703.

Coss, R. G. 1991. Context and animal behavior III. The relationship between early development and evolutionary persistence of ground squirrel antisnake behavior. *Ecological Psychology* 3: 277–315.

Coss, R. G., and D. H. Owings. 1985. Restraints on ground squirrel antipredator behavior: adjustments over multiple time scales. In T. D. Johnston and A. T. Pietrewicz, eds., *Issues in the Ecological Study of Learning*, pp. 167–200. Lawrence Erlbaum, Hillsdale, NJ.

Cothran, R. D. 2008. Direct and indirect fitness consequences of female choice in a crustacean. *Evolution* 62: 1666–1675.

Covert, J. B., and W. W. Reynolds. 1977. Survival value of fever in fish. *Nature* 267: 43–45.

Cox, C. J., P. G. Foster, R. P. Hirt, S. R. Harris, and T. M. Embley. 2008. The archaebacterial origin of eukaryotes. *Proceedings of the National Academy of Sciences of the United States of America* 105: 20356–20361.

Coyne, J. A., and H. A. Orr. 2004. *Speciation*. Sinauer Associates, Sunderland, MA.

Cracraft, J. 1973. Continental drift, paleoclimatology and the evolution and biogeography of birds. *Journal of Zoology* 169: 455–545.

Cracraft, J. 1974a. Continental drift and vertebrate distribution. *Annual Review of Ecology and Systematics* 5: 215–261.

Cracraft, J. 1974b. Phylogeny and the evolution of ratite birds. *Ibis* 116: 494–521.

Cracraft, J. 1989. Speciation and its ontology. In D. Otte and J. Endler, eds., *Speciation and Its Consequences*, pp. 28–59. Sinauer Associates, Sunderland, MA.

Creel, S. 2001. Cooperative hunting and sociality in African wild dogs, *Lycaon pictus*. In L. Dugatkin, ed., *Model Systems in Behavioral Ecology*, pp. 466–490. Princeton University Press, Princeton.

Crick, F. 1981. *Life Itself: Its Origin and Nature*. Simon and Schuster, New York.

Crooks, K. 1994. Demography and status of the island fox and the island spotted skunk on Santa-Cruz Island, California. *Southwestern Naturalist* 39: 257–262.

Crow, J., and M. Kimura. 1965. Evolution in sexual and asexual populations. *American Naturalist* 99: 439–450.

Cummings, C. L., H. M. Alexander, A. A. Snow, L. H. Rieseberg, M. J. Kim, and T. M. Culley. 2002. Fecundity selection in a sunflower crop-wild study: can ecological data predict crop allele changes? *Ecological Applications* 12: 1661–1671.

Curcio, M. J., and K. M. Derbyshire. 2003. The outs and ins of transposition: from mu to kangaroo. *Nature Reviews Molecular Cell Biology* 4: 865–877.

Currie, C. R., U. G. Mueller, and D. Malloch. 1999a. The agricultural pathology of ant fungus gardens. *Proceedings of the National Academy of Sciences of the United States of America* 96: 7998–8002.

Currie, C. R., J. A. Scott, R. C. Summerbell, and D. Malloch. 1999b. Fungus-growing ants use antibiotic-producing bacteria to control garden parasites. *Nature* 398: 701–704.

Daeschler, E. B., N. H. Shubin, and F. A. Jenkins. 2006. A Devonian tetrapod-like fish and the evolution of the tetrapod body plan. *Nature* 440: 757–763.

Dahl, E. L., and P. J. Rosenthal. 2008. Apicoplast translation, transcription and genome replication: targets for antimalarial antibiotics. *Trends in Parasitology* 24: 279–284.

Daly, M. J., J. D. Rioux, S. F. Schaffner, T. J. Hudson, and E. S. Lander. 2001. High-resolution haplotype structure in the human genome. *Nature Genetics* 29: 229–232.

Danforth, B. 2007. Bees. *Current Biology* 17: R156–R161.

Darwin, C. 1859. *On the Origin of Species*. John Murray, London.

Darwin, C. 1868. *The Variation of Animals and Plants under Domestication*, 1st Ed. John Murray, London.

Darwin, C. 1871. *The Descent of Man and Selection in Relation to Sex*. John Murray, London.

Darwin, C. 1872. *The Origin of Species*. John Murray, London.

Darwin, C. 1875. *The Variation of Animals and Plants under Domestication*, 2nd Ed. John Murray, London.

Darwin, E. 1796. *Zoonomia*. J. Johnson, London.

Darwin Project 2010. Entry-7471. Accessed December 16, 2010. www.darwinproject.ac.uk/home.

Daszak, P., L. Berger, A. A. Cunningham, A. D. Hyatt, D. E. Green, and R. Speare. 1999. Emerging infectious diseases and amphibian population declines. *Emerging Infectious Diseases* 5: 735–748.

Davies, N. B. 1992. *Dunnock Behavior and Social Evolution*. Oxford University Press, Oxford.

Dawkins, R., and J. R. Krebs. 1978. Animal signals: information or manipulation? In J. R. Krebs and N. B. Davies, eds., *Behavioural Ecology*, pp. 282–315. Sinauer Associates, Sunderland, MA.

Dawkins, R., and J. Krebs. 1979. Arms races between and within species. *Proceedings of the Royal Society B: Biological Sciences* 205: 489–511.

DeBeer, G. 1930. *Embryology and Evolution*. Claredon Press, Oxford.

DeBeer, G. 1940. *Embryos and Ancestors*. Claredon Press, Oxford.

Del Castillo, R. C., J. Núñez-Farfán, and Z. Cano-Santana. 1999. The role of body size in mating success of *Sphenarium purpurascens* in Central Mexico. *Ecological Entomology* 24: 146–155.

Delsuc, F., H. Brinkmann, and H. Philippe. 2005. Phylogenomics and the reconstruction of the tree of life. *Nature Reviews Genetics* 6: 361–375.

Dempster, W. 1996. *Natural Selection and Patrick Matthew: Evolutionary Concepts in the Nineteenth Century*. Pentland Press, Edinburgh.

Denison, R. F. 2000. Legume sanctions and the evolution of symbiotic cooperation by rhizobia. *American Naturalist* 156: 567–576.

Denison, R. F., E. T. Kiers, and S. A. West. 2003. Darwinian agriculture: when can humans find solutions beyond the reach of natural selection? *The Quarterly Review of Biology* 78: 145–168.

Denoel, M., P. Joly, and H. H. Whiteman. 2005. Evolutionary ecology of facultative paedomorphosis in newts and salamanders. *Biological Reviews* 80: 663–671.

De Oliveira, T., O. Pybus, A. Rambaut, M. Salemi, S. Cassol, M. Ciccozzi, G. Rezza, G. Gattinara, R. D'Arrigo, M. Amicosante, L. Perrin, V. Colizzi, C. Perno, and Benghazi Study Group. 2006. Molecular epidemiology: HIV-1 and HCV sequences from Libyan outbreak. *Nature* 444: 836–837.

De Queiroz, K. 2007. Species concepts and species delimitation. *Systematic Biology* 56: 879–886.

Desponts, M., and J.-P. Simon. 1987. *Structure et variabilité génétique de populations d'épinette noire (Picea mariana [Mill] B.S.P.) dans la zone hémiarctique du Nouveau-Québec. Canadian Journal of Forest Research* 17: 1006–1012.

De Vasser, J. and S. F. Elena. 2007. The evolution of sex: empirical insights into the roles of epistasis and drift. *Nature Reviews Genetics* 8: 139–149.

Diamond, J. 1984a. Historic extinctions: a Rosetta stone for understanding prehistoric extinctions. In P. Martin and R. Klein, eds., *Quaternary Extinctions: A Prehistoric Revolution*, pp. 824–862. University of Arizona Press, Tucson.

Diamond, J. 1984b. "Normal" extinctions of isolated populations. In M. Nitecki, ed., *Extinctions*, pp. 191–246. University of Chicago Press, Chicago.

Dice, L. R. 1947. Effectiveness of selection by owls of deer mice (*Peromyscus maniculatus*) which contrast in color with their background. *Contributions from the Laboratory of Vertebrate Biology, University of Michigan* 34: 1–20.

Dice, L. R., and P. Blossom. 1937. Studies of mammalian ecology in southwestern North America with special attention to the colors of desert mammals. *Publications of the Carnegie Institute* 485: 1–129.

Dietl, G. P. 2003a. Interaction strength between a predator and dangerous prey: *Sinistrofulgur* predation on *Mercenaria. Journal of Experimental Marine Biology and Ecology* 289: 287–301.

Dietl, G. P. 2003b. Coevolution of a marine gastropod predator and its dangerous bivalve prey. *Biological Journal of the Linnean Society* 80: 409–436.

Dietrich, M. R. 1994. The origins of the neutral theory of molecular evolution. *Journal of the History of Biology* 27: 21–59.

Dobzhansky, T. 1937. *Genetics and the Origin of Species.* Columbia University Press, New York.

Dobzhansky, T. 1958. Species after Darwin. In S. Barnett, ed., *A Century of Darwin*, pp. 19–55. Harvard University Press, Cambridge.

Dobzhansky, T. 1970. *Genetics of the Evolutionary Process.* Columbia University Press, New York.

Dobzhansky, T. 1973. Nothing in biology makes sense except in the light of evolution. *American Biology Teacher* 35: 125–129.

Dobzhansky, T., and O. Pavlovsky. 1957. An experimental study of interaction between genetic drift and natural selection. *Evolution* 11: 311–319.

Domingo, E., A. Grande-Perez, and V. Martin. 2008. Future prospects for the treatment of rapidly evolving viral pathogens: insights from evolutionary biology. *Expert Opinion on Biological Therapy* 8: 1455–1460.

Donley, J. M., C. A. Sepulveda, P. Konstantinidis, S. Gemballa, and R. E. Shadwick. 2004. Convergent evolution in mechanical design of lamnid sharks and tunas. *Nature* 429: 61–65.

Donoghue, P. C. J., and M. J. Benton 2007. Rocks and clocks: calibrating the tree of life using fossils and molecules. *Trends in Ecology & Evolution* 22: 424–431.

Doolittle, W. F. 2000. Uprooting the tree of life. *Scientific American* 2000: 90–95.

Doolittle, W. F., and C. Sapienza. 1980. Selfish genes, the phenotype paradigm and genome evolution. *Nature* 284: 601–603.

Dorus, S., E. J. Vallender, P. D. Evans, J. R. Anderson, S. L. Gilbert, M. Mahowald, G. J. Wyckoff, C. M. Malcom, and B. T. Lahn. 2004. Accelerated evolution of nervous system genes in the origin of *Homo sapiens. Cell* 119: 1027–1040.

dos Reis, M., L. Wernisch, and R. Savva. 2003. Unexpected correlations between gene expression and codon usage bias from microarray data for the whole *Escherichia coli* K-12 genome. *Nucleic Acids Research* 31: 6976–6985.

Dougherty, E. 1955. Comparative evolution and the origin of sexuality. *Systematic Zoology* 4: 145–169.

Drea, S. C., N. T. Lao, K. H. Wolfe, and T. A. Kavanagh. 2006. Gene duplication, exon gain and neofunctionalization of OEP16-related genes in land plants. *Plant Journal* 46: 723–735.

Dubnau, D. 1999. DNA uptake in bacteria. *Annual Review of Microbiology* 53: 217–244.

Dugatkin, L. A. 1997. *Cooperation Among Animals: An Evolutionary Perspective.* Oxford University Press, New York.

Dugatkin, L. A. 2004. *Principles of Animal Behavior*, 1st Ed. W.W. Norton, New York.

Dugatkin, L. A. 2009. *Principles of Animal Behavior*, 2nd Ed. W.W. Norton, New York.

Dugatkin, L. A., and H. K. Reeve. 1994. Behavioral ecology and "levels of selection": dissolving the group selection controversy. *Advances in The Study of Behaviour* 23: 101–133.

Dugatkin, L. A., M. Perlin, and R. Atlas. 2003. The evolution of group-beneficial traits in the absence of between-group selection: a model. *Journal of Theoretical Biology* 220: 67–74.

Dugatkin, L. A., M. Perlin, and R. Atlas. 2005. Antibiotic resistance and the evolution of group-beneficial traits. II: a metapopulation model. *Journal of Theoretical Biology* 236: 392–396.

Dugatkin, L. A., M. Perlin, J. S. Lucas, and R. Atlas. 2004. Group-beneficial traits, frequency-dependent selection and genotypic diversity: an antibiotic resistance paradigm. *Proceedings of the Royal Society B: Biological Sciences* 272: 79–83.

Dunn, L. C., A. B. Beasley, and H. Tinker. 1958. Relative fitness of wild house mice heterozygous for a lethal allele. *American Naturalist* 92: 215–220.

Dunn, L., and D. Bennett. 1967. Maintenance of gene frequency of a male sterile, semi-lethal *t*-allele in a confined population of wild mice. *American Naturalist* 101: 535–538.

Du Rietz, G. 1930. The fundamental units of biological taxonomy. *Svensk Botanisk Tidskrift* 24: 333–428.

Duvick, D., and K. Cassmann. 1999. Post-green trends in yield potential of temperate maize in the north-central United States. *Crop Science* 39: 1622–1630.

Dybdahl, M. F., and C. M. Lively. 1998. Host–parasite coevolution: evidence for rare advantage and time-lagged selection in a natural population. *Evolution* 52: 1057–1066.

Dyson, F. 1985. *Origins of Life.* Cambridge University Press, Cambridge.

Eberhard, W. G. 2009. Postcopulatory sexual selection: Darwin's omission and its consequences. *Proceedings of the National Academy of Sciences of the United States of America* 106: 10025–10032.

Egea, R., S. Casillas, and A. Barbadilla. 2008. Standard and generalized McDonald–Kreitman test: a website to detect selection by comparing different classes of DNA sites. *Nucleic Acids Research* 36: W157–W162.

Eigen, M., W. Gardiner, P. Schuster, and R. Winkler-Oswatitsch. 1981. The origin of genetic information. *Scientific American* 244: 88–118.

Eigen, M., and P. Schuster. 1977. The hypercycle: a principle of natural selection organization. Part A: emergence of the hypercycle. *Naturwissenschaften* 58: 465–523.

Eldredge, N., and S. J. Gould. 1972. Punctuated equilibrium: an alternative to phyletic gradualism. In T. J. M. Schopf, ed., *Models of paleobiology*, pp. 82–115. Freeman, Cooper and Company, San Francisco.

Elena, S. F., and R. E. Lenski. 1997. Test of synergistic interactions among deleterious mutations in bacteria. *Nature* 390: 395–398.

Elena, S. F., and R. E. Lenski. 2003. Evolution experiments with microorganisms: the dynamics and genetic bases of adaptation. *Nature Reviews Genetics* 4: 457–469.

Elgar, M., and N. Pierce. 1988. Mating success and fecundity in an ant-tended lycaenid butterfly. In T. Clutton-Brock, ed., *Reproductive Success*, pp. 59–75. University of Chicago Press, Chicago.

Elias, I. and J. Lagergren. 2007. Fast computation of distance estimators. Accessed January 18, 2011. www.biomedcentral.com/1471-2105/8/891.

Elliot, S. L., S. Blanford, and M. B. Thomas. 2002. Host–pathogen interactions in a varying environment: temperature, behavioural fever and fitness. *Proceedings of the Royal Society B: Biological Sciences* 269: 1599–1607.

Emerson, R., and E. M. East. 1913. Inheritance of quantitative characters in maize. *Nebraska Agriculture Experimental Station Research Bulletin* 2: 1–120.

Enard, W., and S. Pääbo. 2004. Comparative primate genomics. *Annual Review of Genomics and Human Genetics* 5: 351–378.

Endler, J. 1986. *Natural Selection in the Wild*. Princeton University Press, Princeton.

Endler, J. 1995. Multiple trait co-evolution and environmental gradients in guppies. *Trends in Ecology & Evolution* 10: 22–29.

Endler, J., and T. McLellan. 1988. The process of evolution: toward a newer synthesis. *Annual Review of Ecology and Systematics* 19: 395–421.

Engelstädter, J. 2008. Constraints on the evolution of asexual reproduction. *BioEssays* 30: 1138–1150.

Engelstädter, J., and G. D. D. Hurst. 2009. The ecology and evolution of microbes that manipulate host reproduction. *Annual Review of Ecology, Evolution and Systematics* 40: 127–149.

Enne, V. I., D. M. Livermore, P. Stephens, and L. M. Hall, 2001. Persistence of sulphonamide resistance in *Escherichia coli* in the UK despite national prescribing restriction. *Lancet* 357: 1325–1328.

Eppinger, M., C. Baar, G. Raddatz, D. H. Huson, and S. C. Schuster. 2004. Comparative analysis of four Campylobacterales. *Nature Reviews Microbiology* 2: 872–885.

Erwin, D. H. 2008. Extinction as the loss of evolutionary history. *Proceedings of the National Academy of Sciences of the United States of America* 105: 11520–11527.

Erwin, D. H., and R. L. Anstey. 1995. Speciation in the fossil record. In *New Approaches to Speciation in the Fossil Record*, pp. 11–38. Columbia University Press, New York.

Eshet, Y., M. R. Rampino, and H. Visscher. 1995. Fungal event and palynological record of the ecological crisis and recovery across the Permian–Triassic boundary. *Geology* 23: 967–970.

Esser, C., and W. Martin. 2007. Supertrees and symbiosis in eukaryote genome evolution. *Trends in Microbiology* 15: 435–437.

Evans, D. T., and R. C. Desrosiers. 2001. Immune evasion strategies of the primate lentiviruses. *Immunological Reviews* 183: 141–158.

Evart, J. C. 1921. The nestling feathers of the mallard with observations on the composition, origin and history of feathers. *Proceedings of the Zoological Society of London* 1921: 609–642.

Exploratorium. 2010. Mutant fruit flies. Accessed November 18. 2010. www.exploratorium.edu/exhibits/mutant_flies.

Eyre-Walker, A. 2006. The genomic rate of adaptive evolution. *Trends in Ecology & Evolution* 21: 569–575.

Fahrenholz, H. 1909. *Aus dem Myobien-Nachlass des Herrn Poppe. Abhandlungen des Naturwissenschaftlichen Vereins Zu Bremen* 19: 359–370.

Fairbanks, A., ed. 1898. *The First Philosophers of Greece*. Paul, Trench and Trubner, London.

Faith, D. P. 1992. Conservation evaluation and phylogenetic diversity. *Biological Conservation* 61: 1–10.

Fan, Y., E. Linardopoulou, C. Friedman, E. Williams, and B. J. Trask. 2002. Genomic structure and evolution of the ancestral chromosome fusion site in 2q13–2q14.1 and paralogous regions on other human chromosomes. *Genome Research* 12: 1651–1662.

Farrelly, F., and R. A. Butow. 1983. Rearranged mitochondrial genes in the yeast nuclear genome. *Nature* 301: 296–301.

Feder, J. L., and K. E. Filchak. 1999. It's about time: the evidence for host plant-mediated selection in the apple maggot fly, *Rhagoletis pomonella*, and its implications for fitness trade-offs in phytophagous insects. *Entomologia Experimentalis Et Applicata* 91: 211–225.

Feder, J. L., S. B. Opp, B. Wlazlo, K. Reynolds, W. Go, and S. Spisak. 1994. Host fidelity is an effective premating barrier between sympatric races of the apple maggot fly. *Proceedings of the National Academy of Sciences of the United States of America* 91: 7990–7994.

Feder, J. L., J. B. Roethele, B. Wlazlo, and S. H. Berlocher. 1997. Selective maintenance of allozyme differences among sympatric host races of the apple maggot fly. *Proceedings of the National Academy of Sciences of the United States of America* 94: 11417–11421.

Felsenstein J. 1965. The effect of linkage on directional selection. *Genetics* 52: 349–363.

Felsenstein, J. 1974. Evolutionary advantage of recombination. *Genetics* 78: 737–756.

Felsenstein, J. 1978. Cases in which parsimony or compatibility methods will be positively misleading. *Systematic Zoology* 27: 401–410.

Felsenstein, J. 1985. Phylogenies and the comparative method. *American Naturalist* 125: 1–15.

Felsenstein, J. 1988. Sex and the evolution of recombination. In R. E. Michod and B. Levin, eds., *The Evolution of Sex*, pp. 74–86. Sinauer Associates, Sunderland, MA.

Felsenstein, J. 2004. *Inferring Phylogenies*. Sinauer Associates, Sunderland, MA.

Felsenstein, J., and S. Yokoyama. 1976. Evolutionary advantage of recombination. 2. Individual selection for recombination. *Genetics* 83: 845–859.

Feng, D. F., G. Cho, and R. F. Doolittle. 1997. Determining divergence times with a protein clock: update and reevaluation. *Proceedings of the National Academy of Sciences of the United States of America* 94: 13028–13033.

Feng, D. F., and R. F. Doolittle. 1987. Progressive sequence alignment as a prerequisite to correct phylogenetic trees. *Journal of Molecular Evolution* 25: 351–360.

Fey, P., A. S. Kowal, P. Gaudet, K. E. Pilcher, and R. L. Chisholm. 2007. Protocols for growth and development of *Dictyostelium discoideum*. *Nature Protocols* 2: 1307–1316.

Fiedler, K., and U. Maschwitz. 1988. Functional analysis of the myrmecophilous relationships between ants (Hymenoptera: Formicidae) and lycaenids (Lepidotera: Lycaenidae). *Oecologia* 75: 204–206.

Fiers, W., R. Contreras, F. Duerinck, G. Haegeman, D. Iserentant, J. Merregaert, W. Minjou, F. Molemans, A. Raeymaekers, A. Vandenberghe, G. Volckaert, and M. Ysebaert. 1976. Complete nucleotide sequence of bacteriophage MS2 RNA: primary and secondary structure of replicase gene. *Nature* 260: 500–507.

Filchak, K. E., J. B. Roethele, and J. L. Feder. 2000. Natural selection and sympatric divergence in the apple maggot *Rhagoletis pomonella*. *Nature* 407: 739–742.

Fisher, P. R. 1997. Genetics of phototaxis in a model eukaryote, *Dictyostelium discoideum*. *BioEssays* 19: 397–407.

Fisher, R. A. 1915. The evolution of sexual preference. *Eugenics Review* 7: 184–192.

Fisher, R. 1918. The correlation between relatives on the supposition of Mendelian inheritance. *Transactions of the Royal Society of Edinburgh* 52: 399–433.

Fisher, R. A. 1930. *The Genetical Theory of Natural Selection*. Dover, New York.

Fisher, R. A. 1958. *The Genetical Theory of Natural Selection*, 2nd Rev. Ed. Dover, New York.

Fitch, W. M. 1971. Toward defining the course of evolution: minimum change for a specified tree topology. *Systematic Zoology* 20: 406–416.

Fleischmann, R. D., M. D. Adams, O. White, R. A. Clayton, E. F. Kirkness, A. R. Kerlavage, C. J. Bult, J. F. Tomb, B. A. Dougherty, J. M. Merrick, K. McKenney, G. Sutton, W. FitzHugh, C. Fields, J. D. Gocayne, J. Scott, R. Shirley, L. Liu, A. Glodek, J. M. Kelley, J. F. Weidman, C. A. Phillips, T. Spriggs, E. Hedblom, M. D. Cotton, T. R. Utterback, M. C. Hanna, D. T. Nguyen, D. M. Saudek, R. C. Brandon, L. D. Fine, J. L. Fritchman, J. L. Fuhrmann, N. S. M. Geoghagen, C. L. Gnehm, L. A. McDonald, K. V. Small, C. M. Fraser, H. O. Smith, and J. C. Venter. 1995. Whole-genome random sequencing and assembly of *Haemophilus influenzae* Rd. *Science* 269: 496–512.

Foerster, K., T. Coulson, B. C. Sheldon, J. Pemberton, and T. Clutton-Brock. 2007. Sexually antagonistic genetic variation for fitness in red deer. *Nature* 447: 1107–1111.

Foote, M. 2003. Origination and extinction through the Phanerozoic: a new approach. *Journal of Geology* 111: 125-148.

Forterre, P. 2006. The origin of viruses and their possible roles in major evolutionary transitions. *Virus Research* 117: 5–16.

Fowler, N. L., and D. A. Levin. 1984. Ecological constraints on the establishment of a novel polyploid in competition with its diploid progenitor. *American Naturalist* 124: 703–711.

Fox, S. W., and K. Dose. 1977. *Molecular Evolution and the Origin of Life*. Marcel Dekker, New York.

Frankham, R. 1995. Effective population size/adult population size ratios in wildlife: a review. *Genetical Research* 66: 95–107.

Franklin, L. R. 2007. Bacteria, sex, and systematics. *Philosophy of Science* 74: 69–95.

Franks, S. J., S. Sims, and A. E. Weis. 2007. Rapid evolution of flowering time by an annual plant in response to a climate fluctuation. *Proceedings of the National Academy of Sciences of the United States of America* 104: 1278–1282.

Frary, A., T. C. Nesbitt, A. Frary, S. Grandillo, E. van der Knaap, B. Cong, J. P. Liu, J. Meller, R. Elber, K. B. Alpert, and S. D. Tanksley. 2000. fw2.2: a quantitative trait locus key to the evolution of tomato fruit size. *Science* 289: 85–88.

Freeberg, T. M. 2004. Social transmission of courtship behavior and mating preferences in brown-headed cowbirds, *Molothrus ater*. *Learning & Behavior* 32: 122–130.

Freedberg, S., and M. Wade. 2001. Cultural inheritance as a mechanism for sex-ratio bias in reptiles. *Evolution* 55: 1049–1055.

Frentiu, F., and A. Briscoe. 2008. A butterfly's eye view of birds. *BioEssays* 30: 1151–1162.

Fricke, C., J. Perry, T. Chapman, and L. Rowe. 2009. The conditional economics of sexual conflict. *Biology Letters* 5: 671–674.

Frohlich, M. W., and M. W. Chase. 2007. After a dozen years of progress the origin of angiosperms is still a great mystery. *Nature* 450: 1184–1189.

Fry, B. G. 2003a. Isolation of a neurotoxin (alpha colubritoxin) from a nonvenomous colubrid: evidence for early origin of venom in snakes. *Journal of Molecular Evolution* 57: 446–452.

Fry, B. G. 2003b. Molecular evolution and phylogeny of elapid snake venom three-finger toxins. *Journal of Molecular Evolution* 57: 110–129.

Fry, B. G., N. Vidal, J. A. Norman, F. J. Vonk, H. Scheib, S. F. R. Ramjan, S. Kuruppu, K. Fang, S. B. Hedges, M. K. Richardson, W. C. Hodgson, V. Ignjatovic, R. Summerhayes, and E. Kochva. 2006. Early evolution of the venom system in lizards and snakes. *Nature* 439: 584–588.

Funes, S., E. Davidson, A. Reyes-Prieto, S. Magallon, P. Herion, M. P. King, and D. Gonzalez-Halphen. 2002. A green algal apicoplast ancestor. *Science* 298: 2155.

Funes, S., A. Reyes-Prieto, X. Perez-Martinez, and D. Gonzalez-Halphen. 2004. On the evolutionary origins of apicoplasts: revisiting the rhodophyte vs. chlorophyte controversy. *Microbes and Infection* 6: 305–311.

Futuyma, D. J. 2010. Evolutionary constraint and ecological consequences. *Evolution* 64: 1865–1884.

Gage, M. J. G. 2003. Evolutionary biology: scramble for the eggs. *Nature* 426: 22–23.

Galef, B. G., and K. N. Laland. 2005. Social learning in animals: empirical studies and theoretical models. *Bioscience* 55: 489–499.

Galef, B. G., and E. E. Whiskin. 2006. Increased reliance on socially acquired information while foraging in risky situations? *Animal Behaviour* 72: 1169–1176.

Galef, B., and S. Wigmore. 1983. Transfer of information concerning distant foods: a laboratory investigation of the "information-centre" hypothesis. *Animal Behaviour* 31: 748–758.

Galtier, N., G. Piganeau, D. Mouchiroud, and L. Duret. 2001. GC-content evolution in mammalian genomes: the biased gene conversion hypothesis. *Genetics* 159: 907–911.

Gamache, I., J. P. Jaramillo-Correa, S. Payette, and J. Bousquet. 2003. Diverging patterns of mitochondrial and nuclear DNA diversity in subarctic black spruce: imprint of a founder effect associated with postglacial colonization. *Molecular Ecology* 12: 891–901.

Ganapathy, R. 1980. A major meteorite impact on the earth 65 million years ago: evidence from the Cretaceous–Tertiary boundary clay. *Science* 209: 921–923.

Ganti, T. 1978. Chemical systems and super-systems. 3. Models of self-reproducing chemical super-systems—Chemotons. *Acta Chimica Academiae Scientiarum Hungaricae* 98: 265–283.

Gardner, M. J., N. Hall, E. Fung, O. White, M. Berriman, R. W. Hyman, J. M. Carlton, A. Pain, K. E. Nelson, S. Bowman, I. T. Paulsen, K. James, J. A. Eisen, K. Rutherford, S. L. Salzberg, A. Craig, S. Kyes, M. S. Chan, V. Nene, S. J. Shallom, B. Suh, J. Peterson, S. Angiuoli, M. Pertea, J. Allen, J. Selengut, D. Haft, M. W. Mather, A. B. Vaidya, D. M. A. Martin, A. H. Fairlamb, M. J. Fraunholz, D. S. Roos, S. A. Ralph, G. I. McFadden, L. M. Cummings, G. M. Subramanian, C. Mungall, J. C. Venter, D. J. Carucci, S. L. Hoffman, C. Newbold, R. W. Davis, C. M. Fraser, and B. Barrell. 2002. Genome sequence of the human malaria parasite *Plasmodium falciparum*. *Nature* 419: 498–511.

Garrard, A. 1999. Charting the emergence of cereal and pulse domestication in Southeast Asia. *Environmental Archaeology* 4: 67–86.

Gaucher, E., S. Govindarajan, and O. Ganesh. 2007. Palaeotemperature trend for Precambrian life inferred from resurrected proteins. *Nature* 451: 704–708.

Gaut, B. S., B. R. Morton, B. C. McCaig, and M. T. Clegg. 1996. Substitution rate comparisons between grasses and palms: synonymous rate differences at the nuclear gene *Adh* parallel rate differences at the plastid gene *rbcL*. *Proceedings of the National Academy of Sciences of the United States of America* 93: 10274–10279.

Genereux, D., and C. T. Bergstrom. 2005. Evolution in action: understanding antibiotic resistance. In J. Cracraft and R. Bybee, eds., *Evolutionary Science and Society*, pp. 145–153. AIBS, Washington, DC.

Genome Center of Wisconsin. 2011. Illustrated map of the *E. coli* O157:H7 genome being studied in Fred Blattner's research lab at the Genome Center of Wisconsin. Board of Regents of the University of Wisconsin System. Accessed March 17, 2011. www.news.wisc.edu/newsphotos/images/Ecoli_o157_genome01.jpg.

Gerrish, P. J., and R. E. Lenski 1998. The fate of competing beneficial mutations in an asexual population. *Genetica* 103: 127–144.

Ghedin, E., N. A. Sengamalay, M. Shumway, J. Zaborsky, T. Feldblyum, V. Subbu, D. J. Spiro, J. Sitz, H. Koo, P. Bolotov, D. Dernovoy, T. Tatusova, Y. Bao, K. St. George, J. Taylor, D. J. Lipman, C. M. Fraser, J. K. Taubenberger, and S. L. Salzberg. 2005. Large-scale sequencing of human influenza reveals the dynamic nature of viral genome evolution. *Nature* 437: 1162–1166.

Ghiselin, M. T. 1974. *The Economy of Nature and the Evolution of Sex*. University of California Press, Berkeley.

Gibbons, J. 1979. A model of sympatric speciation in *Megarhyssa* (Hymenoptera: Ichneumonidae): competitive speciation. *American Naturalist* 114: 719–741.

Gibson, D. G., J. I. Glass, C. Lartigue, V. N. Noskov, R. Y. Chuang, M. A. Algire, G. A. Benders, M. G. Montague, L. Ma, M. M. Moodie, C. Merryman, S. Vashee, R. Krishnakumar, N. Assad-Garcia, C. Andrews-Pfannkoch, E. A. Denisova, L. Young, Z. Q. Qi, T. H. Segall-Shapiro, C. H. Calvey, P. Parmar, C. A. Hutchison III, H. O. Smith, and J. C. Venter 2010. Creation of a bacterial cell controlled by a chemically synthesized genome. *Science* 329: 52–56.

Gil, R., A. Latorre, and A. Moya. 2004a. Bacterial endosymbionts of insects: insights from comparative genomics. *Environmental Microbiology* 6: 1109–1122.

Gil, R., F. J. Silva, J. Pereto, and A. Moya. 2004b. Determination of the core of a minimal bacterial gene set. *Microbiology and Molecular Biology Reviews* 68: 518–537.

Gilbert, W. 1986. The RNA world. *Nature* 319: 618.

Gilbert, W. 1987. The exon theory of genes. *Cold Spring Harbor Symposia on Quantitative Biology* 52: 901–905.

Gilmour, J. S. 1937. A taxonomic problem. *Nature* 139: 1040–1042.

Giovannoni, S. J., S. Turner, G. J. Olsen, S. Barns, D. J. Lane, and N. R. Pace. 1988. Evolutionary relationships among cyanobacteria and green chloroplasts. *Journal of Bacteriology* 170: 3584–3592.

Glick, T. F., ed. 1974. *The Comparative Reception of Darwinism*. University of Texas Press, Austin.

Glick, T. F., ed. 1988. *The Comparative Reception of Darwinism* [with a new preface]. University of Chicago Press, Chicago.

Gluckman, P., and N. Hansen, N. 2006. *Mismatch: Why Our World No Longer Fits Our Bodies*. Oxford University Press, Oxford.

Goddard, M. R., H. C. J. Godfray, and A. Burt. 2005. Sex increases the efficacy of natural selection in experimental yeast populations. *Nature* 434: 636–640.

Godfrey-Smith, P. 2006. The strategy of model-based science. *Biology & Philosophy* 21: 725–740.

Goffeau, A., B. G. Barrell, H. Bussey, R. W. Davis, B. Dujon, H. Feldmann, F. Galibert, J. D. Hoheisel, C. Jacq, M. Johnston, E. J. Louis, H. W. Mewes, Y. Murakami, P. Philippsen, H. Tettelin, and S. G. Oliver. 1996. Life with 6000 genes. *Science* 274: 546–567.

Gojobori, T., E. Moriyama, and M. Kimura. 1990. Molecular clock of viral evolution and the neutral theory. *Proceedings of the National Academy of the Sciences of the United States of America* 87: 10015–10018.

Goldsby, R. A., T. J. Kindt, and B. A. Osborne. 2000. *Kuby Immunology*, 4th Ed. Freeman, New York.

Goldschmidt, R. 1938. *Physiological Genetics*. McGraw-Hill, New York.

Goldschmidt, R. 1940. *The Material Basis of Evolution*. Yale University Press, New Haven, CT.

Goldsmith, T. H. 1990. Optimization, constraint and history in the evolution of eyes. *The Quarterly Review of Biology* 65: 281–322.

Golubovsky, M. D., and Y. M. Gall. 2003. R. Goldschmidt and J. Huxley: creative parallelisms. *Zhurnal Obshchei Biologii* 64: 510–518.

Gonzalez, E. R., and L. Watling. 2002. Redescription of *Hyalella azteca* from its type locality, Vera Cruz, Mexico (Amphipoda: Hyalellidae). *Journal of Crustacean Biology* 22: 173–183.

González-Andrés, F., J. Chávez, G. Montáñez, and J. L. Ceresuela. 1999. Characterisation of woody *Medicago* (sect. *Dendrotelis*) species, on the basis of seed and seedling morphometry. *Genetic Resources and Crop Evolution* 46: 505–519.

Gottlieb, L. D. 1973. Genetic differentiation, sympatric speciation, and origin of a diploid species of *Stephanomeria*. *American Journal of Botany* 60: 545–553.

Gottlieb, L. D., S. I. Warwick, and V. S. Ford. 1985. Morphological and electrophoretic divergence between *Layia discoidea* and *Layia glandulosa*. *Systematic Botany* 10: 484–495.

Gould, S. J. 1974. Size and shape: the immutable laws of design set limits on all organisms. *Natural History*, January 1974.

Gould, S. J. 1977. *Ontogeny and Phylogeny*. Harvard University Press, Cambridge.

Gould, S. J. 1979. Mickey Mouse meets Konrad Lorenz. *Natural History* 88: 30–36.

Gould, S. J. 1985. The paradox of the first tier: an agenda for paleobiology. *Paleobiology* 11: 2–12.

Gould, S. J. 1987. Justice Scalia's misunderstanding. *Natural History* 96: 14–21.

Gould, S. J. 1991. Fall in the house of Ussher. *Natural History* 100: 12–20.

Gould, S. J. 2002. *The Structure of Evolutionary Theory*. Harvard University Press, Cambridge.

Gould, S., and N. Eldredge. 1993. Punctuated equilibrium comes of age. *Nature* 366: 223–227.

Gould, S. J., and E. Vrba. 1982. Exaptation: a missing term in science of form. *Paleobiology* 8: 4–15.

Grafen, A. 1984. Natural selection, kin selection and group selection. In J. Krebs and N. Davies, eds., *Behavioural Ecology: An Evolutionary Approach*, pp. 62–84. Blackwell, London.

Grafen, A. 1990. Biological signals as handicaps. *Journal of Theoretical Biology* 144: 517–546.

Graham, J. B., and K. A. Dickson. 2000. The evolution of thunniform locomotion and heat conservation in scombrid fishes: new insights based on the morphology of *Allothunnus fallai*. *Zoological Journal of the Linnean Society* 129: 419–466.

Graham, J. B., and K. A. Dickson. 2004. Tuna comparative physiology. *Journal of Experimental Biology* 207: 4015–4024.

Grant, B. R., and P. R. Grant. 1996. Cultural inheritance of song and its role in the evolution of Darwin's finches. *Evolution* 50: 2471–2487.

Grant, P. R., and B. R. Grant. 1994. Phenotypic and genetic effects of hybridization in Darwin's finches. *Evolution* 48: 297–316.

Grant, P. R., and B. R. Grant. 1997. Hybridization, sexual imprinting, and mate choice. *American Naturalist* 149: 1–28.

Grant, V. 1992. Floral isolation between ornithophilous and sphingophilus species of *Ipomopsis* and *Aquilegia*. *Proceedings of the National Academy of Sciences of the United States of America* 89: 11828–11831.

Grant, V. 1994. Modes and origins of mechanical and ethological isolation in angiosperms. *Proceedings of the National Academy of Sciences of the United States of America* 91: 3–10.

Gray, D. A., and W. H. Cade. 1999. Quantitative genetics of sexual selection in the field cricket, *Gryllus integer*. *Evolution* 53: 848–854.

Gray, M. W., G. Burger, and B. F. Lang. 1999. Mitochondrial evolution. *Science* 293: 1476–1482.

Green, R. E., J. Krause, A. W. Briggs, T. Maricic, U. Stenzel, M. Kircher, N. Patterson, H. Li, W. W. Zhai, M. H. Y. Fritz, N. F. Hansen, E. Y. Durand, A. S. Malaspinas, J. D. Jensen, T. Marques-Bonet, C. Alkan, K. Prufer, M. Meyer, H. A. Burbano, J. M. Good, R. Schultz, A. Aximu-Petri, A. Butthof, B. Hober, B. Hoffner, M. Siegemund, A. Weihmann, C. Nusbaum, E. S. Lander, C. Russ, N. Novod, J. Affourtit, M. Egholm, C. Verna, P. Rudan, D. Brajkovic, Z. Kucan, I. Gusic, V. B. Doronichev, L. V. Golovanova, C. Lalueza-Fox, M. de la Rasilla, J. Fortea, A. Rosas, R. W. Schmitz, P. L. F. Johnson, E. E. Eichler, D. Falush, E. Birney, J. C. Mullikin, M. Slatkin, R. Nielsen, J. Kelso, M. Lachmann, D. Reich, and S. Pääbo. 2010. A draft sequence of the Neandertal genome. *Science* 328: 710–722.

Gregory, T. R. 2001. Coincidence, coevolution, or causation? DNA content, cell size, and the C-value enigma. *Biological Reviews* 76: 65–101.

Gregory, T. R. 2005. Synergy between sequence and size in large-scale genomics. *Nature Reviews Genetics* 6: 699–708.

Gregory, T. R. 2008. Evolutionary trends. *Evolution: Education and Outreach* 1: 259–273.

Gregory, T. R. 2011. Genomes large and small: the evolution of genome size in eukaryotes. In S. Gilles and S. Hewitt, eds., *Biology on the Cutting Edge: Concepts, Issues, and Canadian Research around the Globe*, pp. 107–111. Pearson, Toronto.

Gregory, T. R., and R. DeSalle. 2005. Comparative genomics in prokaryotes. In T. R. Gregory, ed., *The Evolution of the Genome*, pp. 521–583. Elsevier, San Diego.

Greig, D., R. H. Borts, and E. J. Louis. 1998. The effect of sex on adaptation to high temperature in heterozygous and homozygous yeast. *Proceedings of the Royal Society B: Biological Sciences* 265: 1017–1023.

Grenfell, B. T., O. G. Pybus, J. R. Gog, J. L. N. Wood, J. M. Daly, J. A. Mumford, and E. C. Holmes. 2004. Unifying the epidemiological and evolutionary dynamics of pathogens. *Science* 327–332.

Gribaldo, S., and C. Brochier-Armanet. 2006. The origin and evolution of Archaea: a state of the art. *Philosophical Transactions of the Royal Society of London Series B—Biological Sciences* 361: 1007–1022.

Grosberg, R. K., and R. R. Strathmann. 2007. The evolution of multicellularity: a minor major transition? *Annual Review of Ecology and Systematics* 38: 621–654.

Gross, M. R. 1982. Sneakers, satellites, and parentals: polymorphic mating strategies in North American sunfishes. *Zeitschrift fur Tierpsychologie* 60: 1–26.

Gross, M. 1985. Disruptive selection for alternative life histories in salmon. *Nature* 313: 47–48.

Gross, M., and R. Charnov. 1980. Alternative male life histories in bluegill sunfish. *Proceedings of the National Academy of Sciences of the United States of America* 77: 6937–6940.

Grunstein, M., P. Schedl, and L. Kedes. 1976. Sequence analysis and evolution of sea urchin (*Lytechinus pictus* and *Strongylocentrotus purpuratus*) histone H4 messenger-RNAs. *Journal of Molecular Biology* 104: 3513–69.

Gschloessl, B., Y. Guermeur, and J. M. Cock. 2008. HECTAR: a method to predict subcellular targeting in heterokonts. *BMC Bioinformatics* 9: 393.

Guerrier-Takada, C., K. Gardiner, T. Marsh, N. Pace, and S. Altman. 1983. The RNA moiety of ribonuclease P is the catalytic subunit of the enzyme. *Cell* 35: 849–857.

Gupta, R. S. 1998. Protein phylogenies and signature sequences: a reappraisal of evolutionary relationships among archaebacteria, eubacteria, and eukaryotes. *Microbiology and Molecular Biology Reviews* 62: 1435–1491.

Haddrath, O., and A. J. Baker. 2001. Complete mitochondrial DNA genome sequences of extinct birds: ratite phylogenetics and the vicariance biogeography hypothesis. *Proceedings of the Royal Society B: Biological Sciences* 268: 939–945.

Hahn, M. W., and G. A. Wray. 2002. The G-value paradox. *Evolution & Development* 4: 73–75.

Haig, D. 2010. Games in tetrads: segregation, recombination, and meiotic drive. *American Naturalist* 176: 404–413.

Hain, T., and B. Neff. 2006. Promiscuity drives self-referent recognition. *Current Biology* 16: 1807–1811.

Haldane, J. B. S. 1922. Sex ratio and unisexual sterility in animals. *Journal of Genetics* 12: 101–109.

Haldane, J. B. S. 1927. A mathematical theory of natural and artificial selection. Part V. Selection and mutation. *Proceedings of the Cambridge Philosophical Society* 23: 838.

Haldane, J. B. S. 1928. *Possible Worlds*. Harper Brothers, New York.

Haldane, J. B. S. 1932. *The Causes of Evolution*. Longmans Green, London. Reprinted in 1990 with a new afterword by E. G. Leigh, Jr. Princeton University Press, Princeton. Page reference is to the 1990 edition.

Haldane, J. B. S. 1941. *New Paths in Genetics*. George Allen & Unwin, London.

Haldane, J. B. S. 1957. The cost of natural selection. *Journal of Genetics* 55: 511–524.

Hale, M. L., M. H. Verduijn, A. P. Moller, K. Wolff, and M. Petrie. 2009. Is the peacock's train an honest signal of genetic quality at the major histocompatibility complex? *Journal of Evolutionary Biology* 22: 1284–1294.

Hallam, A., and P. B. Wignall. 1997. *Mass Extinctions and Their Aftermath*. Oxford University Press, Oxford.

Hamilton, W. D. 1963. The evolution of altruistic behavior. *American Naturalist* 97: 354–356.

Hamilton, W. D. 1964. The genetical evolution of social behaviour. I and II. *Journal of Theoretical Biology* 7: 1–52.

Hamilton, W. D. 1967. Extraordinary sex ratios. *Science* 156: 477–487.

Hamilton, W. D. 1971. Geometry for the selfish herd. *Journal of Theoretical Biology* 31: 295–311.

Hamilton, W. D. 1980. Sex versus non-sex versus parasite. *Oikos* 35: 282–290.

Hamilton, W. D., R. Axelrod, and R. Tanese. 1990. Sexual reproduction as an adaptation to resist parasites. *Proceedings of the National Academy of Sciences of the United States of America* 87: 3566–3573.

Hamilton, W. D., and M. Zuk. 1982. Heritable true fitness and bright birds: a role for parasites. *Science* 218: 384–387.

Harcourt, A. H., P. Harvey, S. Larson, and R. Short. 1981. Testis weight, body size and breeding system in primates. *Nature* 293: 55–57.

Hardin, G. 1968. The tragedy of the commons. *Science* 162: 1243–1248.

Hardy, G. H. 1908. Mendelian proportions in a mixed population. *Science* 28: 49–50.

Harris, H. 1966. Enzyme polymorphisms in man. *Proceedings of the Royal Society B: Biological Sciences* 164: 298–310.

Harris, R. N. 1987. Density-dependent pedomorphosis in the salamander, *Notophthalmus viridescens dorsalis*. *Ecology* 68: 705–712.

Harrison, R. G., and D. M. Rand. 1989. Mosaic hybrid zones and the nature of species boundaries. In D. Otte and J. Endler, eds. *Speciation and Its Consequences*, pp. 111–133. Sinauer Associates, Sunderland, MA.

Hartl, D. L. 1972. Population dynamics of sperm and pollen killers. *Theoretical and Applied Genetics* 42: 81–88.

Hartl, D. L., and A. G. Clark 2007. *Principles of Population Genetics*, 4th Ed. Sinauer Associates, Sunderland, MA.

Hartl, D. L., Y. Hiraizumi, and J. Crow. 1967. Evidence for sperm dysfunction as the mechanism of segregation distortion in *Drosophila melanogaster*. *Proceedings of the National Academy of Sciences of the United States of America* 58: 2240–2245.

Hartman, H., and A. Fedorev. 2002. The origin of the eukaryotic cell: a genomic investigation. *Proceedings of the National Academy of Sciences of the United States of America* 99: 1420–1425.

Hasday, J. D., and I. S. Singh. 2000. Fever and the heat shock response: distinct, partially overlapping processes. *Cell Stress and Chaperones* 5: 471–480.

Hasday, J. D., K. D. Fairchild, and C. Shanholtz. 2000. The role of fever in the infected host. *Microbes and Infection* 2: 1891–1904.

Haskell, D. 1994. Experimental evidence that nestling begging behavior incurs a cost due to nest predation. *Proceedings of the Royal Society B: Biological Sciences* 257: 161–164.

Haug, G. H., R. Tiedemann, and L. D. Keigwin. 2004. How the Isthmus of Panama put ice in the Arctic: drifting continents open and close gateways between oceans and shift Earth's climate. Woods Hole Oceanographic Institution. Accessed February 25, 2011. www.whoi.edu/oceanus/viewArticle.do?id=2508.

Hauser, L., G. J. Adcock, P. J. Smith, J. H. B. Ramirez, and G. R. Carvalho. 2002. Loss of microsatellite diversity and low effective population size in an overexploited population of New Zealand snapper (*Pagrus auratus*). *Proceedings of the National Academy of Sciences of the United States of America* 99: 11742–11747.

Hawley, G. J., and D. H. Dehayes. 1994. Genetic diversity and population structure of red spruce (*Picea rubens*). *Canadian Journal of Botany-Revue Canadienne de Botanique* 72: 1778–1786.

Hazkani-Covo, E., R. Sorek, and D. Graur. 2003. Evolutionary dynamics of large numts in the human genome: rarity of independent insertions and abundance of post-insertion duplications. *Journal of Molecular Evolution* 56: 169–174.

He, X. L., W. F. Qian, Z. Wang, Y. Li, and J. Z. Zhang. 2010. Prevalent positive epistasis in *Escherichia coli* and *Saccharomyces cerevisiae* metabolic networks. *Nature Genetics* 42: 272–276.

Healy, J. F., ed. 1991. *Natural History, a Selection by Pliny (the Elder)*. Penguin Classics, London.

Hebert, P. D. N. 1974. Enzyme variability in natural populations of *Daphnia magna* 3. Genotypic frequencies in intermittent populations. *Genetics* 77: 335–341.

Hebert, P. D. N., and T. J. Crease. 1980. Clonal coexistence in *Daphnia pulex*—another planktonic paradox. *Science* 207: 1363–1365.

Held, L. I., Jr. 2009. *Quirks of Human Anatomy: An Evo–Devo Look at the Human Body*. Cambridge University Press, New York.

Hellmann, I., and R. Nielson. 2008. Human evolutionary genomics. In M. Pagel and A. Pomiankowski, eds., *Evolutionary Genomics and Proteomics*, Chapter 31. Sinauer Associates, Sunderland, MA.

Helyer, B., and J. Howie. 1963a. Renal disease associated with positive lupus erythematosus tests in a cross-bred strain of mice. *Nature* 197: 197.

Helyer, B., and J. Howie. 1963b. Spontaneous auto-immune disease in NZB/BL mice. *British Journal of Haematology* 9: 119–131.

Henig, R. 2001. *The Monk in the Garden: The Lost and Found Genius of Gregor Mendel, the Father of Genetics*. Mariner Books, Boston.

Henikoff, S., K. Ahmad, and H. S. Malik. 2001. The centromere paradox: stable inheritance with rapidly evolving DNA. *Science* 293: 1098–1102.

Hennig, W. 1966. *Phylogenetic Systematics*. University of Illinois Press, Urbana.

Herndon, W., and J. W. Weik. 1893. *Abraham Lincoln: The Story of a Great Life*. Samson Low, London.

Hero, J. M., and G. R. Gillespie. 1997. Epidemic disease and amphibian declines in Australia. *Conservation Biology* 11: 1023–1025.

Herron, M. D., and R. E. Michod. 2008. Evolution of complexity in the volvocine algae: transitions in individuality through Darwin's eye. *Evolution* 62: 436–451.

Herron, M., J. Hackett, F. Aylward, and R. Michod. 2009. Triassic origin and early radiation of multicellular volvocine algae. *Proceedings of the National Academy of Sciences of the United States of America* 16: 3254–3258.

Hewitt, G. 1989. The subdivision of species by hybrid zones. In D. Otte and J. Endler, eds. *Speciation and Its Consequences*, pp. 85–110. Sinauer Associates, Sunderland, MA.

Hewitt, G. M. 1996. Some genetic consequences of ice ages, and their role in divergence and speciation. *Biological Journal of the Linnean Society* 58: 247–276.

Hewitt, G. M. 2000. The genetic legacy of the Quaternary ice ages. *Nature* 405: 907–913.

Hickey, D. A. 1993. Molecular symbionts and the evolution of sex. *Journal of Heredity* 84: 410–414.

Higgins, D. G., and P. M. Sharp. 1988. Clustal: a package for performing multiple sequence alignment on a microcomputer. *Gene* 73: 237–244.

Hildebrand, A. R., G. T. Penfield, D. A. Kring, M. Pilkington, A. Camargo, S. B. Jacobsen, and W. V. Boynton. 1991. Chicxulub crater: a possible Cretaceous/Tertiary boundary impact crater on the Yucatán Peninsula, Mexico. *Geology* 19: 867–871.

Hill, C., and N. Pierce. 1989. The effect of adult diet on the biology of butterflies, 1: the common imperial blue, *Jalmenus evagoras. Oecologia* 81: 249–257.

Hill, K., and A. M. Hurtado. 1996. *Aché Life History: The Ecology and Demography of a Foraging People.* Transaction Publishers Rutgers—The State University of New Jersey, Piscataway, NJ.

Hilu, K. W. 1993. Polyploidy and the evolution of domesticated plants. *American Journal of Botany* 80: 1494–1499.

Hinde, R. A. 1973. Nobel recognition for ethology. *Nature* 245: 346.

Hobolth, A., O. F. Christensen, T. Mailund, and M. H. Schierup. 2007. Genomic relationships and speciation times of human, chimpanzee, and gorilla inferred from a coalescent hidden Markov model. *PLoS Genetics* 3: e7.

Hodges, S. A., M. Fulton, J. Y. Yang, and J. B. Whittall. 2004. Verne Grant and evolutionary studies of *Aquilegia. New Phytologist* 161: 113–120.

Hoekstra, H. E. 2006. Genetics, development and evolution of adaptive pigmentation in vertebrates. *Heredity* 97: 222–234.

Hoekstra, H. E., K. E. Drumm, and M. W. Nachman. 2004. Ecological genetics of adaptive color polymorphism in pocket mice: geographic variation in selected and neutral genes. *Evolution* 58: 1329–1341.

Hoekstra, H. E., R. J. Hirschmann, R. A. Bundey, P. A. Insel, and J. P. Crossland. 2006. A single amino acid mutation contributes to adaptive beach mouse color pattern. *Science* 313: 101–104.

Hoekstra, H. E., J. M. Hoekstra, D. Berrigan, S. N. Vignieri, A. Hoang, C. E. Hill, P. Beerli, and J. G. Kingsolver. 2001. Strength and tempo of directional selection in the wild. *Proceedings of the National Academy of Sciences of the United States of America* 98: 9157–9160.

Hoelzel, A. R. 1999. Impact of population bottlenecks on genetic variation and the importance of life history: a case study of the northern elephant seal. *Biological Journal of the Linnean Society* 68: 23–39.

Hoelzel, A. R., J. Halley, S. J. O'Brien, C. Campagna, T. Ambom, B. Le Boeuf, K. Rails, and G. Dover. 1993. Elephant seal genetic variation and the use of simulation models to investigate historical population bottlenecks. *Journal of Heredity* 84: 443–449.

Hoelzel, A. R., B. LeBoeuf, J. Reiter, and C. Campagna. 1999a. Alphamale paternity in elephant seals. *Behavioral Ecology and Sociobiology* 46: 298–306.

Hoelzel, A. R., J. C. Stephens, and S. J. O'Brien. 1999b. Molecular genetic diversity and evolution at the MHC DQB locus in four species of pinnipeds. *Molecular Biology and Evolution* 16: 611–618.

Hoffmann, A. 1999. Laboratory and field heritabilities: some lessons from *Drosophila.* In T. Mousseau, B. Sinervo, and J. Endler, eds., *Adaptive Genetic Variation in the Wild*, pp. 200–218. Oxford University Press, New York.

Hoglund, J., and R. Alatalo. 1995. *Leks.* Princeton University Press, Princeton.

Holland, L. Z., and N. D. Holland. 2001. Evolution of neural crest and placodes: amphioxus as a model for the ancestral vertebrate? *Journal of Anatomy* 199: 85–98.

Holland, N. D., G. Panganiban, E. L. Henyey, and L. Z. Holland. 1996. Sequence and developmental expression of AmphiDII, an amphioxus Distal-less gene transcribed in the ectoderm, epidermis and nervous system: insights into evolution of craniate forebrain and neural crest. *Development* 122: 2911–2920.

Holldobler, B., and E. O. Wilson. 1990. *The Ants.* Harvard University Press, Cambridge.

Holloway, A. K., D. C. Cannatella, H. C. Gerhardt, and D. M. Hillis. 2006. Polyploids with different origins and ancestors form a single sexual polyploid species. *American Naturalist* 167: E88–E101.

Holmes, S. 2003. Bootstrapping phylogenetic trees: theory and methods. *Statistical Science* 18: 241–255.

Hone, D. W., and M. Benton. 2005. The evolution of large size: how does Cope's Rule work? *Trends in Ecology & Evolution* 20: 4–6.

Horder, T. 2006. Gavin Rylands de Beer: how embryology foreshadowed the dilemmas of the genome. *Nature Reviews Genetics* 7: 892–898.

Hori, M. 1993. Frequency-dependent natural selection in the handedness of scale-eating cichlid fish. *Science* 260: 216–219.

Horiike, T., K. Hamada, S. Kanaya, and T. Shinozawa. 2001. Origin of eukaryotic cell nuclei by symbiosis of Archaea in Bacteria is revealed by homology-hit analysis. *Nature Cell Biology* 3: 210–214.

Horiike, T., K. Hamada, and T. Shinozawa. 2002. Origin of eukaryotic cell nuclei by symbiosis of Archaea in Bacteria supported by the newly clarified origin of functional genes. *Genes & Genetic Systems* 77: 369–376.

Houde, A. E. 1997. *Sex, Color and Mate Choice in Guppies.* Princeton University Press, Princeton.

Houde, A. E., and J. A. Endler. 1990. Correlated evolution of female mating preference and male color pattern in the guppy, *Poecilia reticulata. Science* 248: 1405–1408.

Houlahan, J. E., C. S. Findlay, B. R. Schmidt, A. H. Meyer, and S. L. Kuzmin. 2000. Quantitative evidence for global amphibian population declines. *Nature* 404: 752–755.

Houtman, A. M., and J. B. Falls. 1994. Negative assortative mating in the white-throated sparrow, *Zonotrichia albicollis*: the role of mate choice and intrasexual competition. *Animal Behaviour* 48: 378–383.

Huang, C. Y., M. A. Ayliffe, and J. N. Timmis. 2003. Direct measurement of the transfer rate of chloroplast DNA into the nucleus. *Nature* 422: 72–76.

Huang, C. Y., M. A. Ayliffe, and J. N. Timmis. 2004. Simple and complex nuclear loci created by newly transferred chloroplast DNA in tobacco. *Proceedings of the National Academy of Sciences of the United States of America* 101: 9710–9715.

Huang, J. L. 2004. Phylogenomic evidence supports past endosymbiosis, intracellular and horizontal gene transfer in *Cryptosporidium parvum. Genome Biology* 5: R88.

Huber, C., W. Eisenreich, S. Hecht, and G. Wachtershauser. 2003. A possible primordial peptide cycle. *Science* 301: 938–940.

Huber, C., and G. Wachtershauser. 1997. Activated acetic acid by carbon fixation on (Fe,Ni)S under primordial conditions. *Science* 276: 245–247.

Hudson, R. R. 1990. Gene geneologies and the coalescent process. *Oxford Surveys in Evolutionary Biology* 7: 1–44.

Huey, R. B., and P. E. Hertz. 1984. Is a jack-of-all-temperatures a master of none? *Evolution* 38: 441–444.

Hughes, A. L. 2002. Adaptive evolution after gene duplication. *Trends in Genetics* 18: 433–434.

Hughes, L., B. Chang, D. Wagner, and N. Pierce. 2000. Effects of mating history on ejaculate size, fecundity and copulation duration in the ant-tended lycaenid butterfly, *Jalmenus evagoras. Behavioral Ecology and Sociobiology* 47: 119–128.

Hughes, W. O. H., B. P. Oldroyd, M. Beekman, and F. L. W. Ratnieks. 2008. Ancestral monogamy shows kin selection is key to the evolution of eusociality. *Science* 320: 1213–1216.

Hulbert, A. J., R. Pamplona, R. Buffenstein, and W. A. Buttemer. 2007. Life and death: metabolic rate, membrane composition, and life span of animals. *Physiological Reviews* 87: 1175–1213.

Hurst, G. D. D., and J. H. Werren. 2001. The role of selfish genetic elements in eukaryotic evolution. *Nature Reviews Genetics* 2: 597–606.

Hurst, L. 2002. The Ka/Ks ratio: diagnosing the form of sequence evolution. *Trends in Genetics* 18: 486–487.

Hutton, J. 1795. *Theory of the Earth*. Creech, Edinburgh.

Huxley, J. 1938. Darwin's theory of sexual selection and the data subsumed by it, in light of recent research. *American Naturalist* 72: 416–433.

Huxley, J. 1942. *Evolution: The Modern Synthesis*. Harper, New York.

Huxley, T. H. 1863. *Evidence as to Man's Place in Nature*. D. Appleton, New York.

IDEALS. 2001. Data and publications from the Illinois long-term selection experiment for oil and protein in corn. University of Illinois. Accessed March 8, 2001. www.ideals.illinois.edu/handle/2142/3525.

Ikemura, T. 1981a. Correlation between the abundance of *Escherichia coli* transfer RNAs and the occurrence of the respective codons in its protein genes. *Journal of Molecular Biology* 146: 1–21.

Ikemura, T. 1981b. Correlation between the abundance of *Escherichia coli* transfer RNAs and the occurrence of the respective codons in its protein genes: a proposal for a synonymous codon choice that is optimal for the *Escherichia coli* translational system. *Journal of Molecular Biology* 151: 389–409.

Ikemura, T. 1982. Correlation between the abundance of yeast transfer RNAs and the occurrence of the respective codons in its protein genes: differences in synonymous choice patterns of yeast and *Escherichia coli* with reference to the abundance of isoaccepting transfer RNAs. *Journal of Molecular Biology* 158: 573–597.

Ingman, M., H. Kaessmann, S. Pääbo, and U. Gyllensten. 2000. Mitochondrial genome variation and the origin of modern humans. *Nature* 408: 708–713.

Ingram, G. J., and K. R. McDonald. 1993. An update on the decline of Queensland's frogs. In D. Klunney and D. Ayers, eds., *Herpetology in Australia: A Diverse Discipline*, pp. 297–303. Royal Zoological Society of New South Wales, New South Wales.

International Human Genome Sequencing Consortium. 2001. Initial sequencing and analysis of the human genome. *Nature* 409: 860–921.

Ioannou, C. C., C. R. Tosh, L. Neville, and J. Krause. 2008. The confusion effect—from neural networks to reduced predation risk. *Behavioral Ecology* 19: 126–130.

Irwin, D. E., J. H. Irwin, and T. D. Price. 2001. Ring species as bridges between microevolution and speciation. *Genetica* 112: 223–243.

Isack, H. A., and H.-U. Reyer. 1989. Honeyguides and honey gatherers: interspecific communication in a symbiotic relationship. *Science* 243: 1343–1346.

Isozaki, Y., J. X. Yao, T. Matsuda, H. Sakai, Z. S. Ji, N. Shimizu, N. Kobayashi, H. Kawahata, H. Nishi, M. Takano, and T. Kubo. 2004. Stratigraphy of the Middle–Upper Permian and Lowermost Triassic at Chaotian, Sichuan, China: record of Late Permian double mass extinction event. *Proceedings of the Japan Academy Series B—Physical and Biological Sciences* 80: 10–16.

Issac, R., and J. Chmielewski. 2002. Approaching exponential growth with a self-replicating peptide. *Journal of the American Chemical Society* 124: 6808–6809.

IUCN. 2001. IUCN list of categories and criteria. The World Conservation Union, Gland, Switzerland.

Izett, G. A. 1991. Tektites in Cretaceous–Tertiary boundary rocks on Haiti and their bearing on the Alvarez impact extinction hypothesis. *Journal of Geophysical Research, Planets* 96: 20879–20905.

Jablonski, D. 1997. Body-size evolution in Cretaceous molluscs and the status of Cope's rule. *Nature* 385: 250–252.

Jablonski, D. 2001. Lessons from the past: evolutionary impacts of mass extinctions. *Proceedings of the National Academy of Sciences of the United States of America* 98: 5393–5398.

Jablonski, D. 2002. Survival without recovery after mass extinctions. *Proceedings of the National Academy of Sciences of the United States of America* 99: 8139–8144.

Jablonski, D., and R. A. Lutz. 1983. Larval ecology of marine benthic invertebrates: paleobiological implications. *Biological Reviews of the Cambridge Philosophical Society* 58: 21–89.

Jackson, J. B. C., and A. H. Cheetham. 1999. Tempo and mode of speciation in the sea. *Trends in Ecology & Evolution* 14: 72–77.

Jackson, M. E., and R. D. Semlitsch. 1993. Pedomorphosis in the salamander *Ambystoma talpoideum*: effects of a fish predator. *Ecology* 74: 342–350.

Jacobs, H. T., J. W. Posakony, J. W. Grula, J. W. Roberts, J. H. Xin, R. J. Britten, and E. H. Davidson. 1983. Mitochondrial-DNA sequences in the nuclear genome of *Strongylocentrotus purpuratus*. *Journal of Molecular Biology* 165: 609–632.

Jaenike, J. 1978. An hypothesis to account for the maintenance of sex within populations. *Evolutionary Theory* 3: 191–194.

Jansson, R. 2003. Global patterns in endemism explained by past climatic change. *Proceedings of the Royal Society B: Biological Sciences* 270: 583–590.

Jaramillo-Correa, J. P., and J. Bousquet. 2003. New evidence from mitochondrial DNA of a progenitor–derivative species relationship between black spruce and red spruce (Pinaceae). *American Journal of Botany* 90: 1801–1806.

Jaramillo-Correa, J. P., J. Bousquet, J. Beaulieu, N. Isabel, M. Perron, and M. Bouille. 2003. Cross-species amplification of mitchondrial DNA sequence-tagged-site markers in conifers: the nature of polymorphism and variation within and among species in *Picea*. *Theoretical and Applied Genetics* 106: 1353–1367.

Javaux, E. J., C. P. Marshall, and A. Bekker. 2010. Organic-walled microfossils in 3.2-billion-year-old shallow-marine siliciclastic deposits. *Nature* 463: 934–938.

Jehl, J., and W. Everett. 1985. History and status of the avifauna of Isla Guadalupe, Mexico. *Transactions of the San Diego Society of Natural History* 20: 313–336.

Jennings, P., and J. de Jesus. 1968. Studies on competition in rice I. Competition in mixtures of varieties. *Evolution* 22: 119–124.

Jessen, T. H., R. E. Weber, G. Fermi, J. Tame, and G. Braunitzer. 1991. Adaptation of bird hemoglobins to high altitudes: demonstration of molecular mechanism by protein engineering. *Proceedings of the National Academy of Sciences of the United States of America* 88: 6519–6522.

Jesson, L. K., and S. C. H. Barrett. 2002. The genetics of mirror-image flowers. *Proceedings of the Royal Society B: Biological Sciences* 269: 1835–1839.

Ji, Q., P. J. Currie, M. A. Norell, and S.-A. Ji. 1998. Two feathered dinosaurs from northeastern China. *Nature* 393: 753–761.

Jin, Y. G., Y. Wang, W. Wang, Q. H. Shang, C. Q. Cao, and D. H. Erwin. 2000. Pattern of marine mass extinction near the Permian–Triassic boundary in South China. *Science* 289: 432–436.

Johanson, D. 2001. Origins of modern humans: multiregional or out of Africa? American Institute of Biological Science. Accessed June 21, 2011. www.actionbioscience.org/evolution/johanson.html.

Johnson, D. O. 1997. *An English Translation of Claudius Aelianus' Varia historia.* Edwin Mellen Press, Lewiston, NY.

Jokela, J., M. F. Dybdahl, and C. M. Lively. 2009. The maintenance of sex, clonal dynamics, and host–parasite coevolution in a mixed population of sexual and asexual snails. *American Naturalist* 174: S43–S53.

Jokela, J., C. M. Lively, M. F. Dybdahl, and J. A. Fox. 1997. Evidence for a cost of sex in the freshwater snail *Potamopyrgus antipodarum. Ecology* 78: 452–460.

Jones, K. E., A. Purvis, and J. L. Gittleman. 2003. Biological correlates of extinction risk in bats. *American Naturalist* 161: 601–614.

Jorba, J., R. Campagnoli, L. De, and O. Kew. 2008. Calibration of multiple poliovirus molecular clocks covering an extended evolutionary range. *Journal of Virology* 82: 4429–4440.

Jordan, M. A., R. L. Hammond, H. L. Snell, H. M. Snell, and W. C. Jordan. 2002. Isolation and characterization of microsatellite loci from Galápagos lava lizards (*Microlophus* spp.). *Molecular Ecology Notes* 2: 349–351.

Jordan, M. A., and H. L. Snell. 2008. Historical fragmentation of islands and genetic drift in populations of Galápagos lava lizards (*Microlophus albemarlensis* complex). *Molecular Ecology* 17: 1224–1237.

Joyce, G. F. 2002. The antiquity of RNA-based evolution. *Nature* 418: 214–221.

Jukes, T. H., and M. Kimura. 1984. Evolutionary constraints and the neutral theory. *Journal of Molecular Evolution* 21: 90–92.

Kadereit, B., P. Kumar, W.-J. Wang, D. Miranda, E. L. Snapp, N. Severina, I. Torregroza, T. Evans, and D. L. Silver. 2008. Evolutionarily conserved gene family important for fat storage. *Proceedings of the National Academy of Sciences of the United States of America* 105: 94–99.

Kameda, Y. C., A. Kawakita, and M. Kato. 2007. Cryptic genetic divergence and associated morphological differentiation in the arboreal land snail *Satsuma* (*Luchuhadra*) *largillierti* (Camaenidae) endemic to the Ryukyu Archipelago, Japan. *Molecular Phylogenetics and Evolution* 45: 519–533.

Kameda, Y., A. Kawakita, and M. Kato. 2009. Reproductive character displacement in genital morphology in *Satsuma* land snails. *American Naturalist* 173: 689–697.

Kandler, O. 1994a. Cell wall biochemistry in Archaea and its phylogenetic implications. *Journal of Biological Physics* 20: 165–169.

Kandler, O. 1994b. The early diversification of life. In S. Bengtson, ed., *Nobel Symposium No. 84. Early Life on Earth*, pp. 152–160. Columbia University Press, New York.

Kaplan, H. S., K. R. Hill, J. B. Lancaster, and A. M. Hurtado. 2000. A theory of human life history evolution: diet, intelligence, and longevity. *Evolutionary Anthropology* 9: 156–185.

Kappeler, P. M., and C. van Schaik, eds. 2004. *Sexual Selection in Primates: New and Comparative Perspectives.* Cambridge University Press, Cambridge.

Karlin, S., L. Brocchieri, A. Campbell, M. Cyert, and J. Mrázek. 2005. Genomic and proteomic comparisons between bacterial and archaeal genomes and related comparisons with the yeast and fly genomes. *Proceedings of the National Academy of Sciences of the United States of America* 102: 7309–7314.

Kasting, J. F. 1993. Earth's early atmosphere. *Science* 259: 920–926.

Kastner, M., F. Asaro, H. V. Michel, W. Alvarez, and L. W. Alvarez. 1984. The precursor of the Cretaceous–Tertiary boundary clays at Stevns Klint, Denmark, and DSDP Hole 465a. *Science* 226: 137–143.

Kato, M., A. Takimura, and A. Kawakita. 2003. An obligate pollination mutualism and reciprocal diversification in the tree genus *Glochidion* (Euphorbiaceae). *Proceedings of the National Academy of Sciences of the United States of America* 100: 5264–5267.

Katz, S. L. 2002. Design of heterothermic muscle in fish. *Journal of Experimental Biology* 205: 2251–2266.

Kaufman, D. W. 1974. Adaptive coloration in *Peromyscus polionotus*: experimental selection by owls. *Journal of Mammalogy* 55: 271–283.

Kawakita, A., and M. Kato. 2006. Assessment of the diversity and species specificity of the mutualistic association between *Epicephala* moths and Glochidion trees. *Molecular Ecology* 15: 3567–3581.

Kawakita, A., A. Takimura, T. Terachi, T. Sota, and M. Kato. 2004. Cospeciation analysis of an obligate pollination mutualism: have Glochidion trees (Euphorbiaceae) and pollinating *Epicephala* moths (Gracillariidae) diversified in parallel? *Evolution* 58: 2201–2214.

Keightley, P. D., and A. Eyre-Walker. 2010. What can we learn about the distribution of fitness effects of new mutations from DNA sequence data? *Philosophical Transactions of the Royal Society of London Series B—Biological Sciences* 365: 1187–1193.

Keller, G., T. Adatte, W. Stinnesbeck, M. Rebolledo-Vieyra, J. Urrutia Fucugauchi, U. Kramar, and D. Stuben. 2004a. Chicxulub impact predates the K–T boundary mass extinction. *Proceedings of the National Academy of Sciences of the United States of America* 101: 3753–3758.

Keller, G., T. Adatte, W. Stinnesbeck, D. Stuben, Z. Berner, U. Kramar, and M. Harting. 2004b. More evidence that the Chicxulub impact predates the K/T mass extinction. *Meteoritics & Planetary Science* 39: 1127–1144.

Kellogg, V. L. 1896. New Mallophaga, I—with special reference to a collection made from maritime birds of the Bay of Monterey, California. *Proceedings of the California Academy of Sciences* 6: 31–196.

Kendler, K. S., M. Gatz, C. O. Gardner, and N. L. Pedersen. 2006. A Swedish national twin study of lifetime major depression. *American Journal of Psychiatry* 163: 109–114.

Kerr, B., and P. Godfrey-Smith. 2002. On individualist and multilevel perspectives on selection in structured populations. *Biology and Philosophy* 17: 477–517.

Kessin, R. H. 2001. *Dictyostelium: Evolution, Ecology and Development of Multicellularity.* Cambridge University Press, Cambridge.

Kessin, R. H., G. Gundersen, V. Zaydfudim, M. Grimson, and R. Blanton. 1996. How cellular slime molds evade nematodes. *Proceedings of the National Academy of Sciences of the United States of America* 93: 4857–4861.

Khaitovich, P., I. Hellmann, W. Enard, K. Nowick, M. Leinweber, H. Franz, G. Weiss, M. Lachmann, and S. Pääbo. 2005. Parallel patterns of evolution in the genomes and transcriptomes of humans and chimpanzees. *Science* 309: 1850–1854.

Khaitovich, P., W. Enard, M. Lachmann, and S. Pääbo. 2006. Evolution of primate gene expression. *Nature Reviews Genetics* 7: 693–702.

Kiers, E. T., and R. F. Denison. 2008. Sanctions, cooperation, and the stability of plant–rhizosphere mutualisms. *Annual Review of Ecology Evolution and Systematics* 39: 215–236.

Kiers, E. T., R. A. Rousseau, and R. F. Denison. 2006. Measured sanctions: legume hosts detect quantitative variation in rhizobium cooperation and punish accordingly. *Evolutionary Ecology Research* 8: 1077–1086.

Kiers, E. T., R. A. Rousseau, S. A. West, and R. F. Denison. 2003. Host sanctions and the legume–rhizobium mutualism. *Nature* 425: 78–81.

Killian, J. K., T. R. Buckley, N. Stewart, B. L. Munday, and R. L. Jirtle. 2001. Marsupials and Eutherians reunited: genetic evidence for the Theria hypothesis of mammalian evolution. *Mammalian Genome* 12: 513–517.

Kilner, R. M., D. G. Noble, and N. B. Davies. 1999. Signals of need in parent–offspring communication and their exploitation by the common cuckoo. *Nature* 397: 667–672.

Kim, S. T., M. J. Yoo, V. A. Albert, J. S. Farris, P. S. Soltis, and D. E. Soltis. 2004. Phylogeny and diversification of B-function MADS-box genes in angiosperms: evolutionary and functional implications of a 260-million-year-old duplication. *American Journal of Botany* 91: 2102–2118.

Kimball, J. W. 2011. Telomeres. Accessed March 17, 2011. http://users.rcn.com/jkimball.ma.ultranet/BiologyPages/T/Telomeres.html.

Kimura, M. 1961. Natural selection as the process of accumulating genetic information in adaptive evolution. *Genetical Research 2*: 127–140.

Kimura, M. 1968. Evolutionary rates at the molecular level. *Nature* 217: 214–216.

Kimura, M. 1977. Preponderance of synonymous changes as evidence for the neutral theory of molecular evolution. *Nature* 267: 275–276.

Kimura, M. 1979. The neutral theory of molecular evolution. *Scientific American* 241: 94–104.

Kimura, M. 1983. *The Neutral Theory of Molecular Evolution*. Cambridge University Press, Cambridge.

Kimura, M. 1993. Retrospective of the last quarter-century of the neutral theory. *Japanese Journal of Genetics* 68: 521–528.

King-Hele, D. 1998. The 1997 Wilkins lecture: Erasmus Darwin, the lunaticks and evolution. *Notes and Records of the Royal Society of London* 52: 153–180.

Kingman, J. F. C. 1982. On the genealogy of large populations. *Journal of Applied Probability* 19A: 27–43.

Kingsolver, J. G., and M. A. R. Koehl. 1985. Aerodynamics, thermoregulation and the evolution of insect wings: differential scaling and evolutionary change. *Evolution* 39: 488–504.

Kingsolver, J. G., and D. W. Pfennig. 2004. Individual-level selection as a cause of Cope's Rule of phyletic size increase. *Evolution* 58: 1608–1612.

Kirk, D. L. 2005. A twelve-step program for evolving multicellularity and a division of labor. *BioEssays* 27: 299–310.

Kirk, M. M., K. Stark, S. M. Miller, W. Muller, B. E. Taillon, H. Gruber, R. Schmitt, and D. L. Kirk. 1999. RegA, a Volvox gene that plays a central role in germ–soma differentiation, encodes a novel regulatory protein. *Development* 126: 639–647.

Kirkpatrick, M., and M. Ryan. 1991. The evolution of mating preferences and the paradox of the lek. *Nature* 350: 33–38.

Kirkwood, T. B. L. 1977. Evolution of ageing. *Nature* 270: 301–304.

Kirkwood, T. B. L., and S. N. Austad. 2000. Why do we age? *Nature* 408: 233–238.

Kishimoto, R., and T. Kawamichi. 1996. Territoriality and monogamous pairs in a solitary ungulate, the Japanese serow, *Capricornis crispus*. *Animal Behaviour* 52: 673–682.

Kistler, A. L., D. R. Webster, S. Rouskin, V. Magrini, J. J. Credle, D. P. Schnurr, H. A. Boushey, E. R. Mardis, H. Li, and J. L. DeRisi. 2007. Genome-wide diversity and selective pressure in the human rhinovirus. *Virology Journal* 4: 40.

Klassen, G. J. 1992. Coevolution: a history of the macroevolutionary approach to studying host–parasite associations. *Journal of Parasitology* 78: 573–587.

Klein, R. 2003. Whither the Neanderthals? *Science* 299: 1525–1527.

Kluger, M. J., W. Kozak, C. A. Conn, L. R. Leon, and D. Soszynski. 1996. The adaptive value of fever. *Infectious Disease Clinics of North America* 10: 1–20.

Kluger, M. J., W. Kozak, C. A. Conn, L. R. Leon, and D. Soszynski. 1998. Role of fever in disease. *Annals of the New York Academy of Sciences* 856: 224–233.

Knauth, L. P., and D. R. Lowe. 2003. High Archean climatic temperature inferred from oxygen isotope geochemistry of cherts in the 3.5 Ga Swaziland Supergroup, South Africa. *Geological Society of America Bulletin* 115: 566–580.

Knell, R. J., and K. M. Webberley. 2004. Sexually transmitted diseases of insects: distribution, evolution, ecology and host behaviour. *Biological Reviews* 79: 557–581.

Knoll, A. H. 2006. Eukaryotic organisms in Proterozoic oceans. *Philosophical Transactions of the Royal Society of London Series B—Biological Sciences* 361: 1023–1038.

Knowlton, J. L., C. J. Donlan, G. W. Roemer, A. Samaniego-Herrera, B. S. Keritt, B. Wood, A. Aguirre-Munoz, K. R. Faulkner, and B. R. Tershy. 2007. Eradication of non-native mammals and the status of insular mammals on the California Channel Islands, USA, and Pacific Baja California Peninsula Islands, Mexico. *Southwestern Naturalist* 52: 528–540.

Knowlton, N. 1993. Sibling species in the sea. *Annual Review of Ecology and Systematics* 24: 189–216.

Knowlton, N., and L. A. Weigt. 1998. New dates and new rates for divergence across the Isthmus of Panama. *Proceedings of the Royal Society B: Biological Sciences* 265: 2257–2263.

Knowlton, N., L. A. Weigt, L. A. Solorzano, D. K. Mills, and E. Bermingham. 1993. Divergence in proteins, mitochondrial-DNA, and reproductive compatibility across the Isthmus of Panama. *Science* 260: 1629–1632.

Koch, N., B. Lynch, and R. Rochette. 2007. Trade-off between mating and predation risk in the marine snail, *Littorina plena*. *Invertebrate Biology* 126: 257–267.

Koch, U., E. von Elert, and D. Straile. 2009. Food quality triggers the reproductive mode in the cyclical parthenogen *Daphnia* (Cladocera). *Oecologia* 159: 317–324.

Kodric-Brown, A., and J. H. Brown. 1984. Truth in advertising: the kinds of traits favored by sexual selection. *American Naturalist* 124: 309–323.

Koh, D., A. Armugam, and K. Jeyaseelan. 2006. Snake venom components and their applications in biomedicine. *Cellular and Molecular Life Sciences* 63: 3030–3041.

Kojima, K., and H. Schaffer. 1967. Survival process of linked mutant genes. *Evolution* 21: 518–531.

Kokko, H., R. Brooks, M. D. Jennions, and J. Morley. 2003. The evolution of mate choice and mating biases. *Proceedings of the Royal Society B: Biological Sciences* 270: 653–664.

Kokko, H., M. D. Jennions, and R. Brooks. 2006. Unifying and testing models of sexual selection. *Annual Review of Ecology, Evolution, and Systematics* 37: 43–66.

Kokko, H., W. J. Sutherland, J. Lindstrom, J. D. Reynolds, and A. Mackenzie. 1998. Individual mating success, lek stability, and the neglected limitations of statistical power. *Animal Behaviour* 56: 755–762.

Koltes, J. E., B. P. Mishra, D. Kumar, R. S. Kataria, L. R. Totir, R. L. Fernando, R. Cobbold, D. Steffen, W. Coppieters, M. Georges, and J. M. Reecy. 2009. A nonsense mutation in cGMP-dependent type II protein kinase (PRKG2) causes dwarfism in American Angus cattle. *Proceedings of the National Academy of Sciences of the United States of America* 106: 19250–19255.

Kondrashov, A. 1982. Selection against harmful mutations in large sexual and asexual populations. *Genetical Research* 40: 325–332.

Kondrashov, A. 1988. Deleterious mutations and the evolution of sexual reproduction. *Nature* 336: 435–440.

Kondrashov, A. S. 1993. Classification of hypotheses on the advantage of amphimixis. *Journal of Heredity* 84: 372–387.

Kondrashov, A. 2001. Sex and U. *Trends in Genetics* 17: 75–77.

Kondrashov, F. A., and A. S. Kondrashov. 2010. Measurements of spontaneous rates of mutations in the recent past and the near future. *Philosophical Transactions of the Royal Society of London B—Biological Sciences* 365: 1169–1176.

Koonin, E. V. 2000. How many genes can make a cell: the minimal-gene-set concept. *Annual Review of Genomics and Human Genetics* 1: 99–116.

Koonin, E. V. 2003. Horizontal gene transfer: the path to maturity. *Molecular Microbiology* 50: 725–727.

Koonin, E. V. 2006. The ancient virus world and the evolution of cells. *Molecular Microbiology* 50: 725–727.

Koonin, E. V., K. S. Makarova, and L. Aravind. 2001. Horizontal gene transfer in prokaryotes: quantification and classification. *Annual Review of Microbiology* 55: 709–742.

Kozur, H. W. 1998. Some aspects of the Permian–Triassic boundary (PTB) and of the possible causes for the biotic crisis around this boundary. *Palaeogeography Palaeoclimatology Palaeoecology* 143: 227–272.

Krama, T., and I. Krams. 2005. Cost of mobbing call to breeding pied flycatcher, *Ficedula hypoleuca*. *Behavioral Ecology* 16: 37–40.

Krams, I., T. Krama, K. Igaune, and R. Mand. 2008. Experimental evidence of reciprocal altruism in the pied flycatcher. *Behavioral Ecology and Sociobiology* 62: 599–605.

Krause, E. 1879. Contribution to the history of the descent theory. *Kosmos*.

Krause, J., L. Orlando, D. Serre, B. Viola, K. Prüfer, M. P. Richards, J.-J. Hublin, C. Hänni, A. P. Derevianko and S. Pääbo, 2007. Neanderthals in central Asia and Siberia. *Nature* 449: 902–904.

Krebs, J. R., and R. Dawkins. 1984. Animal signals: mind-reading and manipulation? In J. R. Krebs and N. B. Davies, eds., *Behavioural Ecology*, pp. 380–401. Sinauer Associates, Sunderland, MA.

Kreitman, M. 2000. Methods to detect selection in populations with applications to the human. *Annual Review of Genomics and Human Genetics* 1: 539–559.

Kring, D. A. 2010. Chicxulub impact event: understanding the K–T boundary. Accessed December 17, 2010. www.lpi.usra.edu/science/kring/epo_web/impact_cratering/Chicxulub/Chicx_title.html.

Krings, M., A. Stone, R. W. Schmitz, H. Krainitzki, M. Stoneking, and S. Pääbo. 1997. Neandertal DNA sequences and the origin of modern humans. *Cell* 90: 19–30.

Krizek, B. A., and J. C. Fletcher. 2005. Molecular mechanisms of flower development: an armchair guide. *Nature Reviews Genetics* 6: 688–698.

Kruger, K., P. Grabowski, A. Zaug, J. Sands, D. Gottschling, and T. Cech. 1982. Self-splicing RNA: autoexcision and autocyclization of the ribosomal RNA intervening sequence of *Tetrahymena*. *Cell* 31: 147–157.

Kuchta, S. R. 2005. Experimental support for aposematic coloration in the salamander *Ensatina eschscholtzii xanthoptica*: implications for mimicry of Pacific newts. *Copeia* 2005: 265–271.

Kuchta, S. R., A. H. Krakauer, and B. Sinervo. 2008. Why does the yellow-eyed *ensatina* have yellow eyes? Batesian mimicry of Pacific newts (genus *Taricha*) by the salamander *Ensatina eschscholtzii xanthoptica*. *Evolution* 62: 984–990.

Kuchta, S. R., D. S. Parks, R. L. Mueller, and D. B. Wake. 2009. Closing the ring: historical biogeography of the salamander ring species *Ensatina eschscholtzii*. *Journal of Biogeography* 36: 982–995.

Kuhn, T. 1962. *The Structure of Scientific Revolutions*. University of Chicago Press, Chicago.

Kumar, S. 2006. Molecular clocks: four decades of evolution. *Nature Genetics* 6: 654–662.

Kumar, S., and S. Subramanian. 2002. Mutation rates in mammalian genomes. *Proceedings of the National Academy of Sciences of the United States of America* 99: 803–808.

Kuppers, B., and M. Sumper. 1975. Minimal requirements for template recognition by bacteriophage Q-Beta replicase: approach to general RNA-dependent RNA synthesis. *Proceedings of the National Academy of Sciences of the United States of America* 72: 2640–2643.

Kutzbach, J., R. Gallimore, S. Harrison, P. Behling, R. Selin, and F. Laarif. 1998. Climate and biome simulations for the past 21,000 years. *Quaternary Science Reviews* 17: 473–506.

Kuzdzal-Fick, J. J., K. R. Foster, D. C. Queller, and J. E. Strassmann. 2007. Exploiting new terrain: an advantage to sociality in the slime mold *Dictyostelium discoideum*. *Behavioral Ecology* 18: 433–437.

Kvenvolden, K., J. Lawless, K. Pering, E. Peterson, J. Flores, C. Ponnamperuma, I. R. Kaplan, and C. Moore. 1970. Evidence for extraterrestrial amino-acids and hydrocarbons in the Murchison meteorite. *Nature* 228: 923–926.

Kyte, F. T., Z. Zhou, and J. T. Wasson. 1980. Siderophile-enriched sediments from the Cretaceous–Tertiary boundary. *Nature* 288: 651–656.

Lacey, R. W. 1973. Genetic basis, epidemiology, and future significance of antibiotic resistance in *Staphylococcus aureus*: a review. *Journal of Clinical Pathology* 26: 899–913.

Lachlan, R. F., and M. R. Servedio. 2004. Song learning accelerates allopatric speciation. *Evolution* 58: 2049–2063.

Lachmann, M., S. Szamado, and C. Bergstrom. 2001. Cost and conflict in animal signals and human language. *Proceedings of the National Academy of Sciences of the United States of America* 98: 13189–13194.

Lahn, B. T., N. M. Pearson, and K. Jegalian. 2001. The human Y chromosome, in the light of evolution. *Nature Reviews Genetics* 2: 207–216.

Laitman, J. T., and J. S. Reidenberg. 1993. Specializations of the human upper respiratory and upper digestive systems as seen through comparative and developmental anatomy. *Dysphagia* 8: 318–325.

Lamarck, J.-B. 1801. *Système des animaux sans vertèbres*. Agasse, Paris.

Lamarck, J.-B. 1809. *Zoological Philosophy*. Dentu, Paris.

Lande, R. 1980. Sexual dimorphism, sexual selection and adaptation in polygenic characters. *Evolution* 34: 292–307.

Lane, C. E., and J. M. Archibald. 2008. The eukaryotic tree of life: endosymbiosis takes its TOL. *Trends in Ecology & Evolution* 23: 268–275.

Laroche, J., P. Li, L. Maggia, and J. Bousquet. 1997. Molecular evolution of angiosperm mitochondrial introns and exons. *Proceedings of the National Academy of Sciences of the United States of America* 94: 5722–5727.

Larsen, C. S. 2008. *Our Origins: Discovering Physical Anthropology*. W.W. Norton, New York.

Laurance, W. F., K. R. McDonald, and R. Speare. 1996. Epidemic disease and the catastrophic decline of Australian rain forest frogs. *Conservation Biology* 10: 406–413.

Law, J. H., and B. J. Crespi. 2002. Recent and ancient asexuality in Timema walkingsticks. *Evolution* 56: 1711–1717.

Lawrence, J. G., and H. Ochman. 1998. Molecular archaeology of the *Escherichia coli* genome. *Proceedings of the National Academy of Sciences of the United States of America* 95: 9413–9417.

Layzell, D. B., R. M. Rainbird, C. A. Atkins, and J. S. Pate. 1979. Economy of photosynthate use in nitrogen-fixing legume nodules—observations on 2 contrasting symbioses. *Plant Physiology* 64: 888–891.

Lazcano, A. 2006. The origins of life. *Natural History* 115: 36–43.

Lee, M. S. Y., and M. W. Caldwell. 1998. Anatomy and relationships of *Pachyrhachis problematicus*, a primitive snake with hindlimbs. *Philosophical Transactions of the Royal Society of London Series B—Biological Sciences* 353: 1521–1552.

Lee, S., C. Parr, Y. Hwang, D. Mindell, and J. Choe. 2003. Phylogeny of magpies (genus *Pica*) inferred from mtDNA data. *Molecular Phylogenetics and Evolution* 29: 250–257.

Leech, S. M., and M. L. Leonard. 1997. Begging and the risk of predation in nestling birds. *Behavioral Ecology* 8: 644–646.

Lehmann, L., L. Keller, S. West, and D. Roze. 2007. Group selection and kin selection: two concepts but one process. *Proceedings of the National Academy of Sciences of the United States of America* 104: 6736–6739.

Leitner, T., and J. Albert. 1999. The molecular clock of HIV-1 unveiled through analysis of a known transmission history. *Proceedings of the National Academy of Sciences of the United States of America* 96: 10752–10757.

Lenski, R. E. 2004. Phenotypic and genomic evolution during a 20,000-generation experiment with the bacterium *Escherichia coli*. *Plant Breeding Reviews* 24: 225–265.

Lenski, R. E., and M. Travisano. 1994. Dynamics of adaptation and diversification—a 10,000-generation experiment with bacterial populations. *Proceedings of the National Academy of Sciences of the United States of America* 91: 6808–6814.

Levene, H. 1953. Genetic equilibrium when more than one ecological niche is available. *American Naturalist* 87: 331–333.

Levin, B. R., and C. Bergstrom. 2000. Bacteria are different: Observations, interpretations, speculations, and opinions about the mechanisms of adaptive evolution in prokaryotes. *Proceedings of the National Academy of Sciences of the United States of America* 97: 6981–6985.

Levin, B. R., V. Perrot, and N. Walker, 2000. Compensatory mutations, antibiotic resistance and the population genetics of adaptive evolution in bacteria. *Genetics* 154: 985–997.

Levin, D. 2002. *The Role of Chromosomal Change in Plant Evolution*. Oxford University Press, New York.

Levins, R., and R. Lewontin. 1987. *The Dialectical Biologist*. Harvard University Press, Cambridge.

Lewis, E. B. 1978. A gene complex controlling segmentation in *Drosophila*. *Nature* 276: 556–570.

Lewis, S., S. M. J. Searle, N. Harris, M. Gibson, V. Iyer, J. Richter, C. Wiel, L. Bayraktaroglu, E. Birney, M. Crosby, J. Kaminker, B. B. Matthews, S. Prochnik, C. D. Smith, J. L. Tupy, G. Rubin, S. Misra, C. Mungall, and M. E. Clamp. 2002. Apollo: a sequence annotation editor. *Genome Biology* 3: e12.

Lewontin, R., and J. Hubby. 1966. A molecular approach to the study of genic heterozygosity in natural populations. II. Amount of variation and degree of heterozygosity in natural populations of *Drosophila pseudoobscura*. *Genetics* 54: 595–609.

Li, H. H., U. B. Gyllensten, X. F. Cui, R. K. Saiki, H. A. Erlich, and N. Arnheim. 1988. Amplification and analysis of DNA sequence in single human sperm and diploid cells. *Nature* 335: 414–417.

Li, J. W., M. Ishaq, M. Prudence, X. Xi, T. Hu, Q. Z. Liu, and D. Y. Guo. 2009. Single mutation at the amino acid position 627 of PB2 that leads to increased virulence of an H5N1 avian influenza virus during adaptation in mice can be compensated by multiple mutations at other sites of PB2. *Virus Research* 144: 123–129.

Liem, K. F. 1988. Form and function of lungs: the evolution of air breathing mechanisms. *American Zoologist* 28: 739–759.

Lighthill, M. J. 1969. Hydromechanics of aquatic animal propulsion—a review. *Annual Review of Fluid Mechanics* 1: 413–446.

Lilley, D. M. J. 2003. The origins of RNA catalysis in ribozymes. *Trends in Biochemical Sciences* 28: 495–501.

Lim, L., and G. I. McFadden. 2010. The evolution, metabolism and functions of the apicoplast. *Philosophical Transactions of the Royal Society of London B—Biological Sciences* 365: 749–763.

Lincoln, T. A., and G. F. Joyce. 2009. Self-sustained replication of an RNA enzyme. *Science* 323: 1229–1232.

Lindner, A. B., R. Madden, A. Demarez, E. J. Stewart, and F. Taddei. 2008. Asymmetric segregation of protein aggregates is associated with cellular aging and rejuvenation. *Proceedings of the National Academy of Sciences of the United States of America* 105: 3076–3081.

Liow, L. H., M. Fortelius, K. Lintulaakso, H. Mannila, and N. C. Stenseth. 2009. Lower extinction risk in sleep-or-hide mammals. *American Naturalist* 173: 264–272.

Liston, A., and S. McColl. 2003. Subversion of the chemokine world by microbial pathogens. *BioEssays* 25: 478–488.

Little, A. E. F., T. Murakami, U. G. Mueller, and C. R. Currie. 2003. The infrabuccal pellet piles of fungus-growing ants. *Naturwissenschaften* 90: 558–562.

Little, A. E. F., T. Murakami, U. G. Mueller, and C. R. Currie. 2006. Defending against parasites: fungus-growing ants combine specialized behaviours and microbial symbionts to protect their fungus gardens. *Biology Letters* 2: 12–16.

Lively, C. M. 1987. Evidence from a New Zealand snail for the maintenance of sex by parasitism. *Nature* 328: 519–521.

Lockhart, A. B., P. H. Thrall, and J. Antonovics. 1996. Sexually transmitted diseases in animals: ecological and evolutionary implications. *Biological Reviews of the Cambridge Philosophical Society* 71: 415–471.

Loewe, L., and W. G. Hill. 2010. The population genetics of mutations: good, bad and indifferent. *Philosophical Transactions of the Royal Society of London Series B—Biological Sciences* 365: 1153–1167.

Long, T. A. F. 2005. The influence of mating system on the intensity of parent–offspring conflict in primates. *Journal of Evolutionary Biology* 18: 509–515.

Looy, C., R. Twitchet, C. Dilcher, H. Van Konijnenburg-Van Cittert, and H. Visscher. 2001. Life in the end-Permian dead zone. *Proceedings of the National Academy of Sciences of the United States of America* 98: 7879–7883.

Lorenz, K. 1966. *On Aggression*. Harcourt, Brace and World, New York.

Luria, S. E., and M. Delbrück. 1943. Mutations of bacteria from virus sensitivity to virus resistance. *Genetics* 28: 491–511.

Lush, J. 1948. *The Genetics of Populations*. Iowa State University, Special report 94 of the College of Agriculture, Ames.

Lutz, B., H. Lu, G. Eichele, D. Miller, and T. Kaufman. 1997. Rescue of *Drosophila* labial null mutant by the chicken ortholog Hoxb-1 demonstrates that the function of Hox genes is phylogenetically conserved. *Genes and Development* 10: 176–184.

Lutzoni, F., and M. Pagel. 1997. Accelerated evolution as a consequence of transitions to mutualism. *Proceedings of the National Academy of Sciences of the United States of America* 94: 11422–11427.

Lyell, C. 1830. *Principles of Geology: Being an Attempt to Explain the Former Changes of the Earth's Surface by Reference to Causes Now in Operation*. John Murray, London.

Lynch, M. 2007. *The Origins of Genome Architecture*. Sinauer Associates, Sunderland, MA.

Lynch, M. 2010a. Evolution of the mutation rate. *Trends in Genetics* 26: 345–352.

Lynch, M. 2010b. Rate, molecular spectrum, and consequences of human mutation. *Proceedings of the National Academy of Sciences of the United States of America* 107: 961–968.

Lynch, M., and J. S. Conery. 2003. The origins of genome complexity. *Science* 302: 1401–1404.

Lynch, M., W. Sung, K. Morris, N. Coffey, C. Landry, E. Dopman, W. Dickinson, K. Okamoto, S. Kulkarni, D. Hartl, and W. Thomas. 2008. A genome-wide view of the spectrum of spontaneous mutations in yeast. *Proceedings of the National Academy of Sciences of the United States of America* 105: 9272–9279.

Lyttle, T. W. 1991. Segregation distorters. *Annual Review of Genetics* 25: 511–557.

MacAndrew, A. 2003. A process for human/chimpanzee divergence. Alec's Evolution Pages. Accessed February 1, 2011. www.evolutionpages.com/homo_pan_divergence.htm.

Mace, G. M., J. L. Gittleman, and A. Purvis. 2003. Preserving the Tree of Life. *Science* 300: 1707–1709.

MacFadden, B. 1992. *Fossil Horses: Systematics, Paleobiology and Evolution of the Family Equidae.* Cambridge University Press, New York.

MacFadden, B. J. 2005. Fossil horses—evidence for evolution. *Science* 307: 1728–1730.

Mackie, G. O. 1995. On the visceral nervous system of *Ciona. Journal of the Marine Biological Association of the United Kingdom* 75: 141–151.

Mackowiak, P. A. 1994. Fever: blessing or curse? A unifying hypothesis. *Annals of Internal Medicine* 120: 1037–1041.

MacPhee, R., ed. 1999. *Extinctions in Near Time: Causes, Contexts and Consequences.* Kluwer, New York.

Maddison, W. 1997. Gene trees in species trees. *Systematic Biology* 46: 523–536.

Magurran, A. E. 2005. *Evolutionary Ecology: The Trinidadian Guppy.* Oxford University Press, Oxford.

Magurran, A. E., and B. H. Seghers. 1991. Variation in schooling and aggression amongst guppy (*Poecilia reticulata*) populations in Trinidad. *Behaviour* 118: 214–234.

Magurran, A. E., B. H. Seghers, G. R. Carvalho, and P. W. Shaw. 1992. Behavioural consequences of an artificial introduction of guppies (*Poecilia reticulata*) in N. Trinidad: evidence for the evolution of anti-predator behaviour in the wild. *Proceedings of the Royal Society B—Biological Sciences* 248: 117–122.

Magurran, A. E., B. H. Seghers, P. W. Shaw, and G. R. Carvalho. 1995. The behavioral diversity and evolution of guppy, *Poecilia reticulata*, populations in Trinidad. *Advances in the Study of Behavior* 24: 155–202.

Malicki, J., K. Schughart, and W. McGinnis. 1990. Mouse HOX-2.2 specifies thoracic segmental identity in *Drosophila* embryo and larvae. *Cell* 63: 961–967.

Malik, H. S., and S. Henikoff. 2001. Adaptive evolution of Cid, a centromere-specific histone in *Drosophila. Genetics* 157: 1293–1298.

Malik, H. S., and S. Henikoff. 2002. Conflict begets complexity: the evolution of centromeres. *Current Opinion in Genetics & Development* 12: 711–718.

Malik, S., A. W. Pightling, L. Stefaniak, A. Schurko, and J. Logsdon. 2008. An expanded inventory of conserved meiotic genes provides evidence for sex in *Trichomonas vaginalis.* PLoS ONE 3: e2879.

Malthus, T. 1798. *An Essay on the Principle of Population, As It Affects the Future Improvement of Society.* J. Johnson, London.

Mangone, D. M., and C. R. Currie. 2007. Garden substrate preparation behaviours in fungus-growing ants. *Canadian Entomologist* 139: 841–849.

Mann, A., and M. Weiss. 1996. Hominoid phylogeny and taxonomy: a consideration of the molecular and fossil evidence in an historical perspective. *Molecular Phylogenetics and Evolution* 5: 169–181.

Manne, L. L., T. M. Brooks, and S. L. Pimm. 1999. Relative risk of extinction of passerine birds on continents and islands. *Nature* 399: 258–261.

Margoliash, E. 1963. Primary structure and evolution of Cytochrome c. *Proceedings of the National Academy of Sciences of the United States of America* 50: 672–679.

Margulis, L. 1970. *Origin of Eukaryotic Cells.* Yale University Press, New Haven, CT.

Margulis, L., M. Dolan, and R. Guerrero. 2000. The chimeric eukaryote: origin of the nucleus from the karyomastigont in amitoichondriate protists. *Proceedings of the National Academy of Sciences of the United States of America* 97: 6954–6959.

Marler, P., and D. R. Griffin. 1973. The 1973 Nobel prize for physiology or medicine. *Science* 182: 464–466.

Martin, A., and S. R. Palumbi. 1993. Body size, metabolic rate, generation time, and the molecular clock. *Proceedings of the National Academy of Sciences of the United States of America* 90: 4087–4091.

Martin, N., J. Jaubert, P. Gounon, E. Salido, G. Haase, M. Szatanik, and J. L. Guenet. 2002. A missense mutation in *Tbce* causes progressive motor neuronopathy in mice. *Nature Genetics* 32: 443–447.

Martin, P., and R. Kelin, eds. 1984. *Quaternary Extinctions: A Prehistoric Revolution.* University of Arizona Press, Tucson.

Martin, R. 2004. *Missing Links: Evolutionary Concepts and Transitions Through Time.* Jones and Bartlett Publishers, Boston.

Martin, S. 2006. Lipid droplets: a unified view of a dynamic organelle. *Nature Reviews Molecular Cell Biology* 7: 373–378.

Martin, W., J. Baross, D. Kelley, and M. J. Russell. 2008. Hydrothermal vents and the origin of life. *Nature Reviews Microbiology* 6: 805–814.

Martin, W., A. Z. Mustafa, K. Henze, and C. Schnarrenberger. 1996. Higher-plant chloroplast and cytosolic fructose-1,6-bisphosphatase isoenzymes: origins via duplication rather than prokaryote–eukaryote divergence. *Plant Molecular Biology* 32: 485–491.

Matthew, P. 1831. *On Naval Timber and Arboriculture.* Longmans, London.

Mavelli, F., and K. Ruiz-Mirazo. 2007. Stochastic simulations of minimal self-reproducing cellular systems. *Philosophical Transactions of the Royal Society of London Series B—Biological Sciences* 362: 1789–1802.

Maxwell, W. D. 1992. Permian and Early Triassic extinction of non-marine tetrapods. *Paleontology* 135: 571–583.

May, M., and D. R. Brown. 2009. Diversifying and stabilizing selection of sialidase and N-acetylneuraminate catabolism in *Mycoplasma synoviae. Journal of Bacteriology* 191: 3588–3593.

Maynard Smith, J. 1966. Sympatric speciation. *American Naturalist* 100: 637–650.

Maynard Smith, J. 1971. The origin and maintenance of sex. In G. Williams, ed., *Group Selection*, pp. 163–175. Aldine, Chicago.

Maynard Smith, J. 1978. *The Evolution of Sex.* Cambridge University Press, Cambridge.

Maynard Smith, J. 1982. *Evolution and the Theory of Games.* Cambridge University Press, Cambridge.

Maynard Smith, J. 1988. The evolution of recombination. In R. E. Michod and B. Levin, eds., *The Evolution of Sex*, pp. 107–125. Sinauer Associates, Sunderland, MA.

Maynard Smith, J. 1989. *Evolutionary Genetics*. Oxford University Press, Oxford.

Maynard Smith, J., and J. Haigh. 1974. The hitch-hiking effect of a favourable gene. *Genetical Research* 23: 23–35.

Maynard Smith, J., and G. Price. 1973. The logic of animal conflict. *Nature* 246: 15–18.

Maynard Smith, J., N. H. Smith, M. O'Rourke, and B. G. Spratt. 1993. How clonal are bacteria? *Proceedings of the National Academy of Sciences of the United States of America* 90: 4384–4388.

Maynard Smith, J., and E. Szathmary. 1997. *The Major Transitions in Evolution*. Oxford University Press, New York.

Maynard Smith, J., and E. Szathmary. 1999. *The Origins of Life*. Oxford University Press, New York.

Mayr, E. 1942. *Systematics and the Origin of Species*. Columbia University Press, New York.

Mayr, E. 1970. *Populations, Species and Evolution*. Harvard University Press, Cambridge.

Mayr, E. 1982. *The Growth of Biological Thought*. Harvard University Press, Cambridge.

Mayr, E. 1983. How to carry out the adaptationist program. *American Naturalist* 121: 324–334.

Mayr, E. 1991. *One Long Argument*. Harvard University Press, Cambridge.

Mayr, E. 2002. Ernst Mayr through time on the biological species concept—a conceptual analysis—comments by Ernst Mayr. *Theory in Biosciences* 121: 99–100.

Mays, H. L., and G. E. Hill. 2004. Choosing mates: good genes versus genes that are a good fit. *Trends in Ecology & Evolution* 19: 554–559.

McArthur, E. D., D. C. Freeman, J. H. Graham, H. Wang, S. C. Sanderson, T. A. Monaco, and B. N. Smith. 1998. Narrow hybrid zone between two subspecies of big sagebrush (*Artemisia tridentata*: Asteraceae). VI. Respiration and water potential. *Canadian Journal of Botany-Revue Canadienne de Botanique* 76: 567–574.

McCallum, M. L. 2007. Amphibian decline or extinction? Current declines dwarf background extinction rate. *Journal of Herpetology* 41: 483–491.

McChesney, G. J., and B. R. Tershy. 1998. History and status of introduced mammals and impacts to breeding seabirds on the California Channel and Northwestern Baja California islands. *Colonial Waterbirds* 21: 335–347.

McCracken, K. G., C. P. Barger, and M. D. Sorenson. 2010. Phylogenetic and structural analysis of the HbA (α^A/β^A) and HbD (α^D/β^A) hemoglobin genes in two high-altitude waterfowl from the Himalayas and the Andes: bar-headed goose (*Anser indicus*) and Andean goose (*Chloephaga melanoptera*). *Molecular Phylogenetics and Evolution* 56: 649–658.

McCutcheon, J. P., B. R. McDonald, and N. A. Moran. 2009. Convergent evolution of metabolic roles in bacterial co-symbionts of insects. *Proceedings of the National Academy of Sciences of the United States of America* 106: 15394–15399.

McCutcheon, J. P., and N. A. Moran. 2010. Functional convergence in reduced genomes of bacterial symbionts spanning 200 million years of evolution. *Genome Biology and Evolution* 2: 708–718.

McDonald, J. H., and M. Kreitman. 1991. Adaptive protein evolution at the *Adh* locus in *Drosophila*. *Nature* 351: 652–654.

McGinnis, N., M. A. Kuziora, and W. McGinnis. 1990. Human Hox4.2 and *Drosophila deformed* encode similar regulatory specificities in *Drosophila* embryos and larvae. *Cell* 63: 969–976.

McInerney, J. O., and M. Wilkinson. 2005. New methods ring changes for the tree of life. *Trends in Ecology & Evolution* 20: 105–107.

McLean, M. J., K. H. Wolfe, and K. M. Devine. 1998. Base composition skews, replication orientation, and gene orientation in 12 prokaryote genomes. *Journal of Molecular Evolution* 47: 691–696.

McPherron, S. P., Z. Alemseged, C. W. Marean, J. G. Wynn, D. Reed, D. Geraads, R. Bobe, and H. A. Bearat. 2010. Evidence for stone-tool-assisted consumption of animal tissues before 3.39 million years ago at Dikika, Ethiopia. *Nature* 466: 857–860.

McShea, D. W. 1994. Mechanisms of large-scale evolutionary trends. *Evolution* 48: 1747–1763.

McShea, D. W. 1998. Possible largest-scale trends in organismal evolution: eight "live" hypotheses. *Annual Review of Ecology and Systematics* 29: 293–318.

Meckel, J. F. 1821. *System der vergleichenden Anatomie*. Rengersche Buchhandlung, Halle, Germany.

Medawar, P. B. 1952. *An Unsolved Problem of Biology*. H. K. Lewis, London.

Mehdiabadi, N. J., H. K. Reeve, and U. G. Mueller. 2003. Queens versus workers: sex-ratio conflict in eusocial Hymenoptera. *Trends in Ecology & Evolution* 18: 88–93.

Meirmans, S., and M. Neiman. 2006. Methodologies for testing a pluralist idea for the maintenance of sex. *Biological Journal of the Linnean Society* 89: 605–613.

Meissner, M., K. Stark, B. Cresnar, D. L. Kirk, and R. Schmitt. 1999. Volvox germline-specific genes that are putative targets of RegA repression encode chloroplast proteins. *Current Genetics* 36: 363–370.

Merrill, C., L. Bayraktaroglu, A. Kusano, and B. Ganetzky. 1999. Truncated RanGAP encoded by the *Segregation Distorter* locus of *Drosophila*. *Science* 283: 1742–1745.

Mesterton-Gibbons, M., and L. A. Dugatkin. 1992. Cooperation among unrelated individuals—evolutionary factors. *The Quarterly Review of Biology* 67: 267–281.

Meyer, A., J. M. Morrissey, and M. Schartl. 1994. Recurrent origin of a sexually selected trait in *Xiphophorus* fishes inferred from a molecular phylogeny. *Nature* 368: 539–542.

Meyer, A., and R. Zardoya. 2003. Recent advances in the (molecular) phylogeny of vertebrates. *Annual Review of Ecology, Evolution, and Systematics* 34: 311–338.

MHHE. 2010. A great deal is being learned about the dynamics of extinction. Accessed December 17, 2010. www.mhhe.com/biosci/genbio/olc_linkedcontent/j_enhancement/raven_27-2.html.

Michener, C. D., and R. Sokol. 1957. A quantitative approach to a problem in classification. *Evolution* 11: 130–162.

Michod, R. E. 1997. Cooperation and conflict in the evolution of individuality. 1. Multilevel selection of the organism. *American Naturalist* 149: 607–645.

Michod, R. E. 2007. Evolution of individuality during the transition from unicellular to multicellular life. *Proceedings of the National Academy of Sciences of the United States of America* 104: 8613–8618.

Michod, R. E., H. Bernstein, and A. M. Nedelcu. 2008. Adaptive value of sex in microbial pathogens. *Infection Genetics and Evolution* 8: 267–285.

Michad, R. E., and B. Levin, eds. 1988. *The Evolution of Sex*. Sinauer Associates, Sunderland, MA.

Mikkelsen, T. S., L. W. Hillier, E. E. Eichler, M. C. Zody, D. B. Jaffe, S. P. Yang, W. Enard, I. Hellmann, K. Lindblad-Toh, T. K. Altheide, N. Archidiacono, P. Bork, J. Butler, J. L. Chang, Z. Cheng, A. T. Chinwalla, P. deJong, K. D. Delehaunty, C. C. Fronick, L. L. Fulton, Y. Gilad, G. Glusman, S. Gnerre, T. A. Graves, T. Hayakawa, K. E. Hayden, X. Q. Huang, H. K. Ji, W. J. Kent, M. C. King, E. J. Kulbokas, M. K. Lee, G. Liu, C. Lopez-

Otin, K. D. Makova, O. Man, E. R. Mardis, E. Mauceli, T. L. Miner, W. E. Nash, J. O. Nelson, S. Pääbo, N. J. Patterson, C. S. Pohl, K. S. Pollard, K. Prufer, X. S. Puente, D. Reich, M. Rocchi, K. Rosenbloom, M. Ruvolo, D. J. Richter, S. F. Schaffner, A. F. A. Smit, S. M. Smith, M. Suyama, J. Taylor, D. Torrents, E. Tuzun, A. Varki, G. Velasco, M. Ventura, J. W. Wallis, M. C. Wendl, R. K. Wilson, E. S. Lander, and R. H. Waterston. 2005. Initial sequence of the chimpanzee genome and comparison with the human genome. *Nature* 437: 69–87.

Milberg, P., and T. Tyrberg. 1993. Native birds and noble savages—a review of man-caused prehistoric extinctions of island birds. *Ecography* 16: 229–250.

Milinski, M. 1979. Can an experienced predator overcome the confusion of swarming prey more easily? *Animal Behaviour* 27: 1122–1126.

Milinski, M., and T. Bakker. 1990. Female sticklebacks use male coloration in mate choice and hence avoid parasitized sticklebacks. *Nature* 344: 330–333.

Miller, S. 1953. A production of amino acids under possible primitive Earth conditions. *Science* 117: 528–529.

Mills, D. K., R. Peterson, and S. Spiegelman. 1967. An extracellular Darwinian experiment with a self-replicating nucleic acid model. *Proceedings of the National Academy of Sciences of the United States of America* 58: 217–220.

Mira, A., H. Ochman, and N. A. Moran. 2001. Deletional bias and the evolution of bacterial genomes. *Trends in Genetics* 17: 589–596.

Mirsky, A. E., and H. Ris. 1951. The desoxyribonucleic acid content of animal cells and its evolutionary significance. *Journal of General Physiology* 34: 451–462.

Mitchell, W. A., and T. J. Valone. 1990. Commentary: the optimization approach—studying adaptations by their function. *The Quarterly Review of Biology* 65: 43–52.

Mittlebach, G. 1984. Group size and feeding rate in bluegills. *Copeia* 1984: 998–1000.

Mock, D. 1980. White–dark polymorphism in herons. In D. L. Drawe, ed., *Proceedings of the First Welder Wildlife Symposium*, pp. 145–161. Welder Wildlife Foundation, Sinton, TX.

Mock, D., and G. Parker. 1997. *The Evolution of Sibling Rivalry*. Oxford University Press, New York.

Møller, A., and M. Jennions. 2001. How important are direct benefits of sexual selection. *Naturwissenschaften* 88: 401–415.

Monagas, W. R., and R. E. Gatten. 1983. Behavioural fever in the turtles *Terrapene carolina* and *Chrysemys picta*. *Journal of Thermal Biology* 8: 285–288.

Moore, J. E., and A. J. Read. 2008. A Bayesian uncertainty analysis of cetacean demography and bycatch mortality using age-at-death data. *Ecological Applications* 18: 1914–1931.

Moore, W. S. 1977. Evaluation of narrow hybrid zones in vertebrates. *The Quarterly Review of Biology* 52: 263–277.

Moose, S. P., J. W. Dudley, and T. R. Rocheford. 2004. Maize selection passes the century mark: a unique resource for 21st century genomics. *Trends in Plant Sciences* 9: 358–364.

Moran, N., and P. Baumann. 1994. Phylogenetics of cytoplasmically inherited microorganisms of Arthropods. *Trends in Ecology & Evolution* 9: 15–20.

Morgan, T. H. 1934. *Embryology and Genetics*. Columbia University Press, New York.

Morgan, T. H., C. Bridges, and A. Sturtevant. 1925. The genetics of *Drosophila*. *Bibliographica Genetica* 2: 1–262.

Morris, S. W. 1994. Fleeming Jenkin and "The Origin of Species": a reassessment. *British Journal for the History of Science* 27: 313–343.

Morse, D. H. 1970. Ecological aspects of some mixed species foraging flocks of birds. *Ecological Monographs* 4: 119–168.

Mourier, T., A. J. Hansen, E. Willerslev, and P. Arctander. 2001. The human genome project reveals a continuous transfer of large mitochondrial fragments to the nucleus. *Molecular Biology and Evolution* 18: 1833–1837.

Mousseau, T. A., and D. A. Roff. 1987. Natural selection and the heritability of fitness components. *Heredity* 59: 181–197.

Mousseau, T., B. Sinervo, and J. Endler, eds. 1999. *Adaptive Genetic Variation in the Wild*. Oxford University Press, New York.

Moya-Sola, S., D. M. Alba, S. Almecija, I. Casanovas-Vilar, M. Kohler, S. De Esteban-Trivigno, J. M. Robles, J. Galindo, and J. Fortuny. 2009. A unique Middle Miocene European hominoid and the origins of the great ape and human clade. *Proceedings of the National Academy of Sciences of the United States of America* 106: 9601–9606.

Mueller, U. G., and C. Rabeling. 2008. A breakthrough innovation in animal evolution. *Proceedings of the National Academy of Sciences of the United States of America* 105: 5287–5288.

Muir, W. M. 1996. Group selection for adaptation to multiple-hen cages: selection program and direct responses. *Poultry Science* 75: 447–458.

Mullen, L. M., S. N. Vignieri, J. A. Gore, and H. E. Hoekstra. 2009. Adaptive basis of geographic variation: genetic, phenotypic and environmental differences among beach mouse populations. *Proceedings of the Royal Society B—Biological Sciences* 276: 3809–3818.

Müller, F. 1879. *Ituna* and *Thyridia*: a remarkable case of mimicry in butterflies. *Transactions of the Entomological Society of London* 1879: xx–xxix.

Muller, H. J. 1925. Why polyploidy is rarer in animals than in plants. *American Naturalist* 59: 346–353.

Muller, H. J. 1932. Genetic aspects of sex. *American Naturalist* 66: 118–138.

Muller, H. J. 1942. Isolating mechanisms, speciation, and temperature. *Biology Symposium* 6: 71–125.

Muller, H. J. 1964. The relation of recombination to mutational advance. *Mutation Research* 1: 2–9.

Muniesa, M., F. Lucena, and J. Jofre. 1999. Comparative survival of free Shiga toxin 2-encoding phages and *Escherichia coli* strains outside the gut. *Applied and Environmental Microbiology* 65: 5615–5618.

Murphy, W. J., E. Eizirik, W. E. Johnson, Y. P. Zhang, O. A. Ryder, and S. J. O'Brien. 2001. Molecular phylogenetics and the origins of placental mammals. *Nature* 409: 614–618.

Myerowitz, R. 1997. Tay-Sachs disease–causing mutations and neutral polymorphisms in the *Hex A* gene. *Human Mutation* 9: 195–208.

Myers, N. 1988. Threatened biotas: "hot spots" in tropical forests. *Environmentalist* 8: 187–208.

Myers, S., L. Bottolo, C. Freeman, G. McVean, and P. Donnelly. 2005. A fine-scale map of recombination rates and hotspots across the human genome. *Science* 310: 321–324.

Nachman, M. W. 2005. The genetic basis of adaptation: lessons from concealing coloration in pocket mice. *Genetica* 123: 125–136.

Nachman, M. W., H. E. Hoekstra, and S. L. D'Agostino. 2003. The genetic basis of adaptive melanism in pocket mice. *Proceedings of the National Academy of Sciences of the United States of America* 100: 5268–5273.

National Geographic. 2010. Dinosaur extinction. National Geographic Society. Accessed December 17, 2010. http://science.national geographic.com/science/prehistoric-world/dinosaur-extinction.html.

National Research Council. 2000. *The Future Role of Pesticides in U.S. Agriculture*. National Academies Press, Washington, DC.

Nedelcu, A., and R. Michod. 2006. The evolutionary origin of an altruistic gene. *Molecular Biology and Evolution* 23: 1460–1464.

Nee, S., and R. M. May. 1997. Extinction and the loss of evolutionary history. *Science* 278: 692–694.

Neff, B. D., P. Fu, and M. R. Gross. 2003. Sperm investment and alternative mating tactics in bluegill sunfish (*Lepomis macrochirus*). *Behavioral Ecology* 14: 634–641.

Nei, M., and T. Gojobori. 1986. Simple methods for estimating the numbers of synonymous and nonsynymous nucleotide substitutions. *Molecular Biology and Evolution* 3: 418–426.

Neiman, M., G. Hehman, J. T. Miller, J. M. Logsdon, and D. R. Taylor. 2010. Accelerated mutation accumulation in asexual lineages of a freshwater snail. *Molecular Biology and Evolution* 27: 954–963.

Neiman, M., J. Jokela, and C. M. Lively. 2005. Variation in asexual lineage age in *Potamopyrgus antipodarum*, a New Zealand snail. *Evolution* 59: 1945–1952.

Neiman, M., S. Meirmans, and P. G. Meirmans. 2009. What can asexual lineage age distribution tell us about the maintenance of sex? *Annals of the New York Academy of Sciences* 1168: 185–200.

Nelson, D. A., H. Khanna, and P. Marler. 2001. Learning by instruction or selection: implications for patterns of geographic variation in birdsong. *Behaviour* 138: 1137–1160.

Nesse, R. M. 2001. The smoke detector principle: natural selection and the regulation of defensive responses. *Annals of the New York Academy of Sciences* 935: 75–85.

Nesse, R. M. 2005. Maladaptation and natural selection. *The Quarterly Review of Biology* 80: 62–70.

Nesse, R. M., S. C. Stearns, and G. S. Omenn. 2006. Medicine needs evolution. *Science* 311: 1071.

Nesse, R. M., and G. C. Williams. 1994. *Why We Get Sick: The New Science of Darwinian Medicine*. Times Books, New York.

Ng, M., and M. F. Yanofsky. 2001. Function and evolution of the plant MADS-box gene family. *Nature Reviews Genetics* 2: 186–195.

Nielsen, R. 2005. Molecular signatures of natural selection. *Annual Review of Genetics* 39: 197–218.

Nikaido, H. 2009. Multidrug resistance in bacteria. *Annual Review of Biochemistry* 78: 119–146.

Nilsson, D. E., and S. Pelger. 1994. A pessimistic estimate of the time required for an eye to evolve. *Proceedings of the Royal Society B—Biological Sciences* 256: 53–58.

Nilsson, L. A., L. Jonsson, L. Ralison, and E. Randrianjohany. 1987. Angrecoid orchids and hawkmoths in central Madagascar—specialized pollination systems and generalist foragers. *Biotropica* 19: 310–318.

Nilsson-Ehle, H. 1908. *Einige Ergebnisse von Kreuzungen bei Hafer und Weizen. Botaniska Notiser*: 257–294.

Nisbet, R. M., and W. W. Murdoch. 1995. Final report: Framework for predicting the effects of environmental change on populations. Accessed October 3, 2011. http://cfpub.epa.gov/ncer_abstracts/index.ctm/fuseaction/display.abstractDetail/abstract/484/report/F.

Nonacs, P. 1986. Ant reproductive strategies and sex allocation theory. *The Quarterly Review of Biology* 61: 1–21.

Nordborg, M. 2007. Coalescent theory. In D. J. Balding, M. J. Bishop, and C. Cannings, eds., *Handbook of Statistical Genetics*, 3rd Ed., pp. 843–877. Wiley and Sons, Chichester, UK.

Normark, B. B., O. P. Judson, and N. A. Moran. 2003. Genomic signatures of ancient asexual lineages. *Biological Journal of the Linnean Society* 79: 69–84.

Novick, L. R., and K. M. Catley, 2007. Understanding phylogenies in biology: the influence of a Gestalt perceptual principle. *Journal of Experimental Psychology: Applied* 13: 197–223.

Nüsslein-Volhard, C., and E. Wieschaus. 1980. Mutations affecting segment number and polarity in *Drosophila*. *Nature* 287: 795–801.

Oakley, T. H., and M. S. Pankey. 2008. Opening the "Black Box": the genetic and biochemical basis of eye evolution. *Evolution: Education and Outreach* 1: 390–402.

O'Brien, A. D., J. W. Newland, S. F. Miller, R. K. Holmes, H. W. Smith, and S. B. Formal. 1984. Shiga-like toxin-converting phages from *Escherichia coli* strains that cause hemorrhagic colitis or infantile diarrhea. *Science* 226: 694–696.

O'Donald, P. 1980. *Genetic Models of Sexual Selection*. University of Cambridge Press, Cambridge.

Ohno, S. 1970. *Evolution by Gene Duplication*. Springer-Verlag, New York.

Ohta, T. 1992. The nearly neutral theory of molecular evolution. *Annual Review of Ecology and Systematics* 23: 263–286.

Orgel, L. E. 1963. The maintenance of the accuracy of protein synthesis and its relevance to aging. *Proceedings of the National Academy of Sciences of the United States of America* 49: 517–521.

Orgel, L. E. 2004. Prebiotic chemistry and the origin of the RNA world. *Critical Reviews in Biochemistry and Molecular Biology* 39: 99–123.

Orgel, L. E., and F. H. C. Crick. 1980. Selfish DNA: the ultimate parasite. *Nature* 284: 604–607.

Oro, J. 1961. Comets and the formation of biochemical compounds on the primitive earth. *Nature* 190: 389–390.

Orr, H. A. 1990. Why polyploidy is rarer in animals than in plants revisited. *American Naturalist* 136: 759–770.

Orr, H. A. 2009. Testing natural selection. *Scientific American* 300: 44–50.

Orr, H. A. 2010. The population genetics of beneficial mutations. *Philosophical Transactions of the Royal Society of London Series B—Biological Sciences* 365: 1195–1201.

Orth, C. J., J. S. Gilmore, J. D. Knight, C. L. Pillmore, R. H. Tschudy, and J. E. Fassett. 1981. An iridium abundance anomaly at the palynological Cretaceous–Tertiary boundary in northern New Mexico. *Science* 214: 1341–1343.

Ortiz-Monasterio, J., K. Sayre, S. Rajaram, and M. McMahon. 1997. Genetic progress in wheat yield and nitrogen efficiency under four nitrogen rates. *Crop Science* 37: 898–904.

Osborn, H. F. 1894. *From the Greeks to Darwin: An Outline of the Development of the Evolution Idea*. MacMillan, London.

Ostrom, J. H. 1974. *Archaeopteryx* and the origin of flight. *The Quarterly Review of Biology* 49: 27–47.

Othman, R. (GC Biologi). 2010. Human digestive system. Accessed October 3, 2011. http://cikgurozaini.blogspot.com/2010/08/human-digestive-system.html.

Otte, D., and J. A. Endler, eds. 1989. *Speciation and Its Consequences*. Sinauer Associates, Sunderland, MA.

Otto, S. P. 2009. The evolutionary enigma of sex. *American Naturalist* 174: S1–S14.

Otto, S. P., and T. Lenormand. 2002. Resolving the paradox of sex and recombination. *Nature Reviews Genetics* 3: 252–261.

Otto, S., and M. Whitlock. 2005. Fixation probabilities and times. *Encyclopedia of Life Sciences*. Wiley and Sons, New York.

Otto, S. P., and J. Whitton. 2000. Polyploid incidence and evolution. *Annual Review of Genetics* 34: 401–437.

Owings, D. H., and R. G. Coss. 1977. Snake mobbing by California ground squirrels: adaptive variation and ontogeny. *Behaviour* 62: 50–69.

Owings, D. H., and D. W. Leger. 1980. Chatter vocalization of California ground squirrels: predator- and social-role specificity. *Zeitschrift fur Tierpsychologie* 54: 163–184.

Pääbo, S. 1999. Human evolution (Reprinted from *Trends in Biochemical Science*, Volume 12, December, 1999). *Trends in Cell Biology* 9: M13–M16.

Page, R. B., M. A. Boley, J. J. Smith, S. Putta, and S. R. Voss. 2010. Microarray analysis of a salamander hopeful monster reveals transcriptional signatures of paedomorphic brain development. *BMC Evolutionary Biology* 10: 199.

Page, R. D. M. 1994. Parallel phylogenies: reconstructing the history of host–parasite assemblages. *Cladistics* 10: 155–173.

Page, R. D. M., ed. 2003. *Tangled Trees: Phylogeny, Cospeciation, and Coevolution.* University of Chicago Press, Chicago.

Pal, C., B. Papp, and M. J. Lercher. 2005. Adaptive evolution of bacterial metabolic networks by horizontal gene transfer. *Nature Genetics* 37: 1372–1375.

Pal, C., B. Papp, M. J. Lercher, P. Csermely, S. G. Oliver, and L. D. Hurst. 2006. Chance and necessity in the evolution of minimal metabolic networks. *Nature* 440: 667–670.

Palaima, A. 2007. The fitness cost of generalization: present limitations and future possible solutions. *Biological Journal of the Linnean Society* 90: 583–590.

Paland, S., and M. Lynch. 2006. Transitions to asexuality result in excess amino acid substitutions. *Science* 311: 990–992.

Paley, W. 1802. *Natural Theology,* 2nd Ed. R. Paulder Publishing, London.

Park, N. H., I. H. Song, and Y. H. Chung. 2006. Chronic hepatitis B in hepatocarcinogenesis. *Postgraduate Medical Journal* 82: 507–515.

Parker, G. 1970. Sperm competition and its evolutionary consequences in insects. *Biological Reviews* 45: 525–567.

Parker, G. A. 1979. Sexual selection and sexual conflict. In M. Blum and N. Blum, eds., *Sexual Selection and Reproductive Competition in Insects,* pp. 123–166. Academic Press, New York.

Parker, G. A., N. Royle, and I. Hartley. 2002. Intrafamilial conflict and parental investment: a synthesis. *Philosophical Transactions of the Royal Society B—Biological Sciences* 357: 295–307.

Parker, G. A., V. G. F. Smith, and R. R. Baker. 1972. Origin and evolution of gamete dimorphism and male–female phenomenon. *Journal of Theoretical Biology* 36: 529–553.

Partridge, L., and N. H. Barton. 1993. Optimality, mutation, and the evolution of ageing. *Nature* 362: 305–311.

Paul, N., and G. F. Joyce. 2002. A self-replicating ligase ribozyme. *Proceedings of the National Academy of Sciences of the United States of America* 99: 12733–12740.

Paul, N., and G. F. Joyce. 2004. Minimal self-replicating systems. *Current Opinion in Chemical Biology* 8: 634–639.

Paul, N., G. Springsteen, and G. F. Joyce. 2006. Conversion of a ribozyme to a deoxyribozyme through in vitro evolution. *Chemistry & Biology* 13: 329–338.

Pauling, L., H. A. Itano, S. J. Singer, and I. C. Wells, 1949. Sickle cell anemia, a molecular disease. *Science* 110: 543–548.

Payne, J. L., and S. Finnegan. 2007. The effect of geographic range on extinction risk during background and mass extinction. *Proceedings of the National Academy of Sciences of the United States of America* 104: 10506–10511.

Payne, J. L., A. G. Boyer, J. H. Brown, S. Finnegan, M. Kowalewski, R. A. Krause, S. K. Lyons, C. R. McClain, D. W. McShea, P. M. Novack-Gottshall, F. A. Smith, J. A. Stempien, and S. C. Wang. 2009. Two-phase increase in the maximum size of life over 3.5 billion years reflects biological innovation and environmental opportunity. *Proceedings of the National Academy of Sciences of the United States of America* 106: 24–27.

Pearson, J. C., D. Lemons, and W. McGinnis. 2005. Modulating Hox gene functions during animal body patterning. *Nature Reviews Genetics* 6: 893–904.

Pellmyr, O. 2003. Yuccas, yucca moths, and coevolution: a review. *Annals of the Missouri Botanical Garden* 90: 35–55.

Peris, J. B., P. Davis, J. M. Cuevas, M. R. Nebot, and R. Sanjuan. 2010. Distribution of fitness effects caused by single-nucleotide substitutions in bacteriophage f1. *Genetics* 185: 603–609.

Perna, N. T., G. Plunkett III, V. Burland, B. Mau, J. D. Glasner, D. J. Rose, G. F. Mayhew, P. S. Evans, J. Gregor, H. A. Kirkpatrick, G. Posfai, J. Hackett, S. Klink, A. Boutin, Y. Shao, L. Miller, E. J. Grotbeck, N. W. Davis, A. Lim, E. T. Dimalanta, K. D. Potamousis, J. Apodaca, T. S. Anantharaman, J. Y. Lin, G. Yen, D. C. Schwartz, R. A. Welch, and F. R. Blattner. 2001. Genome sequence of enterohaemorrhagic *Escherichia coli* O157:H7. *Nature* 409: 529–533. (published erratum appears in *Nature* 410: 240).

Perron, M., A. G. Gordon, and J. Bousquet. 1995. Species-specific RAPD fingerprints for the closely-related *Picea mariana* and *P. rubens.* *Theoretical and Applied Genetics* 91: 142–149.

Perron, M., D. J. Perry, C. Andalo, and J. Bousquet. 2000. Evidence from sequence-tagged-site markers of a recent progenitor–derivative species pair in conifers. *Proceedings of the National Academy of Sciences of the United States of America* 97: 11331–11336.

Petrie, M. 1994. Improved growth and survival of offspring of peacocks with more elaborate trains. *Nature* 371: 598–599.

Petrie, M., T. Halliday, and C. Sanders. 1991. Peahens prefer peacocks with elaborate trains. *Animal Behaviour* 41: 323–331.

Pfeffer, S., M. Zavolan, F. A. Grässer, M. Chien, J. J. Russo, J. Ju, B. John, A. J. Enright, D. Marks, C. Sander, and T. Tuschl. 2004. Identification of virus-encoded microRNAs. *Science* 304: 734–736.

Pfennig, D. W. 1995. Absence of joint nesting advantage in desert seed harvester ants: evidence from a field experiment. *Animal Behaviour* 49: 567–575.

Phelps, C. B., R. R. Wang, S. Choo, and R. Gaudet. 2010. Differential regulation of TRPV1, TRPV3, and TRPV4 sensitivity through a conserved binding site on the Ankyrin repeat domain. *Journal of Biological Chemistry* 285: 731–740.

Philipp, D., and M. Gross. 1994. Genetic evidence for cuckoldry in bluegill *Lepomis macrochirus.* *Molecular Ecology* 3: 563–569.

Piatigorsky, J., and G. Wistow. 1989. Enzyme/crystallins: gene sharing as an evolutionary strategy. *Cell* 57: 197–199.

Piddock, L. J. V., C. A. Hart, A. M. Johnston, and D. Taylor. 1998. Review of the literature on antibiotic resistance in foodborne pathogens. *ASM News* 64: 311–312.

Piel, F. B., A. P. Patil, R. E. Howes, O. A. Nyangiri, P. W. Gething, T. N. Williams, D. J. Weatherall, and S. I. Hay. 2010. Global distribution of the sickle cell gene and geographical confirmation of the malaria hypothesis. *Nature Communications* 1: 104.

Pierce, B. 2010. *Genetics Essentials.* Freeman, New York.

Pierce, N., R. Kitchling, R. Buckley, M. Talor, and K. Benbow. 1987. The costs and benefits of cooperation between the Australian lycaenid butterfly, *Jalmenus evagoras,* and its attendant ants. *Behavioral Ecology and Sociobiology* 21: 237–248.

Pierce, N. E., M. F. Braby, A. Heath, D. J. Lohman, J. Mathew, D. B. Rand, and M. A. Travassos. 2002. The ecology and evolution of ant association in the Lycaenidae (Lepidoptera). *Annual Review of Entomology* 47: 733–771.

Pijanowska, J., and G. Stolpe. 1996. Summer diapause in *Daphnia* as a reaction to the presence of fish. *Journal of Plankton Research* 18: 1407–1412.

Pimentel, D., and H. Lehman, eds. 1991. *The Pesticide Question: Environment, Economics and Ethics*. Chapman & Hall, New York.

Pimm, S. L., M. P. Moulton, L. J. Justice, N. J. Collar, D. Bowman, and W. J. Bond. 1994. Bird extinctions in the central Pacific. *Philosophical Transactions of the Royal Society of London Series B—Biological Sciences* 344: 27–33.

Pimm, S., M. Moulton, and L. Justice. 1995. Bird extinctions in the central Pacific. In J. Lawton and R. May, eds., *Extinction Rates*, pp. 75–87. Oxford University Press, Oxford.

Pimm, S., P. Raven, A. Peterson, C. H. Sekercioglu, and P. R. Ehrlich. 2006. Human impacts on the rates of recent, present, and future bird extinctions. *Proceedings of the National Academy of Sciences of the United States of America* 103: 10941–10946.

Pinto, J. P., G. R. Gladstone, and Y. L. Yung. 1980. Photochemical production of formaldehyde in earth's primitive atmosphere. *Science* 210: 183–184.

Pisani, D., J. A. Cotton, and J. O. McInerney. 2007. Supertrees disentangle the chimerical origin of eukaryotic genomes. *Molecular Biology and Evolution* 24: 1752–1760.

Pitcher, T. J., 1986. Functions of shoaling behaviour in teleost. In T. J. Pitcher, ed., *The Behaviour of Teleost Fishes*, pp. 294–337. Johns Hopkins University Press, Baltimore.

Pitcher, T. J., and C. Wyche. 1983. Predator-avoidance behaviour in sand-eel schools: why do schools seldom split. In D. Noakes, D. Lindquist, G. Helfman, and J. Ward, eds., *Predator and Prey in Fishes*, pp. 193–204. Junk, The Hague.

Pizzari, T., C. Cornwallis, H. Leville, S. Jakobsson, and T. Birkhead. 2003. Sophisticated sperm allocation in male fowl. *Nature* 426: 70–74.

Pomiankowski, A. 1988. The evolution of female mate preferences for male genetic quality. In P. Harvey and L. Partridge, eds., *Oxford Surveys in Evolutionary Biology*, pp. 136–184. Oxford University Press, Oxford.

Ponder, W., and D. Lindberg. 1997. Towards a phylogeny of gastropod molluscs: an analysis using morphological characters. *Zoological Journal of the Linnean Society* 119: 83–265.

Powell, J. F. F., S. M. Reska-Skinner, M. O. Prakash, W. H. Fischer, M. Park, J. E. Rivier, A. G. Craig, G. O. Mackie, and N. M. Sherwood. 1996. Two new forms of gonadotropin-releasing hormone in a protochordate and the evolutionary implications. *Proceedings of the National Academy of Sciences of the United States of America* 93: 10461–10464.

Powner, M. W., B. Gerland, and J. D. Sutherland. 2009. Synthesis of activated pyrimidine ribonucleotides in prebiotically plausible conditions. *Nature* 459: 239–242.

Price, K. 1998. Benefits of begging for yellow-headed blackbird nestlings. *Animal Behaviour* 56: 571–577.

Price, M. V., and N. M. Waser. 1982. Population structure, frequency dependent selection and the maintenance of sexual reproduction. *Evolution* 36: 35–43.

Price, T., and D. Schulter. 1991. On the low heritability of life-history traits. *Evolution* 45: 853–861.

Price, T., D. Schulter, and N. E. Heckman. 1993. Sexual selection when the female benefits directly. *Biological Journal of the Linnean Society* 48: 187–211.

Proctor, H., and I. Owens. 2000. Mites and birds: diversity, parasitism and coevolution. *Trends in Ecology & Evolution* 15: 358–364.

Prothero, D. 2003. *Bringing Fossils to Life: An Introduction to Paleobiology*. McGraw-Hill Science, New York.

Provine, W. 1986. *Sewall Wright and Evolutionary Biology*. University of Chicago Press, Chicago.

Prum, R. O., and A. Brush. 2002. The evolutionary origin and diversification of feathers. *The Quarterly Review of Biology* 77: 261–295.

Pryke, S., S. G. E. Andersson, and M. Lawes. 2001. Sexual selection of multiple handicaps in the red-collared widowbird: female choice of tail length but not carotenoid display. *Evolution* 55: 1452–1463.

Ptak, S. E., D. A. Hinds, K. Koehler, B. Nickel, N. Patil, D. G. Ballinger, M. Przeworski, K. A. Frazer, and S. Pääbo. 2005. Fine-scale recombination patterns differ between chimpanzees and humans. *Nature Genetics* 37: 429–434.

Pulido, F., P. Berthold, G. Mohr, and U. Querner. 2001. Heritability of the timing of autumn migration in a natural bird population. *Proceedings of the Royal Society B—Biological Sciences* 268: 953–959.

Pulliam, R. 1973. On the advantages of flocking. *Journal of Theoretical Biology* 38: 419–422.

Pulliam, R., and T. Caraco. 1984. Living in groups: is there an optimal group size? In J. Krebs and N. Davies, eds., *Behavioral Ecology*, pp. 122–148. Sinauer Associates, Sunderland, MA.

Purvis, A. 2008. Phylogenetic approaches to the study of extinction. *Annual Review of Ecology Evolution and Systematics* 39: 301–319.

Queller, D. C. 1992. Quantitative genetics, inclusive fitness and group selection. *American Naturalist* 139: 540–558.

Queller, D. C. 2003. Theory of genomic imprinting conflict in social insects. *BMC Evolutionary Biology* 3: 15.

Raby, P. 2001. *Alfred Russel Wallace: A Life*. Princeton University Press, Princeton.

Raff, R., and T. C. Kaufman. 1983. *Embryos, Genes and Evolution*. MacMillan, New York.

Ralph, S. A., G. G. van Dooren, R. F. Waller, M. J. Crawford, M. J. Fraunholz, B. J. Foth, C. J. Tonkin, D. S. Roos, and G. I. McFadden. 2004. Metabolic maps and functions of the *Plasmodium falciparum* apicoplast. *Nature Reviews Microbiology* 2: 203–216.

Ramírez, S., B. Gravendeel, R. Singer, C. Marshall, and N. E. Pierce. 2007. Dating the origin of the Orchidaceae from a fossil orchid with its pollinator. *Nature* 448: 1042–1045.

Rampino, M. R. 1999. Evidence for abrupt latest Permian mass extinction of foraminifera: results of tests for the Signor–Lipps effect: reply. *Geology* 27: 383–384.

Rampino, M. R., and A. C. Adler. 1998. Evidence for abrupt latest Permian mass extinction of foraminifera: results of tests for the Signor–Lipps effect. *Geology* 26: 415–418.

Ramsey, J., and D. W. Schemske. 1998. Pathways, mechanisms, and rates of polyploid formation in flowering plants. *Annual Review of Ecology and Systematics* 29: 467–501.

Randerson, J. P., and L. D. Hurst. 2001. The uncertain evolution of the sexes. *Trends in Ecology & Evolution* 16: 571–579.

Raper, K. B. 1935. *Dictyostelium discoideum*: a new species of slime mold from decaying forest leaves. *Journal of Agricultural Research* 50: 135–147.

Rashevsky, N. 1938. *Mathematical Biophysics*. University of Chicago Press, Chicago.

Ratnieks, F. L. W., and H. Helantera. 2009. The evolution of extreme altruism and inequality in insect societies. *Philosophical Transactions of the Royal Society of London Series B—Biological Sciences* 364: 3169–3179.

Raup, D. M. 1986. Biological extinction in earth history. *Science* 231: 1528–1533.

Raup, D. M. 1992. Large-body impact and extinction in the Pharenozoic. *Paleobiology* 18: 80–88.

Raup, D. M., and D. Jablonski. 1993. Geography of end-Cretaceous marine bivalve extinctions. *Science* 260: 971–973.

Raup, D. M., and J. J. Sepkoski, Jr. 1979. Size of the Permo–Triassic bottleneck and its evolutionary implications. *Science* 206: 217–218.

Recker, D. 1990. There's more than one way to recognize a Darwinian–Lyell Darwinism. *Philosophy of Science* 57: 459–478.

Reddy, V., J. P. Emery, M. J. Gaffey, W. F. Bottke, A. Cramer, and M. S. Kelley. 2009. Composition of 298 Baptistina: implications for the K/T impactor link. *Meteoritics & Planetary Science* 44: 1917–1927.

Redfield, R. J. 2001. Do bacteria have sex? *Nature Reviews Genetics* 2: 634–639.

Reed, J. L., T. D. Vo, C. H. Schilling, and B. O. Palsson. 2003. An expanded genome-scale model of *Escherichia coli* K-12 (iJR904 GSM/GPR). *Genome Biology* 4: R54.1–R54.12.

Reeve, H. K., and B. Holldobler. 2007. The emergence of a superorganism through intergroup competition. *Proceedings of the National Academy of Sciences of the United States of America* 104: 9736–9740.

Reeve, H. K., and P. W. Sherman. 1993. Adaptation and the goals of evolutionary research. *The Quarterly Review of Biology* 68: 1–32.

Reich, D., R. E. Green, M. Kircher, J. Krause, N. Patterson, E. Y. Durand, B. Viola, A. W. Briggs, U. Stenzel, P. L. F. Johnson, T. Maricic, J. M. Good, T. Marques-Bonet, C. Alkan, Q. Fu, S. Mallick, H. Li, M. Meyer, E. E. Eichler, M. Stoneking, M. Richards, S. Talamo, M. V. Shunkov, A. P. Derevianko, J.-J. Hublin, J. Kelso, M. Slatkin, and S. Pääbo. 2010. Genetic history of an archaic hominin group from Denisova Cave in Siberia. *Nature* 468: 1053–1060.

Reichard, M., M. Ondrackova, A. Bryjova, C. Smith, and J. Bryja. 2009. Breeding resource distribution affects selection gradients on male phenotypic traits: experimental study of lifetime reproductive success in the bitterling fish (*Rhodeus amarus*). *Evolution* 63: 377–390.

Reinke, V., H. E. Smith, J. Nance, J. Wang, C. Van Doren, R. Begley, S. J. M. Jones, E. B. Davis, S. Scherer, S. Ward, and S. K. Kim. 2000. A global profile of germline gene expression in *C. elegans*. *Molecular Cell* 6: 605–616.

Repcheck, J. 2003. *The Man Who Found Time: James Hutton and the Discovery of Earth's Antiquity*. Perseus Publishing, New York.

Retallack, G. J. 1999. Postapocalyptic greenhouse paleoclimate revealed by earliest Triassic paleosols in the Sydney Basin, Australia. *Bulletin of the Geological Society of America* 111: 52–70.

Retallack, G. J., J. Veevers, and R. Morante. 1996. Global coal gap between Permian–Triassic extinction and Middle Tertiary recovery of peat-forming plants. *Bulletin of the Geological Society of America* 108: 195–207.

Reynolds, W. M., M. E. Casterlin, and J. B. Covert. 1976. Behavioural fever in teleost fishes. *Nature* 259: 41–42.

Reznick, D. 1996. Life history evolution in guppies: a model system for the empirical study of adaptation. *Netherlands Journal of Zoology* 46: 172–190.

Reznick, D., H. Bryga, and J. A. Endler. 1990. Experimentally induced life-history evolution in a natural population. *Nature* 346: 357–359.

Ribeiro, S., and G. B. Golding. 1998. The mosaic nature of the eukaryotic nucleus. *Molecular Biology and Evolution* 15: 779–788.

Rice, J., D. A. Warner, C. D. Kelly, M. P. Clough, and J. T. Colbert. 2010. The theory of evolution is not an explanation for the origin of life. *Evolution: Education and Outreach* 3: 141–142.

Rice, W. R. 1994. Degeneration of a nonrecombining chromosome. *Science* 263: 230–232.

Rice, W. R. 2002. Experimental tests of the adaptive significance of sexual recombination. *Nature Reviews Genetics* 3: 241–251.

Ridley, M. 2004. *Evolution*, 3rd Ed. Blackwell Scientific Publishing, Malden, MA.

Rieseberg, L. H. 2001. Chromosomal rearrangements and speciation. *Trends in Ecology & Evolution* 16: 351–358.

Rissing, S., and G. Pollock. 1986. Social interaction among pleometric queens of *Veromessor pergandei* during colony foundation. *Animal Behaviour* 34: 226–234.

Rissing, S., and G. Pollock. 1987. Queen aggression, pleometric advantage and brood raiding in the ant *Veromessor pergandei*. *Animal Behaviour* 35: 975–982.

Rissing, S., and G. Pollock. 1991. An experimental analysis of pleometric advantage in *Messor pergandei*. *Insectes Sociaux* 63: 205–211.

Rissing, S., G. Pollock, M. Higgins, R. Hagen, and D. Smith. 1989. Foraging specialization without relatedness or dominance among cofounding ant queens. *Nature* 338: 420–422.

Rivera, M. C., R. Jain, J. E. Moore, and J. A. Lake. 1998. Genomic evidence for two functionally distinct gene classes. *Proceedings of the National Academy of Sciences of the United States of America* 95: 6239–6244.

Rivera, M. C., and J. A. Lake. 2004. The ring of life provides evidence for a genome fusion origin of eukaryotes. *Nature* 431: 152–155.

Robert, F., and M. Chaussidon. 2006. A palaeotemperature curve for the Precambrian oceans based on silicon isotopes in cherts. *Nature* 443: 969–972.

Roberts, R., and 113 fellow Nobel Laureates. 2006. An open letter to Colonel Muammar al-Gaddafi. *Nature* 444: 146.

Robertson, M. P., and S. L. Miller. 1995a. An efficient prebiotic synthesis of cytosine and uracil. *Nature* 375: 772–774.

Robertson, M. P., and S. L. Miller. 1995b. Prebiotic synthesis of 5-substituted uracils: a bridge between the RNA world and the DNA protein world. *Science* 268: 702–705.

Robichaux, R. H., G. D. Carr, M. Liebman, and R. W. Pearcy. 1990. Adaptive radiation of the Hawaiian silversword alliance (Compositae Madiinae): ecological, morphological and physiological diversity. *Annals of the Missouri Botanical Garden* 77: 64–72.

Robinson, G. E., C. M. Grozinger, and C. W. Whitfield. 2005. Sociogenomics: social life in molecular terms. *Nature Reviews Genetics* 6: 257–271.

Robinson, R. 1993. Expressivity of the *Manx* gene in cats. *Journal of Heredity* 84: 170–172.

Robson, A. J., C. T. Bergstrom, and J. K. Pritchard. 1999. Risky business: sexual and asexual reproduction in variable environments. *Journal of Theoretical Biology* 197: 541–556.

Rocha, J. L., J. O. Sanders, D. M. Cherbonnier, T. J. Lawlor, and J. F. Taylor. 1998. Blood groups and milk and type traits in dairy cattle: after forty years of research. *Journal of Dairy Science* 81: 1663–1680.

Rodriguez, D. J. 1996. A model for the establishment of polyploidy in plants. *American Naturalist* 147: 33–46.

Rodriguez-Trelles, F., R. Tarrio, and F. J. Ayala. 2006. Origins and evolution of spliceosomal introns. *Annual Review of Genetics* 40: 47–76.

Roemer, G. W., T. J. Coonan, D. K. Garcelon, J. Bascompte, and L. Laughrin. 2001a. Feral pigs facilitate hyperpredation by golden eagles and indirectly cause the decline of the island fox. *Animal Conservation* 4: 307–318.

Roemer, G. W., D. A. Smith, D. K. Garcelon, and R. K. Wayne. 2001b. The behavioural ecology of the island fox (*Urocyon littoralis*). *Journal of Zoology* 255: 1–14.

Roemer, G. W., C. J. Donlan, and F. Courchamp. 2002. Golden eagles, feral pigs, and insular carnivores: how exotic species turn native predators into prey. *Proceedings of the National Academy of Sciences of the United States of America* 99: 791–796.

Rogaev, E. I., Y. K. Moliaka, B. A. Malyarchuk, F. A. Kondrashov, M. V. Derenko, I. Chumakov, and A. Grigorenko. 2006. Complete mitochondrial genome and phylogeny of Pleistocene mammoth *Mammuthus primigenius*. *PLoS Biology* 4: e73.

Roger, J. 1997. *Buffon: A Life in Natural History*. Cornell University Press, Ithaca.

Rogers, D., and T. Tanimoto. 1960. A computer program for classifying plants. *Science* 132: 1115–1118.

Rohwer, S. 1977. Status signaling in Harris sparrows—some experiments in deception. *Behaviour* 61: 107–129.

Ronquist, F. 1995. Reconstructing the history of host–parasite associations using generalized parsimony. *Cladistics* 11: 73–89.

Roos, D. S., M. J. Crawford, R. G. K. Donald, M. Fraunholz, O. S. Harb, C. Y. He, J. C. Kissinger, M. K. Shaw, and B. Striepen. 2002. Mining the *Plasmodium* genome database to define organellar function: what does the apicoplast do? *Philosophical Transactions of the Royal Society of London Series B—Biological Sciences* 357: 35–46.

Rosenblum, E. B., H. Rompler, T. Schoneberg, and H. E. Hoekstra. 2010. Molecular and functional basis of phenotypic convergence in white lizards at White Sands. *Proceedings of the National Academy of Sciences of the United States of America* 107: 2113–2117.

Rosenzweig, M. 1978. Competitive speciation. *Biological Journal of the Linnean Society* 10: 275–289.

Rowe, L., and T. Day. 2006. Detecting sexual conflict and sexually antagonistic coevolution. *Philosophical Transactions of the Royal Society of London Series B—Biological Sciences* 361: 277–285.

Rudan, I., N. Smolej-Narancic, H. Campbell, A. Carothers, A. Wright, B. Janicijevic, and P. Rudan. 2003. Inbreeding and the genetic complexity of human hypertension. *Genetics* 163: 1011–1021.

Ruse, M. 1980. Charles Darwin and group selection. *Annals of Science* 37: 615–630.

Russell, B. 1952. *The Impact of Science on Society*. George Allen & Unwin, London.

Russell, D., and W. Tucker. 1971. Supernovae and the extinction of dinosaurs. *Nature* 229: 553–554.

Ruta, M., M. I. Coates, and D. L. J. Quicke. 2003. Early tetrapod relationships revisited. *Biological Reviews* 78: 251–345.

Ruvinsky, A., and J. A. Marshall Graves, eds. 2005. *Mammalian Genomics*. CABI Publishing, Wallingford, UK.

Ryan Lab. 2011. I. Tungara frog calls. Accessed September 23, 2011. www.sbs.utexas.edu/ryan/multi_media.html.

Ryan, M. J. 1990. Sexual selection, sensory systems and sensory exploitation. *Oxford Surveys in Evolutionary Biology* 7: 157–195.

Ryan, M. J., J. H. Fox, W. Wilczynski, and A. S. Rand. 1990. Sexual selection for sensory exploitation in the frog *Physalaemus pustulosus*. *Nature* 343: 66–67.

Ryan, M. J., and A. S. Rand. 2003. Mate recognition in túngara frogs: a review of some studies of brain, behavior, and evolution. *Acta Zoologica Sinica* 49: 713–726.

Ryti, R. T., and T. J. Case. 1984. Spatial arrangement and diet overlap between colonies of desert ants. *Oecologia* 62: 401–404.

Saladino, R., C. Crestini, G. Costanzo, and E. DiMauro. 2004. Advances in the prebiotic synthesis of nucleic acids bases: implications for the origin of life. *Current Organic Chemistry* 8: 1425–1443.

Salathe, M., R. D. Kouyos, and S. Bonhoeffer. 2008. The state of affairs in the kingdom of the Red Queen. *Trends in Ecology & Evolution* 23: 439–445.

Salser, W., S. Bowen, D. Browne, F. Eladli, N. Fedoroff, K. Fry, H. Heindell, G. Paddock, R. Poon, B. Wallace, and P. Whitcome. 1976. Investigation of organization of mammalian chromosomes at DNA sequence level. *Federation Proceedings* 35: 23–35.

Salvini-Plawen, L., and E. Mayr. 1977. *On the Evolution of Photoreceptors and Eyes*. Plenum, New York.

Sanger, F., A. R. Coulson, T. Friedmann, G. Air, B. Barrell, N. L. Brown, J. C. Fiddes, J. C. Hutchison, P. M. Slocombe, and M. Smith. 1978. The nucleotide sequence of bacteriophage φX 174. *Journal of Molecular Biology* 125: 225–246.

Santagati, F., and F. M. Rijli. 2003. Cranial neural crest and the building of the vertebrate head. *Nature Reviews Neuroscience* 4: 806–818.

Santorelli, L. A., C. R. L. Thompson, E. Villegas, J. Svetz, C. Dinh, A. Parikh, R. Sucgang, A. Kuspa, J. E. Strassmann, D. C. Queller, and G. Shaulsky. 2008. Facultative cheater mutants reveal the genetic complexity of cooperation in social amoebae. *Nature* 451: 1107–1110.

Sarich, V. M., and A. C. Wilson. 1967. Immunological time scale for hominid evolution. *Science* 158: 1200–1203.

Schidlowski, M. 2001. Carbon isotopes as biogeochemical recorders of life over 3.8 Ga of Earth history: evolution of a concept. *Precambrian Research* 106: 117–134.

Schneider, J. G. 1862. *Ten Books*. H. G. Bohn, London.

Schneider, R. A., and J. A. Helms. 2003. The cellular and molecular origins of beak morphology. *Science* 299: 565–568.

Schneiker, S., O. Perlova, O. Kaiser, K. Gerth, A. Alici, M. O. Altmeyer, D. Bartels, T. Bekel, S. Beyer, E. Bode, H. B. Bode, C. J. Bolten, J. V. Choudhuri, S. Doss, Y. A. Elnakady, B. Frank, L. Gaigalat, A. Goesmann, C. Groeger, F. Gross, L. Jelsbak, L. Jelsbak, J. Kalinowski, C. Kegler, T. Knauber, S. Konietzny, M. Kopp, L. Krause, D. Krug, B. Linke, T. Mahmud, R. Martinez-Arias, A. C. McHardy, M. Merai, F. Meyer, S. Mormann, J. Munoz-Dorado, J. Perez, S. Pradella, S. Rachid, G. Raddatz, F. Rosenau, C. Ruckert, F. Sasse, M. Scharfe, S. C. Schuster, G. Suen, A. Treuner-Lange, G. J. Velicer, F. J. Vorholter, K. J. Weissman, R. D. Welch, S. C. Wenzel, D. E. Whitworth, S. Wilhelm, C. Wittmann, H. Blocker, A. Puhler, and R. Mueller. 2007. Complete genome sequence of the myxobacterium *Sorangium cellulosum*. *Nature Biotechnology* 25: 1281–1289.

Schrag, S. J., and V. Perrot. 1996. Reducing antibiotic resistance. *Nature* 381: 120–121.

Schrag, S. J., V. Perrot, and B. R. Levin. 1997. Adaptation to the fitness costs of antibiotic resistance in *Escherichia coli*. *Proceedings of the Royal Society B—Biological Sciences* 264: 1287–1291.

Schrödinger, E. 1944. *What Is Life?* Cambridge University Press, Cambridge.

Schultz, T. R., and S. G. Brady. 2008. Major evolutionary transitions in ant agriculture. *Proceedings of the National Academy of Sciences of the United States of America* 105: 5435–5440.

Schurko, A. M., M. Neiman, and J. M. Logsdon. 2009. Signs of sex: what we know and how we know it. *Trends in Ecology & Evolution* 24: 208–217.

Schwilk, D. W., and D. D. Ackerly. 2001. Flammability and serotiny as strategies: correlated evolution in pines. *Oikos* 94: 326–336.

Scotese, C. R. 2003. Paleomap project. Accessed January 18, 2011. www.scotese.com/.

Searcy, W. A., and S. Nowicki. 2005. *The Evolution of Animal Communication*. Princeton University Press, Princeton.

Secord, J. A. 2000. *Victorian Sensation: The Extraordinary Publication, Reception and Secret Authorship of* Vestiges of the Natural History of Creation. University of Chicago Press, Chicago.

Seeley, T. 1985. *Honeybee Ecology: A Study of Adaptation in Social Life.* Princeton University Press, Princeton.

Seger, J. 1985. Intraspecific resource competition as a cause of sympatric speciation. In P. Greenwood, P. Harvey, and M. Slatkin, eds., *Evolution: Essays in Honor of John Maynard Smith,* pp. 43–53. Cambridge University Press, Cambridge.

Seger, J. 1989. All for one, one for all, that is our device. *Nature* 338: 374–375.

Seghers, B. H. 1973. *An Analysis of Geographic Variation in the Antipredator Adaptations of the Guppy, Poecilia reticulata.* University of British Columbia, Vancouver.

Semlitsch, R. D. 1987. Pedomorphosis in *Ambystoma talpoideum:* effects of density, food, and pond drying. *Ecology* 68: 994–1002.

Seppälä, H., T. Klaukka, J. Vuopio-Varkila, A. Muotiala, H. Helenius, K. Lager, and P. Huovinen. 1997. The effect of changes in the consumption of macrolide antibiotics on erythromycin resistance in group A streptococci in Finland. *New England Journal of Medicine* 337: 441–446.

Serre, D., A. Langaney, M. Chech, M. Teschler-Nicola, M. Paunovic, P. Mennecier, M. Hofreiter, G. Possnert, and S. Pääbo. 2004. No evidence of Neandertal mtDNA contribution to early modern humans. *PLoS Biology* 2: 313–317.

Sfakiotakis, M., D. M. Lane, and J. B. C. Davies. 1999. Review of fish swimming modes for aquatic locomotion. *IEEE Journal of Oceanic Engineering* 24: 237–252.

Shadwick, R. E. 2005. How tunas and lamnid sharks swim: an evolutionary convergence. *American Scientist* 93: 524–531.

Sharpton, V. L., G. B. Dalrymple, L. E. Marin, G. Ryder, B. C. Schuraytz, and J. Urrutia-Fucugauchi. 1992. New links between the Chicxulub impact structure and the Cretaceous/Tertiary boundary. *Nature* 359: 819–821.

Shaw, P. W., G. R. Carvalho, B. H. Seghers, and A. E. Magurran. 1992. Genetic consequences of an artificial introduction of guppies (*Poecilia reticulata*) in N. Trinidad. *Proceedings of the Royal Society B—Biological Sciences* 248: 111–116.

Shimeld, S. M., and P. W. H. Holland. 2000. Vertebrate innovations. *Proceedings of the National Academy of Sciences of the United States of America* 97: 4449–4452.

Shin, Y., A. Hiraishi, and J. Sugiyama. 1993. Molecular systematics of the genus *Zoogloea* and emendation of the genus. *International Journal of Systematic Bacteriology* 43: 826–831.

Short, M. B., C. A. Solari, S. Ganguly, T. R. Powers, J. Kessler, and R. E. Goldstein. 2006. Flows driven by flagella of multicellular organisms enhance long-range molecular transport. *Proceedings of the National Academy of Sciences of the United States of America* 103: 8315–8319.

Shubin, N. 2009. *Your Inner Fish: A Journey into the 3.5-Billion-Year History of the Human Body.* Vintage, New York.

Shubin, N. H., E. B. Daeschler, and F. A. Jenkins. 2006. The pectoral fin of *Tiktaalik roseae* and the origin of the tetrapod limb. *Nature* 440: 764–771.

Shute, P. 1951. *Culex molestus. Transactions of the Royal Entomological Society of London* 102: 380–382.

Signor, P., and J. Lipps. 1982. Sampling bias, gradual extinction patterns, and catastrophes in the fossil record. In L. Silver and P. Schultz, eds., *Geological Implications of Large Asteroids and Comets on the Earth,* pp. 291–296. Geological Society of America, Boulder, CO.

Sigurdsson, H., S. Dhondt, M. A. Arthur, T. J. Bralower, J. C. Zachos, M. Vanfossen, and J. E. T. Channell. 1991. Glass from the Cretaceous/ Tertiary Boundary in Haiti. *Nature* 349: 482–487.

Simmons, L. W. 2001. *Sperm Competition and Its Evolutionary Consequences.* Princeton University Press, Princeton.

Simmons, L. W., D. J. Emlen, and J. L. Tomkins. 2007. Sperm competition games between sneaks and guards: a comparative analysis using dimorphic male beetles. *Evolution* 61: 2684–2692.

Simmons, L. W., and F. Garcia-González. 2008. Evolutionary reduction in testes size and competitive fertilization success in response to experimental removal of sexual selection in dung beetles. *Evolution* 62: 2580–2591.

Simpson, G. 1953. *The Major Features of Evolution.* Columbia University Press, New York.

Simpson, G. 1961. *Principles of Animal Taxonomy.* Columbia University Press, New York.

Sjölund, M., E. Tano, M. J. Blaser, D. I. Andersson, and L. Engstrand. 2005. Persistence of resistant *Staphylococcus epidermidis* after single course of clarithromycin. *Emerging Infectious Diseases* 11, 1389–1393.

Sjölund, M., K. Wreiber, D. I. Andersson, M. J. Blaser, and L. Engstrand. 2003. Long-term persistence of resistant *Enterococcus* species after antibiotics to eradicate *Helicobacter pylori. Annals of Internal Medicine* 139: 483–487.

Skyrms, B. 2010. The flow of information in signaling games. *Philosophical Studies* 147: 155–165.

Slabbekoorn, H., and T. B. Smith. 2002. Birdsong, ecology and speciation. *Philosophical Transactions of the Royal Society of London Series B—Biological Sciences* 357: 493–503.

Slonczewski, J. L., and J. W. Foster. 2011. *Microbiology: An Evolving Science,* 2nd Ed. W.W. Norton, New York.

Slusarczyk, M. 1999. Predator-induced diapause in *Daphnia* magna may require two chemical cues. *Oecologia* 119: 159–165.

Smit, J., and J. Hertogen. 1980. An extraterrestrial event at the Cretaceous–Tertiary boundary. *Nature* 285: 198–200.

Smit, J., and F. T. Kyte. 1984. Siderophile-rich magnetic spheroids from the Cretaceous–Tertiary boundary in Umbria, Italy. *Nature* 310: 403–405.

Smit, J., A. Montanari, N. H. M. Swinburne, W. Alvarez, A. R. Hildebrand, S. V. Margolis, P. Claeys, W. Lowrie, and F. Asaro. 1992. Tektite-bearing, deep-water clastic unit at the Cretaceous–Tertiary boundary in northeastern Mexico. *Geology* 20: 99–103.

Smith, A. B., A. S. Gale, and N. E. A. Monks. 2001. Sea-level change and rock-record bias in the Cretaceous: a problem for extinction and biodiversity studies. *Paleobiology* 27: 241–253.

Smith, F. D. M., R. May, R. Pellew, T. H. Johnson, and K. S. Walter. 1993. Estimating extinction rates. *Nature* 364: 494–496.

Smith, R. L., ed. 1984. *Sperm Competition and the Evolution of Animal Mating Systems.* Academic Press, New York.

Smith, R. M. H., and P. Ward. 2001. Pattern of vertebrate extinctions across an event bed at the Permian–Triassic boundary in the Karoo Basin of South Africa. *Geology* 29: 1147–1150.

Smith, S., and M. Donoghue. 2008. Rates of molecular evolution are linked to life history in flowering plants. *Science* 322: 86–89.

Sneath, P. H. A. 1995. Thirty years of numerical taxonomy. *Systematic Biology* 44: 281–298.

Snoke, M. S., and D. E. L. Promislow. 2003. Quantitative genetic tests of recent senescence theory: age-specific mortality and male fertility in *Drosophila melanogaster. Heredity* 91: 546–556.

Sobel, J. M., G. F. Chen, L. R. Watt, and D. W. Schemske. 2010. The biology of speciation. *Evolution* 64: 295–315.

Sober, E. 1984. *The Nature of Selection: Evolutionary Theory in Philosophical Focus*. Bradford/MIT, Cambridge.

Sober, E. 1987. What is adaptationism? In J. Dupre, ed., *The Latest on the Best: Essays on Evolution and Optimality*, pp. 105–118. MIT Press, Cambridge.

Sober, E., and D. S. Wilson. 1998. *Unto Others*. Harvard University Press, Cambridge.

Sokol, R. 1985. The principles of numerical taxonomy: twenty-five years later. In M. Goodfellow, D. Jones, and F. Priest, eds., *Computer-Assisted Bacterial Systematics*, pp. 1–20. Academic Press, London.

Sokol, R., and P. H. A. Sneath. 1963. *Principles of Numerical Taxonomy*. Freeman, London.

Solari, C. A., J. O. Kessler, and R. E. Michod. 2006a. A hydrodynamics approach to the evolution of multicellularity: flagellar motility and germ–soma differentiation in volvocalean green algae. *American Naturalist* 167: 537–554.

Solari, C. A., S. Ganguly, J. O. Kessler, R. E. Michod, and R. E. Goldstein. 2006b. Multicellularity and the functional interdependence of motility and molecular transport. *Proceedings of the National Academy of Sciences of the United States of America* 103: 1353–1358.

Soltis, D. E., P. S. Soltis, and J. A. Tate. 2004. Advances in the study of polyploidy since plant speciation. *New Phytologist* 161: 173–191.

Sompayrac, L. 2008. *How the Immune System Works*, 3rd Ed. Wiley-Blackwell, Malden, MA.

Sordahl, T. A. 1990. The risks of avian mobbing and distraction behavior: an anecdotal review. *Wilson Bulletin* 102: 349–352.

Speare, R. 1994. *Preliminary Study of Diseases in Australian Wet Tropics Amphibians: Deaths of Rainforest Frogs at O'Keefe Creek, Big Tableland*. Queensland Department of Environment and Heritage, Brisbane.

Spiegelman, S. 1970. Extracellular evolution of replicating molecules. In F. Schmitt, ed., *The Neurosciences: A Second Study Program*, pp. 927–945. Rockefeller University Press, New York.

Spinney, L. 2010. Evolution: Dreampond revisited. *Nature* 446: 174–175.

Stadler, K., V. Masignani, M. Eickmann, S. Becker, S. Abrignani, H. D. Klenk, and R. Rappuoli. 2003. SARS—beginning to understand a new virus. *Nature Reviews Microbiology* 1: 209–218.

Stanley, S. M. 1973. An explanation for Cope's rule. *Evolution* 27: 1–26.

Stanley, S. M., and X. Yang. 1994. A double mass extinction at the end of the Paleozoic era. *Science* 266: 1340–1344.

Stebbins, G. L. 1938. Cytological characteristics associated with the different growth habitats in the dicotyledons. *American Journal of Botany* 25: 189–198.

Stebbins, R. C. 1949. Speciation and salamanders of the plethodontid genus *Ensatina*. *University of California Publications in Zoology* 48: 377–526.

Steiner, C. C., J. N. Weber, and H. E. Hoekstra. 2007. Adaptive variation in beach mice produced by two interacting pigmentation genes. *PLoS Biology* 5: 1880–1889.

Sternfeld, J., and C. N. David. 1981. Oxygen gradients cause pattern orientation in *Dictyostelium* cell clumps. *Journal of Cell Science* 50: 9–17.

Storfer, A. 2003. Amphibian declines: future directions. *Diversity and Distributions* 9: 151–163.

Stover, S. 2003. The epidemiology of thoroughbred racehorse injuries. *Clinical Techniques in Equine Practice* 2: 312–322.

Strait, D., F. E. Grine, and J. Fleagle. 2007. Analyzing hominid phylogeny. In W. Henke and I. Tattersall, eds., *Handbook of Paleoanthropology*, pp. 1781–1806. Springer, New York.

Strong, D. R. 1973. *Amphipod amplexus*: significance of ecotypic variation. *Ecology* 54: 1383–1388.

Stuben, D., U. Kramar, Z. Berner, W. Stinnesbeck, G. Keller, and T. Adatte. 2002. Trace elements, stable isotopes, and clay mineralogy of the Elles II K–T boundary section in Tunisia: indications for sea level fluctuations and primary productivity. *Palaeogeography, Palaeoclimatology, Palaeoecology* 178: 321–345.

Studentreader.com. 2011. Exon shuffling. Accessed March 17, 2011. http://studentreader.com/exon-shuffling/.

Stumpf, M. P. H., and G. A. T. McVean. 2003. Estimating recombination rates from population-genetic data. *Nature Reviews Genetics* 4: 959–968.

Sturtevant, A. H. 1939. On the subdivision of the genus *Drosophila*. *Proceedings of the National Academy of Sciences of the United States of America* 25: 137–141.

Sultan, S. E., and F. A. Bazzaz. 1993. Phenotypic plasticity in *Polygonum persicaria*. I. Diversity and uniformity in genotypic norms of reaction to light. *Evolution* 47: 1009–1031.

Sumper, M., and M. Luce. 1975. Evidence of de novo production of self replicating and environmentally adapted RNA structures by bacteriophage QBeta replicase. *Proceedings of the National Academy of Sciences of the United States of America* 72: 162–166.

Sundqvist, M., P. Geli, D. I. Andersson, M. Sjolund-Karlsson, A. Runehagen, H. Cars, K. Abelson-Storby, O. Cars, and G. Kahlmeter. 2010. Little evidence for reversibility of trimethoprim resistance after a drastic reduction in trimethoprim use. *Journal of Antimicrobial Chemotherapy* 65: 350–360.

Surridge, A. K., D. Osorio, and N. I. Mundy. 2003. Evolution and selection of trichromatic vision in primates. *Trends in Ecology & Evolution* 18: 198–205.

Suwa, G., B. Asfaw, R. T. Kono, D. Kubo, C. O. Lovejoy, and T. D. White. 2009. The *Ardipithecus ramidus* skull and its implications for hominid origins. *Science* 326:69.

Swanson, W. J., A. G. Clark, H. M. Waldrip-Dail, M. F. Wolfner, and C. F. Aquadro. 2001. Evolutionary EST analysis identifies rapidly evolving male reproductive proteins in *Drosophila*. *Proceedings of the National Academy of Sciences of the United States of America* 98: 7375–7379.

Swisher, C. C., J. M. Grajalesnishimura, A. Montanari, S. V. Margolis, P. Claeys, W. Alvarez, P. Renne, E. Cedillopardo, F. Maurrasse, G. H. Curtis, J. Smit, and M. O. McWilliams. 1992. Coeval Ar-40/Ar-39 ages of 65.0 million years ago from Chicxulub crater melt rock and Cretaceous–Tertiary boundary tektites. *Science* 257: 954–958.

Symons, G. 1888. *The Eruption of Krakatoa and Subsequent Phenomena*. Krakatoa Committee of the Royal Society, London.

Szathmary, E., F. Jordan, and C. Pal. 2001. Molecular biology and evolution: can genes explain biological complexity? *Science* 292: 1315–1316.

Szathmary, E., and J. Maynard Smith. 1995. The major evolutionary transitions. *Nature* 374: 227–232.

Taft, R., M. Pheasant, and J. Mattick. 2007. The relationship between non-protein-coding DNA and eukaryotic complexity. BioEssays 29: 288–299.

Tajima, F. 1989. Statistical method for testing the neutral mutation hypothesis by DNA polymorphism. *Genetics* 123: 585–595.

Takata, T., S. Miyaishi, Y. Yamamoto, S. Inagaki, K. Yoshitome, T. Ishikawa, and H. Ishizu. 2002. Allele frequencies of single nucleotide polymorphisms in the second exon of the myoglobin gene among the Japanese. *Human Biology* 74: 317–320.

Tatar, M., D. E. L. Promislow, A. A. Khazaeli, and J. W. Curtsinger. 1996. Age-specific patterns of genetic variance in *Drosophila melanogaster*. II. Fecundity and its genetic covariance with age-specific mortality. *Genetics* 143: 849–858.

Tatar, M., A. Bartke, and A. Antebi. 2003. The endocrine regulation of aging by insulin-like signals. *Science* 1346–1351.

Taubes, G. A. 2010. The fruitful fruit fly: discovering the homeobox. Howard Hughes Medical Institute. Accessed November 18, 2010. www.hhmi.org/genesweshare/b120.html.

Taylor, J. S., and J. Raes. 2004. Duplication and divergence: the evolution of new genes and old ideas. *Annual Review of Genetics* 38: 615–643.

Thanukos, A. 2009. How the adaptation got its start. *Evolution: Education and Outreach* 2: 612–616.

Thelander, C. G. 1994. *Life on the Edge: A Guide to California's Endangered Natural Resources.* Ten Speed Press, Berkeley.

Theofilopoulos, A., and F. Dixon. 1985. Murine models of systemic lupus erythematosus. *Advances in Immunology* 37: 269–358.

Thieben, G., and H. Saedler. 2001. Plant biology: floral quartets. *Nature* 409: 469–471.

Thomas, A. L. R., G. K. Taylor, R. B. Srygley, R. L. Nudds, and R. J. Bomphrey. 2004. Dragonfly flight: free-flight and tethered flow visualizations reveal a diverse array of unsteady lift-generating mechanisms, controlled primarily via angle of attack. *Journal of Experimental Biology* 207: 4299–4323.

Thomas, C. M., and K. M. Nielsen. 2005. Mechanisms of, and barriers to, horizontal gene transfer between bacteria. *Nature Reviews Microbiology* 3: 711–721.

Thomas, M. B., and S. Blanford. 2003. Thermal biology in insect–parasite interactions. *Trends in Ecology and Evolution* 18: 344–350.

Thompson, D. A. 1917. *On Growth and Form.* Cambridge University Press, Cambridge.

Thompson, J. N. 1982. *Interaction and Coevolution.* John Wiley and Sons, New York.

Thompson, J. N. 1994. *The Coevolutionary Process.* University of Chicago Press, Chicago.

Thompson, J. N. 1999. The evolution of species interactions. *Science* 284: 2116–2118.

Thompson, J. N. 2005. *The Geographic Mosaic of Coevolution.* University of Chicago Press, Chicago.

Thompson, J. N. 2009. The coevolving web of life. *American Naturalist* 173: 125–140.

Thompson, J. N. 2010. Four central points about coevolution. *Evolution: Education and Outreach* 3: 7–13.

Thompson, J. N., and B. M. Cunningham. 2002. Geographic structure and dynamics of coevolutionary selection. *Nature* 417: 735–738.

Thornhill, R. 1976. Sexual selection and nuptial feeding behavior in *Bittacus apicalis. American Naturalist* 110: 529–548.

Tibbetts, E. A., and J. Dale. 2004. A socially enforced signal of quality in a paper wasp. *Nature* 432: 218–222.

Tibbetts, E. A., and A. Izzo. 2010. Social punishment of dishonest signalers caused by mismatch between signal and behavior. *Current Biology* 20: 1637–1640.

Tibbetts, E. A., and R. Lindsay. 2008. Visual signals of status and rival assessment in *Polistes dominulus* paper wasps. *Biology Letters* 4: 237–239.

Tilford, C. A., T. Kuroda-Kawaguchi, H. Skaletsky, S. Rozen, L. G. Brown, M. Rosenberg, J. D. McPherson, K. Wylie, M. Sekhon, T. A. Kucaba, R. H. Waterston, and D. C. Page. 2001. A physical map of the human Y chromosome. *Nature* 409: 943–945.

Timmis, J., M. Ayliffe, C. Huang, and W. Martin. 2004. Endosymbiotic gene transfer: organelle genomes forge eukaryotic chromosomes. *Nature Reviews Genetics* 5: 123–135.

Tompkins, R., and J. K. Townsend. 1977. Control of metamorphic events in a neotenous urodele, *Ambystoma mexicanum. Journal of Experimental Zoology* 200: 191–196.

Tourmen, Y., O. Baris, P. Dessen, C. Jacques, Y. Malthiery, and P. Reynier. 2002. Structure and chromosomal distribution of human mitochondrial pseudogenes. *Genomics* 80: 71–77.

Track and Field News. 2010. Men's world records. Accessed June 1, 2011. www.trackandfieldnews.com.

Train, J. 1845. *A Historical and Statistical Account of the Isle of Man.* J. Lumsden and Sons, Glasgow.

Trainor, P. A., K. R. Melton, and M. Manzanares. 2003. Origins and plasticity of neural crest cells and their roles in jaw and craniofacial evolution. *International Journal of Developmental Biology* 47: 541–553.

Travasso, M., and N. Pierce. 2000. Acoustics, context and function of vibrational signalling in a lycaenid butterfly–ant mutualism. *Animal Behaviour* 60: 13–26.

Travers, K., and M. Barza. 2002. Morbidity of infections caused by antimicrobial-resistant bacteria. *Clinical Infectious Diseases* 34: S131–S134.

Trindade, S., L. Perfeito, and I. Gordo. 2010. Rate and effects of spontaneous mutations that affect fitness in mutator *Escherichia coli. Philosophical Transactions of the Royal Society of London Series B—Biological Sciences* 365: 1177–1186.

Trivers, R. L. 1971. The evolution of reciprocal altruism. *The Quarterly Review of Biology* 46: 189–226.

Trivers, R. L. 1974. Parent–offspring conflict. *American Zoologist* 14: 249–265.

Trivers, R. 1985. *Social Evolution.* Benjamin Cummings, Menlo Park, CA.

Trivers, R., and H. Hare. 1976. Haplo-diploidy and the evolution of the social insects. *Science* 191: 249–263.

Tupler, R., G. Perini, and M. R. Green. 2001. Expressing the human genome. *Nature* 409: 832–833.

Twitchet, R., C. Looy, R. Morante, H. Visscher, and P. B. Wignall. 2001. Rapid and synchronous collapse of marine and terrestrial ecosystems during the end-Permian biotic crisis. *Geology* 29: 351–354.

Tyler, M. 1991. Declining amphibian populations—a global phenomenon? An Australian perspective. *Alytes* 9: 43–50.

Ueshima, R., and T. Asami. 2003. Single-gene speciation by left–right reversal—a land-snail species of polyphyletic origin results from chirality constraints on mating. *Nature* 425: 679–679.

Understanding Evolution. 2008. Evolution in the fast lane? The University of California Museum of Paleontology, Berkeley, and the Regents of the University of California. Accessed March 8, 2011. http://evolution.berkeley.edu.evolibrary/news/080101_recenthumanevo.

Understanding Evolution. 2010. Reading trees: a quick review. University of California Museum of Paleontology. Accessed November 10, 2010. http://evolution.berkeley.edu/evolibrary/article/phylogenetics_02.

University of Illinois. 2011. Integrative Biology 335: systematics of plants, molecular systematics. Accessed January 18, 2011. www.life.illinois.edu/ib/335/MolSyst.html.

Urrutia-Fucugauchi, J., L. Marin, and A. Trejo-Garcia. 1996. UNAM scientific drilling program of Chicxulub impact structure—Evidence for a 300 kilometer crater diameter. *Geophysical Research Letters* 23: 1565–1568.

Vaglia, J. L., and K. K. Smith. 2003. Early differentiation and migration of cranial neural crest in the opossum, *Monodelphis domestica. Evolution & Development* 5: 121–135.

Valentine, J. W., J. Collins, and C. Meyer. 1994. Morphological complexity increase in metazoans. *Paleobiology* 20: 131–142.

Van der Pijl, L., and C. Dodson. 1966. *Orchid Flowers: Their Pollination and Evolution.* University of Miami Press, Coral Gables, FL.

Van Valen, L. 1973. A new evolutionary law. *Evolutionary Theory* 1: 1–30.

Varki, A., D. H. Geschwind, and E. E. Eichler. 2008. Explaining human uniqueness: genome interactions with environment, behaviour and culture. *Nature Reviews Genetics* 9: 749–763.

Vaughn, L. K., H. A. Bernheim, and M. J. Kluger. 1974. Fever in the lizard *Dipsosaurus dorsalis*. *Nature* 252: 473–474.

Vawter, L., and W. M. Brown. 1986. Nuclear and mitochondrial DNA comparisons reveal extreme rate variation in the molecular clock. *Science* 234: 194–196.

Velicer, G. J. 2003. Social strife in the microbial world. *Trends in Microbiology* 11: 330–337.

Vidal, N. 2002. Colubroid systematics: evidence for an early appearance of the venom apparatus followed by extensive evolutionary tinkering. *Journal of Toxicology-Toxin Reviews* 21: 21–41.

Viereck, L. A., and W. F. Johnston. 1990. *Picea mariana* (Mill.) B. S. P. In R. M. Burns and B. H. Honkala, eds., *Silvics of North America: 1. Conifers; 2. Hardwoods*, U.S. Department of Agriculture, Forest Service, Washington, DC. Accessed January 25, 2011. www.na.fs.fed.us/pubs/silvics_manual/Volume_1/picea/mariana.htm.

Vignieri, S., H. Larson, and H. E. Hoekstra. 2010. The selective advantage of crypsis in mice. *Evolution* 64: 2153–2158.

Voight, B., S. Kudaravalli, X. Wen, and J. Pritchard. 2006. A map of recent positive selection in the human genome. *PLoS Biology* 4: e72.

von Baer, K. 1828. *Entwicklungsgeschichte der Thiere: Beobachtung und Reflexion*. Schmitzdorf, St. Petersburg.

Vonk, F. J., J. F. Admiraal, K. Jackson, R. Reshef, M. A. G. de Bakker, K. Vanderschoot, I. van den Berge, M. van Atten, E. Burgerhout, A. Beck, P. J. Mirtschin, E. Kochva, F. Witte, B. G. Fry, A. E. Woods, and M. K. Richardson. 2008. Evolutionary origin and development of snake fangs. *Nature* 454: 630–633.

Vos, M. 2009. Why do bacteria engage in homologous recombination? *Trends in Microbiology* 17: 226–232.

Voyles, J., S. Young, L. Berger, C. Campbell, W. F. Voyles, A. Dinudom, D. Cook, R. Webb, R. A. Alford, L. F. Skerratt, and R. Speare. 2009. Pathogenesis of chytridiomycosis, a cause of catastrophic amphibian declines. *Science* 326: 582–585.

Voytek, S. B., and G. F. Joyce. 2007. Emergence of a fast-reacting ribozyme that is capable of undergoing continuous evolution. *Proceedings of the National Academy of Sciences of the United States of America* 104: 15288–15293.

Vrijenhoek, R. C., R. M. Dawley, C. J. Cole, and J. P. Bogart. 1989. A list of known unisexual vertebrates. In R. C. Dawley and J. P. Bogart, eds., *Evolution and Ecology of Unisexual Vertebrates*, pp. 19–23. New York State Museum, Albany.

Wada, H., and N. Satoh. 2001. Patterning the protochordate neural tube. *Current Opinion in Neurobiology* 11: 16–21.

Wade, M. J. 2007. The co-evolutionary genetics of ecological communities. *Nature Reviews Genetics* 8: 185–195.

Wagner, G. P., C. Amemiya, and F. Ruddle. 2003. Hox cluster duplications and the opportunity for evolutionary novelties. *Proceedings of the National Academy of Sciences of the United States of America* 100: 14603–14606.

Wagner, P. L., and M. K. Waldor. 2002. Bacteriophage control of bacterial virulence I. *Infection and Immunity* 70: 3985–3993.

Wake, D. B. 1997. Incipient species formation in salamanders of the Ensatina complex. *Proceedings of the National Academy of Sciences of the United States of America* 94: 7761–7767.

Wake, D. B., and V. T. Vredenburg. 2008. Are we in the midst of the sixth mass extinction? A view from the world of amphibians. *Proceedings of the National Academy of Sciences of the United States of America* 105: 11466–11473.

Wake, D. B., and K. P. Yanev. 1986. Geographic variation in allozymes in a ring species, the Plethodontid salamander, *Ensatina eschscholtzii* of North America. *Evolution* 40: 702–715.

Wake, D. B., K. P. Yanev, and C. W. Brown. 1986. Intraspecific sympatry in a ring species, the Plethodontid salamader, *Ensatina eschscholtzii*, in Southern California. *Evolution* 40: 866–868.

Wakeley, J. 2008. *Coalescent Theory*. Roberts and Company Publishers, Greenwood Village, CO.

Wallace, A. R. 1855. On the law which has regulated the introduction of new species. *The Annals and Magazine of Natural History* 16: 184–196.

Wallace, A. R. 1891. *Natural Selection and Tropical Nature; Essays on Descriptive and Theoretical Biology*. Macmillan, London and New York.

Waller, R. F., P. J. Keeling, G. G. van Dooren, and G. I. McFadden. 2003. Comment on "A green algal apicoplast ancestor." *Science* 301: 49.

Walsh, B. D. 1864. On phytophagous varieties and phytophagous species. *Proceedings of the Entomological Society of Philadelphia* 3: 403–430.

Walsh, B. D. 1867. The apple worm and the apple maggot. *Journal of Horticulture* 2: 338–343.

Wang, H., D. W. Byrd, J. L. Howard, E. D. McArthur, J. H. Graham, and D. C. Freeman. 1998. Narrow hybrid zone between two subspecies of big sagebrush (*Artemisia tridentata*: Asteraceae). V. Soil properties. *International Journal of Plant Sciences* 159: 139–147.

Wang, H., E. D. McArthur, and D. C. Freeman. 1999. Narrow hybrid zone between two subspecies of big sagebrush (*Artemisia tridentata*: Asteraceae). IX. Elemental uptake and niche separation. *American Journal of Botany* 86: 1099–1107.

Wang, H., E. McArthur, S. Sanderson, J. Graham, and D. Freeman. 1997. Narrow hybrid zone between two subspecies of big sagebrush. *Evolution* 51: 95–102.

Wang, M. B., X. Y. Bian, L. M. Wu, L. X. Liu, N. A. Smith, D. Isenegger, R. M. Wu, C. Masuta, V. B. Vance, J. M. Watson, A. Rezaian, E. S. Dennis, and P. M. Waterhouse. 2004. On the role of RNA silencing in the pathogenicity and evolution of viroids and viral satellites. *Proceedings of the National Academy of Sciences of the United States of America* 101: 3275–3280.

Wang, Y. Z., M. B. Slade, A. A. Gooley, B. J. Atwell, and K. L. Williams. 2001. Cellulose-binding modules from extracellular matrix proteins of *Dictyostelium discoideum* stalk and sheath. *European Journal of Biochemistry* 268: 4334–4345.

Wardenburg, J. B., W. A. Williams, and D. Missiakas. 2006. Host defenses against *Staphylococcus aureus* infection require recognition of bacterial lipoproteins. *Proceedings of the National Academy of Sciences of the United States of America* 103: 13831–13836.

Weber, D. S., B. S. Stewart, J. C. Garza, and N. Lehman. 2000. An empirical genetic assessment of the severity of the northern elephant seal population bottleneck. *Current Biology* 10: 1287–1290.

Weber, D. S., B. S. Stewart, J. Schienman, and N. Lehman. 2004. Major histocompatibility complex variation at three class II loci in the northern elephant seal. *Molecular Ecology* 13: 711–718.

Weber, R. E., T. H. Jessen, H. Malte, and J. Tame. 1993. Mutant hemoglobins (Alpha 119 and Beta 55- Ser): functions related to high-altitude respiration in geese. *Journal of Applied Physiology* 75: 2646–2655.

Wedell, N., M. J. G. Gage, and G. A. Parker. 2002. Sperm competition, male prudence and sperm-limited females. *Trends in Ecology & Evolution* 17: 313–320.

Weigensberg, I., and D. Roff. 1996. Natural heritabilities: can they be reliably estimated in the laboratory? *Evolution* 50: 2149–2157.

Weiner, J. 1995. *The Beak of the Finch: A Story of Evolution in Our Time.* Vintage Books, New York.

Weinstock, J., E. Willerslev, A. Sher, W. F. Tong, S. Y. W. Ho, D. Rubenstein, J. Storer, J. Burns, L. Martin, C. Bravi, A. Prieto, D. Froese, E. Scott, X. L. Lai, and A. Cooper. 2005. Evolution, systematics, and phylogeography of Pleistocene horses in the New World: a molecular perspective. *PLoS Biology* 3: 1373–1379.

Welch, J. J. 2006. Estimating the genomewide rate of adaptive protein evolution in *Drosophila. Genetics* 173: 821–837.

Welch J. J., and L. Bromham. 2005. Molecular dating when rates vary. *Trends in Ecology & Evolution* 20: 320–327.

Wells, K. D., and J. Schwartz. 2007. The behavioral ecology of anuran communication. *Springer Handbook of Auditory Research* 28: 44–86.

West, S. A., C. M. Lively, and A. F. Read. 1999. A pluralist approach to sex and recombination. *Journal of Evolutionary Biology* 12: 1003–1012.

West, S. A., E. T. Kiers, I. Pen, and R. F. Denison. 2002a. Sanctions and mutualism stability: when should less beneficial mutualists be tolerated? *Journal of Evolutionary Biology* 15: 830–837.

West, S. A., E. T. Kiers, E. L. Simms, and R. F. Denison. 2002b. Sanctions and mutualism stability: why do rhizobia fix nitrogen? *Proceedings of the Royal Society B—Biological Sciences* 269: 685–694.

West-Eberhard, M. J. 1979. Sexual selection, social competition and evolution. *Proceedings of the American Philosophical Society* 123: 222–234.

West-Eberhard, M. J. 1981. Intragroup selection and the evolution of insect societies. In R. D. Alexander and D. W. Tinkle, eds., *Natural Selection and Social Behavior*, pp. 3–17. Chiron Press, New York.

Wetherill, G. W. 1979. Apollo objects. *Scientific American* 240: 54–65.

Wheatcroft, D. J., and T. D. Price. 2008. Reciprocal cooperation in avian mobbing: playing nice pays. *Trends in Ecology & Evolution* 23: 416–419.

Wheeler, J., and S. Rissing. 1975. Natural history of *Veromessor pergandei* I. The nest. *Pan-Pacific Entomologist* 51: 205–216.

Whiteman, H. 1994. Evolution of facultative paedomorphosis in salamanders. *The Quarterly Review of Biology* 69: 205–221.

Whittall, J., and S. Hodges. 2007. Pollinator shifts drive increasingly long nectar spurs in columbine flowers. *Nature* 447: 706–709.

Whittle, C. A., and M. O. Johnston. 2003. Broad-scale analysis contradicts the theory that generation time affects molecular evolutionary rates in plants. *Journal of Molecular Evolution* 56: 223–233.

Widmer, A., C. Lexer, and S. Cozzolino. 2009. Evolution of reproductive isolation in plants. *Heredity* 102: 31–38.

Wikimedia Commons. (Lec CRP1) 2006. The ancestry of King Charles II of Spain (1661–1700). Accessed November 15, 2010. http://commons.wikimedia.org/wiki/File:Carlos_segundo80.png.

Wilbur, H. M., and J. P. Collins. 1973. Ecological aspects of amphibian metamorphosis. *Science* 182: 1305–1314.

Wiley, E. O. 1978. The evolutionary species concept reconsidered. *Systematic Zoology* 27: 17–26.

Wiley, E. 1988. Vicariance biogeography. *Annual Review of Ecology and Systematics* 19: 513–542.

Wilkinson, G. S., and P. R. Reillo. 1994. Female choice response to artificial selection on an exaggerated male trait in a stalk-eyed fly. *Proceedings of the Royal Society B—Biological Sciences* 255: 1–6.

Williams, G. C. 1957. Pleiotropy, natural selection, and the evolution of senescence. *Evolution* 11: 398–411.

Williams, G. 1966. *Adaptation and Natural Selection.* Princeton University Press, Princeton.

Williams, G. C. 1975. *Sex and Evolution.* Princeton University Press, Princeton.

Williams, G. C., and J. B. Mitton. 1973. Why reproduce sexually? *Journal of Theoretical Biology* 39: 545–554.

Wilson, A. C., S. S. Carlson, and T. J. White. 1977. Biochemical evolution. *Annual Review of Biochemistry* 46: 573–639.

Wilson, D. S. 1980. *The Natural Selection of Populations and Communities.* Benjamin Cummings, Menlo Park, CA.

Wilson, D. S., and E. O. Wilson. 2007. Rethinking the theoretical foundation of sociobiology. *The Quarterly Review of Biology* 82: 327–348.

Wilson, E. O., and B. Holldobler. 2005a. Eusociality: origin and consequences. *Proceedings of the National Academy of Sciences of the United States of America* 102: 13367–13371.

Wilson, E. O., and B. Holldobler. 2005b. The rise of the ants: a phylogenetic and ecological explanation. *Proceedings of the National Academy of Sciences of the United States of America* 102: 7411–7414.

Wilson, M. A., and K. D. Makova. 2009. Evolution and survival on eutherian sex chromosomes. *PLoS Genetics* 5: e1000568.

Witter, M. S. 1990. Evolution in the Madiinae—evidence from enzyme electrophoresis. *Annals of the Missouri Botanical Garden* 77: 110–117.

Woese, C. R. 1998a. A manifesto for microbial genomics. *Current Biology* 8: R781–R783.

Woese, C. R. 1998b. The universal ancestor. *Proceedings of the National Academy of Sciences of the United States of America* 95: 6854–6859.

Woese, C. R. 2000. Interpreting the universal phylogenetic tree. *Proceedings of the National Academy of Sciences of the United States of America* 97: 8392–8396.

Woese, C. R. 2002. On the evolution of cells. *Proceedings of the National Academy of Sciences of the United States of America* 99: 8742–8747.

Woese, C. R., O. Kandler, and M. L. Wheelis. 1990. Towards a natural system of organisms: proposal for the domains archaea, bacteria and eucarya. *Proceedings of the National Academy of Sciences of the United States of America* 87: 4576–4579.

Woese, C. R., G. J. Olsen, M. Ibba, and D. Soll. 2000. Aminoacyl-tRNA synthetases, the genetic code, and the evolutionary process. *Microbiology and Molecular Biology Reviews* 64: 202–236.

Wolfram Demonstrations Project. 2011. Coalescent gene genealogies. Accessed January 25, 2011. http://demonstrations.wolfram.com/CoalescentGeneGenealogies.

Wood, B., and N. Lonergan. 2008. The hominin fossil record: taxa, grades and clades. *Journal of Anatomy* 212: 354–376.

Wood, T. E., N. Takebayashi, M. S. Barker, I. Mayrose, P. B. Greenspoon, and L. H. Rieseberg. 2009. The frequency of polyploid speciation in vascular plants. *Proceedings of the National Academy of Sciences of the United States of America* 106: 13875–13879.

Wood, V., R. Gwilliam, M. A. Rajandream, M. Lyne, R. Lyne, A. Stewart, J. Sgouros, N. Peat, J. Hayles, S. Baker, D. Basham, S. Bowman, K. Brooks, D. Brown, S. Brown, T. Chillingworth, C. Churcher, M. Collins, R. Connor, A. Cronin, P. Davis, T. Feltwell, A. Fraser, S. Gentles, A. Goble, N. Hamlin, D. Harris, J. Hidalgo, G. Hodgson, S. Holroyd, T. Hornsby, S. Howarth, E. J. Huckle, S. Hunt, K. Jagels, K. James, L. Jones, M. Jones, S. Leather, S. McDonald, J. McLean, P. Mooney, S. Moule, K. Mungall, L. Murphy, D. Niblett, C. Odell, K. Oliver, S. O'Neil, D. Pearson, M. A. Quail, E. Rabbinowitsch, K. Rutherford, S. Rutter, D. Saunders, K. Seeger, S. Sharp, J. Skelton, M. Simmonds, R. Squares, S. Squares, K. Stevens, K. Taylor, R. G. Taylor, A. Tivey, S. Walsh, T. Warren, S. Whitehead,

J. Woodward, G. Volckaert, R. Aert, J. Robben, B. Grymonprez, I. Weltjens, E. Vanstreels, M. Rieger, M. Schafer, S. Muller-Auer, C. Gabel, M. Fuchs, C. Fritzc, E. Holzer, D. Moestl, H. Hilbert, K. Borzym, I. Langer, A. Beck, H. Lehrach, R. Reinhardt, T. M. Pohl, P. Eger, W. Zimmermann, H. Wedler, R. Wambutt, B. Purnelle, A. Goffeau, E. Cadieu, S. Dreano, S. Gloux, V. Lelaure, S. Mottier, F. Galibert, S. J. Aves, Z. Xiang, C. Hunt, K. Moore, S. M. Hurst, M. Lucas, M. Rochet, C. Gaillardin, V. A. Tallada, A. Garzon, G. Thode, R. R. Daga, L. Cruzado, J. Jimenez, M. Sanchez, F. del Rey, J. Benito, A. Dominguez, J. L. Revuelta, S. Moreno, J. Armstrong, S. L. Forsburg, L. Cerrutti, T. Lowe, W. R. McCombie, I. Paulsen, J. Potashkin, G. V. Shpakovski, D. Ussery, B. G. Barrell, and P. Nurse. 2002. The genome sequence of *Schizosaccharomyces pombe*. *Nature* 415: 871–880.

World Conservation Monitoring Center. 1992. *Global Biodiversity: Status of the Earth's Living Resources*. Chapman & Hall, London.

World Masters Athletics. 2011. Track and field world records. Accessed June 1, 2011. www.world-masters-athletics.org.

Wray, G. A., M. W. Hahn, E. Abouheif, J. P. Balhoff, M. Pizer, M. V. Rockman, and L. A. Romano. 2003. The evolution of transcriptional regulation in eukaryotes. *Molecular Biology and Evolution* 20: 1377–1419.

Wright, S. 1931. Evolution in Mendelian populations. *Genetics* 16: 97–159.

Wright, S. 1932. The roles of mutation, inbreeding crossbreeding and selection in evolution. *International Congress of Genetics* 1: 356–366.

Wright, S. 1938. Size of population and breeding structure in relation to evolution. *Science* 87: 430–431.

Wright, S. 1969. *Evolution and Genetics of Populations: The Theory of Gene Frequencies*. University of Chicago Press, Chicago.

Wu, C. I., and W. H. Li. 1985. Evidence for higher rates of nucleotide substitution in rodents than in man. *Proceedings of the National Academy of Sciences of the United States of America* 82: 1741–1745.

Wu, C., and C. T. Ting. 2004. Genes and speciation. *Nature Reviews Genetics* 5: 114–122.

Wynne-Edwards, V. C. 1962. *Animal Dispersion in Relation to Social Behavior*. Oliver & Boyd, Edinburgh.

Wynne-Edwards, V. C. 1986. *Evolution through Group Selection*. Blackwell, Oxford.

Wynne-Edwards, V. C. 1993. A rationale for group selection. *Journal of Theoretical Biology* 162: 1–22.

Xiao, S. H., and M. Laflamme. 2009. On the eve of animal radiation: phylogeny, ecology and evolution of the Ediacara biota. *Trends in Ecology & Evolution* 24: 31–40.

Xu, X., Z. Zhou, and R. Prum. 2001. Branched integumental structures in *Sinornithosaurus* and the origin of feathers. *Nature* 410: 200–204.

Xu, X., X. Zheng, and H. You. 2009. A new feather type in a nonavian theropod and the evolution of feathers. *Proceedings of the National Academy of Sciences of the United States of America* 106: 832–834.

Xu, X., X. Zheng, and H. You. 2010. Exceptional dinosaur fossils show ontogenetic development of early feathers. *Nature* 464: 1338–1341.

Yamamoto, M. 1977. Some aspects of behavior of the migrating slug of the cellular slime-mold *Dictyostelium discoideum*. *Development Growth & Differentiation* 19: 93–102.

Yanai, I., Y. Wolf, and E. V. Koonin. 2002. Evolution of gene fusions: horizontal versus independent events. *Genome Biology* 3: 1–13.

Yu, J. C., J. L. Borke, and G. Zhang. 2004. Brief synopsis of cranial sutures: optimization by adaptation. *Seminars in Pediatric Neurology* 11: 249–255.

Yule, G. U. 1902. Mendel's laws and their probable relations to intra-racial heredity. *New Phytologist* 1: 193–207, 222–238.

Yunis, J. J., and O. Prakash. 1982. The origin of man: a chromosomal pictorial legacy. *Science* 215: 1525–1530.

Zaaijer, H. L., F. J. van Hemert, M. H. Koppelman, and V. V. Lukashov. 2007. Independent evolution of overlapping polymerase and surface protein genes of hepatitis B virus. *Journal of General Virology* 88: 2137–2143.

Zahavi, A. 1977. The cost of honesty (further remarks on the handicap principle). *Journal of Theoretical Biology* 67: 603–605.

Zenisek, S. F. M., E. J. Hayden, and N. Lehman. 2007. Genetic exchange leading to self-assembling RNA species upon encapsulation in artificial protocells. *Artificial Life* 13: 279–289.

Zeyl, C., and G. Bell. 1997. The advantage of sex in evolving yeast populations. *Nature* 388: 465–468.

Zhang, J. Z. 2003. Evolution by gene duplication: an update. *Trends in Ecology & Evolution* 18: 292–298.

Zhang, J. Z., Y. P. Zhang, and H. Rosenberg. 2002. Adaptive evolution of a duplicated pancreatic ribonuclease gene in leaf-eating monkeys. *Nature Genetics* 30: 411–415.

Zhang, P., and D. B. Wake. 2009. Higher-level salamander relationships and divergence dates inferred from complete mitochondrial genomes. *Molecular Phylogenetics and Evolution* 53: 492–508.

Zhang, Y.-P., and O. A. Ryder. 1994. Phylogenetic relationships of the bears (the Ursidae) inferred from mitochondrial DNA sequences. *Molecular Phylogenetics and Evolution* 3: 351–359.

Zhao, M. X., and J. L. Bada. 1989. Extraterrestrial amino acids in the Cretaceous/Tertiary sediments at Stevn's Klint, Denmark. *Nature* 339: 463–465.

Zohary, D., and M. Hopf. 2000. *Domestication of Plants in the Old World: The Origins and Spread of Cultivated Plants in West Africa, Europe and the Nile Valley*. Oxford University Press, Oxford.

Zuckerkandl, E., and L. Pauling. 1962. Molecular disease, evolution and genic heterogeneity. In M. Kasha and B. Pullman, eds., *Horizons in Biochemistry*, pp. 189–225. Academic Press, New York.

Zwaan, B., R. Bulsma, and R. F. Hoekstra. 1995. Direct selection on life span in *Drosophila melanogaster*. *Evolution* 49: 649–659.

CREDITS

with permission. Figure 5.37: From "A Devonian tetrapod-like fish and the evolution of the tetrapod body plan" by Daeschler, E. B., N. H. Shubin, and F. A. Jenkins. *Nature* 440:757–763. Copyright © 2006 Nature Publishing Group. Reprinted with permission. Figure 5.38: From "Palaeontology: A firm step from water to land" by Ahlberg and Clark. *Nature* 440, 747–749, April 6, 2006. Copyright © 2006 Nature Publishing Group. Reprinted with permission. Figure 5.49: From "Flammability and Serotiny as Strategies: Correlated Evolution in Pines" by Dylan W. Schwilk and David D. Ackerly. *Oikos* Vol. 94, No. 2, Aug. 2001, pp. 326–336. Reprinted with permission.

Chapter 6

Photos: p. 177: Frans Lanting/Corbis; p. 178: Frans Lanting/Corbis; p. 202: Frans Lanting/Corbis; p. 204 (top, both): Courtesy of John Brew; (center): Sarah Leen/National Geographic Stock; (top left): Lehmann, K. PNAS October 14, 2003 vol. 100 No. 21 12277–12282. © National Academy of Sciences; (center left): Chisnikov/Dreamstime.com; p. 216 (A): Nachman. *Genetica*. Volume 123, Numbers 1–2, 125–136 © 2005, Springer Netherlands; (B): Nachman et al. PNAS April 29, 2003 vol. 100 No. 9 5268–5273. © National Academy of Sciences; p. 223 (top to bottom): Courtesy of The Jackson Laboratory; p. 223: Courtesy of The Jackson Laboratory; Redmond Durrell/Alamy; p. 225: Asami et al. *The American Naturalist* Vol. 152, No. 2, pp. 225–236 © 1998 The University of Chicago Press; p. 227 (A): Keiji Iwai/Alamy; (B): Radius Images/Alamy; p. 235: Bill Hilton Jr. www.hiltonpond.org.

Drawn art: Figure 6.15: From *Genetics Essentials: Concepts and Connections* by Benjamin Pierce. Reprinted by permission of W.H. Freeman & Company. Figure 6.18: From "Evolution of the Mutation Rate" by Michael Lynch. *Trends in Genetics* vol. 26, Issue 8, 345–352, July 1, 2010. Copyright © 2010 Elsevier Ltd. All rights reserved. Reprinted with permission. Figure 6.19: From "Distribution of Fitness Effects Caused by Single-Nucleotide Substitutions in Bacteriophage f1" by J.B. Peris et al. *Genetics* 185, June 2010. Copyright © 2010 by the Genetics Society of America. Reprinted with permission.

Chapter 7

Drawn art: Figure 7.32: From "Estimating human inbreeding coefficients: comparison of genealogical and marker heterozygosity approaches" by Carothers et al. *Annals of Human Genetics*, Vol. 70, No. 5, Sept. 2006, pp. 666–76. Reprinted by permission of Blackwell Publishers, Ltd.

Chapter 8

Photos: p. 242: Frans Lanting/Corbis; p. 244 (B): "Cats and All About Them" by Frances Simpson. (c) 1902, Frederick H. Stokes, Company, Publishers, p. 141 (C): *Isle of Man: From Earliest Times to Present Date* by Joseph Train © 1845 M.A. Quiggen; London, Simpkin Marshal & Co. North Quay, p. 21; p. 249 (A): Brian Skerry/National Geographic/Getty Images; p. 253: Courtesy of Brian Gratwicke/flickr; p. 255: Jmjm/Dreamstime.com; p. 266: Courtesy of Michael L. Baird, bairdphotos.com/Flickr; p. 270 (A): Chris Schenk/Foto Natura/Minden/Getty Images; p. 278 (A): Courtesy of Jeff Hardin, University of Wisconsin; p. 279: Phelps, C.B. et al. *The Journal of Biological Chemistry*, 285, 731–740 © 2010, by the American Society for Biochemistry and Molecular Biology.

Drawn art: Figure 8.35 (B): From "The neutral theory of molecular evolution" by M. Kimura. *Scientific American* 241:94–104. Reprinted with permission. Figure 8.37: From "Differential Regulation of TRPV1, TRPV3, and TRPV4 Sensitivity through a Conserved

Binding Site on the Ankyrin Repeat Domain" by Phelps et al. *The Journal of Biological Chemistry* vol. 285 (1):731–740. Jan. 1, 2010. Copyright © 2010 by The American Society for Biochemistry and Molecular Biology, Inc. Reprinted with permission. Figure 8.39: From "In Vitro Selection of Functional Nucleic Acids" by David S. Wilson and Jack W. Szostak. *Annual Review of Biochemistry*, Vol. 68: 611–647. Reprinted with permission. Figure 8.44: Figure 5 from "Mutation rates in mammalian genomes" by Kumar, S. and S. Subramanian, *PNAS* vol. 99 (2), 803–808, Jan 15, 2002. Copyright © 2002 National Academy of Sciences, U.S.A. Reprinted with permission. Figure 8.45: From "The Molecular Clock and the Relationship between Population Size and Generation Time" by Lin Chao and David E. Carr. *Evolution* Vol. 47, No. 2, Apr. 1993, pp. 688–690. Reprinted with permission.

Chapter 9

Photos: p. 290: Frans Lanting/Corbis; p. 315 (all): Joshua L. Puhn; p. 317 (A): Rssfhs/Dreamstime; (B): Pierivb/Dreamstime.com; p. 318: Lew Robertson/Corbis; p. 323: Imagebroker/Alamy.

Chapter 10

Photos: p. 334: Frans Lanting/Corbis; p. 339: Courtesy of Dr. T. Ryan Gregory/University of Guelph; p. 344: Bentley et al. *Nature* 417, 141–147 (9 May 2002) © 2002, Nature Publishing Group; p. 346: Professor Stanley N. Cohen/Photo Researchers, Inc; p. 349: Dennis Kunkel Microscopy, Inc; p. 356: © 2009 Arakawa et al; licensee BioMed Central Ltd. BMC Bioinformatics 2009, 10:31; p. 368 (A): G. D. Carr; (B): A.C. Medeiros; (C–D): G. D. Carr; p. 369 (top): Martin Beebee/Alamy; 369 (center): May & Brown *Journal of Bacteriology* June 2009, pp. 3588–3593, Vol. 191, No. 11. © 2009, American Society for Microbiology; (bottom): Courtesy of Forest and Kim Starr.

Drawn art: Figure 10.3: Figure 2.1 from "The Origins of Genome Architecture" by M. Lynch. *Journal of Heredity* vol. 98 No. 6, 2007. Reprinted by permission of Sinauer Associates, Inc. Figure 10.11: The E. Coli 0157: H7 Genome. From The Genome Center of Wisconsin. Figure courtesy of Frederick R. Blattner. Figure 10.14: From "Comparative genomics in prokaryotes" in *The Evolution of the Genome* by T. Ryan Gregory. Copyright © 2005 Elsevier. Reprinted by permission of the publisher. Figure 10.15: From "Comparative genomics in prokaryotes" in *The Evolution of the Genome* by T. Ryan Gregory. Copyright © 2005 Elsevier. Reprinted by permission of the publisher. Figure 10.19: From "Complete genome sequence of the myxobacterium Sorangium cellulosum" by S. Scneiker, et al. *Nature Biotechnology* 25:1281–128. Copyright © 2007 by Nature Publishing Group. Reprinted with permission. Figure 10.20: From "Unexpected correlations between gene expression and codon usage bias from microarray data for the whole *Escherichia coli* K-12 genome" by Dos Reis et al. *Nucleic Acids Research* vol. 31 No. 23, Dec. 1, 2003. Copyright © 2003 Oxford University Press. Reprinted with permission. Figure 10.25: Figure 2 from "Molecular archaeology of the Escherichia coli genome" by Lawrence, J. G. and H. Ochman, *PNAS* vol. 95 (16), 9413–9417. Copyright © 1998 National Academy of Sciences, U.S.A. Reprinted with permission. Figure 10.32: From "Apollo: A Sequence Annotation Editor" by Lewis et al., *Genome Biology* vol. 3, No. 12. Dec. 23, 2002. Copyright © 2002 BioMed Central.

Chapter 11

Photos: p. 377: Frans Lanting/Corbis; p. 378: Frans Lanting/Corbis; p. 380: Courtesy of E. Javaux; p. 382: NOAA; p. 385: G. Thomas

Bishop/Newscom; p. 388: Courtesy of Feng Guo and Thomas R. Cech; p. 398 (A): SPL/Photo Researchers, Inc; (B): David M. Phillips/Photo Researchers, Inc.

Drawn art: Figure 11.11 (B): From "The antiquity of RNA-based evolution" by Gerald F. Joyce. *Nature* 418, 214–221, July 11, 2002. Copyright © 2002 by Nature Publishing Group. Reprinted with permission.

Chapter 12

Photos: p. 404: Frans Lanting/Alamy; p. 406: M.J. Grimson & R.L. Blanton/Biological Sciences Electron Microscopy Laboratory, Texas Tech University; p. 409: FR Images/Alamy; p. 411 (ciliate): Peter Arnold, Inc/Alamy; (radiolarians): Getty Images/Visuals Unlimited; (euglenids): medicalpicture/Alamy; (amoeba): Melba Photo Agency/Alamy; (horse): Bjakko/Dreamstime.com; (fungi): Pkemp/Dreamstime.com; (land plants): Sandyprints/Dreamstime.com; (red algae): Biodisc/Visuals Unlimited/Alamy; (micrograph): Knoll, et al. Phil. Trans. R. Soc. B 29 June 2006 vol. 361 No. 1470 1023–1038. © The Royal Society; p. 419 (top): Dictyostelium Aggregation January 2008 Bruno in Columbus http://en.wikipedia.org/wiki/Public_domain; (bottom, all): Kessin et al. PNAS May 14, 1996 vol. 93 No. 10 4857–4861 © National Academy of Sciences; p. 420: Dr M.Schleicher, ABI/Cell Biology, LMU Munich; p. 422: C. Solari University of Arizona; p. 424: Piotr Naskrechi/Minden Pictures/FLPA; p. 425: Christopher Boesch; p. 428: Frans Lanting/Corbis; p. 429: Carolina Biological/Visuals Unlimited/Corbis.

Drawn art: Figure 12.2 (A): From "Two-phase increase in the maximum size of life over 3.5 billion years reflects biological innovation and environmental opportunity" by Payne et al. *PNAS* 106 (1) 24–27. Copyright © 2009 National Academy of Sciences, U.S.A. Reprinted with permission. Figure 12.2 (B): From "Morphological Complexity Increase in Metazoans" by James W. Valentine, Allen G. Collins and C. Porter Meyer. *Paleobiology*, Vol. 20, No. 2, Spring 1994, pp. 131–142. Reprinted with permission. Figure 12.7: From "Evolutionary relationships among cyanobacteria and green chloroplasts" by Giovannoni et al. *Journal of Bacteriology*, Aug. 1988, pp. 3584–3592. Copyright © 1988 American Society for Microbiology. Reprinted with permission. Figure 12.11: From "The Evolution of Multicellularity: A Minor Major Transition?" by Grosberg and Strathmann. *Annual Review of Ecology, Evolution, and Systematics* Vol. 38: 621–654. Reprinted with permission. Figure 12.24 (B): From "Group Size and Ectoparasitism Affect Daily Survival Probability in a Colonial Bird" by Charles R. Brown and Mary Bomberger Brown. *Behavioral Ecology and Sociobiology* vol. 56, No. 5, Sep., 2004, pp. 498–511. Copyright © 2004 Springer Science + Business Media. Reprinted with permission. Figure 12.24 (C): From "Group Size and Ectoparasitism Affect Daily Survival Probability in a Colonial Bird" by Charles R. Brown and Mary Bomberger Brown. *Behavioral Ecology and Sociobiology* vol. 56, No. 5, Sep., 2004, pp. 498–511. Copyright © 2004 Springer Science + Business Media. Reprinted with permission. Figure 12.25 (B): From "Between-group transmission dynamics of the swallow bug, Oeciacus vicarious" by Brown and Brown. *Journal of Vector Ecology.* Vol. 30, No. 1, June 2005, pp. 137–43. Reprinted with permission.

Chapter 13

Photos: p. 432: Frans Lanting/Corbis; p. 438 (A): Photoshot Holdings Ltd/Alamy; (B): Gerold and Cynthia Merker/Visuals Unlimited/Getty Images; p. 439: Brown. PNAS: 1997 November 25; 94(24):

13011–13016. Copyright © 1997, The National Academy of Sciences of the USA; p. 443: Malicki et al, Mouse Hox-2.2 Specifies Thoracic Segmental Identity in Drosophila Embryos and Larvae, *Cell* 1990, Vol 63, Issue 5; p. 444 (top down): Custom Life Science Image/Alamy; David Bagnall/Alamy; Marcos Veiga/Alamy; Visuals Unlimited, Inc./Wally Eberhart/Getty Images; p. 448: Arco Images GmbH/Alamy; p. 450: Reproduced with permission from The International Journal of Developmental Biology. Trainor et al. 47: 541–553 (2003); p. 451: Schneider & Helms. *Science* 24 January 2003: Vol. 299 No. 5606 pp. 565–568 © AAAS.

Drawn art: Figure 13.11: From "Regulating Evolution: How Gene Switches Make Life" by Carroll et al. *Scientific American*, May 5, 2008. Copyright © 2008 Tolpa Studios, Inc. Reprinted by permission of Tami Tolpa. Figure 13.12: From "Regulating Evolution: How Gene Switches Make Life" by Carroll et al. *Scientific American*, May 5, 2008. Copyright © 2008 Tolpa Studios, Inc. Reprinted by permission of Tami Tolpa. Figure 13.15: From "Gene duplication, exon gain and neofunctionalization of OEP16-related genes in land plants" by Drea et al. *The Plant Journal*, Jun. 2006, Vol. 46, No. 5: 723–35. Reprinted by permission of Blackwell Publishing, Ltd.

Chapter 14

Photos: p. 454: Frans Lanting/Corbis; p. 456: Liam Craik-Horan; p. 459 (A): O. Digoit/Alamy; (B): blickwinkel/Alamy; (C): Steffen Hauser/botanikfoto/Alamy; p. 464: WaterFrame/Alamy; p. 465 (A-1): tbkmedia.de/Alamy; (A-2): Peter Arnold, Inc./Alamy; (B-1): All Canada Photos/Alamy; (B-2): Danita Delimont/Alamy; p. 468: Robert Clay/Alamy; p. 470: Philip Scalia/Alamy; p. 473 (A): Bill Beatty/Visuals Unlimited, Inc; (B): Colombo4956/Dreamstime; (C): Worldfoto/Dreamstime; p. 476 (A–C): Whittall & Hodges. *Nature* 447, 706–709 (7 June 2007) © 2007 Nature Publishing Group; p. 482: Richard Wong/Alamy; p. 485: Harry Taylor/Dorling Kindersley/Getty Images; p. 487 (A–B): Brunet, M. et al. *Nature* 418, 145–151 (11 July 2002). © 2002 Nature Publishing Group; p. 488 (top A): Suwa, G. et al. *Science* 2 October 2009: Vol. 326 No. 5949 pp. 68, 68e1–68e7. Copyright © 2009, American Association for the Advancement of Science; (top B): Image © Tim White. Suwa, G. et al. *Science* 2 October 2009: Vol. 326 No. 5949 pp. 68, 68e1–68e7. Copyright © 2009, American Association for the Advancement of Science; (center A): Fred Spoor/AFP/Newscom; (center B): Vilem Bischof/AFP/Getty Images; (center C): David L. Brill; (bottom A): © Australian Museum; (bottom B–C): David L. Brill; p. 489: (top A): John Reader/Photo Researchers, Inc; (top B): Deco Images/Alamy; (center, all): © Australian Museum, photo by Carl Bento; p. 492 (A): Pascal Goetgheluck/Photo Researchers, Inc; (B): Philippe Plailly/Photo Researchers, Inc; (C): Time Life Pictures/Getty Images.

Chapter 15

Photos: p. 498: Frans Lanting/Corbis; p. 500 (A): Mark Hallett Paleoart/Photo Researchers, Inc; (B): Michael R Long/The Natural History Museum, London; (C): North Wind Picture Archives; (D): Tom McHugh/Photo Researchers, Inc; p. 503 (A): AP Photo; (B): Schmidta, A.R. et al. Cretaceous African Life Captured in Amber. *PNAS* April 20, 2010 vol. 107 No. 16, 7329–7334; (bottom): The Natural History Museum, London; p. 505: © 2008 Elsevier Ltd. All rights reserved. *Trends in Ecology & Evolution*, Volume 24, Issue 1, 31–40, 27 October 2008; p. 509 (top down): Richard Coss (3); (left to right): Unclejay/Dreamstime.com; (c) H. Clarke/VIREO; (c) R.L. Pitman/VIREO; Oceanodroma Macrodactyla, lithograph

(Euphorbiaceae) and Pollinating Epicephala Moths (Gracillariidae) Diversified in Parallel?" by Atsushi Kawakita, Atsushi Takimura, Toru Terachi, Teiji Sota and Makoto Kato. *Evolution* vol. 58, No. 10, Oct. 2004, pp. 2201–2214. Copyright © 2004 John Wiley & Sons, Inc. Reprinted with permission. Figure 19.12: From *Tangled trees: phylogeny, cospeciation, and coevolution* by Roderic D. M. Page. Reprinted by permission of the University of Chicago Press. Figure 19.14: Figure 2 a&b from "Why does the yellow-eyed Ensatina have yellow eyes? Batesian mimicry of Pacific newts (genus Taricha) by the salamander Ensatina eschscholtzii xanthoptica" by Kuchta, Krakauer and Sinervo. *Evolution*, Vol. 62, No. 4, April 2008. Reprinted by permission of Blackwell Publishers, Ltd. Figure 19.19: Figure 6 from "Cultural Inheritance of Song and Its Role in the Evolution of Darwin's Finches" by B. Rosemary Grant and Peter R. Grant, *Evolution* Vol. 50, No. 6 (Dec., 1996), pp. 2471–2487. Copyright © 1996. Reprinted by permission of Blackwell Publishing, Ltd.

Chapter 20

Photos: p. 646: Frans Lanting/Corbis; p. 649: Joe McDonald/Visuals Unlimited/Getty Images; p. 664 (A–B): Laitman & Reidenberg. *Dysphagia* Volume 8, Number 4, 318–325, Sept. 1, 1993 © 1993 Springer New York; p. 673: Yves Brun, Indiana University.

Drawn art: Figure 20.6: From "Unifying the Epidemiological and Evolutionary Dynamics of Pathogens" by Grenfell et al. *Science* 16 January 2004: Vol. 303, No. 5656, pp. 327–332. Reprinted by permission of the American Association for the Advancement of Science. Figure 20.7: From "Unifying the Epidemiological and Evolutionary Dynamics of Pathogens" by Grenfell et al. *Science* 16 January 2004: Vol. 303, No. 5656, pp. 327–332. Reprinted by permission of the American Association for the Advancement of Science. Figure 20.15: From "Life and death: metabolic rate, membrane composition, and life span of animals" by Hulbert, Pamplona, Buffenstein and Buttemer. *Physiological Reviews*, Oct. 2007; 87(4):1175–213. Reprinted with permission. Figure 20.23: From "Senescence in a Bacterium with Asymmetric Division" by Martin Ackermann, Stephen C. Stearns, and Urs Jenal. *Science,* 20 June 2003: 1920. Reprinted by permission of the American Association for the Advancement of Science. Review Question 6: Figure 2 from "Cross-protective immunity can account for the alternating epidemic pattern of dengue virus serotypes" *PNAS*, Vol.103, No. 38, pp. 14234–14239. Copyright © 2006 National Academy of Sciences, U.S.A. Reprinted with permission.

INDEX

Note: Material in figures or tables is indicated by italic page numbers.